Environmental Science and Engineering

More information about this series at http://www.springer.com/series/7487

Pravat Kumar Shit · Hamid Reza Pourghasemi ·
Pulakesh Das · Gouri Sankar Bhunia
Editors

Spatial Modeling in Forest Resources Management

Rural Livelihood and Sustainable Development

Springer

Editors
Pravat Kumar Shit
PG Department of Geography
Raja Narendra Lal Khan Women's College
(Autonomous), Vidyasagar University
Midnapore, West Bengal, India

Hamid Reza Pourghasemi
Department of Natural Resources
and Environmental Engineering
College of Agriculture, Shiraz University
Shiraz, Iran

Pulakesh Das
Department of Remote Sensing & GIS
Vidyasagar University
Midnapore, West Bengal, India

Gouri Sankar Bhunia
TPF Getinsa Euroestudios S.L.
Gurgaon, Haryana, India

ISSN 1863-5520 ISSN 1863-5539 (electronic)
Environmental Science and Engineering
ISBN 978-3-030-56541-1 ISBN 978-3-030-56542-8 (eBook)
https://doi.org/10.1007/978-3-030-56542-8

This Springer imprint is published by the registered company Springer Nature Switzerland AG
The registered company address is: Gewerbestrasse 11, 6330 Cham, Switzerland

Dedicated to beloved teachers and parents

Foreword

It is a great pleasure to pen the Foreword of the book *Spatial Modelling in Forest Management: Rural Livelihood and Sustainable Development* edited by Dr. Pravat Kumar Shit, Dr. Hamid Reza Pourghasemi, Dr. Pulakesh Das, Dr. Gouri Sankar Bhunia. This book is comprised of 28 empirical research articles contributed by dedicated researchers from various disciplines.

This book has been divided into three parts. The first part comprises nine articles related with forest resource measurements, monitoring systems and mapping techniques. Nine articles of the second part deal with modelling, risk assessment and vulnerability. The final part with ten articles throws light on rural livelihood and sustainable management.

These 28 research papers encompass in-depth scientific analysis of various socio-economic perspectives of forest resources. This publication volume critically analyses the recent trend of forest resource utilisation with particular reference to micro and macro-level issues. Moreover, this book gives emphasis to rural livelihood for the sustainable management of forest. Concisely, this book covers almost all the emerging forest-related issues of the present era.

This book would be a piece of extreme appreciation for researchers, conservationists and social workers. I wish all the very best for its wide circulation and admiration.

July 2020

Malay Mukhopadhyay
Professor and Former Head
Department of Geography
Visva-Bharati
Santiniketan, India

Preface

Climate change is one of the leading ecological, economical and geopolitical issues of the twenty-first century. According to the United Nations Framework Convention on Climate Change (UNFCCC 1992 and IPCC 2007), natural climate variability along with the ever-increasing anthropogenic disturbances (via land-use change, deforestation, urban and cropland expansion, increase of artificial surface, use of fossil fuels, exercise of agro-chemicals, etc.) has led to increasing 'greenhouse gas' concentrations in the atmosphere and uprising the average global temperature. Forest ecosystem, as a source of huge carbon pool, plays key role in reducing the greenhouse gases and maintaining the water and energy fluxes and regulating the atmospheric processes. To mitigate greenhouse gas effects while maintain the forest ecosystem services, it is essential to provide managers and policymakers with accurate information on the current state, dynamics and spatial distribution of carbon sources and sinks point in forest area.

This book has considered 28 chapters associated to spatial modelling in forest resources issues, management and researches on forest health, forest biomass, carbon stocks and climate change studies, preferably. Currently, there are various challenges and uncertainties due of climate change and man-made interferences, which imposes great difficulties in adopting the appropriate decision. On the other hand, suitable management activities and policies for forest conservation have gained much attention globally. The latest advances in geo-spatial (remote sensing [RS] and GIS) technology, data processing platforms and modelling approaches have proven the potentially in developing sustainable and climate adaptive management plans. The integration of various geo-spatial and non-spatial data via advance data processing packages enables to generate reliable data for decision making. Various studies on forest health, forest conservation, carbon stock assessment, non-timber forest resources and climate change mitigation using the evident-based research and supplemented by the latest satellite data and data processing platforms will contribute to the development of appropriate policy and action plans for sustainable livelihoods.

We are very much thankful to all the authors who have meticulously completed their documents on a short announcement and paid in building this a very edifying and beneficial publication. We do believe that this will be a very convenient book for the geographers, ecologists, forest scientists and others working in the field of forest resources management including the research scholars, environmentalists and policymakers.

Midnapore, West Bengal, India

Pravat Kumar Shit
Hamid Reza Pourghasemi
Pulakesh Das
Gouri Sankar Bhunia

Acknowledgements

We are obliged to a chunk of experts for their kind support and valuable time to evaluate the chapters succumbed for insertion in this book. We are very much thankful to our respected teacher's Prof. Maloy Mukhopadhya, Prof. Sunando Bandyopadhyay, Prof. L. N. Satpati, Prof. Ashis Kumar Paul, Prof. Ramkrishna Maiti, Prof. Soumendu Chatterjee, Prof. Nilanjana Das Chatterjee, Prof. Dilip Kr. Pal, Prof. Sanat Kumar Guchait, Prof. N. C. Jana, Dr. Jatishankar Bandopadhay and Dr. Ratan Kumar Samanta for sharing their experiences, useful suggestions, continuous encouragement and immense support throughout the work.

We would like to thank the anonymous reviewers, acted as independent referees. Their input was consistently constructive and has substantially improved the quality of the final product.

We would also like to thank Ranita and Debjani, whose love, encouragement and support kept us motivated up to the final shape of the book. Finally, the book has been several years in the making and we therefore want to thank family and friends for their continuous support.

Dr. Pravat Kumar Shit would like to thank Dr. Jayasree Laha, Principal, Raja N. L Khan Women's College (Autonomous), Midnapore for her administrative support to carry on this project. We also acknowledge the Department of Geography, Raja N. L. Khan Women's College (Autonomous) for providing the logistic support and infrastructure facilities.

Dr. Hamid Reza Pourghasemi would like to thank the Shiraz University, College of Agriculture and Watershed Management Society of Iran for kind supports during preparation of this book.

This work would not have been possible without constant inspiration from my students, lessons from my teachers, enthusiasm from my colleagues and collaborators, and support from my family.

Midnapore, West Bengal, India

Pravat Kumar Shit
Hamid Reza Pourghasemi
Pulakesh Das
Gouri Sankar Bhunia

Contents

About the Editors

Pravat Kumar Shit Assistant Professor has been working in the Postgraduates Department of Geography, Raja N. L. Khan Women's College (Autonomous), Gope Palace, Midnapore, West Bengal, India. He received his Ph.D. in Geography from Vidyasagar University (India) in 2013, M.Sc. in Geography and Environment Management from Vidyasagar University in 2005 and PG Diploma in Remote Sensing & GIS from Sambalpur University in 2015. His main fields of research are soil erosion spatial modelling, non-timber forest and rural livelihood, forest biomass and carbon stocks, water resources and natural resources mapping and environmental modelling and published more than 50 international and national research articles in various renowned journals. His research work has been funded by the University Grants Commission (UGC), Government of India and Department of Science and Technology (DST), Government of West Bengal. He is an associate editor and on the editorial boards member of three international journal in geography, and earth environment science. He has served as a reviewer for many international journals. He published seven (07) books. Also, he published five books in Springer. He is currently the editor of the GIScience and Geo-environmental Modelling (GGM) Book Series, Springer-Nature.

Hamid Reza Pourghasemi is an Associate Professor of Watershed Management Engineering in the College of Agriculture, Shiraz University, Iran. He has a B.Sc. in Watershed Management Engineering from the University of Gorgan (2004), Iran; an M.Sc. in Watershed Management Engineering from Tarbiat Modares University (2008), Iran; and a Ph.D. in Watershed Management Engineering from the same University (February 2014). His main research interests are GIS-based spatial modelling using machine learning/data mining techniques in different fields such as landslide, flood, gully erosion, forest fire, land subsidence, species distribution modelling and groundwater/hydrology. Also, he works on multi-criteria decision-making methods in natural resources and environment. He has published more than 130 peer-reviewed papers in high-quality journals and three books in Springer and Elsevier as editor. Also, he is an active reviewer in more of 60 international journals.

Pulakesh Das is currently work in World Resources Institute India (WRII), New Delhi, India. Previously, he was teaching as an Assistant Professor in the Department of Remote Sensing & GIS, Vidyasagar University, Midnapore, West Bengal, India. He has received his Ph.D. degree from the Indian Institution of Technology (IIT) Kharagpur, India in July 2019. He completed his M.Sc. (2012) in Remote Sensing & GIS and B.Sc. (2010) in Physics from the Vidyasagar University, Midnapore, West Bengal, India. His primary research area includes land use forest cover (LUFC) modelling, hydrological modelling, forest cover dynamics and climate change, digital image processing, microwave remote sensing for soil moisture and forest biomass estimation, plant biophysical characterisation, etc. He has published more than 13 research articles in reputed peer-reviewed journals.

Gouri Sankar Bhunia received his Ph.D. from the University of Calcutta, India, in 2015. His Ph.D. dissertation work focused on disease transmission modelling using geospatial technology. His research interests include health geography, environmental modelling, risk assessment, data mining, urban planning and information retrieval using geospatial technology. He is an Associate Editor and on the editorial boards of three international journal in Health GIS and Geosciences. Currently, he is involved various smart city planning programme in India. He is also working as a visiting faculty in a private university of West Bengal. He has worked as a 'Resource Scientist' in Bihar Remote Sensing Application Centre, Patna (Bihar, India). He is the recipient of the Senior Research Fellow (SRF) from Rajendra Memorial Research Institute of Medical Sciences (ICMR, India) and has contributed to multiple research programs kala-azar disease transmission modelling, development of customised GIS software for kala-azar 'risk' and 'non-risk' area, and entomological study. He has published more than 60 paper in in reputed peer-reviewed national and international journal and three books in Springer. He is currently the editor of the GIScience and Geo-environmental Modelling (GGM) Book Series, Springer-Nature.

Part I
Forest Resources Measurement, Monitoring and Mapping

Chapter 1
Forest Management with Advance Geoscience: Future Prospects

Gouri Sankar Bhunia and Pravat Kumar Shit

Abstract The creation and implementation, involving key stakeholders, of context-specific forest management practices plays a significant role in the achievements of sustainable forest management. A number of site-growth modelling studies have been funded in recent years with the goal of developing quantitative relations between the site Index and specific biophysical indicators. With considerable time period, the role of forests in meeting the requirements for minor resources and ecological services has been recognized beyond the mere supply of forest. Present chapter describes advance geoscience application in forest management and also suggesting present research work to be adopted in future forest management plan. Counter-measures and recommendations were suggested on different forest management aspects, including developing consolidated structured data sets, designing top-ranking model monitoring and analysis and creating a multi-scenario decision support network. Finally, we proposed the main field of research in forestry research by incorporating and developing the participatory method, crowd sourcing, crisis mapping models and simulation systems and by linking data integration framework of geospatial technology, evaluation system and decision support system, to enhance forestry management by systematically and efficiently.

Keywords Forestry · Geospatial science · Crowd sourcing · Crisis mapping · Participatory mapping · Sustainable management

---

G. S. Bhunia (✉)
Department of Geography, Nalini Prabha Deo Prasad Roy Arts & Commerce College, Jorapara, Ashok Nagar, Sarkanda, Bilaspur 495006, Chhattisgarh, India
e-mail: rsgis6gouri@gmail.com

P. K. Shit
PG Department of Geography, Raja N.L. Khan Women's College (Autonomous), Gope Palace, Vidyasagar University, Paschim Medinipur, West Bengal 721102, India

P. K. Shit et al. (eds.), *Spatial Modeling in Forest Resources Management*, Environmental Science and Engineering,
https://doi.org/10.1007/978-3-030-56542-8_1

1.1 Introduction

Forest resource modelling and its status are very important economically and ecologically. Scientific community and policy makers are becoming increasingly aware of the fact that sustainable forest management is affected by several factors linked to global change. Between 1999 and 2012, the rapid rise in population contributed to the increase of €1 trillion, leading to an over 7 trillion people worldwide that need to be maintained by Earth resources. Forests are important to humanity because they provide a broader range of critical ecosystem resources, but the increasing depletion of forest cover means that the need for an ever-shrinking resource must be met with increased demand (Brockerhoff et al. 2013). Today, forest cover is about 31 percent of the land area or 4 billion ha. About half of the Earth's largest forests have been destroyed from land development, with the remaining 16 million hectares losing annually. At the same time, forests became more and more popular as sources of water and food, drugs, wood goods, and other leisure, economic, artistic, and spiritual advantages. Forests have been adequately or abusively exploited, but more effort to make sustainable use of them has been made. Forests are assessed annually at the global and country levels in terms of their scale, quality, usage and importance. In fact, as trees greatly add to the Earth's carbon balance. International interest in the identification of biomass is closely related to the protection of trees, photosynthetic development, and other carbon cycle processes and climatic variation (Houghton et al. 2009). In fact, the inventory of forests gives information on forest activities, conservation of forests and associated decision-making. Forest canopy and booth information was retrieved mainly by remote sensing and space-borne technologies (Tomppo et al. 2008) for wide areas of the world. Other environmental changes caused by human beings, such as increase in low ozone levels, deposition of nitrogenic contaminants, introduction of exotic insect pests and pathagogens, the fragmentation of ecosystems, and increased destruction such as fire may worsen these consequences (Bernier and Schöne 2009). Forestry can also have other consequences of climate change.

Some forest surveillance currently relies on data on development, i.e. improvements to forest cover, and two methods are used (DeVries and Herold 2013). Most environmental regulations are currently focussed on details on environmental operations, i.e. changes in land cover, and two methods are applied: top-down and bottom-up. The top-down approach utilizes satellite systems, while the bottom-up approach employs ground observation by government agencies, community-based surveillance (CBM), participatory surveillance or voluntary data (Danielsen et al. 2009). Satellite data provide global coverage and improved acquisition speed at a low cost, necessary for near-real-time forest surveillance (NRT) (Lynch et al. 2013). The scientific community is now generally recognized as major factors to latest increases in greenhouse gasses in the atmosphere and changes in the global hydrological cycle deforestation and degradation of forests (Hansen et al. 2013). The sample plot data reference data is still obtained largely through manual measurements while significant work is underway for, for example, terrestrial laser scanning, mobile laser scanning (MLS). The characteristics calculated in the inventories of operating forests are primarily the number of trees, tree types and breast height diameters (BHDs). Satellite Observations (SOs)—including Earth Observation (EOs) surveillance of the Earth's home planet; International Space Station (ISS) calculation, experiment and photo surveys; observations of the space science (SS); and Global Satellite Navigation System (GNSS) observations—are the basis for research to better understand our atmosphere and our environment.

The globalization of global trading networks and an increase in the volume of traded goods have been a contributing factor to population growth (Hulme 2009). Climate change can exacerbate invasions and impacts of forest pests. Climate change may, for example, promote the spread of both native pests and exotic ones (insects and pathogens) or affect tree pest resistance (Jactel et al. 2012) and there is growing evidence of an increasingly widely used phenomenon (Anderegg et al. 2015). Trumbore et al. (2015) described invasive species and diseases, as well as climate change, and deforestation as the major stressors in today's world's forests. The ongoing escalation and mechanization of forest management, which has increased forest vulnerability to biological invasion, climate change and other stressors, is an additional contributory to the forest health issue (Seidl et al. 2011). There have been several shifts in the forests in recent global warming (Lucier et al. 2009). Climate change effects can be beneficial in certain areas for certain tree species. In some areas, the growth of trees is increasing in longer growing seasons, hotter temperatures and higher CO_2 rates. Many of the expected climate changes and their indirect impacts are likely to adversely impact forests. Observed changes in vegetation (Lenoir et al. 2010) or increased mortality from drought and heat in forests around the world (Allen et al. 2010) may not be caused by climate change triggered by human beings but may demonstrate the potential consequences of the rapid environment. However, the vulnerability of tropical humid forests has been discussed recently (Huntingford et al. 2013) and temperate forests may be at greater threat in areas subject to a more extreme climate (Choat et al. 2012). A variety of viewpoints are available to consider adjusting to these changing and unpredictable future circumstances (McEvoy et al. 2013). Forest management would have to prepare on a variety of spatial and time levels in order to resolve potential problems and implement more flexible and collaborative management strategies. The tacit belief that local climate conditions will continue to be constant is often the basis of local forest activities (Guariguata et al. 2008). Additional social and economic developments in forest management will also continue to push transition (Ince et al. 2011). A growing global population, rapid economic growth, and increased wealth, for example, are driving demand in multiple developing countries for food and fiber crops and forest conversions into agriculture (Gibbs et al. 2010). The goals of climate change mitigation are to raise demand for biomass-based bioenergy and biomass in construction and manufacturing systems. Increasing urbanization shifts the essence of social demands for forests, and a reduction in rural communities decreases the supply of labor and capacity for intensive initiatives on forest management.

Regional climate, biodiversity, domestic environmental protection, even global environmental changes are strongly affected by the change in forestry area. The further squandering of ecological building space and the greater demands for strict forest red line expansion and industrialisation. In current land use and ecological cultures, how forest resource management is optimized and improved by complex monitoring for change in forest areas has become an extremely significant and urgent mission. Dynamic surveillance needs database tools have been developed to direct forest area surveillance capacity building (Gillis et al. 2005). For dynamic monitoring, a georeferenced digital database is commonly used as a basis of capacity building monitoring work. As an indicator of changes to forest area (Illera et al. 1996), temporary evolution of vegetation indices may be possible. With the exponential growth of earth observation technology (EO) and the continuing launch of remote sensing satellites, the Earth observation data resolution is growing and the number and range of data are rising as well, indicating that EO data are increasingly entering the age of big data (Xia et al. 2018). The Earth Observation Satellites Committee's (CEOS) figures indicate that over the last half century in 500 EO satellites were deployed, and more than 150 satellites will be launched in the next 12 years (Guo 2017). The Big Earth Observation Data (BEOD) slowly supported the growth of the world industries, research institutions, and application sectors which had a profound effect upon the Earth system science, contributing to human activities, environmental monitoring, and climate change (Yao et al. 2020). Furthermore, Web-based Geographic Information Systems (WGIS) is accessible in order to allow access to digital maps and geographic models. The reports are available publicly. It is a significant step in democratizing exposure for various users to geographical information. There are no spatial analysis resources available in current WGIS programs for this area, which use specific data sources, and easy access to reports with maps, graphs, text and table data.

In particular for data collection, for the production of a technical model and for research platform construction there remain many defects and limitations in technology and capacity building. Failure to coordinate forest land knowledge resulted from a lack of inventory requirements (Managi et al. 2019). Forest region adjustments are difficult to incorporate details based on the different forest land inventory and land grading requirements. There are significant issues with the uniformity of reporting practices in the reporting implementation process. Further development is required in conjunction with an integrated research model and a dynamic monitoring of forestry change. Systematic analysis for forestry area changes includes comprehensive data bases and models, and a system development tool is also required to help the conversion process from data analysis to application decision-making study. Spatial data items for the Multi-Period and Multi-Scenario Forest Region must be routinely analyzed and planned. The current state and rising forest area patterns combine with environmental and socio-economic influences interacting with each other. By using methods of GIS and Space Economics, the findings of the analyzes can be better adapted to natural change and better decision-making data for the optimal management of forest lands and other land types, using methodological methodologies. The findings of such research are focused on the effects of the natural changes. In addition, it is necessary for forest change to be improved in prediction and dynamic analysis, for development paths to be framed and regional development objectives to be recognized. The basic project to improve the study of forest dynamic change is the perfecting of data integration—model study—policy modelling Integrated development and the realization of scenarios for forest land changes under various systems. Restrictive factors influencing the growth of forestry areas must be identified, forest growth adaptive management measures examined and transition and strategies for development based on different stages promoted. The essential pre-conditions under the current climate change, urbanization growth, industrial system transformation, environmental protection and so on are the identification of the major contradictions in the cycle of forest growth and the main factors restricting development and implementing adaptive management.

1.2 Geosciences to Improve Forest Assessment

Through technical and statistical advancement, the processing of forestry data and their analyzes have steadily progressed (Kleinn 2002). Of starters, field dimensions, such as diameter or height scales, usually measured using tape or wood compasses and relascopes are now being improved with the use of emerging technology, including laser scope discoverers. In addition, the technology of remote sensing has been used rapidly to enhance soil sampling (Maniatis and Mollicone 2010), to measure improvements in vegetation and areas and to monitor other value variables, including forest fires, rodents and trees outside forest (Barducci et al. 2002). The usage, along with ground-based observations, of remote sensed data has gained considerable interest in estimating greenhouse gas emissions and forest-related removals, especially in the context of REDD+ (GFOI 2014). Recently a free Landsat satellite sample has been used by Food and Agriculture Organization of the United Nations (UN-FAO) to record forest land and area changes figures for the period 1990–2005 (FAO and JRC 2012) for woodland, other forested land and other ecosystem services. Therefore, a specific challenge for enhancing forest cover projections, carbon reserves and complexities is to efficiently integrate numerous top-down and ground-up strategies, a suggestion issued by the United Nations Framework Convention on Climate Change in the sense of emission reduction from deforestation and forest loss (REDD+) (UNFCCC 2009).

In the last few years, major changes have been made in LiDAR's systems leading to a boost in LiDAR position precision and surface density. LiDAR technology applies to a vast range of laser measurement devices, three primary approaches to the sensing of forest structures being terrestrial, airborne and space-borne approaches (Yao et al. 2011). Terrestrial laser scanning (TLS) has the ability to estimate tree diameters, height of the tree, tree volume and thus biomass in a structured and automatic manner (Hosoi et al. 2013). There is still an overview of these massive, three-dimensional datasets, but many ongoing methodological advances will make this technology useful soon. A digital elevation model (DEM) can be created from the point-cloud data created with LiDAR from the points reaching the ground and a canopy height-model from those intercepted by the upper canopy can be made. LiDAR's precision, combined with high spatial and point density, makes airborne LiDAR systems an enticing data acquisition method for estimating a large array of tree and forest parameters (Laes et al. 2011) like tree height (Detto et al. 2013), tree biomass (Li et al. 2008), leaf area index (Morsdorf et al. 2006) or stem volume (Heurich and Thoma 2008). Spaceborne data like LiDAR enables forest structures to be mapped globally with a vertical structure (e.g. by the Geoscience Laser Altimeter System (GLAS)) (Simard et al. 2011). In 2018, a similar sensor, ICESat2, has a smaller footprint than the previous GLAS instrument. Finally, the Global Ecosystem Dynamics Investigation (GEDI) project aims to make a high-resolution observation of a forest vertical structure at the global level using a LiDAR-backed instrument embarked on the International Space Station (https://science.nasa.gov/missions/gedi/). Moreover, a system called Synthetic Aperture Radar (SAR) is used to improve resolution beyond physical antenna aperture limits in order to achieve a high radar spatial resolution. For example, since it has a wavelength (5–6 cm), a C-band SAR signal is known to quick saturate with forest biomass (Thurner et al. 2014). In April 2014, Sentinel-1A was successfully launched with C-Band Radar as part of the European Space Agency's Copernicus Mission (ESA). Nonetheless, a loss of sensitivity at values greater than 100–150 Mg ha^{-1}, sometimes interpreted as signal saturation, was also observed in several studies (Woodhouse et al. 2012). Mermoz et al. (2015) have shown recently that L-band scatters appear to attenuate, rather than saturate, over and above this biomass threshold which could result in new opportunities in the mapping of L-band SARs. Currently, the L-band ALOS PALSAR is the single, wavelength radar sensor for monitoring the structure of forests, and in 2014, its sequel—ALOS2—was launched. In the case of forest-carbon evaluation, LiDAR, radar, textural and stereo-photogrammetry analysis have made considerable progress and allow the measurement (LVG 2012), over a significant shorter duration than conventional field sampling campaigns, of several variables of interests—for example, the diameter of tree, the height of tree and crown size (Table 1.1).

However, it is currently little understood how precise additional forestry characteristics such as timber volume per hectare are modellable by high-resolution data (almost 1.0 m and < 5 m) and high-resolution satellite stereo data (<1.0 m). For forestry survey methods, such as the extraction of quantitative information on canopy structure and forest biomass estimates, also in a setting of high biomass (Bastin et al. 2014). Therefore, it was possible for researchers to study ecologic structures with far greater detail than those provided by the start of high-resolution satellite sensors such as CARTOSAT (Spatial Resolution: 2.5 m), IKONOS, (spatial resolution in MS: 4 m), Quickbird (spatial resolution in MS: 2.88 m), and OrbView-3, (spatial resolution in the MS: 4 m),GEOEYE (Straub et al. 2013; Goward et al. 2003; Gibbs et al. 2007). For the calculation of the heights of individual pine trees and lading stands at Appomattox-Buckingham State Forest in Virginia, USA Popescu and Wynne (2004) used LiDAR and ATlAS multi-spectral (visible, near-IR and mid-IR) optical data with spatial resolution of 4 m. Table 1.2 illustrated the high spatial resolution satellite data in forestry mapping and monitoring. They showed that combined multi-spectrum imaging and LiDAR data can reliably predict forest inventory and evaluation tree heights of value. Nagendra (2001) assessed 'remote sensing capacity for determining the diversity of the ecosystem.' He concluded that a decade ago it was

Table 1.1 Satellite data and methods for forestry resources mapping and monitoring

Satellite/sensor	Aims	Methods	Reference
Synthetic aperture radar (SAR) and/or LiDAR	To detect and map forest degradation; Estimates above ground biomass	Spectral fractions, unmixing or classification	Mitchell et al. (2017)
Airborne X-band SAR data	To enhance discriminability of the forest types and features	Leaf Area Index; Spatial textural analysis	Roy et al. (1994)
Japanese Earth Resource Satellite (JERS)-1 Synthetic Aperture Radar (SAR)	Assesses the feasibility of forest cover mapping and the delineation of deforestation	Multi-image segmentation, post-classification detection	Thiel et al. (2006)
JERS-1, ERS-1 SAR and RADARSAT	Objectives are biomass estimation, forest and land-cover-type recognition in boreal forests	Textural measures, multitemporal approach, mixed pixel approach	Kurvonen et al. (2002)
Passive Microwave Remote Sensing (C-band, L-band and X-bands)	To compute the emissivity e of forests	Radiative transfer theory, matrix doubling algorithm	Ferrazzoli and Guerriero (1996)
Synthetic aperture radar (SAR); airborne and terrestrial LiDAR	Degradation and forest change assessment	Random forest (RF), REDD+ mechanism	Calders et al. (2020)
Synthetic Aperture Radar (SAR)	Quantification of spatial and temporal changes in forest cover	Random Forests, Extremely Randomised Trees	Devaney et al. (2015)

not yet possible to delineate a large number of species with spectral data. However, a 2-m spatial resolution was launched in 2009 for WorldView-2 (WV2) (Coastal, Blue, Green, Red, Red-Edge, near infrared (NIR)—1 and NIR—2) with a high resolution of 0.5 m (Coastal, Blue, Green, Yellow and NIR). Several studies in recent years have used WV2 images for the study of tree habitats. The accuracy of mapping six species/groups of trees improved with WV2 imagery by 16–18% compared to IKONOS satellite images. The research, however, covered trees/groups with sparse vegetation and not in a forest, within a dense urban area. Carter's (2013) use of multitemporal data from June and September 2010 in a multi-temporal forest mix in Upstate New York from two WV2 images for classifying ash, maple, oak, beech, evergreen and 6 other tree classifications. This would also promote the grouping of tree species into mixed near-nature, natural, urban forests with a wide variety of tree species. Very high spectral resolution imaging often mounted on aerial systems offers important, unreviewed eye information on forest function with a greater number

Table 1.2 Use of high spatial resolution satellite data in forest mapping and monitoring

Satellite/sensor	Region of study	Aims	Methods	Outcome	Reference
IKONOS II	Italy	Forest Inventory and Mapping	Supervised classification, Object-based approach	Forest cover density and dominant tree species composition	Giannetti et al. (2003)
IKONOS; QuickBird	Costa Rica, Central America	To evaluate tree death rates	Calibration factor	Calculated a landscape-scale annual exponential death rate	Clark et al. (2004)
QuickBird	Tully, New York	Tree identification and tree crown delineation	Rule-based classification; segmentation algorithm	Classification trees were built and results were evaluated using a cross-validation approach; spectral metrics, texture, elevation features, and geometric features were calculated for each image object	Ke and Quackenbush (2007)
WorldView-2, LiDAR	Ljubljana	Tree species inventory	Object-based image analysis; Digital elevation model (DEM); Principal component analysis (PCA)	The accuracy of the proportions of individual tree species that form the forest stand canopy was lower than in some other studies. The distinction between deciduous and coniferous tree species was the most reliable	Verlič et al. (2014)
QuickBird, IKONOS	Nepal	Forest Condition Monitoring	Geographic object-based image analysis (GEOBIA)	Tree crown detection, delineation, and change assessment	Uddin et al. (2015)

(continued)

Table 1.2 (continued)

Satellite/sensor	Region of study	Aims	Methods	Outcome	Reference
LiDAR, WorldView-2	Victoria, Australia	Characterisation and classification of forest communities	k-means clustering algorithm; TreeVaW algorithm; support vector machines (SVMs) and decision trees	Identified individual trees, including locations and crown sizes identification of Myrtle Beech and adjacent tree species—notably at individual tree level	Zhang (2017)
WorldView-2	Long Island, New York (US)	Assessment of forest fire	Multiple Endmember Spectral Mixture Analysis (MESMA) fraction; spectral indices	Forest burn severity mapping from VHR data	Meng et al. (2017)
Unmanned aerial vehicles (UAVs), Pléiades	Czech Republic	Estimation of basic tree attributes, such as tree height, crown diameter, diameter at breast height (DBH), and stem volume	Structure from motion (SfM) algorithms; spectral Correlation	Predict tree characteristics with high accuracy (i.e., crown projection, stem volume, cross-sectional area (CSA), and height)	Abdollahnejad et al. (2018)
Airborne LiDAR	Peru	measure and monitor carbon stocks and emissions; measurements of top-of-canopy height	Random forest machine learning regression	Aboveground carbon stocks and emissions	Csillik et al. (2019)

of narrow spectral bands (up to 200 or more contiguous spectral bands). Imaging spectroscopy, for example, may relay valuable information on variability in canopy chemistry (Baccini and Asner 2013) and thus provide direct information on the functioning of the Ecosystem. The taxonomic and functional structure of canopy trees can also be described in a highly successful way.

For research and development, the bulk of the above technological methods are still considered. Technology development, modifying and implementing existing systems in accordance with country circumstances, has the possibility, as necessary, of improving field measurement alertness, reducing time and expense of field sampling campaigns and improving forest extrapolation estimates over broad spatial scales including remote or conflict areas. The implementation of transparent national forest surveillance systems can also be assisted by new technologies. However, national and subnational corporations, private businesses, research and academic institutions, NGOs and civil society face a great many constraints in implementing, adapting and activating these technologies. Of these, minimal technical skills are possibly the most critical when using these new technologies; thus, training and capacity building are necessary and must be expected.

1.3 Cloud Computing and Forest Management

The rapid advancement of cloud computing technology in recent years provides strong computing power, especially for the efficiency of big geospatial data management and processing, which makes it possible to perform complex simulations on a global scale. Cloud computing is used as a framework to allow users to access a common community of computational tools that is configurable and can easily be supplied and published with minimal management effort and/or interference between service providers (Li and Huang 2017). Cloud computing has transformed the conventional information technology model entirely by offering at least three types of services: infrastructure as a service (IaaS), platform as a service (PaaaS) and software as a service (SaaS). In order to address persistent spatial data model problems spatial cloud computing (SCC), a data layer as a service (DaaS) was proposed (Yang et al. 2011). The discrete global grid systems (DGGS) have had a fairly flawless theoretical statistical history and basic functions over the last two decades (Zhao et al. 2016). DGGS is known as an Earth reference system (ERS) which uses cells to divide and address the globe (Bauer-Marschallinger et al. 2014). The DGGS Standards Working Group was set up in 2014 and an international specification was adopted by the Open Geospatial Consortium.

Cloud technologies, and particularly in the field of data storage, have begun to infiltrate all facets of life. Cloud computing cannot make use of itself explicitly for the visualization and maintenance of forest resources, because it essentially lacks the functionality of the spatial data collection. This effort aims to address four strength issues in the geospace, namely data, machine, competitor and space-time. After several years of growth, cloud-related computing technology and forest observation tools are also increasingly being developed, for example Google Earth Engine and Esri Geospatial Software (Yao et al. 2020).

Figure 1.1 shows that cloud computing provides some services for forest observation mapping and monitoring including spatial data infrastructure (SDI), EO data resource, algorithm or model library, processing and computation, systems and applications. The easiest approach is to provide a wide variety of nodes, computers, and servers which can provide customers with on-demand network resources such as the AWS, Google Cloud or Aliyun space storage network. The second is to provide forest

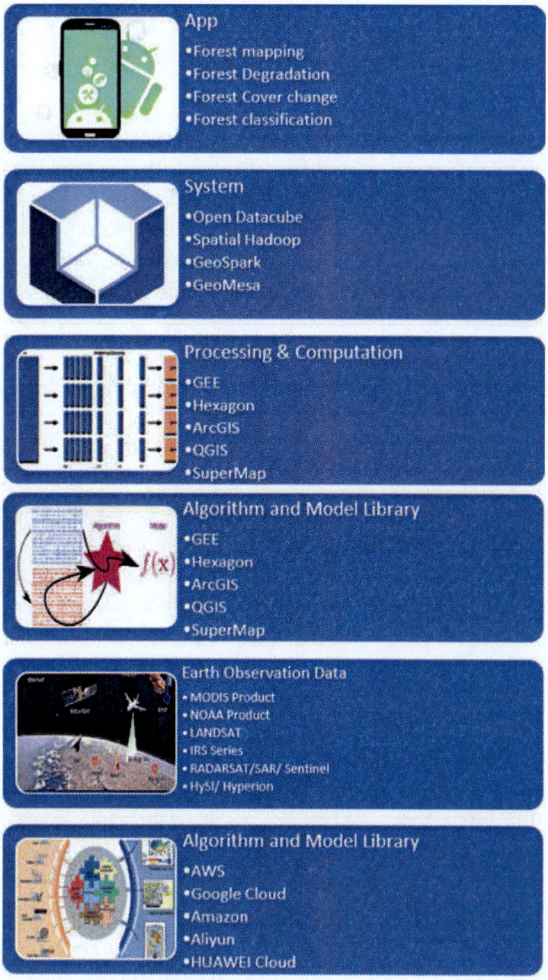

Fig. 1.1 Cloud computing for forest observation mapping and monitoring

resource mapping and tracking computer services for Earth observation, known as the EO computer cloud. The data cloud is actually the most sophisticated and simple form of cloud computing. The Global Earth Observation System of Systems (GEOSS) for example has developed a scalable platform for regional and multidisciplinary data exchange in EO sector, with its cloud-based exploration and access solutions. An algorithm or a software database, processing and computational power is given for the third and fourth versions. The two pieces are comparatively more professional and are only open to a few study teams or industrial firms. They are the most growing trends for structures and applications.

1.4 Integration of Participatory Approach and Geospatial Technology

Rambaldi et al. (2006) states that PGIS "combines a variety of geospatial knowledge and methods, such as maps of drawings, participatory 3D modeling, community-based air and visual analysis of satellite images, GPS transect walks and cognitive GIS mapping." GIS (Participatory GIS) was widely used to support group mapping to ensure subsistence sources and their cultural areas (such as holy sites, historical sites, ancestor routes) Corbett et al. (2006) stated, "this participatory concept implies a degree of control over decision-making, managerial authority and accountability by the group at all levels." In the last two decades the number of companies and advocacy groups interested in natural resource management has dramatically grown. In promoting engagement preparation systems, ICT has the ability to play a significant role. Engagement forms include either one-on-one (in an anonymous interview) or group interaction (individual forums such as public juries, round tables, research circles, and collective advisory groups). The ability of the local people to participate largely depends on the sort of opportunity they may anticipate (Robiglio and Mala 2005). Learning and sharing experiences with local community organizations have helped us recognize the conditions required for effective measurement and reporting at local level. Measuring, documenting and testing involves cost-effective and accurate testing methods. We researched conditions for a fairly easy and cost-effective method to gather information concerning land use (LU) and land cover changes (LCC) using remote sensing and geographical information systems. We have identified land use and drivers of deforestation and forest destruction, and their effects, using satellite imagery analysis and knowledge from local people (Fig. 1.2). GIS technology and the maps remain primarily focused on characterising, evaluating characteristics of places instead of communities and livelihoods given these efforts and the rapid growth of new research areas Participatory GIS (PGIS) and Public Participation GIS (PPGIS). Sulistyawan et al. (2018) demonstrate how participatory mapping results can be integrated into Spatial Planning Regulation. The integration of PGIS and Space Planning Control is being carried out in three stages. Throughout the first step, the group and district governments formed a shared vision and dedicated all

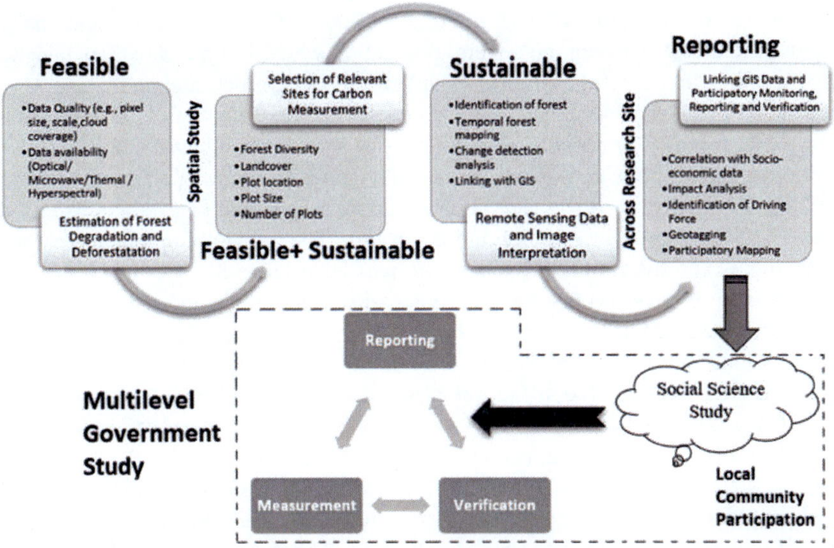

Fig. 1.2 Local Community participation in forest research analysis and participatory mapping. Deforestation and forest degradation scales, still using satellite data and space analyzes, and the knowledge provided by local communities to select appropriate locations for the estimation of carbon stocks and change drivers for forest cover. This figure illustrates how the social science team conducted its investigations in each local context (Modified after Boissière et al. 2014)

parties to embracing the end results of mapping for use in the future planning phase. The second step was to promote the involvement of GIS by the Community and to incorporate the appropriate community areas in the regulation on spatial planning. A clear evaluation of their strengths and limitations for the different applications is required to combine participatory and GIS-mapping approaches and is important for carving practitioners, designers and community members alike (Vajjhala 2005). Within this multidisciplinary approach we incorporated biophysical information and information (carbon stock estimates), social science information and remote sensing information (cards using satellite images and knowledge of the local population). In this multidisciplinary approach. When it is the only tool for biomass assessment, the use of remote sensing is limited. Participatory mapping can also help local people draw maps based on their experiences in the land cover (Mapedza et al. 2003). Vegetation forms for rising land cover can be established in local communities (Abraao et al. 2008). Ground inspections are also needed to validate the discrepancies in remote sensing maps. Without local community engagement, remote sensing and GIS cannot provide too much knowledge about the drivers of transition. They offer a full picture of the changes in forest cover (Mapedza et al. 2003) and thus the reasons for variations in space and time in carbon stocks. Remote sensing experts may also provide information on sensitive areas that need careful attention when working together with local communities for monitoring and plot-measuring (e.g.

high environmental value areas (Balram et al. 2004). New ICT technologies help to close the divide between the general population who will now take their insight into the policy process more efficiently, and experts, academics, and politicians who take action every day on behalf of the country. The social network is a significant piece of knowledge for social capital analysis as it focuses on how social networks promote and restrict incentives, attitudes and cognition (Paletto et al. 2010). Social media enables users to access digital information, upload and distribute content as well. They can not only access digital information. The growth of mobile telephone technologies and the resulting decline in prices have made it easier to access the internet. The Internet has been revolutionized in social media and turned from an information source into a communication platform. The media that expressing our thoughts, emotions and general mental state through social networks such as Twitter, Inc, Facebook, Inc and Reddit Inc. are also a part of our daily lives (Wongkoblap et al. 2017). This website has also become a powerful data bank of advertisers and analysts, who can examine consumer practices, social content and related knowledge, as well as other attitudes and habits to define their interests and tastes. The most popular social media enables knowledge and views to be exchanged by forums or by Wikis in a forum, or more sophisticated ways of gathering information such as MySpace or Facebook (Sweeney 2009).

1.5 Mobile Application in Forest Management

The software, Web App, Online Download, iPhone app or smartphone application may be named a mobile device. Mobile apps are generally available through native distribution platforms, known as app stores, run by mobile operating system owners. The most rising smartphone apps in India include Whatsapp, biking, Instagram, Bookmyshow, Paytm, Indian Rail App Disha. The essential mobile app for forestry is listed below.

1.5.1 Hejje (Pug Mark)

Hejje is an indigenously developed Android-based application. It co-ordinates foot patrolling of forest staff apart from providing the range forest officers live update of their respective anti-poaching activities such as patrol time, water level in lakes, suspicious activities, tree population and forest fires. The staff using the mobile application can take photographs.

1.5.2 Urban Forest Cloud Tree Inventory App

This app allows user to inventory trees in an easy-to-use web map and export the data to an ESRI shapefile or MS Excel/Access file for use in other software applications. It serves as a tool for individuals without tree inventory software and as supplemental, highly accessible tool for those with inventory software.

1.5.3 Tree Sense

The app allows user to quantify and qualify the benefits of trees, including air quality, electricity savings and storm water reduction. Users can also figure out the best placement for future tree plantings in order to maximize their benefits.

1.5.4 Timber Tracker

Timber Tracker, was developed by App Pros, LLC of Springfield, Missouri. Timber Tracker was designed by loggers for loggers. This app allows you to estimate timber harvest, price lumber, and prepare and send a quote PDF to customers.

1.5.5 Leafsnap

This Smartphone App is a North American tree identification guide.

1.5.6 Tree Trails

Tree Trails planned to gather trees knowledge throughout the state of Texas. The app is designed to allow teachers, youth leaders and the general public to trail and learn about these themes and share this knowledge with other people through the corresponding curriculum.

1.5.7 Tree Book

Tree Book is the leading guide to 100 of North America's most famous trees.

1.5.8 Tree Tagger

The Tree Tagger Mobile Forest Health App transforms a smartphone into a device that tracks ill tree and provides scientists and wood management around the world with forest health data.

Nowadays, digital cameras are mostly attached to smartphones. Apps are already used to calculate a forest sample plot or stand using smartphone camera data. Such fast, easy-to-use plot-measurement application for forests has become increasingly popular with foresters and are tested by various forest organisations. To order to consider the impact of constant clearance and irrigation of bushes along the riversides it should better benefit local people including farmers (Pratihast et al. 2012). Under standard boreal forest conditions Melkas et al. (2008) and Vastaranta et al. (2009) were testing a laser camera. It was a digital Canon EOS 400D reflex with an integrated laser line generator from Mitsubishi ML101J27. It measured the diameters of trees without visiting them from the middle of the sampling plot. Forest sample measurements, Vastaranta et al. (2015) tested the TRESTIMATM smartphone software. The software interprets the images from sample images captured on the smartphone using a camera. It then calculates forest inventory attributes including species-specific basal areas (G) as well as basal area medium-tree diameter (DgM) and height (HgM). The smartphone app seeks to assist forest conservationists, government officials and stakeholders in creating, amending and applying an efficacious community-based conservation program that preserves forest protection in real time updates on the status of forests and associated habitats (Fig. 1.3). Satellite images and videos are used by the iOS device to provide in real time information on the situation in certain areas of woodland habituating. India's Natural Resources Department has a free mobile program called "Indiana DNR," for Android as well as Apple iOS. "Forest Watcher" is a US-based charity that monitors changes in forest cover to allow offline

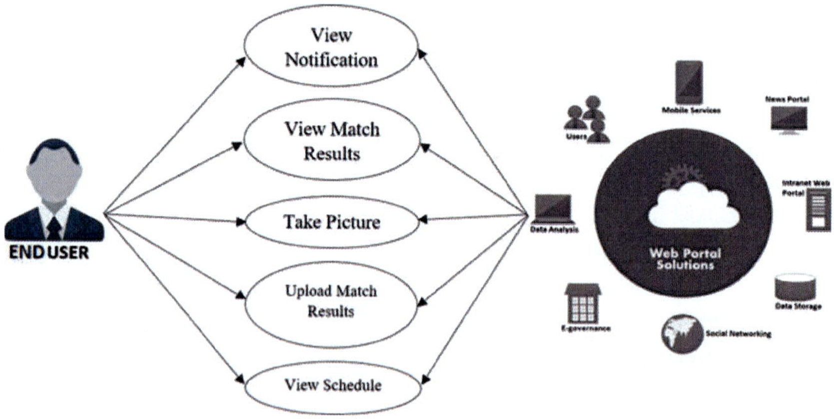

Fig. 1.3 Mobile application use case diagram

access to real-time Satellite Maps and data gathered. The app shows forest shifts on mobile devices via their Global Positioning System (GPS) system, which depends not on internet connection. Moreover, the Forest Survey of India (FSI) adds importance to fire warnings in the field of forests, attaches attributes to a list, prepares map-based items, etc. for example. Fire areas are MODIS 1 km grid centres. The KML file, Google compatible format and sent to the registered end users within 2 h of satellite overpass, is created with forest fire alerts from the active fire locations. Emails and SMS can be used to send warnings to registered users.

Forests used computer technology to adopt best practices. Incorporating computer technology into forestry has reduced the problems of information, expertise and data sharing in order to increase decision-making. It strengthens the openness of forest science. This lead to successful policy making around the world to safeguard the forest and increase the development of forest products through environmental conservation. As the role of computer technology is important for the education of forest scientists and the management of forest science, it is vital that information technology is required to educate forests scientists. This can be done easily through forest information technology.

1.6 Near Real Time Monitoring of the Forest-Sensitive Zones

Thanks to the local community's involvement in the land, forest change may be indicated to the municipality, date, scale, and proximity drivers in NRT transition (deforestation, forest destruction, or reforestation). Digital tools such as smartphones for mobile correspondence ease data storage and delivery activities (Pratihast et al. 2016). The incorporation of community-based monitoring (CBM) data into national forest monitoring system (NFMS) has, however, caused some problems, including: (1) lack of trust in the method of data collection, (2) incoherently controlled size, (3) restricted geographical coverage, (4) variable data quality and (5) a lack of confidence in data providers. Recent advances in technology like cloud 2.0 have provided potential solutions to such problems such as GIS, remote sensing, big data analytics, smart apps and social media (Conrad and Hilchey 2011; Skarlatidou et al. 2011). Given this ability, for many reasons there are currently a lack of successful implementation of the integrated NRT forest monitoring program. Secondly, forestry transition research results from multisource data sources (i.e. satellite and CBM in NRT) cannot be managed by organizational processes. Second, no device can archive, envision and provide local actors with information on forest change across the Internet. Fourth, there is the absence of the quest capabilities for spatial and temporal forest transition. Ultimately, in the sense that individuals do not provide input about the information submitted the interaction among users and the program is usually "passive." The integrated forest surveillance network is shown by Fig. 1.4. The architecture is scalable and extensible, enabling the application to easily add additional functionality or map

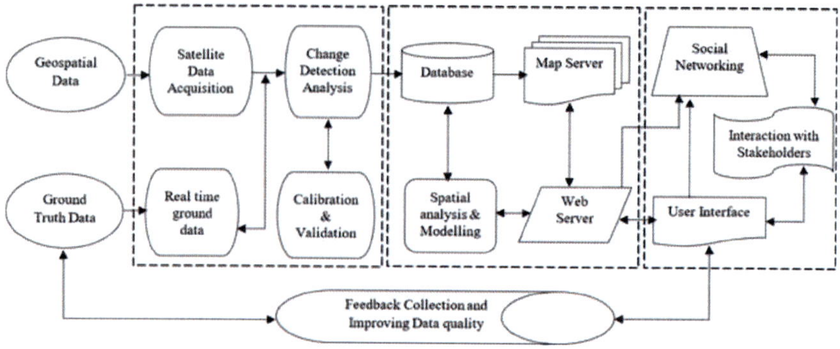

Fig. 1.4 Interactive web-based near real-time forest monitoring

layers of external data sources. The server conducts the necessary spatial analysis and provides input, interactions and visualization of the findings for the client.

The GIS virtual support program is accessible on the Internet. The key goal is to make the forest information on the spatial and non-spatial datasets on land cover and vegetation type and other forestry layers in the country accessible freely available to existing, accurate and reliable land resource users. NSDI is intended to facilitate the compilation, aggregation and dissemination by different mapping agencies of geographic datasets on various issues in a shared set of specified standards and formats. The national forest fire program is coordinated to locate fire areas through its "Easy Reading Out" service. Modest Resolution Imaging Spectroradiometer (MODIS) Data transmission and analysis is performed (Tang et al. 2019). The stakeholders are sent within 90 min to designated emergency areas (hot spots). The Indian Space Research Organization (ISRO) along with Forest Survey of India (FSI) has been conducted spatial scale and patterns in forest cover shifts in India using multi-source and multi-date data (1930–2013). This research has evaluated the spatial scale and patterns in forest cover shifts in India using multi-source and multi-date data (1930–2013). As a guide for classifying the other four periods (1975, 1985, 1995, 2005), visual interpretations have been used to evaluate the forest cover map created from the Resourcesat-2 AWiFS 2013 image for transition between forest and non-forest cover. For time series evaluation and analysis of trends in forest distribution (1930–1975, 1975–1985, 1985–1995, 1995–2005 and 2005–2013), a grid cell measuring 5 km by 5 km were developed. The e-planning program of these two American federal agencies aims to offer advanced, immersive, internet-based planning papers with similarly smart backend technologies for public commentary delivery Pratihast et al. (2016). developed an interactive web-based near real-time (NRT) forest monitoring system in Ethiopia. The functionality of the program comprises (1) the download, store, and analysis of NRT forestry changes identification by means of Landsat time series images; (2) the possibility to submit land observations and collection positions for forest change on request; (3) the ability to monitor and display the hotspot for forest changes in time. The spatial database framework was built to allow various types of data to be processed, managed and accessed via structured query language (SQL), including basic geographic data, ground observation data, and distance sensing information. This problem is tackled by a new, 80-year-old technology—radar—the Monitoring of the Andean Amazon Project (MAAP), with the use of remote sensing data to identify hotspots in the West Amazon and the activities triggered by forest depletion (https://maaproject.org/en/). In order to allow its team to track deforestation during Peru's entire year in almost real-time, MAAP now integrates high-resolution optical data with the capacity for radar imagery. Inspired by the DETER and SAD forest surveillance programs protecting Brazil's Amazon, the FORMA project originated at the World Development Center and entered the Data Lab at the World Resources Institute (https://www.globalforestwatch.org/). Such devices easily generate maps of hotspots for forest destruction. These also allowed police, civil society groups and the media to respond quickly to criminal crime and to reduce the deforestation rate in Brazil. The FORMA network consists of several components: wildfire extreme and fire data from the MODIS instrument on NASA Terra satellite, NOAA weather data and forest clearance historical data. In relation to dryness or other seasonal variability—see Fig. 1.5—a mathematical model uses increasing pixel background to identify relevant signs of the failure for wood cover. The end result is a map displaying regions of interest based on the new satellite images. The Indonesian version of Mongabay.com (mongabay.co.id) launched recently over the coming month aims to create a pilot program to investigate some deforestation hotspots in Indonesia, found by the GloF-DAS in local correspondents. The ground-based project, which is witnessing high rates of mining deforestation, the conversion to the oil palm assets and to pulp and paper plantations and agriculture, could boost forest transparency in Indonesia (Fig. 1.5). The GloF-DAS is based on a new NASA Moderate Resolution Imaging Spectroradiometer (MODIS) satellite data product.

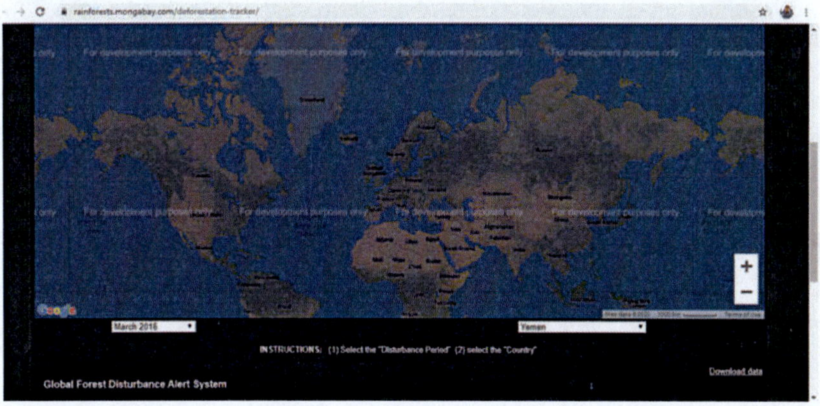

Fig. 1.5 The Global Forest Disturbance Alert System (GloFDAS) provides data on forest disturbance globally on a quarterly basis. GloF-DAS is freelyavailable at rainforests.mongabay.com/deforestation-tracker/

1.7 Crowd Sourcing in Forest Management

Jeff Howe and Mark Robinson, editors of the Wired Journal, coined the word 'crowdsourcing' in 2005 for the purpose of representing the activities of a global band of people for information, ideas and services (World Meteorological Organization 2017). The terms "crowd" and "outsource" were a mixture of phrases. Crowdsourcing is the type of participatory online operation that provides a community of people with diverse skills, complexity and number with ample resources to volunteer their work through a scalable, accessible request. The generation of geosphere data by volunteer people, who are untrained in astronomy, mapping or related fields is spatial crowdsourcing (Heipke 2010).

Crowdsourcing was introduced in the forest industry to test urban trees even though there have been some concerns about their durability (Fritz et al. 2009). Current technology makes it possible for machines to identify the forest area automatically by means of satellite data and chart most forestry worldwide accurately. The forest cover in a pixel is observed by conventional remote sensing techniques instead of trapping individual trees in the landscape. Within the less dense woodland or in individual trees, as is the essence of the drylands most commonly, the approach can be overlooked. Google Earth receives satellite data from many satellites with various technological capacities and resolutions (Fig. 1.6). The array of Google's dryland satellite imagery from a variety of providers like Digital Globe is especially high, as deserts are cloud-free. While the identification of non-dominant ground cover is difficult for algorithms, the human eyes do not have a problem identifying trees in landscapes.

A program that supports organizations involved in REDD+ and forest discussion was created during the second half of 2015 by the European Space Agency (ESA). The REDD+ is a global initiative to encourage countries to reduce CO_2 emissions, encourage forest restoration and sustainable land management, and increase the sum of land carbon stocks. Emissions from forest destruction is a global initiative. To order to figure out illicit deforestation and to monitor the condition of trees, forest management is important for REDD+. The monitoring of forests from the land, however, requires time. Certain regions, particularly in locations where the resources and records are scarce, are also difficult to protect.

1.8 Crisis Mapping of Forest Cover

The crisis map is a real-time data collection, view and review of a crisis of specific severity of growth, including financial, social and environmental data. Crisis mapping allows a vast range of individuals, even indirectly, to monitor crisis response by sharing information. "Crisis mapping can be defined as the three-part combinations—information compilation, display and analysis—as defined by Patrick Meier (Meyer 2011). All are stored in a dynamic, interactive map. Disaster maps are used for the

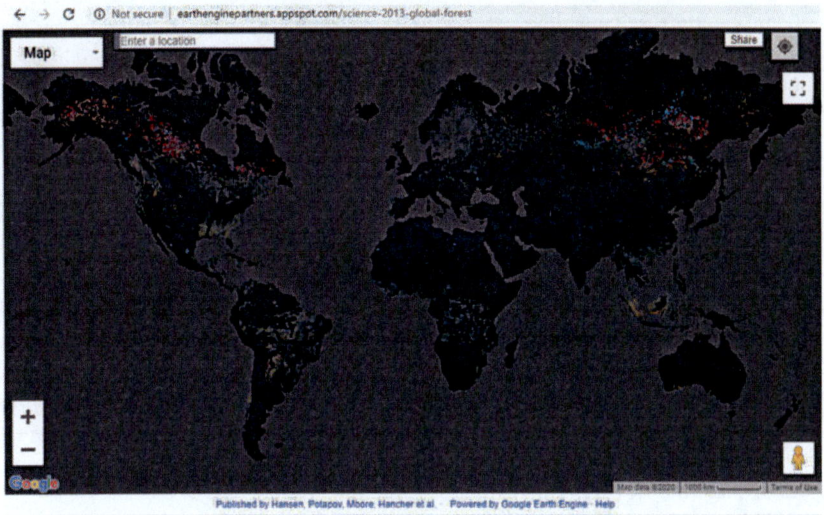

Fig. 1.6 Forest map showing Forest Loss in the World in 2018. Reference 2000 and 2018 imagery are median observations from a set of quality assessment-passed growing season observations. Trees are defined as vegetation taller than 5 m in height and are expressed as a percentage per output grid cell as '2000 Percent Tree Cover'. 'Forest Cover Loss' is defined as a stand-replacement disturbance, or a change from a forest to non-forest state, during the period 2000–2018. 'Forest Cover Gain' is defined as the inverse of loss, or a non-forest to forest change entirely within the period 2000–2012. 'Forest Loss Year' is a disaggregation of total 'Forest Loss' to annual time scales (*Source* https://earthenginepartners.appspot.com/science-2013-global-forest)

development in the field of early warning systems and disaster-response services by means of real-time, crowd-sourced crisis data, satellite imagery, data analytics, computer analysis and web-based applications. A non-profit, open-source development organization, for instance, is building the Ushahidi Web site. Ushahidi will disseminate and gather data on a situation in any region of the world. Users can provide information through email, e-mail or web pages, and the data can be aggregated to form a map or timeline. Google Crisis Response is a Google.org department that "sees to make knowledge critical of natural disasters and humanitarian emergencies more available." Google that respond: updating the disaster area satellite imagery, developing an emergency information and resources web page, hosting a crisis map featuring accurate and crowd-based geographic information, contributing charity to on site organisations, engineering products and information services, such as Google's person-based Finder and landing pages. The humanitarian OpenStreet Map Team (HOT) manages free mapping tools in several locations worldwide to develop, produce and distribute them. The HOT delivers "free, up-to-date maps" as a "critical resource for emergency relief agencies and global emergencies". Although the dataset is available in a format which the technically knowledgeable can most probably use, users will generate derivative charts, views and aggregations.

More than 50 football fields of the forest are damaged every minute, as per the World Community Institute. 20.8 million acres of woodland were overlooked in 2012 alone (over 80,000 square miles).Global Forest Monitor is a program that combines satellite technologies, open data, and crowdfunding to enhance policymaking in real time. Global Forest Watch (https://www.globalforestwatch.org/) was originally founded by the World Bank Institute as an initiative in 2007. This was a way of combining the latest up-to-day technologies with collaborations between various countries to create a global forest surveillance network. In an attempt to foster global forest accessibility, Global Forest Watch merges real-time satellite technologies, web devices, crowd-sourced info, forest maps, protected area maps and on-the-ground networks. For example, Global Forest Watch (GFW) is an immersive global forest monitoring and alert program that offers updates on forests all over the world in real time. Forest image processing and cloud storage capability, GFW using satellite technologies in building exchanging distributed data sets on trees (Fig. 1.7). Forest data collection, it was also developed as a resource for crowd sourcing that allows users to share their own findings directly from the ground.

Rapid growth in ICT (Internet), cloud computing, social networks (including mobile telephony) in recent years has revolutionized people's way of communicating and sharing knowledge with one another. The emergence of smartphones and open Internet connectivity has contributed significantly to the availability of vast information for the public. The ability of crowd-sourced knowledge to provide consumers with a better view of the importance of shopping plans is used by popular webpages such as TripAdvisor, Amazon, eBay and the new e-commerce sites. At the other hand, Wikipedia, Flickr and OpenStreetMap provide an outstanding example of the modern world where crowdsourcing has provided a huge resource of resources for organisations and people all over the globe to continue to use. The concept of the digital world also reveals how important people are as data sources and participants to daily life. The importance of citizen-generated crowd-data is expressed in Digital

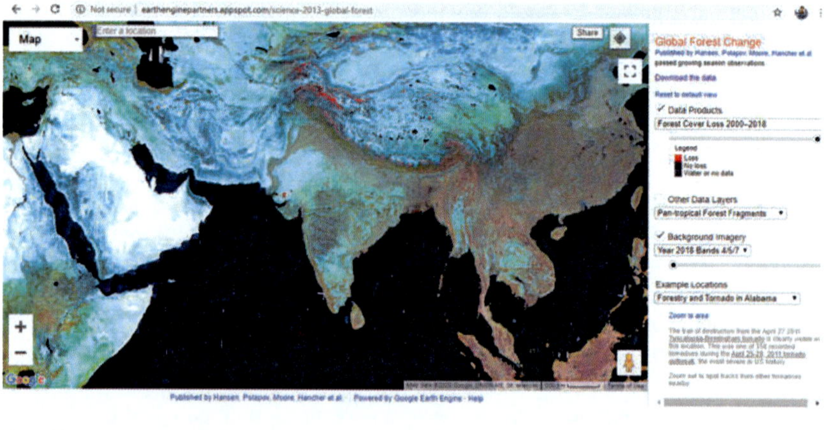

Fig. 1.7 Forest change of Indian sub-continent (*Source* Hansen et al. 2013)

World (Craglia et al. 2012). The Digital Earth vision addresses the key government, science and social forces that make it possible to locate, imagine and interpret vast quantities of data from a "DigitalEarth" as a multi-resolution, three-dimensional image of the earth.

1.9 Conclusion

Forests and forest ecosystems are evolving more as a result of natural and human transition. And if individuals will not participate, they are still subject to tangible improvements. Conservation importance can be established by the use of multiple ecological landscape criteria, as well as stringent soil monitoring for areas with biological value, highlights, warm spots and hot specks. In global efforts to mitigate greenhouse gas emissions, the protection of carbon stocks, particularly in peat bog forests and mangrove areas, should be a key priority. In the current phase of global economic growth, afforestation, reforestation and the restoration of the natural environment will concurrently provide jobs. Specific challenges are compounded by the geographical and temporal aspects of sustainable forest management. A global and multidisciplinary initiative is required to minimize habitat degradation and to boost public consciousness, increase forestry law enforcement initiatives, increase support for conservation areas and introduce environmental protections in tandem with construction activities. Nevertheless, for the broader application of sustainable forestry management, the relative weakness of the forest legislation to tackle equity concerns remains a major problem. The roles of forest habitats and the use of wood in Anthropocene are evolving. Land managers need new method to understand and change management strategies using revised information. The rapid speed of transition will intensify the need in both natural and managed forests to identify ecological responses over broad scales. There is a commitment to track forest conditions in near real time that substantial progress is being made with the remote sensing-based change detection and Tracking system; it is, however, unclear if this tool is sufficient for quantification and review of the efficacy of management activities. New insights must also be easily incorporated into management decisions and user-friendly predictive model. The level of experiential awareness, insights and data collection can be increased by emerging approaches to gathering and dissemination of information such as citizen science networks and 'crowd sourcing' methods in order to complement written information. Eventually, over the past few decades, upgrades to data acquisition, storage, and access processes have provided vast volumes of readily accessible data, allowing large-scale "big data" biogeochemical pattern analyses. For potential control of forest resources, forest ecosystem services and mixed natural-human processes will be considered for trade-offs. With the purposes of prevention and transition to the adverse impacts of natural and anthropogenic disturbances that will escalate in Anthropocene, current best management practices (BMPs) should have been reviewed.

Acknowledgements We are grateful to the PG Department of Geography, Raja N. L. Khan Women's College (Autonomous), affiliated to Vidyasagar University, Midnapore, West Bengal, India for supporting this research. The author (P. K. Shit) acknowledged West Bengal DSTBT for financial support through R&D Research Project Memo no. 104(Sanc.)/ST/P/S&T/10G-5/2018).

References

Abdollahnejad A, Panagiotidis D, Surový P (2018) Estimation and extrapolation of tree parameters using spectral correlation between UAV and Pléiades data. Forests 9(2):85. https://doi.org/10.3390/f9020085

Abraao MB, Nelson BW, Baniwa JC, Yu DW, Shephard GH Jr (2008) Ethnobotanical ground-truthing: indigenous knowledge, floristic inventories and satellite imagery in the upper Rio Negro, Brazil. J Biogeogr 35:2237–2248

Allen CD, Macalady AK, Chenchouni H, Bachelet D, McDowell N, Vennetier M, Kitzberger T, Rigling A, Breshears DD, Hogg EH, Gonzalez P, Fensham R, Zhang Z, Castro J, Demidova N, Lim J-H, Allard G, Running SW, Semerci A, Cobb N (2010) A global overview of drought and heat-induced tree mortality reveals emerging climate change risks for forests. For Ecol Manag 259:660–684

Anderegg WR, Hicke JA, Fisher RA, Allen CD, Aukema J, Bentz B et al (2015) Tree mortality from drought, insects, and their interactions in a changing climate. New Phytol 208:674–683

Baccini A, GP Asner (2013) Improving pantropical forest carbon maps with airborne LiDAR sampling. Carbon Manag 4

Balram S, Dragićević S, Meredith T (2004) A collaborative GIS method for integrating local and technical knowledge in establishing biodiversity conservation priorities. Biodivers Conserv 13:1195–1208

Barducci A, Guzzi D, Marcoionni P, Pippi I (2002) Infrared detection of active fires and burnt areas: theory and observations. Infrared Phys Technol 43:119–125

Bastin J-F, Barbier N, Couteron P, Adams B, Shapiro A, Bogaert J, De Cannière C (2014) Above-ground biomass mapping of African forest mosaics using canopy texture analysis: toward a regional approach. Ecol Appl 24:1984–2001

Bauer-Marschallinger B, Sabel D, Wagner W (2014) Optimisation of global grids for high-resolution remote sensing data. Comput. Geosci. 72:84–93

Bernier P, Schöne D (2009) Adapting forests and their management to climate change: an overview. Unasylva 60:5–11

Boissière M, Beaudoin G, Hofstee C, Rafanoharana S (2014) Participating in REDD+ measurement, reporting, and verification (PMRV): opportunities for local people? Forests 5(8):1855–1878

Brockerhoff EG, Jactel H, Parrotta JA, Ferraz SF (2013) Role of eucalypt and other planted forests in biodiversity conservation and the provision of biodiversity-related ecosystem services. For Ecol Manage 301:43–50

Calders K, Jonckheere I, Nightingale J, Vastaranta M (2020) Remote sensing technology applications in forestryand REDD+. Forests 11:188. https://doi.org/10.3390/f11020188

Carter N (2013) An assessment of worldview-2 imagery for the classification of a mixed deciduous forest. Rochester Institute of Technology; College of Science: Thomas H. Gosnell School of Life Sciences; Program of Environmental Science, Rochester, NY, p 61

Choat B, Jansen S, Brodribb TJ, Cochard H, Delzon S, Bhaskar R, Bucci SJ, Feild TS, Gleason SM, Hacke UG, Jacobsen AL, Lens F, Maherali H, Martinez-Vilalta J, Mayr S, Mencuccini M, Mitchell PJ, Nardini A, Pittermann J, Pratt RB, Sperry JS, Westoby M, Wright IJ, Zanne AE (2012) Global convergence in the vulnerability of forests to drought. Nature 491:752–755

Clark DB, Castro CS, Alvarado LDA, Read JM (2004) Quantifying mortality of tropical rain forest trees using high-spatial-resolution satellite data. Ecol Lett 7(1):52–59

Conrad CC, Hilchey KG (2011) A review of citizen science and community-based environmental monitoring: issues and opportunities. Environ Monit Assess 176(1):273–291

Corbett J, Keller P, Kyem PAK, Rambaldi G, Weiner D, Olson R, Muchemi J, McCall M, Chambers R (2006) Mapping for change: practice, technologies and communication. Participatory Learn Action 54:1–13

Craglia M, de Bie K, Jackson D, Pesaresi M, Remetey-Fülöpp G, Wang C, Annoni A, Bian L, Campbell F, Ehlers M et al (2012) Digital Earth 2020: towards the vision for the next decade. Int J Digit Earth 5:4–21

Csillik O, Kumar P, Mascaro J et al (2019) Monitoring tropical forest carbon stocks and emissions using Planet satellite data. Sci Rep 9:17831. https://doi.org/10.1038/s41598-019-54386-6

Danielsen F, Burgess ND, Balmford A, Donald PF, Funder M, Jones JPG et al (2009) Local participation in natural resource monitoring: a characterization of approaches. Conserv Biol 23(1):31–42

Detto M, Muller-Landau HC, Mascaro J, Asner GP (2013) Hydrological networks and associated topographic variation as templates for the spatial organization of tropical forest vegetation. PLoS ONE 8:e76296

Devaney J, Barrett B, Barrett F, Redmond J, O Halloran J (2015) Forest cover estimation in Ireland using radar remote Sensing: a comparative analysis of forest cover assessment methodologies. PLoS One 10(8):e0133583. Published 2015 Aug 11. https://doi.org/10.1371/journal.pone.0133583

DeVries B, Herold M (2013) The science of Measuring, Reporting and Verication (MRV). In: Lyster R, MacKenzie C, McDermott C (eds) Law, Tropical forests and carbon: the case of REDD+. Cambridge University Press, Cambridge, pp 151–183

FAO, JRC (2012) Global forest land-use change 1990–2005. In: Lindquist EJ, D'Annunzio R, Gerrand A, MacDicken K, Achard F, Beuchle R, Brink A, Eva HD, Mayaux P, San-Miguel-Ayanz J, Stibig H-J (eds) FAO Forestry Paper No 169. Food and Agriculture Organization of the United Nations, European Commission Joint Research Centre. FAO, Rome

Ferrazzoli P, Guerriero L (1996) Passive microwave remote sensing of forests: a model investigation. IEEE Trans Geosci Remote Sens 34(2):433–443

Fritz S, McCallum I, Schill C, Perger, C, Grillmayer R, Achard F, Kraxner F, Obersteiner M (2009) Geo-Wiki.Org: the use of crowdsourcing to improve global land cover. Remote Sens 1:345–354

Giannetti F, Gottero F, Terzuolo PG (2003) Use of high resolution satellite images in the forest inventory and mapping of Piemonte Region (Italy). In: Corona P, Köhl M, Marchetti M (eds) Advances in forest inventory for sustainable forest management and biodiversity monitoring. Forestry sciences, vol 76. Springer, Dordrecht

Gibbs HK, Brown S, Niles JO, Foley JA (2007) Monitoring and estimating tropical forest carbon stocks: making REDD a reality. Environ Res Lett 2(4):045023

Gibbs HK, Ruesch AS, Achard F, Clayton MK, Holmgren P, Ramankutty N, Foley JA (2010) Tropical forests were the primary sources of new agricultural land in the 1980s and 1990s. Proc Natl Acad Sci 107:16732–16737

Gillis M, Omule A, Brierley T (2005) Monitoring Canada's forests: the national forest inventory. The Forestry Chronicle 81(2):214–221

Global Forest Observations Initiative (2014) Integrating remote-sensing and ground-based observations for estimation of emissions and removals of greenhouse gases in forests: methods and guidance from the Global Forest Observations Initiative. Group on Earth Observations, Geneva

Goward SN, Davis PE, Fleming D, Miller L, Townshend JRG (2003) Empirical comparison of Landsat 7 and IKONOS multispectral measurements for selected Earth Observation System (EOS) validation sites, Remote Sens Environ 88:196–209

Guariguata MR, Cornelius JP, Locatelli B, Forner C, Sánchez-Azofeifa GA (2008) Mitigation needs adaptation: tropical forestry and climate change. Mitig Adapt Strateg Glob Chang 13:793–808

Guo H (2017) Big earth data: a new frontier in earth and information sciences. Big Earth Data 1:4–20

Hansen MC, Potapov PV, Moore R, Hancher M, Turubanova SA, Tyukavina A, Thau D, Stehman SV, Goetz SJ, Loveland TR, Kommareddy A, Egorov A, Chini L, Justice CO, Townshend JRG (2013) High-resolution global maps of 21st-century forest cover change. Science 342(6160):850–853

Heipke C (2010) Crowdsourcing geospatial data. ISPRS-J Photogramm Remote Sens 65:550–557

Heurich M, Thoma F (2008) Estimation of forestry stand parameters using laser scanning data in temperate, structurally rich natural European beech (Fagus sylvatica) and Norway spruce (Picea abies) forests. Forestry 81(5):645–661. https://doi.org/10.1093/forestry/cpn038

Hosoi F, Nakai Y, Omasa K (2013) 3-D voxel-based solid modeling of a broad-leaved tree for accurate volume estimation using portable scanning lidar. ISPRS J Photogrammetry Remote Sens 82:41–48

Houghton RA, Hall F, Goetz SJ (2009) Importance of biomass in the global carbon cycle. J Geophys Res 114:2156–2202

Hulme PE (2009) Trade, transport and trouble: managing invasive species pathways in an era of globalization. J Appl Ecol 46:10–18

Huntingford C, Zelazowski P, Galbraith D, Mercado LM, Sitch S, Fisher R, Lomas M, Walker AP, Jones CD, Booth BBB, Malhi Y, Hemming D, Kay G, Good P, Lewis SL, Phillips OL, Atkin OK, Lloyd J, Gloor E, Zaragoza-Castells J, Meir P, Betts R, Harris PP, Nobre C, Marengo J, Cox PM (2013) Simulated resilience of tropical rainforests to CO_2-induced climate change. Nat Geosci 6:268–273

Illera P, Fernandez A, Delgado J (1996) Temporal evolution of the NDVI as an indicator of forest fire danger. Int J Remote Sens 17(6):1093–1105

Ince PJ, Kramp AD, Skog KE, Yoo DI, Sample VA (2011) Modeling future U.S. forest sector market and trade impacts of expansion in wood energy consumption. J For Econ 17:142–156

Jactel H, Petit J, Desprez-Loustau ML, Delzon S, Piou D, Battisti A et al (2012) Drought effects on damage by forest insects and pathogens: a meta-analysis. Glob Change Bio. 18:267–276

Ke Y, Quackenbush LJ (2007) Forest species classification and tree crown delineation using Quick-Bird imagery. In: ASPRS 2007 Annual ConferenceTampa, Florida. Available at: https://www.asprs.org/wp-content/uploads/2011/01/0037.pdf

Kleinn C (2002) New technologies and methodologies for national forest inventories. Unasylva 210

Kurvonen L, Pulliainen J, Hallikainen M (2002) Active and passive microwave remote sensing of boreal forests. Acta Astronaut 51(10):707–713

Laes D, Reutebuch SE, McGaughey RJ, Mitchell B (2011) Guidelines to estimate forest inventory parameters from lidar and field plot, companion document to the advanced lidar applications—forest inventory modeling class, pp 1–22. Available at: https://pdfs.semantic-scholar.org/8979/90a39c23a868fb07787b022e8ee290f06878.pdf

Lenoir J, Gegout JC, Dupouey JL, Bert D, Svenning JC (2010) Forest plant community changes during 1989–2007 in response to climate warming in the Jura Mountains (France and Switzerland). J Veg Sci 21:949–964

Li G, Huang Z (2017) Data infrastructure for remote sensing big data: integration, management and on-demand service. JisuanjiYanjiu Yu Fazhan/Comput Res Dev 54:267–283

Li YZ, Anderson H-E, McGaughey R (2008) A comparison of statistical methods for estimating forest biomass from light detection and ranging data. West J Appl For 23:223–231

Lucier A, Ayres M, Karnosky D, Thompson I, Loehle C, Percy K, Sohngen B (2009) Forest responses and vulnerabilities to recent climate change. In: Seppälä R, Buck A, Katila P (eds) Adaptation of forests and people to climate change: a global assessment report, vol World Series Volume 22. IUFRO Helsinki, pp 29–52

LVG (2012) DigitaleGeländemodelle (DGM). Product information of the Bavarian Office for Surveying and Geographic Information (LandesamtfürVermessung und Geoinformation Bayern). https://vermessung.bayern.de/file/pdf/1614/download_faltblatt-dgm09.pdf. Accessed on 11 Dec 2012

Lynch J, Maslin M, Balzter H, Sweeting M (2013) Choose satellites to monitor deforestation. Nature 496(7445):293–294
Managi S, Wang J, Zhang L (2019) Research progress on monitoring and assessment of forestry area for improving forest management in China. For Econ Rev 1(1):57–70
Maniatis D, Mollicone D (2010) Options for sampling and stratification for national forest inventories to implement REDD + under the UNFCCC. Carb Bal Manag 5:1–9
Mapedza E, Wright J, Fawcett R (2003) An investigation of land cover change in Mafungautsi Forest, Zimbabwe, using GIS and participatory mapping. Appl Geogra 23:1–21
McEvoy D, Fünfgeld H, Bosomworth K (2013) Resilience and climate change adaptation: the importance of framing. Plan Pract Res 28:280–293
Meier P (2011) What is crisis mapping? An update on the field and looking ahead. iRevolutions.irevolution.net/2011/01/20/what-is-crisis-mapping/
Melkas T, Vastaranta M, Holopainen M, Hill R, Rosette J, Suárez J (2008) In accuracy and efficiency of the laser-camera. In: Proceedings of SilviLaser 2008, 8th international conference onLiDAR applications in forest assessment and inventory, Heriot-Watt University, Edinburgh, UK, 17–19 September 2008; SilviLaser 2008 Organizing Committee: Edinburgh, UK, pp 315–324.
Meng R, Wu J, Schwager KL, Zhao F, Dennison PE, Cook BD, Brewster K, Green TM, Serbin SP (2017) Using high spatial resolution satellite imagery to map forest burn severity across spatial scales in a Pine Barrens ecosystem. Remote Sens Environ 191:95–109
Mermoz S, Réjou-Méchain M, Villard L, Le Toan T, Rossi V, Gourlet-Fleury S (2015) Decrease of L-band SAR backscatter with biomass of dense forests. Remote Sens Environ 159:307–317
Mitchell AL, Rosenqvist A, Mora B (2017) Current remote sensing approaches to monitoring forest degradation in support of countries measurement, reporting and verification (MRV) systems for REDD+. Carbon Balance Manage 12:9. https://doi.org/10.1186/s13021-017-0078-9
Morsdorf F, Koetz B, Meier E, Itten KI, Allgöwer B (2006) Estimation of LAI and fractional cover from small footprint airborne laser scanning data based on gap fraction. Remote Sens Environ 104:50–61
Nagendra H (2001) Using remote sensing to assess biodiversity. Int J Remote Sens 22:2377–2400
Paletto A, De Meo I, Ferretti F (2010) Social network analysis to support the forest landscape planning: an application in arci-grighine, Sardinia (Italy). Forestry Ideas 16 1(39):28–35
Popescu SC, Wynne RH (2004) Seeing the trees in the forest: using lidar and multispectral data fusion with local filtering and variable window size for estimating tree height. Photogramm Eng Remote Sens 70:589–604
Pratihast AK, Souza Jr CM, Herold M, Ribbe L (2012) Application of mobile devices for community-based forest monitoring. Sensing a Changing World, 1–6
Pratihast AK, DeVries B, Avitabile V, de Bruin S, Herold M, Bergsma A (2016) Design and implementation of an interactive web-based near real-time forest monitoring system. PLoS One 11(3):e0150935. Published 2016 Mar 31. https://doi.org/10.1371/journal.pone.0150935
Rambaldi G, McCall M, Weiner D, Kyem PAK (2006) Participatory spatial information management and communication in developing countries. Electron J Inf Syst Dev Countries 2006: 1, 2, 6,
Robiglio V, Mala WA (2005) Integrating local and expert knowledge using participatory mapping and GIS to implement integrated forest management options in Akok, Cameroon. The Forestry Chronicle 81(3):392–397
Roy PS, Diwakar PG, Singh IJ et al (1994) Evaluation of microwave remote sensing data for forest stratification and canopy characterisation. J Ind Soc Remote Sens 22:31–44. https://doi.org/10.1007/BF03015118
Seidl R, Rammer W, Lexer MJ (2011) Adaptation options to reduce climate change vulnerability of sustainable forest management in the Austrian Alps. Can J For Res 41(4):694–706. https://doi.org/10.1139/x10-235
Simard M, Pinto N, Fisher JB, Baccini A (2011) Mapping forest canopy height globally with spaceborne lidar. J Geophys Res 116:G04021. https://doi.org/10.1029/2011jg001708

Skarlatidou A, Haklay M, Cheng T (2011) Trust in Web GIS: the role of the trustee attributes in the design of trustworthy Web GIS applications. Int J Geogr Inf Sci 25(12):1913–1930. https://doi.org/10.1080/13658816.2011.557379

Straub C, Tian J, Seitz R, Reinartz P (2013) Assessment of Cartosat-1 and WorldView-2 stereo imagery in combination with a LiDAR-DTM for timber volume estimation in a highly structured forest in Germany. For Int J For Res 86(4):463–473. https://doi.org/10.1093/forestry/cpt017

Sulistyawan BS, Verweij PA, Boot RGA, Purwanti B, Rumbiak W, Wattimena MC, Rahawarin P, Adzan G (2018) Integrating participatory GIS into spatial planning regulation: the case of Merauke District, Papua, Indonesia. Int J Commons 12(1):26–59

Sweeney RW (2009) There's no 'I' in Youtube: socialmedia, networked identity and art education. Int J Educ through Art 5(2 and 3):201–212

Tang X, Bullock EL, Olofsson P, Estel S, Woodcock CE (2019) Near real-time monitoring of tropical forest disturbance: new algorithms and assessment framework. Remote Sens Environ 224:202–218

Thiel C, Drezet P, Weise C, Quegan S, Schmullius C (2006) Radar remote sensing for the delineation of forest cover maps and the detection of deforestation. For Int J For Res 79(5):589–597

Thurner M, Beer C, Santoro M, Carvalhais N, Wutzler T, Schepaschenko D, Shvidenko A, Kompter E, Ahrens B, Levick SR, Schmullius C (2014) Carbon stock and density of northern boreal and temperate forests. Glob Ecol Biogeogr 23:297–310

Tomppo E, Olsson H, Ståhl G, Nilsson M, Hagner O, Katila M (2008) Combining national forest inventory field plots and remote sensing data for forest databases. Remote Sens Environ 112:1982–1999

Trumbore S, Brando P, Hartmann H (2015) Forest health and global change. Science 349:814–818

Uddin K, Gilani H, MurthyMSR, Kotru R, Qamer FM (2015) Forest condition monitoring using very-high-resolution satellite imagery in a remote mountain watershed in Nepal. Mountain Res Dev 35(3):264–277

UNFCCC (2009) 4/CP.15 Methodological guidance for activities relating to reducing emissions from deforestation and forest degradation and the role of conservation, sustainable management of forests and enhancement of forest carbon stocks in developing countries. FCCC/CP/2009/11/Add1. Report of the Conference of the Parties on its fifteenth session, held in Copenhagen from 7 to 19 December 2009

Vajjhala SP (2005) Integrating GIS and participatory Mapping in community development planning. ESRI International User Conference, Sustainable Development and Humanitarian Affairs Track, San Diego, CA, July 2005. Available at: https://proceedings.esri.com/library/-userconf/proc05/papers/pap1622.pdf

Vastaranta M, Melkas T, Holopainen M, Kaartinen H, Hyyppä J, Hyyppä H (2009) Laser-basedfield measurements in tree-level forest data acquisition. Photogramm J Finl 21:51–61

Vastaranta M, Latorre EG, Luoma V, Saarinen N, Holopainen M, Hyyppä J (2015) Evaluation of a smartphone App for forest sample plot measurements. Forests 6:1179–1194. https://doi.org/10.3390/f6041179

Verlič A, Đurić N, Kokalj Z, Marsetič A, Simončič P, Oštir K (2014) Tree species classification using WorldView-2 Satellite images and laser scanning data in a natural urban forest. Preliminary CommunicationŠumarski List 9–10:477–488

Wongkoblap A, Vadillo MA, Curcin V (2017) Researching mental health disorders in the era of social media: systematic review. J Med Internet Res 19(6):e228

Woodhouse IH, Mitchard ET, Brolly M, Maniatis D, Ryan CM (2012) Radar backscatter is not a'directmeasure'of forest biomass. Nat Clim Chang 2:556–557

World Meteorological Organization (2017) Integrated flood management tools series-crisis mapping and crowdsourcing in flood management, No. 26 version 1.0. Available at: https://www.floodmanagement.info/publications/tools/APFM-_Tool_26_e.pdf

Xia J, Yang C, Li Q (2018) Building a spatiotemporal index for earth observation big data. Int J Appl Earth Obs 73:245–252

Yang C, Goodchild M, Huang Q, Nebert D, Raskin R, Xu Y, Bambacus M, Fay D (2011) Spatial cloud computing: how can the geospatial sciences use and help shape cloud computing? Int J Digit Earth 4:305–329

Yao T, Yang X, Zhao F, Wang Z, Zhang Q, Jupp D, Lovell J, Culvenor D, Newnham G, Ni-Meister W, Schaaf C, Woodcock C, Wang J, Li X, Strahler A (2011) Measuring forest structure and biomass in New England forest stands using Echidna ground-based lidar. Remote Sens Environ 115:2965–2974. https://doi.org/10.1016/j.rse.2010.03.019

Yao X, Li G, Xia J, Ben J, Cao Q, Long Zhao L, Ma Y, Zhang L, Zhu D (2020) Enabling the big Earth observation data via cloud computing and DGGS: opportunities and challenges. Remote Sens 12(1):62

Zhang Z (2017) Adoption of airborne LiDAR data and high spatial resolution satellite imagery for characterisation and classification of forest communities: tests applications in Australian cool temperate rainforest environment. Thesis. https://doi.org/10.4225/03/5897df9948341

Zhao X, Ben J, Sun W, Tong X (2016) Overview of the research progress in the earth tessellation grid. CehuiXuebao/Acta Geodaetica et CartographicaSinica 45:1–14

Chapter 2
Estimation of Net Primary Productivity: An Introduction to Different Approaches

Pramit Kumar Deb Burman

Abstract The net primary productivity (*NPP*) is defined as the net carbon gain by plants in natural and agricultural ecosystems, which is computed by subtracting the autotrophic respiration from the gross photosynthetic carbon uptake by the ecosystems. It acts as the indicators of carbon sequestration, ecosystem health, and agricultural yield which are important in the context of climate change, its impact and mitigation, and food security. The *NPP* can be estimated in multiple ways including the direct and indirect measurements and modelling. The various direct *NPP* measurements are ground-based in situ observations of ecosystem-atmosphere carbon flux such as the micrometeorological flux-gradient method, eddy covariance, flux chamber measurements etc. The indirect measurements of *NPP* include the satellite-derived *NPP* estimates which are computed from the directly measured spectral reflectances, using different biophysical relations such as the light use efficiency model etc. However the accuracy of these products varies geospatially and largely depends on the retrieval of input parameters and representativeness of underlying model parameterization. There are two major modelling approaches to estimate the *NPP* namely bottom-up and top-down estimates. The bottom-up models compute the *NPP* from the directly recorded variables such as temperature, precipitation, radiation, wind, atmospheric CO_2 concentration etc. using the biome-specific functional relations due to which these are also known as the process-based models. The top-down or inverse models use the matrix inversion method to predict the sources and sinks of CO_2 emission in a region from the directly measured concentrations by the surface stations and/or satellites and thus the *NPP* of that region. The *NPP* estimates from measurements and models are used to calculate the carbon budgets at different scales from ecosystem-level to global scale. However significant

P. K. Deb Burman (✉)
Centre for Climate Change Research, Indian Institute of Tropical Meteorology, Ministry of Earth Sciences, Pune 411008, India
e-mail: pramit.cat@tropmet.res.in; pkdebburman@gmail.com

Department of Atmospheric and Space Sciences, Savitribai Phule Pune University, Pune 411007, India

P. K. Shit et al. (eds.), *Spatial Modeling in Forest Resources Management*, Environmental Science and Engineering,
https://doi.org/10.1007/978-3-030-56542-8_2

uncertainties exist in such estimates due to insufficient surface measurements, under-representation of several regions and ecosystems, imperfect boundary conditions and parameterizations in models. While the direct measurements provide more accurate estimates of *NPP*, these require to be carried over for long duration using multiple different instruments which are prone to errors and data-loss whereas the models can provide large-scale estimates of *NPP* but need to be validated against realistic in situ measurements across an wide array of ecosystems. The aforementioned aspects of *NPP* estimation are discussed in detail in the present chapter.

Keywords Carbon cycle · Terrestrial ecosystems · Gross primary productivity · Net ecosystem productivity · Eddy covariance · Ecosystem models · Inverse models · Vegetation indices

2.1 Introduction

Since the industrial revolution the atmospheric CO_2 concentration (c_a in ppm) has risen from ~280 to 416 ppm (till the time of writing this book chapter) in an unprecedented rate (https://www.esrl.noaa.gov/gmd/ccgg/trends/mlo.html). Due to the increased energy demand the carbon stored in fossil fuel deposits has been burnt and released into the atmosphere. To meet the food and fibre demands of an increasing population the rapid agricultural expansion has taken place at the cost of natural ecosystems such as forests. The increased amount of atmospheric greenhouse gases of which CO_2 is a major component, is changing the Earth's radiation feedback resulting in global warming and paving the way for climate change (IPCC 2013).

Such climate change is predicted to have adverse effects on the Earth such as abrupt changes in atmospheric and oceanic circulation patterns, polar ice cap melt, sea-level rise, shifting treeline, increased frequency and intensity of extreme events etc. In order to devise the climate change mitigation strategies the sources and sinks of atmospheric CO_2 need to be identified and their strengths and patterns need to be characterised. The terrestrial ecosystems play a regulatory role in the Earth's radiation budget due to their roles in determining the surface albedo and photosynthetic carbon uptake (Betts 2000). According to the global carbon budget 2019 (Friedlingstein et al. 2019) the terrestrial ecosystems were the largest sink of atmospheric CO_2 in the latest decade during 2009–2018 with a sink-strength of 3.2 ± 0.6 GtC y^{-1}. Several land-based mitigation strategies are designed based on these ecosystems such as afforestation and reforestation, biochar, bioenergy with carbon capture and storage etc. (Minx et al. 2018). Moreover proper quantification of the carbon cycles of these ecosystems is important to estimate the intended nationally determined contribution (INDC) of the nations in compliance with the Paris climate accord.

The changes in c_a, trends of air temperature and precipitation due to climate change will modify the capacity and pattern of the photosynthetic carbon uptake by terrestrial ecosystems (IPCC 2019). The response of the ecosystems to such changes remains uncertain which is required for the climate change impact assessment. For this purpose the long-term observations of terrestrial carbon cycle are required. It is predicted that in changed environmental conditions the indigenous plant species are at high risk to be replaced by the more sturdy invasive species, thus resulting in the extinction of species and loss of biodiversity (Bongaarts 2019). The accelerated c_a is also predicted to cause forest dieback in several regions (Cox et al. 2004).

In this contest, the present chapter is aimed at the estimation of net primary productivity (*NPP*), a component of the carbon cycle and its significance. The different observation and modelling techniques to achieve this are discussed in the subsequent sections. In compliance with the theme of this book, the contents of this chapter are restricted to the forest ecosystems. It is to be noted that *NPP* estimation of aquatic ecosystems such as marine phytoplankton is a different topic and not discussed here. Also the agricultural ecosystems are not included as those are not considered as natural ecosystems. Considering the wide span of the subject matter and a vast amount of literature existing on the different aspects of the same, I have not tried to make this chapter as a comprehensive review article but as an indicative document on the progress done in this field and its present status. The future directions from here are also discussed briefly towards the end. This topic being an interdisciplinary one, care has been taken in the formulation of the chapter to make it apprehensible to the potential readers who belong to different academic and professional backgrounds.

2.2 Data and Modelling

2.2.1 The Carbon Cycle Components

The flux of any variable is defined as the amount of that variable exchanged across a unit surface per unit time. In this regard the vertical CO_2 flux (F_c) between the ecosystem and atmosphere is the measure of carbon exchanged between the ecosystem and atmosphere, also known as the net ecosystem exchange (*NEE*). It is a resultant of the photosynthetic uptake and respirative loss of carbon which are defined as the gross primary productivity (*GPP*) and total ecosystem respiration (*TER*) respectively. The *TER* is comprised of respired carbon fluxes by the autotrophs and heterotrophs which are defined as autotrophic respiration (R_A) and heterotrophic respiration (R_H) respectively. As per the meteorological convention the negative and positive values of *NEE* stand for the carbon gain and loss by the canopy, which is opposite to the convention followed in ecology. The net ecosystem productivity (*NEP*) is defined as the negative of *NEE*. According to these definitions (Chapin et al. 2006),

$$GPP = NPP + R_A \tag{2.1}$$

$$NPP = NEE + R_H \tag{2.2}$$

and,

$$TER = R_A + R_H \tag{2.3}$$

All the variables used this book chapter are described in Table 2.1.

2.2.2 *In Situ Measurements*

The *NEE* can be directly estimated from the observations in contrast with *GPP* and *TER*. Hence the *GPP* and *TER* are estimated from the *NEE* measurements using a set of ecophysiological relations among the ecosystem and environmental variables. Following are the different methodologies for measuring *NEE*.

2.2.2.1 Eddy Covariance Measurements

The eddy covariance (EC) method is probably the most accurate technique for estimating the biosphere–atmosphere scalar and energy fluxes using the direct measurements of wind parameters and scalars. It has been widely used across the globe for continuous monitoring of long-term CO_2 exchange by the ecosystems in different geographical location, altitudes and terrains (Baldocchi 2003; Deb Burman et al. 2020a, b). Several continental, regional and national networks exist comprising the dense arrays of such towers (Baldocchi et al. 2001; Beringer et al. 2016; Deb Burman et al. 2017; Deb Burman et al. 2018; Rebmann et al. 2018). A comprehensive global map of such active and past EC flux towers can be found in Burba (2019).

Any variable in the atmosphere is exchanged and mixed among the layers by the random turbulent wind motions, also known as the eddies. In EC method these eddies of different temporal scales are sampled by the fast (5, 10 or 20 Hz) measurements of wind velocity components and gas concentrations. Finally the contributions of all such eddies are summed up to compute the net fluxes using the Reynolds averaging technique (Reynolds 1895; Deb Burman et al. 2018). The ecosystem-atmosphere CO_2 flux (F_c in μmol m^{-2} s^{-1}) is computed from the vertical component of wind velocity (w in m s^{-1}) and atmospheric CO_2 molar concentration (c in μmol m^{-3}) which can be expressed as follows,

$$F_c = \overline{w'c'} \tag{2.4}$$

and

$$X' = X - \overline{X} \tag{2.5}$$

Table 2.1 Variables used in the present study, listed alphabetically

Symbol	Definition	Unit	Reference
A_n	Net CO_2 uptake by leaves	$\mu mol\ m^{-2}\ s^{-1}$	Collatz et al. (1991)
APAR	Absorbed Photosynthetically Active Radiation	$W\ m^{-2}$	Running et al. (1999)
α_{nir}	Surface reflectance in the near-infrared range of electromagnetic spectrum	–	Carlson and Ripley (1997)
α_{vis}	Surface reflectance in the visible range of electromagnetic spectrum	–	Carlson and Ripley (1997)
b	Regression parameter	$m\ s^{-1}$	Collatz, et al. (1991)
B	Bowen ratio	–	Stull (1988)
β	Proportionality constant	–	Businger and Oncley (1990)
c	Atmospheric CO_2 molar concentration	$\mu mol\ m^{-3}$	–
$c(z)$	c measured at height z	$\mu mol\ m^{-3}$	–
c_a	Atmospheric CO_2 concentration	ppm	–
c_s	CO_2 concentration at leaf surface	ppm	Collatz et al. (1991)
c_u	Updraft CO_2 molar concentration	$\mu mol\ m^{-3}$	Businger and Oncley (1990)
c_d	Downdraft CO_2 molar concentration	$\mu mol\ m^{-3}$	Businger and Oncley (1990)
EVI	Enhanced vegetation index	–	Jiang et al. (2008)
ε	Actual light use efficiency	$gC\ MJ^{-1}$	Monteith (1972)
ε_F	ε analogous factor for *SIF*		Guanter et al. (2014)
ε_{max}	Theoretically maximum light use efficiency	$gC\ MJ^{-1}$	Monteith (1972)
f	Ratio of ε_{max} and ε	–	Monteith (1972)
fAPAR	Fraction of Absorbed Photosynthetically Active Radiation	–	Running et al. (1999)
F_c	Vertical CO_2 flux	$\mu mol\ m^{-2}\ s^{-1}$	Aubinet et al. (2012)
F_q	Vertical water vapour flux	$mmol\ m^{-2}\ s^{-1}$	Stull (1988)
F_s	Storage flux of CO_2	$\mu mol\ m^{-2}\ s^{-1}$	Aubinet et al. (2012)
GPP	Gross Primary Productivity	$gC\ m^{-2}\ y^{-1}$	Chapin et al. (2006)
h	Measurement height	m	Chapin et al. (2006)
h_s	Relative humidity at leaf surface	–	Collatz et al. (1991)

(continued)

Table 2.1 (continued)

Symbol	Definition	Unit	Reference
H	Sensible heat flux	W m^{-2}	Stull (1988)
K_c	Eddy diffusivity factor for CO_2	m^2 s^{-1}	Lee (2018)
K_q	Eddy diffusivity factor for water vapour	m^2 s^{-1}	Lee (2018)
LAI	Leaf Area Index	–	Watson (1947)
LE	Latent heat flux	W m^{-2}	Stull (1988)
λ	Measurement wavelength	nm	–
m	Regression parameter	–	Collatz et al. (1991)
NDVI	Normalized Difference Vegetation Index	–	Carlson and Ripley (1997)
NEE	Net Ecosystem Exchange	μmol m^{-2} s^{-1}	Chapin et al. (2006)
NEP	Net Ecosystem Productivity	gC m^{-2} y^{-1}	Chapin et al. (2006)
NPP	Net Primary Productivity	gC m^{-2} y^{-1}	Chapin et al. (2006)
PAR	Photosynthetically Active Radiation	W m^{-2}	Alados et al. (1996)
q	Atmospheric water vapour molar concentration	mmol m^{-3}	–
R_A	Autotrophic Respiration	μmol m^{-2} s^{-1}	Chapin et al. (2006)
R_H	Heterotrophic Respiration	μmol m^{-2} s^{-1}	Chapin et al. (2006)
$r_{a,c}$	Aerodynamic resistance to CO_2 transfer	m^{-1} s	Lee (2018)
r_b	Leaf boundary layer resistance	m^{-1} s	Lee (2018)
r_c	Canopy resistance	m^{-1} s	Lee (2018)
r_s	Stomatal resistance	m^{-1} s	Lee (2018)
SAVI	Soil-Adjusted Vegetation Index	–	Huete (1988)
SIF	Solar-Induced Fluorescence	W m^{-2} sr^{-1} μm^{-1}	Meroni et al. (2009)
σ_w	Standard deviation of the vertical component of wind velocity	m s^{-1}	–
TER	Total Ecosystem Respiration	gC m^{-2} y^{-1}	Reichstein (2005)
$u*$	Friction velocity	m s^{-1}	Foken (2008)
VI	Vegetation Index	–	Huete et al. (2002)
VPD	Vapour Pressure Deficit	hPa	–
w	Vertical component of wind velocity	m s^{-1}	–

(continued)

Table 2.1 (continued)

Symbol	Definition	Unit	Reference
LSWI	Land-Surface Water Index	–	Fensholt and Sandholt (2003)

where X' stands for the instantaneous fluctuations in the measured values of variable X from its mean over the averaging period ($\overline{X}$). The overbar denotes the temporal averaging which is usually done every 30 or 60 min.

A typical EC measurement setup for CO_2 includes sonic anemometer and infrared gas analyser for wind velocity and CO_2 concentration measurements respectively, which are usually placed at a single height above the ecosystem canopy or at several recommended heights within the canopy (Deb Burman et al. 2019). These measurement heights determine the footprint or representativeness of the EC measurement (Kormann and Meixner 2001; Kljun et al. 2015). In addition the EC tower is instrumented at various levels and depths with a set of associated meteorological, radiation and soil sensors (Aubinet et al. 2012). The fast measurements by EC sensors are prone to various errors such as the random spikes, faulty measurement values due to inaccurate sensor geometry, density fluctuation of the ambient air due to the presence of moisture, obstruction of the wind by the sensors etc. The raw EC data is rigorously filtered for removing such errors following a set of recommended pre-processing techniques such as despiking (Vickers and Mahrt 1997; Mauder et al. 2013), detrending, coordinate rotations (Kaimal and Finnigan 1994), angle of attack correction (Kaimal and Finnigan 1994), Webb-Pearman-Leuning correction (Mauder et al. 2013), low (Moncrieff et al. 1997) and high-pass noise filtering (Moncrieff et al. 2004), time-lag between velocity and concentration measurements (Burba 2013) etc. Such a flux tower is shown in Fig. 2.1 which is installed at the Pichavaram mangrove ecosystem as part of the MetFlux India network in Tamil Nadu, India (Deb Burman et al., 2017; Gnanamoorthy et al. 2019; Chakraborty et al. 2020; Gnanamoorthy et al. 2020).

While implementing these corrections the resulting half-hourly NEE values are flagged according to a 10-point scale from 0 to 9 with increasing order suggesting reduced confidence (Foken et al. 2004). Such classification takes into account the atmospheric conditions including non-stationary and non-integral turbulences. The choice of best quality data depends on the requirement. Usually for developing the functional relationships between carbon flux and environment, the data values not exceeding flag 2 are used while in some cases such strict quality-control results in 60–65% of data loss, mostly during nighttime rendering the remaining data heavily biased towards daytime. In such cases the flags are gradually relaxed for more uniform representation of day and nighttime values in the final data record. The friction velocity ($u*$) is a measure of the atmospheric turbulence (Foken 2008). At very low values of turbulence the fundamental principles of EC measurement are violated. To avoid this condition $u*$-filtering is done in which the *NEE* values corresponding to $u*$ below a certain threshold are rejected. This $u*$-threshold determination is crucial for the accurate estimation of *NEE* and several recipes exists to determine this

Fig. 2.1 A surface flux tower instrumented at multiple levels with eddy covariance and other associated measurement sensors at the Pichavaram mangroves, Tamil Nadu, India; this tower is part of the MetFlux India network

(Gu et al. 2005; Barr et al. 2013); Wutzler et al. 2018) . Apart from these a certain amount of CO_2 is trapped in the canopy that does not mix with the atmosphere by turbulent mixing. This is termed as the storage flux (Baldocchi 2003). The storage flux between ground and measurement height h (F_s) is computed from the measured CO_2 concentration (c) time series at height h as follows,

$$F_s = \int_0^h \frac{\partial c}{\partial t} dz \tag{2.6}$$

Finally the *NEE* is computed as ($F_c + F_s$). All of these measures are documented in several books such as Burba and Anderson (2007) and Aubinet et al. (2012). The various quality-control measures result gaps in the data. Gaps also occur due to the instrument malfunctioning at several times. However to account for the NEP of any ecosystem, continuous measurement record is required which is achieved by filling the gaps in data. Several recipes of gap-filling exist in literature e.g. mean diurnal variation (MDV), marginal distribution sampling (MDS), look-up table (LUT) etc. (Moffat et al. 2007; Reichstein 2005). Selection of any particular recipe for gap-filling depends on various factors such as the extent and severity of data loss, environmental conditions, local climatology, availability of supporting measurements etc. These recipes have evolved over years owing to active research by several groups and are documented elsewhere (Falge et al. 2001).

Although the EC measurement offers an unprecedented advantage of real time monitoring of *NEP* of any ecosystem it has its own limitations (Massman and Lee 2002). Advective fluxes remain difficult to be separated from the vertical exchange (Paw et al. 2000; Etzold et al. 2010). The fluxes measured over mountainous, undulated terrains are laced with lots of measurement uncertainty (Geissbühler and Siegwolf 2000). During low-turbulence conditions, mostly in the nocturnal periods fluxes are severely undermined (Aubinet et al. 2010). The EC method is still under active research and evolving fast. A large number of scientific publications and references exist on this technique, its development and adaptations not all of which are possible to be included in the limited span of the present chapter. Interested readers however are suggested to read several such articles appearing in the meteorology, forestry and agricultural journals.

2.2.2.2 Gradient Flux Measurements

In the absence of direct EC measurements F_c can be estimated from the spatial gradient of mean CO_2 molar concentration ($\overline{c}$ in μmol m^{-3}) assuming a diffusive model. In this model $\overline{c}$ is assumed to vary slowly and the Reynolds averaged covariance F_c is parameterized as the function of vertical gradient of $\overline{c}$ as follows,

$$F_c = -K_c \frac{\partial \overline{c}}{\partial z} \tag{2.7}$$

where, K_c is defined as the eddy diffusivity factor for carbon dioxide. It has the unit of $m^2\ s^{-1}$. Due to the turbulent diffusion CO_2 is transported from high concentration zone to low concentration zone. Hence the negative sign is introduced in Eq. (2.7) to maintain conformity with the flux convention described earlier. In a simplistic setup with two-level measurements Eq. (2.7) can be reformulated as,

$$F_c = -K_c \frac{\overline{c(2)} - \overline{c(1)}}{z(2) - z(1)} \tag{2.8}$$

where, $\overline{c(1)}$ and $\overline{c(2)}$ stand for the mean CO_2 concentrations at lower ($z(1)$) and upper ($z(2)$) measurement heights respectively (Lee 2018).

Although convenient in the absence of fast measurements, the gradient flux measurement technique is heavily criticised for several reasons. In the gradient flux formulation the spatial heterogeneity of CO_2 source strength at the horizontal surfaces of concentration measurement is not considered. Moreover the vertical variation in CO_2 source strength is overlooked (Dyer 1974). This problem is partially circumvented by introducing more concentration measurements in the intermediate levels and fitting a vertical profile to the measured values for computing the gradient. $K_c$has a strong functional dependency on atmospheric stability for which a set of empirically determined stability correction factors are introduced for better estimation of F_c (Dyer and Hicks 1970; Businger et al. 1971). In practical applications the gradient flux method is mostly used in conjunction with the primary EC measurement with iterative determination of $K_c$from the latter set of measurements.

2.2.2.3 Resistance Methods

Assuming a constant flux layer, where F_c does not vary with height, Eq. (2.7) can be reformulated as,

$$\overline{c} = -F_c \int_0^h \frac{1}{K_c} dz \tag{2.9}$$

or,

$$F_c = -\overline{c} \frac{1}{\mathrm{r_{a,c}}} \tag{2.10}$$

where,

$$r_{a,c} = \int_0^h \frac{1}{K_c} dz \quad (2.11)$$

is defined as the aerodynamic resistance to CO_2 transfer, in analogy to the electrical resistance as defined in the Ohm's law (Lee 2018). The $r_{a,c}$ increases with increasing stability and depth of the diffusion layer. Increased turbulence in the atmosphere decreases $r_{a,c}$.

In addition to aerodynamic resistance, the plant-atmosphere photosynthetic CO_2 exchange pathway involves leaf boundary layer and stomatal resistances, connected in series as the CO_2 molecules transport through these sequentially. The layer of atmosphere in close vicinity of the leaves is usually thin but unperturbed. The exchange of CO_2 through this layer only takes place through molecular diffusion which gives rise to the leaf boundary layer resistance (r_b).

The stomatal resistance (r_s in s m^{-1}) is defined as the resistance faced by CO_2 molecules while escaping to the atmosphere from the stomatal cavity. The plant-atmosphere CO_2 and water vapour exchanges by photosynthesis and transpiration respectively, are coupled by the stomatal opening and closure mechanism (Farquhar and Sharkey 1982). There are two major schemes to parameterize r_s. According to the Jarvis-Stewart formulation (Jarvis 1976; Stewart 1988), r_s is empirically determined from radiation, leaf temperature, vapour pressure deficit (*VPD*) and soil moisture content.

In another approach, introduced by Ball et al. (1987) and Collatz et al. (1991) r_s is expressed as a function of net CO_2 uptake (A_n) and relative humidity and CO_2 concentration at the leaf surface (h_s and c_s, respectively) as follows,

$$r_s = \frac{1}{m\frac{A_n \cdot h_s}{cs} + b} \quad (2.12)$$

where, m and b are the linear regression parameters derived experimentally. These models have been widely used by several researchers and adopted according to different climate types (Leuning 1990; Tuzet et al. 2003; Whitley et al. 2008; Ye and Yu 2008). The inverse of any resistance parameter is defined as the corresponding conductance. In simplified bigleaf models where the entire ecosystem is considered to behave like a single leaf (Monteith et al. 1965), the canopy resistance (r_c) is used to compute the ecosystem-atmosphere fluxes. However, the interpretation of r_c is not trivial. In a simplistic formulation where all the leaves in the canopy can be considered as individual resistors with stomatal resistance r_s connected in parallel, $r_c$can be expressed as r_c/n where n is the number of leaves in the canopy.

2.2.2.4 Modified Bowen Ratio Method

The Bowen ratio (B) is defined as the ratio of sensible (H in W m^{-2}) and latent (LE in W m^{-2}) heat fluxes. It has been widely used to estimate LE from the energy flux measurements in the absence of water vapour measurement assuming a perfect closure of the surface energy budget known as the Bowen ratio method (Stull 1988). This technique is modified to estimate F_c from the vertical flux of water vapour (F_q in mmol m^{-2} s^{-1}) in the absence of fast measurement of c from the two level slow measurements of $\overline{q}$ and $\overline{c}$ as follows,

$$F_c = \frac{\overline{c(2)} - \overline{c(1)}}{\overline{q(2)} - \overline{q(1)}} \cdot F_q \tag{2.13}$$

where, $\overline{c(1)}$ and $\overline{q(1)}$ are the values of c and atmospheric water vapour molar concentration (q in mmol m^{-3}) at lower measurement height (z1) and $\overline{c(2)}$ and $\overline{q(2)}$ are the values of c and q at upper measurement height (z2). This technique assumes the eddy diffusivities of CO_2 and water vapour transport (i.e. K_c and K_q, respectively) are equal to each other and is known as the modified Bowen ratio method (Meyers et al. 1996).

2.2.2.5 Associated Micro-meteorological Methods

There have been several micrometeorological methods proposed to measure F_c in the absence of fast EC measurements, using the available slow gas concentration measurements as described below.

Disjunct Eddy Covariance

The disjunct eddy covariance (DEA) is a modification of EC method first proposed by Rinne et al. (2001) for measuring the fluxes of volatile organic compounds (VOC). In this method the turbulence is assumed to be fully developed and hence the time series of w and gas concentration are sampled at a much coarser temporal scale than EC method. This method has been used by few researchers to estimate F_c as well with good confidence (Hörtnagl et al. 2010; Baghi et al. 2012).

Eddy Accumulation

The eddy accumulation (EA) method is a modified version of the EC method where air samples are stored in two separate containers based on updraft and downdraft (Hicks and McMillen 1984). The collection time is proportional to the strength of updraft or downdraft i.e. the magnitude of w. After the data is collected for 30 or

60 min the average CO_2 concentration in both the collection volumes is measured and subtracted from each other for computing F_c (the detailed mathematical formulation is similar to the EC method). Instead of sampling all the eddies separately as done in the EC method (within the practically limiting smallest and largest time scales, as decided by the sampling frequency and averaging time), in EA method all the eddies are augmented according to upward or downward motions and the mean concentration for both of these segments are computed. This method was first proposed by Desjardins (1972).

Relaxed Eddy Accumulation

The relaxed eddy accumulation (REA) is a modification of the EA technique where the air volumes are sampled separately at constant flow rate for updrafts and downdrafts, with the separate measurements of CO_2 concentrations for both the volumes. Finally the difference between updraft and downdraft averages of CO_2 molar concentrations ($\overline{c_u}$ and $\overline{c_d}$ respectively) is multiplied by the standard deviation of vertical velocity (σ_w) during the entire duration of updraft or downdraft event and an empirically determined dimensionless factor of proportionality (β) to compute F_c as follows,

$$F_c = \beta \cdot \sigma_w \cdot (\overline{c_u} - \overline{c_d}) \tag{2.14}$$

This method had been first proposed by Businger and Oncley (1990) who also proposed the value of β to be 0.6. However the choice of β is a major source of uncertainty of this method which was later shown to vary within 0.40–0.63 under different experimental conditions by several researchers (Baker et al. 1992; Milne et al. 1999). Additionally the accurate segregation of air samples in updraft and downdraft periods and precise measurement of $\overline{c_u}$ and $\overline{c_d}$ are essential which are difficult to be achieved in field conditions (Pattey et al. 1993). The REA has been mostly used in the measurements of trace gas, aerosol, VOCs and isotopic fluxes (Guenther et al. 1996; Valentini et al. 1997; Myles et al. 2007) for which fast concentration measurements are not available, as required by the EC technique. However, several inter comparison studies have shown the REA to perform well in estimating F_c under strict measurement control as compared to EC method in agricultural and forest ecosystems (Pattey et al. 1993; Gaman et al. 2004). There have been several modifications of REA since its inception such as the hyperbolic relaxed eddy accumulation (HREA) by Bowling et al. (1999) etc. which I am not going to discuss in detail in the limited span of the present chapter. The interested readers are suggested to read the relevant literature for the detailed information on these techniques.

Chamber-Based Measurements

In the chamber-based methods enclosure chambers are used to isolate a part of the atmosphere within which the CO_2 concentration is allowed to change only by the respiration or photosynthesis by the plant, part of plant or soil enclosed by these chambers. The F_c is computed from the change in c over the sampling time interval. The walls of the chambers are impermeable and do not allow the CO_2 inside to interact with the ambient atmosphere through diffusion. In static chamber methods the ambient air flow is restricted (Wang et al. 2010) whereas in dynamic chamber methods the ambient air is allowed to flow through the chamber at a constant base flow rate (Ohkubo et al. 2007). Several different geometries of the chambers exist (Kutsch et al. 2010).

The operation of flux chambers can be manual or automatic. These are portable and have less stringent requirements than the EC technique. However they significantly alter the microclimate within the chamber volume e.g. the opaque chambers increase the probability of dark respiration, static chambers hinder the turbulent mixing etc. Despite the ease of installation of the flux chambers, these methods have small footprint, are not suitable for long term canopy-scale CO_2 flux measurements in tall forest ecosystems and mostly used for grasslands with small canopy height or leaf, bole and soil respiration components (Law et al. 1999).

Once *NEE* is estimated by the above-mentioned methods, it is partitioned into *GPP* and *TER* using respiration (Reichstein 2005), light-use efficiency (Lasslop et al. 2010), isotopic fractionation (Bowling et al. 2001) or statistical correlation based methods (Skaggs et al. 2018).

2.2.3 *Satellite Measurements*

In situ measurements by far provide the most realistic estimates of the carbon cycle of terrestrial ecosystems. However, these have limited footprints. A dense network of ground-based towers instrumented with multiple sensors is required to be deployed for estimating the *NPP* of any region, as outlined in the previous section. Also the measurements need to be continued seamlessly at least for several years before a reasonable estimation can be achieved with the seasonal, intra-seasonal and inter-annual variabilities (Baldocchi 2001). Establishing such networks are challenging in remote and inaccessible locations. Moreover the ground-based measurement systems are marred by severe data loss due to power shortage, adverse weather conditions etc. Maintaining such systems required long-term dedicated effort, human involvement and have high establishment and operating costs. The data collection, processing and interpretation can be tedious for difficult terrains such as mountainous regions, dense rainforests etc. In view of these *NPP* can be remotely monitored using satellite and other remote sensing techniques as described below.

2.2.3.1 Vegetation Indices

Vegetation indices (*VI*s) are spectral transformations of the reflectances recorded by satellite sensors in multiple bands of the electromagnetic spectrum (Huete et al. 2002). These are dimensionless and used for monitoring the vegetation health and photosynthetic activities at different spatio-temporal scales. Different *VI*s are formulated such as the normalized difference vegetation index (*NDVI*), leaf area index (*LAI*), enhanced vegetation index (*EVI*), land-surface water index (*LSWI*), soil-adjusted vegetation index (*SAVI*) etc. which are used in combination with the meteorological measurements as input to the bottom-up models for predicting the *NPP* (Liu et al. 1997) or *GPP* (Deb Burman et al. 2017). The *LAI* is defined as the total one-sided leaf surface area relative to per unit ground area (Watson 1947) which can be estimated from the ground-based or satellite measurements (Bréda 2003; Deng et al. 2006). It is closely connected with *NDVI* which is estimated from the surface reflectances as defined below,

$$NDVI = \frac{\alpha_{nir} - \alpha_{vis}}{\alpha_{nir} + \alpha_{vis}} \tag{2.15}$$

where, α_{nir} and α_{vis} are the averaged surface reflectances in the visible and near infrared regions of the electromagnetic radiation spectrum, respectively (Carlson and Ripley 1997). The *EVI* is an adaptation of *NDVI* corrected for the atmospheric and canopy background noises which is more sensitive towards dense canopies (Jiang et al. 2008). On a similar note, the *LSWI* is a modification of *NDVI* that takes care of the effect of vegetation leaf structure, moisture contents in leaf and soil on the spectral reflectances (Fensholt and Sandholt 2003; Xiao et al. 2004). The *SAVI* is a modification of *NDVI* adjusted for the soil brightness effect as defined by Huete (1988).

Different adaptations of this methodology exists in literature where the *VI* had either been estimated from the satellites such as LANDSAT (Ganguly et al. 2012), Moderate Resolution Imaging Spectroradiometer (MODIS) (Demarty et al. 2007), Advanced Very High Resolution Radiometer (AVHRR) (Buermann et al. 2001) etc. or the ground-based measurements (Deb Burman et al. 2017) and the meteorological variables are either taken from the surface measurements (Bao et al. 2016; Deb Burman et al. 2017), remote sensing observations (Sims et al. 2008), model predictions (Yan et al. 2016) or a suitable assimilation of these (Demarty et al. 2007). The functioning of bottom-up models is described in Sect. 2.2.4.1 of this chapter. The development of *VIs* from reflectances measured by the satellites is challenging due to multiple constraints, such as atmospheric scattering, vegetation structure, leaf inclination, albedo etc. (Knyazikhin et al. 1998) and hence remains to be an active area of research.

2.2.3.2 Light Use Efficiency Approach

According to Monteith (1972, 1977) the photosynthetic yield of any ecosystem is directly proportional to the amount of solar radiation absorbed by the canopy in absence of any water or nutrient stress in soil. The plants absorb solar radiation in the photosynthetically active radiation (*PAR*) or visible range of the electromagnetic spectrum (wavelength varying within 400–700 nm) (Alados et al. 1996). The amount of *PAR* absorbed by the plants is known as the absorbed photosynthetically active radiation (*APAR*) which is related to *PAR* as follows,

$$APAR = fAPAR * PAR \tag{2.16}$$

where, *fAPAR* is known as the fraction of photosynthetically active radiation. This is the basis of light use efficiency (*LUE*) approach to estimate *NPP* of any ecosystem formulated as,

$$NPP = fAPAR * PAR * \varepsilon \tag{2.17}$$

where, ε is the effective *LUE* of the ecosystem in the measured environmental condition that varies widely across the ecosystems depending on their geographical location, species type, canopy structure, presence of enzymes, evaporative demand, temperature stress, availabilities of moisture and nutrients etc. It is expressed as a downscaled fraction of the theoretical maximum *LUE*, $\varepsilon_{\max}$ as,

$$\varepsilon = \varepsilon_{\max} * f \tag{2.18}$$

where the factor f accounts for the deviation of ε from $\varepsilon_{\max}$ owing to the non-optimal environmental conditions (Monteith 1972). Initially $\varepsilon_{\max}$ was largely thought be constant at 0.405 gC MJ^{-1} (Potter et al. 1993), which was later found to vary between a wider range (Ahl et al. 2004; Yu et al. 2009). Multiple models use this methodology to predict *NPP* such as the Carnegie-Ames-Stanford Approach (CASA) (Potter et al. 1993), Vegetation Photosynthesis Model (VPM) (Xiao et al. 2004), Physiological Principles for Predicting Growth (3-PG) (Coops et al. 2005) etc.

2.2.3.3 Derived Biophysical Products

Using the *LUE* approach a daily *NPP* product MOD17 is developed from the Moderate Resolution Imaging Spectroradiometer (MODIS) observations. It was shown by several researchers that compared to *NPP*, *GPP* has a better correlation with *APAR* (Raymond Hunt 1994; Prince and Goward 1995). Hence while developing the MOD17 product, the *GPP* and respiration components (including growth respiration *GR* and maintenance respiration *MR*) are calculated separately and subtracted to produce *NPP* (Running et al. 1999). The *GPP* is calculated daily directly from the *fAPAR* measured by MODIS and *PAR* from an assimilated data product whereas the *GR* and *MR* are calculated from the carbon allometric equations of plants and *LAI* estimates by MODIS namely MOD15 (Myneni et al. 1999). The biome-specific plant-physiological parameters governing the carbon allometric equations are derived from a bottom-up model Biome-BGC (Running and Hunt 1993). The functioning of such models is discussed in 2.2.4.1 of the present chapter. The MOD17 product is validated across several in situ measurements such as FLUXNET (Running et al. 1999; Zhao et al. 2005; Turner et al. 2006).

2.2.3.4 Solar Induced Fluorescence

The Solar-Induced Fluorescence (*SIF*) stands for the part of energy released from the chlorophyll-a (expressed in the unit of W m^{-2} sr^{-1} μm^{-1}) after absorbing the *PAR*, in the light pathway of photosynthesis, apart from the electron transport and thermal energy dissipation. It has a typical wavelength range of 650–800 nm, which is larger than the absorbed radiation. The strongest peak in *SIF* spectrum occurs at around740 nm, in the far-red or near-IR regime with the second strongest peak appearing at around 685 nm, in the red zone (Meroni et al. 2009; Mohammed et al. 2019). This can be attributed to the fact that in optimum condition PSI photosystem emits preferably in near-IR whereas the PSII photosystem emits in both red and near-IR (Govindje 1995).

Globally, the *SIF* is seen to have linear dependencies with *GPP* estimates from top-down and bottom-up approaches (Zhang et al. 2014; Cui et al. 2017). In a simplistic model (Guanter et al. 2014) the *SIF* observed from space can be related to *GPP* in the following way, similar to the LUE formulation described in (2.17) as,

$$GPP = SIF(\lambda) * \frac{\varepsilon}{\varepsilon_F} \tag{2.19}$$

where, SIF(λ) corresponds to the *SIF* measurement at wavelength λ (usually taken as either of 685 or 740 nm) and ε_F is an efficiency factor for *SIF*, analogous to ε. In present times, a plethora of SIF estimates are available from different satellite-based sensors e.g. ENVironmental SATellite (ENVISAT) (Guanter et al. 2007), Greenhouse gases Observing SATellite (GOSAT) (Guanter et al. 2012), Global Ozone Monitoring Experiment-2 (GOME-2) (Joiner et al. 2013), SCanning Imaging Absorption SpectroMeter for Atmospheric CHartographY (SCIAMACHY) (Wolanin et al. 2015), Orbiting Carbon Observatory-2 (OCO-2) (Sun et al. 2017), TROPOspheric Monitoring Instrument (TROPOMI) (Köhler et al. 2018) etc. and have been used extensively in *GPP* estimation (Wagle et al. 2016; He et al. 2020). Different algorithms have been designed based the wavelength, atmospheric condition and canopy type. It is worthwhile to note that in a recent work Patel et al. (2018) found out the exponential relationship between cropland *NPP*, estimated from ground-based harvesting estimates and 740 nm *SIF*, measured by GOME-2 over the Indo-Gangetic Plain in India which can be further exploited for *NPP* estimation and harvest yield prediction.

2.2.3.5 Other Remote Sensing Measures

In conjunction with the ground-based and satellite measurements, aircrafts (Cihlar et al. 1992; Desjardins et al. 1992; Macpherson et al. 1992; Chou et al. 2002) and unmanned aerial vehicles (Pirk et al. 2017) have been used for CO_2 flux measurements. The instrumentation of such campaigns include vertical concentration profile, EC (Oechel et al. 1998), *VI* and *SIF* (Zarco-Tejada et al. 2013) measurements. The SIF measurements have also been deployed on the ground (Grossmann et al. 2018). The working principle of these measurements remain the same as discussed earlier but additional quality control measures are implemented to account for the complexities arising in such measurements.

2.2.4 Modelling

The observations provide real time diagnostic information about the *NPP*. However for prediction of the ecosystem health and carbon sequestration in response to the changes in climate and environmental conditions prognostic information is required, which can be achieved by the models (Prentice et al. 2001; Levy et al. 2004). The models are a set of mathematical equations, algebraic, differential, integral or a combination of these to describe the coupled meteorological and ecophysiological processes. These equations are built upon the theories as well as the phenomenological relations derived from the observations (Bonan et al. 2011) and can be solved analytically or numerically. These models are tested by comparing the predicted variables against their direct observations for a certain measurement interval. This process is known as the calibration. Further a calibrated model is run again for a different time interval and the simulated output is checked against the measurement, thus validating the model (Enting and Pearman 1986; Friend et al. 2007). In a sense the models act as a bridge between the observations and predictions. There are two major modelling approaches as described below.

2.2.4.1 Bottom-Up Approach

As the name suggests, in bottom-up approach the leaf or canopy scale measurements are upscaled to local, national, regional or global scales using the biogeophysical and biogeochemical processes. Hence these models are also known as the process-based models (Ito and Oikawa 2002). The vegetation growth and decay in process-based models depend on climate forcing and environmental conditions which are improvement over the traditional land-surface models where the vegetation is prescribed with no dynamical change (Foley et al. 1996; Clark et al. 2011; Lawrence et al. 2011). Due to this property the process-based models can be used for ecosystem growth prediction under changed climate such as crop production in a water-stressed condition (Gervois et al. 2004). Hence the process-based models are known as the dynamic global vegetation models, abbreviated as DGVM (Sitch et al. 2003; Prentice and Cowling 2013). These models can be simulated at a point or grid scale as a standalone model or a constituent module of a couple climate, general circulation or Earth system model (Zeng et al. 2002; Bonan and Levis 2006; Kato et al., 2009). Either of surface measurements, satellite products, reanalysis datasets or a suitable combination of these are used as input variables to the process-based models. The input data can be provided at multi-temporal time scales e.g. half-hourly, hourly, daily etc. (Williams et al. 1996; Deb Burman et al. 2017).

The process-based models have different components i.e. soil, ecosystem, atmosphere, hydrology etc. linked to each other by different complex feedback mechanisms which can be broadly classified into two categories namely, biogeochemistry and biogeophysics (El-Masri et al. 2013). Input variables of these models vary among different formulations. However, most of these models require meteorological parameters such as the shortwave and longwave radiations, air temperature, pressure, humidity, wind speed, precipitation, c_a etc. as input. In addition, *VI*s such as *LAI* are required by several models.

The photosynthesis in the models is parameterized according to the pathways i.e. C3 (Farquhar et al. 1980; Collatz et al. 1991), C4 (Collatz et al. 1992), CAM (Cortázar and Nobel 1990; Kluge and Ting 2012) etc. where the carbon assimilation is governed by the availability of *PAR* (Sellers et al. 1992), RuBisCO activity (Bernacchi et al. 2001), phosphophenolpyruvate (PEP)-carboxylase enzyme functionality (Vidal and Chollet 1997), electron transport capacity, leaf nitrogen content (Kattge et al. 2009) etc. In addition the air temperature, moisture demand and soil moisture also affect the carbon uptake by stomatal opening and closure resulting from heat and water stresses. The amount of *PAR* absorbed by the plants depends on leaf structure, orientation and area, incident *PAR* and albedo (Knyazikhin et al. 1998; Dai et al. 2004; Fensholt et al. 2004). The incident *PAR* can be directly measured or computed from the incoming solar radiation (Deb Burman et al. 2020a, b) whereas the leaf area is computed from *LAI*. As the carbon and water cycles are interlinked with nutrient cycles e.g. nitrogen, phosphorus etc. such models also need information regarding soil bulk density, organic matter content, texture and nutrient profiles for spin-up (Thornton and Rosenbloom 2005; Oleson et al. 2013).

A part of the total photosynthetic carbon uptake or gross primary productivity (*GPP*) is lost to the atmosphere by respiration. Different respiration components include growth and maintenance respirations of leaves, stems and roots (Barman et al. 2014a, b) which are computed from the direct temperature dependence of respiration (Ryan 1991) or a temperature dependent activation factor (Arora and Boer 2005). This approach is slightly different from the observational approach where the autotrophic and heterotrophic respirations are clubbed together as total ecosystem respiration (*TER*) and computed from the nighttime temperature dependence of F_c

by statistical regression (Lloyd and Taylor 1994). Finally the GPP is computed as the sum of *TER* and F_c.

Next, the fixed carbon is allocated among different pools i.e. root, leaf, stem, bole, litter etc. using the experimentally determined parameter values e.g. leaf area per unit carbon, leaf nitrogen content etc. (Sitch et al. 2003). A schematic of such a process-based model showing different compartments of carbon allocation and their interrelation is shown in Fig. 2.2, reprinted with permission from El-Masri et al. (2013). The plants are not modelled individually in the process-based models, as that would require much detailed parameterization and more computational resources, rather the plants are broadly categorized into several plat functional types (PFTs). Such clubbing of plants is implemented based on their photosynthesis pathway, geographical and climatological distributions such as tropical broadleaf deciduous, boreal coniferous evergreen, shrub, tundra, pasture, grassland etc. (Poulter et al.

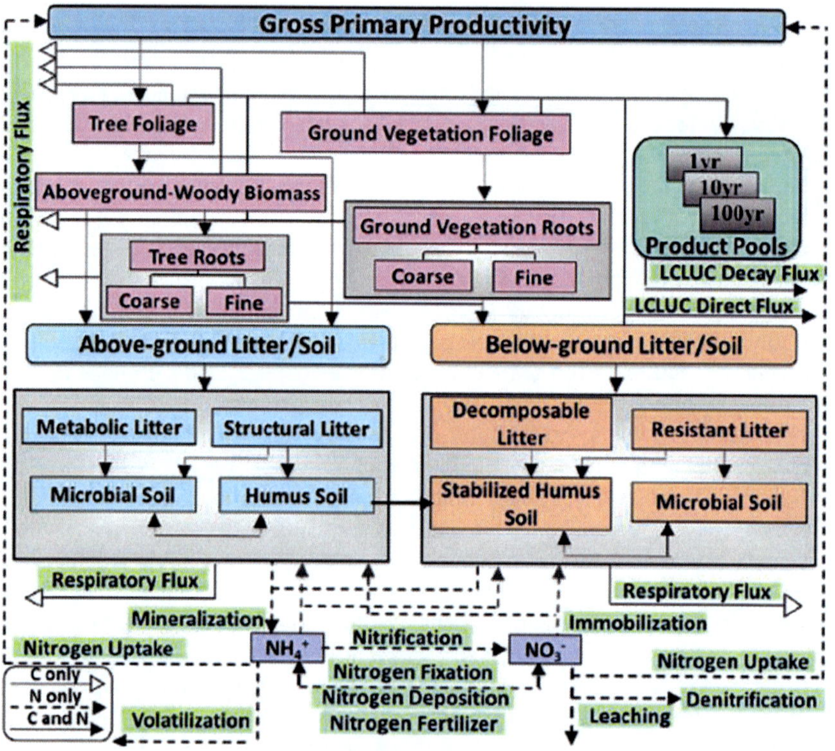

Fig. 2.2 A schematic diagram of the different carbon pools and processes in a process-based ecosystem model Integrated Science Assessment Model (ISAM) (reprinted from El-Masri et al. (2013) with permission)

2011). It is worthwhile to note here that several ecosystems remain poorly or under-represented in such models due to the lack of understanding of their ecophysiological processes owing to a lack of adequate experimental evidences or complexity in modelling those such as mangrove, rice paddy etc. (Langerwisch et al. 2018; Kumar and Scheiter 2019). For regional or global scale applications the vegetation map is prescribed to the bottom-up models as land cover and use map (Ramankutty and Foley 1999; Klein Goldewijk et al. 2011). The representativeness and accuracy of these maps remain a crucial control of uncertainty of the *NPP* estimation by these models (Arora and Boer 2010).

2.2.4.2 Top-Down Approach

The alternate approach to model the Earth-atmosphere trace gas fluxes is the top-down approach. As evident in its name, in this method the atmospheric trace gases concentrations are observed at multiple spatial and temporal scales and the sources and sinks responsible for these distributions are computed by a backward modelling approach (Gurney et al. 2004; Rayner et al. 2005). This methodology is also known as the inverse modelling which is opposite in approach to the bottom-up modelling where a forward scheme is implemented. The variables (ψ) and observations (χ) matrices are connected by the following relation,

$$G \cdot \psi = \chi \tag{2.20}$$

where G is the coefficient matrix. The basic problem in inverse modelling lies in finding the inverse of G which is mostly calculated by Bayesian inversion (Heimann and Kaminski 1999). The observed spatio-temporal distribution of trace gases are apportioned into potential sources and sinks using atmospheric transport models (Kaminski et al. 1999) which can be solved in Lagrangian (Pisso et al. 2019), Eularian (Pillai et al. 2012) or hybrid schemes (Siqueira et al. 2000). The terrestrial sources and sinks of atmospheric CO_2 include the natural and agricultural ecosystems which are our elements of interest in the present chapter. A good knowledge of the biospheric-atmospheric CO_2 exchange, its seasonal variation and controlling parameters is required for the proper evaluation of *NEP* using inverse methods. The limited scope of the present chapter will not allow me to go into the mathematical and technical details of these techniques but the interested readers are suggested to consult the available vast scientific literature for more insight (Bousquet et al. 1999; Gurney et al. 2002; Peylin et al. 2013).

This apparently simple problem is not very straightforward in reality as I discuss next. A plethora of different sets of available data are used in inverse modelling. These include surface observations (Pickett-Heaps et al. 2011), aircraft measurements (Pisso et al. 2019), ship measurements (Bousquet et al. 1999), satellite products (Houweling et al. 2003, 2015) etc. The measurements of atmospheric traces gases concentrations and fluxes are not spread uniformly across the globe. While some regions are well-mapped with dense observation networks with measurements carried at high temporal resolution (Bousquet et al. 1999) such as north America, some regions do not have adequate observation stations or high frequency measurements (Heimann and Kaminski 1999) such as India (Nalini et al. 2019). Aircraft measurements are costly and still limited in number like the ship observations (Patra et al. 2011). The satellite observations are often contaminated with scattering at different levels, boundary-layer dynamics and presence of clouds which are more prominent over the tropical regions (Rayner et al. 2002; Pandey et al. 2016). Together these result in the data gaps in χ.

In such cases the inverse problem becomes ill-posed with the number of observations being less than the number of control variables and an unique solution to the inverse problem ceases to exist (Heimann and Kaminski 1999). Moreover the errors in χ result in inaccurate estimation of the inverse of G. In case of Bayesian inversion the directly measured surface fluxes (Chevallier et al. 2006) or bottom-up model outputs (Dargaville et al. 2002; Patra et al. 2011) are provided to constrain the problems as a priori information of source and sink strengths. Subsequently the source and sink strength are predicted from the inverse models. These a posteriori estimates are subsequently compared with the a priori information. To reduce the uncertainty in this process a cost function is defined between the observed and simulated concentrations and a priori and a posteriori flux estimates (Kadygrov et al. 2009) which is subsequently minimized by several optimization methods used in data assimilation such as Kalman filter, 4D-var etc. (Liu et al. 2016). Apart from these the errors in transport model formulation propagate in the source and sink patterns by the inverse models (Bousquet et al. 1996; Schuh et al. 2019). The success of inverse modelling of any trace gas requires a proper understanding of the trace gas species chemistry. It has been mostly successful for long-lived trace gases such as CO_2, CH_4 etc. which take part in the long-range transport. Often to reduce the computational cost the Earth surface is divided into several regions (typically less than 100) and the average flux for each of these zones are modelled by the inverse models. A schematic of different regions used in the TRANSCOM inversion is shown in Fig. 2.3 (reprinted with permission from Houweling et al. (2015)). However such averaging reduces the spatial heterogeneity and increases the probability of misrepresentation of any region. For example it has been argued (Valsala et al. 2013) that the terrestrial biospheric CO_2 fluxes in the Carbon Tracker dataset (Peters et al. 2007) is misrepresented over the Indian subcontinent as no CO_2 measurement from this region had been used as a priori flux in the Carbon Tracker dataset due to which the flux over this region is highly biased by the emissions from neighbouring regions such as China, Korea, Kazakhstan, Indonesia etc. which belong to the same averaged spatial domains as India namely Eurasian Temperate and Asia Tropical.

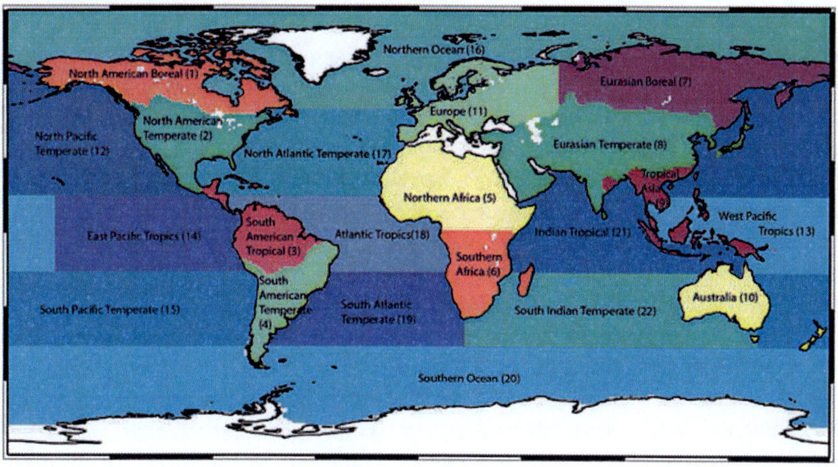

Fig. 2.3 A schematic showing the different land (and ocean) zones used in the TRANSCOM inversion (reprinted with permission from Houweling et al. (2015))

2.3 Discussion and Conclusions

Several studies have reported the different components of carbon cycle including *NPP* at global, regional and ecosystem levels using different techniques discussed above or a combination of these (Beer et al. 2010; Jung et al. 2011; Barman et al. 2014b; Tramontana et al. 2016). According to the multi-model study by Cramer et al. (1999) the global *NPP* ranges within 44.4–66.3 Gt C y^{-1}. This is shown in Fig. 2.4 which is

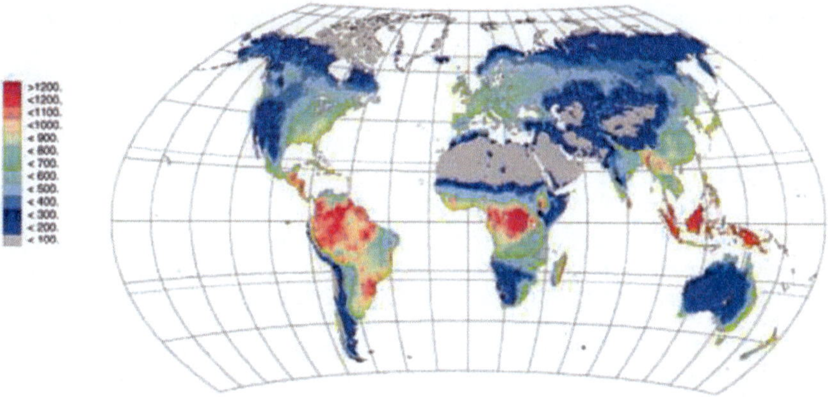

Fig. 2.4 A global map of the average annual *NPP* (gC m^{-2} y^{-1}) estimated by an ensemble of models (reprinted with permission from Cramer et al. (1999)).

reprinted from Cramer et al. (1999) with permission. Based on the MODIS measurements the average global *NPP* is 56 Gt C y^{-1}(Zhang et al. 2009). According to the combined bottom-up and top-down CO_2 data assimilation approach by Rayner et al. (2005) the global *NPP* is approximately 40.5 Gt C y^{-1}. Such studies are important to know about the sources and sinks of atmospheric CO_2 and their annual patterns. These are required not only to assess their roles in global climate change but also to predict the effect of future climate change on these ecosystems e.g. increased air temperature, increased c_a etc. However, due to the sparsity of measurements such estimates have often been riddled with lots of uncertainties both at global and regional levels (Graven et al. 2013; Patra et al. 2013; Cervarich et al. 2016). Moreover, still significant differences exist among observed, bottom-up and top-down modelled estimates of ecosystem carbon uptake (Bastos et al. 2020). In order to devise the climate change mitigation strategies the relations among plant carbon uptake and environmental variables need to be accurately known and represented in the assessment models. The terrestrial carbon cycle components predicted by the different models taking part in coupled modelintercomparison project (CMIP) (Meehl et al. 2000) differ significantly from each other due to the apparent inconsistency in the underlying land carbon cycle formulations (Anav et al. 2013; Friedlingstein et al. 2014). Such uncertainties can be reduced by improved parameterizations which can derived from the long-term measurements in the diverse ecosystem types using an optimally designed denser network of surface measurements.

Acknowledgements I gratefully acknowledge the constant encouragement and support by the Director, Indian Institute of Tropical Meteorology (IITM), Pune. The IITM is an autonomous institute of the Ministry of Earth Sciences (MoES), the Government of India. I thank Supriyo Chakraborty and other colleagues of the MetFlux India team for all possible helps. I also take this opportunity to thank Vinod Kumar Gaur, Martin Heimann, Philippe Ciais, Marc Aubinet, Fortunat Joos, George Burba, KT Paw U, Hartmut Bösch, Paul Palmer, Christof Ammann, Atul K. Jain, Prabir K. Patra, Bimal K. Bhattacharya, Grant Allen, Eiko Nemitz, Somnath Baidya Roy and Sabine Fuss for the valuable discussions I had with them during different scientific meetings at different stages of my career.

References

Ahl DE, Gower ST, Mackay DS, Burrows SN, Norman JM, Diak GR (2004) Heterogeneity of light use efficiency in a northern Wisconsin forest: Implications for modeling net primary production with remote sensing. Remote Sens Environ 93(1–2):168–178. https://doi.org/10.1016/j.rse.2004.07.003

Alados I, Foyo-Moreno I, Alados-Arboledas L (1996) Photosynthetically active radiation: measurements and modelling. Agric For Meteorol 78(1–2):121–131. https://doi.org/10.1016/0168-1923(95)02245-7

Anav A, Friedlingstein P, Kidston M, Bopp L, Ciais P, Cox P, Jones C, Jung M, Myneni M, Zhu Z (2013) Evaluating the land and ocean components of the global carbon cycle in the CMIP5 earth system models. J Clim 26(1s8):6801–6843. https://doi.org/10.1175/JCLI-D-12-00417.1

Arora VK, Boer GJ (2005) A parameterization of leaf phenology for the terrestrial ecosystem component of climate models. Glob Change Biol 11:39–59. https://doi.org/10.1111/j.1365-2486.2004.00890.x

Arora VK, Boer GJ (2010) Uncertainties in the 20th century carbon budget associated with land use change. Glob Change Biol 27:3310–3348. https://doi.org/10.1111/j.1365-2486.2010.02202.x

Aubinet M, Feigenwinter C, Heinesch B, Bernhofer C, Canepa E, Lindroth A, Montagnani L, Rebmann C, Sedlak P, Van Gorsel E (2010) Direct advection measurements do not help to solve the night-time CO_2 closure problem: evidence from three different forests. Agric For Meteorol 150(5):655–664. https://doi.org/10.1016/j.agrformet.2010.01.016

Aubinet M, Vesala T, Papale D (2012) Eddy covariance: a practical guide to measurement and data analysis. Springer.

Baghi R, Durand P, Jambert C, Jarnot C, Delon C, Serça D, Keravec P (2012) A new disjunct eddy-covariance system for BVOC flux measurements—validation on CO_2 and H_2O fluxes. Atmos Meas Tech 5:3119–3132. https://doi.org/10.5194/amt-5-3119-2012

Baker JM, Norman JM, Bland WL (1992) Field-scale application of flux measurement by conditional sampling. Agric For Meteorol 62:31–52. https://doi.org/10.1016/0168-1923(92)90004-N

Baldocchi DD (2001) Assessing ecosystem carbon balance: problems and prospects of the eddy covariance technique. Ann Rev Ecol Syst

Baldocchi DD (2003) Assessing the eddy covariance technique for evaluating carbon dioxide exchange rates of ecosystems: past, present and future. Global Change Biol 9(4):479–492. https://doi.org/10.1046/j.1365-2486.2003.00629.x

Baldocchi D, Falge E, Gu L, Olson R, Hollinger D, Running S, Wofsy S (2001) FLUXNET: a new tool to study the temporal and spatial variability of ecosystem-scale carbon dioxide, water vapor, and energy flux densities. Bull Am Meteor Soc 82(February):2415–2434. https://doi.org/10.1175/1520-0477(2001)082%3c2415:FANTTS%3e2.3.CO;2

Ball JT, Woodrow IE, Berry JA (1987) A model predicting stomatal conductance and its contribution to the control of photosynthesis under different environmental conditions. In: Progress in photosynthesis research. Martinus Nijhoff Publishers, Dordrecht. https://doi.org/10.1007/978-94-017-0519-6_48

Bao G, Bao Y, Qin Z, Xin X, Bao Y, Bayarsaikan S, Chuntai B (2016) Modeling net primary productivity of terrestrial ecosystems in the semi-arid climate of the Mongolian Plateau using LSWI-based CASA ecosystem model. Int J Appl Earth Obs Geoinf 46:84–93. https://doi.org/10.1016/j.jag.2015.12.001

Barman R, Jain AK, Liang M (2014a) Climate-driven uncertainties in modeling terrestrial gross energy and water fluxes: A site level to global-scale analysis. Glob Change Biol 20(6):1885–1900

Barman R, Jain AK, Liang M (2014b) Climate-driven uncertainties in modeling terrestrial gross primary production: A site level to global-scale analysis. Glob Change Biol 20(5):1394–1411. https://doi.org/10.1111/gcb.12474

Barr AG, Richardson AD, Hollinger DY, Papale D, Arain MA, Black TA, Schaeffer K (2013) Use of change-point detection for friction-velocity threshold evaluation in eddy-covariance studies. Agric For Meteorol 171:31–45. https://doi.org/10.1016/j.agrformet.2012.11.023

Bastos A, O'Sullivan M, Ciais P, Makowski D, Sitch S, Friedlingstein P, Chevallier F, Rödenbeck C, Pongratz J, Luijkx IT, Patra PK, Peylin P, Canadell JG, Lauerwald R, Li W, Smith NE, Peters W, Goll DS, Jain AK, Kato E, Lienert S, Lombardozzi DL, Haverd V, Nabel JEMS, Poulter B, Tian H, Walker AP, Zaehle S (2020) Sources of uncertainty in regional and global terrestrial CO_2 exchange estimates. Glob Biogeochem Cycles 34(2):e2019GB006393. https://doi.org/10.1029/2019GB006393

Beer C, Reichstein M, Tomelleri E, Ciais P, Jung M, Carvalhais N, Papale D (2010) Terrestrial gross carbon dioxide uptake: Global distribution and covariation with climate. Science 329:834–838. https://doi.org/10.1126/science.1184984

Beringer J, Hutley LB, McHugh I, Arndt SK, Campbell D, Cleugh HA, Cleverly J, de Dios VS, Eamus D, Evans B, Ewenz C, Grace P, Griebel A, Haverd V, Hinko-Najera N, Huete A, Isaac P,

Kanniah K, Leuning R, Liddell MJ, Macfarlane C, Meyer W, Moore C, Pendall E, Phillips A, Phillips RL, Prober SM, Restrepo-Coupe N, Rutledge S, Schroder I, Silberstein R, Southall P, Yee MS, Tapper NJ, van Gorsel E, Vote C, Walker J, Wardlaw T (2016) An introduction to the Australian and New Zealand flux tower network—OzFlux. Biogeosciences 13(21). https://doi.org/10.5194/bg-13-5895-2016

Bernacchi CJ, Singsaas EL, Pimentel C, Portis AR, Long SP (2001) Improved temperature response functions for models of Rubisco-limited photosynthesis. Plant Cell Environ 24:2. https://doi.org/10.1046/j.1365-3040.2001.00668.x

Betts RA (2000) Offset of the potential carbon sink from boreal forestation by decreases in surface albedo. Nature 408:187–190. https://doi.org/10.1038/35041545

Bonan GB, Levis S (2006) Evaluating aspects of the community land and atmosphere models (CLM3 and CAM3) using a dynamic global vegetation model. J Clim 19(11):2290–2301. https://doi.org/10.1175/JCLI3741.1

Bonan GB, Lawrence PJ, Oleson KW, Levis S, Jung M, Reichstein M, Lawrence DM, Swenson SC (2011) Improving canopy processes in the community land model version 4 (CLM4) using global flux fields empirically inferred from FLUXNET data. J Geophys Res 116. https://doi.org/10.1029/2010jg001593

Bongaarts J (2019) IPBES, 2019. Summary for policymakers of the global assessment report on biodiversity and ecosystem services of the Intergovernmental Science-Policy Platform on Biodiversity and Ecosystem Services. Popul Dev Rev. https://doi.org/10.1111/padr.12283

Bousquet P, Ciais P, Monfray P, Balkanski Y, Ramonet M, Tans P (1996) Influence of two atmospheric transport models on inferring sources and sinks of atmospheric CO_2. Tellus, Ser B Chem Phys Meteorol 48B:568–582. https://doi.org/10.3402/tellusb.v48i4.15932

Bousquet P, Ciais P, Peylin P, Ramonet M, Monfray P (1999) Inverse modeling of annual atmospheric CO_2 sources and sinks 1. Method and control inversion. J Geophys Res Atmos 104(D21):26161–26178. https://doi.org/10.1029/1999JD900342

Bowling DR, Delany AC, Turnipseed AA, Baldocchi DD, Monson RK (1999) Modification of the relaxed eddy accumulation technique to maximize measured scalar mixing ratio differences in updrafts and downdrafts. J Geophys Res Atmos 104:9121–9133. https://doi.org/10.1029/1999JD900013

Bowling DR, Tans PP, Monson RK (2001) Partitioning net ecosystem carbon exchange with isotopic fluxes of CO_2. Glob Change Biol 7:127–145. https://doi.org/10.1046/j.1365-2486.2001.00400.x

Bréda NJJ (2003) Ground-based measurements of leaf area index: a review of methods, instruments and current controversies. J Exp Bot 54(392):2403–2417. https://doi.org/10.1093/jxb/erg263

Buermann W, Dong J, Zeng X, Myneni RB, Dickinson RE (2001) Evaluation of the utility of satellite-based vegetation leaf area index data for climate simulations. J Clim 14:3536–3550. https://doi.org/10.1175/1520-0442(2001)014%3c3536:EOTUOS%3e2.0.CO;2

Burba G (2013) Eddy covariance method-for scientific, industrial, agricultural, and regulatory applications. Book

Burba G (2019) Illustrative maps of past and present eddy covariance measurement locations: II. High-resolution images. https://doi.org/10.13140/RG.2.2.33191.70561

Burba G, Anderson D (2007) Introduction to the eddy covariance method: general guidelines and conventional workflow. Li-Cor Biosci 141

Businger JA, Oncley SP (1990) Flux measurement with conditional sampling. J Atmos Oceanic Technol 7:349–352. https://doi.org/10.1175/1520-0426(1990)007%3c0349:fmwcs%3e2.0.co;2

Businger JA, Wyngaard JC, Izumi Y, Bradley EF (1971) Flux-profile relationships in the atmospheric surface layer. J Atmos Sci 28(2):181–189. https://doi.org/10.1175/1520-0469(1971)028%3c0181:fprita%3e2.0.co;2

Carlson TN, Ripley DA (1997) On the relation between NDVI, fractional vegetation cover, and leaf area index. Remote Sens Environ 62(3):241–252. https://doi.org/10.1016/S0034-4257(97)00104-1

Cervarich M, Shu S, Jain AK, Arneth A, Canadell J, Friedlingstein P, Houghton RA, Kato E, Koven C, Patra P, Poulter B, Sitch S, Stocker B, Viovy N, Wiltshire A, Zeng N (2016). The terrestrial carbon budget of South and Southeast Asia. Environ Res Lett 11(10):105006. https://doi.org/10.1088/1748-9326/11/10/105006

Chakraborty S, Tiwari YK, Deb Burman PK, Baidya Roy S, Valsala V (2020) Observations and modeling of GHG concentrations and fluxes over India. In: Krishnan R, Sanjay J, Gnanaseelan C, Mujumdar M, Kulkarni A, Chakraborty S (eds) Assessment of climate change over the Indian region. Springer, Singapore. https://doi.org/10.1007/978-981-15-4327-2_4

Chapin FS, Woodwell GM, Randerson JT, Rastetter EB, Lovett GM, Baldocchi DD, Clark DA, Harmon ME, Schimel DS, Valentini R, Schulze ED (2006) Reconciling carbon-cycle concepts, terminology, and methods. Ecosystems. https://doi.org/10.1007/s10021-005-0105-7

Chevallier F, Viovy N, Reichstein M, Ciais P (2006) On the assignment of prior errors in Bayesian inversions of CO_2 surface fluxes. Geophys Res Lett 33(L13802). https://doi.org/10.1029/2006GL026496

Chou WW, Wofsy SC, Harriss RC, Lin JC, Gerbig C, Sachse GW (2002) Net fluxes of CO_2 in Amazonia derived from aircraft observations. J Geophys Res Atmos 107(D22):4614. https://doi.org/10.1029/2001JD001295

Cihlar J, Caramori PH, Schuepp PH, Desjardins RL, Macpherson JI (1992) Relationship between satellite-derived vegetation indices and aircraft-based CO_2 measurements. J Geophys Res 97(D17):18515–18521. https://doi.org/10.1029/92jd00655

Clark DB, Mercado LM, Sitch S, Jones CD, Gedney N, Best MJ, Pryor M, Rooney GG, Essery RLH, Blyth E, Boucher O, Harding RJ, Huntingford C, Cox PM (2011) The joint UK land environment simulator (JULES), model description—Part 2: Carbon fluxes and vegetation dynamics. Geosci Model Dev 4(3):701–722. https://doi.org/10.5194/gmd-4-701-2011

Collatz GJ, Ball JT, Grivet C, Berry JA (1991) Physiological and environmental-regulation of stomatal conductance, photosynthesis and transpiration—a model that includes a laminar boundary-layer. Agric For Meteorol 54(2–4):107–136. https://doi.org/10.1016/0168-1923(91)90002-8

Collatz GJ, Ribas-Carbo M, Berry JA (1992) Coupled photosynthesis-stomatal conductance model for leaves of C4 plants. Aust J Plant Physiol 19(1139):519–539. https://doi.org/10.1071/PP9920519

Coops NC, Waring RH, Law BE (2005) Assessing the past and future distribution and productivity of ponderosa pine in the Pacific Northwest using a process model, 3-PG. Ecol Model 183(1):107–124. https://doi.org/10.1016/j.ecolmodel.2004.08.002

Cox PM, Betts RA, Collins M, Harris PP, Huntingford C, Jones CD (2004) Amazonian forest dieback under climate-carbon cycle projections for the 21st century. Theoret Appl Climatol 78:137–156. https://doi.org/10.1007/s00704-004-0049-4

Cramer W, Kicklighter DW, Bondeau A, Iii BM, Churkina G, Nemry B, Ruimy A, Schloss AL, ThE Participants OF ThE Potsdam NpP Model Intercomparison (1999). Comparing global models of terrestrial net primary productivity (NPP): overview and key results. Glob Change Biol 5:1–15. https://doi.org/10.1046/j.1365-2486.1999.00009.x

Cui Y, Xiao X, Zhang Y, Dong J, Qin Y, Doughty RB, Zhang G, Wang J, Wu X, Qin Y, Zhou S (2017) Temporal consistency between gross primary production and solar-induced chlorophyll fluorescence in the ten most populous megacity areas over years. Sci Rep 7:14963. https://doi.org/10.1038/s41598-017-13783-5

Dai Y, Dickinson R, Wang Y (2004) A two-big-leaf model for canopy temperature, photosynthesis, and stomatal conductance. J Clim 17:2281–2299. https://doi.org/10.1175/1520-0442(2004)017%3c2281:ATMFCT%3e2.0.CO;2

Dargaville R, McGuire AD, Rayner P (2002) Estimates of large-scale fluxes in high latitudes from terrestrial biosphere models and an inversion of atmospheric CO_2 measurements. Clim Change 55:273–285. https://doi.org/10.1023/A:1020295321582

Deb Burman PK, Sarma D, Williams M, Karipot A, Chakraborty S (2017) Estimating gross primary productivity of a tropical forest ecosystem over north-east India using LAI and meteorological variables. J Earth Syst Sci 126(7):1–16. https://doi.org/10.1007/s12040-017-0874-3

Deb Burman PK, Prabha TV, Morrison R, Karipot A (2018) A case study of turbulence in the nocturnal boundary layer during the Indian summer monsoon. Bound-Layer Meteorol 169(1):115–138

Deb Burman PK, Sarma D, Morrison R, Karipot A, Chakraborty S (2019) Seasonal variation of evapotranspiration and its effect on the surface energy budget closure at a tropical forest over north-east India. J Earth Syst Sci 128(5):1–21. https://doi.org/10.1007/s12040-019-1158-x

Deb Burman PK, Sarma D, Chakraborty S, Karipot A, Jain AK (2020a) The effect of Indian summer monsoon on the seasonal variation of carbon sequestration by a forest ecosystem over North-East India. SN Appl Sci 2(2):154. https://doi.org/10.1007/s42452-019-1934-x

Deb Burman PK, Shurpali NJ, Chowdhuri S, Karipot A, Chakraborty S, Lind SE, Prabha TV (2020b) Eddy covariance measurements of CO_2 exchange from agro-ecosystems located in subtropical (India) and boreal (Finland) climatic conditions. J Earth Syst Sci 129(1):43. https://doi.org/10.1007/s12040-019-1305-4

Demarty J, Chevallier F, Friend AD, Viovy N, Piao S, Ciais P (2007) Assimilation of global MODIS leaf area index retrievals within a terrestrial biosphere model. Geophys Res Lett 34(15):L15402. https://doi.org/10.1029/2007GL030014

Deng F, Chen JM, Plummer S, Chen M, Pisek J (2006) Algorithm for global leaf area index retrieval using satellite imagery. IEEE Trans Geosci Remote Sens 44(8):2219–2229. https://doi.org/10.1109/TGRS.2006.872100

Desjardins RL (1972) A study of carbon dioxide and sensible heat fluxes using the eddy correlation technique. Cornell University

Desjardins RL, Hart RL, Macpherson JI, Schuepp PH, Verma SB (1992) Aircraft- and tower-based fluxes of carbon dioxide, latent, and sensible heat. J Geophys Res 97(D17):18477–18485. https://doi.org/10.1029/92jd01625

Dyer AJ (1974) A review of flux-profile relationships. Bound-Layer Meteorol 7(3):363–372. https://doi.org/10.1007/BF00240838

Dyer AJ, Hicks BB (1970) Flux-gradient relationships in the constant flux layer. Q J R Meteorol Soc 96(410):715–721. https://doi.org/10.1002/qj.49709641012

El-Masri B, Barman R, Meiyappan P, Song Y, Liang M, Jain AK (2013) Carbon dynamics in the Amazonian Basin : integration of eddy covariance and ecophysiological data with a land surface model. Agric For Meteorol 182–183:156–167. https://doi.org/10.1016/j.agrformet.2013.03.011

Enting IG, Pearman GI (1986) The use of observations in calibrating and validating carbon cycle models. In: The changing carbon cycle. Springer, New York. https://doi.org/10.1007/978-1-4757-1915-4_21

Etzold S, Buchmann N, Eugster W (2010) Contribution of advection to the carbon budget measured by eddy covariance at a steep mountain slope forest in Switzerland. Biogeosciences 7(8):2461–2475. https://doi.org/10.5194/bg-7-2461-2010

Falge E, Baldocchi D, Olson R, Anthoni P, Aubinet M, Bernhofer C et al (2001) Gap filling strategies for defensible annual sums of net ecosystem exchange. Agric For Meteorol 107(1):43–69. https://doi.org/10.1016/S0168-1923(00)00225-2

Farquhar GD, Sharkey TD (1982) Stomatal conductance and photosynthesis. Ann Rev Plant Physiol 33(1):317–345. https://doi.org/10.1146/annurev.pp.33.060182.001533

Farquhar GD, Von Caemmerer S, Berry JA (1980) A biochemical model of photosynthetic CO_2 assimilation in leaves of C_3 species. Planta 149:78–90. https://doi.org/10.1007/BF00386231

Fensholt R, Sandholt I (2003) Derivation of a shortwave infrared water stress index from MODIS near- and shortwave infrared data in a semiarid environment. Remote Sens Environ 87:111–121. https://doi.org/10.1016/j.rse.2003.07.002

Fensholt R, Sandholt I, Rasmussen MS (2004) Evaluation of MODIS LAI, fAPAR and the relation between fAPAR and NDVI in a semi-arid environment using in situ measurements. Remote Sens Environ 91:490–507. https://doi.org/10.1016/j.rse.2004.04.009

Foken T (2008) Micrometeorology. Micrometeorology. https://doi.org/10.1007/978-3-540-74666-9

Foken T, Göckede M, Mauder M, Mahrt L, Amiro B, Munger W (2004) Post-field data quality control. In: Handbook of micrometeorology, pp 181–208. https://doi.org/10.1007/1-4020-2265-4_9

Foley JA, Prentice IC, Ramankutty N, Levis S, Pollard D, Sitch S, Haxeltine A (1996) An integrated biosphere model of land surface processes, terrestrial carbon balance, and vegetation dynamics. Global Biogeochem Cycles 10(4):603–628. https://doi.org/10.1029/96GB02692

Friedlingstein P, Meinshausen M, Arora VK, Jones CD, Anav A, Liddicoat SK, Knutti R (2014) Uncertainties in CMIP5 climate projections due to carbon cycle feedbacks. J Clim 27(2):511–526. https://doi.org/10.1175/JCLI-D-12-00579.1

Friedlingstein P, Jones MW, Sullivan M, Andrew RM, Hauck J, Peters GP, Peters W, Pongratz J, Sitch S, Le Quéré C, Bakker DCE, Canadell JG, Ciais P, Jackson RB, Anthoni P, Barbero L, Bastos A, Bastrikov V, Becker M, Bopp L, Buitenhuis E, Chandra N, Chevallier F, Chini LP, Currie KI, Feely RA, Gehlen M, Gilfillan D, Gkritzalis T, Goll DS, Gruber N, Gutekunst S, Harris I, Haverd V, Houghton RA, Hurtt G, Ilyina T, Jain AK, Joetzjer E, Kaplan JO, Kato E, Goldewijk KK, Korsbakken JI, Landschützer P, Lauvset SK, Lefèvre N, Lenton N, Lienert S, Lombardozzi D, Marland G, McGuire PC, Melton JR, Metzl N, Munro DR, Nabel JEMS, Nakaoka S-I, Neill C, Omar AM, Ono T, Peregon A, Pierrot D, Poulter B, Rehder G, Resplandy L, Robertson E, Rödenbeck C, Séférian R, Schwinger J, Smith N, Tans PP, Tian H, Tilbrook B, Tubiello FN, van der Werf GR, Wiltshire AJ, Zaehle S (2019) Global carbon budget 2019. Earth Syst Sci Data 11(4):1783–1838. https://doi.org/10.5194/essd-11-1783-2019

Friend AD, Arneth A, Kiang NY, Lomas M, Ogee J, Rödenbeck C, Running SW, Santaren JD, Sitch S, Viovy N, Ian Woodward F (2007) FLUXNET and modelling the global carbon cycle. Global Change Biol 13(3):610–633. https://doi.org/10.1111/j.1365-2486.2006.01223.x

Gaman A, Rannik Ü, Aalto P, Pohja T, Siivola E, Kulmala M, Vesala T (2004) Relaxed eddy accumulation system for size-resolved aerosol particle flux measurements. J Atmos Oceanic Technol 21:933–943. https://doi.org/10.1175/1520-0426(2004)021%3c0933:REASFS%3e2.0.CO;2

Ganguly S, Nemani RR, Zhang G, Hashimoto H, Milesi C, Michaelis A, Wang W, Votava P, Samanta A, Melton F, Dungan JL, Vermote E, Gao F, Knyazikhin Y, Myneni RB (2012) Generating global Leaf Area Index from Landsat: algorithm formulation and demonstration. Remote Sens Environ 122:185-2-2. https://doi.org/10.1016/j.rse.2011.10.032

Garcia de Cortázar V, Nobel PS (1990) Worldwide environmental productivity indices and yield predictions for a cam plant, Opuntia ficus-indica, including effects of doubled CO_2 levels. Agric For Meteorol 49:261–279. https://doi.org/10.1016/0168-1923(90)90001-M

Geissbühler P, Siegwolf R, Eugster W (2000) Eddy covariance measurements on mountain slopes: the advantage of surface-normal sensor orientation over a vertical set-up. Bound-Layer Meteorol 96:371–392. https://doi.org/10.1023/A:1002660521017

Gervois S, de Noblet-Ducoudré N, Viovy N, Ciais P, Brisson N, Seguin B, Perrier A (2004) Including croplands in a global biosphere model: methodology and evaluation at specific sites. Earth Interact. https://doi.org/10.1175/1087-3562(2004)8%3c1:iciagb%3e2.0.co;2

Gnanamoorthy P, Selvam V, Ramasubramanian R, Nagarajan R, Chakraborty S, Deb Burman PK, Karipot A (2019) Diurnal and seasonal patterns of soil CO_2 efflux from the Pichavaram mangroves, India. Environ Monit Assess 191(4)

Gnanamoorthy P, Selvam V, Deb Burman PK, Chakraborty S, Karipot A, Nagarajan R, Ramasubramanian R, Song Q, Zhang Y, Grace J (2020) Seasonal variations of net ecosystem (CO_2) exchange in the Indian tropical mangrove forest of Pichavaram. Estuar, Coast Shelf Sci 243:106828

Govindje (1995) Sixty-three years since Kautsky: chlorophyll a fluorescence. Funct Plant Biol 22:131–160. https://doi.org/10.1071/pp9950131

Graven HD, Keeling RF, Piper SC, Patra PK, Stephens BB, Wofsy SC, Welp LR, Sweeney C, Tans PP, Kelley JJ, Daube BC, Kort EA, Santoni GW, Bent JD (2013) Enhanced seasonal exchange of CO_2 by Northern ecosystems since 1960. Science 341(6150):1085–1089. https://doi.org/10.1126/science.1239207

Grossmann K, Frankenberg C, Magney TS, Hurlock SC, Seibt U, Stutz J (2018) PhotoSpec: a new instrument to measure spatially distributed red and far-red Solar-Induced Chlorophyll Fluorescence. Remote Sens Environ 216:311–327. https://doi.org/10.1016/j.rse.2018.07.002

Guanter L, Alonso L, Gómez-Chova L, Amorós-López J, Vila J, Moreno J (2007) Estimation of solar-induced vegetation fluorescence from space measurements. Geophys Res Lett 34:L08401. https://doi.org/10.1029/2007GL029289

Guanter L, Frankenberg C, Dudhia A, Lewis PE, Gómez-Dans J, Kuze A, Suto H, Grainger RG (2012) Retrieval and global assessment of terrestrial chlorophyll fluorescence from GOSAT space measurements. Remote Sens Environ 12:236–251. https://doi.org/10.1016/j.rse.2012.02.006

Guanter L, Zhang Y, Jung M, Joiner J, Voigt M. Berry JA, Frankenberg C, Huete AR, Zarco-Tejada P, Lee J-E, Moran M-S, Ponce-Campos G, Beer C, Camps-Valls G, Buchmann N, Gianelle D, Klumpp K, Cescatti A, Bake JM, Griffis TJ (2014) Global and time-resolved monitoring of crop photosynthesis with chlorophyll fluorescence. Proc Natl Acad Sci U S A 111(14):E1327–E1333. https://doi.org/10.1073/pnas.1320008111

Guenther A, Baugh W, Davis K, Hampton G, Harley P, Klinger L, Vierling L, Zimmerman P, Allwine W, Dilts S, Lamb B, Westberg H, Baldocchi D, Geron C, Pierce T (1996) Isoprene fluxes measured by enclosure, relaxed eddy accumulation, surface layer gradient, mixed layer gradient, and mixed layer mass balance techniques. J Geophys Res Atmos 101:18555–18567. https://doi.org/10.1029/96jd00697

Gurney, K. R., Law, R. M., Denning, A. S., Rayner, P. J., Baker, D., Bousquet, P., Bruhwiler L, Chen Y-H, Ciais P, Fan S, Fung IY, Gloor M, Heimann M, Higuchi K, John K, Maki T, Maksyutov S, Masarie K, Peylin P, Prather M, Pak BC, Randerson J, Sarmiento J, Taguchi S, Takahashi T, Yuen CW (2002) Towards robust regional estimates of CO_2 sources and sinks using atmospheric transport models. Nature 415:626–629. https://doi.org/10.1038/415626a

Gurney KR, Law RM, Denning AS, Rayner PJ, Pak BC, Baker D, Bousquet P, Bruhwiler L, Chen Y-H, Ciais P, Fung IY, Heimann M, John J, Maki T, Maksyutov S, Peylin P, Prather M, Taguchi, S. (2004). Transcom 3 inversion intercomparison: Model mean results for the estimation of seasonal carbon sources and sinks. Global Biogeochem Cycles 18(1):GB1010. https://doi.org/10.1029/2003gb002111

Gu L, Falge EM, Boden T, Baldocchi DD, Black TA, Saleska SR, Suni T, Verma SB, Vesala T, Wofsy SC, Xu L (2005) Objective threshold determination for nighttime eddy flux filtering. Agric For Meteorol 128(3–4):179–197. https://doi.org/10.1016/j.agrformet.2004.11.006

Heimann M, Kaminski T (1999) Inverse modelling approaches to infer surface trace gas fluxes from observed atmospheric mixing ratios. In Approaches to scaling a trace gas fluxes in ecosystems. Elsevier B.V., p. 2770295. https://doi.org/10.1016/S0167-5117(98)80035-9

He L, Magney T, Dutta D, Yin Y, Köhler P, Grossmann K, Stutz J, Dold C, Hatfield J, Guan K, Peng B, Frankenberg C (2020) From the ground to space: using solar-induced chlorophyll fluorescence to estimate crop productivity. Geophys Res Lett 47(7):e2020GL087474. https://doi.org/10.1029/2020GL087474

Hicks BB, McMillen RT (1984) A simulation of the Eddy accumulation method for measuring pollutant fluxes. J Climate Appl Meteorol 23:637–643. https://doi.org/10.1175/1520-0450(1984)023%3c0637:ASOTEA%3e2.0.CO;2

Hörtnagl L, Clement R, Graus M, Hammerle A, Hansel A, Wohlfahrt G (2010) Dealing with disjunct concentration measurements in eddy covariance applications: a comparison of available approaches. Atmos Environ 44:2024–2032. https://doi.org/10.1016/j.atmosenv.2010.02.042

Houweling S, Breon F-M, Aben I, Rödenbeck C, Gloor M, Heimann M, Ciais P (2003) Inverse modeling of CO_2 sources and sinks using satellite data: a synthetic inter-comparison of measurement techniques and their performance as a function of space and time. Atmos Chem Phys Discuss 4:523–538. https://doi.org/10.5194/acpd-3-5237-2003

Houweling S, Baker D, Basu S, Boesch H, Butz A, Chevallier F, Deng F, Dlugokencky EJ, Feng L, Ganshin A, Hasekamp O (2015). An intercomparison of inverse models for estimating sources and sinks of CO_2 using GOSAT measurements. J Geophys Res 120:5253–5266. https://doi.org/10.1002/2014JD022962

Huete AR (1988) A soil-adjusted vegetation index (SAVI). Remote Sens Environ 25:295–309. https://doi.org/10.1016/0034-4257(88)90106-X

Huete A, Didan K, Miura T, Rodriguez EP, Gao X, Ferreira LG (2002) Overview of the radiometric and biophysical performance of the MODIS vegetation indices. Remote Sens Environ 83:195–213. https://doi.org/10.1016/S0034-4257(02)00096-2

IPCC (2013) Climate Change 2013: The Physical Science Basis. Contribution of Working Group I to the Fifth Assessment Report of the Intergovernmental Panel on Climate Change [Stocker TF, Qin D, Plattner G-K, Tignor M, Allen SK, Boschung J, Nauels A, Xia Y, Bex V, Midgley PM (eds)]. Cambridge University Press, Cambridge, United Kingdom and New York, NY, USA, 1535 pp

IPCC (2019) Climate Change and Land: an IPCC special report on climate change, desertification, land degradation, sustainable land management, food security, and greenhouse gas fluxes in terrestrial ecosystems [Shukla PR, Skea J, Calvo Buendia E, Masson-Delmotte V, Pörtner H-O, Roberts DC, Zhai P, Slade R, Connors S, van Diemen R, Ferrat M, Haughey E, Luz S, Neogi S, Pathak M, Petzold J, Portugal Pereira J, Vyas P, Huntley E, Kissick K, Belkacemi M, Malley J (eds)]. In press

Ito A, Oikawa T (2002) A simulation model of the carbon cycle in land ecosystems (Sim-CYCLE): A description based on dry-matter production theory and plot-scale validation. Ecol Model 151:143–176. https://doi.org/10.1016/S0304-3800(01)00473-2

Jarvis PG (1976) The interpretation of the variations in leaf water potential and stomatal conductance found in canopies in the field. Philos Trans R Soc Lond B, Biol Sci 273(927):593–610. https://doi.org/10.1098/rstb.1976.0035

Jiang Z, Huete AR, Didan K, Miura T (2008) Development of a two-band enhanced vegetation index without a blue band. Remote Sens Environ 112(10):3833–3845. https://doi.org/10.1016/j.rse.2008.06.006

Joiner J, Guanter L, Lindstrot R, Voigt M, Vasilkov AP, Middleton EM, Huemmrich KF, Yoshida Y, Frankenberg C (2013) Global monitoring of terrestrial chlorophyll fluorescence from moderate-spectral-resolution near-infrared satellite measurements: methodology, simulations, and application to GOME-2. Atmos Meas Tech 6:2803–2823. https://doi.org/10.5194/amt-6-2803-2013

Jung M, Reichstein M, Margolis HA, Cescatti A, Richardson AD, Arain MA, Arneth A, Bernhofer C, Bonal D, Chen J, Gianelle D, Gobron N, Kiely G, Kutsch W, Lasslop G, Law BE, Lindroth A, Merbold L, Montagnani L, Moors EJ, Papale D, Sottocornola M, Vaccari F, Williams C (2011) Global patterns of land-atmosphere fluxes of carbon dioxide, latent heat, and sensible heat derived from eddy covariance, satellite, and meteorological observations. J Geophys Res Biogeosci 116(3):2–4. https://doi.org/10.1029/2010JG001566

Kadygrov N, Maksyutov S, Eguchi N, Aoki T, Nakazawa T, Yokota T, Inoue G (2009) Role of simulated GOSAT total column CO_2 observations in surface CO_2 flux uncertainty reduction. J Geophys Res: Atmos 114(D21)

Kaimal JC, Finnigan JJ (1994) Atmospheric boundary layer flows: their structure and measurement. Oxford University Press

Kaminski T, Heimann M, Giering R (1999) A coarse grid three-dimensional global inverse model of the atmospheric transport 1. Adjoint model and Jacobian matrix. J Geophys Res Atmos 104(D15):18535–18553. https://doi.org/10.1029/1999JD900147

Kato T, Ito A, Kawamiya M (2009) Multiple temporal scale variability during the twentieth century in global carbon dynamics simulated by a coupled climate—terrestrial carbon cycle model. Clim Dyn 32:901–923. https://doi.org/10.1007/s00382-009-0548-1

Kattge J, Knorr W, Raddatz T, Wirth C (2009) Quantifying photosynthetic capacity and its relationship to leaf nitrogen content for global-scale terrestrial biosphere models. Glob Change Biol 15:976–991. https://doi.org/10.1111/j.1365-2486.2008.01744.x

Klein Goldewijk K, Beusen A, Van Drecht G, De Vos M (2011) The HYDE 3.1 spatially explicit database of human-induced global land-use change over the past 12,000 years. Global Ecol Biogeograp 20(1):73–86. https://doi.org/10.1111/j.1466-8238.2010.00587.x

Kljun N, Calanca P, Rotach MW, Schmid HP (2015) A simple two-dimensional parameterisation for Flux Footprint Prediction (FFP). Geosci Model Dev 8(11):3695–3713. https://doi.org/10.5194/gmd-8-3695-2015

Kluge M, Ting IP (2012) Crassulacean acid metabolism: analysis of an ecological adaptation, 30th edn. Springer, Berlin. https://doi.org/10.1007/978-3-642-67038-1

Knyazikhin Y, Martonchik JV, Myneni RB, Diner DJ, Running SW (1998) Synergistic algorithm for estimating vegetation canopy leaf area index and fraction of absorbed photosynthetically active radiation from MODIS and MISR data. J Geophys Res Atmos 103:32257–32275. https://doi.org/10.1029/98JD02462

Köhler P, Frankenberg C, Magney TS, Guanter L, Joiner J, Landgraf J (2018) Global retrievals of solar-induced chlorophyll fluorescence with TROPOMI: first results and intersensor comparison to OCO-2. Geophys Res Lett 45(19):10456–10463. https://doi.org/10.1029/2018GL079031

Kormann R, Meixner FX (2001) An analytical footprint model for non-neutral stratification. Bound-Layer Meteorol 99(2):207–224. https://doi.org/10.1023/A:1018991015119

Kumar D, Scheiter S (2019) Biome diversity in South Asia—how can we improve vegetation models to understand global change impact at regional level? Sci Total Environ 671:1001–1016. https://doi.org/10.1016/j.scitotenv.2019.03.251

Kutsch WL, Bahn M, Heinemeyer A (2010) Soil carbon dynamics: an integrated methodology. https://doi.org/10.1017/CBO9780511711794

Langerwisch F, Vaclavik T, Von Bloh W, Vetter T, Thonicke K (2018) Combined effects of climate and land-use change on the provision of ecosystem services in rice agro-ecosystems. Environ Res Lett 13:015003. https://doi.org/10.1088/1748-9326/aa954d

Lasslop G, Reichstein M, Papale D, Richardson AD, Arneth A, Barr A, Stoy P, Wohlfahrt G (2010) Separation of net ecosystem exchange into assimilation and respiration using a light response curve approach : critical issues and global evaluation. Global Change Biol 16(1):187–208. https://doi.org/10.1111/j.1365-2486.2009.02041.x

Law BE, Ryan MG, Anthoni PM (1999) Seasonal and annual respiration of a ponderosa pine ecosystem. Glob Change Biol 5(2):169–182

Lawrence DM, Oleson KW, Flanner MG, Thornton PE, Swenson SC, Lawrence PJ, Slater AG (2011) Parameterization improvements and functional and structural advances in Version 4 of the Community Land Model. J Adv Model Earth Syst 3(1). https://doi.org/10.1029/2011ms000045

Lee X (2018) Fundamentals of boundary-layer meteorology. Springer, Cham, Switzerland. https://doi.org/10.1007/978-3-319-60853-2

Leuning R (1990) Modelling stomatal behaviour and and photosynthesis of *Eucalyptus grandis*. Funct Plant Biol 17(2):159–175. https://doi.org/10.1071/pp9900159

Levy PE, Cannell MGR, Friend AD (2004) Modelling the impact of future changes in climate, CO_2 concentration and land use on natural ecosystems and the terrestrial carbon sink. Global Environ Change 14:21–30. https://doi.org/10.1016/j.gloenvcha.2003.10.005

Liu J, Chen JM, Cihlar J, Park WM (1997) A process-based boreal ecosystem productivity simulator using remote sensing inputs. Remote Sens Environ 62:158–175. https://doi.org/10.1016/S0034-4257(97)00089-8

Liu J, Bowman KW, Lee M (2016) Comparison between the local ensemble transform kalman filter (LETKF) and 4D-Var in atmospheric CO_2 flux inversion with the goddard earth observing system-chem model and the observation impact diagnostics from the LETKF. J Geophys Res 121:13066–13087. https://doi.org/10.1002/2016JD025100

Lloyd J, Taylor JA (1994) On the temperature dependence of soil respiration. Funct Ecol 8(3):315. https://doi.org/10.2307/2389824

Macpherson JI, Grossman RL, Kelly RD (1992) Intercomparison results for FIFE flux aircraft. J Geophys Res 97(D17):18499–18514. https://doi.org/10.1029/92jd00272

Massman WJ, Lee X (2002) Eddy covariance flux corrections and uncertainties in long-term studies of carbon and energy exchanges. Agric For Meteorol 113(1–4):121–144. https://doi.org/10.1016/S0168-1923(02)00105-3

Mauder M, Cuntz M, Drüe C, Graf A, Rebmann C, Schmid HP, Schmidt M, Steinbrecher R (2013) A strategy for quality and uncertainty assessment of long-term eddy-covariance measurements. Agric For Meteorol 169:122–135. https://doi.org/10.1016/j.agrformet.2012.09.006

Meehl GA, Boer GJ, Covey C, Latif M, Stouffer RJ (2000) The coupled model intercomparison project (CMIP). Bull Am Meteor Soc 81(2):313–318. https://doi.org/10.1175/1520-0477(2000)081%3c0313:tcmipc%3e2.3.co;2

Meroni M, Rossini M, Guanter L, Alonso L, Rascher U, Colombo R, Moreno J (2009) Remote sensing of solar-induced chlorophyll fluorescence: review of methods and applications. Remote Sens Environ 113:2037–2051. https://doi.org/10.1016/j.rse.2009.05.003

Meyers TP, Hall ME, Lindberg SE, Kim K (1996) Use of the modified Bowen-ratio technique to measure fluxes of trace gases. Atmos Environ 30(19):3321–3329. https://doi.org/10.1016/1352-2310(96)00082-9

Milne R, Beverland IJ, Hargreaves K, Moncrieff JB (1999) Variation of the β coefficient in the relaxed eddy accumulation method. Bound-Layer Meteorol 93:221–225. https://doi.org/10.1023/A:1002061514948

Minx JC, Lamb WF, Callaghan MW, Fuss S, Hilaire J, Creutzig F, Amann T, Beringer T, de Oliveira Garcia W, Hartmann J, Khanna T, Lenzi D, Luderer G, Nemet GF, Rogelj J, Smith P, Vicente JLV, Wilcox J, Del Mar Zamora Dominguez M (2018) Negative emissions—Part 1: Research landscape and synthesis. Environ Res Lett 13:063001. https://doi.org/10.1088/1748-9326/aabf9b

Moffat AM, Papale D, Reichstein M, Hollinger DY, Richardson AD, Barr AG, Beckstein C, Braswell BH, Churkina G, Desai AR, Falge E (2007) Comprehensive comparison of gap-filling techniques for eddy covariance net carbon fluxes. Agric For Meteorol 147(3):209–232. https://doi.org/10.1016/j.agrformet.2007.08.011

Mohammed GH, Colombo R, Middleton EM, Rascher U, van der Tol C, Nedbal L, Goulas Y, Pérez-Priego O, Damm A, Meroni M, Joiner J, Cogliati S, Verhoef W, Malenovský Z, Gastellu-Etchegorry J-P, Miller JR, Guanter L, Moreno J, Moya I, Berry JA, Frankenberg C, Zarco-Tejada PJ (2019) Remote sensing of solar-induced chlorophyll fluorescence (SIF) in vegetation: 50 years of progress. Remote Sens Environ. https://doi.org/10.1016/j.rse.2019.04.030

Moncrieff JB, Massheder JM, De Bruin H, Elbers J, Friborg T, Heusinkveld B, Kabat P, Scott S, Søgaard H, Verhoef A (1997) A system to measure surface fluxes of momentum, sensible heat, water vapour and carbon dioxide. J Hydrol 188:589–611.

Moncrieff J, Clement R, Finnigan J, Meyers T (2004) Averaging, detrending, and filtering of eddy covariance time series. In: Handbook of micrometeorology, pp 7–31. https://doi.org/10.1007/1-4020-2265-4_2

Monteith JL et al (1965) Evaporation and environment. In: Symposia of the society for experimental biology, vol 19, p 4

Monteith JL (1972) Solar radiation and productivity in tropical ecosystems. J Appl Ecol 9(3):747–766. https://doi.org/10.2307/2401901

Monteith JL (1977) Climate and the efficiency of crop production in Britain. *Philosophical Transactions of the Royal Society of London. B, Biological Sciences, 281*(980), 277–294. https://doi.org/10.1098/rstb.1977.0140

Myles L, Meyers TP, Robinson L (2007) Relaxed eddy accumulation measurements of ammonia, nitric acid, sulfur dioxide and particulate sulfate dry deposition near Tampa, FL, USA. Environ Res Lett 2(3):034004. https://doi.org/10.1088/1748-9326/2/3/034004

Myneni RB, Knyazikhin Y, Privette JL, Running SW, Nemani R, Zhang Y, et al (1999) MODIS Leaf Area Index (LAI) and Fraction of Photosynthetically Active Radiation absorbed by vegetation (FPAR) product. Algorithm Theoretical Basis Document. https://eospso.gsfc.nasa.gov/atbd/modistables.html

Nalini K, Sijikumar S, Valsala V, Tiwari YK, Ramachandran R (2019) Designing surface CO2 monitoring network to constrain the Indian land fluxes. Atmos Environ 218:117003. https://doi.org/10.1016/j.atmosenv.2019.117003

Oechel WC, Vourlitis GL, Brooks S, Crawford TL, Dumas E (1998) Intercomparison among chamber, tower, and aircraft net CO_2 and energy fluxes measured during the Arctic System Science

Land-Atmosphere-Ice Interactions (ARCSS-LAII) Flux Study. J Geophys Res 103(D22):28993–29903. https://doi.org/10.1029/1998JD200015

Ohkubo S, Kosugi Y, Takanashi S, Mitani T, Tani M (2007) Comparison of the eddy covariance and automated closed chamber methods for evaluating nocturnal CO_2 exchange in a Japanese cypress forest. Agric For Meteorol 142:50–65. https://doi.org/10.1016/j.agrformet.2006.11.004

Oleson KW, Lawrence DM, Bonan GB, Drewniak B (2013) Technical description of version 4.5 of the Community Land Model (CLM). https://doi.org/10.5065/D6RR1W7M

Pandey S, Houweling S, Krol M, Aben I, Chevallier F, Dlugokencky EJ, Gatti LV, Gloor E, Miller JB, Detmers R, Machida T, Röckmann T (2016) Inverse modeling of GOSAT-retrieved ratios of total column CH_4 and CO_2 for 2009 and 2010. Atmos Chem Phys 16:5043–5062. https://doi.org/10.5194/acp-16-5043-2016

Patel NR, Padalia H, Devadas R, Huete A, Senthil Kumar A, Krishna Murthy YVN (2018) Estimating net primary productivity of croplands in Indo-Gangetic Plains using GOME-2 sun-induced fluorescence and MODIS NDVI. Curr Sci 114(6):1333–1337. https://doi.org/10.18520/cs/v114/i06/1333-1337

Patra PK, Niwa Y, Schuck TJ, Brenninkmeijer CAM, MacHida T, Matsueda H, Sawa Y (2011) Carbon balance of South Asia constrained by passenger aircraft CO_2 measurements. Atmos Chem Phys 11(9):4163–4175. https://doi.org/10.5194/acp-11-4163-2011

Patra PK, Canadell JG, Houghton RA, Piao SL, Oh NH, Ciais P, Manjunath KR, Chhabra A, Wang T, Bhattacharya T, Bousquet P, Hartman J, Ito A, Mayorga E, Niwa Y, Raymond PA, Sarma VVSS, Lasco R (2013) The carbon budget of South Asia. Biogeosciences 10(1):513–527. https://doi.org/10.5194/bg-10-513-2013

Pattey E, Desjardins RL, Rochette P (1993) Accuracy of the relaxed eddy-accumulation technique, evaluated using CO_2 flux measurements. Bound-Layer Meteorol 66:341–355. https://doi.org/10.1007/BF00712728

Paw UKT, Baldocchi DD, Meyers TP, Wilson KB (2000) Correction of eddy-covariance measurements incorporating both advective effects and density fluxes. Bound-Layer Meteorol 97:487–511. https://doi.org/10.1023/A:1002786702909

Peters W, Jacobson AR, Sweeney C, Andrews AE, Conway TJ, Masarie K, Miller JB, Bruhwiler LMP, Petron G, Hirsch AI, Worthy DEJ, van der Werf GR, Randerson JT, Wennberg PO, Krol MC, Tans PP (2007) An atmospheric perspective on North American carbon dioxide exchange: CarbonTracker. Proc Natl Acad Sci U S A 104(48):18925–18930. https://doi.org/10.1073/pnas.0708986104

Peylin P, Law RM, Gurney KR, Chevallier F, Jacobson AR, Maki T, Niwa Y, Patra PK, Peters W, Rayner PJ, Rödenbeck C, van der Laan-Luijkx IT, Zhang X (2013) Global atmospheric carbon budget: results from an ensemble of atmospheric CO_2 inversions. Biogeosciences 10:6699–6720. https://doi.org/10.5194/bg-10-6699-2013

Pickett-Heaps CA, Rayner PJ, Law RM, Ciais P, Patra PK, Bousquet P, Peylin P, Maksyutov S, Marshall J, Rödenbeck C, Langenfelds RL, Steele LP, Francey RJ, Tans P, Sweeney C (2011) Atmospheric CO_2 inversion validation using vertical profile measurements: analysis of four independent inversion models. J Geophys Res Atmos 116(D12305):1–17. https://doi.org/10.1029/2010JD014887

Pillai D, Gerbig C, Kretschmer R, Beck V, Karstens U, Neininger B, Heimann M (2012) Comparing Lagrangian and Eulerian models for CO_2 transport-a step towards Bayesian inverse modeling using WRF/STILT-VPRM. Atmos Chem Phys 12:8979–8991. https://doi.org/10.5194/acp-12-8979-2012

Pirk N, Sievers J, Mertes J, Parmentier FJW, Mastepanov M, Christensen TR (2017) Spatial variability of CO_2 uptake in polygonal tundra: Assessing low-frequency disturbances in eddy covariance flux estimates. Biogeosciences 14(12):3157–3169. https://doi.org/10.5194/bg-14-3157-2017

Pisso I, Patra P, Takigawa M, Machida T, Matsueda H, Sawa Y (2019) Assessing Lagrangian inverse modelling of urban anthropogenic CO_2 fluxes using in situ aircraft and ground-based measurements in the Tokyo area. Carbon Balance Manage 14(6):1–23. https://doi.org/10.1186/s13021-019-0118-8

Potter CS, Randerson JT, Field CB, Matson PA, Vitousek PM, Mooney HA, Klooster SA (1993) Terrestrial ecosystem production: a process model based on global satellite and surface data. Global Biogeochem Cycles 7(4):811–841. https://doi.org/10.1029/93GB02725

Poulter B, Ciais P, Hodson E, Lischke H, Maignan F, Plummer S, Zimmermann NE (2011) Plant functional type mapping for earth system models. Geosci Model Dev 4:993–1010. https://doi.org/10.5194/gmd-4-993-2011

Prentice IC, Cowling SA (2013). Dynamic global vegetation models. In: Encyclopedia of biodiversity: Second Edition. https://doi.org/10.1016/B978-0-12-384719-5.00412-3

Prentice I, Farquhar G, Fasham, M. (2001). The carbon cycle and atmospheric carbon dioxide. In: Climate change 2001: the scientific basis. https://doi.org/10.1256/004316502320517344

Prince SD, Goward SN (1995) Global primary production: a remote sensing approach. J Biogeogr 22(4):815–835

Ramankutty N, Foley JA (1999) Estimating historical changes in global land cover: croplands from 1700 to 1992. Global Biogeochem Cycles 13(4):997–1027. https://doi.org/10.1029/1999GB900046

Raymond Hunt E (1994) Relationship between woody biomass and PAR conversion efficiency for estimating net primary production from NDVI. Int J Remote Sens 15(8):1725–1729. https://doi.org/10.1080/01431169408954203

Rayner PJ, Law RM, O'Brien DM, Butler TM, Dilley AC (2002) Global observations of the carbon budget 3. Initial assessment of the impact of satellite orbit, scan geometry, and cloud on measuring CO_2 from space. J Geophys Res Atmos 107(D21):4557. https://doi.org/10.1029/2001JD000618

Rayner PJ, Scholze M, Knorr W, Kaminski T, Giering R, Widmann H (2005) Two decades of terrestrial carbon fluxes from a carbon cycle data assimilation system (CCDAS). Global Biogeochem Cycles 19:GB2026. https://doi.org/10.1029/2004GB002254

Rebmann C, Aubinet M, Schmid H, Arriga N, Aurela M, Burba, G., Clement R, De Ligne A, Fratini G, Gielen B, Grace J, Graf A, Gross P, Haapanala S, Herbst M, Hörtnagl L, Ibrom A, Joly L, Kljun N, Kolle O, Kowalski A, Lindroth A, Loustau D, Mammarella I, Mauder M, Merbold L, Metzger S, Mölder M, Montagnani L, Papale D, Pavelka M, Peichl M, Roland M, Serrano-Ortiz P, Siebicke L, Steinbrecher R, Tuovinen J-P, Vesala T, Wohlfahrt G, Franz D (2018) ICOS eddy covariance flux-station site setup: a review. Int Agrophys 32(4):471–494. https://doi.org/10.1515/intag-2017-0044

Reichstein M (2005) On the separation of net ecosystem exchange into assimilation and ecosystem respiration: review and improved algorithm. Glob Change Biol 11:1–16

Reynolds O (1895) IV. On the dynamical theory of incompressible viscous fluids and the determination of the criterion. Philos Trans R Soc Lond A 186:123–164. https://doi.org/10.1098/rsta.1895.0004

Rinne HJI, Guenther AB, Warneke C, De Gouw JA, Luxembourg SL (2001) Disjunct eddy covariance technique for trace gas flux measurements. Geophys Res Lett 28:3139–3142. https://doi.org/10.1029/2001GL012900

Running SW, Hunt ER (1993) Generalization of a Forest Ecosystem Process Model for Other Biomes, BIOME-BGC, and an Application for Global-Scale Models. Scal Physiol Process. https://doi.org/10.1016/b978-0-12-233440-5.50014-2

Running SW, Investigator P, Nemani R, Investigator A, Glassy JM, Engineer S, Thornton PE (1999) Modis daily photosynthesis (PSN) and annual net primary production (NPP) product. Algorithm theoretical basis Document

Ryan MG (1991) Effects of climate change on plant respiration. Ecol Appl 1(2):157–167. https://doi.org/10.2307/1941808

Schuh AE, Jacobson AR, Basu S, Weir B, Baker D, Bowman K, Chevallier F, Crowell S, Davis KJ, Deng F, Denning S, Feng L, Jones D, Liu J, Palmer PI (2019) Quantifying the impact of atmospheric transport uncertainty on CO_2 surface flux estimates. Global Biogeochem Cycles 33:484–500. https://doi.org/10.1029/2018GB006086

Sellers PJ, Berry JA, Collatz GJ, Field CB, Hall FG (1992) Canopy reflectance, photosynthesis, and transpiration. III. A reanalysis using improved leaf models and a new canopy integration scheme. Remote Sens Environ 42(3):187–216. https://doi.org/10.1016/0034-4257(92)90102-P

Sims DA, Rahman AF, Cordova VD, El-Masri BZ, Baldocchi DD, Bolstad PV, Flanagan L, Goldstein A, Hollinger D, Xu L (2008) A new model of gross primary productivity for North American ecosystems based solely on the enhanced vegetation index and land surface temperature from MODIS. Remote Sens Environ 112:1633–1646. https://doi.org/10.1016/j.rse.2007.08.004

Siqueira M, Lai CT, Katul G (2000) Estimating scalar sources, sinks, and fluxes in a forest canopy using Lagrangian, Eulerian, and hybrid inverse models. J Geophys Res Atmos 105(D24):29475–29488. https://doi.org/10.1029/2000JD900543

Sitch S, Smith B, Prentice IC, Arneth A, Bondeau A, Cramer W, Kaplan JO, Levis S, Lucht W, Sykes MT, Thonicke K, Venevsky S (2003) Evaluation of ecosystem dynamics, plant geography and terrestrial carbon cycling in the LPJ dynamic global vegetation model. Global Change Biol 9:161–185. https://doi.org/10.1046/j.1365-2486.2003.00569.x

Skaggs TH, Anderson RG, Alfieri JG, Scanlon TM, Kustas WP (2018) Fluxpart: open source software for partitioning carbon dioxide and water vapor fluxes. Agric For Meteorol 253–254:218–224. https://doi.org/10.1016/j.agrformet.2018.02.019

Stewart JB (1988) Modelling surface conductance of pine forest. Agric For Meteorol 43(1):19–35. https://doi.org/10.1016/0168-1923(88)90003-2

Stull RB (1988) An introduction to boundary layer meteorology, vol 13. Springer, Dordrecht. https://doi.org/10.1007/978-94-009-3027-8

Sun Y, Frankenberg C, Wood JD, Schimel DS, Jung M, Guanter L, Drewry DT, Verma M, Porcar-Castell A, Griffis TJ, Gu L, Magney TS, Köhler P, Evans B, Yuen K (2017) OCO-2 advances photosynthesis observation from space via solar-induced chlorophyll fluorescence. Science 358(6360):eaam5747. https://doi.org/10.1126/science.aam5747

Thornton PE, Rosenbloom NA (2005) Ecosystem model spin-up: Estimating steady state conditions in a coupled terrestrial carbon and nitrogen cycle model. Ecol Model 189:25–48. https://doi.org/10.1016/j.ecolmodel.2005.04.008

Tramontana G, Jung M, Schwalm CR, Ichii K, Camps-Valls G, Ráduly B., Reichstein M, Altaf Arain M, Cescatti A, Kiely G, Merbold L, Serrano-Ortiz P, Sickert S, Wolf S, Papale D (2016) Predicting carbon dioxide and energy fluxes across global FLUXNET sites with regression algorithms. Biogeosciences 13:4291–4313. https://doi.org/10.5194/bg-13-4291-2016

Turner DP, Ritts WD, Cohen WB, Gower ST, Running SW, Zhao M, Costa MH, Kirschbaum AA, Ham JM, Saleska SR, Ahl DE (2006) Evaluation of MODIS NPP and GPP products across multiple biomes. Remote Sens Environ 102:282–292. https://doi.org/10.1016/j.rse.2006.02.017

Tuzet A, Perrier A, Leuning R (2003) A coupled model of stomatal conductance, photosynthesis and transpiration. Plant, Cell Environ 26:1097–1116. https://doi.org/10.1046/j.1365-3040.2003.01035.x

Valentini R, Greco S, Seufert G, Bertin N, Ciccioli P, Cecinato A, Brancaleoni E, Frattoni M (1997) Fluxes of biogenic VOC from Mediterranean vegetation by trap enrichment relaxed eddy accumulation. Atmos Environ 31:229–238. https://doi.org/10.1016/S1352-2310(97)00085-X

Valsala V, Tiwari YK, Pillai P, Roxy M, Maksyutov S, Murtugudde R (2013) Intraseasonal variability of terrestrial biospheric CO_2 fluxes over India during summer monsoons. J Geophys Res Biogeosci 118(2):752–769. https://doi.org/10.1002/jgrg.20037

Vickers D, Mahrt L (1997) Quality control and flux sampling problems for tower and aircraft data. J Atmos Oceanic Technol 14:512–526. https://doi.org/10.1175/1520-0426(1997)014%3c0512:QCAFSP%3e2.0.CO;2

Vidal J, Chollet R (1997) Regulatory phosphorylation of C4 PEP carboxylase. Trends Plant Sci 2(6):230–237. https://doi.org/10.1016/S1360-1385(97)89548-9

Wagle P, Zhang Y, Jin C, Xiao X (2016) Comparison of solar-induced chlorophyll fluorescence, light-use efficiency, and process-based GPP models in maize. Ecol Appl 26(4):1211–1222. https://doi.org/10.1890/15-1434

Wang M, Guan DX, Han SJ, Wu JL (2010) Comparison of eddy covariance and chamber-based methods for measuring CO2 flux in a temperate mixed forest. Tree Physiol 30:149–163. https://doi.org/10.1093/treephys/tpp098

Watson DJ (1947) Comparative physiological studies on the growth of field crops: I. Variation in net assimilation rate and leaf area between species and varieties, and within and between years. Ann Botany 11(41):41–76. https://doi.org/10.1093/oxfordjournals.aob.a083148

Whitley R, Zeppel M, Armstrong N, Macinnis-Ng C, Yunusa I, Eamus D (2008) A modified Jarvis-Stewart model for predicting stand-scale transpiration of an Australian native forest. Plant Soil 305:35–47. https://doi.org/10.1007/s11104-007-9399-x

Williams M, Rastetter EB, Fernandes DN, Goulden ML (1996) Modelling the soil-plant-atmosphere continuum in a Quercus-Acer stand at Harvard Forest: the regulation of stomatal conductance by light, nitrogen and soil/plant hydraulics. Plant, Cell Environ 19:911–927. https://doi.org/10.1111/j.1365-3040.1996.tb00456.x

Wolanin A, Rozanov VV, Dinter T, Noël S, Vountas M, Burrows JP, Bracher A (2015) Global retrieval of marine and terrestrial chlorophyll fluorescence at its red peak using hyperspectral top of atmosphere radiance measurements: Feasibility study and first results. Remote Sens Environ 166:243–261. https://doi.org/10.1016/j.rse.2015.05.018

Wutzler T, Lucas-Moffat A, Migliavacca M, Knauer J, Sickel K, Šigut L, Menzer O, Reichstein M (2018) Basic and extensible post-processing of eddy covariance flux data with REddyProc. Biogeosciences 15(16):5015–5030. https://doi.org/10.5194/bg-15-5015-2018

Xiao X, Hollinger D, Aber J, Goltz M, Davidson EA, Zhang Q, Moore B (2004) Satellite-based modeling of gross primary production in an evergreen needleleaf forest. Remote Sens Environ 89:519–534. https://doi.org/10.1016/j.rse.2003.11.008

Yan M, Tian X, Li Z, Chen E, Li C, Fan W (2016) A long-term simulation of forest carbon fluxes over the Qilian Mountains. Int J Appl Earth Obs Geoinf 52:512–526. https://doi.org/10.1016/j.jag.2016.07.009

Ye ZP, Yu Q (2008) A coupled model of stomatal conductance and photosynthesis for winter wheat. Photosynthetica 46(4):637–640. https://doi.org/10.1007/s11099-008-0110-0

Yu D, Shi P, Shao H, Zhu W, Pan Y (2009) Modelling net primary productivity of terrestrial ecosystems in East Asia based on an improved CASA ecosystem model. Int J Remote Sens 30(18):4851–4866. https://doi.org/10.1080/01431160802680552

Zarco-Tejada PJ, Morales A, Testi L, Villalobos FJ (2013) Spatio-temporal patterns of chlorophyll fluorescence and physiological and structural indices acquired from hyperspectral imagery as compared with carbon fluxes measured with eddy covariance. Remote Sens Environ 133:102–115. https://doi.org/10.1016/j.rse.2013.02.003

Zeng X, Shajkh M, Dai Y, Dickinson RE, Myneni R (2002) Coupling of the common land model to the NCAR community climate model. J Clim 15:1832–1854. https://doi.org/10.1175/1520-0442(2002)015%3c1832:COTCLM%3e2.0.CO;2

Zhang Y, Xu M, Chen H, Adams J (2009) Global pattern of NPP to GPP ratio derived from MODIS data: effects of ecosystem type, geographical location and climate. Glob Ecol Biogeogr 18:280–290. https://doi.org/10.1111/j.1466-8238.2008.00442.x

Zhang Y, Guanter L, Berry JA, Joiner J, van der Tol C, Huete A, Gitelson A, Voigt M, Köhler P (2014) Estimation of vegetation photosynthetic capacity from space-based measurements of chlorophyll fluorescence for terrestrial biosphere models. Global Change Biol 20:3727–3742. https://doi.org/10.1111/gcb.12664

Zhao M, Heinsch FA, Nemani RR, Running SW (2005) Improvements of the MODIS terrestrial gross and net primary production global data set. Remote Sens Environ 95:164–176. https://doi.org/10.1016/j.rse.2004.12.011

Chapter 3
Assessing Forest Health using Geographical Information System Based Analytical Hierarchy Process: Evidences from Southern West Bengal, India

Shyamal Dutta, Sufia Rehman, Mehebub Sahana, and Haroon Sajjad

Abstract Vegetation plays an important role in sustaining the ecological biota and maintaining the equilibrium of environment. Thus, assessing vegetation status includes analyzing ecological dynamism, enough soil nutrients and vegetation health. Nearly 24% area of the India is under forest providing a range of resources to local communities. Bankura district has substantial forest cover comprising three divisions i.e., north Bankura division, south Bankura division and Panchet division. Nearly 1463.56 km^2 territorial extent of the district comes under forest jurisdiction constituting 21.27% of the total geographical area of the district. Per capita availability of forest in this district is 0.046 ha which is lower than the other districts of south western districts of West Bengal. Therefore, it is essential to analyze the health of vegetation in this region. Present study aims to analyze the forest health using different indices namely Normalized Difference Vegetation Index (NDVI), Enhanced Vegetation Index (EVI), Greenness Index (GI), Shadow Index (SI), Normalized Difference Bareness Index (NDBaI), Normalized Difference Built-up Index (NDBI) Perpendicular Vegetation Index (PVI), and Normalized Difference Moisture Index (NDMI) during 1990 and 2019 using Landsat 5 TM (1990) and Landsat 8 data (2019) under the model of Analytical Hierarchy Process (AHP). Results revealed that forest health

S. Dutta (✉)
Department of Geography, University of Burdwan, Bardhaman, WB, India
e-mail: duttashyamal1987@gmail.com

S. Rehman
Department of Geography, Jamia Millia Islamia, New Delhi, India
e-mail: srfpinky@gmail.com

M. Sahana
School of Environment, Education & Development, University of Manchester, Manchester, UK
e-mail: mehebubsahana@gmail.com

H. Sajjad
Faculty of Natural Sciences, Department of Geography, Jamia Millia Islamia, New Delhi, India
e-mail: haroon.geog@gmail.com

P. K. Shit et al. (eds.), *Spatial Modeling in Forest Resources Management*, Environmental Science and Engineering,
https://doi.org/10.1007/978-3-030-56542-8_3

was largely affected during the study period due to land transformation and disturbances created by anthropogenic activities. Findings of the study based on this 29-year spatio-temporal vegetation dynamics will ameliorate the local stake holders for managing and maintaining the health of vegetation in the study area.

Keywords Analytical hierarchy process · Indices · Forest health · Southern West Bengal

3.1 Introduction

Forest constitutes 31% area of the planet andis considered an essential part of the environment which helps in sustaining the fragile ecosystems and maintaining the equilibrium of various ecological functions (FSI 2017). Nearly 13.2 million people globally rely on forest for their livelihood and another 41 million people on its relative sectors (WWF 2020). Monitoring and assessment of forest health is thus, imperative to protect and maintain the potentiality of these resources for long-term benefits (Smith 2002). These resources are sensitive to various climatic and human-induced stressors which results in continuous deterioration of such resources (Simula 2009; Jain and Sajjad 2016). Forest health is also affected by anthropogenic activities; grazing, deforestation, land use transformation and climate change (Solberg et al. 2004). Intergovernmental Panel on Climate Change (IPCC) in its special report on 'Climate Change and Land' highlighted that deforestation may result increase in the concentration of greenhouse gases (GHGs) which may only be reduced through forest restoration and protection (IPCC 2019). It is also emphasized that impacts of forest cover change on local environment is more severe and significant than the global effects. Around 30% area in Madhya Pradesh, Maharashtra and Rajasthan has been deforested and degraded due to soil erosion, over grazing, cultivation and deterioration of Wetlands (Business Standard 2019).

Forest health is determined by the various components including plant growth, organism, climate, insects and human. Forest health is defined by the Society of American Forests as '*the perceived condition of a forest derived from concerns about such factors as its age, structure, composition, function, vigor, presence of unusual levels of insects or disease, and resilience to disturbance*'. Forest health concerns with the complete elimination of the practices that disrupts its potentiality and functioning including pests, diseases and lessening the impacts of climate induced disturbances (forest fires). Forest serves uncountable ecological functions and thus, it is important to understand the various stressors affecting forest health and their interaction with forest structure, type and forest history. Effective coping strategies and management practices can help in lessening the threats and maintaining the forest health (Lausch et al. 2017). Forest health monitoring been carried out at individual level by private owners, locals and administrators at various spatial scales. These assessments were intensive and rely on the suitable indicators. Remote sensing has provided a way in overcoming these limitations in assessing forest health at different spatio-temporal scales (Lausch et al. 2016). In the end of World War-II, forest health was determined using aerial sketch mapping in USA and Canada. Colour infrared and colour aerial photographs have also help in various forest assessments. High resolution satellite imagery such as Lidar and aerial videography are being utilized for forest health monitoring (Ciesla 2000). In last few decades, increase in the types and numbers of remote sensing instruments have enhanced the image processing capabilities (Wulder et al. 2006).

Various scholars have utilized several methods for assessing forest health (Cammarano et al. 2014; Yengohet al. 2015; Jain et al. 2016; Xue and Su 2017; Acharya et al. 2018). Malik et al. (2019) utilized analytical hierarchy process on a set of indices to analyze the vegetation status of Sali River basin in West Bengal. They emphasized that integration of analytical hierarchy process (AHP) and Landsat 8 data provide better assessment of forest health. Chitalea et al. (2014) analyzed the forest vulnerability in Uttarakhand as a function of exposure, sensitivity and adaptation Ellison (2015). Examined the vulnerability of mangroves to climate change and sea level rise. Sharma et al. (2015) analyzed the inherent forest vulnerability using pair wise comparison method of Western Ghats forest in India. Dash et al. (2017) analyzed the forest health using very high resolution unmanned aerial vehicles imagery. They highlighted that spatial data being consistent is advantageous over traditional aerial photographs, field surveys and provide more accurate forest health assessment. Mensah et al. (2019) examined the vegetation dynamics using normalized difference index and supervised classification. Analytical hierarchy process (AHP) introduced by Saaty (1980a, b) is an effective tool of multi-criteria decision-making analysis. In recent decades, AHP has been widely used for forest assessment and land use planning due to effectiveness and handling the critical decision-making problems (Wolfslehner and Vacik 2008; Jalilova et al. 2012). It is based on the principle of decomposition and forming a set of pair wise comparisons. It is an essential tool for assigning priorities to indicators and the criteria.

Forest covers 24% area of the total geographical area of India. Nearly 100 million tribal people in rural areas lives in the forest depend directly and indirectly on forest resources (Biswas 1993; Mongabay 2018). Various policies were formulated to ensure the rights of the forest communities over the forest goods. Forest Right Act (FRA 2006) was enacted to assure the rights of forest dependent communities specially schedule tribes over the forest land. National Forest Policy (1988) was came into effect to conserve the forest resources for maintaining the stability and sustainability of the environment. The main objectives of these policies were to enhance the potentiality and sustainability of the forest resources. Bankura district has substantial forest cover which provides sustenance to a large tribal population. In this backcloth, this study aims to analyze the forest health between 1990 and 2019. The study has three main objectives. Firstly, to assess the forest health as well as dynamics over the period of last 29 years (between 1990 and 2019). Secondly, to identify theme wise spatio temporal dynamics of vegetation status through positive indices Normalized Difference Vegetation Index (NDVI), Normalized Difference Moisture Index (NDMI), Enhanced Vegetation Index (EVI), Greenness Index (GI), Shadow Index

(SI), Perpendicular Vegetation Index (PVI), as well as negative indices like Normalized Difference Bareness Index (NDBaI), Normalized Difference Built up Index (NDBI). Moisture, vegetation cover, bare areas and soil condition are essential to analyze the vegetation health thus, the suitable indices were chosen to examine the vegetation health. Lastly, AHP was used to provide weightage to the different indices and to assess forest health in the study area.

3.2 Methods and Database

3.2.1 Study Area

Bankuradistrict (administratively Bardhaman Division of state of west Bengal) is in the South-western part of the West Bengaland geographically known as "*Rarh*" (O'Malley 1908). The study area lies between 22°38′ and 23°38′ North latitudes and 86°36′and 87°46′ East longitudes (Fig. 3.1). River Damodar serves as a natural boundary separating Bankura from Paschim and Purba Bardhaman Districts. It is bounded by Puruliya district in the west, Paschim Medinipur district in the south and Hugli district in the south-east. People in the district are engaged in agricultural activities. The district is enormously influenced by the climatic conditions (e.g., temperature, rainfall and humidity), complex topography (Plateau fringe as well as alluvial plains), geo-hydrology and edhaphic condition. Two major agro-climatic zones covers whole Bankura district viz., Undulating Red with lateritic terrain and Vindhyan Alluvial Zone. Undulating Red zone is found in Sonamukhi, Joypur, Bishnupur, Ranibandh, Gangajalghati, Barjora, Saltora, Hirbandh, Onda, Simlapal, Mejia, Taldangra, Raipur, Sarenga, Chhatna, Indpur, Khatra, Bankura-I and Bankura-II blocks.

Vindhyan Alluvial Zone covering the three blocks namely Patrasayer, Indus and Kotulpur blocks. Groundwater extraction is difficult to rugged topography thus, rainfed agriculture is being practiced here specially during late monsoon (Das and Paul 2015). The common species in the study area are Sal (*Shorea robusta*), Mahua (*Madhuca latifolia*), Teak (*Techtona grnadis*), Asan (*Terminalia tomentosa*), Kendu (*Diospyros melanoxylon*), Kusum (*Schleicheratrijuga*), Long leave *Acacia* (*Acacia latifolia*), Palash (*Butea monosperma*) and Arjun (*Terminalia arjuna*). Among the Rarh Bengal districts, Bankura district has substantial forest cover comprising three divisions i.e., north Bankura division, south Bankura division and Panchet division (Fig. 3.1c). Of the total area of the district (6882 km^2), forest occupies nearly 1285.5 km^2 comprising open forest (667.98 km^2), moderate dense forest (395.27 km^2) and very dense forest (222.33 km^2). Nearly 28.59 km^2 area is found under scrub land (FSI 2019). Thus, forest health assessment is imperative to manage and increase the potentiality of forest in this district.

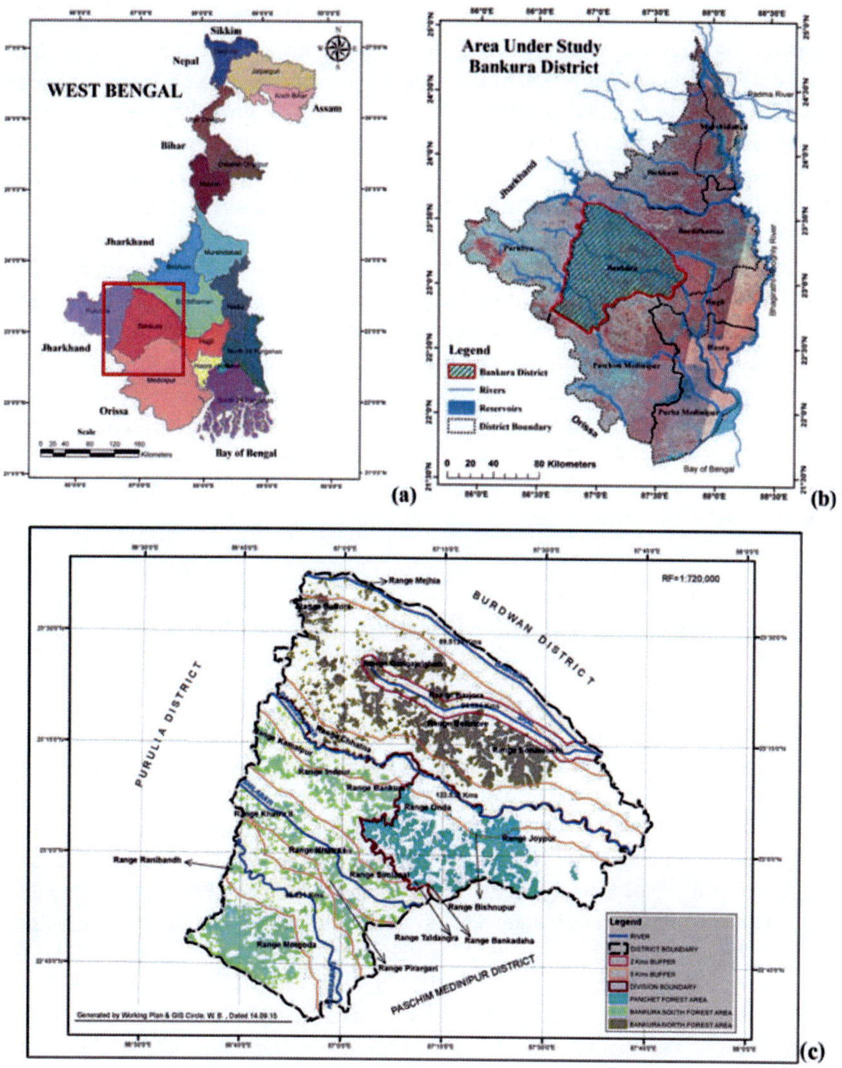

Fig. 3.1 Location and the forest divisions of the study area: Bankura District in South Western Part of West Bengal (**a**), Geographical Setup of the Study area in Rarh Bengal (**b**), Forest Divisions of Bankura District (**c**)

3.2.2 Database and Methods

In this study Landsat Thematic Mapper image (Path-139, Row-044) of 23rd December 1990 and Landsat 8 OLI data (Path-139, Row-044) of 5th January 2019 were used to analyze the forest health dynamics. The images were downloaded from United States Geological Survey- USGS (https://earthexplorer.usgs.gov/) for assessing the vegetation status through different indices. The study area is characterized by tropical deciduous forest category mainly dominated by *sal* (*Shorea robusta*). These trees shed their leaves during June–August. Therefore, images between December and January have been collected for analysis, when vegetation is fully flourished with broad leaves. Moderate resolution (i.e. 30 m pixel size) land satellite data Landsat TM as well as Landsat 8-OLI sensor has been chosen for assessment tools because these become very useful source for working in multi-faceted academic fields from to geology to regional planning as well as forest-mapping to identification of global LULC dynamics (Zhou et al. 2014). Vegetation indices like NDVI, EVI, NDBI, SI and NDMI were apply through Raster Calculator tool for forest health assessment. The eight indices were prepared from these images and were categorized into five classes i.e. very high, high, moderate, low and very low based on quartile classification in Arc GIS 10.5. Saaty's analytical hierarchy process (AHP) was used to give the weightage to different indices. The accuracy of the derived results was checked through field verification following stratified random sampling. Detailed methodology has been presented in Fig. 3.2.

3.2.3 Normalized Difference Vegetation Index (NDVI)

NDVI is an important index used to measure the various aspects of vegetation at spatio-temporal scale. It has been considered as an important tool to assess the vegetation status spatially (Anderson et al. 1993; Myneni and Williams 1994; Paruelo et al. 1997; Moleele et al. 2001; Verbesselt et al. 2006; Gu et al. 2007; Avtar et al. 2014). NDVI was initially used by Rouse et al. (1974) calculated through band ratio among the Red and near-infrared bands. This index is the outcome of some normal ratio function of absorption by visible Red (R) and reflection of near infrared (NIR) bands. Theindex values range between -1 and $+1$ (one) where all negative values upto zero represents water body, moist surface and values starting from $+ 1$ represents the completely flourished green space. Therefore, high values of index indicate more proliferated green vegetation and vice versa. As NDVI is not an elemental measure of vegetation rather it is used for ordinary spatio temporal change recognition (Malmstrom et al. 1997; Zhou et al. 2001; Geerken and Ilaiwi 2004; Palmer and Fortescue 2004; Kurtz et al. 2016; Wang et al. 2017), sparse vegetation patches along with grazing activities (Blanco et al. 2008; Kunwar et al. 2018), canopy cover identification based on LAI (Turner et al. 1999; Bajwa et al. 2008; González et al. 2008, Gamon et al. 1995; Wellens 1997; Fuller 1998, Kunwar et al. 2018) and base layer for forest fragmentation analysis (Dutta et al. 2017; Sahana et al. 2015). The index is calculated as:

$$Normalized\ Difference\ Vegetation\ Index\ (NDVI) = \frac{NIR - RED}{NIR + RED}$$

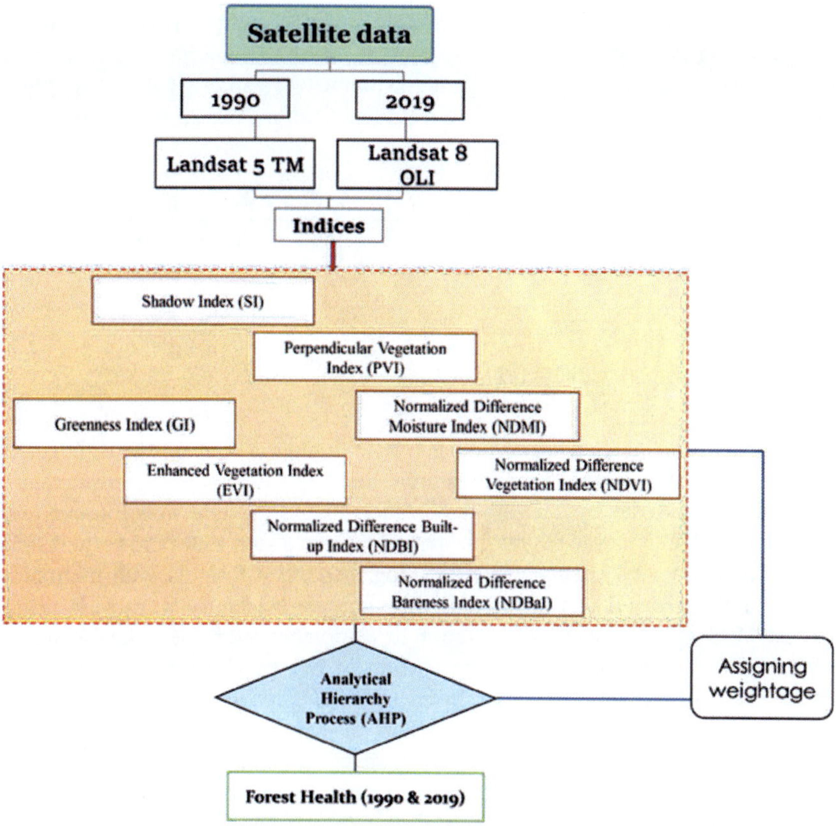

Fig. 3.2 Methodological framework

3.2.4 Enhanced Vegetation Index (EVI)

EVI has been considered as modified NDVI with high capability for vegetation monitoring and enhanced sensitivity to biomass region. It also incorporates some crucial issues like reduction of the control of atmospheric abnormalities and canopy as well as leaf area along with plant phenology more judiciously than NDVI (Miura et al. 2001; Huang et al. 2002, 2013; Matsushita et al. 2007; Lillesand et al. 2007). Use of Blue band of satellite images (band 1 in case of Landsat TM and Band 2 in case of Landsat 8) has become the main shortcoming of this index as this band is affected by noise (Jiang et al. 2008). EVI has been calculated by using Landsat5-TM and 8-OLI Data as follows: Here, band numbers 1, 3, 4 and 2, 4, 5 representing blue, red and NIR bands along with the correction coefficient of atmospheric abnormalities are C1, C2 and L (i.e., aerosol resistance) has been used respectively (Landscape Tool Box 2019). Similarly, like NDVI, the positive values from '0' progressing to '1' indicates from water surface, and value 1 represents good quality vegetation in case of EVI. EVI has proved to be an efficient tool covering multi-dimensional aspects like measuring evapotranspiration (Nagler et al. 2005, 2007), large area vegetation change detection, structural change identification of vegetation (Huete et al. 1997, 2002; Gurung et al. 2009) and canopy cover identification (Gao et al. 2000). EVI is calculated as:

$$Enhanced\ Vegetation\ Index\ (EVI) = \left(\frac{(2.5 * Red - Green)}{Red + 6 * Green - 7.5 * TIR + 1)}\right)$$

3.2.5 Greenness Index (GI)

GI was firstly used by Kauth and Thomas (1976) through performing an orthogonal conversion of the Landsat MSS data to examine the vegetation status. Various scholars have modified GI index by deriving the coefficients from visible, near infrared and middle infrared for Landsat TM satellite data (Crist and Kauth 1986; Crist et al. 1986; Jensen and Lulla 1987). Though by applying theoretical formula with minimum regional adjustment GI can be used in different space identity worldwide using Landsat MSS orTM. We used this index to comprehend the forest health status as biomass play an essential role in forest health assessment (Jensen and Lulla 1987; Mondal et al. 2013; Huete et al. 1985). Like NDVI and EVI here also higher the values of GI indicate the presence of more amounts of green biomass with chlorophyll content and vice versa. Presently GI has been widely used at different scales to identify gross primary production-GPP (Westergaard-Nielsen et al. 2013) and crop health (Wiegand et al. 1991; Cohen et al. 2003). GI was calculated as:

$$\begin{aligned} Greenness\ Index\ (GI) = {} & (0.2728 * Blue) - (0.2174 * Green) \\ & - (0.5508 * Red) + (0.7221 * NIR) \\ & + (0.0733 * SWIR1) - (0.1648 * SWIR2) \end{aligned}$$

3.2.6 Perpendicular Vegetation Index (PVI)

Richardson and Wiegand (1977), Jackson et al. (1980) and Bannari et al. (1995) used perpendicular distance from demarcation line of soil areal coverage and a sign of plant growth in which they estimated the perpendicular distance between the vegetation patch above soil surface using the NIR and red bands. Using linear measurement by formulating equation of the red and NIR bands was also done by Richardson and Wiegand (1977); Hueteet al. (1985), Major et al. (1990), Baret and Guyot (1991) and Cyr (1994). PVI concept was developed on the basic fact of remote sensing i.e., vegetation has addedretort in near-infrared bands and inferiorretort in red band and their reflectance. This fact is enthusiastically prolonged to adaptable band spaces whereby the spectral signatures of vegetation resemble a 3D ‘tasseled cap’ shape determined by a greenness alignment, a soil level surface, and a yellowness alignment (plant) or a wetness alignment depending on the bands used (Crist and Kauth 1986; Bannari et al. 1995). High values PVI indicates to the good-quality vegetation while lower values depict the bare surfaces. PVI was calculated as:

$$Perpendicular\ VegetationIndex\ (PVI) = (\sin(a) \times NIR)(\cos(a) \times RED)$$

3.2.7 Normalized Difference Moisture Index (NDMI)

Moisture content and vegetation structure are closely related to each other. Thus, assessment of moisture content is essential to examine vegetation health (Vogelmann and Rock 1988; Sader 1989; Fiorella and Ripple 1995, b). SWIR bands correction in Landsat images are mostly used in this analysis to enrich the correlation between Leaf Area Index (LAI) and NDVI (Horler and Ahern 1986; Nemani et al. 1993). Scholars have also used NDMI to understand the vegetation status (Turner et al. 1999, Hunt and Rock 1989; Hardisky et al. 1983; Wilson and Sader 2002; Goodwin et al. 2008). NDMI is highly correlated to water content of canopy as well as changes in plant biomass also closely tracked superiorly than the NDVI and other indices in vegetation studies. So, to understand the moisture condition of vegetation NDMI was used here. It is calculated as:

$$Normalized\ Difference\ Moisture\ Index\ (NDMI) = \frac{NIR - SWIR}{NIR + SWIR}$$

3.2.8 Shadow Index (SI)

Pattern of shadow modify spectral responses due to change in crown arrangement of the forest land depending of the height of any tree or a group of trees. Shadow index was developed to identify the canopy cover and association of vegetation to the ground (Liu and Yamazaki 2012). Following the two broad categories for assessment (i.e., model-based methods and feature-based methods) feature-based approaches are found more consistent for evaluation as these show the difference between non-shadowed and shadowed regions. The model-based method is more complex and time consuming (Polidorio et al. 2003; Liu et al. 2017; Nakajimaet al. 2002; Zhan et al. 2005; Liu et al. 2017). SI has been widely used to estimate vegetation condition (Pal et al. 2018), vegetation gap identification, its situation & classification

(Ono et al. 2010) and examining forest canopy density (Poggi et al. 2004; Andersen et al. 2005). SI was calculated using the following equation:

$$Shadow\,Index\,(SI) = [(1 - Red) * (1 - Green) * (1 - Blue)]^{1/3}$$

3.2.9 Normalized Difference Bareness Index (NDBaI)

NDBaI is used for the identification of urban and open surface. Since bare soil plays an essential role in sustaining its capacity in any type of ecosystem (Zhao and Chen 2005; Li and Chen 2014) thus, in this study, our main focus to use this index was to identify the bare surface (i.e. area devoid of any vegetation cover) which is negatively correlated to the vegetative cover (Roy et al. 1997; Rikimaru and Miyatake 1997; Sahana et al. 2015, 2019). NDBaI reflects spectral signature based on different background properties of soil and its association (Pal et al. 2018). Indicating degree of bareness higher values of NDBaI shows completely bare surface and built up inurban areas whereas the areas with vegetation cover came under lower values. Apart from separating vegetation form its backgrounds, this index also helps to understand the impact of landform, topography along with anthropogenic impacts (Zhao and Chen 2005, Chen et al. 2006; Poggi et al. 2004; Andersen et al. 2005). NDBI was calculated as:

$$Normalized\,Difference\,Bareness\,Index\,(NDBal) = \frac{SWIR - Thermal\,Infrared}{SWIR + Thermal\,Infrared}$$

3.2.9.1 Normalized Difference Built up Index (NDBI)

Normalized Difference Built-up Index (NDBI) is one of the most common indices for analyzing the built-up areas (Sahana et al. 2019). The build-up areas and bare soil reflects more SWIR than NIR and water body doesn't reflect on Infrared spectrum. In case of vegetated surface, reflection of NIR is higher than SWIR spectrum. So, we incorporated NDBI as a negative correlated indicator ofgreen space in the study area. The values of NDBI lies between -1 and $+\,1$. Negative value of NDBI represents water bodies whereas higher values represent build-up areas. For vegetation NDBI reflects lower values. The index is calculated as:

$$Normalized\,Difference\,Built\,up\,Index\,(NDBI) = (SWIR - NIR)/(SWIR + NIR)$$

3.2.9.2 Multi-criteria Decision Analysis

Analytical hierarchy process (AHP) is an extensively used technique of multi-criteria decision-makinganalysisintroduced by Saaty in 1970s (Saaty 1980a, b; de Jong 1984; Chakrabortty et al. 2018). This technique is widely used by scholars for assessing waste management in urban areas (Morrissey and Browne 2004; Contreras et al. 2008), identification of groundwater potential (Dunning et al. 2000; Flug et al. 2000; Joubert et al. 2003; Das et al. 2018), mining & landslide susceptibility analysis (Pourghasemi et al. 2012), revegetation policy implication in Riparian region (Qureshi and Harrison 2003) and analyzing the agricultural land suitability (Jamil et al. 2018; Mandal et al. 2020). AHP aims at making comparative judgement, disintegration and assigning priorities to the criteria (Podvezko 2009). In this study, we used AHP to assign weightage to the indices for assessing forest health in the study area. Priority matrix was prepared to comparative assessment of the indices and consistency ratio was kept to less than 10 forlogical analysis of the overall work. Assigning weights of the theme considering its suborders (Rao and Briz-Kishore 1991), the entire theme based-layers are integrated into single GIS platform to demarcate the forest health status by applying the following equation:

$$
\begin{aligned}
Vegetation\,Status\,(VS) = [&(NDVIW)*(NDVIWI)+(EVIW)\\
&*(EVIWI)+(GIW)*(GIWI)\\
&+(PVIW)*(PVIWI)+NDBaIW\\
&*(NDBaIWI)+(NDMIW)*(NDMIWI)\\
&+(SIW)*(SIWI)+(NDBIW)*(NDBIWI)]
\end{aligned}
$$

where; VS = vegetation status, NDVI = Normalized Difference Vegetation Index, EVI = Enhanced Vegetation Index, GI = Greenness Index, PVI = Perpendicular Vegetation Index, NDBaI = Normalized Difference Bareness Index, NDMI = Normalized Difference Moisture Index, SI = Shadow Index, NDBI = Normalized difference Built up Index. Subscripts 'WI' and 'W' indicate the normalized weights of the individual feature of a theme and normalized weights of a theme, respectively. Values of VS helped to understand the vegetation status in the study area. VS values were classified into five categories starting from very low to very high.

3.3 Result and Discussions

3.3.1 Normalized Difference Vegetation Index (NDVI)

In the Study area NDVI values ranged from 0 to 1 for 1990 and 2019 but varied spatially. Thus, it is classified into five categories based on Quartiles classification namely very low (0–0.49 in 1990 and 0–0.38 in 2019), low (0.5–0.52 in 1990 and

0.39–0.42 in 2019), moderate (0.53–0.56 in 1990 and 0.43–0.45 in 2019), high (0.57–0.63 in 1990 and 0.46–0.49 in 2019) and very high (greater than 0.63 in 1990 and greater than 0.49 in 2019). Very low zone represents mainly water body as well as other moist surfaces without any vegetation (mainly in the South-western and northern Bankura) whereas low NDVI was found in the bare and concrete surfaces. Grassland and agricultural tracts were found under the moderate category while fragmented forest patches were identified under high category of NDVI (Fig. 3.3). High values of NDVI mainly found in the Sal forest of central and southern part of

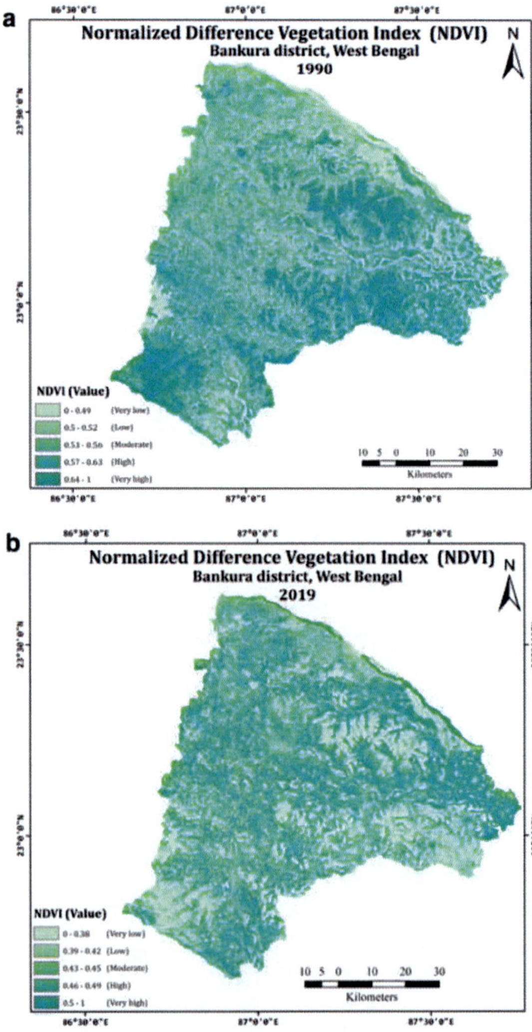

Fig. 3.3 Normalized Difference Vegetation Index (NDVI) in 1990 (**a**) and 2019 (**b**)

the study area covering the Beliatore, Joypur and Motgodaranges in 1990 (Fig. 3.3). But after 20 years mainly the NDVI values of Beliatore and Joypur ranges reduces significantly in 2019. They lost their position from very high in 1990 to high category in 2019.

3.3.2 *Enhanced Vegetation Index (EVI)*

Normalized Difference Vegetation Index (NDVI) is chlorophyll sensitive while EVI is more responsive to canopy structural variations. In the Study area EVI values ranged from 0 to 1 after normalization for 1990 and 2019 but with varying differences in spatial coverage in the study area. EVI values for both the years were classified into five categories based on Quartiles classification. These categories are very low (0–0.47 in 1990 and 0–0.32 in 2019), low (0.48–0.51 in 1990 and 0.33–0.43 in 2019), moderate (0.52–0.54 in 1990 and 0.44–0.51 in 2019), high (0.55–0.56 in 1990 and 0.52–0.62 in 2019) and very high (greater than 0.56 in 1990 and greater than 0.62 in 2019). High EVI was found in the Sal forest of central as well as southern part covering the *Beliatore, Joypur* and *Motgoda* ranges in 1990 (Fig. 3.4). Very low values were found around the water body and other moist surface devoid of vegetative cover (mainly in the South-western and northern Bankura). It has been identified that EVI values of Beliatore and Joypur ranges reduced significantly in 2019. The vegetation from high category in 1990 came to moderate category in 2019.

3.3.3 *Greenness Index (GI)*

GI values for both the years 1990 and 2019 showed significant spatial differences though the value ranges between 0 and 1. The lower values represent very low biomass and chlorophyll content while the higher values signify good quality of vegetation biomass (Fig. 3.5). GI was also classified for 1990 and 2019 into five categories based on Quartiles classification. These are very low (0–0.37 in 1990 and 0–0.50 in 2019), low (0.38–0.40 in 1990 and 0.51–0.56 in 2019), moderate (0.41–0.43 in 1990 and 0.57–0.62 in 2019), high (0.44–0.48 in 1990 and 0.63–0.71 in 2019) and very high (greater than 0.48 in 1990 and greater than 0.71 in 2019). High values of GI were found in central and south-western parts of the study area in 1990 which came under moderate category in 2019.

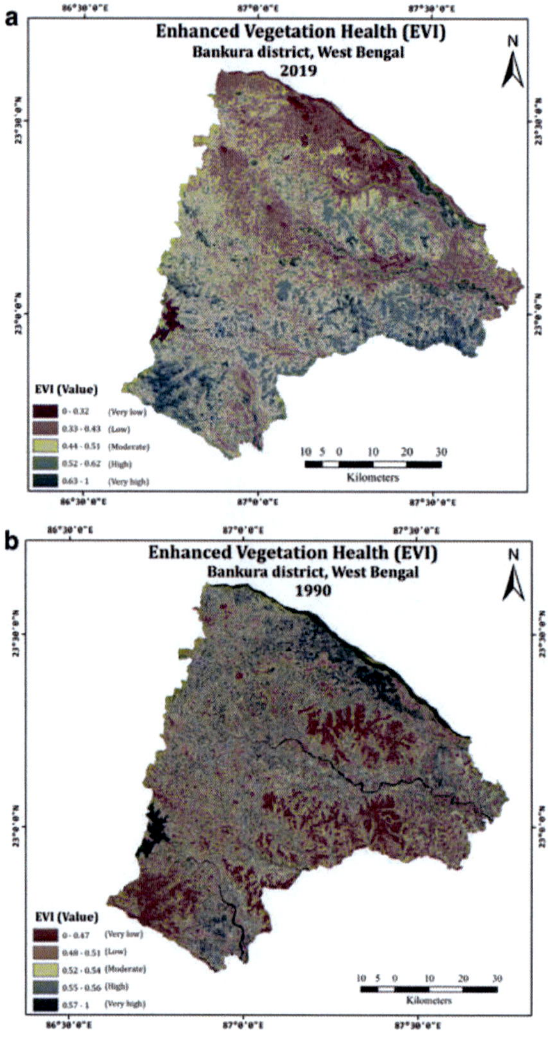

Fig. 3.4 Enhanced Vegetation Index (EVI) in 1990 (**a**) and 2019 (**b**)

3.3.4 Perpendicular Vegetation Index (PVI)

PVI values in the study areas how immense changes during 29 years. The values of PVI were classified for 1990 and 2019 into five categories based on Quartiles classification. These are very low (0–0.31 in 1990 and 0–0.33 in 2019), low (0.32–0.35 in 1990 and 0.34–0.36 in 2019), moderate (0.36–0.38 in 1990 and 0.37–0.4 in 2019), high (0.39–0.42 in 1990 and 0.41–0.44 in 2019) and very high (greater than 0.42 in 1990 and greater than 0.44 in 2019). High PVI was found in the forest range

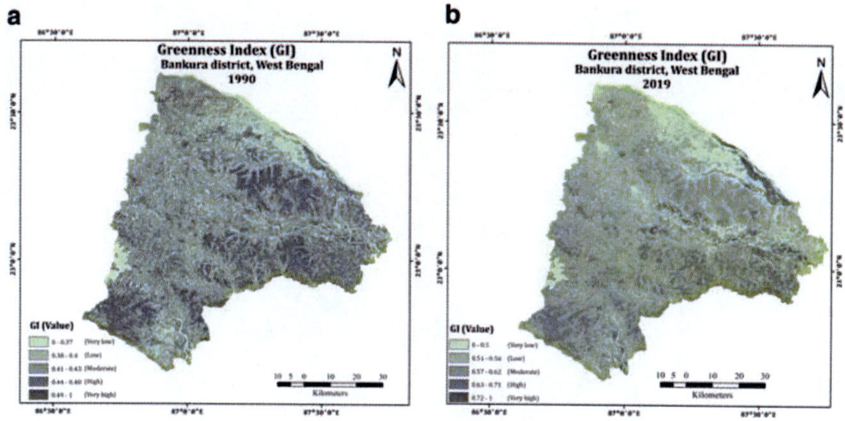

Fig. 3.5 Greenness Index (GI) in 1990 (**a**) and 2019 (**b**)

of Simlapal and Joypur in the southern and south-eastern part of the study area in both years of 1990 and 2019 (Fig. 3.6). Conversion of high PVI to low to very high PVI category mostly found in the northern part of the district covering the forest ranges of Beliatore, Sonamukhi and Gangajalghati.

3.3.5 Shadow Index (SI)

SI was developed to recognize the canopy cover and alliance of vegetation to the ground. SI values were classified following the Quartiles classification method for both years. Very low SI was found over the water body of Damodar, river channels, dams in the south eastern part of the study area and along the sandy riverbed of the district (Fig. 3.7). Low SI class is mainly associated with the barren land, settled area to water-logged moist agricultural land. In 1990 high and very high shadow index was found in Sonamukhi Joypur and Simlapal forest ranges which was found under moderate category in 1990 and low in 2019.

3.3.6 Normalized Difference Bareness Index (NDBaI)

NDBaI indicates the urban and open surfaces devoid of any vegetative cover. In the study area, NDBaI value range significantly changed spatially in the year 29 years. All the values NDBaI were classified on the basis of quartiles classification method i.e., very low (0–0.51 in 1990 and 0–0.31 in 2019), low (0.52–0.59 in 1990 and 0.32–0.44 in 2019), moderate (0.6–0.64 in 1990 and 0.45–0.51 in 2019), high (0.65–0.69 in 1990 and 0.52–0.63 in 2019) and very high (greater than 0.69 in 1990 and greater

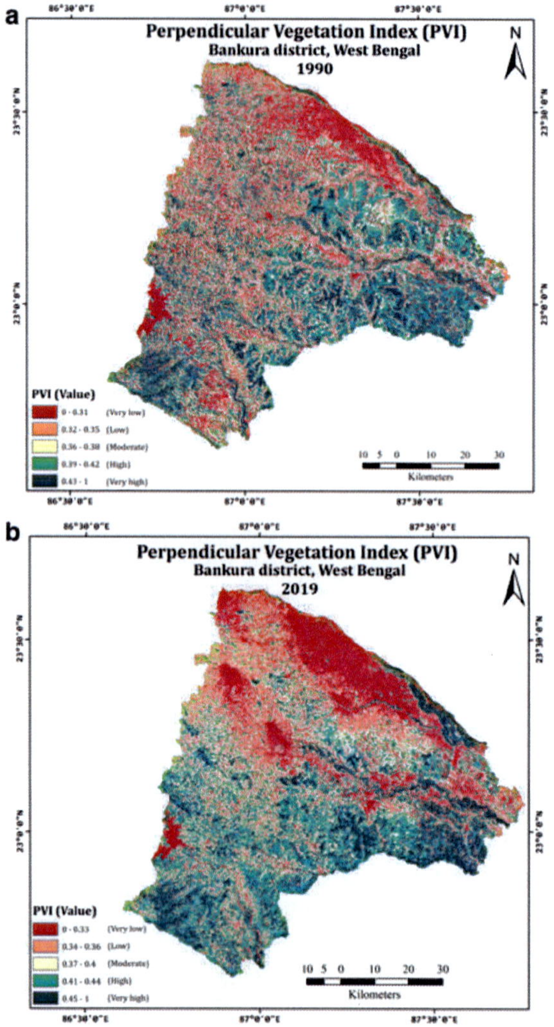

Fig. 3.6 Perpendicular Vegetation Index (PVI) in 1990 (**a**) and 2019 (**b**)

than 0.63 in 2019). High NDBaI values were found in the central and eastern part of the study area (mainly associated with shrubs and fallow agricultural land) became sparse after 29 years (Fig. 3.8). Northern part of the study area and central forested region was found under low NDBaI zone.

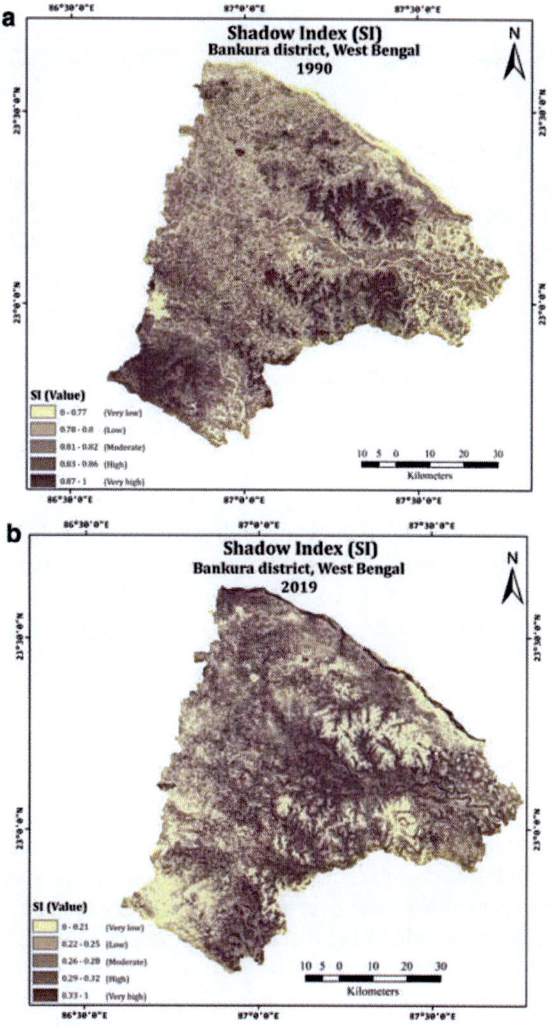

Fig. 3.7 Shadow Index (SI) in 1990 (**a**) and 2019 (**b**)

3.3.7 Normalized Difference Built-Up Index (NDBI)

Normalized difference built up index (NDBI) is a useful tool to identify the urban impervious land where the positive values represent the built-up areas and negative values represent vegetation and water body. The index was also divided into five categories based on Quartiles classification method namely very low (0–0.68 in 1990 and 0–0.4 in 2019, low (0.69–0.77 in 1990 and 0.41–0.44 in 2019), moderate (0.78–0.81 in 1990 and 0.45–0.48 in 2019), high (0.82–0.84 in 1990 and 0.49–0.54

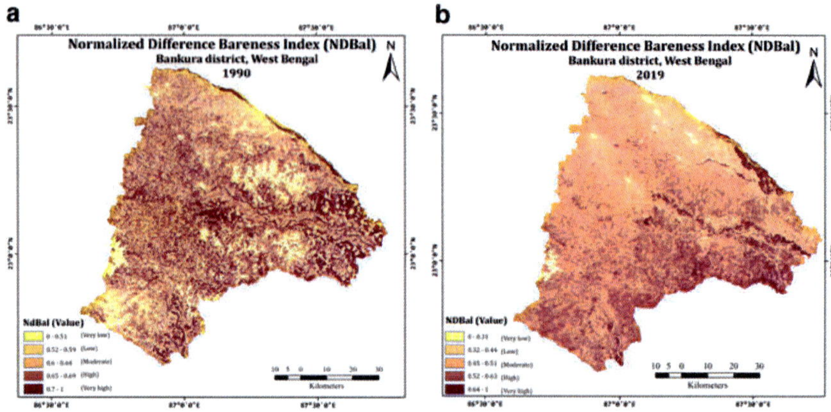

Fig. 3.8 Normalized Difference Bareness Index (NDBaI) in 1990 (**a**) and 2019 (**b**)

in 2019) and very high (greater than 0.84 in 1990 and greater than 0.54 in 2019). Low NDBI values are found in the central, south-western and in small of eastern parts of Bankura district in 1990 which are converted mostly into moderate zone as a result of agricultural proliferation in 29 years. High NDBI areas significantly increased in 29 years (Fig. 3.9).

3.3.8 *Normalized Difference Moisture Index (NDMI)*

NDMI provides valuable information about moisture condition, water stress for irrigation decisions and used to assess the drought conditions. In the study area, NDMI mean values significantly decreases over time. After classifying the NDMI following Quartiles classification we identified five different categories namely very low (0–0.16 in 1990 and 0–0.47 in 2019, low (0.17–0.19 in 1990 and 0.48–0.53 in 2019), moderate (0.2–0.24 in 1990 and 0.54–0.56 in 2019), high (0.25–0.33 in 1990 and 0.57–0.6 in 2019) and very high (greater than 0.33 in 1990 and greater than 0.6 in 2019). High NDMI was found in the central and south-western part and in small of eastern Bankura district in 1990 which later converted mostly into moderate zone due to increase in agricultural activities in 29 years (Fig. 3.10).

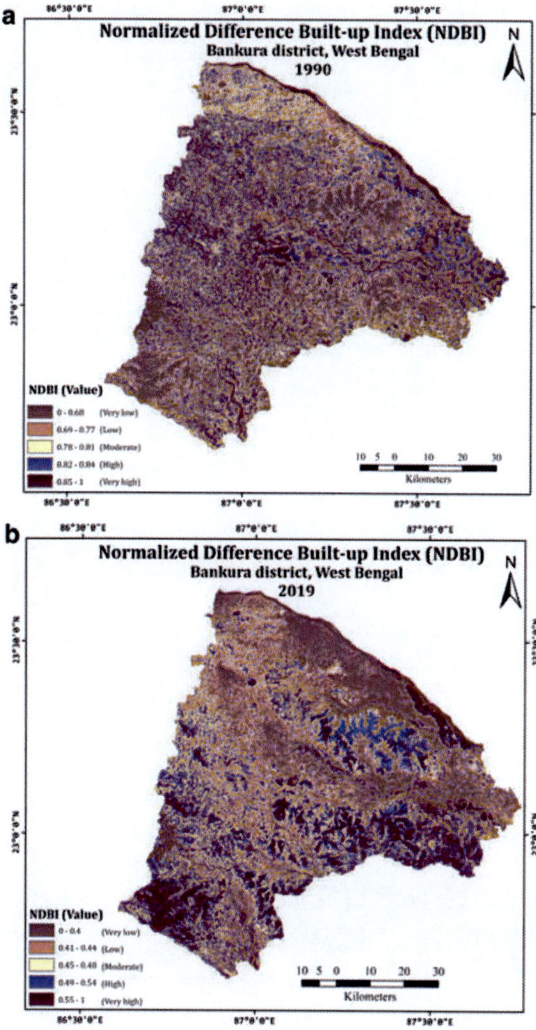

Fig. 3.9 Normalized Difference Built-up Index (NDBI) in 1990 (**a**) and 2019 (**b**)

3.4 Discussion

3.4.1 Vegetation Status Identification Through AHP

Eight indices namely NDVI, EVI, GI, SI, NDBaI, NDBI, PVI and NDMIwere used to assess the forest health in the Bankura district. After brief literature review, it has been identified that single index is not enough to analyze the health. To understand all the aspects of forest, these indices were used in order to better interpret the forest

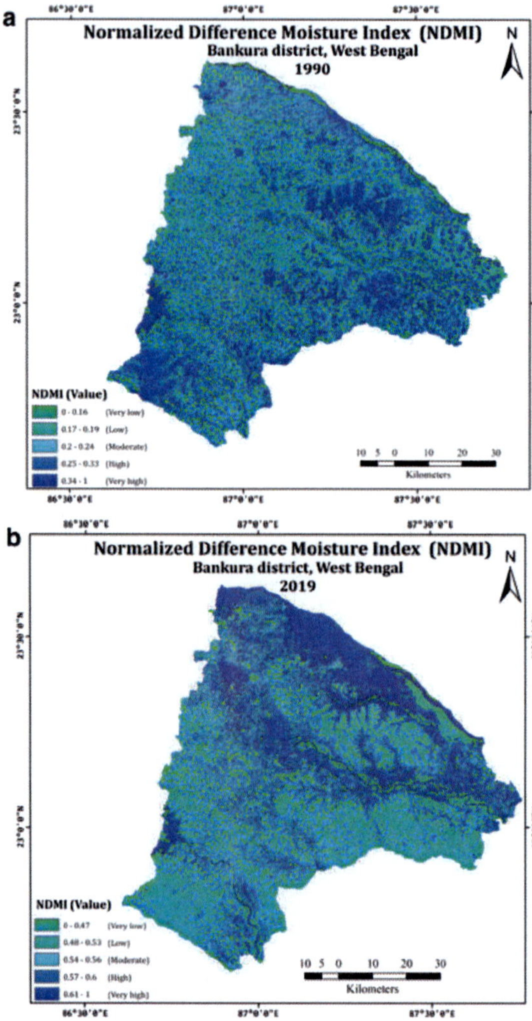

Fig. 3.10 Normalized Difference Moisture Index (NDMI) in 1990 (**a**) and 2019 (**b**)

health in the study area. NDVI has been identified an important index for analyzing vegetation spatially thus, high priority was given to NDVI during AHP calculation followed by EVI. GI was used to analyze the impact of biomass and chlorophyll content in forest. Through the application of PVI, perpendicular distance from the bare soil has been extracted. This dimension was included in the analysis as the stud area is least affected by urbanization except some spot like northern mining areas, Bankura and Sonamukhi municipalities. Shadow Index (SI) helped in assessing the agree of bareness while NDBaI and NDBI manifested the degree of built up stress

Table 3.1 Pair wise comparison matrix

Indices	NDVI	GI	NDBaI	NDBI	NDMI	EVI	PVI	SI
NDVI	1	1	1	1	1	2	1	7
GI	1	1	1	1	1	1	1	7
NDBaI	1	1	1	3	5	5	4	4
NDBI	1	1	0.33	1	6	5	2	5
NDMI	1	1	0.2	0.17	1	1	1	6
EVI	0.5	1	0.2	0.2	1	1	1	6
PVI	1	1	0.25	0.5	1	1	1	7
SI	0.14	0.14	0.25	0.2	0.17	0.17	0.14	1

Number of comparisons = 28
Consistency Ratio CR = 9.4%
Principal Eigen value = 8.927
Eigenvector solution: 7 iterations, delta = 6.2E−8

on forest cover (inverse way). Lastly, monitoring the vegetation status through close positive relation with high density forest with moisture content, NDMI was used.

All the factors found significant in the present study identified through Spearman rho Rank test (99% significant level) for both the year of 1990 and 2019 and showed highest accuracy of the indices. The indices used in the study were categorized based on Quartiles classification into five categories namely very high, high, moderate, low and very low. Weightage to classes were assigned based on positive (NDVI, EVI, SI, PVI etc.) and negative consideration (NDBI, NDBaI) in relation to vegetation (Table 3.1). NDVI, EVI, GI, SI, PVI, NDMI were provided rank positively with descending order (very high-5, high-4, moderate-3, low-2 and very low-1) and considered as positive parameter to assess the forest health whereas NDBI and NDBaI has been assigned weightages inversely (very low-5, low-4, moderate-3, high-2, very high-1). NDBI and NDBaI were considered negative associated with vegetation (Tables 3.1 and 3.2). Thereafter, analytical hierarchy process (AHP) was applied for assigning weights to different indices. Pair-wise comparison matrix was used for normalizing the weights and for comparing the indices (Saaty 1990). Vegetation health was spatially represented for the year 1990 & 2019 (Fig. 3.11). The analysis of vegetation health map revealed that vegetation healthvaries at spatio-temporal scale.

In 1990, area under very low vegetation health was very nominal i.e., 1% to total district which is reduced to 0.002% after 29 years which indicates full reduction of this category (Table 3.3). It was found around water body, moist surfaces and built up areas. Low vegetation status was found along the low-lying agricultural land with water body covering an area of 41.009 km^2 (37.6%) in 1990 and found mainly in the northern as well as north eastern part of the district along the bank of river *Damodar*. This category also significantly reduced to 24.65% to total district area with a net loss of 14.108 km^2 green cover in 29 years (1990–2019). Moderate vegetation health was found over agricultural field with some grass-covered area representing an area of about 41.613 km^2 (38.1%) in 1990 (Table 3.3). This category increased up to

Table 3.2 Normalized rank of the indices

Indices	Class	Assigned weight	Normalized weight
SI	Very high	5	0.33
	High	4	0.27
	Moderate	3	0.20
	Low	2	0.13
	Very low	1	0.07
GI	Very high	5	0.33
	High	4	0.27
	Moderate	3	0.20
	Low	2	0.13
	Very low	1	0.07
NDVI	Very high	5	0.33
	High	4	0.27
	Moderate	3	0.20
	Low	2	0.13
	Very low	1	0.07
NDMI	Very High	5	0.33
	High	4	0.27
	Moderate	3	0.20
	Low	2	0.13
	Very low	1	0.07
EVI	Very high	5	0.33
	High	4	0.27
	Moderate	3	0.20
	Low	2	0.13
	Very low	1	0.07
PVI	Very high	5	0.33
	High	4	0.27
	Moderate	3	0.20
	Low	2	0.13
	Very low	1	0.07
NDBI	Very high	1	0.07
	High	2	0.13
	Moderate	3	0.20
	Low	4	0.27
	Very low	5	0.33
NDBal	Very high	1	0.07
	High	2	0.13

(continued)

Table 3.2 (continued)

Indices	Class	Assigned weight	Normalized weight
	Moderate	3	0.20
	Low	4	0.27
	Very low	5	0.33

Source Based on authors' calculations

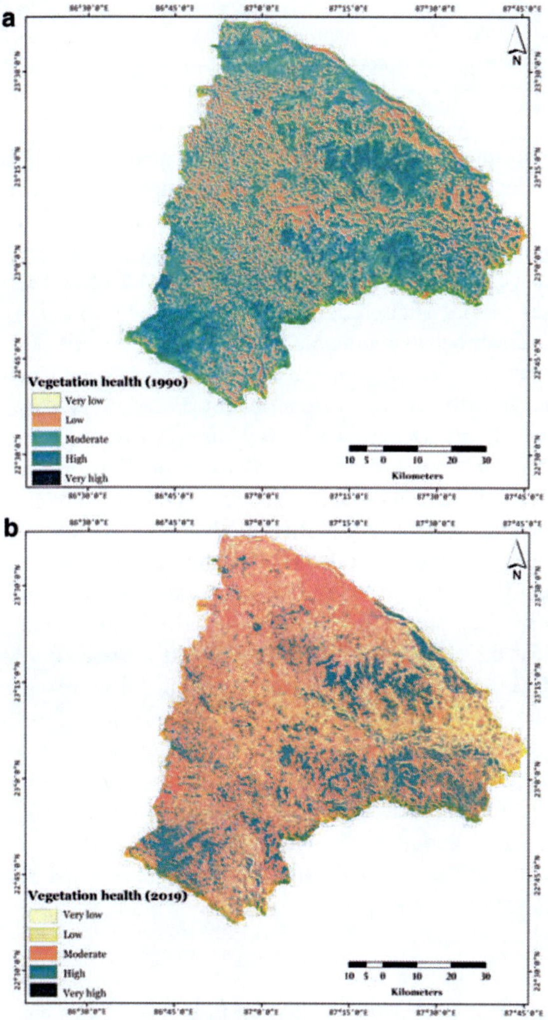

Fig. 3.11 Vegetation health status in 1990 (**a**) and 2019 (**b**)

Table 3.3 Area under different classes

Forest Health	Area in 1990		%	Area in 2019		%
	Hectares	sq. km		Hectares	sq. km	
Very low	109.6	1.096	1.0	0.27	0.0027	0.002
Low	4100.9	41.009	37.6	2690.1	26.901	24.657
Moderate	4161.3	41.613	38.1	5692.1	56.921	52.173
High	2538.3	25.383	23.3	2525.9	25.259	23.152
Very high	0.09	0.0009	0.0	1.8	0.018	0.016
	10,910.19	109.1019	100.0	10,910.19	109.1019	100.00

52.17% with a net increase of around 11 km^2 in green space. Remaining forest health categories i.e., high and very high, are representing relatively good vegetation condition with area coverage of 26 km^2 (23.5%) in 1990. This category mainly belonged to fragmented forest patches within protected forest boundary and dense contiguous forest cover located in the forest ranges of Beliatore, Joypur, Sonamukhi and Simlapal (Fig. 3.11). Though they are least affected forest categories, but their peripheral forest cover became vulnerable to loss (1 km^2) in 29 years.

Overall, the forest cover of Bankura was about 18.68% and most of the forest cover is associated with *Sal* forest reported by Forest Survey Report from FSI (2019) and District Forest Department of Bankura (2018). Therefore, present study presented a well forest health assessment in Bankura district. Sub-tropical deciduous forest (mainly *sal*) were mainly found concentrate in the district and considered for forest health assessment here LULC map produced by National Remote Sensing Centre (NRSC) Bhuvan, India revealed 609.01 km^2 area under deciduous forest and 211.48 km^2 under plantation covering 12% area of the district in the year 2011–12. In present study we found more than 23% (1990 and 2019) good quality forest cover which came under the open forest categories. Forest representing scrubs forest, in 2011–12 (NRSC) is only under 2% to the total area. In our study we found 1% forest under very low-quality forest in 1990 which is almost abolished in 2019 (Fig. 3.12). Such type of vegetation health zone is required most attention as the periphery of this patches converted into fragmented small patches. As the good quality of the forest of this district is under national average thus, priority should be given to this category for maintain the forest condition (Laman 2011; Wilkie et al. 2011; Fetriyuna and Fiantis 2017). Therefore, this area requires immediate actions to be implemented for improving the forest condition.

3.5 Conclusion and Policy Implication

Present study has analyzed vegetation health in Bankura district using suitable indices between 1990 and 2019. Vegetation health was determined using Landsat TM and L-8 OLI satellite database using Analytical hierarchy process (AHP) and overlay

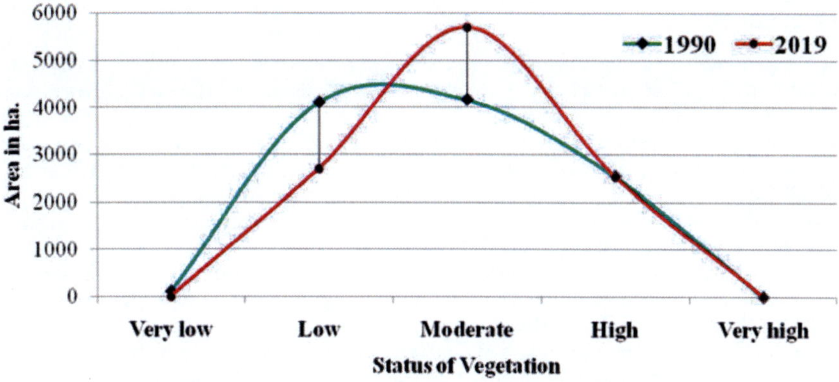

Fig. 3.12 Change in different vegetation vealth categories in 29 years (1990–2019)

analysis in the study area. We examined the vegetation health using eight indices namely NDVI, NDMI, NDBI, NDBal, SI, GI, EVI and PVI. The reason behind using such indices is to deeply analyze the vegetation health. These indices were found effective in examining the vegetation health. All the indices have shown high degree of positive correlated and statistically significant with each other. Vegetation health was found deteriorating in the north-western and central parts of the district. These areas were recorded high to moderate vegetation health during 1990 which became very low to low in 2019.

With the initiation of Joint Forest management under National Forest Policy (1988) the forest covers significantly improved which is cleared from forest health status of 1990. But after the 29 years the scenario significantly altered with high level of forest transformation to patches into other land use. This drastic change depicted from the forest health in 2019 in which only protected forest regions in southern and eastern Bankura were found unaltered. However, these were found patchy and fragmented along deciduous forest mainly *Sal*. Effective policy measures are required to manage and maintain the potentiality of the forest in the study area. Collaborative forest management practices including stakeholders and forest community are essential to increase the sustainability of forest. Controlling forest clearance for forest goods and land conversion will also help in managing the long-term potentiality of the forest. Findings of the study based on these 29 years spatio-temporal vegetation dynamics will increase the understanding of the local stake holders for managing and maintaining the health of vegetation in the study area.

References

Acharya TD, Subedi A, Lee DH (2018) Evaluation of water indices for surface water extraction in a Landsat 8 scene of Nepal. Sensors 18(8):2580

Anderson GL, Hanson JD, Haas RH (1993) Evaluating Landsat Thematic Mapper derived vegetation indices for estimating above-ground biomass on semiarid rangelands. Remote Sens Environ 45(2):165–175

Andersen HE, McGaughey RJ, Reutebuch SE (2005) Estimating forest canopy fuel parameters using LIDAR data. Remote Sens Environ 94(4):441–449

Avtar R, Herath S, Saito O, Gera W, Singh G, Mishra B et al (2014) Application of remote sensing techniques toward the role of traditional water bodies with respect to vegetation conditions. Environ Dev Sustain 16(5):995–1011

Bajwa SG, Gowda PH, Howell TA, Leh M (2008) Comparing artificial neural network with least square regression techniques for LAI retrieval from remote sensing data. Pecora. Retrieved from https://www.researchgate.net/profile/Prasanna_Gowda2/publication/256908657_Comparing_artificial_neural_network_and_leastsquare_regressiontechniquesfor_LAI_retrieval_from_remote_sensing_data/links/54cbf6b80cf29ca810f489a8.pdf. Accessed 18 Jan 2019

Bankura District Forest Department (2018) Govt. of West Bengal. https://bankuraforest.in/publication/. Accessed 18 Jan 2020.

Bannari A, Morin D, Bonn F, Huete AR (1995) A review of vegetation indices. Remote Sens Rev 13(1–2):95–120

Baret F, Guyot G (1991) Potentials and limits of vegetation indices for LAI and APAR assessment. Remote Sens Environ 35(2–3):161–173. https://doi.org/10.1016/0034-4257(91)90009-U

Biswas PK (1993) Forestry-based sustainable development: The social dimensions. Indian J Public Adm XXXIX(3)

Blanco LJ, Aguilera MO, Paruelo JM, Biurrun FN (2008) Grazing effect on NDVI across an aridity gradient in Argentina. J Arid Environ 72(5):764–776

Business Standard (2019) As India hosts desertification meet, 30% of its land is already degraded. https://www.business-standard.com/article/current-affairs/as-india-hosts-desertification-meet-30-of-its-land-is-already-degraded-119090200088_1.html. Accessed on 22 March 2020

Cammarano D, Fitzgerald GJ, Casa R, Basso B (2014) Assessing the robustness of vegetation indices to estimate wheat N in Mediterranean environments. Remote Sens 6(4):2827–2844

Chakrabortty R, Pal SC, Malik S, Das B (2018) Modeling and mapping of groundwater potentiality zones using AHP and GIS technique: a case study of Raniganj Block, Paschim Bardhaman, West Bengal. Model Earth Syst Environ 4:1–26

Chen PY, Fedosejevs G, Tiscareno-Lopez M, Arnold JG (2006) Assessment of MODIS-EVI, MODIS-NDVI and VEGETATION-NDVI composite data using agricultural measurements: an example at corn fields in western Mexico. Environ Monit Assess 119(1–3):69–82

Chitalea VS, Shresthaa HL, Agrawala NK, Choudhurya D, Gilania H, Dhonjua HK, Murthya MSR (2014). Forest climate change Vulnerability and Adaptation Assessment in Himalayas. Int Arch Photogramm Remote Sens Spat Inf Sci 8

Ciesla WM (2000) Remote sensing in forest health protection (No. 3). US Department of Agriculture, Forest Service, Forest Health Technology Enterprise Team

Cohen WB, Maiersperger TK, Gower ST, Turner DP (2003) An improved strategy for regression of biophysical variables and Landsat ETM+ data. Remote Sens Environ 84(4):561–571

Contreras F, Hanaki K, Aramaki T, Connors S (2008) Application of analytical hierarchy process to analyze stakeholders preferences for municipal solid waste management plans, Boston, USA. Resour Conserv Recycl 52(7):979–991

Crist EP, Kauth R J (1986b) The Tasseled Cap de-mystified. **(transformations of MSS and TM data)**

Crist EP, Laurin R, Cicone RC (1986a) Vegetation and soils information contained in transformed Thematic Mapper data. In Proceedings of IGARSS'86 symposium, pp 1465–1470. European Space Agency Publications Division, Parisa

Cyr L (1994) Apport des indices de vegetation pourl'evaluation de la couverture du sol envued'unemodelisationspatiale de l'erosion

Das B, Pal SC, Malik S, Chakrabortty R (2018) Modeling groundwater potential zones of Puruliya district, West Bengal, India using remote sensing and GIS techniques. Geol Ecol Landscapes 3(3):1–15

Das K, Paul PK (2015) Soil moisture retrieval model by using RISAT-1, C-band data in tropical dry and sub-humid zone of Bankura district of India. Egypt J Remote Sens Space Sci 18(2):297–310

Dash JP, Watt MS, Pearse GD, Heaphy M, Dungey HS (2017) Assessing very high resolution UAV imagery for monitoring forest health during a simulated disease outbreak. ISPRS J Photogramm Remote Sens 131:1–14

de Jong P (1984) A statistical approach to Saaty's scaling method for priorities. J Math Psychol 28(4):467–478

Dunning DJ, Ross QE, Merkhofer MW (2000) Multiattribute utility analysis for addressing Section 316 (b) of the Clean Water Act. Environ Sci Policy 3:7–14

Dutta S, Sahana M, Guchhait SK (2017) Assessing anthropogenic disturbance on forest health based on fragment grading in Durgapur Forest Range, West Bengal, India. Spat Inf Res 25(3):501–512

Ellison JC (2015) Vulnerability assessment of mangroves to climate change and sea-level rise impacts. Wetlands Ecol Manage 23(2):115–137

Fetriyuna H, Fiantis D (2017) Impact of land-use changes on Kuranji River basin functions chapter 8. In: Shivakoti G, Pradhan U, Helmi H (eds) Redefining diversity and dynamics of naturalresources management in Asia, vol 4, pp 105–114. Elsevier. https://doi.org/10.1016/B978-0-12-805451-2.00008-9. Accessed 18 Jan 2020

Fiorella M, Ripple WJ (1995) Analysis of conifer forest regeneration using landsat thematic mapper data

Flug M, Seitz HL, Scott JF (2000) Multicriteria decision analysis applied to Glen Canyon Dam. J Water Resour Plan Manag 126(5):270–276

Forest Rights Act-FRA (2006) Act, rules and guidelines. Ministry of Tribal Affairs, Government of India. https://tribal.nic.in/FRA/data/FRARulesBook.pdf

Forest Survey of India (2019) West Bengal. https://fsi.nic.in/isfr19/vol2/isfr-2019-vol-ii-west-bengal.pdf. Accessed on 18 April 2020

FSI-Forest Survey of India (2017) Forest cover. https://fsi.nic.in/isfr2017/isfr-forest-cover-2017.pdf. Accessed on 18 April 2020

Fuller DO (1998) Trends in NDVI time series and their relation to rangeland and crop production in Senegal, 1987–1993. Int J Remote Sens 19(10):2013–2018

Gamon JA, Field CB, Goulden ML, Griffin KL, Hartley AE, Joel G et al (1995) Relationships between NDVI, canopy structure, and photosynthesis in three Californian vegetation types. Ecol Appl 5(1):28–41

Gao X, Huete AR, Ni W, Miura T (2000) Optical–biophysical relationships of vegetation spectra without background contamination. Remote Sens Environ 74(3):609–620

Geerken R, Ilaiwi M (2004) Assessment of rangeland degradation and development of a strategy for rehabilitation. Remote Sens Environ 90(4):490–504

González-Sanpedro MC, Le Toan T, Moreno J, Kergoat L, Rubio E (2008) Seasonal variations of leaf area index of agricultural fields retrieved from Landsat data. Remote Sens Environ 112(3):810–824. https://doi.org/10.1016/J.RSE.2007.06.018

Goodwin NR, Coops NC, Wulder MA, Gillanders S, Schroeder TA, Nelson T (2008) Estimation of insect infestation dynamics using a temporal sequence of Landsat data. Remote Sens Environ 112(9):3680–3689. https://doi.org/10.1016/J.RSE.2008.05.005

Gu Y, Brown JF, Verdin JP, Wardlow B (2007) A five-year analysis of MODIS NDVI and NDWI for grassland drought assessment over the central great plains of United States. Geophys Res Lett 34(L06407):1–6

Gurung RB, Breidt FJ, Dutin A, Ogle SM (2009) Predicting Enhanced Vegetation Index (EVI) curves for ecosystem modeling applications. Remote Sens Environ 113(10):2186–2193

Hardisky MA, Klemas V, Smart RM (1983) The influence of soil salinity, growth form, and leaf moisture on the spectral radiance of Spartina alterniflora canopies. Photogram Eng Remote Sens 49:77–83

Horler DNH, Ahern FJ (1986) Forestry information content of Thematic Mapper data. Int J Remote Sens 7(3):405–428. https://doi.org/10.1080/01431168608954695

Huang C, Wylie B, Yang L, Homer C, Zylstra G (2002) Derivation of a tasselled cap transformation based on Landsat 7 at-satellite reflectance. Int J Remote Sens 23(8):1741–1748

Huang N, He JS et al (2013) Estimating the spatial pattern of soil respiration in Tibetan alpine grasslands using Landsat TM images and MODIS data. Ecol Ind 26:117–125

Huete AR, Jackson RD, Post DF (1985) Spectral response of a plant canopy with different soil backgrounds. Remote Sens Environ 17(1):37–53

Huete AR, Liu H, van Leeuwen WJ (1997) The use of vegetation indices in forested regions: issues of linearity and saturation. In: Geoscience and remote sensing, 1997. IGARSS'97. Remote sensing—ascientific vision for sustainable development, 1997 IEEE international, vol 4, pp 1966–1968. IEEE

Huete A, Didan K, Miura T, Rodriguez EP, Gao X, Ferreira LG (2002) Overview of the radiometric and biophysical performance of the MODIS vegetation indices. Remote Sens Environ 83(1–2):195–213

Hunt ER, Rock BN (1989) Detection of changes in leaf water content using near- and middle infrared reflectance. Remote Sens Environ 30(1):43–54. https://doi.org/10.1016/0034-4257(89)90046-1

IPCC (2019) Climate change and land. An IPCC Special report on climate change, desertification, land degradation, sustainable land management, food security, and greenhouse gas fluxes in terrestrial ecosystems, Summary for Policymakers. https://www.ipcc.ch/site/assets/uploads/2019/08/4.-SPM_Approved_Microsite_FINAL.pdf. Accessed on 22 March 2020

Jackson RD, Pinter Jr PJ, Reginato RJ, Idso SB (1980) Hand-held radiometry. Agricultural Reviews and Manuals W-19. US Dept. of Agriculture. Science and Education Admin, Oakland, CA

Jain P, Sajjad H (2016) Household dependency on forest resources in the Sariska Tiger Reserve (STR), India: implications for management. J Sustain For 35(1):60–74

Jain P, Ahmed R, Sajjad H (2016) Assessing and monitoring forest health using a forest fragmentation approach in Sariska Tiger Reserve, India. NorskGeografiskTidsskrift-Norwegian J Geogr 70(5):306–315

Jalilova G, Khadka C, Vacik H (2012) Developing criteria and indicators for evaluating sustainable forest management: a case study in Kyrgyzstan. For Policy Econ 21:32–43

Jamil M, Ahmed R, Sajjad H (2018) Land suitability assessment for sugarcane cultivation in Bijnor district, India using geographic information system and fuzzy analytical hierarchy process. GeoJournal 83(3):595–611

Jensen JR, Lulla K (1987) Introductory digital image processing: a remote sensing perspective. Geocarto Int 2(1):65. https://doi.org/10.1080/10106048709354084

Jiang Z, Huete AR, Didan K, Miura T (2008) Development of a two-band enhanced vegetation index without a blue band. Remote Sens Environ 112(10):3833–3845

Joubert A, Stewart TJ, Eberhard R (2003) Evaluation of water supply augmentation and water demand management options for the City of Cape Town. J Multi-Criteria Decision Anal 12(1):17–25

Kauth RJ, Thomas GS (1976) The tasselled cap—a graphic description of the spectral-temporal development of agricultural crops as seen by Landsat. In: LARS symposia, p 159

Kunwar RM, Evans A, Mainali J, Ansari AS, Rimal B, Bussmann RW (2018) Change in forest and vegetation cover influencing distribution and uses of plants in the Kailash Sacred Landscape. Nepal. Environment, Development and Sustainability, pp 1–16

Kurtz DB, Asch F, Giese M, Hülsebusch C, Goldfarb MC, Casco JF (2016) High impact grazing as a management tool to optimize biomass growth in northern Argentinean grassland. Ecol Ind 63:100–109

Laman T (2011) State of the world's forests. Retrieved from https://www.fao.org/docrep/013/i2000e/i2000e.pdf

Landscape Tool Box (2019) https://wiki.landscapetoolbox.org/doku.php/remote_sensing_methods:enhanced_vegetatioindex. Accessed 27 Jan 2020

Lausch A, Erasmi S, King DJ, Magdon P, Heurich M (2016) Understanding forest health with remote sensing-part I—A review of spectral traits, processes and remote-sensing characteristics. Remote Sens 8(12):1029

Lausch A, Erasmi S, King D, Magdon P, Heurich M (2017) Understanding forest health with remote sensing-Part II—A review of approaches and Data models. Remote Sens 9(2):129. https://doi.org/10.3390/rs9020129

Li S, Chen X (2014) A new bare-soil index for rapid mapping developing areas using LANDSAT 8 data. Int Arch Photogramm Remote Sensi Spat Inf Sci 40(4):139

Lillesand T, Kiefer RW, Chipman J (2007) Remote sensing and image interpretation, 5th edn. Wiley

Liu W, Yamazaki F (2012) Object-based shadow extraction and correction of high-resolution optical satellite images. IEEE J Select Topics Appl Earth Obs RemoteSens 5(4):1296–1302

Liu X, Hou Z, Shi Z, Bo Y, Cheng J (2017) A shadow identification method using vegetation indices derived from hyperspectral data. Int J Remote Sens 38(19):5357–5373

Major DJ, Baret F, Guyot G (1990) A ratio vegetation index adjusted for soil brightness. Int J Remote Sens 11(5):727–740

Malik S, Pal SC, Das B, Chakrabortty R (2019) Assessment of vegetation status of Sali River basin, a tributary of Damodar River in Bankura District, West Bengal, using satellite data. Environ Dev Sustain, 1–35

Malmström CM, Thompson MV, Juday GP, Los SO, Randerson JT, Field CB (1997) Inter-annual variation in global-scale net primary production: Testing model estimates. Global Biogeochem Cycles 11(3):367–392

Mandal VP, Rehman S, Ahmed R, Masroor M, Kumar P, Sajjad H (2020) Land suitability assessment for optimal cropping sequences in Katihar district of Bihar, India using GIS and AHP. Spat Inf Res, 1–11

Matsushita B, Yang W, Chen J, Onda Y, Qiu G (2007) Sensitivity of the enhanced vegetation index (EVI) and normalized difference vegetation index (NDVI) to topographic effects: a case study in high-density cypress forest. Sensors 7(11):2636–2651

Mensah AA, Sarfo DA, Partey ST (2019) Assessment of vegetation dynamics using remote sensing and GIS: A case of Bosomtwe Range Forest Reserve, Ghana. Egypt J Remote Sens Space Sci 22(2):145–154

Miura T, Huete AR, Yoshioka H, Holben BN (2001) An error and sensitivity analysis of atmospheric resistant vegetation indices derived from dark target-based atmospheric correction. Remote Sens Environ 78(3):284–298

Moleele N, Ringrose S, Arnberg W, Lunden B, Vanderpost C (2001) Assessment of vegetation indexes useful for browse (forage) prediction in semi-arid rangelands. Int J Remote Sens 22(5):741–756

Mondal I, Bandyopadhyay J, Kumar JM (2013) Mangrove zonation and succession pattern of Fazergange and Bakkhali area at Sundarban, W. B., India using remote sensing & GIS techniques. Indian Cartographer 33:311–315

Mongabay (2018) State of Forest Report says that India's forest and tree cover has increased by 1 percent. https://india.mongabay.com/2018/02/state-of-forest-report-says-that-indias-forest-and-tree-cover-has-increased-by-1-percent/. Accessed on 18 April 2020

Morrissey AJ, Browne J (2004) Waste management models and their application to sustainable waste management. Waste Manage 24(3):297–308

Myneni RB, Williams DL (1994) On the relationship between FAPAR and NDVI. Remote Sens Environ 49:200–211

Nagler PL, Cleverly J, Glenn E, Lampkin D, Huete A, Wan Z (2005) Predicting riparian evapotranspiration from MODIS vegetation indices and meteorological data. Remote Sens Environ 94(1):17–30

Nagler PL, Glenn EP, Kim H, Emmerich W, Scott RL, Huxman TE et al (2007) Relationship between evapotranspiration and precipitation pulses in a semiarid rangeland estimated by moisture flux towers and MODIS vegetation indices. J Arid Environ 70(3):443–462

Nakajima T, Tao G, Yasuoka Y (2002) Simulated recovery of information in shadow areas on IKONOS image by combing ALS data. In: Proceeding of Asian conference on remote sensing (ACRS)

National Forest Policy (1988) Government of India Ministry of Environment and Forests New Delhi. http://asbb.gov.in/Downloads/National%20Forest%20Policy.pdf. Accessed on 18 April 2020

Nemani R, Pierce L, Running S, Goward S (1993) Developing satellite-derived estimates of surface moisture status. J Appl Meteorol 32(3):548–557

O'Malley LS (1908) Bengal District Gazeteers Bankura. Government of West Bengal, India

Ono A, Kajiwara K, Honda Y (2010) Development of new vegetation indexes, shadow index (SI) and water stress trend (WST). Intern. Arch Photogram, Remote Sens Spat Inf Sci 38:710–714

Pal SC, Chakrabortty R, Malik S, Das B (2018) Application of forest canopy density model for forest cover mapping using LISS-IV satellite data: a case study of Sali watershed, West Bengal. Model Earth Syst Environ 4(2):853–865

Palmer AR, Fortescue A (2004) Remote sensing and change detection in rangelands. Afr J Range Forage Sci 21(2):123–128

Paruelo JM, Epstein HE, Lauenroth WK, Burke IC (1997) ANPP estimates from NDVI for the central grassland region of the United States. Ecology 78(3):953–958

Podvezko V (2009) Application of AHP technique. J Bus Econ Manag 10(2):181–189

Poggi D, Porporato A, Ridolfi L, Albertson JD, Katul GG (2004) The effect of vegetation density on canopy sub-layer turbulence. Bound-Layer Meteorol 111(3):565–587

Polidorio AM, Flores FC, Imai NN, Tommaselli AM, Franco C (2003) Automatic shadow segmentation in aerial color images. In XVI Brazilian symposium on computer graphics and imageprocessing, 2003. SIBGRAPI 2003, pp 270–277. IEEE.

Pourghasemi HR, Pradhan B, Gokceoglu C (2012) Application of fuzzy logic and analytical hierarchy process (AHP) to landslide susceptibility mapping at Haraz watershed, Iran. Nat Hazards 63(2):965–996

Qureshi ME, Harrison SR (2003) Application of the analytic hierarchy process to riparian revegetation policy options. Small-Scale For Econ Manag Policy 2(3):441

Rao BV, Briz-Kishore BH (1991) A methodology for locating potential aquifers in a typical semiarid region in India using resistivity and hydrogeological parameters. Geoexploration 27(1–2):55–64

Richardson AJ, Wiegand CL (1977) Distinguishing vegetation from soil background information. Photogramm Eng Remote Sens 43(12):1541–1552

Rikimaru A, Miyatake S (1997) Development of forest canopy density mapping and monitoring model using indices of vegetation, bare soil and shadow, presented paper for the 18th ACRS. Kuala Lumpur, Malaysia

Rouse Jr J, Haas RH, Schell JA, Deering DW (1974) Monitoring vegetation systems in the Great Plains with ERTS, pp 309–317. NASA, Washington. https://ntrs.nasa.gov/search.jsp?R=19740022614. Accessed 5 Jan 2019

Roy PS, Miyatake S, Rikimaru A (1997) Biophysical spectral response modelling approach for forest density stratification. In: Proceedings of the 18th Asian conference on remote sensing

Saaty T (1980a) The analytic process: planning, priority setting, resources allocation. McGraw, New York

Saaty TL (1980b) The analytical hierarchy process. MCGraw-Hill, New York

Saaty TL, Decision HTMA (1990) The analytic hierarchy process. Eur J Oper Res 48:9–26

Sader SA (1989) Multispectral and seasonal characteristics of northern hardwood and boreal forest types in Maine. Image Process 89:109–116

Sahana M, Dutta S, Sajjad H (2019) Assessing land transformation and its relation with land surface temperature in Mumbai city, India using geospatial techniques. Int J Urban Sci 23(2):205–225

Sahana M, Sajjad H, Ahmed R (2015) Assessing spatio-temporal health of forest cover using forest canopy density model and forest fragmentation approach in Sundarban reserve forest, India. Model Earth Syst Environ 1(4):49

Sharma J, Chaturvedi RK, Bala G, Ravindranath NH (2015) Assessing "inherent vulnerability" of forests: a methodological approach and a case study from Western Ghats, India. Mitig Adapt Strat Glob Change 20(4):573–590

Simula, M. (2009). Towards defining forest degradation: comparative analysis of existing definitions. *Forest Resources Assessment Working Paper, 154.*

Smith WB (2002) Forest inventory and analysis: a national inventory and monitoring program. Environ Pollut 116:S233–S242

Solberg S, Næsset E, Lange H, Bollandsås OM (2004) Remote sensing of forest health. Int Arch Photogramm Remote Sens Spat Inf Sci 36(Part 8):W2

Turner DP, Cohen WB, Kennedy RE, Fassnacht KS, Briggs JM (1999) Relationships between Leaf Area Index and Landsat TM spectral vegetation indices across three temperate zone sites. Remote Sens Environ 70(1):52–68. https://doi.org/10.1016/S0034-4257(99)00057-7

United States Geological Survey (USGS) site. https://earthexplorer.usgs.gov/. Accessed on 17 Sept 2019

Verbesselt J, Jonsson P, Lhermitte S, Van Arad J, Coppin P (2006) Evaluating satellite and climate data-derived indices as fire risk indicators in savanna ecosystems. IEEE Trans Geosci Remote Sens 44(6):1622–1632

Vogelmann JE, Rock BN (1988) Assessing forest damage in high-elevation coniferous forests in Vermont and New Hampshire using thematic mapper data. Remote Sens Environ 24(2):227–246. https://doi.org/10.1016/0034-4257(88)90027-2

Wang H, Chen F, Zhang R, Qin L (2017) Seasonal dynamics of vegetation of the central Loess Plateau (China) based on tree rings and their relationship to climatic warming. Environ Dev Sustain 19(6):2535–2546

Wellens J (1997) Rangeland vegetation dynamics and moisture availability in Tunisia: an investigation using satellite and meteorological data. J Biogeogr 24(6):845–855

Westergaard-Nielsen A, Lund M, Hansen BU, Tamstorf MP (2013) Camera derived vegetation greenness index as proxy for gross primary production in a low Arctic wetland area. ISPRS J Photogramme Remote Sens 86:89–99

Wiegand CL, Richardson AJ, Escobar DE, Gerbermann AH (1991) Vegetation indices in crop assessments. Remote Sens Environ 35:105–119

Wilkie DS, Bennett EL, Peres CA, Cunningham AA (2011) The empty forest revisited. Ann N Y Acad Sci 1223(1):120–128

Wilson EH, Sader SA (2002) Detection of forest harvest type using multiple dates of Landsat TM imagery. Remote Sens Environ 80(3):385–396. https://doi.org/10.1016/S0034-4257(01)00318-2

Wolfslehner B, Vacik H (2008) Evaluating sustainable forest management strategies with the analytic network process in a pressure-state-response framework. J Environ Manage 88(1):1–10

Wulder MA, Dymond CC, White JC, Leckie DG, Carroll AL (2006) Surveying mountain pine beetle damage of forests: a review of remote sensing opportunities. For Ecol Manage 221(1–3):27–41

WWF-World Wildlife Fund (2020) Threats: deforestation and forest degradation. https://www.worldwildlife.org/threats/deforestation-and-forest-degradation. Accessed on 18 April 2020

Xue J, Su B (2017) Significant remote sensing vegetation indices: a review of developments and applications. J Sens

Yengoh GT, Dent D, Olsson L, Tengberg AE, Tucker III CJ (2015) Use of the Normalized Difference Vegetation Index (NDVI) to assess Land degradation at multiple scales: current status, future trends, and practical considerations. Springer

Zhan Q, Shi W, Xiao Y (2005) Quantitative analysis of shadow effects in high-resolution images of urban areas. Int Arch Photogramm Remote Sens 36(8/W27)

Zhao H, Chen X (2005) Use of normalized difference bareness index in quickly mapping bare areas from TM/ETM+. In: Geoscience and Remote sensing symposium, 2005. IGARSS'05. Proceedings. 2005 IEEE international, vol 3, pp 1666–1668. IEEE

Zhou L, Tucker CJ, Kaufmann RK, Slayback D, Shabanov NV, Myneni RB (2001) Variations in northern vegetation activity inferred from satellite data of vegetation index during 1981 to 1999. J Geophys Res Atmos 106(D17):20069–20083

Zhou Y, Yang G, Wang S, Wang L, Wang F (2014) A new index for mapping built-up and bare land areas from Landsat-8 OLI data. Remote Sens. Retrieved from https://www.tandfonline.com/doi/abs/10.1080/2150704X.2014.973996. Accessed 5 Jan 2019

Chapter 4
Ecological Determinants of Woody Plant Species Richness in the Indian Himalayan Forest

Mahanand Swapna, Tamang Deeke Doma, Sikder Arunima, Gudasalamani Ravikanth, Muneeswaran Mariappan, and Ganesan Rengaian

Abstract The ecological importance of woody plant species richness is well known. The role of abiotic ecological determinants on structuring the vegetation has been well studied. The present study evaluated the independent and integrated strength of the abiotic and biotic determinants in explaining species richness of woody plants in the Indian Himalayan forest. The primary field inventory data was collected using nested quadrat method (tree species at 10×10 m^2, shrub species at 5×5 m^2, and herb species at 1×1 m^2 quadrats) for different life forms and for the abundance estimation within each 1 km transect. Each transect was laid in a 6.3×6.3 km^2 grid on the study site. The biotic determinants included diameter at breast height (d.b.h.) and tree height, whereas the abiotic determinants were temperature, precipitation, soil moisture, relative humidity and elevation. A total of 302 woody plant species (233 genera and 53 families) were recorded from the field inventory. The woody plant species richness was found to range from 1 to 54 per ha at transect level. Structural Equation Model (SEM) evaluated different combinations of ecological determinants for woody plant species richness. The abiotic or biotic determinants were non-significant if considered independently; however, the integration of both resulted in a significant relation with woody plant species richness. The best combination of ecological determinants include density d.b.h. ≥ 2.5 cm, tree height, relative humidity, and elevation ($R^2 = 0.53$). Overall, the integration of biotic and abiotic determinants better explained woody plant species richness in the Indian Himalayan forest.

M. Swapna (✉) · T. D. Doma · S. Arunima · G. Ravikanth · M. Mariappan · G. Rengaian
Ashoka Trust for Research in Ecology and the Environment, Royal Enclave, Srirampura, Jakkur, Bangalore 560064, India
e-mail: swapna.mahanand@atree.org

T. D. Doma · S. Arunima
Manipal Academy of Higher Education, Manipal, Udupi 576104, India

M. Mariappan
International Union for Conservation of Nature(IUCN), Rue Mauverney 28, 1196 Gland, Switzerland

P. K. Shit et al. (eds.), *Spatial Modeling in Forest Resources Management*, Environmental Science and Engineering,
https://doi.org/10.1007/978-3-030-56542-8_4

Keywords Ecological variability · Species richness · Structural equation model · Himalayan forest

4.1 Introduction

Ecological variability (biodiversity and environmental heterogeneity) is an important indicator of ecosystem function and ecosystem health (Gaston and Spicer 2013). In particular, understanding the dynamics of species diversity is an important aspect of many habitats. As the absolute number of species can change over time, due to climatic variability, speciation, extinction, or dispersal events and rapid land-use change, the persistence of predictable patterns tell us that such events and their consequences are somehow geographically constrained (Groom et al. 2006; Latham and Ricklefs 1993; Mahanand and Behera 2019). Therefore, factors governing species compositions, structure, and patterns are of primary interest to ecologists.

Woody plant species are usually characterized by trees and shrubs with the diameter at breast height (d.b.h.) of ≥1 cm (Condit 1995), ≥2.5 cm (Gentry 1982) and ≥10 cm (Gentry 1988; Smith et al. 1998). Canopy patches beneath woody plants and the inter canopy patches experience heterogeneous patterns of energy, water, and biogeochemistry, and the level of this heterogeneity depends on the architecture of woody plant canopies (Breshears 2006; Pandey et al. 2018). Thus, the higher richness of woody plant species in a habitat effectively maintains the microenvironment in terms of reduced soil erosion and air temperatures, wind speed and radiation (Blondel et al. 2010; Pickett et al. 2009). Consequently, it enhances the soil moisture content and relative humidity (Fine 2015). Alternately, increased radiation is often associated with reduced water availability, resulting in low woody plant species richness (Pausas and Austin 2001; Vereecken et al. 2015). The relationship of plant species richness to abiotic determinants often account for >80% of the spatial variation in richness; however, there is no global consistency due to scale variant characters (Asner et al. 2017; Francis and Currie 2003; Li et al. 2019).

Several complex and interacting abiotic determinants such as topography (Körner 2004), edaphic factors (Hawkins et al. 2003; Russell-Smith 1991), precipitation (Hawkins et al. 2006), soil moisture (van der Molen et al. 2011), natural disturbances (Sinha et al. 2018), and anthropogenic disturbances (Gogoi and Sahoo 2018; Lakshminarasimhan and Paul 2018) have been identified as predictors of woody-plant species richness in tropical forests. Moreover, the optimized range of these interacting abiotic determinants (temperature, radiation, soil moisture) supports higher woody plant species richness (Blondel et al. 2010; O'Brien 1998). Many of these predictors of species richness are region-specific due to the unique evolutionary, geographic, and land-use histories of each region (Pennington et al. 2009; Whittaker et al. 2001). The Indian Himalayan forest is the part of a global biodiversity hotspot and encompasses a large fraction of global climate and plant diversity as well (Bookhagen and Burbank 2010; Fick and Hijmans 2017). The major forest types observed are tropical moist deciduous forests along the foothills of the Himalaya, while the temperate broad-leaved forests are found between 1500 and 3000 m elevation in the eastern Himalaya. Moreover, the region also harbors broad-leaved hill forests, subtropical forests, temperate mixed forests and wet/dry evergreen forest (Roy et al. 2015). In a phytogeographical sense, forests of this region are species rich, and harbor a number of phylogenetically primitive plant species and hence regarded as a "treasure trove" of ancient and unique vegetation (Champion and Seth 1968). However, climatic variability has clearly affected plant speciation, extinction, and dispersal events in the Himalaya (Manish and Pandit 2019). All these issues need to be addressed in order to protect and conserve the unique biodiversity of the Indian Himalayan forest (Banerjee et al. 2019; Basnett et al. 2019; Mehta et al. 2020).

The abiotic determinant, namely temperature was found to be a limiting factor for the woody plant species richness in the temperate forests (Currie 1991), whereas studies in tropical regions emphasize the importance of moisture and related factors (Brown 1990). Moreover, a study in the subtropical region compared different abiotic determinants and found that relative air humidity and soil moisture significantly explained the plant species richness (Drissen et al. 2019). A study in east Nepal confirmed the significance of precipitation and moisture in explaining woody plant species richness (Bhattarai and Vetaas 2003). Although, species richness and distribution patterns of plants are largely regulated by elevation and other environmental factors (Saikia et al. 2017), it has been pointed out that many components of climate and local environment (e.g., temperature, precipitation, seasonality and disturbance regime) vary along the elevation gradients and ultimately create the variation in species richness (Lomolino 2001). The declining trend of species richness along elevation gradient has been recently reported for the Indian Himalayan forests (Bhutia et al. 2019; Malhi et al. 2010). Studies focused on deforestation and land use effect along the elevational range of Sikkim Himalaya, summarised that the primary forests at higher elevational range are subject to low deforestation compared to the broadleaf forest ranging from low to higher elevational range (Kanade and John 2018).

The biotic determinants have been identified or documented based on qualitative assessment in the Indian Himalayan forest (Chatterjee et al. 2006; Singh and Sanjappa 2011; Tambe and Rawat 2010). For Sikkim Himalaya, the different plant structural components (tree species richness, density d.b.h., basal area and distribution range) were evaluated along with the elevational range and summarized as highest tree density and richness at 1000 m elevation of Sikkim (Pandey et al. 2018). Shooner et al. (2018), report an inverse relationship between elevation and species richness along an elevation gradient in Arunachal Pradesh. Sharma et al. (2019), in Sikkim Himalayas, report a hump shaped richness pattern along elevation gradient. Other explorations utilized the phenological traits of woody species and found significance for the abiotic determinants (day length and temperature), while the role of elevation was not significant in Sikkim Himalaya (Basnett et al. 2019; Ranjitkar et al. 2013). Bhutia et al. (2019) report an inverse 'J' shaped species richness emphasizing the role of abiotic determinants and habitat filtering to explain this pattern.

The distribution of species is partly determined by ecological gradients of water, nutrients, heat and radiation, and partly by biotic determinants, stochastic events or disturbances caused by natural events or forest management (Austin and Smith 1990). Pau et al. (2012), analysed the influence of abiotic and biotic factors in determining woody plant species richness and productivity, and concluded the direct effect of precipitation (0.72) and its mediation through plant structure (0.74), in the dry forests of Hawaii island. However, there are studies that have only evaluated either abiotic or biotic determinants to understand the woody plant species richness in Indian Himalayan forest and hardly any work have considered both these factors together (Dutta and Devi 2013; Gogoi and Sahoo 2018; Sinha et al. 2018). Therefore, it is essential to consider both abiotic and biotic determinants together to reach a better understanding of the composition of woody plants in the Indian Himalayan forest.

This is a maiden attempt to evaluate the independent as well as integrated effect of ecological determinants (i.e., biotic and abiotic) to explain the richness of woody plant species in the Indian Himalayan forest. The geography greatly challenges the relationship between woody plant species richness and abiotic determinants and yet to be generalized. On the other hand, woody plant structure plays an important role to modify the microenvironment. The present study aimed to explore (i) the significance of a best individual ecological determinant, and (ii) the best combinations of ecological determinants to explain the woody plant species richness in Indian Himalayan forest.

4.2 Methods

4.2.1 Study Area

We selected the following states and respective districts in India for this study that run along the foothills of Himalaya, i.e., Assam (Dhubri, Kokrajhar, Bongaigaon, Barpeta, Goalpara, Nalbari), Sikkim (East, West, North, South), and West Bengal (Malda, Uttar Dinajpur, Dakshin Dinajpur, Darjeeling, Jalpaiguri, Coochbehar) (Fig. 4.1). The spatial extent of the study area ranges between 87.79°–91.74 °E and 24.69°–28.09 °N and encapsulates a geographic area of 44,617.6 km^2. The varied topography and climatic conditions account for highly diverse flora (Champion and Seth 1968). The vegetation changes from tropical to subtropical upwards through middle hills with much coniferous and Oak forests of temperate character to the higher slopes with subalpine scrub and alpine "meadows" generally related to higher alpine flora of the north temperate zone (Singh and Sanjappa 2011).

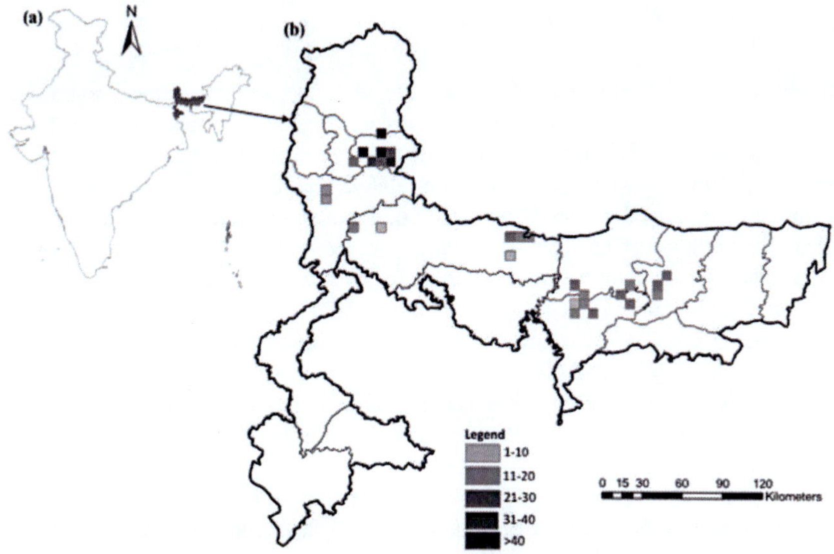

Fig. 4.1 **a** Map showing the study area along the foothills of Indian Himalaya, **b** the spatial distribution of sampled grids and the woody plant species richness ranging from 1 to 54

4.2.2 Biotic Determinants

The primary geo-tagged plants used for this study were collected under the project 'Bioresources and Sustainable Livelihoods in Northeast India' funded by the Department of Biotechnology, New Delhi, India. The entire study site was divided into 6.3 $\times$ 6.3 km^2 grid. The field inventory was carried out in the foothills of Himalaya that holds the diverse floristic composition and forest types. Moreover, the grid sampled for this study falls within the elevational range of 500–2200 m. The sampling of flowering plants were performed in each of the 6.3 $\times$ 6.3 km^2 grid using two belt transects of 10 m $\times$ 500 m. Tree species were enumerated in the entire belt transect and the shrub and herb layers were enumerated at the beginning and end of the transects by laying sub plots of 5 $\times$ 5 m^2 for shrubs and 1 $\times$ 1 m^2 for herbs. Girth at breast height (GBH) and height of the trees were measured and voucher specimens were collected for identification purposes. Further geographic coordinates (latitude and longitude), suite of environmental variables (elevation, temperature, rainfall, moisture, slope, and aspect) and habitat information (canopy cover, canopy height, litter cover, litter depth, invasive species, soil temperature, moisture, disturbance, human use, etc.) were also recorded from each transect as part of metadata collection.

4.2.3 Abiotic Determinants

The available global data on the four essential climatic variables and one topographic variable were used as representative of abiotic determinants, i.e., mean annual surface air temperature (°C), mean annual precipitation (mm), mean annual volumetric soil moisture (m^3/m^3), mean annual surface air relative humidity (%), and elevation (m). Precipitation is the accumulated liquid and frozen water, including rain and snow, that falls to the earth's surface. The determinants do not include fog, dew or the precipitation that evaporates in the atmosphere before it lands at the surface of the earth. Surface air temperature is the temperature of the air at 2 m above the surface of land. The temperature at 2 m level from the ground is calculated by interpolating between the lowest model level and the earth's surface, taking account of the atmospheric conditions. Volumetric soil moisture is the volume of water in soil layer 1 (0–7 cm, the surface is at 0 cm). The volumetric soil water is associated with the soil texture (or classification), soil depth, and the underlying groundwater level. Surface air relative humidity is the ratio of the partial pressure of water vapour to the equilibrium vapour pressure of water at the same temperature near the surface. The four essential climatic variables were downloaded ERA5 climate reanalysis from the Copernicus data portal (ecmwf.int/en/forecasts/datasets/reanalysis-datasets/era5). The downloaded climatic data were the monthly mean of each variable at the spatial resolution of $0.25° \times 0.25°$ for the year 2019. The elevation data was downloaded from Shuttle Radar Topography Mission (SRTM) digital elevation model at a spatial resolution of 90 m. The abiotic determinants mentioned henceforth include temperature, precipitation, soil moisture, relative humidity and elevation.

4.2.4 Data Preparation

A total of 27 transects were laid in the study site as one transect per 6.3×6.3 km^2 grid. For each transect, the g.b.h. was converted to d.b.h. by dividing the g.b.h. value with pi or 3.14. The woody species counts were made and tabulated for the d.b.h. $\geq$ 2.5 cm and d.b.h. $\geq$ 10 cm in each transect. Similarly, the tree heights were averaged for the total woody species in the transects for d.b.h. $\geq$ 2.5 cm. For this study, three biotic variables d.b.h. $\geq$ 2.5 cm, d.b.h. $\geq$ 10 cm and tree height were selected. Also, a record was created for the woody plants genera and family. On the other hand, the downloaded global ERA5 data of the four essential climatic variables were masked using the shapefile of the study sites in QGIS software. The data were normalized, averaged and calculated for each essential climatic variable. Zonal statistics analysis was performed using the essential climatic variables and integrated to each sampled grid of 6.3×6.3 km^2 grid.

4.2.5 *Statistical Analysis*

Spearman rank correlation (Rho) is a non-parametric correlation test that evaluates the multicollinearity of ecological determinants through the correlation matrix. We utilised both the ecological determinants (abiotic and biotic) against woody plant species richness using CRAN package “GGally” in R software. Two ecological determinants having Rho ≥ 0.7 are considered as co-linear. The subsequent analysis considered one out of those collinear determinants in the model, to avoid biases and overestimation. Thereafter, the linear regression model (LM) was performed between the ecological determinants and woody plant species richness.

The structural equation model (SEM) was used to evaluate the relationship between ecological determinants and woody plant species richness. SEM is a complex and robust regression model widely used in ecological analysis (Panda et al. 2017; Rana et al. 2019). The general methodology of SEM involves the specification of a multivariate dependence model that can be statistically tested against field data. In essence, the specification of a hypothesized model yields an expected covariance matrix that can be tested against the actual covariance matrix. Also, SEM allows the estimation of latent variables using multiple observed indicators and reports the measurement errors. This multivariate technique not only assists in selecting models among alternatives but also provides an efficient and simultaneous solution to a series of overlapping regression relationships (Daou and Shipley 2019).

The analysis was performed in R software using the packages namely “lavaan”, “haven”, and “semPlot” from CRAN repository.We compared each ecological determinant against woody plant species richness. Thereafter, we tested the combination of two, three, and four ecological determinants with the woody plant species richness. We restricted the SEM evaluation with four variables as thereafter the degree of freedom was found to be zero. The model performance was evaluated using the beta coefficient, R^2 value, root mean square error. Also, the covariance among the variables were evaluated.

4.3 Results

The primary woody plant data was collected from 27 transects sampled in 27 respective grids of size 6.3×6.3 km^2 in the Himalayan forests which are contiguous in Sikkim, West Bengal and Assam states, India. The Indian Himalayan forest has high spatial heterogeneity and hence from the foothills to an elevation of 2200 m harbors varied forest types and transition zones. The forest types encountered during this study are tropical, subtropical, temperate, along with the elevation ranges in ascending manner. These could also be classified as deciduous, mixed-deciduous, warm broadleaf, semi-evergreen and evergreen forest based on the vegetation types. The transect level woody plant species richness was found to range from 1 to 54/ha among the sampled transects from the study site (Fig. 4.1b). A total of 302 unique woody plants belonging to 233 genera were identified and recorded from the 27 sampled transects in the study site. The most abundant family to which the woody plants belonged are *Fagaceae, Theaceae, Dipterocarpaceae, Symplocaceae, Cupressaceae.*

The five most abundant woody plants such as *Shorea robusta* (561), *Terminalia bellirica* (167), *Tectona grandis* (140), *Mallotus philippensis* (128), *Melia azadirachta* (112) were recorded in the sampled transects from Assam. A total of 107 unique woody plants were recorded from the study sites of Assam. Similarly, the five most abundant woody plants *Castanopsis sp.* (310), *Viburnum erubescens* (231), *Symplocos sp.* (158), *Eurya cavinervis* (141), *Cryptomeria japonica* (116) were recorded in the sampled transects from Sikkim. A total of 151 unique woody plants were recorded from the study sites of Sikkim. The five most abundant woody plants *Shorea robusta* (99), *Machilus edulis* (79), *Magnolia champaca* (77), *Aglaia spectabilis* (67), *Tectona grandis* (41) were recorded in the sampled transects from West Bengal. A total of 82 unique woody plants were recorded from the study sites of West Bengal.

Among the abiotic determinants, mean annual temperature ranged from −5.5 to 25.5 °C in the study site for the year 2019 (Fig. 4.2a). The low to medium range of temperature pattern was observed in Sikkim, whereas higher range of temperature was recorded for forests in the Assam and West Bengal states (Fig. 4.2a). The annual mean precipitation was found to range from 584 to 7650 mm. The range of medium

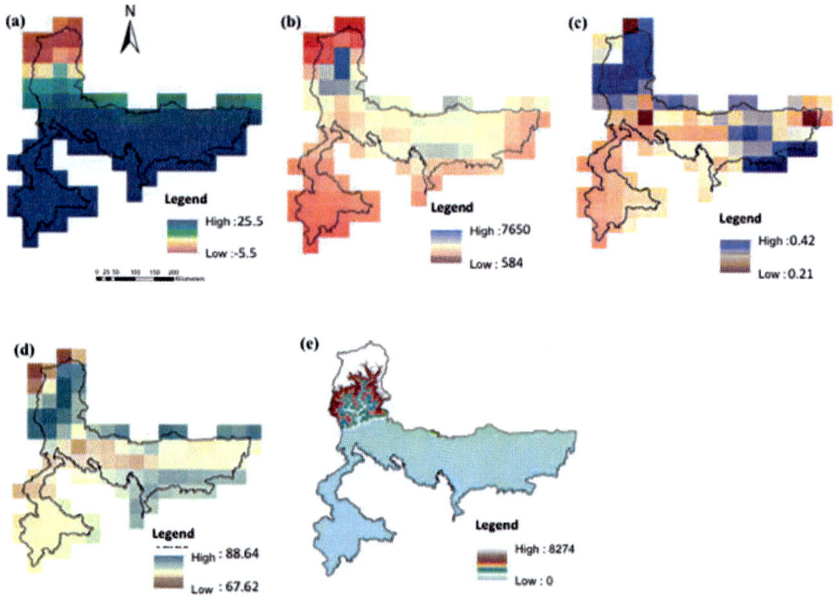

Fig. 4.2 Representing the abiotic determinants selected for this study, i.e., **a** temperature (°C), **b** precipitation (mm), **c** soil moisture in m^3/m^3, **d** relative humidity (%), and **e** elevation (m)

to high precipitation was recorded for forests in Sikkim and Assam states, except the snow covered area of Sikkim. The low to medium level precipitation was recorded in the forests of West Bengal state (Fig. 4.2b). The volumetric soil moisture was found to range from 0.21 to 0.42 m^3/m^3. The soil in the forests of Sikkim and Assam states had high soil moisture holding capacity compared to the forests of West Bengal (Fig. 4.2c). The relative air humidity recorded for the study site was found to be ranging from 67.62 to 88.64%. Similar to the pattern of soil moisture, the relative air humidity was higher for Sikkim and Assam, while it was low to medium in the forests of West Bengal state (Fig. 4.2d). The mean elevation of the grids for the study site was found to be ranging from 0 to 8274 m. The majority of higher elevational variation was recorded in the Sikkim region, whereas the Assam and West Bengal regions varied from moderate elevation to plain (Fig. 4.2e).

The Spearman rank correlation test was performed among and within the ecological determinants and woody plant richness. The resulting correlation matrix was found to be ranging from -1 to $+1$ (Fig. 4.3). The woody plant species richness was

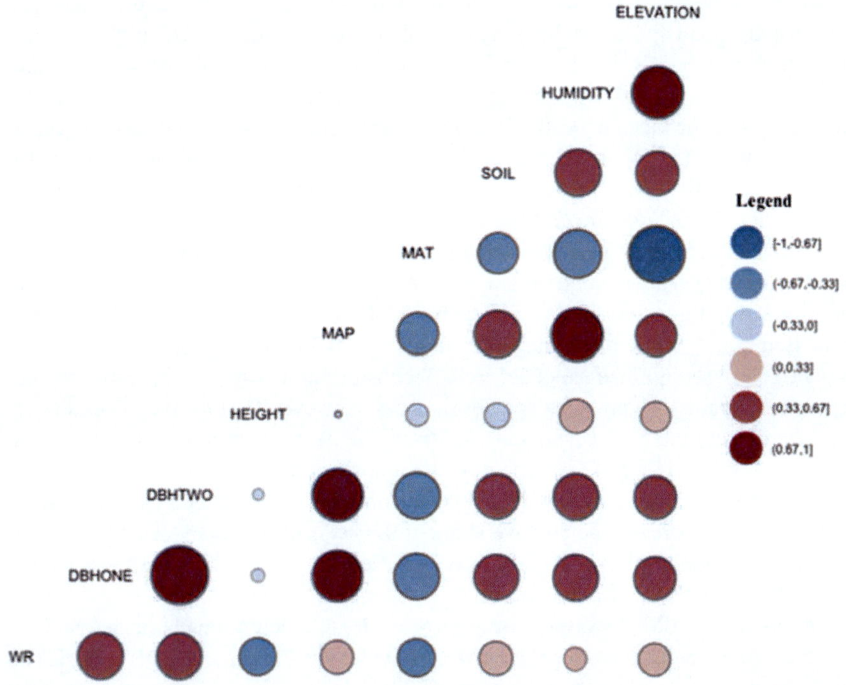

Fig. 4.3 The result from Spearman rank correlation performed between and among woody plant species richness and ecological determinants are represented as a correlation matrix and the range varies from -1 to $+1$. (WR—woody plant species richness, DBHONE—density d.b.h. $\geq$ 2.5 cm, DBHTWO—density d.b.h. $\geq$ 10 cm, HEIGHT—Tree height, MAP—Mean annual precipitation, MAT—mean annual temperature, SOIL—soil moisture, HUMIDITY—relative humidity, ELEVATION—elevation)

found to be moderately correlated with the two biotic determinants namely density d.b.h. ≥ 2.5 cm and density d.b.h. ≥ 10 cm (Fig. 4.3). However, the correlation between woody plant species richness and tree height was observed to be moderately negative. Among the abiotic determinants precipitation, soil moisture, relative humidity and elevation have resulted in low positive correlation with woody plant species richness. However, the woody plant species richness was negatively correlated with temperature (Fig. 4.3). The ecological determinants were found to be positively correlated among each other, except tree height and temperature. Due to high collinearity (Rho ≥ 0. 7) between density d.b.h. ≥ 2.5 cm and density d.b.h. ≥ 10 cm, one out of these determinants may be considered for better model representation (Fig. 4.3).

A comparison was made between LM and SEM to evaluate the relationship between woody plant species richness and individual ecological determinants. It was observed that the SEM outputs were statistically significant and the obtained root mean square error (RMSE) was <0.08 for each tested model using SEM (Table 4.1). Among the biotic determinants, the density d.b.h. ≥ 2.5 cm and density d.b.h. ≥ 10 cm individually resulted in highest significance ($R^2 = 0.4$) and slope (0.6) with woody plant species richness (Fig. 4.4b, d). However, the tree height showed the significance of ($R^2 = 0.15$) and negative slope (−0.34) with the woody plant species richness (Fig. 4.4f). Among the abiotic determinants, temperature resulted in highest significance ($R^2 = 0.15$) and a negative slope (−0.39) with woody plant species richness (Fig. 4.4h). However, the remaining abiotic determinants exhibited a significance of $R^2 \leq 0.1$ and positive slope ≤0.31 with woody plant species richness in the Indian Himalayan forest (Table 4.1; Fig. 4.4j–p).

SEM was performed using two ecological determinants (density d.b.h. ≥ 2.5 cm and tree height) against woody plant species richness exhibited highest significance of $R^2 = 0.49$ (Table 4.1; Fig. 4.5a). The model-I showed positive slope of 0.59 with density d.b.h. ≥ 2.5 cm and negative slope of −0.39 with tree height. Also, the model-J showed significance of $R^2 = 0.49$ considering density d.b.h. ≥ 10 cm and tree height against woody plant species richness. However, the observed slopes were 0.58 for density d.b.h. ≥ 10 cm and −0.36 for tree height with woody plant species richness (Table 4.1; Fig. 4.5b). The model-K considered the density d.b.h. ≥ 2.5 cm and precipitation against woody plant species richness and resulted in a significance of $R^2 = 0.37$. However, the positive slope of 0.63 was observed by density d.b.h. ≥ 2.5 cm and negative slope of −0.04 was observed for precipitation with the woody plant species richness (Table 4.1; Fig. 4.5c).

In model-O, SEM performed using three ecological determinants (density d.b.h. ≥ 2.5 cm, tree height, and precipitation) exhibited highest significance ($R^2 = 0.53$) in explaining woody plant species richness (Table 4.1; Fig. 4.6a). Moreover, a positive slope of 0.72 was observed for density d.b.h. ≥ 2.5 cm, while tree height (−0.44), and precipitation (−0.25) exhibited negative slope (Fig. 4.6a). Similarly, the model-P (density d.b.h. ≥ 2.5 cm, tree height, relative humidity) and model-Q (density d.b.h. ≥ 2.5 cm, tree height, elevation) resulted in significance of $R^2 = 0.51$ and $R^2 = 0.50$, respectively (Table 4.1; Fig. 4.6b, c). The model-T, analysed a combination of four ecological determinants (density d.b.h. ≥ 2.5 cm, tree height, elevation, relative

Table 4.1 Representing the significant outputs (root mean square error <0.08) from the structural equation model (SEM), where different combinations of ecological determinants were tested against woody plant species richness (WR)

Model	Combinations of ecological determinants	R^2
WR versus one ecological determinant		
A	WR ~ density d.b.h. ≥ 2.5 cm	0.4
B	WR ~ density d.b.h. ≥ 10 cm	0.4
C	WR ~ tree height	0.15
D	WR ~ temperature	0.15
E	WR ~ precipitation	0.10
F	WR ~ soil moisture	0.10
G	WR ~ relative humidity	0.10
H	WR ~ elevation	0.02
WR versus two ecological determinants		
I	WR ~ density d.b.h. ≥ 2.5 cm + tree height	0.49
J	WR ~ density d.b.h. ≥ 10 cm + tree height	0.49
K	WR ~ density d.b.h. ≥ 10 cm + precipitation	0.37
L	WR ~ density d.b.h. ≥ 2.5 cm + density d.b.h. ≥ 10 cm	0.36
M	WR ~ tree height + temperature	0.34
WR versus three ecological determinants		
N	WR ~ density d.b.h. ≥ 2.5 cm + tree height + precipitation	0.53
O	WR ~ density d.b.h. ≥ 2.5 cm + tree height + relative humidity	0.51
P	WR ~ density d.b.h. ≥ 2.5 cm + tree height + elevation	0.50
Q	WR ~ density d.b.h. ≥ 2.5 cm + tree height + soil moisture	0.50
R	WR ~ density d.b.h. ≥ 2.5 cm + tree height + temperature	0.50
S	WR ~ density d.b.h. ≥ 2.5 cm + density d.b.h. ≥ 10 cm + tree height	0.49
WR versus four ecological determinants		
T	WR ~ density d.b.h. ≥ 2.5 cm + tree height + relative humidity + elevation	0.56
U	WR ~ density d.b.h. ≥ 2.5 cm + tree height + relative humidity + temperature	0.55
V	WR ~ density d.b.h. ≥ 2.5 cm + tree height + precipitation + Elevation	0.54

(continued)

Table 4.1 (continued)

Model	Combinations of ecological determinants	R^2
W	WR ~ density d.b.h. ≥ 2.5 cm + tree height + precipitation + temperature	0.54
X	WR ~ density d.b.h. ≥ 2.5 cm + tree height + precipitation + relative humidity	0.53
Y	WR ~ density d.b.h. ≥ 2.5 cm + tree height + precipitation + soil moisture	0.53

humidity) using SEM can be considered as the best model ($R^2 = 0.55$) to explain woody plant species richness of the Indian Himalayan forest (Table 4.1; Fig. 4.7a). The positive slope was observed for density d.b.h. ≥ 2.5 cm (0.67) and elevation (0.36), whereas negative slope was observed for tree height (−0.30) and relative humidity (−0.45; Fig. 4.7a). The model-U resulted in the significance of $R^2 = 0.54$ in explaining woody plant species richness of the Indian Himalayan forest (Table 4.1; Fig. 4.7b).

4.4 Discussion

The ecological determinants tested individually against woody plant species richness observed that biotic determinants ($R^2 \leq 0.4$) are more significant than abiotic determinants in explaining the patterns ($R^2 \leq 0.15$; Fig. 4.4). Among all the abiotic determinants, temperature significantly explained the woody plant species richness; that corroborates with the study performed by Basnett et al. (2019) to evaluate phenological traits of woody plant species. Moreover, they defined no significant role of elevation. Similarly, our study also observed a poor fit ($R^2 \leq 0.1$) between woody plant species richness and elevation (Fig. 4.4n). In contrast, Pandey et al. (2018) found the highest woody plant species richness at an elevation of 1000 m and summarized elevation as a significant abiotic determinant. Probably, scale dependent characteristics of ecological determinant must be considered to compare the observation made for woody plant species richness (Asner et al. 2017; Li et al. 2019). Moreover, our field inventory made up to the mean elevation of 2200 m out of 8274 m. Therefore, considering the woody species occurrence from the whole elevation range can better explain the variation along elevation. The remaining abiotic determinants such as precipitation, relative humidity and soil moisture tested individually against woody plant species richness exhibited a poor fit ($R^2 \leq 0.1$) with a positive slope.

The best combination of two ecological determinants were density d.b.h. ≥ 2.5 cm and tree height, although these two are biotic determinants. This summarised that the significance of two biotic determinants is higher than any other combinations (Table 4.1). On the other hand, the combinations of three ecological determinants such as density d.b.h. ≥ 2.5 cm, tree height and precipitation is observed to be the best model compared to other combinations. It can be summarised that the combined effect of

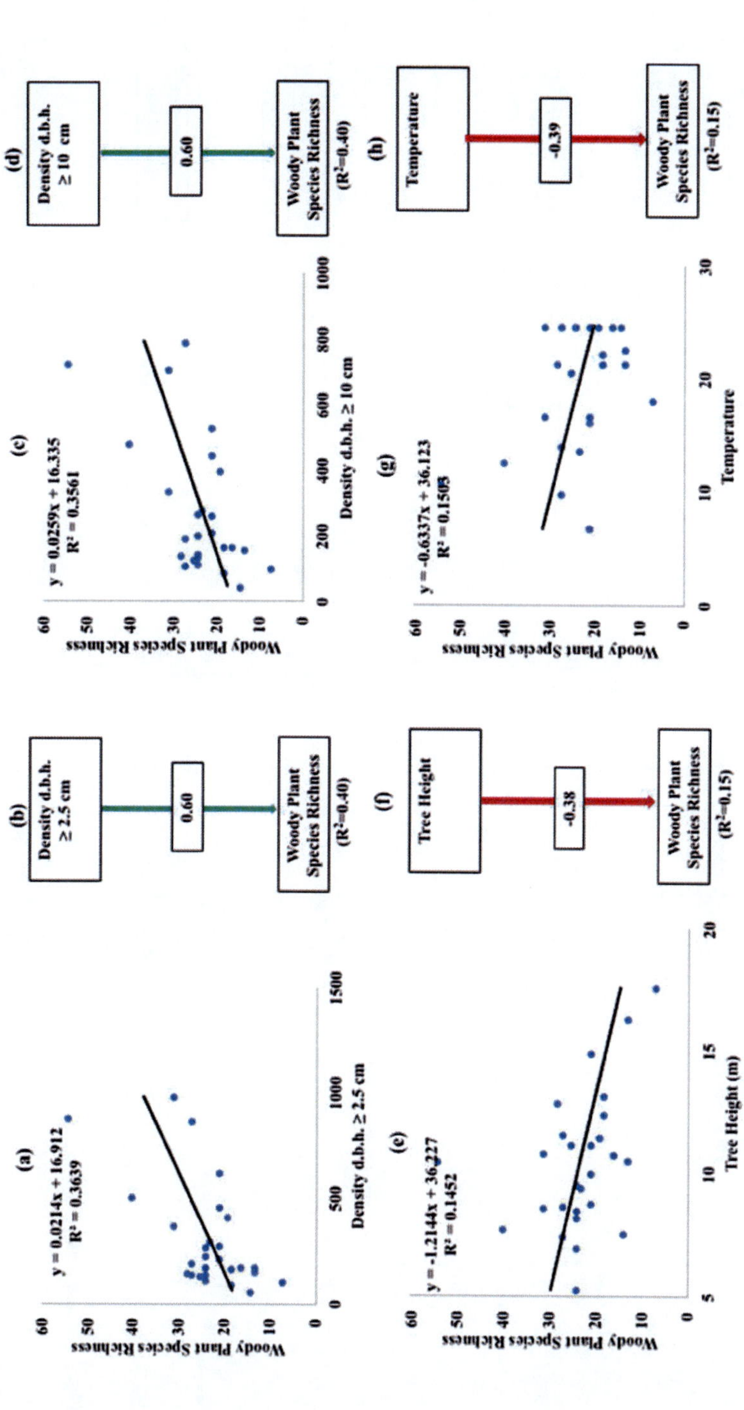

Fig. 4.4 Comparison of linear regression model (**a, c, e, g, i, k, m, o**) and structural equation model-SEM (**b, d, f, h, j, l, n, p**) by evaluating the significance (slope, R^2) of woody plant species richness against individual ecological determinants

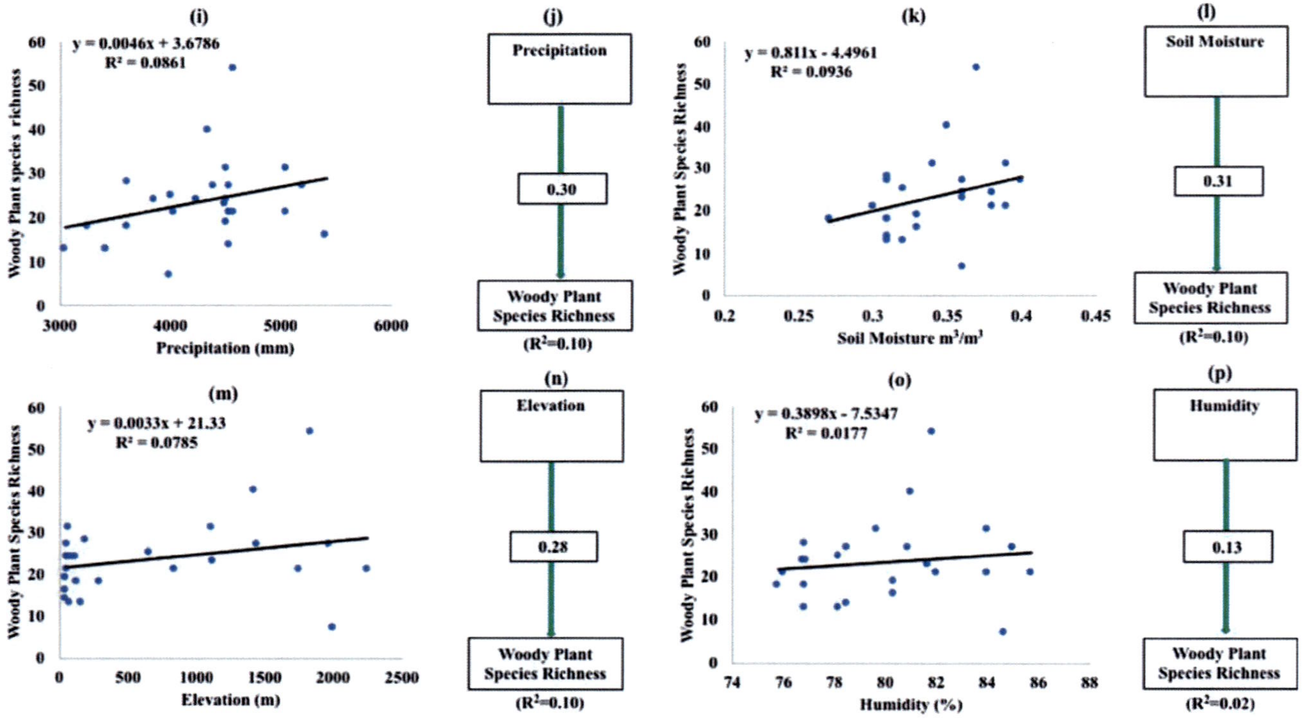

Fig. 4.4 (continued)

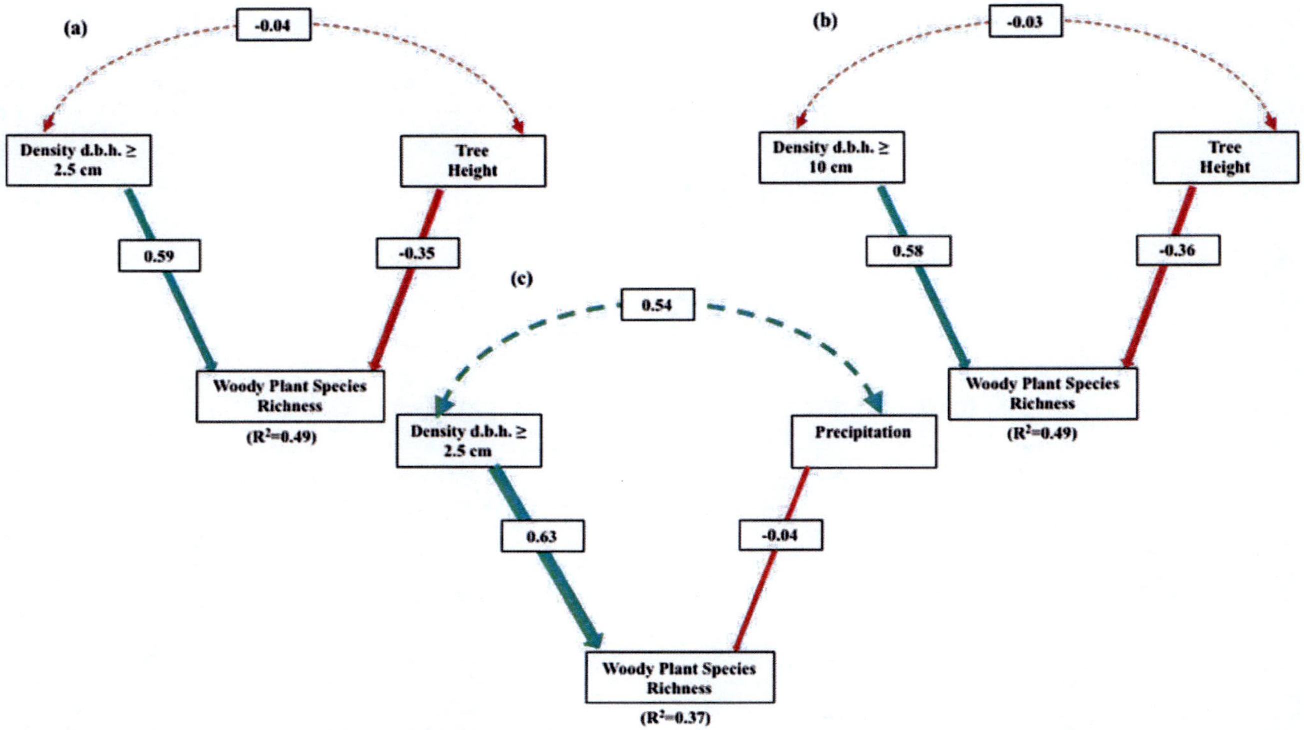

Fig. 4.5 SEM output to evaluate the significance of two ecological determinants, **a** density d.b.h. ≥ 2.5 cm and tree height (two biotic determinants), **b** density d.b.h. ≥ 10 cm and tree height (two biotic determinants), **c** density d.b.h. ≥ 2.5 cm and precipitation (one biotic and one abiotic determinants) in explaining woody plant species richness of the Indian Himalayan forest (where, solid line-slope dotted line-covariance, red colour-negative slope/covariance, green colour-positive slope or covariance)

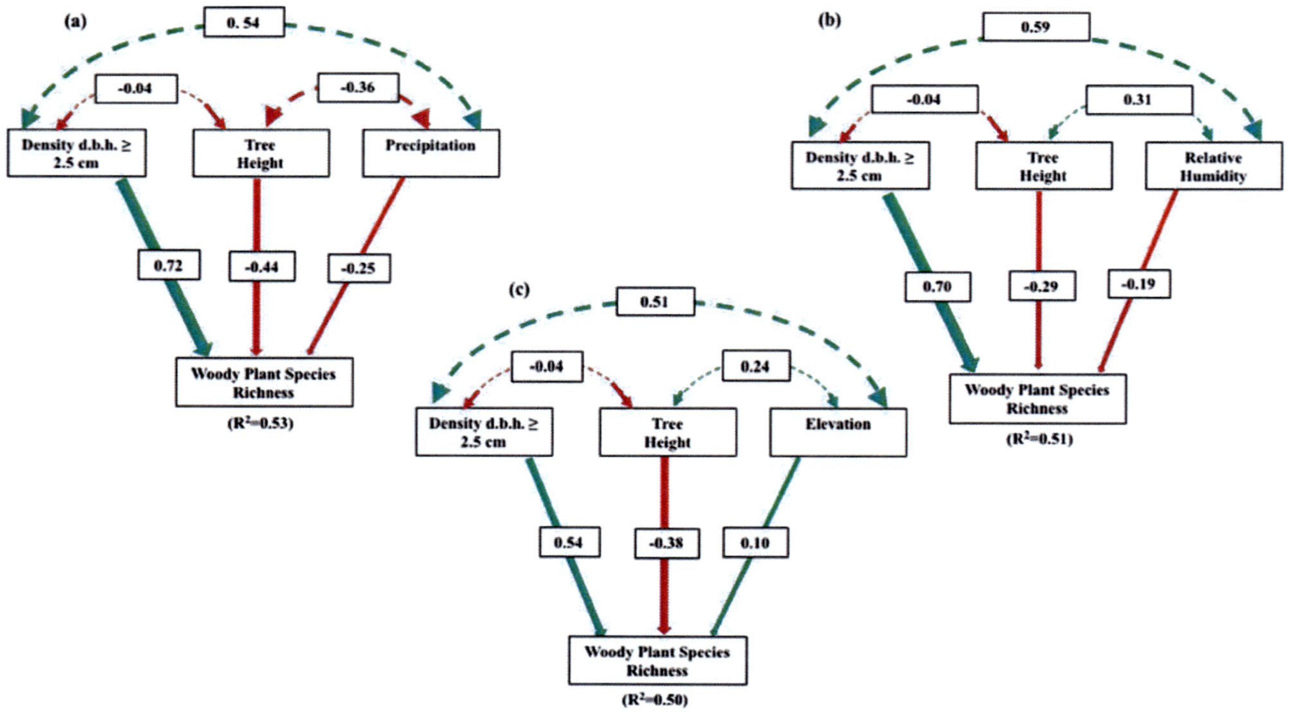

Fig. 4.6 SEM output to evaluate the significance of three ecological determinants, i.e., **a–c** utilized two biotic and one abiotic determinants against the woody plant species richness of the Indian Himalayan forest

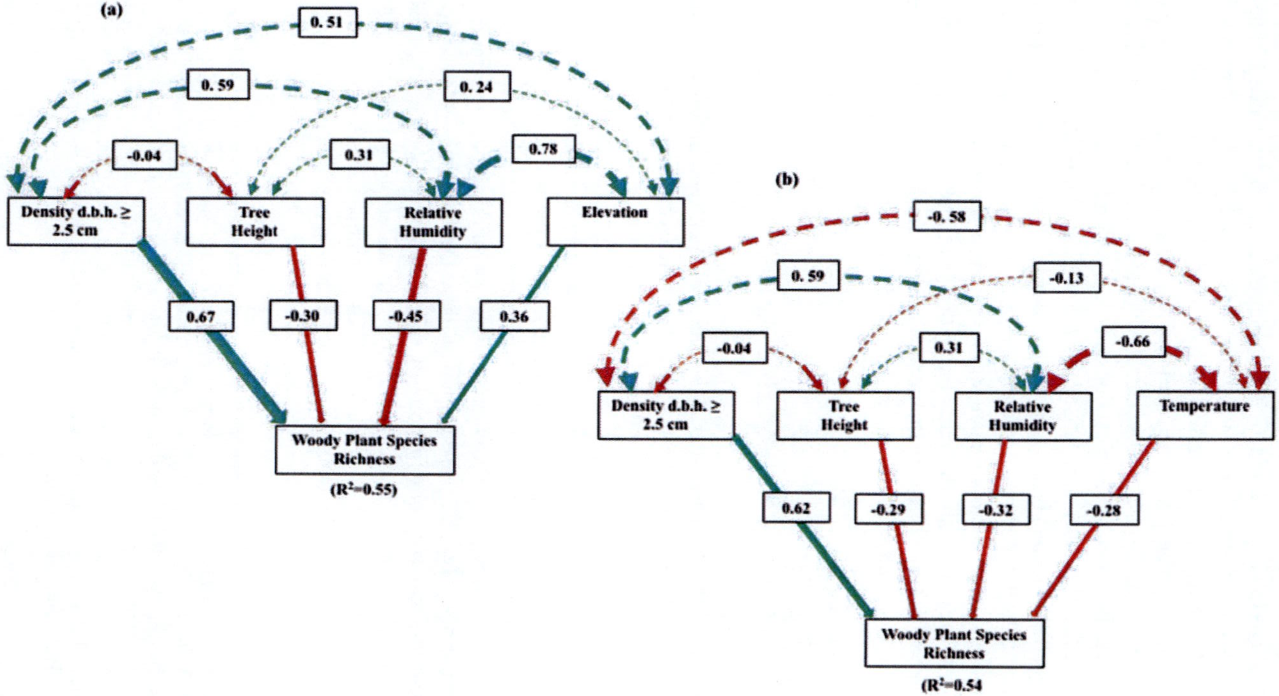

Fig. 4.7 SEM output to evaluate the significance of four ecological determinants (two biotic and two abiotic) as showin in **a** and **b** to explain woody plant species richness of the Indian Himalayan forest

both biotic and abiotic determinants is a good explanation. However, the combination of four ecological determinants d.b.h. $\geq$ 2.5 cm, tree height, relative humidity and elevation emerged as the best combination to explain woody plant species richness in the Indian Himalayan forest. These observations can be validated with inferences made by Austin and Smith (1990) and Pau et al. (2012) that distribution of woody plant species richness can be determined partly by abiotic and partly by biotic determinants. Also, the similar observations were made by Saikia et al. (2017) that the high species richness can be explained by elevation and other ecological determinants and not by any sole factor. The study summarised that the combination of three ecological determinants can be the second best approach to understand the woody plant species richness. However, the best combination includes four ecological determinants (both biotic and abiotic) to explain woody plant species richness for the study site in the Indian Himalaya forest. Overall, the study infers that both biotic and abiotic determinants combined together better explain the woody plant species richness of Indian Himalayan forest, than individual abiotic and biotic determinants independently.

4.5 Conclusions

The main objective of the study was to evaluate the ecological determinants (abiotic, biotic) for woody plant species richness in the Indian Himalayan forest. Using SEM, the tested ecological determinants showed that, integration of biotic and abiotic determinants could outperform in explaining woody plant species richness. Probably, more field inventories for biotic determinants and higher resolution of abiotic determinants could enhance their significance to explain the woody plant species richness. However, field inventory in Indian Himalayan forest is subjected to various challenges (economic, time, labour), whereas high resolution abiotic determinants are not accessible publicly. The present study utilized the best available resources and summarizes the suitable combinations of ecological determinants, i.e., d.b.h. $\geq$ 2.5 cm, tree height, relative humidity, elevation for understanding woody plant species richness of the Indian Himalayan forest. Also, it emphasizes the strength of both biotic and abiotic determinants together, which can be considered to carry out similar studies in future.

Acknowledgements This work was supported by the research grants from Department of Biotechnology, Government of India for the research Project Bioresource and Sustainable Development in Northeast India (BT/01/17/NE/TAX Dt 29 March 2018). The State Forest Departments of Assam, Sikkim and West Bengal are acknowledged for providing research permits in the respective forest areas. We also thank the State Biodiversity Boards of the above States and the National Biodiversity Authority of India for the support to carry out the research project. The assistance provided by Mr. Sonam Dorjee Bhutia and Mr. Passang Tamang is kindly acknowledged. We thank Dr. Sarala Khaling for facilitating the research work.

References

Asner GP et al (2017) Scale dependence of canopy trait distributions along a tropical forest elevation gradient. New Phytol 214:973–988

Austin M, Smith T (1990) A new model for the continuum concept. In: Progress in theoretical vegetation science. Springer, Berlin, pp 35–47

Banerjee A, Chakraborty P, Bandopadhyay R (2019) Urgent conservation needs in the Sikkim Himalaya biodiversity hotspot. Biodiversity 20:88–97

Basnett S, Nagaraju SK, Ravikanth G, Devy SM (2019) Influence of phylogeny and abiotic factors varies across early and late reproductive phenology of Himalayan Rhododendrons. Ecosphere 10:e02581

Bhattarai KR, Vetaas OR (2003) Variation in plant species richness of different life forms along a subtropical elevation gradient in the Himalayas, east Nepal. Glob Ecol Biogeogr 12:327–340

Bhutia Y, Gudasalamani R, Ganesan R, Saha S (2019) Assessing forest structure and composition along the altitudinal gradient in the state of Sikkim, Eastern Himalayas, India. Forests 10:633

Blondel J, Aronson J, Bodiou J-Y, Boeuf G (2010) The Mediterranean region: biological diversity in space and time. Oxford University Press

Bookhagen B, Burbank DW (2010) Toward a complete Himalayan hydrological budget: Spatiotemporal distribution of snowmelt and rainfall and their impact on river discharge. J Geophys Res Earth Surf 115

Breshears DD (2006) The grassland–forest continuum: trends in ecosystem properties for woody plant mosaics? Front Ecol Environ 4:96–104

Brown JH (1990) Species diversity. In: Analytical biogeography. Springer, Berlin, pp 57–89

Champion H, Seth S (1968) A revised survey of the forest types of India. Govt India Publication, Delhi

Chatterjee S, Saikia A, Dutta P, Ghosh D, Worah S (2006) Review of biodiversity in Northeast India Background Paper 13

Condit R (1995) Research in large, long-term tropical forest plots. Trends Ecol Evol 10:18–22

Currie DJ (1991) Energy and large-scale patterns of animal-and plant-species richness. Am Nat 137:27–49

Daou L, Shipley B (2019) The measurement and quantification of generalized gradients of soil fertility relevant to plant community ecology. Ecol 100:e02549

Drissen T, Faust C, Stadtmann R, Treitler JT, Zerbe S, Mantilla-Contreras J (2019) Plant composition and diversity in a semi-natural Mediterranean island landscape: the importance of environmental factors. Plant Biosyst Int J Deal Aspects Plant Biol 153:756–766

Dutta G, Devi A (2013) Plant diversity, population structure, and regeneration status in disturbed tropical forests in Assam, northeast India. J For Res 24:715–720

Fick SE, Hijmans RJ (2017) WorldClim 2: new 1-km spatial resolution climate surfaces for global land areas. Int J Climatol 37:4302–4315

Fine PV (2015) Ecological and evolutionary drivers of geographic variation in species diversity. Ann Rev Ecol Evol Systemat 46:369–392

Francis AP, Currie DJ (2003) A globally consistent richness-climate relationship for angiosperms. Am Nat 161:523–536

Gaston KJ, Spicer JI (2013) Biodiversity: an introduction. Wiley, New York

Gentry AH (1982) Patterns of neotropical plant species diversity. In: Evolutionary biology. Springer, Berlin, pp 1–84

Gentry AH (1988) Tree species richness of upper Amazonian forests. Proc Natl Acad Sci 85:156–159

Gogoi A, Sahoo UK (2018) Impact of anthropogenic disturbance on species diversity and vegetation structure of a lowland tropical rainforest of eastern Himalaya, India. J Mountain Sci 15:2453–2465

Groom MJ, Meffe GK, Carroll CR, Andelman SJ (2006) Principles of conservation biology (vol Sirsi) i9780878935185. Sinauer Associates Sunderland

Hawkins BA et al (2003) Energy, water, and broad-scale geographic patterns of species richness. Ecology 84:3105–3117

Hawkins BA, Diniz-Filho JAF, Jaramillo CA, Soeller SA (2006) Post-Eocene climate change, niche conservatism, and the latitudinal diversity gradient of New World birds. J Biogeogr 33:770–780

Kanade R, John R (2018) Topographical influence on recent deforestation and degradation in the Sikkim Himalaya in India; implications for conservation of East Himalayan broadleaf forest. Appl Geogr 92:85–93

Körner C (2004) Mountain biodiversity, its causes and function. AMBOCX 33:11–17

Lakshminarasimhan P, Paul T (2018) Plant diversity in the Biodiversity Hotspot of Indian Eastern Himalaya and its conservation. Bishen Singh Mahendra Pal Singh, pp 565–584

Latham RE, Ricklefs RE (1993) Continental comparisons of temperate-zone tree species diversity. In: Species diversity in ecological communities: historical and geographical perspectives, pp 294–314

Li Y et al (2019) Relationships between plant diversity and biomass production of alpine grasslands are dependent on the spatial scale and the dimension of biodiversity. Ecol Eng 127:375–382

Lomolino MV (2001) Elevation gradients of species-density: historical and prospective views. Glob Ecol Biogeogr 10:3–13

Mahanand S, Behera MD (2019) Understanding the Indian mainland–island biogeography through plant dispersal mechanism. Biodivers Conserv 28:2063–2084

Malhi Y, Silman M, Salinas N, Bush M, Meir P, Saatchi S (2010) Introduction: elevation gradients in the tropics: laboratories for ecosystem ecology and global change research. Glob Change Biol 16:3171–3175

Manish K, Pandit MK (2019) Identifying conservation priorities for plant species in the Himalaya in current and future climates: a case study from Sikkim Himalaya, India. Biol Conserv 233:176–184

Mehta P et al (2020) Conservation and prioritization of threatened plants in Indian Himalayan Region. Biodivers Conserv 1–23

O'Brien E (1998) Water-energy dynamics, climate, and prediction of woody plant species richness: an interim general model. J Biogeogr 25:379–398

Panda RM, Behera MD, Roy PS, Biradar C (2017) Energy determines broad pattern of plant distribution in Western Himalaya. Ecol Evol 7:10850–10860

Pandey A, Rai S, Kumar D (2018) Changes in vegetation attributes along an elevation gradient towards timberline in Khangchendzonga National Park, Sikkim. Trop Ecol 59:259–271

Pau S, Gillespie TW, Wolkovich EM (2012) Dissecting NDVI–species richness relationships in Hawaiian dry forests. J Biogeogr 39:1678–1686

Pausas JG, Austin MP (2001) Patterns of plant species richness in relation to different environments: an appraisal. J Veg Sci 12:153–166

Pennington RT, Lavin M, Oliveira-Filho A (2009) Woody plant diversity, evolution, and ecology in the tropics: perspectives from seasonally dry tropical forests. Ann Rev Ecol Evol Syst 40:437–457

Pickett S, Cadenasso M, Meiners S (2009) Ever since clements: from succession to vegetation dynamics and understanding to intervention. Appl Veg Sci 12:9–21

Rana SK, Price TD, Qian H (2019) Plant species richness across the Himalaya driven by evolutionary history and current climate. Ecosphere 10:e02945

Ranjitkar S, Luedeling E, Shrestha KK, Guan K, Xu J (2013) Flowering phenology of tree rhododendron along an elevation gradient in two sites in the Eastern Himalayas. Int J Biometeorol 57:225–240

Roy PS et al (2015) New vegetation type map of India prepared using satellite remote sensing: comparison with global vegetation maps and utilities. Int J Appl Earth Obs Geoinf 39:142–159

Russell-Smith J (1991) Classification, species richness, and environmental relations of monsoon rain forest in northern Australia. J Veg Sci 2:259–278

Saikia P et al (2017) Plant diversity patterns and conservation status of eastern Himalayan forests in Arunachal Pradesh, Northeast India. For Ecosyst 4:28

Sharma N, Behera MD, Das AP, Panda RM (2019) Plant richness pattern in an elevation gradient in the Eastern Himalaya. Biodivers Conserv 28:2085–2104

Shooner S, Davies TJ, Saikia P, Deka J, Bharali S, Tripathi OP, Singha L, Latif Khan M, Dayanandan S (2018) Phylogenetic diversity patterns in Himalayan forests reveal evidence for environmental filtering of distinct lineages. Ecosphere 9:1–13

Singh P, Sanjappa M (2011) Flowering plants of Sikkim-an analysis biodiversity of Sikkim-exploring and conserving a global hotspot. In: Arrawatia ML, Tambe S (eds) Department of information and public relations. Government of Sikkim, Gangtok, pp 65–88

Sinha S, Badola HK, Chhetri B, Gaira KS, Lepcha J, Dhyani PP (2018) Effect of altitude and climate in shaping the forest compositions of Singalila National Park in Khangchendzonga Landscape, Eastern Himalaya, India. J Asia-Pacific Biodivers 11:267–275

Smith DN, Killeen TJ, Dallmeier F, Comiskey JA (1998) A comparison of the structure and composition of montane and lowland tropical forest in the Serranía Pilón Lajas, Beni, Bolivia. Man Bio Series 21:681–700

Tambe S, Rawat G (2010) The alpine vegetation of the khangchendzonga landscape, Sikkim Himalaya. Mt Res Dev 30:266–274

van der Molen MK et al (2011) Drought and ecosystem carbon cycling. Agric For Meteorol 151:765–773

Vereecken H et al (2015) Soil hydrology: recent methodological advances, challenges, and perspectives. Water Resour Res 51:2616–2633

Whittaker RJ, Willis KJ, Field R (2001) Scale and species richness: towards a general, hierarchical theory of species diversity. J Biogeogr 28:453–470

Chapter 5
Multivariate Analysis of Soil-Vegetation Interaction and Species Diversity in a Natural Environment of *Rhus coriaria* L. (Case Study: Bideskan Habitat, Southern Khorasan, Iran)

Sh. Ghollasimod, H. Memarian, and J. Shamshiri

Abstract To manage rangeland ecosystems, the first step is to determine effective factors on species distribution and diversity. The prediction models of species distribution determine the most effective factors for any plant species and examine behavior of the species interacted with environmental variables and also accompanying species. In this work, to study ecological characteristics and to determine the most important environmental factors affecting the Sumac (*Rhus coriaria* L.) species, its range was mapped using a randomly-systematic approach to take 30 plots of 10 m^2. The soil samples were taken from a depth of 0 to 30 cm. The evenness and richness indices were computed based on species frequency in each plot and each community, i.e. witness and *Rhus coriaria* L. The independent samples t-test, Principal Component Analysis (PCA) and Canonical Correspondence Analysis (CCA) were employed for comparing natural Sumac habitat with control area (without the presence of *Rhus coriaria* L.). According to the Shannon-Weiner diversity index, Sumac habitat was more diverse and based on the evenness index of 0.717, it showed more uniform distribution compared to control area. The student's t-test of independent samples in two areas demonstrated a significant higher amount (between 30 and 140%) of electrical conductivity, saturated electrical conductivity, potassium, organic matter in Sumac habitat, as compared with control area. Finally, the relationship analysis between soil factors and vegetation using the multivariate techniques of PCA and CCA showed that the soil characteristics, saturation moisture percentage, electrical conductivity, nitrogen, organic matter, lime, potassium, silt and acidity had the most impact on separation of two regions and distribution of Sumac species.

Sh. Ghollasimod · H. Memarian (✉)
Department of Rangeland and Watershed Management, Faculty of Natural Resources and Environment, University of Birjand, Birjand, Iran
e-mail: hadi_memarian@birjand.ac.ir

J. Shamshiri
MS, Rangeland Management, University of Birjand, Birjand, Iran

P. K. Shit et al. (eds.), *Spatial Modeling in Forest Resources Management*, Environmental Science and Engineering,
https://doi.org/10.1007/978-3-030-56542-8_5

Graphical Abstract

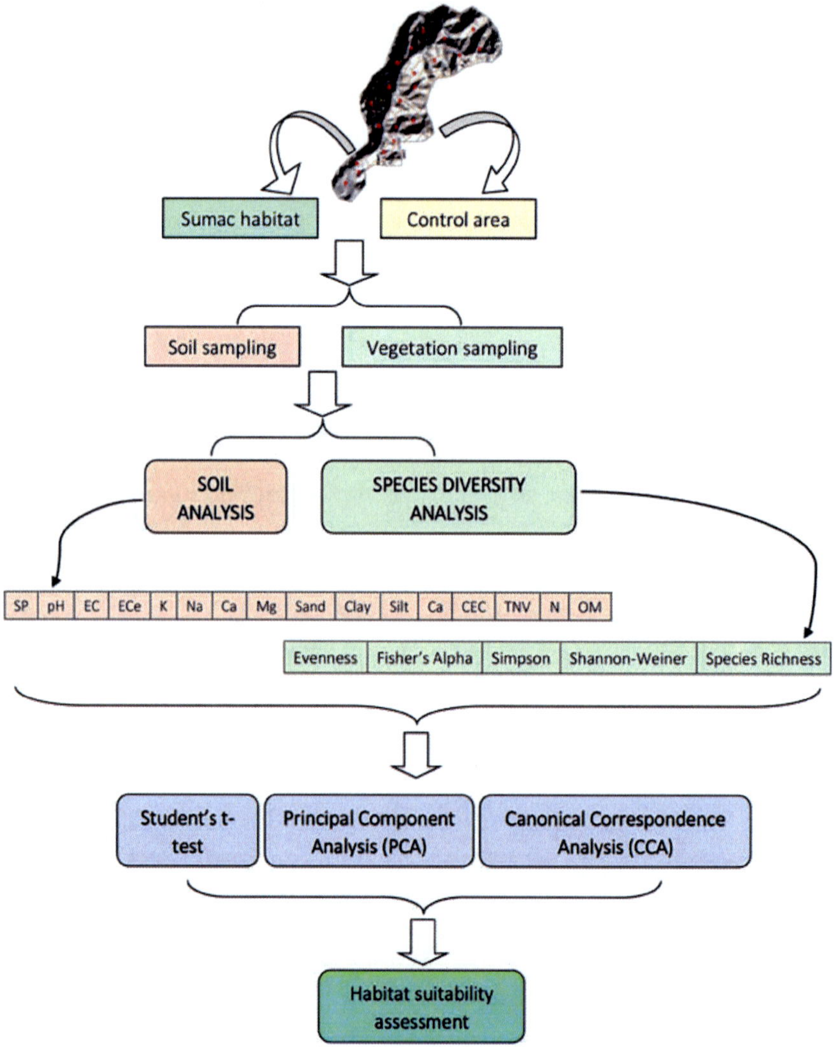

Keywords Species diversity · Sumac habitat · Principal component analysis · Canonical correspondence analysis · Soil characteristics

Abbreviations

2D Two dimensional

A	Annual
C	Carbon
Ca	Calcium
CA	Correspondence Analysis
$CaCo_3$	Calcium carbonate
CCA	Canonical Correspondence Analysis
CEC	Cation exchange capacity
Cl	Chlorine
ECe	Saturated electrical conductivity
F	Forbs
G	Grass
Ge	Geophyte
He	Hemichryptophyte
K	Potassium
m	Meter
$meq\ lit^{-1}$	Milliequivalent per liter
Mg	Magnesium
mm	Millimeter
N	Nitrogen
Na	Sodium
OM	Organic matter
P	Perennial
PC1	First principal component
PC2	Second principal component
PCA	Principal Component Analysis
pH	Acidity
Ph	Phanerophytes
RDA	Redundancy analysis
Sh	Shrub
SP	Saturation Moisture Percentage
T	Therophyte
TNV	Lime
χ^2	Chi-squared

5.1 Introduction

The rangeland ecosystems in arid and semi-arid regions are strongly influenced by their constituent factors due to special governing physical and environmental conditions. Therefore, recognizing the relationships between these factors has a significant function in rangeland management and planning (Jongman et al. 1995). The natural ecosystems of Iran are one of the most important origins of speciation in the world and protecting their diversity is of great importance (Jowkar et al. 2016).

The presence of species and distribution of plant communities in rangeland ecosystems are not accidental; however, climatic, soil, topography, and ecological factors play an important role in their development. Physical and chemical properties of soil in relation to vegetation cover cause a wide geographic distribution of plant species (Ghollasimod et al. 2017; Enright et al. 2005). Investigating the relationship between plant communities and environmental factors has a particular complexity. This means that, firstly, driving forces under study have great variety, second, there are complex actions between environmental and vegetative variables, and thirdly, obtained correlations are often associated with uncertainty (Jongman et al. 1995). Nowadays, what adds to increasing importance of biodiversity is its role in maintaining the sustainability of the ecosystems. In other words, the presence of more species in a region will grant a more complex structure to natural communities, and as a result, these ecosystems will be more adaptive and responsive to the changes and consequently more sustainable, as well (Jenkins and Parker 1998). Burley (2002) stated that the best way to save biodiversity and understand its values is to evaluate and monitor it in different ecosystems. Plant communities are inherently dynamic, and changes in environmental factors such as climate, topography, and soil will change this dynamic behavior (Dirnböck et al. 2002). A simultaneous study of vegetative and environmental factors can lead to more favorable results. Evidences suggest that, on a large scale, such as continent and major environmental regions, the impact of climate is stronger (Jarema et al. 2009); however, in small and local scale, soil factors have a stronger effect than other environmental driving forces (Cui et al. 2009). Goodall et al. (2009) showed that soil characteristics are the major factors in distribution of plant communities, especially in arid regions. Blesky and Canham (1994) declared that characteristics and nutrient reserves of the soil depend closely on plant types. Jin-Tun (2002) explored that distribution of vegetation is mainly a function of climate and soil. Rezaei Pourbaghdar et al. (2014) indicated that among physical and chemical characteristics of the soil, soil texture, lime ($CaCo_3$), electrical conductivity (EC), organic matter (OM), calcium (Ca), magnesium (Mg), chlorine (Cl) and sodium (Na) had the greatest effect in separation of plant communities and distribution of *Rheum ribes* and *Dorema ammoniacum* species. The relationship between vegetation cover and chemical properties of soil can be discussed from two perspectives. The first viewpoint is that distribution of vegetation in a region is a response to soil chemical properties. In the second viewpoint, soil chemical properties are a consequence of the vegetation cover that has been established on it (Li et al. 2012). Farrukh et al. (1994) using the statistical multivariate analysis categorized and analyzed three vegetative communities of Swabi area in Pakistan and determined the factors, soil pH, phosphorus (P) and calcium carbonate ($CaCo_3$) as the most important factors controlling plant distribution in the region. El-Ghani and Amer (2003) established a significant relationship between some soil characteristics, i.e. lime, saturation moisture content, pH and OM with plant distribution in some species of Poaceae, Chenopodiaceae, Fabaceae and Asteraceae. He et al. (2007) also showed the significant impact of silt, clay and OM on distribution of plant species. Yibing et al. (2004) examined the relationship between vegetation distribution and soil factors in China deserts using the Canonical Correspondence Analysis (CCA). The results showed that soil moisture,

OM, salinity and pH had different impacts on the presence of plant species of shrubs and short grasses. Tonggui et al. (2011) considered the role of soil pH as a very important factor in distribution and spread of plants in the coastal zone of China. The relationship between plant species diversity and environmental variables in arid and semi-arid rangelands of the Forg District of South Khorasan Province, Iran was also investigated (Ghollasimod et al. 2017). In this study, the ordination techniques of CCA and Principal Component Analysis (PCA) were employed to determine the effective factors on variations of species diversity. The results indicated that pH, EC, Ca, Na, OM, lime and sand had the most effect, while the topographic factors (slope, altitude and direction) had the least effect on species diversity in the studied area. Considering the performed review, it can be generally concluded that soil factors have a direct and significant effect on vegetation heterogeneity and resilience of plant communities.

Sumac species in Bideskan habitat is considered as one of the most important rangeland by-products that in addition to regional economic prosperity, it provides sustainable employment for villagers and it is considered as an important species for soil conservation. Therefore, by studying environmental conditions and the needs of the Sumac species, it would be possible to judge about its geographic distribution, density and activity in different environments (van der Maarel and Franklin 2012). There are some limited literature and research works on the Sumac interaction with soil characteristics in Iran. For instance, Rezaeipor et al. (2014) demonstrated a significant difference between OM and potassium content (K) (in the depth of 0–10 cm) within the Sumac habitat and outside. They concluded that Sumac species is established on the soils with a high amount of N, OM, K and Ca. Roosta (2015) also introduced K, OM, and N as the most important factors in establishment of the Sumac shrubs.

Thus, according to the above mentioned review, this research is mainly aimed at investigating the relationships between Sumac species with some of the most important physical and chemical properties of soil in the Bideskan habitat of Ferdows district, Iran. The most important soil properties affecting on distribution of plant species in habitat and species diversity in the region are investigated, as well. This information can be finally used and the results are generalized to similar areas to propose reasonable solutions for rehabilitation and distribution development of the Sumac species.

5.2 Materials and Methods

5.2.1 Study Area

The Bideskan habitat with an area of 3685 ha is considered as one of the sub basins of the Lut Desert great basin in Iran. This area is located 24 km from Ferdows city at the longitude of 58° 25′E–58° 29′E and the latitude of 34° 3′N–34° 6′N (Fig. 5.1).

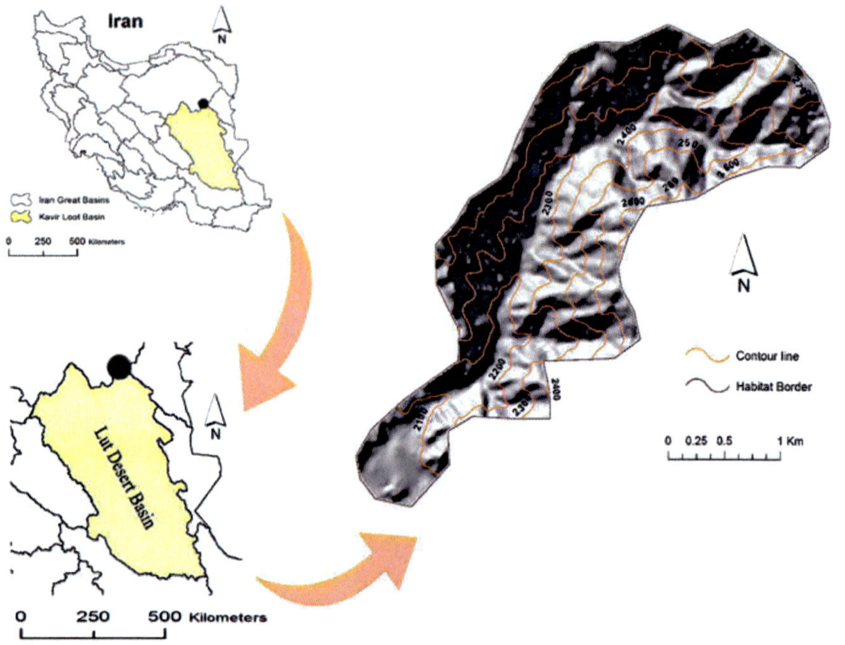

Fig. 5.1 Geographic location of study area

The lowest altitude of the area is 2030 m and the highest is 2734 m above sea level, with the average slope of 18%. Based on a 30-years record of weather parameters, and according to the Emberger Climatogram (Yaghmaei et al. 2009), the climate of the region is dry and cold (mountainous type) with an average annual rainfall of 276 mm. The average annual temperature of the region is 12.3 °C and frosting period is 3.5 months, mostly in December, January and February. According to the observations, the soil of the region is sandy loam with relatively good permeability. The soil depth varies from low to moderate in different slopes of the study area. The geological units of this area belong to the Mesozoic period. The clastic rocks in the study area with a high slope mostly belong to the Jurassic era, that mainly consist of sandstone, quartz sandstone, shale and thin interlayer lime.

5.2.2 *Sumac Species*

Rhus coriaria L. (Sumac) (Fig. 5.2) is a shrub from Anacardiaceae, which involves more than 700 species in 82 genera that are mainly distributed in tropical regions (Mozaffarian 2010). Rhus genus includes over 250 species, distributed in Africa, Asia, Central America, Europe, Madagascar, North America and the Pacific islands.

Fig. 5.2 *Rhus coriaria* L.

In recent years, Sumac has been studied mostly in terms of drug aspect, i.e. anti-oxidant, anti-diabetes, and especially anti-microbial effects. The reproduction in this species is mostly Asexual, using underbrush, seed, scion, stem and rhizomes. The best way of reproduction for *Rhus coriaria* L. is through underbrush (Emad 2001). Due to various issues such as rangeland destruction, cutting and inappropriate harvesting of the plant, Sumac cultivation in most parts of Iran is limited to some remote spots with steep slopes. The average slope of the main habitats of this plant is 30–60%. The plant is established in different slope directions, however it has more distribution and development in northern slopes (Emad 2001). The altitude range for this species in Iran is mostly between 1000 and 1700 m and reaches by 2300 m above sea level in the heights of Shiraz region and up to 2700 m in Khorasan area (Mozaffarian 2010). Due to the nativity of Sumac species in mountainous regions, its limited requirement for water and nutrients and its presence in the soils with moderate and even weak nutrient elements, development of this plant is recommended as an appropriate species for greenery and soil conservation targets (Emad 2001). The lime and marl structures, mainly gravelly and low depth soils with debris deposits, rocky outcrops, as well as clay sandy or loamy clay sandy soils with alkaline pH, are main characteristics of the Sumac habitats in Iran. Sumac does not grow well in shadow; it is also resistant to strong winds, pests and diseases. If it is dried out by dehydration, a little moisture brings the underbrush (Emad 2001).

5.2.3 Research Methodology

Prior to visit the study area, the basic information, i.e. geology, geomorphology, topography, slope, aspect, and land use maps were extracted and information on habitat climate were obtained through the detailed studies on rangeland and watershed management performed by the Department of Natural Resources and Watershed Management of Iran. Then, during May 2016, through field visit, vegetation information and environmental factors were monitored and sampled. The area was divided into two parts, including natural Sumac beds and control area, however with the same geological formation. In each section, 15 plots (total of 30 plots and soil samples) were taken by a randomly-systematic approach. To provide floristic list, the land

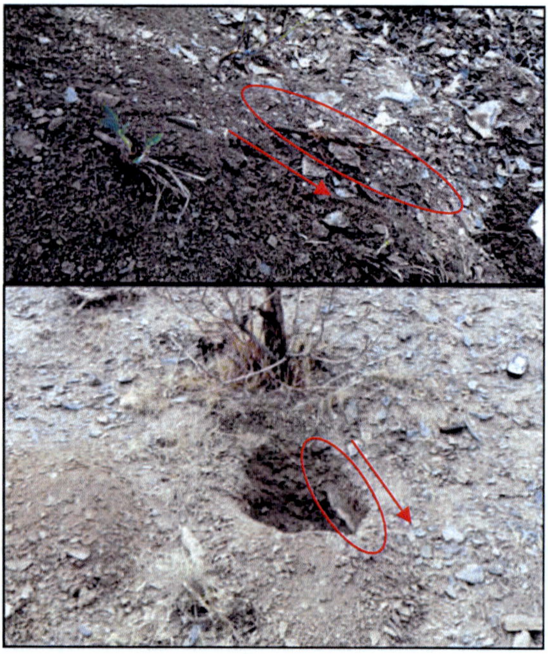

Fig. 5.3 The rootage development of the *Rhus coriaria* L. in direction of the slope gradient

survey method, as one of the methods in regional taxonomic studies, was employed (Schulz et al. 2009). A size of 10 × 10 m for the sampled plots was considered according to the type of plants and their distribution in the area (Schulz et al. 2009). Within each plot, the information related to the plants, including number and type was recorded. During sampling, it was determined that rootage direction of the *Rhus coriaria* L. follows a longitudinal growth in the soil along direction of the gradient (Fig. 5.3), as a result, soil samples were taken from a depth of 0–30 cm. Figure 5.4a shows natural habitat of the Sumac and Fig. 5.4b depicts control area.

In the next step, taken soil samples were dried and passed through a 2 mm sieve in order to be prepared for soil tests. Subsequently, the parameters, soil texture, saturation moisture content, pH, EC, OM, lime, Na, K, Ca, Mg, cation exchange capacity (CEC), and total nitrogen (N) were determined in laboratory. Soil texture was determined through the Bouyoucos hydrometer method (Gee et al. 1986) and saturation moisture content was estimated by providing saturated clay via the weighting approach. After providing saturated extract, soil acidity was determined by pH meter and in order to evaluate soil salinity, electrical conductivity was measured by EC meter. The OM, lime (TNV) and CEC were measured by Walkley and Black (1934), Calcimeter (Hawkins and Kunze 1965) and centrifuge (Bache 1976) approaches, respectively. The Na and K cations were determined by flame photometer (Gee et al. 1986); Ca and Mg were determined via the Complexometry approach (Gee et al. 1986) and the Kjeldahl (1883) method was employed to measure N.

To determine diversity and species richness, the number of species was counted in each plot within two habitats of the Sumac and control area. The Shannon-Weiner Species Diversity Index (1), Simpson (2) and Fisher Alpha (3) (Schulz et al. 2009; Seaby and Henderson 2006) were calculated based on the frequency of plant species using the EstimateS win 9.1 (Colwell 2013), according to the following formulas:

$$H = -\sum_{i=1}^{S} P_i Ln(P_i) \quad (5.1)$$

$$1 - D = 1 - \sum_{i-1}^{s} (P_i)^2 \quad (5.2)$$

$$S = \propto \log\left(1 + \frac{N}{\propto}\right) \quad (5.3)$$

$$\mathrm{E} = \mathrm{H}' / \ln(\mathrm{S}) \quad (5.4)$$

(a)

(b)

Fig. 5.4 **a** Natural Sumac (*Rhus coriaria* L.) habitat; **b** control area

In above equations, H is Shannon-Weiner species diversity, 1-D is Simpson species diversity, S is number of species, N is total number of individuals, or frequency of species, P_i is the ratio of the number of *i*th species to total number of individuals, α is Fisher's alpha diversity index, E is evenness (Pielou 1975), and H′ is the actual degree of evenness (Shannon-Weiner index).

The independent t-test was used in the Statistical Package for Social Sciences (SPSS) software (Green and Salkind 2010) to compare the measured variables in two areas of Sumac habitat and control. To determine the relationship between environmental factors and distribution of Sumac and other plant species, the data ordination techniques, i.e. PCA and CCA were utilized within the PC ORD software environment (McCune and Mefford 2016).

5.2.4 Principal Component Analysis (PCA)

The PCA in mathematical definition is an orthogonal linear transformation that transfers data to a new coordinate system, so that the largest data variance is located on the first coordinate axis, the second largest variance on the second coordinate axis, as well as for the rest. The PCA can be used to reduce the dimension of data by preserving the components of data set that have the greatest impact on variance (Jolliffe 1986). In fact, the number of N plots with M species or environmental factors is converted to N plots with a smaller number of species or environmental factors. For the X^T data matrix with an experimental mean of zero, in which each row is a set of observations, and each column contains the data for a variable, the PCA is defined by Eq. 5.4, as follows:

$$Y^T = X^T W = V \sum \mathrm{W}^{\mathrm{T}} \tag{5}$$

where, $V \sum W^T$ T is decomposition of the individual values of the matrix X^T.

The Broken Stick model is an anticipated technique for approximating the number of statistically significant principal components. The desired scheme is to model N variances by taking a stick of unit length and breaking it into N sections by randomly (and simultaneously) choosing break points from a uniform distribution (Jackson 1993).

5.2.5 Canonical Correspondence Analysis (CCA)

The CCA is the canonical usage of correspondence analysis (CA) (Ter Braak 1986). This method is appropriate for response variables, presenting unimodal distributions and preserves χ^2 (chi-squared) distances between objects. It is calculated from a matrix of χ^2 distances that passes on a type of redundancy analysis (RDA). It employs object marginal sums (row totals) as a weighting factor. The outcome of weighted RDA is the response variables that are maximally related to linear combinations of the descriptive factors, which are ordinated in the Euclidean space. These are then canonical variables. The correlation of the descriptive factors to the final ordination defines their significance (McGarigal et al. 2000).

5.3 Results and Discussion

5.3.1 Vegetation Community

The information of the samples taken via the plots in control area and Sumac habitat indicated that the study region has been occupied by 24 species belonging to 14 families. The largest family is Asteraceae with 8 species, then Apiaceae, Geraniaceae, Poaceae, each with 2 species are the most important families. The families Anacordiaceae, Caryophyllaceae, Chenopodiaceae, Fabaceae, Iridaceae, Lamiaceae, Liliaceae, Polygonaceae, Ranunculaceae, and Schlerophulriaceae, each with one species were observed in the study region. Among them, 11 families and 17 species were found in control area, 9 families and 15 species were explored in Sumac habitat. The Hemichryptophytes were the most important life form with 14 species (54.16%). Then, the Therophytes with seven species (33.33%), Geophytes with two species (8.33%) and Phanerophytes with one species (4.16%) were recognized as the most frequent life form in the study area. The floristic list of the study region is also represented in Table 5.1 and the values of species diversity indices in both natural Sumac habitat and control area are presented in Table 5.2.

In the study area, the Asteraceae species are more frequent than other families. This could be due to a better compatibility of this family with climatic conditions, the evolution and the youngness of the Asteraceae species that gives them a high potential of distribution. However, this degree of incidence indicates a high quantity of vegetation degradation and pressure on the ecosystem, as well. Ghollasimod et al. (2017) argued that an increase in the frequency of Asteraceae species could be resulted from the increase of degradation in the study area due to overgrazing, inappropriate protection and pressure on the ecosystem. Excessive pasture grazing causes the removal of palatable species, belonged to the plant families such as Poaceae and Fabaceae. As a result, invasive species with a high regeneration rate, especially from the Astraceae family, are replaced by palatable species in such pastures. Davis (1965) attributed the high incidence of the Asteraceae species to a wide range of tolerance of these species against unfavorable ecological conditions. The overexploitation of shrubs, the excessive grazing of livestock and soil compaction, the loss of rangelands' natural regeneration, recent droughts, and the presence of plant pests (in particular *Zeuzera pyrina*) can be considered as the important causes for degrading sequence of the study region. Hemicryptophytes involve the most life form of the region; according to Archibold (2012), high frequency of the Hemicryptophytes represents tough living

Table 5.1 Floristic list of the Bideskan habitat

Family	Species	Bio brigade	Life form	Plant form	Usage
Aasterceae	Onopordon	T	P	F	Forage
Anacordiaceae	*Rhus coriaria* L.	Ph	P	Sh	Medicinal-Protector
Apiaceae	*Eryngium billardieri*	He	A	Sh	Protector
Apiaceae	Ferulago	He	P	Sh	Medicinal-Protector
Asteraceae	*Achillea pachycephala*	T	P	F	Medicinal-Protector
Asteraceae	*Artemisia acherui*	He	P	Sh	Medicinal-Protector
Asteraceae	*Gundelia tournefortii*	He	P	F	Forage-Protector
Asteraceae	Polycaria	He	A	F	Medicinal-Protector
Asteraceae	*Scariola orientalis*	He	P	F	Forage
Asteraceae	*Tragopogon graminifolius*	T	P	Sh	Medicinal-Protector
Asteraceae	Amberboa	T	T	F	Protector-Forage
Caryophyllaceae	*Acanthophyllum crasinodum*	He	P	Sh	Protector
Fabaceae	*Astragalus heratensis*	He	P	Sh	Protector
Geraniaceae	*Bibersteinia multifidia*	T	P	F	Medicinal-Protector
Geraniaceae	Geranium	T	A	F	Forage
Iridaceae	*Iris songarica*	G	P	F	Medicinal-Protector
Lamiaceae	*Hymenocrater calycinus*	He	P	Sh	Medicinal-Protector
Liliaceae	*Tulipa montana*	Ge	P	F	Forage
Poaceae	*Bromus tectorum*	He	A	G	Forage
Poaceae	*Stipa barbata*	He	P	F	Protector
Polygonaceae	*Rheum ribes*	T	P	F	Medicinal-Protector
Ranunculaceae	Ranunculus	T	A	F	Forage
Rosaceae	*Sanguisorba minor*	He	P	F	Forage
Schlerophulriaceae	Schlerohachis	He	P	Sh	Medicinal-Protector

T therophyte; *He* hemichryptophyte; *Ge* geophyte; *Ph* phanerophytes; *P* perennial; *A* annual; *F* forbs; *Sh* shrub; *G* grass

Table 5.2 The values of diversity, evenness and richness indices in the studied regions

Region	Evenness	Fisher's Alpha	Simpson	Shannon-Weiner	Species richness
Control area	0.591	3.08	0.137	1.667	17
Rhus coriaria L.	0.717	3.06	0.323	1.942	15

conditions in the region. The presence of 58.33% of plant species of the region in the form of Hemicryptophyte can be attributed to dry/cold climate and mountainous topographic conditions. This finding could be also confirmed through a research work by Hassani et al. (2014). Furthermore, a 33% frequency of plant species in the vegetative form of Therophyte is due to short growing season in the Bideskan habitat. However, low rainfall, recent droughts, unfavorable conditions of enclosure, overgrazing and, consequently the degradation caused by the impact of the pressures from these forces, are among the reasons that affect the abundance of annual plants. Drought mitigation mechanisms of Therophytes enable them to reduce their activities in drought or to complete their life cycle in appropriate humidity conditions (Hassani et al. 2014).

Several indices of species diversity are used in the large amount of literature on biological diversity and ecological monitoring. Understanding the changes in richness of plant species is valuable in the study of global climate change. A commonly used richness indicator is the Shannon-Weiner Index. The values of this index usually alter from 1.5 to 3.5, but in exceptional cases it may even exceed 4.5. The Shannon's index evokes a diversity or variety in a society, and the more varied combination is in the society, the more variety is achieved (Tolera et al. 2008). According to the Shannon-Weiner index, the Sumac habitat with the index of 1.942 is more diverse than control area (with an index of 1.667) (Table 5.2). The higher the evenness index, the distribution of species within the plot or a range is more uniform; therefore, the Sumac habitat with the uniformity equal to 0.717 has more uniform distribution, compared to control area (Table 5.2). According to the Shannon-Weiner and Simpson diversity indices, the Sumac habitat is more diverse, although its number of species (15 species) is less than the number of species in control area (17 species). It has long been known that the richness of vascular plant species stems from many different factors, such as topography, soil, and climate, temperature and other climatic variables, in which soil parameters seem to be the most important factor for species richness (Zhao et al. 2005).

5.3.2 Student's t-test of Independent Samples

The Student's t-test of independent samples was utilized to compare soil characteristics between control area and the habitat of *Rhus coriaria* L. Results showed that there are significant differences in pH, EC, saturated electrical conductivity (ECe), carbon (C), OM, lime (TNV), K and silt between two regions at the confidence level of 95% (Table 5.3). The variables K, EC, OM and TNV in the Sumac habitat showed a significant increase of 37%, 44%, 80% and 140%, respectively compared to control area. These results are consistent with the findings of Roosta (2015), as well. She

Table 5.3 The comparison of soil characteristics between control region and Sumac habitat using t-test

Variable	Region	Mean	t stat
Saturation moisture percentage (SP%)	Control	*25.75*	*0.142 ns*
	Rhus coriaria L.	*25.06*	*0.142 ns*
pH	Control	*8.04*	*0.005**
	Rhus coriaria L.	*8.14*	*0.005**
EC (mmhos cm^{-1})	Control	*220.7*	*0.002**
	Rhus coriaria L.	*318.2*	*0.004**
ECe	Control	*879*	*0.008**
	Rhus coriaria L.	*1119*	*0.009**
Potassium (K) (meq lit^{-1})	Control	*0.87*	*0.011**
	Rhus coriaria L.	*1.19*	*0.011**
Sodium (Na) (meq lit^{-1})	Control	*2.72*	*0.901 ns*
	Rhus coriaria L.	*2.69*	*0.901 ns*
Calcium (Ca) (meq lit^{-1})	Control	*8.25*	*0.999 ns*
	Rhus coriaria L.	*8.25*	*0.999 ns*
Magnesium (Mg) (meq lit^{-1})	Control	*3.7*	*0.337 ns*
	Rhus coriaria L.	*4.96*	*0.338 ns*
Sand (%)	Control	*62*	*0.188 ns*
	Rhus coriaria L.	*58.55*	*0.188 ns*
Clay (%)	Control	*3.7*	*0.824 ns*
	Rhus coriaria L.	*3.6*	*0.824 ns*
Silt (%)	Control	*23.71*	*0.043**
	Rhus coriaria L.	*27.97*	*0.043**
Carbon (C) (%)	Control	*0.21*	*0.002**
	Rhus coriaria L.	*0.37*	*0.002**
CEC (meq/100gr)	Control	*9.41*	*0.461 ns*
	Rhus coriaria L.	*9.97*	*0.463 ns*
Lime percentage (TNV)	Control	*4.7*	*0.002**
	Rhus coriaria L.	*11.3*	*0.002**
N (%)	Control	*0.53*	*0.208 ns*
	Rhus coriaria L.	*0.54*	*0.208 ns*
Organic matter (OM) (%)	Control	*0.36*	*0.002**
	Rhus coriaria L.	*0.65*	*0.002**

n.s. not significant at the level of 5%; * significant at the level of 5%

studied the *Rhus coriaria* L. in the Kakhk district of Gonabad, Iran and explained that the falling of shoot organs on soil surface could be a main reason for the increase of K and OM under the stratum of Sumac plant. Rezaeipor et al. (2014) in a research work on a Sumac habitat in the west of Iran also stated that N, OM, K, Ca and clay content are important elements affecting the distribution of *Rhus coriaria* L. They declared that increase in the content of litters causes an increase of soil porosity, the decrease of bulk density and thus soil gets better permeability conditions. Results confirm the significant role of *Rhus coriaria* L. in soil conservation planning. Therefore, it is necessary to encourage local farmers to preserve this species.

5.3.3 Principal Component Analysis (PCA)

The PCA was applied on 16 soil physical and chemical factors, listed in Table 5.5. According to Table 5.4, the Eigenvalues of the first, second and third components were 4.340, 3.173 and 2.344 respectively, which correspondingly justified 32.131, 24.830 and 19.651%, and totally 76% of the changes in variance. The Broken Stick statistics of the first, second, and third components were smaller than the Eigenvalues of the first, second, and third components; however the Broken Stick figure of the fourth component was higher than its Eigenvalue. As a result, only the first, second and third components had a significant share in the changes of variance (Table 5.4).

According to Table 5.5, the first component had the highest correlation with Ec, ECe, K, silt, Ca, TNV and OM factors. The second component showed the highest correlation with SP, ECe, Na, sand, clay and CEC (Fig. 5.5), while the third component had the highest correlation with SP, K, Mg, silt, Ca and N factors (Table 5.5). The positive and negative sign of the Eigenvectors indicate the positive and negative correlation of the variable with extracted component.

Figure 5.5 depicts a two dimensional (2D) diagram of the first (PC1) and second (PC2) components in PCA. The variables and sampling units, according to the characteristics have a specific position in this diagram. The sampling units close to each axis are influenced by that axis. Closer units are more similar in terms of species composition. According to the Eigenvalues, from left to right, the numerical values of the EC, ECe, K, silt, Ca, TNV and OM increase, while from the bottom of the second axis upwards, the variables ECe, Na and sand decrease, and SP, clay and CEC increase. The control area can be seen in the soils with high OM, Ca, silt, clay and

Table 5.4 The Eigenvalues and the percentage of variance, justified by each component

Component #	Eigenvalue	Variance (%)	Cumulative variance (%)	Broken-Stick Eigenvalue
1	4.341	32.131	32.131	3.381
2	3.173	24.830	56.961	2.381
3	2.344	19.651	76.612	1.881
4	1.264	4.150	80.762	1.547

Table 5.5 The Eigenvector and correlation coefficient of soil variables with each component

Variable	Eigenvector in each principal component			Pearson correlation with each principal component		
	1	2	3	1	2	3
SP	0.2096	0.3579	−0.3696	0.437	0.638	−0.566
pH	0.1967	−0.1853	−0.1127	0.410	−0.330	−0.173
EC	0.4140	−0.0558	−0.2543	**0.863**	−0.099	−0.389
ECe	0.3598	−0.2699	−0.1148	**0.750**	−0.481	−0.176
K	0.2842	−0.2122	−0.2862	**0.592**	−0.378	−0.438
Na	0.1314	−0.2574	0.0070	0.274	−0.459	0.011
Ca	0.0750	0.0912	0.1714	0.156	0.162	0.262
Mg	0.0089	−0.0289	−0.2908	0.019	−0.051	−0.445
Sand	−0.2088	−0.4233	−0.1368	−0.435	−0.754	−0.210
Clay	−0.0311	0.4507	−0.2067	−0.065	0.803	−0.316
Silt	0.2785	0.1512	0.3323	**0.58**	0.269	0.509
Ca	0.3331	0.0369	0.3494	**0.694**	0.066	0.535
CEC	0.2050	0.4207	−0.0142	0.427	0.749	−0.022
TNV	0.3291	−0.0766	−0.2545	**0.686**	−0.136	−0.390
N	−0.1343	0.2255	−0.3037	−0.280	0.402	−0.465
OM	0.3337	0.0288	0.2546	**0.695**	0.051	0.543

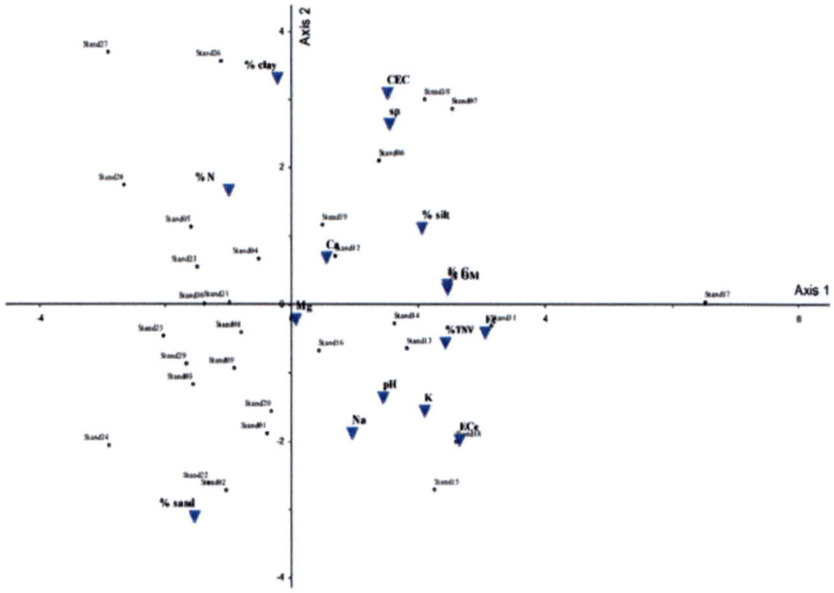

Fig. 5.5 The 2D diagram of the first and second components in PCA

low EC, Mg, N, considering its position in the second and third quarters (in PC1-PC2 diagram) (Fig. 5.6). However, Sumac habitat is distributed in the soils with high OM, Ca, silt, K, EC, ECe and TNV, considering its position in the third and fourth quarters (Fig. 5.7). Therefore, it can be relatively discovered that Sumac habitat is more associated with the variables of EC and TNV on the coarser soils.

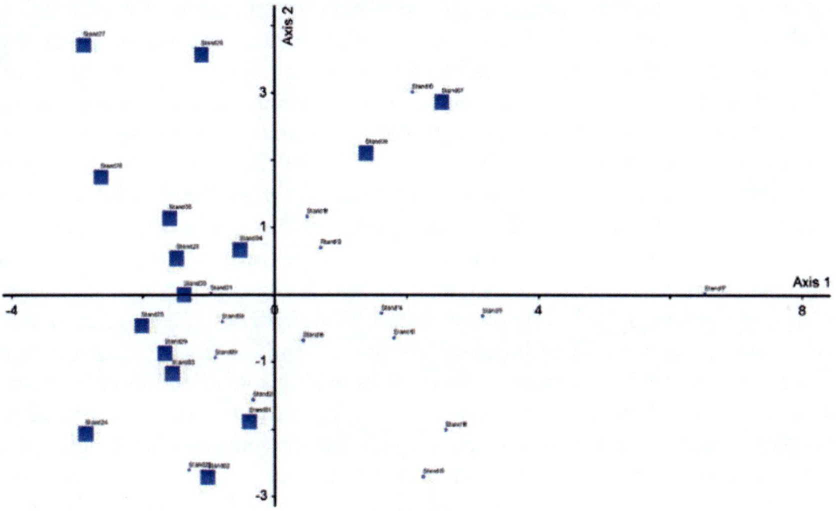

Fig. 5.6 The layout of control area on the 2D diagram of principal components

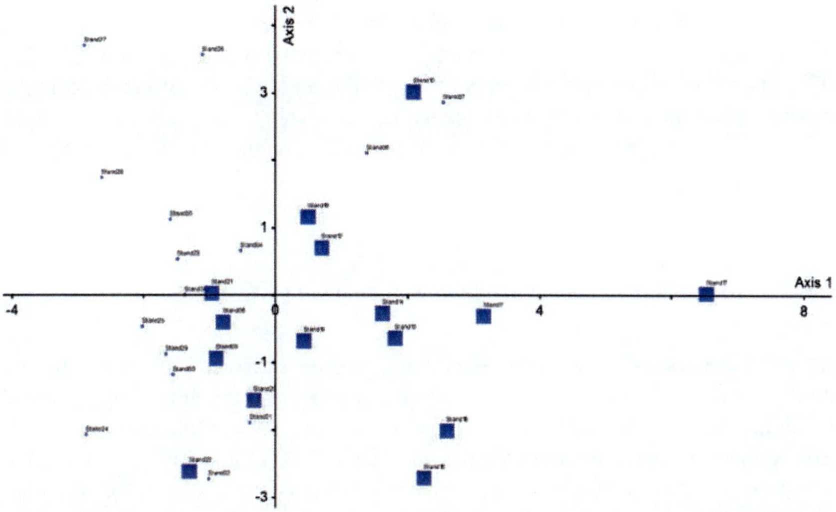

Fig. 5.7 The layout of Sumac habitat (*Rhus coriaria* L.) on the 2D diagram of principal components

According to PCA results, the most important soil factors affecting distinction and distribution of plant species communities in the Bideskan habitat were EC, ECe, OM, Ca, TNV, K and silt content. Many scholars have studied the diversity of plant species in relation with physicochemical properties of soil. Some of these properties include soil structure, water holding capacity, soil fertility and biological activity (Krzic et al. 2003; Härdtle et al. 2003; Akbarlou and Nodehi 2016). In the study region, OM can be considered as an effective factor in distribution of plant species. The biochemical role of OM in the soil is providing a suitable platform for the activity of micro-organisms. It increases nutrients and organic compounds in the soil, which in turn they increase the capacity of absorption and maintenance of nutrients in the soil profile (Biswas and Mukherjee 2001) and consequently affect the separation of ecological species groups (Mirzaei and Karami 2015). Roosta (2015) reported an increase in the elements such as K and OM under the stratum of *Rhus coriaria* L. in the Kalat habitat of Gonabad, Iran as compared to control region. The OM is increased as a positive factor in fertilization that improves soil structure due to vegetation regeneration. It could be one of the reasons for decreasing the alteration range of sodium salts. Following the increase in OM, fertile soil elements such as N and K increase, as well. The silt content also has a significant effect on distribution of species, considering its Eigenvector (0.278). In silty soils, infiltration is carried out moderately and its nutritional elements are sufficiently abundant. Another factor that plays a significant role in separation of plant communities in the region is the K element. This element is often found in the structure of minerals, which after oxidation, it is liberated as potassium ions and enters the soil solution. The amount of K consumption in plants is higher than other elements, except nitrogen. The main reason is the role of K element in photosynthesis regulation, carbohydrate transfer and protein production (Havlin et al. 2005). In addition, the K element facilitates transfer of water and nutrients in the soil profile. The K element also increases root length and, as a result, increases the plant's resistance to drought (Saxena 1985). The EC is one of the important indicators of soil in the region that controls the osmotic pressure and resistance of the plants against higher degree of ion concentration in the soil. Soil salinity in this region is the most important factor affecting the establishment of plant communities.

5.3.4 Canonical Correspondence Analysis (CCA)

The CCA method was used to determine the most effective soil characteristics on the distribution of each plant species. Results showed that the first component had the highest correlation with pH, EC, ECe, K, Mg and TNV, while the second axis showed the highest correlation with pH, K, Ca and CEC (Table 5.6).

Figure 5.8 shows that the plant species of *Bibersteinia multifidia*, *Acanthophyllum crasinodum*, *Onopordon sp.*, *Rheum ribes*, *Scariola orientalis*, *Polycaria*, and *Hymenocrater calycinus* are oriented in the first quarter which involves the soils with low EC, ECe, Mg, TNV and pH. This indicates a reverse relationship with those

Table 5.6 The correlation coefficient of soil variables with the main components of CCA

Variable	Axis#		
	1	2	3
SP	−0.375	−0.098	−0.028
pH	**−0.483**	**−0.436**	**0.289**
EC	**−0.685**	−0.013	−0.086
ECe	**−0.563**	0.022	0.033
K	**−0.621**	**0.378**	0.118
Na	−0.058	0.009	0.109
Ca	0.303	0.148	**0.579**
Mg	**−0.658**	0.039	−0.126
Sand	0.023	0.082	**0.417**
Clay	0.219	−0.105	−0.148
Silt	−0.194	−0.021	**−0.402**
Ca	−0.349	−0.027	0.186
CEC	−0.038	**−0.239**	**−0.538**
TNV	**−0.709**	−0.104	−0.025
N	0.170	0.055	0.095
OM	−0.351	−0.025	0.173

factors. In the second quarter of ordination chart, the plant species *Achillea pachysephala*, *Iris songarica*, *Rhus coriaria* and *Ferulago sp.* are located. This demonstrates a direct relationship with EC, ECe, pH, TNV, K and OM (Fig. 5.9). In the third quarter of ordination chart, the species *Sanguisorba minor*, *Iris songarica* and *Bromus tectorum* are located which have an indirect relationship with the K and a direct relationship with the TNV and pH variables (Fig. 5.10). The plant species *Ranunculus sp.*, *Achillea pachysephala*, *Artemisia aucheri*, *Eryngium billardieri*, *Geranium sp.*, *Tragopogon graminifolius*, *Gundelia tournefortii*, *Amberboa sp.*, *Astragalus heratensis*, *Stipa barbata*, *Tulipa montana* and *Schleohachis sp.* locate in the fourth quarter of ordination chart, which establish indirect relationships with EC, ECe, TNV, K and OM and a direct relationship with Ca (Fig. 5.11).

With regard to the ordination results, it is evident that the vegetation alterations in the region are highly correlated with the soil EC. Therefore, this factor has a large contribution to vegetation changes. Some researchers, such as Carnevale and Torres (1990); Rogel et al. (2001) and El-Ghani and Amer (2003) also showed that soil salinity is one of the most important factors in the establishment of vegetation in arid regions. Salinity and, in general, the concentration of soil salts in the rooting zone, in addition to reducing the plant's available water, can interfere with the balance between ions (Bohra and Doerffling 1993).

In the study area, potassium was one of the factors influencing the distribution of species, especially *Rhus coriaria* L. Rezaeipor et al. (2014) and Enright et al. (2005)

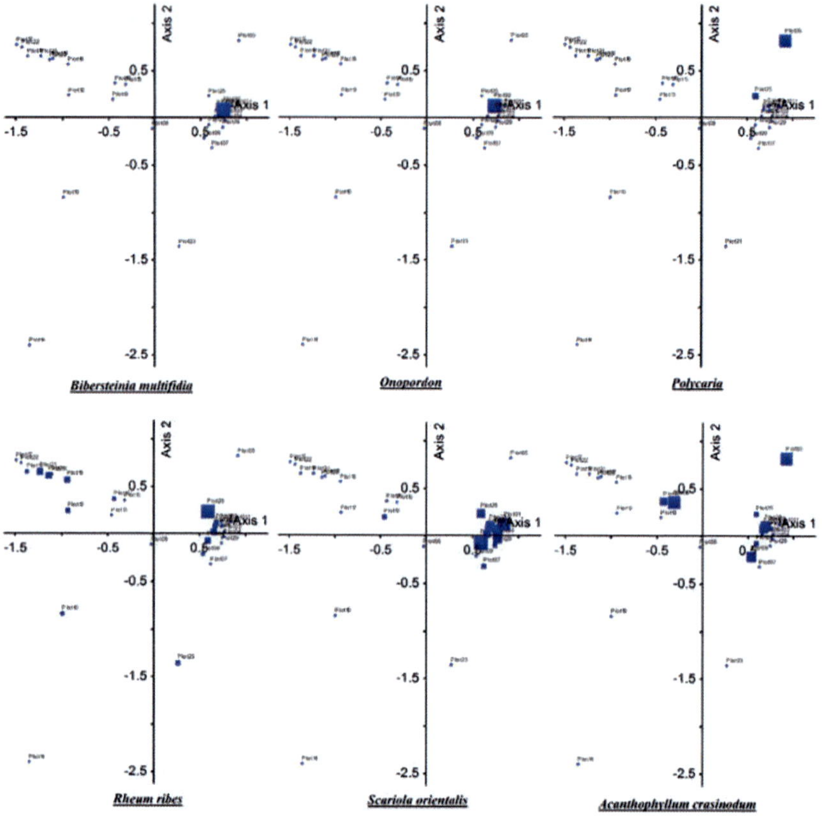

Fig. 5.8 The distribution of plant species on the first quarter of CCA principal components

also suggested that increasing K content in plants plays a very important role in the structure of macromolecules and soil colloids, as well as evapotranspiration of plants, thus playing a major role in the distribution of vegetation in arid areas. Furthermore, Mishra et al. (2003) observed a significant difference between the amount of K under the crown of the plant and outside it. The higher content of the K under the plant's crown was contributed to releasing K from the K minerals or releasing it from litter decomposition.

The amount of TNV is one of the other factors that affect the distribution of vegetation in the studied area. The TNV has low solubility in water and, if it comes in a solution, produces a strong alkaline that limits the growth of plants which need acidic pH. Therefore, lime, except for the plants that grow in calcareous soils, is a growth inhibitor and reduces the ability to use micronutrients such as Zinc and Mg. Some researchers such as Rezaeipor et al. (2014) concluded that TNV was one of the important factors for separating plant communities. The results of a research work by Jafari et al. (2002) showed that among studied topographic and soil factors, pebbles,

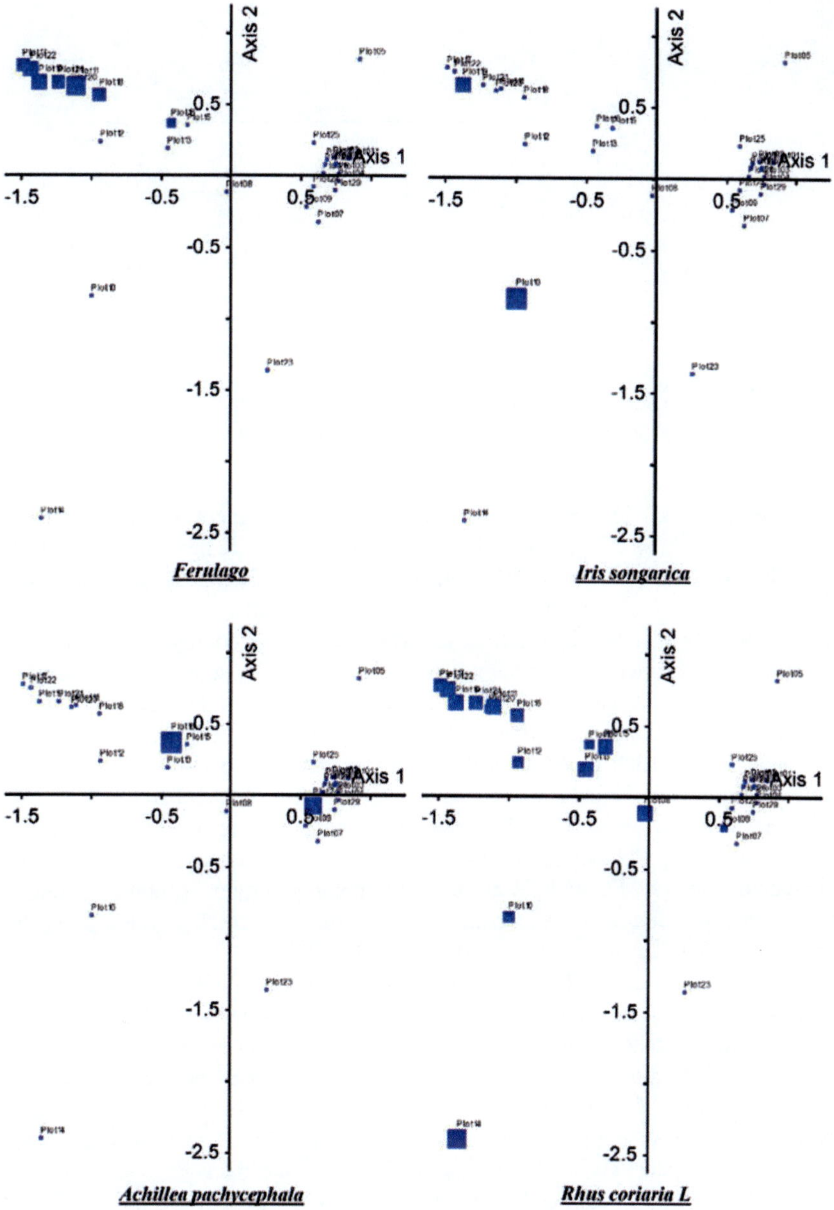

Fig. 5.9 The distribution of plant species on the second quarter of CCA principal components

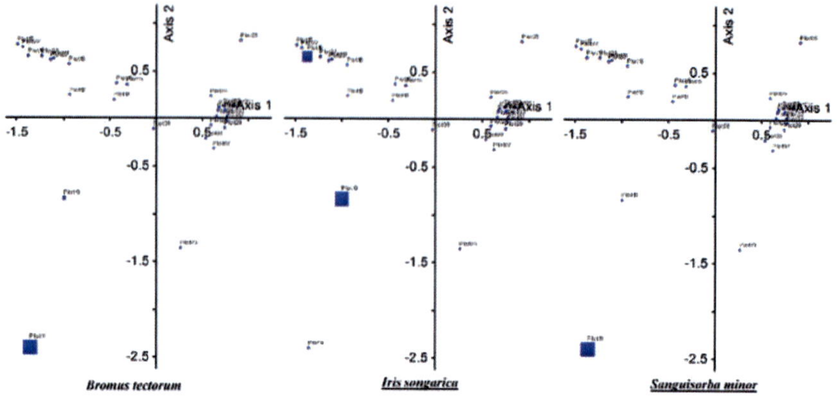

Fig. 5.10 The distribution of plant species on the third quarter of CCA principal components

soil texture, TNV and EC had the greatest impact on the variety of plant species in the region, so that with the reduction of pebbles, soil texture heaviness, increased EC and lime, species diversity in the habitats of *Artemisia sieberi*, *Stipa barbata*, and *Pteropyrum olivieri* gradually changes. Besides, with increasing restrictions, species diversity decrease. In addition, the greater the amount of restrictions, besides reducing diversity, increases the homogeneity. In other words, due to the difficulty of environmental conditions only limited plant species can be established in the area.

Medinski et al. (2010) investigated the relationship between species richness and soil characteristics such as permeability, salinity, clay, silt and pH in southwest Africa and Namibia for five plant species. Results showed that all soil characteristics had a significant impact on species diversity. Furthermore, the results of Yibing et al. (2004) study using PCA in China showed that physical and chemical properties of soil such as OM, SP, EC and pH affect on the homogeneity of habitat and control the pattern of distribution of plant communities in these areas. These findings can be supported by Mirzaei and Karami (2015), as well.

The use of statistical methods could validate the hypothesis of this study regarding the impact of soil factors on presence and establishment of the Sumac (*Rhus coriaria* L.) and showed that the factors EC, TNV, K, pH and OM are the most important factors in the establishment of *Rhus coriaria* species in the study area. The knowledge about the soil necessities of each plant species has an effective role in proposing the species compatible with soil conditions in other similar regions. Therefore, the results of this research can be beneficial for vegetation reclamation and managing rangelands in the Bideskan Sumac habitat and other similar areas.

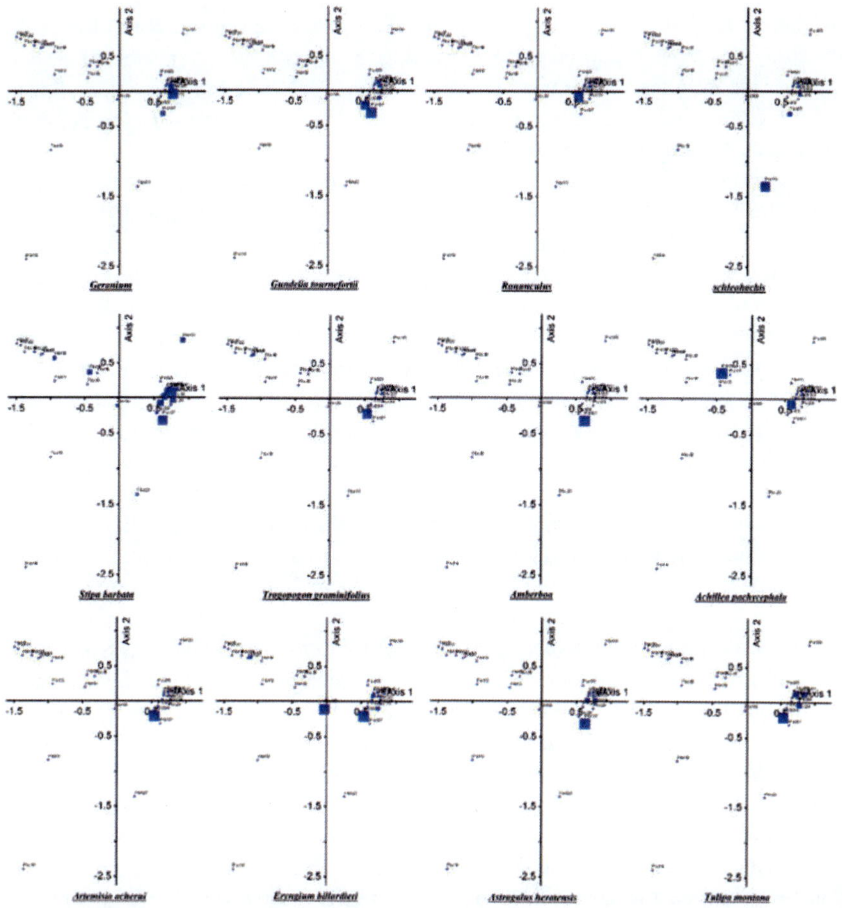

Fig. 5.11 The distribution of plant species on the fourth quarter of CCA principal components

5.4 Conclusion and Recommendation

Awareness of the relationship between soil characteristics and distribution of plant species is vital for the sustainable use of rangelands. Therefore, this study was aimed to determine the effect of soil characteristics in the distribution of vegetation cover, especially *Rhus coriaria* L. in the Bideskan habitat. The sampling results indicated that the most important vegetative form of the region is Hemicryptophytes, belonged to the Asteraceae family, which make up 58.35% of the species population of the area, followed by Throphytes with 33% of frequency in the area. The student's t-test of independent samples on soil data showed that the soil properties, EC, ECe, Ca, OM, TNV, pH, K and silt content had significant differences in two regions, i.e. Sumac habitat and control area. Moreover, four factors of EC, K, OM and TNV in the Sumac habitat showed a significant increase compared to the control area. In this study, the most important factors affecting the distribution of species, identified based on the PCA test results were EC, ECe, OM, Ca, TNV, K and silt content. Based on the CCA analysis, EC, ECe, pH, TNV, K and OM were determined to be effective on the distribution of Sumac plant species.

According to the results of this research, it can be stated that it is possible to establish Sumac (*Rhus coriaria* L.) as a species compatible with arid areas (calcareous soils with higher EC) in the plans to control erosion and sedimentation and improve revival of green spaces in mountainous regions of arid and semi arid areas, especially in Khorasan province and others areas with similar ecological conditions.

Khorasan's large territory in Iran with a variety of climates and soils provides habitats for many vascular plant species. As elsewhere, this diversity is threatened by land use, but increasingly also by climate change. Environmental factors effectively play a significant role in determining the habitats of plants (Escudero et al. 2000; Ghollasimod et al. 2017). Furthermore, species diversity is one of the most important indicators for rangelands evaluation and determining the status of ecosystem (Ghollasimod et al. 2017). In arid regions such as South Khorasan, water availability is the primary limitation to plant productivity and an important factor in plant distribution and diversity in many terrestrial biomes. It is also an ecosystem driver that will be strongly affected in many areas of the world by climate change (Heisler-White et al. 2008). In the future climate change, shifting of habitat and replacement of native species by aggressive and more resilient species are highly predicted. The results of climate change modeling of South Khorasan and Ferdows study area by HADCM3 and ECHO-G general circulation models also show an increase of 3–4 °C in mean temperature by 2050 (Abbasi et al. 2010; Esmaeilnejad and Khashei Siuki 2018). Hazard classification for trend of climate change at the Ferdows station demonstrates a very severe hazard class, as well (Masoudi et al. 2018). Therefore we recommend researchers to investigate bio-diversity of the study area in relation to future climate change to provide a wider visibility and applicability of this work in future works.

Acknowledgements The authors would like to thank the University of Birjand for financial support of this study.

References

Abbasi F, Malbusi S, Babaeian I, Asmari M, Borhani R (2010) Climate change prediction of South Khorasan province during 2010–2039 by using statistical downscaling of ECHO-G data. J Water Soil 24(2):218–233. https://www.sid.ir/en/journal/ViewPaper.aspx?id=177174

Akbarlou M, Nodehi N (2016) Relationship between some environmental factors with distribution of medicinal plants in Ghorkhud Protected Region, Northern Khorasan Province, Iran. J Rangel Sci 6(1):63–72. https://www.rangeland.ir/?_action=articleInfo&article=520586

Archibold OW (2012) Ecology of world vegetation. Springer Science & Business Media, Berlin, 436p. https://www.springer.com/gp/book/9780412442902

Bache BW (1976) The measurement of cation exchange capacity of soils. J Sci Food Agric 27(3):273–280. https://onlinelibrary.wiley.com/doi/10.1002/jsfa.2740270313/full

Biswas TD, Mukherjee SK (2001) Textbook of soil sciences. Tata McGraw-Hill Education. https://www.amazon.com/Textbook-Soil-Sciences-T-D-Biswas/dp/0074620436

Blesky AJ, Canham CD (1994) Forest gaps and isolated savanna trees. An application of patch dynamics in two ecosystems. Bioscience 44(2):77–84. https://sortie-nd.org/research/pubs/Belsky_and_Canham_1994_BioScience_44_77-84.pdf

Bohra JS, Doerffling K (1993) Potassium nutrition of rice (*Oryza sativa* L.) varieties under NaCl salinity. Plant and Soil 152(2):299–303. https://link.springer.com/article/10.1007/BF00029100

Burley J (2002) Forest biological diversity: an overview. UNASYLVA-FAO 53:3–9. https://www.fao.org/docrep/013/i1757e/i1757e03.pdf

Carnevale NJ, Torres PS (1990) The relevance of physical factors on species distributions in inland salt marshes (Argentina). Coenoses 113–120. https://www.jstor.org/stable/43461031?seq=1#page_scan_tab_contents

Colwell RK (2013) EstimateS: statistical estimation of species richness and shared species from samples. Department of Ecology & Evolutionary Biology, University of Connecticut, Storrs. CT 06869-3043. USA. https://viceroy.eeb.uconn.edu/estimates/EstimateSPages/EstSUsersGuide/EstimateSUsersGuide.htm

Cui BS, Zhai HJ, Dong SK, Chen B, Liu SL (2009) Multivariate analysis of the effects of edaphic and topographical factors on plant distribution in the Yilong Lake Basin of Yun-Gui Plateau, China. Can J Plant Sci 89(1):211–221. https://www.nrcresearchpress.com/doi/abs/10.4141/CJPS08013#.WnbCqnwlHIU

Davis PH (1965) Flora of Turkey and the East Aegean Islands, vol 1–9, 140. University of Edinburgh Press, Edinburgh, pp 3–36. https://www.nhbs.com/series/flora-of-turkey-and-the-east-aegean-islands

Dirnböck T, Hobbs RJ, Lambeck RJ, Caccetta PA (2002) Vegetation distribution in relation to topographically driven processes in southwestern Australia. Appl Veg Sci 5(1):147–158. https://www.bioone.org/doi/abs/10.1658/1402-2001(2002)005%5B0147:VDIRTT%5D2.0.CO%3B2

El-Ghani MMA, Amer WM (2003) Soil–vegetation relationships in a coastal desert plain of southern Sinai, Egypt. J Arid Environ 55(4):607–628. https://www.sciencedirect.com/science/article/pii/S014019630200318X

Emad M (2001) Identification of forest and pasture industrial and medicinal herbs and their use. Rural Development of Iran, 112p (in Persian). https://elib.mui.ac.ir/site/catalogue/14576

Enright NJ, Miller BP, Akhter R (2005) Desert vegetation and vegetation-environment relationships in Kirthar National Park, Sindh, Pakistan. J Arid Environ 61(3):397–418. https://www.sciencedirect.com/science/article/pii/S0140196304002095

Escudero A, Iriondo JM, Olano JM, Rubio A, Somolinos RC (2000) Factors affecting establishment of a gypsophyte: the case of *Lepidium subulatum* (Brassicaceae). Am J Bot 87(6):861–871. https://bsapubs.onlinelibrary.wiley.com/doi/full/10.2307/2656894

Esmaeilnejad M, Khashei Siuki A (2018) Modelling climate change impacts on spatial distribution of Saffron Cultivation for future—a case study: South Khorasan. J Saffron Res 6(1):75–88. https://jsr.birjand.ac.ir/article_837_en.html

Farrukh H, Moinuddin A, Ghazala S, Durrani MJ (1994) Phytosociology of the vanishing tropical deciduous forest in district Swabi, Pakistan. II. Ordination. J Bot 26(1):149–160. https://www.pakbs.org/pjbot/PDFs/26(1)/PJB26(1)17.pdf

Gee GW, Bauder JW, Klute A (1986) Particle-size analysis, methods of soil analysis, Part 1. Physical and mineralogical methods, soil science society of America. Inc., Madison, WIS, USA. https://dl.sciencesocieties.org/publications/books/tocs/agronomymonogra/methodsofsoilana

Ghollasimod S, Memarian H, Hoseinpour M, Vahidi MJ (2017) Examination of relationship between species diversity and environmental variables in arid and semi-arid rangelands of Iran. J Environ Sci Manage 20(1):33–46. https://jesam-uplb.org/download_article.php?aid=316-examination-of-relationship-between-species-diversity-and-environmental-variables-in-arid-and-semi-arid-rangelands-of-iran

Goodall DW, Perry RA, Howes KMW (eds) (2009) Arid land ecosystems: volume 2, structure, functioning and management, vol 2. Cambridge University Press. https://books.google.com/books?hl=en&lr=&id=FP2_K1pN22sC&oi=fnd&pg=PR13&dq=+Arid+land+ecosystems:+volume+2,+structure,+functioning+and+management&ots=t3xIw8XbbI&sig=fuI8bUsL6ww4UKAg4taD4UAimZY#v=onepage&q=Arid%20land%20ecosystems%3A%20volume%202%2C%20structure%2C%20functioning%20and%20management&f=false

Green SB, Salkind NJ (2010) Using SPSS for windows and macintosh: analyzing and understanding data. Pearson Prentice Hall Press, 459p. https://www.amazon.com/Using-Windows-Macintosh-Books-Carte/dp/0134319885

Härdtle W, von Oheimb G, Westphal C (2003) The effects of light and soil conditions on the species richness of the ground vegetation of deciduous forests in northern Germany (Schleswig-Holstein). For Ecol Manage 182(1–3):327–338. https://www.sciencedirect.com/science/article/pii/S0378112703000914

Hassani S, Yazdanshenas H, Nazarpoor Fard K, Bassiri R, Pur Rezaee J (2014) Study of physiognomy and origin of plant species in Sarshiv Area of Marivan, Iran. J Rangel Sci 4(4):270–278. https://www.rangeland.ir/article_510002.html

Havlin JL, Beaton JD, Tisdale SL, Nelson WL (2005) Soil fertility and fertilizers: an introduction to nutrient management, vol 515. Pearson Prentice Hall, Upper Saddle River, pp 97–141. https://www.amazon.com/Soil-Fertility-Fertilizers-John-Havlin/dp/013503373X

Hawkins RH, Kunze GW (1965) Phosphate fractions in some Texas Grumusols and their relation to soil weathering and available Phosphorus 1. J Soil Sci Soc Am 29(6):650–656. https://dl.sciencesocieties.org/publications/sssaj/abstracts/29/6/SS0290060650

He MZ, Zheng JG, Li XR, Qian YL (2007) Environmental factors affecting vegetation composition in the Alxa Plateau, China. J Arid Environ 69(3):473–489. https://www.sciencedirect.com/science/article/pii/S0140196306003533

Heisler-White JL, Knapp AK, Kelly EF (2008) Increasing precipitation event size increases aboveground net primary productivity in a semi-arid grassland. Oecologia 158(1):129–140. https://link.springer.com/article/10.1007/s00442-008-1116-9

Jackson DA (1993) Stopping rules in principal components analysis: a comparison of heuristical and statistical approaches. Ecology 74(8):2204–2214. https://www.jstor.org/stable/1939574?seq=1

Jafari M, Zare Chahuki MA, Azarnivand H, Baghestani Meibodi N, Zahedi Amiri G (2002) Relationships between Poshtkouh Rangel Vegetative of Yazd Province and soil physical and chemical characteristics using multivariate analysis methods. J Nat Resour 55(3):419–434. https://www.sid.ir/En/Journal/ViewPaper.aspx?ID=30996

Jarema SI, Samson J, Mcgill BJ, Humphries MM (2009) Variation in abundance across a species' range predicts climate change responses in the range interior will exceed those at the edge: a case study with North American beaver. Global Change Biol 15(2):508–522. https://onlinelibrary.wiley.com/doi/10.1111/j.1365-2486.2008.01732.x/full

Jenkins MA, Parker GR (1998) Composition and diversity of woody vegetation in silvicultural openings of southern Indiana forests. For Ecol Manage 109(1–3):57–74. https://www.sciencedirect.com/science/article/pii/S0378112798002564

Jin-Tun Z (2002) A study on relations of vegetation, climate and soils in Shanxi province, China. Plant Ecology 162(1):23–31. https://link.springer.com/article/10.1023/A:1020355921982

Jolliffe IT (1986) Principal component analysis and factor analysis. In: Principal component analysis. Springer, New York, pp 115–128. https://link.springer.com/chapter/10.1007/978-1-4757-1904-8_7

Jongman RH, Ter Braak CJ, Van Tongeren OF (eds) (1995) Data analysis in community and landscape ecology. Cambridge University Press. https://www.amazon.com/Data-Analysis-Community-Landscape-Ecology/dp/0521475740

Jowkar H, Ostrowski S, Tahbaz M, Zahler P (2016) The conservation of biodiversity in Iran: threats, challenges and hopes. Iran Stud 49(6):1065–1077. https://www.tandfonline.com/doi/abs/10.1080/00210862.2016.1241602

Kjeldahl J (1883) A new method for the determination of nitrogen in organic matter. Z Anal Chem 22(1):366–382. https://books.google.com/books?id=6ePmAAAAMAAJ&pg=PA366#v=onepage&q&f=false

Krzic M, Newman RF, Broersma K (2003) Plant species diversity and soil quality in harvested and grazed boreal aspen stands of northeastern British Columbia. For Ecol Manage 182(1–3):315–325. https://www.sciencedirect.com/science/article/pii/S0378112703000641

Li W, Yan M, Qingfeng Z, Zhikaun J (2012) Effects of vegetation restoration on soil physical properties in the wind–water erosion region of the northern Loess Plateau of China. CLEAN–Soil Air Water 40(1):7–15. https://onlinelibrary.wiley.com/doi/10.1002/clen.201100367/full

Masoudi M, Yousefi M, Behbahani N (2018) Hazard assessment of climate changes in South Khorasan Province, Iran. EQA-Int J Environ Qual 29:29–39. https://eqa.unibo.it/article/view/7910

McCune BJ, Mefford MJ (2016) PC-ORD multivariate analysis of ecological data, Version 7, MjM software design, Glenden Beach, Oregon, USA. https://www.pcord.com/booklet/pinside1.htm

McGarigal K, Cushman S, Stafford S (2000) Multivariate statistics for wildlife and ecology research (No. 04; QH541. 2, M2 2000). Springer, New York, 248p. https://www.springer.com/gp/book/9780387988917

Medinski TV, Mills AJ, Esler KJ, Schmiedel U, Jürgens N (2010) Do soil properties constrain species richness? Insights from boundary line analysis across several biomes in south western Africa. J Arid Environ 74(9):1052–1060. https://www.sciencedirect.com/science/article/pii/S0140196310000844

Mirzaei J, Karami A (2015) Plant diversity and richness in relation to environmental gradient in Zagros ecosystems, West of Iran. J Rangel Sci 5(4):294–302. https://www.rangeland.ir/article_516669_110095.html

Mishra A, Sharma SD, Khan GH (2003) Improvement in physical and chemical properties of sodic soil by 3, 6 and 9 years old plantation of *Eucalyptus tereticornis*: biorejuvenation of sodic soil. For Ecol Manage 184(1–3):115–124. https://www.sciencedirect.com/science/article/pii/S0378112703002135

Mozaffarian V (2010) Trees and Shrubs of Iran. Farhang Moaser Publishing. 1054p. https://www.amazon.com/Trees-Shrubs-Iran-Valiollah-Mozaffarian/dp/9648637032

Pielou EC (1975) Ecological diversity (No. 574.524018 P5). Wiley-Intersection, New York. https://books.google.com/books/about/Ecological_diversity.html?id=12XwAAAAMAAJ

Rezaeipor M, Jahani H, Hoseini SM, Mirzai J, Jafari G (2014) Ecological survey of *Rhus coriaria* L. in west of Iran. J Plant Res (Iran J Biol) 26(4):444–452. https://www.sid.ir/En/Journal/ViewPaper.aspx?ID=411963

Rezaei Pourbaghdar A, Sadeghinia M, Nohegar A, Hakimi MH (2014) Determination of some soil properties on distribution of vegetation types and *Dorema ammoniacum* and Rheum ribes in ranges of Baghedar region in Bafgh city. J Desert Ecosyst Eng 3(4):69–78. https://deej.kashanu.ac.ir/browse.php?a_code=A-10-102-1&sid=1&slc_lang=en

Rogel JÁ, Silla RO, Ariza FA (2001) Edaphic characterization and soil ionic composition influencing plant zonation in a semiarid Mediterranean salt marsh. Geoderma 99(1–2):81–98. https://www.sciencedirect.com/science/article/pii/S0016706100000677

Roosta (2015) Effect of two range plant (*Rhus coriaria* and *Amygdalus scoparia*) on some of soil chemical properties (Case study: Kakhk). M.Sc. thesis, University of Birjand, 89p (in Persian). https://ganj-old.irandoc.ac.ir/articles/850399

Saxena NP (1985) The role of potassium in drought tolerance. Potash Rev 5:1–15. https://oar.icrisat.org/4127/

Schulz B, Bechtold WA, Zarnoch SJ (2009) Sampling and estimation procedures for the vegetation diversity and structure indicator. US Department of Agriculture, Forest Service, Pacific Northwest Research Station, pp 1–53. https://www.fs.fed.us/pnw/pubs/pnw_gtr781.pdf

Seaby RM, Henderson PA (2006) Species diversity and richness version 4. Pisces Conservation Ltd., Lymington, England, 123. https://www.pisces-conservation.com/freestuff.html

Ter Braak CJ (1986) Canonical correspondence analysis: a new eigenvector technique for multivariate direct gradient analysis. Ecology 67(5):1167–1179. https://onlinelibrary.wiley.com/doi/10.2307/1938672/full

Tolera M, Asfaw Z, Lemenih M, Karltun E (2008) Woody species diversity in a changing landscape in the south-central highlands of Ethiopia. Agric Ecosyst Environ 128(1–2):52–58. https://www.sciencedirect.com/science/article/pii/S0167880908001400

Tonggui W, Ming W, Mukui Y, Jianghua X (2011) Plant distribution in relation to soil conditions in Hangzhoubay coastal wetlands, China. Pak J Bot 43(5):2331–2335. https://pdfs.semanticscholar.org/fdb5/44707ea41facbc0381b42a47f08ae1cf4f33.pdf

van der Maarel E, Franklin J (eds) (2012) Vegetation ecology. Wiley, New York. https://www.wiley.com/en-us/Vegetation+Ecology%2C+2nd+Edition-p-9781444338881

Walkley A, Black IA (1934) An examination of the Degtjareff method for determining soil organic matter, and a proposed modification of the chromic acid titration method. J Soil Sci 37(1):29–38. https://journals.lww.com/soilsci/citation/1934/01000/an_examination_of_the_degtjareff_method_for.3.aspx

Yaghmaei L, Soltani S, Khodagholi M (2009) Bioclimatic classification of Isfahan province using multivariate statistical methods. J Climatol 29(12):1850–1861. https://onlinelibrary.wiley.com/doi/10.1002/joc.1835/full

Yibing Q, Zhaoning W, Liyun Z, Qingdong S, Jin J, Lisong T (2004) Impact of habitat heterogeneity on plant community pattern in Gurbantunggut Desert. J Geograph Sci 14(4):447–455. https://link.springer.com/article/10.1007/BF02837488

Zhao CM, Chen WL, Tian ZQ, Xie ZQ (2005) Altitudinal pattern of plant species diversity in Shennongjia Mountains, Central China. J Integr Plant Biol 47(12):1431–1449. https://onlinelibrary.wiley.com/doi/10.1111/j.1744-7909.2005.00164.x/full

Chapter 6
Comparative Assessment of Forest Deterioration through Remotely Sensed Indices—A Case Study in Korba District (Chhattisgarh, India)

Soumen Bramha, Gouri Sankar Bhunia, Sant Ram Kamlesh, and Pravat Kumar Shit

Abstract In various studies, such as meteorology, agriculture and ecology, quantitative estimation of biophysical variables is very important, and thus information about the spatial and temporal distribution of these variables are highly useful. Remote sensing is meanwhile regarded as an important source of knowledge in broad areas for the estimation of fractional vegetation coverage. In the remote sensing, estimation of vegetation characteristics using spectral indices have become very common today, but soil and rocks reflections are also much more than the reflection in these areas of sparse vegetation, which makes it difficult to distinguish plant signals. In this analysis, a variety of spectral indices have been considered to estimate biophysical vegetation parameters to boost vegetation signal in remotely-sensed data and provide an estimated measurement of living green vegetation using Landsat4,5 Thematic Mapper (TM) and Landsat8 Operational Land Imager (OLI) sensor data. To identify the best vegetation index for sparsely vegetated semi-arid and arid region of Chhattisgarh state using four vegetation indexes; Normalized Difference vegetation index (NDVI), transformed normalized difference vegetation index (TNDVI), soil-adjusted vegetation index (SAVI), modified soil-adjusted vegetation index (MSAVI). The TNDVI indices showed the best fractional vegetation cover to estimate the highest precision.

Keywords Landsat data · Forest deterioration · Vegetation indices · Change analysis · Damage assessment

S. Bramha (✉) · G. S. Bhunia
Department of Geography, Nalini Prabha Deo Prasad Roy Arts & Commerce College, Jorapara, Ashok Nagar, Sarkanda, Bilaspur, Chhattisgarh 495006, India
e-mail: brahma6734@gmail.com

S. R. Kamlesh
Government E. Raghvendra Rao Post Graduate Science College, Seepat Road, Sarkanda, Bilaspur, Chhattisgarh 495006, India

P. K. Shit
PG Department of Geography, Raja N.L. Khan Women's College (Autonomous), Gope Palace, Vidyasagar University, Paschim Medinipur, West Bengal 721102, India

P. K. Shit et al. (eds.), *Spatial Modeling in Forest Resources Management*, Environmental Science and Engineering,
https://doi.org/10.1007/978-3-030-56542-8_6

6.1 Introduction

The second largest source of anthropogenic emissions of greenhouse gasses after fossil fuel combustion is forestation and forest deterioration (IPCC 2007). As a result, deforestation and forest degradation have become an important issue regarding climate change mitigation, highlighted in the 2007 based on the Fourth Assessment Report of the Intergovernmental Panel on Climate Change (IPCC). Global programs such as the REDD+ United Nations Framework Convention on Climate Change aim at mitigating climate change by plummeting reductions and degradation in forest cover of tropical forests (Margono et al. 2012). The international forestry community, including the United Nations Forum on Forest (UNFF) has highlighted forest depletion and the Convention on Biological Diversity (CBD) goal in 2010 (Simula 2009). Indeed, it is more difficult to quantify forest degradation compared with deforestation, as deforestation represents a disturbance to stand-replacement, while forest degradation is no improvement in the use of land, and the result still represents forest land by definition (FAO 2007). The use of logging roads, logging courtyards or forest canopy's is evidence of degradation practices, such as the selective logging. The advent of modern remote-sensing methods for collecting information on earth's surface characteristics using photographic cameras and multi-spectra scanners on air and satellite platforms and interactive man-machines, have led to an exciting new era in which the resources investigation methods are revolutionized. They have the capacity to provide broad synoptic coverage of near-real-time data as well as temporary verification, with tremendous potentiality. The currently established remote sensing methods available for forest-relevant surveillance can be grouped into four categories: (1) identification of direct deterioration indicators (canopy coverage percentage) on a single date or on a composite picture (Souza et al. 2003); (2) mapping and assuming such buffering degradation indicators (Potapov et al. 2008) and (3) direct calculation of the forest parameters (system and biomass volumes), now usually performed in radar or LiDAR systems in rainforests (Jubanski et al. 2013). (4) Direct evaluation of forest parameters (SYS).

According to the FAO Global Forest Resource Assessment (FAO 2006) 'forest' is a land spanning more than 0.5 hectare with trees taller than 5 m and a canopy cover of more than 10%, or trees able to reach these thresholds in situ. The trees will have at least 5 m in situ height and a canopy cover of 10%. Land deforestation and degradation contributing more than 17th percent of global carbon dioxide emissions and anthropogenic greenhouse emissions (IPCC 2007). India has 21% of its forest-covered geographic area, hosting almost 88 million indigenous and tribal citizens in 173,000 villages (INCCA 2010). Forest service is projected to be worth 7% of the domestic GDP, which accounts for 57% of rural incomes in India, for fresh water, soil nutrients and non-wood forests. Nonetheless, any change in forest structure and function would have major implications for susceptible societies that are already dependent on forests. Moreover, predicted impacts on India suggest that forests die back, and loss of biodiversity is likely to change in 40–70% of the forested nets in various states under a changing climate (Ravindranath et al. 2006).

Clear identification of forest loss processes applies to shifts in areas and focuses on damage to forest canopies (Herold et al. 2011). Land canopy cover, above-ground biomass and net primary productivity have been a set of measures of land loss identified. Thanks to the ability to measure their spatial and temporal variation using the time series image stack, these measures have been chosen purposely. It has to be recalled however that there are certain limitations in the evaluation of biomass in the saturation phenomenon for biomass due to the inability to penetrate the duct and reflected energy. Remote sensing products were a primary source of knowledge in recent decades for tracking transitions in land cover (Lunetta et al. 2002). However, these items can be seen as the only possible way in which changes in forest cover are continuously tracked over time in large geographic areas (Shimabukuro et al. 2014). The approach uses three spectral enclosure libraries to degrade the percentage of the forest canopy cover for each pixel in Landsat images to three distinct components for each image pixel (Romero-Sanchez and Ponce-Hernandez 2017). The layer photosynthetic vegetation, which consists of 0–100% was used for subsequent study as an equivalent of the cover forest field canopy (0–100%). Souza (2013) points out the variations between future mapping goals such as mapping of the total area affected by the logging (which includes damaged canopy areas, cleared areas or logging facilities and intact forest areas) or mapping of damaged areas (Souza 2013). The availability of medium spatial resolution data including Landsat (30 m × 30 m spatial resolution) is expected to enhance the evaluation of logging areas. The goal of this study is a practical description of what "forest deterioration" means and the calculation and quantification of its various intensity levels and the spatial extent of each landscape.

6.2 Materials and Method

6.2.1 Study Area

Korba District is a central Indian administrative district in the state of Chhattisgarh, extended between 22° 01′–23° 01′ North latitude and 82° 08′–83°09′ East longitude (Fig. 6.1). The district belongs to Bilaspur and is primarily populated by tribes like the Korwas protected tribe. The district of Korba is situated in the northern part of the state, surrounded by Korea, Surguja, Bilaspur, Janjgir-Champa, and is surrounded by the districts. The total area of the district is 7.14.544 ha. The average height of the district Korba is 304.8 m from sea level. The climate is hot and temperate with a warm and dry climate. The district has an annual precipitation of 1506.7 mm. The river Hasdeo and Gagechorai, Tan and Ahiran are the major rivers that flow through Korba. The district's total forest area is 413,787 ha under two Viz forests. Subdivisions of Korba and Katgora. Mahua, Bija, Sagon, Sahaj, Tendu Leaves etc. are the main forest products in the region. The Korba district has 1,206,640 inhabitants according to the 2011 census. The population of the district is 183 people per square kilometer. In

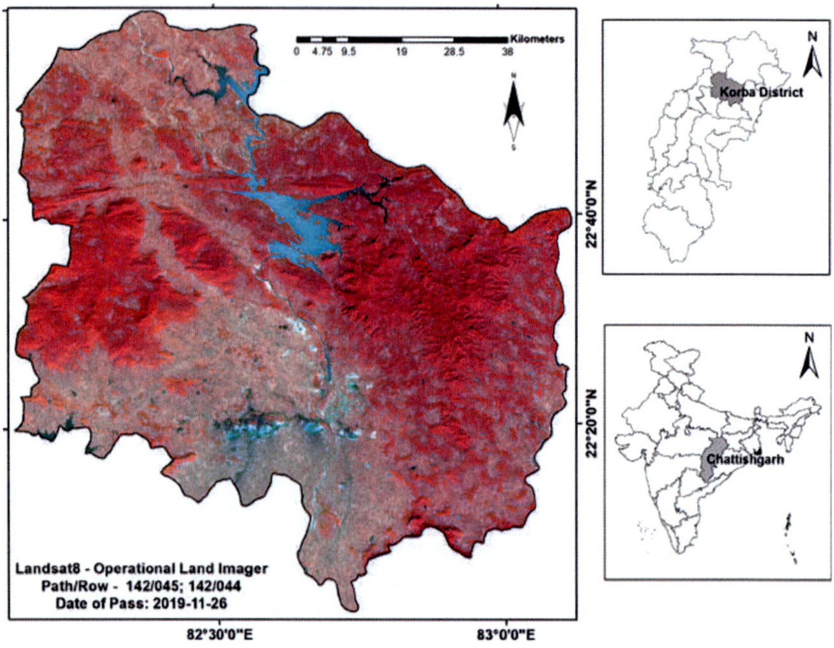

Fig. 6.1 Location map of the Korba district, Chhattisgarh (India)

the decade 2001–2011, population growth was 19.25%. For everyone thousand men, Korba has a sexual relationship of 971 women, and 73.22%.

6.2.2 Data Used

In order to assess mapping and track deforestation intra-annual availability of cloud-free land satellite observations. The Landsat missions are fitted with a multispectral passive optical sensor measuring radiation from the surface of the earth on many arbitrary wide electromagnetic channels known as spectral bands (Table 6.1). TM, OLI and Blue, orange, red, near infrared (NIR), first shortwave infrared ($SWIR_1$), and second shortwave infrared ($SWIR_2$) are identical across all widely used spectral strips (USGS 2017). Imaging from the USGS website was collected with Landsat Thematic Mapper (TM) and the Operational Land Imager (OLI) (https://glovis.usgs.gov). The number of images available each year in the study area, as in the other tropical coverage, is limited; however, at least one cloud free image of Landsat is relatively high per season (Ju and Roy 2008). The images had been corrected geometrically in Level and were atmospherically corrected by the radio-transmission method of the second satellite signal simulation in the Solar Spectrum (6S). Landsat images have been divided by date (i.e. rainy or dry season), and a first cloud filter was added.

Table 6.1 Details of the satellite and sensor characteristics

Satellite	Sensor	Spatial resolution (m)	Spectral resolution (μm)	Temporal resolution	Radiometric resolution
Landsat4,5 (path/row—142/044; 142/045)	Thematic mapper (29/11/1990)	30	0.45–0.52	16 days	8 bit
		30	0.52–0.60		
		30	0.63–0.69		
		30	0.76–0.90		
		30	1.55–1.75		
		120	10.40–12.50		
		30	2.08–2.35		
Landsat8 (path/row—142/044; 142/045)	Operational land imager (26/11/2019)	30	0.433–0.453	16 days	16 bit
		30	0.450–0.515		
		30	0.525–0.600		
		30	0.630–0.680		
		30	0.845–0.885		
		30	1.560–1.660		
		30	2.100–2.300		
		15	0.500–0.680		
		30	1.360–1.390		
		100	10.30–11.30		
		100	11.50–12.50		

Their main benefit over higher-resolution imaging is that it is usable for a fairly long period of time as archival data. Photos with a cloud coverage of more than 10% were avoided and not included in the study (Grecchi et al. 2017; Wulder et al. 2012).

6.2.3 Vegetation Indices

An analysis of a host of vegetation indices permitted an accurate estimate of Aboveground Biomass for those published. In addition, four mainly used VI (Vegetation Index) were initially analyzed (Table 6.2), namely the namely the normalized difference vegetation index (NDVI) (Rouse et al. 1973), modified soil-adjusted vegetation index (MSAVI) (Qi et al. 1994), the normalized difference moisture index (NDMI) (Gao 1996) and normalized burn ratio (NBR) (Key and Benson 2006). The NDVI's theoretical basis lies in the red-NIR distinction of plant spectral reflectance signatures (Rahman et al. 2015). As living, green vegetation grows in one pixel, the chlorophyll absorption causes the red reflectance to decrease while the non-absorption area of NIR spectrums typically increases its basic structure and volume (Baret and Guyot 1991). In order to produce biomass estimates, these indices are used for study and to

Table 6.2 Vegetation indices used in this study

Vegetation index	Equation	Reference
Normalized Difference Vegetation Index (NDVI)	$NDVI = \frac{NIR-Red}{NIR+Red}$	Rouse et al. (1974)
Modified Soil-Adjusted Vegetation Index2 (MSAVI2)	$MSAVI2 = \frac{(2*NIR+1-\sqrt{(2*NIR+1)^2-8*(NIR-RED)}}{2}$	Qi et al. (1994)
Soil Adjusted Vegetation Index (SAVI)	$SAVI = \frac{NIR-RED}{(NIR+RED+L)} * (1+L)$	Huete (1988)
Transformed Normalized Difference Vegetation Index (TNDVI)	$TNDVI = \sqrt{\frac{Infrared-Red}{Infrare+Red}} + 0.5$	Tucker (1979)

classify indicators that have a high correlation with forest biomass (AGB) measurements. The measurement of the vegetation indices used reflectance values at various wavelengths so as to improve information relevant to vegetation while the influences from environmental conditions and shade.

Where RED is the reflection of the red band from the sensor, NIR is the reflectance of the near infrared band and the value of L varies by the amount or cover of green vegetation: in very high vegetation regions, L = 0; and in areas with no green vegetation, $L = 1$.

6.2.4 Change Detection Analysis

Change detection analysis has been performed through image differencing method in which spatially registered imageries are acquired at different times are subtracted pixel by pixel and band by band. The difference between the DN values is calculated by using the formula:

$$DX_{ij}^{k} = X_{ij}^{k}(T_2) - X_{ij}^{k}(T_1)$$

where $X_{ij}^{k}(T_1)$ and $X_{ij}^{k}(T_2)$ is the DN value of pixel 'X' located at row 'i' and column 'j' and 'k' at time 'T_1' and 'T_2'. If the difference between DN value is 0, it does not change. The beliefs become positive or negative as the transition takes place. The issue with this technique is the identification of shift threshold values and no shift in the images. As a criterion for determining the threshold, standard deviation is used. The discrepancy may occur sometimes with the picture discrepancy, even though the shift did not occur over the surface of the earth, because for images acquired at different dates, the correct picture records and perfect radiometric conditions are never obtained.

Finally, using the following formula equation, the rate of change hectare/year and proportion of each class in the time periods studied were calculated.

$$\Delta A(\%) = \frac{At_2 - At_1}{At_1} \times 100$$

where,

$\Delta A(\%)$ Change of percentage between initial time At_1 and time span At_2 forest area.

At_1 Forest area form for first time.

At_2 Forest area form in the end.

The rate of changes in the type of forest cover was calculated using the following formula, according to Abate (2011):

$$R\Delta\left(\frac{ha}{year}\right)\frac{Z - X}{W}$$

where,

RΔ rate of change,

Z recent area of forest cover in ha,

X previous area in forest cover in ha;

W time interval between Z and X in years.

6.2.5 *Accuracy Assessment*

The spatial accuracy of deforestation was assessed based on the count. Note that count-based accuracy does not automatically provide a map accuracy measure (area) in which sample areal representatives are accounted for which this pilot study could not and was not intended. Producer's accuracy (PA), user's accuracy (UA), and overall accuracy (OA) were estimated. OA is the proportion of all samples and all samples that are correctly forecast:

$$OA(\%) = 100 \times \frac{(TP + TN)}{(TP + TN + FP + FN)}$$

where the number of true positive detections is TP, which means the correctly forecast deforestation, TN is the true negative detection number, that means, correctly predicted non-deforestation (no deforestation in reference); FP is the number of false positive detections; i.e. wrongly forecasted non-deforestation. One minus omission error (FN) is a class-specific PA, while a class-specific UA is one minus commission error (FP):

$$UA(\%) = 100\left(\frac{TP}{TP + FP}\right)$$

$$PA(\%) = 100 \times \left(\frac{TP}{(TP + FN)}\right)$$

6.2.6 *Forest Degradation Mapping and Evaluation*

To create a forest degradation image characterizing the variations between the initial and the final status images, use the minus algorithm of spatial analyst tool of QGIS software v3.16. Images of each data form can be single-band images. The difference is determined by removing the initial state image from the final state image (that is, final—original). The gray color of the corresponding pixel images after image registration is subtracted from a subtraction image. The gray value of the picture subtraction indicates the degree of two pictures transition. By selecting the correct threshold values for the gray picture, the modified region and unchanged region will be calculated. The following formula is:

$$Dx_{ij}^{k} = X_{ij}^{k}(t_2) - X_{ij}^{k}(t_1) + c$$

where, i, j as pixel co-ordinates, k for the band, $X_{ij}^{k}(t_1)$ for the pixel (i, j) value of k-band image, t_1 for the pixel (i, j) value of k-band image, t_1, t_2 for the time of the first and the second image, C is constant.

The choice of threshold value determines the ability to detect change for most methods of detection. The option of an acceptable threshold will distinguish the actual changes and the effect areas of random variables as much as possible. A significant number of failed inspections were triggered by the use of a high threshold value and a low threshold value were chosen to inspect zero. One positive change identifies the brighter pixels (end brightness was higher than the initial brightness of status), while one negative change identifies the dimmer pixels. In addition, as an optional pre-processing stage, you can normalize or standardize the images input in a data range from zero to one.

6.3 Results and Discussion

6.3.1 *Forest Cover*

The dynamics for forestry cover in a Chattishgarh state study zone are analysed here because of either deforestation (reversal to other land cover) or forest disturbances over a period of 29 years (1990–2015). We show that although the total area deforested over the years is growing, the amount of forest areas converted to agricultural land in comparison with selective logging disrupting areas have been substantially reduced. Such variations are much more important if we take into account at least once during this time all the places where disturbances have occurred. However, compared to a cycle of deforestation, typically requiring a permanent or long-term transition to other forms of land cover, selective forest area regeneration also happens on a short-term basis, although a portion of these areas is ultimately deforested.

Classification of digital land can be quantitatively demonstrated by the creation and analysis of an error classification matrix. Google Earth images, true and false color images and basic information were the reference data used. The highest OA is calculated as 82.1% for the NDVI, 90.4% for the TNDVI, 85.9% for the SAVI, and 88.7% for MSAVI. The accuracy assessment of TNDVI shows that the achieved product and user accuracy (>85%) is good to very good (>95.8%) for all types of forest coverage.

6.3.2 *Normalized Difference Vegetation Index (NDVI)*

As a vegetation measure the NDVI (Fig. 6.2) is effective because it is sufficiently stable to allow meaningful comparisons of seasonal and interannual changes in the growth and development of vegetation (Chen et al. 2006). In its definition of ratio (Moran et al. 1992), the power of the NDVI reduces several types of multiplicative noise present in several bands (illumination disparity, cloud shadows, atmospheric diminution and other topographical variations) (Chen et al. 2002). In 1991, NDVI value ranges between −0.50 and 0.61. The average NDVI value is calculated as 0.06 with a standard deviation of 0.32. Based on the NDVI value, the study area is classified into five categories, namely (i) <−0.14, (ii) −0.13 to 0.14, (iii) 0.15–0.25, (iv) 0.26–0.36, and (v) >0.37 (Table 6.3). The classification has been done based on the quantile of the NDVI values. Result of the analysis also showed 31.60% (2080.484 km^2) are covered by degraded forest, followed by open forest 28.02% (1844.51 km^2). Most of the dense forest are observed in the central of west and east of the district (Fig. 6.2a). The degraded forest is mainly observed in the surrounding and adjacent region of dense forest cover. Open forest areas are distributed in north of the district. In the central and south of the study area having very less vegetation cover, almost nil.

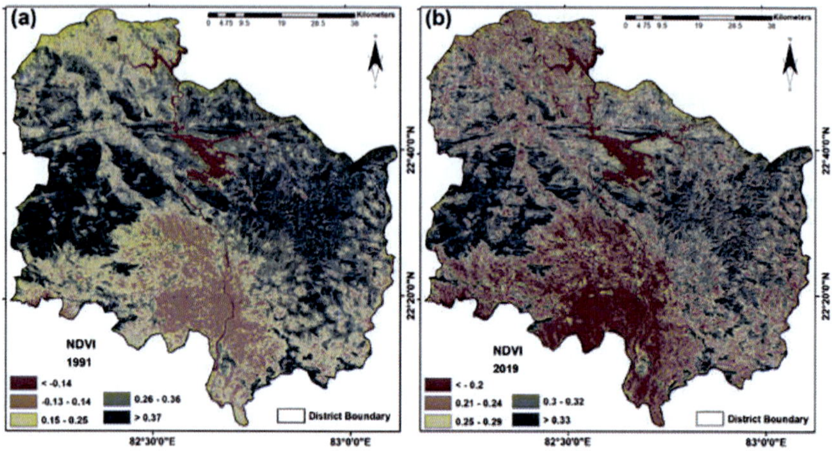

Fig. 6.2 Normalized Difference Vegetation Index (NDVI) map of Korba district in Chhattisgarh, **a** 1991 and **b** 2019

Table 6.3 NDVI characteristics of Korba district in 1991

Parameters	NDVI range	Area (km^2)	Percent	Average NDVI	Std. dev.
Others	<−0.14	152.22	2.31	−0.32	0.11
Scrub/cultivated land	−0.13 to 0.14	1144.03	17.38	0.0001	0.08
Open forest	0.15–0.25	1844.51	28.02	0.21	0.24
Degraded forest	0.26–0.36	2080.484	31.60	0.32	0.03
Dense forest	>0.37	1362.438	20.69	0.49	0.07

In 2019, NDVI value of the Korba district ranges between −0.30 and 0.53 with an average value of 0.11 ± 0.24. Scrub and agricultural fallow land (NDVI: −0.13 to 0.14) covered by 38.98% (2566.30 km^2), followed by open forest 29.82% (1963.11 km^2). The area of others (waterbody/dry fallow/agricultural fallow/built-up) land (NDVI < −0.14) is covered by 0.58 km^2 of the entire study area (Table 6.4) 0.17.41% (1146.36 km^2) of the study area covered by dense forest (NDVI > 0.37). The dense forests are mainly observed in the west of the district, and some small

Table 6.4 NDVI characteristics of Korba district in 2019

Parameters	NDVI ranges	Area (km^2)	Percent	Average NDVI	Std. dev.
Others	<−0.14	0.58	0.009	−0.25	0.03
Scrub/cultivated land	−0.13 to 0.14	2566.30	38.98	0.02	0.13
Open forest	0.15–0.25	1963.11	29.82	0.27	0.02
Degraded forest	0.26–0.36	907.33	13.78	0.31	0.01
Dense forest	>0.37	1146.36	17.41	0.43	0.06

patches of dense forest covers are also observed in the east of the study area. 13.78% of the study area covered by degraded forest, mostly found in the east, north-east and north-west of the study area (Fig. 6.2b).

6.3.3 Transformed Normalized Difference Vegetation Index (TNDVI)

The TNDVI reflects the biomass of plants (Fig. 6.3) and is expressing as the ratio of almost-IR reflection to red reflectance (Tucker 1979). A variety of interrelated effects include: evapotranspirational cooling; solar radiation interception; moisture retention; land-covering; surface energy balance and partial canopy covering depend on the amount of biomass directly and inversely correlated with surface temperature (Yang et al. 2008). TNDVI value of 1991 in Korba district ranges from 0.0004 to 1.05 (mean + S.D. 0.53 ± 0.31). TNVDI value of <0.81 is considered for others land (waterbody/dry fallow/agricultural fallow land/built-up), covered by 26.69% (1627.78 km^2) of the district (Table 6.5). The scrub/cultivated land enclosed by 16.22% (1068.12 km^2). TNDVI value of open forest ranges between 0.86 and 0.91 (mean ± S.D.—0.89 ± 0.01), covered by 26.41% (1738.92 km^2). The degraded forest covered by 17.81% of the district and the TNDVI value ranges from 0.91 to 0.95 (mean ± S.D.—0.91 ± 0.001). The dense forest enclosed by 12.31% (810.77 km^2), having TNDVI value of more than 0.96 (mean ± S.D.—0.96 ± 0.03). Figure 6.4a shows the spatial distribution forest cover in Korba district in 1991. The map shows the central part of west and east of the district have dense forest cover. The degraded forest is mainly distributed in the adjacent areas of dense forest cover, located in the

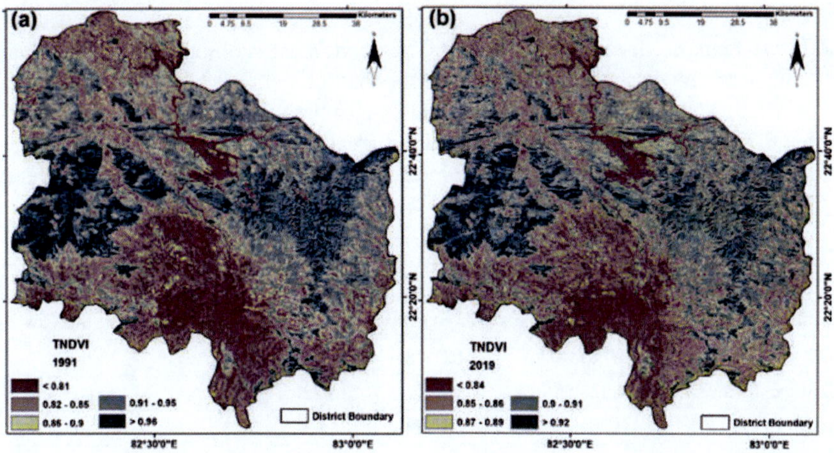

Fig. 6.3 Transformed Normalized Difference Vegetation Index2 (MSAVI2) map of Korba district in Chhattisgarh, **a** 1991 and **b** 2019

Table 6.5 Analysis of areal characteristics of Transformed Normalized Difference Vegetation Index (TNDVI) of Korba district in 1991

Parameters	TNDVI ranges	Area (km^2)	Percent	Average TNDVI	Std. dev.
Others	<0.81	1757.21	26.69	0.65	0.11
Scrub/cultivated land	0.82–0.85	1104.41	16.77	0.85	0.01
Open forest	0.86–0.91	1738.92	26.41	0.89	0.01
Degraded forest	0.91–0.95	1172.57	17.81	0.91	0.001
Dense forest	>0.96	810.77	12.31	0.96	0.03

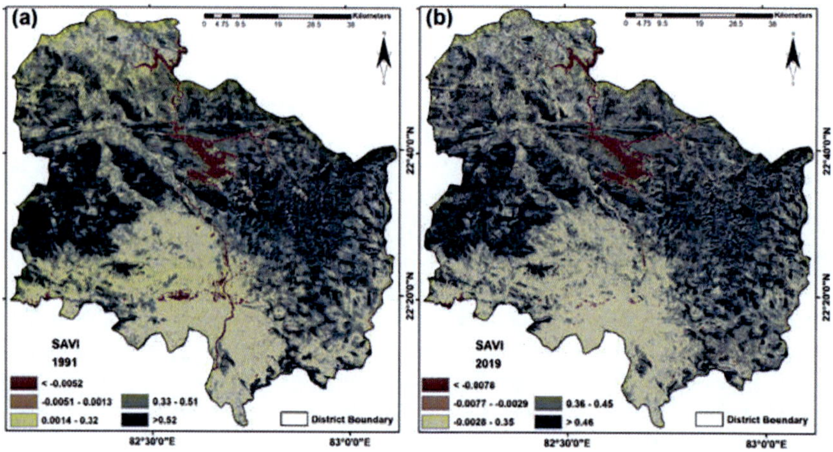

Fig. 6.4 Soil Adjusted Vegetation Index (SAVI) map of Korba district in Chhattisgarh, **a** 1991 and **b** 2019

north-west and north-east of the district. The open forest is observed in the extreme east and extreme north and central part of the district.

In 2019, TNDVI value of the study area ranges between 0.45 and 1.01 (mean + Stdev 0.73 ± 0.16). Table 6.6 illustrated the areal characteristics of various forest cover along with TNDVI value. TNVDI value of <0.81 is considered for others land

Table 6.6 Analysis of areal characteristics of Transformed Normalized Difference Vegetation Index (TNDVI) of Korba district in 2019

Parameters	TNDVI ranges	Area (km^2)	Percent	Average TNDVI	Std. dev.
Others	<0.81	1627.78	24.72	0.4	0.23
Scrub/cultivated land	0.82–0.85	1068.12	16.22	0.83	0.01
Open forest	0.86–0.91	1647.73	25.03	0.88	0.02
Degraded forest	0.91–0.95	1488.41	22.61	0.94	0.01
Dense forest	>0.96	751.79	11.42	1.01	0.03

(waterbody/dry fallow/agricultural fallow land), covered by 24.72% (1627.78 km^2) of the district. The scrub/cultivated land (TNDVI: 0.82–0.85) covered by 16.22% (1068.12 km^2). In open forest, TNDVI value ranges between 0.86 and 0.91 (mean ± S.D.—0.88 ± 0.02), covered by 25.03% (1647.73 km^2). The degraded forest covered by 22.61% of the study area and the TNDVI value ranges from 0.91 to 0.95 (mean ± S.D.—0.94 ± 0.01). The dense forest covered by 11.42% (751.79 km^2), having TNDVI value of more than 0.96 (mean ± S.D.—1.01 ± 0.03). Figure 6.4b shows the density of dense forest cover is decreased in central west and small patches of central of the district. The degraded and open forest covers is also decreased in the north and south-east of the study area.

6.3.4 Soil Adjusted Vegetation Index (SAVI)

In areas where vegetative cover is low and the soil surface is exposed, the reflectance of light in the red and near-infrared spectra can influence vegetation index values (Huete 1988). This is especially problematic when comparisons are being made across different soil types that may reflect different amounts of light in the red and near infrared wavelengths (Huete et al. 2002). The soil-adjusted vegetation index (SAVI) was developed as a modification of the normalized difference vegetation index (NDVI) to correct for the influence of soil brightness when vegetative cover is low (Huete 1988; Lyon et al. 1998). The output of SAVI is a new image layer with values ranging from −1 to 1. The lower the value, the lower the amount/cover of green vegetation. This is especially problematic when comparisons are being made across different soil types that may reflect different amounts of light in the red and near infrared wavelengths (i.e., soils with different brightness values). The soil-adjusted vegetation index was developed as a modification of the Normalized Difference Vegetation Index to correct for the influence of soil brightness when vegetative cover is low.

In 1991, the SAVI value of the Korba district ranges between −0.74 and 0.92 with an average SAVI value of 0.09 ± 0.48. The SAVI value of <−0.0078 is designated as others (waterbody/dry fallow/agricultural fallow land), covered with an area of 3.14% (206.44 km^2) (Table 6.7). The SAVI value of scrub/cultivated land ranges

Table 6.7 Analysis of areal characteristics of Soil Adjusted Vegetation Index (SAVI) of Korba district in 1991

Parameters	SAVI ranges	Area (km^2)	Percent	Average SAVI	Std. dev.
Others	<−0.0078	206.44	3.14	−0.4	0.2
Scrub/cultivated land	−0.0077 to −0.0029	21.74	0.33	−0.01	0.02
Open forest	−0.0028 to 0.35	2302.68	34.98	0.17	0.09
Degraded forest	0.36–0.45	1972.34	29.96	0.42	0.06
Dense forest	>0.46	2080.48	31.60	0.72	0.12

Table 6.8 Analysis of areal characteristics of Soil Adjusted Vegetation Index (SAVI) of Korba district in 2019

Parameters	SAVI ranges	Area (km^2)	Percent	Average SAVI	Std. dev.
Others	<−0.0078	154.48	2.35	−0.26	0.11
Scrub/cultivated land	−0.0077 to −0.0029	24.30	0.37	−0.02	0.03
Open forest	−0.0028 to 0.35	2282.57	34.67	0.19	0.09
Degraded forest	0.36–0.45	2179.52	33.10	0.40	0.03
Dense forest	>0.46	1942.82	29.51	0.62	0.10

from −0.0077 to −0.0029 (mean ± S.D.—0.01 ± 0.02), covered with an area of 0.33% (21.74 km^2). The open forest area covered by 34.98% (2302.68 km^2), and the SAVI value ranges from −0.0028 to 0.35 (mean ± S.D.—0.17 ± 0.09). The SAVI value of degraded forest ranges between 0.36 and 0.45 (mean ± S.D.—0.42 ± 0.06), covered by an area of 29.96% (1972.34 km^2). The dense forest having SAVI value of more than 0.46, covered by 31.60% (2080.48 km^2). Figure 6.4a illustrates the spatial distribution of forest cover derived though SAVI. Results shows eastern and west part of the district having high dense forest cover. Degraded forest is mainly distributed in the north-east and south east of the study area. The small patches of open forest are mainly found in the central north of the study area.

Table 6.8 shows the areal distribution of forest cover estimated through SAVI in 2019. The average SAVI of dense forest is calculated as 0.62 ± 0.1, covered by 29.51% (1942.82 km^2) of the district. The mean SAVI value of degraded forest is estimated as 0.40 ± 0.03 (0.36–0.45), enclosed by an area of 33.10% (2179.52 km^2). The estimated area of open forest (mean ± S.D.—0.19 ± 0.09) in the study area is calculated as 2282.57 km^2 which is 34.67% of the district. Figure 6.4b portrays the spatial coverage of forest of Korba district in 2019. Result of the analysis also shows the degraded open forest cover are increased in the eastern, south-east and northern part of the district.

6.3.5 *Modified Soil Adjusted Vegetation Index2 (MSAVI2)*

The Modified Soil Adjusted Vegetation Index2 (MSAVI2) for the soil is a vegetation that changes the soil to cover areas with a high degree of exposed soil surface and which seeks to resolve some of the limitation of NDVI. The issues with the original SAVI, which is based on the amount of plants in the field studied, was that it was important to define the soil-brightness correction factor (L). It not only prompted the majority of people to use the default L value of 0.5, but also created a circular logic issue that allows them to know the vegetation amount/cover before using SAVI that was supposed to give them details on how much vegetation there was. The MSAVI2 output is a new image layer that represents green vegetation with values from +1 and −1. MSAVI2 has often been used as an input layer for measuring the ground cover or plant groups in a sequence of field study (Chen et al. 2000), biomass index, plots index, and/or leaf area index (Phillips et al. 2009).

The MSAVI2 value of 1991 in the study area ranges between −1.85 and 0.75 (mean ± S.D. −0.55 ± 0.76). The MSAVI2 of dense forest is calculated as more than 0.47, covered with an area of 27.64% (1819.72 km^2). The MSAVI2 value of degraded forest ranges between 0.39 and 0.46 (mean ± S.D. 0.44 ± 0.05), covered by 32.91%. The open forest area covered by 35.60% (2343.63 km^2) with an average MSAVI2 value of 0.21 ± 0.09 (Table 6.9). The MSAVI2 value of scrub/cultivated land varied from −0.0067 to −0.00074 (mean ± S.D.—0.001 ± 0.03), enveloped by an area of 0.44% (28.91 km^2). Figure 6.4a illustrates the spatial distribution of forest cover derived through MSAVI2 in 1991. Results showed the central east and west of the district have high dense forest cover. However, in north and south of the district portrays others land (waterbodies/dry fallow/agricultural fallow land/built-up) with MSAVI2 value of less than −0.0068. In the south-east and north-west, degraded and open forest cover have been found.

The MSAVI2 value of 2019 in the Korba district ranges from −0.84 to 0.69 (mean ± Stdev −0.08 ± 0.44). Areal distribution of forest cover types is represented in Table 6.10. Results showed dense forest area covered by 27.16% (1788.18 km^2), followed by 35.25% (2320.63 km^2) of degraded forest and open forest by 34.59% (2277.17

Table 6.9 Analysis of areal characteristics of Modified Soil Adjusted Vegetation Index2 (MSAVI2) of Korba district in 1991

Parameters	MSAVI2 ranges	Area (km^2)	Percent	Average MSAVI2	Std. dev.
Others	<−0.0068	224.40	3.41	−0.95	0.52
Scrub/cultivated land	−0.0067 to −0.00074	28.91	0.44	−0.001	0.03
Open forest	−0.00073 to 0.38	2343.63	35.60	0.21	0.09
Degraded forest	0.39–0.46	2167.02	32.91	0.44	0.05
Dense forest	>0.47	1819.72	27.64	0.64	0.07

Table 6.10 Analysis of areal characteristics of Modified Soil Adjusted Vegetation Index2 (MSAVI2) of Korba district in 2019

Parameters	MSAVI ranges	Area (km^2)	Percent	Average MSAVI	Std. dev.
Others	<−0.0068	169.16	2.57	−0.46	0.23
Scrub/cultivated land	−0.0067 to −0.00074	28.56	0.43	−0.03	0.02
Open forest	−0.00073 to 0.38	2277.17	34.59	0.20	0.11
Degraded forest	0.39–0.46	2320.63	35.25	0.43	0.03
Dense forest	>0.47	1788.18	27.16	0.58	0.07

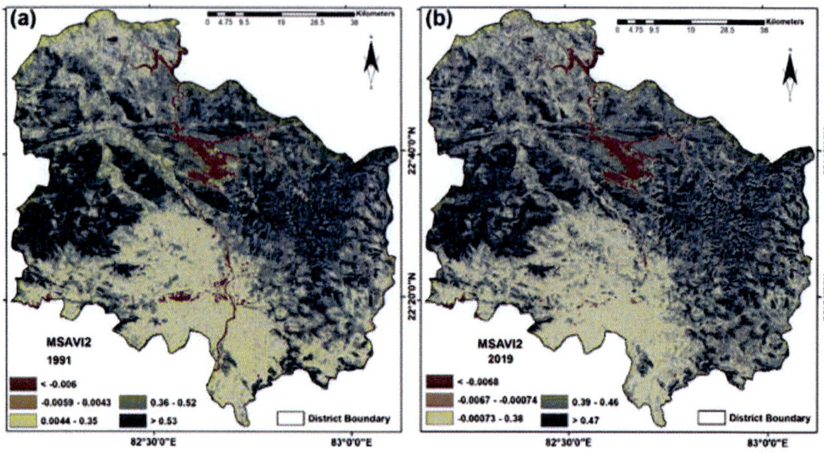

Fig. 6.5 Modified Soil Adjusted Vegetation Index2 (MSAVI2) map of Korba district in Chhattisgarh, **a** 1991 and **b** 2019

km^2). The average MSAVI2 value waterbody/dry fallow/agricultural fallow land is calculated as -0.46 ± 0.23, followed by scrub/cultivated land of -0.03 ± 0.02, open forest of 0.20 ± 0.11. The average MSAVI2 value of degraded forest is calculated as 0.43 ± 0.05, followed by dense forest of 0.64 ± 0.07. The spatial distribution of forest cover is represented by Fig. 6.4b. Results also showed the dense forest cover is portrayed in the west of the district. Although degraded and open forest are mainly observed in north and west of the district and adjacent region of dense forest cover (Fig. 6.5).

6.3.6 Estimation and Spatial Variation of Forest Degradation

The green biomass rates of semiarid grasslands are determined by vegetation indicators derived by remotely sensed data. Nevertheless, the degree to which the knowledge is integrated has a significant effect on the process used. Change detection analysis of forest covered have been performed through NDVI, TNDVI, SAVI and MSAVI2 (Fig. 6.6). Green shades of the maps are represented the positive growth and red shades of the maps are portrays the negative growth of forest cover. The output derived through NDVI, maximum positive changes are observed in the north and south of the district (Fig. 6.6a). Areal analysis of dense forest cover showed 0.03% positive growth for dense forest 0.001 of negative growth during the period between 1991 and 2019 (Table 6.11). The analysis of degraded forest cover shows 0.77% of positive growth and 0.09% of negative growth. The results also showed

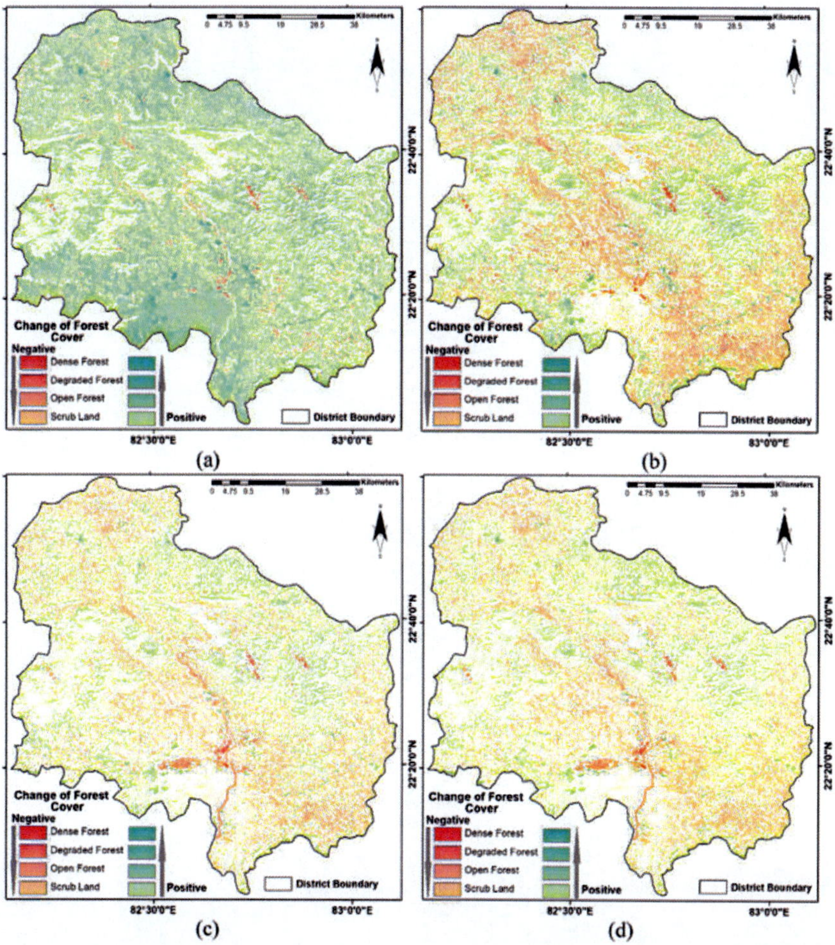

Fig. 6.6 Change of forest coverage calculated through, **a** NDVI, **b** TNDVI, **c** SAVI and **d** MSAVI technique during 1991–2019

11.91% positive growth of open forest and 0.58% of negative growth in the study area. However, the maximum positive growth is observed for the scrub/cultivated land (42.11%) during this study period. Moreover, the negative changes are observed in small pockets of central and west part of the district. The results of TNDVI portrays in negative changes in north, central and south east of the study area. However, the positive growth of forest covers is found in the south-west and north-east of the study area (Fig. 6.6b). The negative growth of dense forest land characteristics is estimated as 0.1% and the estimated positive growth is 0.01%. The estimated negative growth of scrub/cultivated is 17.61%; whereas, the calculated positive growth of scrub/cultivated land cover is 19.45% during the entire study period (1991–2019).

Table 6.11 Change detection analysis of forest cover of Korba district during the period between 1991 and 2019

Growth pattern	Type of forest cover	NDVI			TNDVI			SAVI			MSAVI		
		No. of pixel	Area	Percent	No. of pixel	Area	Percent	No. of pixel	Area	Percent	No. of pixel	Area	Percent
Negative growth	Dense forest	79	0.07	0.001	7045	6.34	0.10	2477	0.99	0.02	2221	2.00	0.02
	Degraded forest	6561	5.90	0.09	33,421	30.08	0.46	6919	2.77	0.06	6219	5.60	0.05
	Open forest	42,457	38.21	0.58	204,007	183.61	2.79	121,217	48.49	1.02	106,309	95.68	0.89
	Scrub/cultivated land	358,154	322.34	4.89	1,289,208	1160.29	17.61	5,560,875	2224.35	46.84	5,509,614	4958.65	46.38
Positive growth	Scrub/cultivated land	3,083,336	2775.00	42.11	1,424,518	1282.07	19.45	1,029,712	411.88	8.67	892,591	803.33	7.51
	Open forest	872,207	784.99	11.91	167,694	150.92	2.29	64,579	25.83	0.54	40,186	36.17	0.34
	Degraded forest	56,301	50.67	0.77	12,848	11.56	0.18	2857	1.14	0.02	2185	1.97	0.02
	Dense forest	2223	2.00	0.03	665	0.60	0.01	248	0.10	0.001	158	0.14	0.001

The results also showed the positive growth of degraded forest cover is 0.46%; while the estimated negative growth is calculated as 0.18%. The negative growth of open forest cover is calculated as 2.79% and the recorded positive growth is 2.29% (Table 6.11).

Figure 6.6c portrays the change detection analysis showed negative changes in central, north and south-east of the study area. The positive growth of forest cover change derived through SAVI, showed in north-east and small pockets of south-west in the study site. Change detection analysis illustrated negative growth of 0.02 and 0.001% of positive growth of dense forest cover. The positive growth of degraded forest cover is calculated as 0.02 and 0.06% of negative growth. The estimated positive growth of open forest is 0.54 and 1.02% of negative growth. However, the positive growth of scrub/cultivated land is estimated as 8.67 and 46.84% of negative growth. The change detection analysis though MSAVI2 is illustrated in Fig. 6.6d. The output showed negative change of forest cover in the south-east, north and central part of the district. Moreover, the positive change of forest cover is illustrated in the north-east and small pockets of south-west of the district. The derived output shows 0.02% negative growth of dense forest cover and 0.001% positive growth. The positive growth of degraded forest cover represents 0.02%, while the estimated negative growth is calculated as 0.05%. The analysis of open forest shows 0.89% of negative growth and 0.34% of positive growth. Moreover, 46.38% negative growth is calculated for the scrub/cultivated land and 7.51% of positive growth is estimated during the study period (Table 6.11).

6.4 Conclusion

Degradation of forests is a severe problem for the climate, culture and the economy. A degraded forest provides products and services from a certain area in a reduced way and preserves only minimal biological diversity. There are many and varying views of forest destruction, depending on the driver and most important products or services. Vegetation indices resulting from remotely sensed data can be calculated on semiarid grasslands with green biomass levels. Nevertheless, the degree to which the knowledge is integrated has a significant effect on the process used. Former-forest land significantly affected by excessive farming, inadequate management, frequent fires, pasture or other troubling uses or land use induced by excessive farming, soil and vegetation prevents or seriously delays recovery after abandonment. This research has examined the relationships between four vegetation indices and a fraction of vegetation. TNDVI is most susceptible to variability in the fraction of vegetation cover. The precision of SAVI and MSAVI2 was very close for the first class of indices. The NDVI is also the least reliable one. The results show that NDVI scaling can produce partial results when dealing with inhomogeneous because it assigns the whole pixel to one of the class features available. Since the indices do not have the strong determination coefficient separately, it is important to boost accuracy after combining them in multivariate mode. Thus, for the fraction estimation of vegetation cover, the established integrated model is recommended in drylands. The accuracy decreases and it is important to be careful to use these indices when the atmosphere in the research area is dry with the use of vegetation indices based on green bands in dry areas.

References

Abate S (2011) Evaluating the land use and land cover dynamics in Borena Woreda of South Wollo highlands, Ethiopia. J Sustain Dev Afr 13(1):87–105

Baret F, Guyot G (1991) Potentials and limits of vegetation indices for LAI and APAR assessment. Remote Sens Environ 35:161–173

Chen J, Chen WJ, Liu J, Cihlar J, Gray S (2000) Annual carbon balance of Canada's forests during 1895–1996, Global Biogeochem. Cycles 14(3):839–849. https://doi.org/10.1029/1999GB001207

Chen WJ, Chen JM, Price DT, Cihlar J (2002) Effects of stand age on net primary productivity of boreal black spruce forests in Ontario, Canada, Can. J For Res 32(5):833–842. https://doi.org/10.1139/x01-165

Chen JM, Deng F, Chen MZ (2006) Locally adjusted cubic-spline capping for reconstructing seasonal trajectories of a satellite-derived surface parameter, IEEE Trans Geosci Remote Sens 44(8):2230–2238. https://doi.org/10.1109/TGRS.2006.872089

FAO (Food and Agricultural Organization) (2006) Global forest resources assessment. 2005 FAO forestry paper no 147. UNFAO, Rome

FAO (Food and Agricultural Organization) (2007) Definitional issues related to reducing emission from deforestation in developing countries. Forests and climate change working paper no 5. UNFAO, Rome

Gao BC (1996) NDWI—a normalized difference water index for remote sensing of vegetation liquid water from space. Remote Sens Environ 58:257–266

Grecchi RC, Beuchle R, Shimabukuro YE, Aragão LEOC, Arai E, Simonetti D, Achard F (2017) An integrated remote sensing and GIS approach for monitoring areas affected by selective logging: a case study in northern Mato Grosso, Brazilian Amazon. Int J Appl Earth Observ Geoinf 61:70–80

Herold M, Román-Cuesta RM, Mollicone D, Hirata Y, Van Laake P, Asner GP, Souza C, Skutsch M, Avitabile V, Macdicken K (2011) Options for monitoring and estimating historical carbon emissions from forest degradation in the context of REDD+. Carbon Balance Manage 6:13

Huete A, Didan K, Miura T, Rodriguez E, Gao X, Ferreira L (2002) Overview of the radiometric and biophysical performance of the MODIS vegetation indices. Remote Sens Environ 83:195–213

Indian Network for Climate Change Assessment (INCCA) (2010) Report 2. Climate Change and India: a 4 × 4 assessment—a sectoral and regional analysis for 2030s. Ministry of Environment and Forests, Government of India, Nov 2010

Huete A (1988) A soil-adjusted vegetation index (SAVI). Remote Sens Environ 25:295–309

Intergovernmental Panel on Climate Change (IPCC) (2007) Climate Change 2007: Mitigation. In: Metz B, Davidson O R, Bosch P R, Dave R and Meyer LA (eds) Contribution of working group III to the fourth assessment report of the intergovernmental panel on climate change. Cambridge University Press, Cambridge. Available at www.ipcc.ch/pdf/assessment-report/ar4/wg3/ar4-wg3-ts.pdf

Ju JC, Roy DP (2008) The availability of cloud-free landsat ETM+ data over the conterminous United States and globally. Remote Sens Environ 112:1196–1211

Jubanski J, Ballhorn U, Kronseder K, Franke J, Siegert F (2013) Detection of large above-ground biomass variability in lowland forest ecosystems by airborne LiDAR. Biogeosciences 10:3917–3930

Key CH, Benson NC (2006) Landscape assessment (LA). FIREMON: fire effects monitoring and inventory system; Gen. Tech. Rep. RMRS-GTR-164-CD, Technical Report; U.S. Department of Agriculture, Forest Service, Rocky Mountain Research Station: Fort Collins, CO, USA

Lunetta RS, Ediriwickrema J, Johnson DM, Lyon JG, McKerrow A (2002) Impacts of vegetation dynamics on the identification of land-cover change in a biologically complex community in North Carolina, USA. Remote Sens Environ 82:258–270

Lyon JG, Yuan D, Lunetta RS, Elvidge CD (1998) A change detection experiment using vegetation indices. Photogramm Eng Remote Sens 64(2):143–150

Margono BA, Turubanova S, Zhuravleva I, Potapov P, Tyukavina A, Baccini A, Goetz S, Hansen MC (2012) Mapping and monitoring deforestation and forest degradation in Sumatra (Indonesia) using Landsat time series data sets from 1990 to 2010. Environ Res Lett 7(3)

Moran MS, Jackson RD, Slater PN, Teillet PM (1992) Evaluation of simplified procedures for retrieval of land surface reflectance factors from satellite sensor output. Remote Sens Environ 41:169–184

Phillips OL, Aragão LEOC, Lewis SL, Fisher JB, Lloyd J, López-González G, Malhi Y, Monteagudo A, Peacock J et al (2009) Drought sensitivity of the Amazon rainforest. Science 323:1344–1347

Potapov P, Yaroshenko A, Turubanova S, Dubinin M, Laestadius L, Thies C, Karpachevskiy M (2008) Mapping the world's intact forest landscapes by remote sensing. Ecol Soc 13(2):51

Qi J, Kerr Y, Chehbouni A (1994) External factor consideration in vegetation index development. In: Proceedings of physical measurements and signatures in remote sensing, ISPRS, pp 723–730

Rahman MM, Khan MNI, Hoque AKF, Ahmed I (2015) Carbon stock in the Sundarbans mangrove forest: spatial variations in vegetation types and salinity zones. Wetlands Ecol Manage 23:269–283. https://doi.org/10.1007/s11273-014-9379-x

Ravindranath NH, Joshi NV, Sukumar R, Saxena A (2006) Impact of climate change on forests in India. Curr Sci 90(3):354–361

Romero-Sanchez ME, Ponce-Hernandez R (2017) Assessing and monitoring forest degradation in a deciduous tropical forest in Mexico via remote sensing indicators. Forests 8:302. https://doi.org/10.3390/f8090302

Rouse JWJ, Hass RH, Schell JA, Deering DW (1974) Monitoring vegetation systems in the Great Plains with erts. NASA Special Publication, Washington, DC, pp 309–317

Rouse JW, Haas RH, Schell JA, Deering DW (1973) Monitoring the vernal advancement and retrogradation (green wave effect) of natural vegetation. Technical report; NASA/GSFC: Greenbelt, MD, USA

Shimabukuro YE, Beuchle R, Grecchi RC, Achard F (2014) Assessment of forest degradation in Brazilian Amazon due to selective logging and fires using time series of fraction images derived from Landsat ETM+ images. Remote Sens Lett 5:773–782

Simula M (2009) Towards defining forest degradation: comparative analysis of existing definitions. Forest resources assessment working paper no 154. UNFAO, Rome

Souza CM Jr (2013) Monitoring of forest degradation: a review of methods in the Amazon Basin. In: Achard F, Hansen MC (eds) Global forest monitoring from Earth observation. CRC Press, Boca Raton

Souza C, Firestone L, Moreira Silva L, Roberts D (2003) Mapping forest degradation in the Eastern Amazon from SPOT 4 through spectral mixture models. Remote Sens Environ 87:494–506

Tucker CJ (1979) Red and photographic infrared linear combinations for monitoring vegetation. Remote Sens Environ 8:127–150

USGS. U.S. Geological Survey. Available online: https://landsat.usgs.gov/what-are-band-designationslandsat-satellites (accessed on 1 June 2017)

Wulder MA, Masek JG, Cohen WB, Loveland TR, Woodcock CE (2012) Opening the archive: how free data has enabled the science and monitoring promise of Landsat. Remote Sens Environ 122:2–10

Yang DM, Li GY, Sun SC (2008) The generality of leaf size versus number trade-off in temperate woody species. Ann Bot 102:623–629. https://doi.org/10.1093/aob/mcn135

Chapter 7
Comparison of Sentinel-2 Multispectral Imager (MSI) and Landsat 8 Operational Land Imager (OLI) for Vegetation Monitoring

Santanu Ghosh, Debabrata Behera, S. Jayakumar, and Pulakesh Das

Abstract The availability of the coarse to moderate resolution no-cost remote sensing data and advances in image processing algorithms have exponentially increased the usage of geo-spatial technology in the last few decades. The latest Sentinel-2 Multispectral Imager (MSI) provides the surface reflectance data in VNIR and SWIR ranges since 2015 at higher spatial, temporal, and spectral resolution compared to the Landsat multispectral sensors, which have been providing such data since 1970s. The symmetry in spectral bands, sensor's spectral response, spatial and radiometric resolution of Sentinel-2 MSI with the Landsat-8 OLI (Operational Land Imager) enables their integrated and complimentary use. In this study, we have compared the surface reflectance and vegetation indices (such as NDVI and EVI) values obtained from MSI and OLI sensors in four homogeneous land use land cover (LULC) features as cropland, agriculture fallow land, dense forest and open forest. The assessment is carried out for pre- and post-monsoon seasons over the Banki sub-division region of Cuttack district, Odisha, India. For all the LULC classes, high similarity is observed in the surface reflectance values in each band except NIR, green and red band. Similarly, for both the vegetation indices derived from Landsat 8 and Sentinel-2 data, high correlation with lower RMSE is observed for all the LULC classes. The correlation (R^2) for cropland varied between 0.87 and 0.96, which varied between 0.56 and 0.97 for agriculture fallow, between 0.58 and 0.9 for dense forest, and between 0.68 and 0.87 for open forest. The surface reflectance pattern obtained for different vegetated features are similar for Landsat 8 and Sentinel-2. However, significantly a higher surface reflectance is observed for Landsat 8 in NIR band followed by red and green bands, where the differences are low for blue, SWIR1 and SWIR2 bands. The comparative assessment of indices suggests a higher correlation in values between Landsat 8 and Sentinel-2 for homogeneous features compared to

S. Ghosh · P. Das (✉)
Department of Remote Sensing and GIS, Vidyasagar University, Midnapore, West Bengal, India
e-mail: das.pulok2011@gmail.com

D. Behera · S. Jayakumar
Department of Ecology and Environmental Sciences, Pondicherry University, Kalapet, Pondicherry, India

P. K. Shit et al. (eds.), *Spatial Modeling in Forest Resources Management*, Environmental Science and Engineering,
https://doi.org/10.1007/978-3-030-56542-8_7

heterogeneous class as forest. Thus, the integrated or complementary use of Landsat 8 and Sentinel-2 for heterogeneous features may induce some biases with a limited accuracy.

Keywords Landsat 8 · Sentinel-2 · Regression · Data integration · Surface reflectance · Vegetation index

7.1 Introduction

The optical remote sensing data from the Sentinel-2 and Landsat series are the two widely used open source medium-to-high spatial resolution multispectral data, which are used in generating various primary and secondary data layers on the land surface features (Hansen and Loveland 2012; Banskota et al. 2014; Gómez et al. 2016; Hill 2013; Hedley et al. 2018). Whereas, the combined use of multi-resolution data obtained from different satellite sensors enable many opportunities including the long-term monitoring and modeling of various natural, anthropogenic and integrated processes. Facing a scenario with an ever-increasing number of the earth observing satellite sensors, the combined application of various remotely sensed data has gain significant attention. The Landsat program by NASA is one of the oldest programs provides the land surface reflectance data in visual and near-infrared (VNIR) and thermal infrared ranges at no-cost from the past five decades (since 1972) and continuing. Since 1982, the Landsat sensors collect the surface reflectance data at 30 m spatial resolution in VNIR range at 24 days intervals with significant radiometric accuracy. Recently, the Sentinel-2 mission by ESA (European Space Agency) with an improved spatial (10, 20 and 60 m spatial resolution) and spectral resolutions (several additional bands in the vegetation red edge zone in comparison to Landsat bands) with 5 days temporal revisit has drawn the attention of the multi-spectral data users. The Sentinel-2A satellite was launched in June 2015, whereas the Sentinel-2B satellite was launch in March 2017, which were boarded with a push broom multi-spectral imager (MSI). The two Sentinel-2 satellites have introduced a significant potential for synergistic use to create unprecedented opportunities for timely and accurate observation of earth surface resources and dynamics (Frampton et al. 2013; Immitzer et al. 2016). With other several advantages, the higher temporal resolution of Sentinel-2 data compared to Landsat data has increased the possibility of getting cloud free satellite images even in tropical regions. Moreover, the opportunity of integrating similar and especially open source remote sensing data from different satellites/sensors is highly useful for long-term and consistent monitoring while enhancing the spatial and temporal resolution. The MSI records the incoming radiations in 13 spectral bands with a swath width of 290 km. The central wavelength and band width of Sentinel-2A and 2B is highly similar, which is analogous to the Landsat 8 (OLI: Operational Land Imager) sensor. This has attracted the researchers worldwide to perform comparative assessment between MSI and OLI for many applications in different landscapes. Several studies have confirmed the potential use of the integrated data, where the data conversion between Sentinel-2 MSI and Landsat 8 OLI is identified as one of the challenging tasks (Mandanici and Bitelli 2016; Lessio et al. 2017; Vuolo et al. 2016). The data standardization or integration between MSI and OLI data requires suitable conversion factors to enable the combined use for time series analysis at much higher spatial and temporal resolutions. This allows better integration and modelling of the satellite data derived parameters and time series ground observations collected at daily or sub-daily scale.

The mapping and monitoring of vegetation at different spatio-temporal scales is one of the prime application areas of earth observing satellite remote sensing. The vegetation spectral response in different wavelength range of the electromagnetic spectrum exhibits the biophysical, structural and biochemical properties of vegetation. The sensors operating in the different spectral ranges, records the radiated energy from vegetation and enables the spatio-temporal characterization of various plant and their attributes integrating the ground observations. Various panchromatic, broadband and narrow band (hyperspectral) VNIR and thermal sensors are used to capture the minor differences in spectral response due to species diversity and their composition and characteristics. In broadband remote sensing, the spectral width and sensor's spectral response determines the wavelength range in which the incoming radiation will be recorded and stored. The high absorption of incident solar radiation in the visual range (400–700 nm; Photosynthetically Active Radiation) by green vegetated features leads to lower emittance or reflection, whereas the majority of the incident energy in near-infrared (NIR) range is reflected by vegetation (through multiple reflection inside mesophyll cells) and leads to higher reflectance values. Moreover, the water absorption bands near to 1400 and 1900 nm induce peak absorption or sharp low reflectance depending on the leaf water content. The surface reflectance captured in sensors operating in blue, red and NIR regions are very sensitive to vegetation greenness, phenology, and structure; whereas, the sensors operate in SWIR region are more sensitive to leaf moisture content and chemical constituents (e.g., proteins, lignin, cellulose, etc.) including the below canopy soil reflectance. The contrasting surface reflectance for vegetation in NIR and red bands produces high gradient known as red edge region, which is used to be exaggerated by applying suitable spectral enhancement techniques. The vegetation indices such as Ratio Vegetation Index or simply Ratio Index (RVI = NIR/red), Difference Vegetation Index (DVI = NIR-red), Normalized Difference Vegetation Index (NDVI), etc. are widely used indices to enhance the vegetation cover in remote sensing imageries and allows better mapping and monitoring (Xie et al. 2010; Krishnaswamy et al. 2009). Enhanced Vegetation Index (EVI) is a modified index over NDVI, incorporates the background soil correction factor and atmospheric resistance through blue band. It has been proved to be more sensitive to higher vegetation density over NDVI, which encounters saturation in high canopy density areas (Matushita et al. 2007). The range of values for both the NDVI and EVI lies between -1 and $+1$, where the higher positive value indicates higher vegetation greenness and density; lower positive value corresponds to less dense or partially green cover area and fallow lands. On the contrary, the negative NDVI value highlights the water body and moist areas in an image. The ratio of spectral bands used in NDVI and EVI reduces the errors due to topographic variations, hill shadow, and makes less sensitive to uncertainties in atmospheric variations and alteration in satellite viewing angle (Li et al. 2014; Anderson et al. 2011; Steven et al. 2003; Teillet and Ren 2008). Such indices have numerous applications in mapping and monitoring vegetated features and their attributes including forest cover type and species diversity, forest density, tree canopy cover, leaf area index (LAI), evapotranspiration, ecosystem productivity, forest fire (FSI 2019; Behera et al. 2018; French et al. 2008; Delegido et al. 2011; Mallinis et al. 2017; Bawa et al. 2002); crop area, type and productivity, intensity (Ghosh et al. 2017; Meng et al. 2013; Lessio et al. 2017; Son et al. 2014; Immitzer et al. 2016); drought monitoring, crop water stress, soil moisture estimation (Zhang and Zhou 2016; Mkhabela et al. 2011), etc. Such a diverse application surge the integration of Landsat and Sentinel optical remote sensing data for long-term monitoring at higher temporal scale with improvement in spatial complexities, most importantly in heterogeneous landscapes and topographically rugged terrain, where the microclimate rapidly varies and leads to highly diverse eco-system (Tucker 1979; Huete et al. 2002).

$$NDVI = (NIR - red)/(NIR + red)$$

$$EVI = G * (NIR - red)/((NIR + C1 * red) - (C2 * blue + L))$$

where G = Gain factor, C1, C2 = Coefficients, L = Canopy background correction factor.

Cross-comparisons of surface reflectance and indices derived from different sensors have been carried out in several studies. Kobayashi et al. (2007) compared multi-sensor reflectance, NDVI and EVI derived from the coarse resolution Moderate Resolution Imaging Spectroradiometer (MODIS) and high-resolution Satellite Pour l'Observation de la Terre Vegetation (SPOT-VGT) data and observed highly similar seasonality. Several studies have been also carried out, where the inter-comparison of sensors from same family or its precedent has been performed. Li et al. (2014) studied the subtle differences in Landsat 8 OLI and Landsat 7 ETM+ sensors while comparing indices (NDVI), Normalized Burned Ratio (NBR), Land Surface Water Index (LSWI), and reported high linear correlation (R = 0.90). Similar studies have been carried out for comparing the Landsat and latest Sentinel-2 optical remote sensing data. Mandanici and Bitelli (2016) studied the homogeneity in surface reflectance values for the homologous bands between Landsat 8 OLI and Sentinel-2 MSI, and derived indices [NDVI, Normalized difference water index (NDWI) and FII (Ferrous iron index)], where they computed the coefficients through linear regression and Pearson correlation coefficient, and endorsed the potency of combined use. Evaluating the Pearson's correlation coefficient, Lessio et al. (2017) tested the synergistic application of Landsat 8 and Sentinel-2 data for improved crop monitoring using the NDVI and NDWI index and observed differences in index values depending on the time lag. While comparing Landsat 8 and Sentinel-2 data over six different sites in Europe, Vuolo et al. (2016) found that all six homologue bands showed a good correlation of 0.90 with RMSE ranging from 0.023 to 0.043. Claverie et al. (2017) developed a methodology to produce Harmonized Landsat-8 Sentinel-2 (HLS) products, i.e., Sentinel 10m (S10), Sentinel 30m (S30), Landsat 30m (L30) to enable global observation at every 2–3 days with medium (<30 m) spatial resolution. Forkuor et al. (2018) compared the land use land cover (LULC) classification accuracy adding the red edge band for Sentinel-2 data and showed that the overall accuracy increases 4–5% compared to Landsat 8 data derived LULC map. Astola et al. (2019) performed comparison between these two sensors in deriving forest variables in a boreal forest and reported the addition of red edge band in Sentinel-2 data outperformed Landsat 8 data. Although a numbers of studies have been carried out evaluating the data similarity between Sentinel-2 and Landsat 8 data, no such study has been reported in any Indian landscape. With this research gap, the present study aims to study band similarities between Landsat 8 OLI and Sentinel-2 MSI sensors and vegetation indices to derive the linear standardization or normalization factors over a sub-tropical region of India.

7.2 Methodology

7.2.1 Study Area

The study was carried out in Banki sub-division of Cuttack district, Odisha state (Fig. 7.1). This region is located between 20.15° N and 20.29° N and between 85.20° E and 85.45° E on the southern bank of Mahanadi river. It has an average elevation of 48 m above msl and the Banki city is located 58 km away from Bhubaneswar, the state capital. In summer, the maximum temperature reaches around 42 °C while minimum temperature is around 11 °C during winter. The average annual rainfall is 1500 mm (https://www.wildlife.odisha.gov.in). Southwest Monsoon is the major source of water in the study area from June to September. The dominant vegetation in the study area mostly comprising of forest, cropland, settlement, and water body (Sahoo et al. 2017; Das et al. 2019). A part of Chandaka-Damapara wildlife sanctuary is situated inside the study area.

7.2.2 Data Acquisition

7.2.2.1 Field Data Collection

We have studied the dominated vegetated land surface features observed in the study area, i.e., dense forest, open forest, cropland and agricultural fallow land to compare the mean surface reflectance between the Landsat 8 and Sentinel-2 data. Field visits were carried out to collect the geo-location data (latitude and longitude) along with

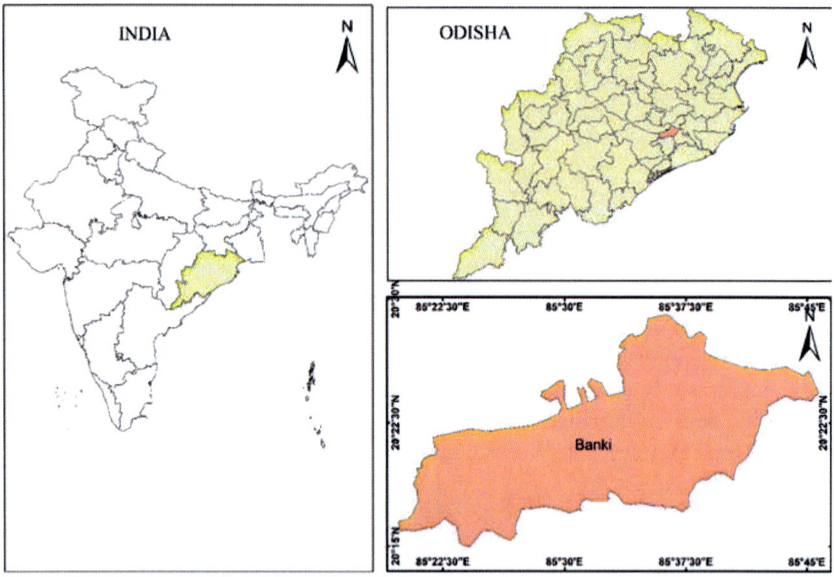

Fig. 7.1 Study area location

field photographs for each LULC category (Fig. 7.2). Moreover, few data points were added through visual image interpretation, which is being considered one of the proven methods of image analysis and feature identification (Roy et al. 2015; Das et al. 2019). The homogeneous land surface patches were chosen for field sampling and visual image interpretation to reduce the errors due to spatial heterogeneity, mix or boundary pixels and differences in spatial resolution.

Fig. 7.2 Field photographs; **a** Cropland, **b** dense forest (Sal forest), **c** open forest (mixed), **d** agricultural fallow land, **e** cropland, **f** open forest (mixed)

Table 7.1 Landsat-8 (OLI) and Sentinel-2 (MSI) images used in this study

Acquisition date		Acquisition time (GMT)		Cloud coverage (%)	
Landsat 8	Sentinel-2	OLI time	MSI time	Landsat 8	Sentinel-2
6th Jan	8th Jan	04.44.05	04.51.51	0.01	0
7th Feb	7th Feb	04.43.53	04.49.31	0.02	0
27th Mar	29th Mar	04.43.30	04.46.51	9.17	0
22th Nov	24th Nov	04.44.06	04.51.21	5.06	0

7.2.2.2 Remote Sensing Data Collection and Pre-processing

The Sentinel-2 level 1C Ortho-rectified satellite data was downloaded from the Copernicus Sentinel data hub (https://scihub.copernicus.eu). The images were masked with the study area vector layer. The image was already pre-processed into a Top of the Atmosphere (ToA) reflectance data. Dark object saturation (DOS) correction was applied using the Semi-automatic plugin in QGIS software. The atmospherically corrected Landsat 8 surface reflectance data were procured from the USGS Earth explorer data portal (https://earthexplorer.usgs.gov). The corresponding and overlapping pair of Landsat 8 and Sentinel-2 images were considered, which were having the lower temporal lag (≤3 days) in imaging date and least cloud cover (≤10%) during the pre-monsoon and post-monsoon of 2017 (Table 7.1). For the geometric accuracy assessment between the Landsat 8 and Sentinel-2 data, the distinct features as river boundaries were matched, over the overlapped regions. The cloud cover regions were masked out from further analysis using the quality assessment band.

7.2.3 Comparison Method Based on Spatial Resolution

The surface reflectance and vegetation indices (EVI and NDVI) derived from Landsat 8 and Sentinel-2 data were compared considering similar ground coverage, i.e., the value represents same ground area. The pure pixels of each land use land cover category were considered, and corresponding pixels from Landsat 8 and Sentinel-2 data were selected based on field observation and visual image interpretation. For each pixel of Landsat 8 data with 30 m spatial resolution, the corresponding and nearest nine pixels of Sentinel-2 data having 10 m spatial resolution (blue, green, red and NIR band) were considered. However, it could be noted that there was a minor separation in corner location in the Landsat and Sentinel data. The average values of nine pixels in Sentinel-2 data was equated with one pixel in Landsat 8 data. On the other hand, for SWIR1 and SWIR2 band, the average value of four pixels in Landsat 8 data (30 m) was compared with nine pixels of Sentinel-2 data (20 m).

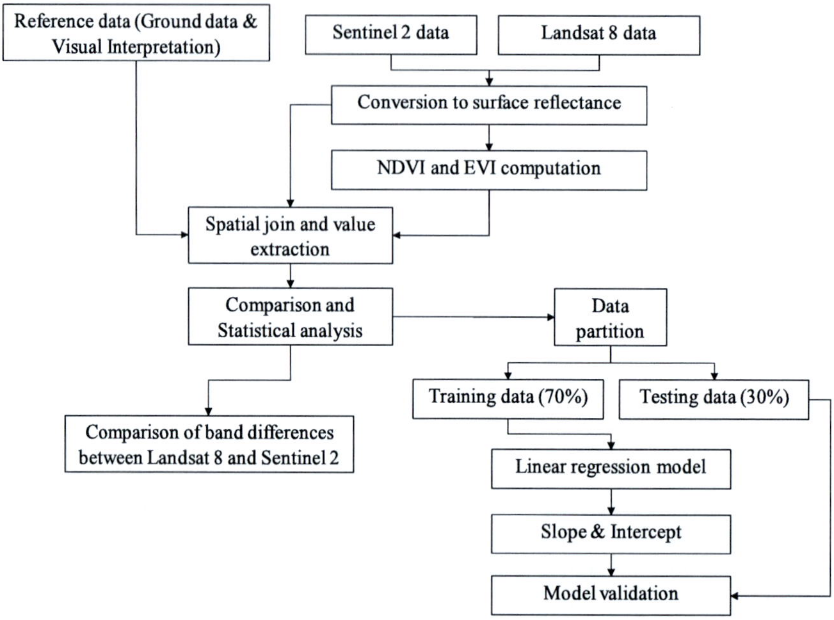

Fig. 7.3 Methodological flow chart

7.2.4 Statistical Analysis

Statistical analysis was carried out using the SPSS software and Microsoft Excel. The total dataset was randomly segregated as training (70%) and testing (30%) using the ArcGIS software. Using training dataset, the linear regression analysis (at 95% confidence interval) was performed to compute parameters (slope and intercept). The derived parameters were then applied in the testing dataset and compared with observed values to demonstrate the accuracy via the coefficient of determination (R^2) and root mean square error (RMSE) values. A methodological flow chart is shown as Fig. 7.3.

7.3 Results and Discussion

7.3.1 Comparison of Landsat-8 OLI and Sentinel-2 Spectral Bands

By discretion, the concurrent surface reflectance bands along with two widely used vegetation indices derived from the Landsat 8 and Sentinel-2 imagery were compared for four LULC types as cropland, agriculture fallow, dense forest and open forest.

The comparison between the average surface reflectance values are graphically represented in Fig. 7.4. The bar diagram shows the similarity between average surface reflectance obtained from the Landsat 8 and Sentinel-2 data for the data acquired on 7th February (similar diagrams were also obtained for other time periods, not shown owing to space limitations). The general pattern indicates a higher average surface reflectance values in Landsat 8 data compared to Sentinel-2 data, except in blue, SWIR1 and SWIR2 band for dense forest and in blue band for open forest and cropland. For cropland, the difference in the average surface reflectance values were minimum or nearly similar in blue, SWIR1 and SWIR bands; whereas, the maximum difference was observed for NIR band (~0.04) followed by green band (~0.03) and red band (~0.02). In case of agriculture fallow, the surface reflectance values in Landsat data were observed higher, where the lower differences were observed (~0.02) in blue, SWIR1 and SWIR2 bands, and highest difference was observed in NIR band (~0.05) followed by green (~0.03) and red (~0.02) band. Similarly, in open and dense forest, the differences in surface reflectance were less in blue, SWIR1 and SWIR2 band (~0.01), which was much higher in NIR band (~0.04) followed by green and red band (~0.02). The result indicates that the average surface reflectance is higher for Landsat 8 compared to Sentinel-2 data for all vegetated features, where the difference is lesser in blue, SWIR1 and SWIR2 bands and highest in NIR band followed by green and red band. If a curve were drawn on the average surface reflectance bar

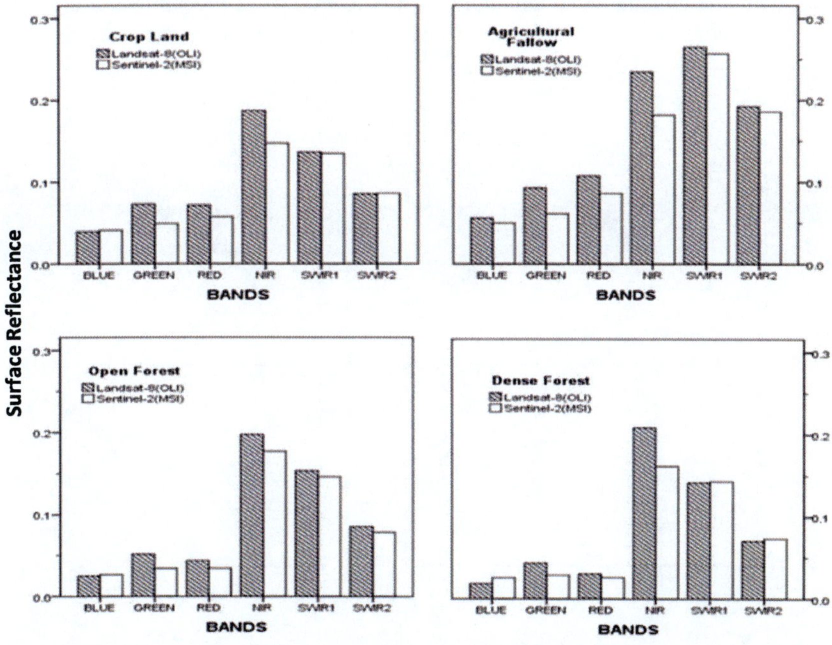

Fig. 7.4 Band to band comparison between the average surface reflectance between Landsat 8 (OLI) and Sentinel-2 (MSI) imagery for four dominated vegetated LULC categories

diagram, it would indicate similar pattern for the both data, indicates the uniform spectral response of the different sensors operating in different wavelength or spectral channel.

7.3.2 Cross-Comparison of Vegetation Indices in Different Land Cover Types

Using the training data points, the scatter diagrams have been drawn to assess the relationship between Landsat 8 and Sentinel-2 for NDVI and EVI (Figs. 7.5 and 7.6). It can be noted that the scatter diagrams have been shown for the data acquired in the month of February, where the rest of diagrams have not been added here due to space limitations. The best-fit regression line has been added in each case to estimate the linear fit equation (slope and intercept) and corresponding similarity (coefficient of determination) was derived. For both sensors, the NDVI values for agriculture fallows varied between 0.22 and 0.7 with the mean value was concentrated around 0.35, whereas few points were seen having NDVI values more than 0.45. The agriculture fallow areas having NDVI values more than 0.45 indicates partial grass cover during the fallow period. In case of cropland, the NDVI values varied between 0.15 and 0.8, where the data points were equally distributed along the NDVI range exhibits adequate data points in each NDVI range or greenness variety. The croplands identified in the study region are Rabi crop areas during the post-monsoon

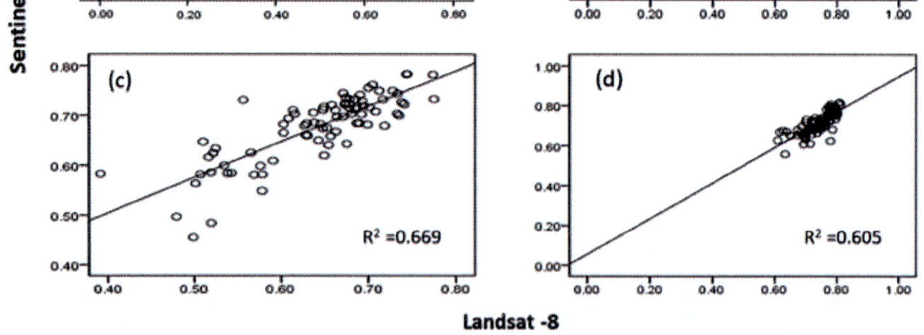

Fig. 7.5 Scatter plot represents NDVI derived from Landsat 8 and Sentinel-2 data for **a** agricultural fallow, **b** crop land, **c** open forest, **d** dense forest

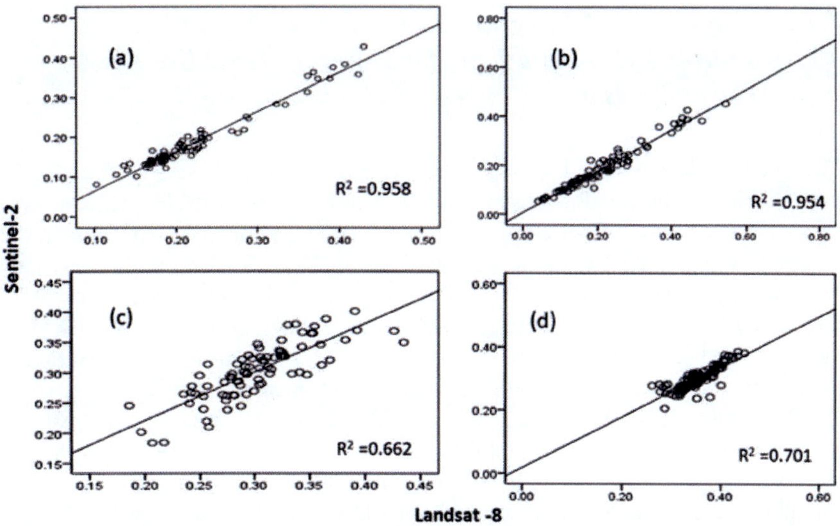

Fig. 7.6 Scatter plot represents EVI derived from Landsat 8 and Sentinel-2 data for **a** agricultural fallow, **b** crop land, **c** open forest, **d** dense forest

season. In case of open forest, the NDVI value ranged between 0.4 and 0.8, which was highly scattered denoting differential greenness identified in the Landsat and Sentinel data. However, the NDVI value range was highly concentrated for dense forest varied between 0.6 and 0.9, denotes the nearly saturated greenness in a pixel owing to the high canopy density in sub-tropical region of the Eastern Ghats. The general pattern of the scatter plots indicates a positive correlation for the four LULC categories, which is obvious. The high R^2 value indicates higher similarity between Landsat 8 and Sentinel-2 data for agriculture fallow ($R^2 = 0.95$) and cropland ($R^2 = 0.96$). However, the relationship was much weaker for the open forest ($R^2 = 0.67$) and dense forest ($R^2 = 0.61$). Similar results were seen for EVI, where high range for agriculture fallow with a concentrated lower value, could be indicating partial grasses during the crop fallow period (Fig. 7.6). The range of EVI value for cropland and open forest was well-distributed, whereas, the range was concentrated for dense forest. The relationship in EVI value between Landsat 8 and Sentinel-2 was highly strong for agriculture fallow ($R^2 = 0.96$) and cropland ($R^2 = 0.95$), which was lower in case of open forest ($R^2 = 0.66$) and dense forest ($R^2 = 0.70$).

7.3.3 *Evaluation of Integral Performance of Vegetation Indices (NDVI and EVI) for Different Land Use Land Cover Types*

The relationship developed through linear regression analysis was verified for its transferability. The relation acquired with 70% data points (training data) were applied in rest of the 30% data points (testing data) and compared with the observed values. The assessment result has been shown in Table 7.2, where the estimated relationship was verified for each month evaluating the R^2 and RMSE values. With NDVI, the R^2 values for cropland varied between 0.87 and 0.96 with the RMSE values varied between 0.03 and 0.08. For agriculture fallow, the R^2 value was higher in January and February (0.95 and 0.96) and lower in March and December (0.71 and 0.66) with the RMSE varied between 0.02 and 0.06. In case of open forest, the R^2 value varied between 0.71 and 0.87 with a RMSE value varied between 0.03 and 0.06 (except 0.16 in December). In case of dense forest, the maximum R^2 value was obtained for January (0.89) followed by March (0.77), December (0.72) and February (0.58) with a comparatively lower RMSE value than open forest varied between 0.02 and 0.03. Nearly similar correlation was observed for EVI for all the land use land cover categories, where the higher R^2 values were observed for cropland and agriculture fallow and a comparatively lower value for open and dense forest.

Table 7.2 Comparison in spectral indices (NDVI and EVI) with testing data points for different land use land cover categories

Land cover	Month	NDVI		EVI	
		R^2	RMSE	R^2	RMSE
Cropland	Jan	0.87	0.05	0.91	0.03
	Feb	0.95	0.03	0.95	0.18
	Mar	0.96	0.08	0.95	0.04
	Dec	0.93	0.06	0.95	0.03
Agricultural fallow	Jan	0.95	0.03	0.92	0.02
	Feb	0.96	0.02	0.97	0.02
	Mar	0.71	0.06	0.56	0.03
	Dec	0.66	0.06	0.73	0.04
Dense forest	Jan	0.89	0.02	0.9	0.02
	Feb	0.58	0.03	0.74	0.02
	Mar	0.77	0.02	0.75	0.01
	Dec	0.72	0.02	0.85	0.01
Open forest	Jan	0.79	0.04	0.79	0.02
	Feb	0.71	0.03	0.68	0.03
	Mar	0.85	0.06	0.76	0.02
	Dec	0.87	0.16	0.82	0.02

The R^2 values were more than 0.91 for cropland, which varied between 0.56 and 0.97 for agriculture fallow. For open forest, the R^2 values varied between 0.68 and 0.82, whereas the range varied between 0.74 and 0.9 for dense forest. The RMSE values for all cases were very low ($\leq$0.04) except for cropland in the month of March (RMSE = 0.18). However, the class and time independent average R^2 values were similar (~0.83) for both NDVI and EVI.

The current study outcome indicates that the surface reflectance values for four different vegetated categories were almost similar in every analogous band with a slightly higher values ($\leq$0.03) in Landsat 8 data compared to Sentinel-2 data, except in NIR band, where the difference was observed very high (0.03–0.05). However, the overall surface reflectance pattern was similar for both the sensors. For vegetation, the surface reflectance difference between Landsat 8 and Sentinel-2 in NIR bands corresponds similar changes in red and green band. This indicates that the use of indices e.g., NDVI and EVI, which utilizes the bands as NIR, red and green band will be similar and comparable. This was corroborated by a range of NDVI values for different land use land cover features, which was almost similar and their strong positive relation. However, in case of open forest, the corresponding NDVI values from Landsat 8 and Sentinel-2 were slightly asymmetric. On the contrary, the relative differences in surface reflectance value between Landsat 8 and Sentinel-2 were more similar in SWIR1, SWIR2 and blue band compared to NIR, red and green band. Thus, indices which use a mixture of bands from two different categories may lead to bias or asymmetry when compared between Landsat 8 and Sentinel-2. In the current study, although the blue band was used along with NIR and red band in the computation of EVI, indicated symmetric values for the different vegetated categories. This could be attributed to the lower contribution from the blue band in EVI computation. Van der Werff and Van der Meer (2016) compared the index values for various indices used for mineral mapping and observed high correlation (0.8) with significant differences in index values from different sensors, which could be indicating the asymmetric differences in surface reflectance from different bands (Table 7.3).

With NDVI, the integration performance between Landsat 8 and Sentinel-2 was the highest for cropland followed by agriculture fallow, open forest and dense forest. For cropland, the high positive correlation during training ($R^2 > 0.95$) and testing ($R^2 > 0.87$) indicates that the NDVI values obtained from Landsat 8 and Sentinel-2 are comparable and can be integrated for combined use. Similarly, in case of agriculture fallow, the high correlation during training ($R^2 > 0.95$) and testing (R^2 varied between 0.66 and 0.96) indicates the compatibility of NDVI values derived from Landsat 8 and Sentinel-2. Comparatively, the lower correlation for open forest ($R^2 = 0.66$ during training and varies between 0.58 and 0.89 during testing) indicates the integrated use of Landsat 8 and Sentinel-2 data could induce some errors. In comparison to open forest, the correlation was higher for dense forest ($R^2 = 0.71$ during training and varies between 0.71 and 0.87), could be indicating lesser chances of errors in the combined or complementarity use of Landsat 8 and Sentinel-2. Nearly similar results were also observed for EVI, where the integration was best modelled for cropland except in February owing to high RMSE value, followed by agriculture fallow, open forest and dense forest. During training the correlation was highest for cropland

Table 7.3 Comparison of Landsat-8 (OLI) and Sentinel-2 (MSI) spectral bands

Landsat-8 (OLI)				Sentinel-2 (MSI)			
Bands name	Central wavelength (μm)	Resolution (m)	Bandwidth (nm)	Bands name	Central wavelength (μm)	Resolution (m)	Bandwidth (nm)
1: Coastal	0.443	30	15.98	1: Coastal aerosol	0.443	60	20
2: Blue	0.482	30	60.04	2: Blue	0.49	10	65
3: Green	0.561	30	57.33	3: Green	0.56	10	35
4: Red	0.655	30	37.47	4: Red	0.665	10	30
5: NIR	0.865	30	28.25	5: Vegetation red edge	0.705	20	15
				6: Vegetation red edge	0.74	20	15
				7: Vegetation red edge	0.783	20	20
				8: NIR	0.842	10	115
				8A: Vegetation red edge	0.865	20	20
6: SWIR1	1.609	30	84.72	11: SWIR	1.61	20	90
7: SWIR2	2.201	30	186.66	12: SWIR	2.19	20	180
9: Cirrus	1.373	30	20.39	10: SWIR-Cirrus	1.375	60	20
8: Pan	0.59	15	172.4				
				9: Water vapour	0.945	60	20
10: TIRS	10.895	100	590				
12: TIRS	12.005	100	1010				

(R^2 = 0.96 during training and varied between 0.91 and 0.95 during testing), which could be indicating the EVI obtained from Landsat 8 and Sentinel-2 is comparable and can be used as a complementary data. For agriculture fallow, the low to high correlation (R^2 = 0.95 during training and varied between 0.56 and 0.97), indicates different chances of errors in corresponding values from Landsat 8 and Sentinel-2 data. In case of dense forest, moderate correlation was observed (R^2 = 0.71 during training and varied between 0.74 and 0.9 during testing), indicates combined use of Landsat 8 and Sentinel-2 data would lead an asymmetric values. Similarly, in case of open forest, the lower correlation (R^2 = 0.66 during training and varied between 0.68 and 0.82) may lead to errors in the values of Landsat 8 and Sentinel-2 when integrated. It can be summarized that the NDVI and EVI values for cropland and agriculture fallow land can be transferred among Landsat 8 and Sentinel-2 with reasonable errors. However, the chances of errors could be much higher for open and dense forest, when the values of Landsat 8 or Sentinel-2 data is transferred to another data and may induce bias during comparative or time-series analysis. Similar studies carried out by Mandanici and Bitelli (2016) for six different land cover types in varied climatic condition and reported similar coefficient of determination values ranging between 0.81 and 0.92 for NDVI and homologous bands. Comparatively a lower correlation was reported by Lessio et al. (2017) obtained the R^2 greater than 0.61 and 0.53 for NDVI and NDWI, respectively. The RMSE value obtained in the study corroborates the values reported by Vuolo et al. (2016) with a correlation of 0.90 with RMSE ranging from 0.023 to 0.043. The results indicate the surface reflectance is higher in NIR, green and red bands for Landsat 8 compared to Sentinel-2 data, which was less in blue, SWIR1 and SWIR2 bands. The study highlights that although, the Landsat 8 and Sentinel-2 data have similar sensors with symmetric spectral response and bandwidth, there are chances of errors when these two are used together. The systematic and minor differences in surface reflectance values in NIR, red and green band would better help in monitoring when vegetation indices are introduced, where chances of errors are minimum for uniform features having low frequency compared to heterogenous features with high frequency.

7.4 Conclusion

In comparison to Landsat, the 5 days revisit of the latest Sentinel-2 satellites with much higher spatial resolution in few important bands (e.g., blue, green, red and NIR) enables improved monitoring the earth surface features and processes, e.g., mapping crop stages and forest fire, where events occur or change at much higher rate. The high spatial resolution of Sentinel-2 enables studies in heterogeneous landscape and high temporal resolution increases chances of getting cloud free scenes and better modelling with ground observed data recorded at daily scale. On the other hand, the Landsat series have global surface reflectance records of past five decades. Thus, the integration bridge between the Landsat inventories and latest Sentinel-2 data would lead to long-term assessment at much higher spatial and temporal resolution. The comparative analysis for the analogous spectral bands operating in nearly similar bandwidth and spectral range indicates high and systematic surface reflectance obtained from Landsat 8 and Sentinel-2. However, the average surface reflectance obtained from Landsat 8 bands indicates higher than that of the Sentinel-2 bands. The difference in surface reflectance is much higher in NIR band followed by red and green band; whereas, the difference is low for blue, SWIR1 and SWIR2 band. The high similarity in surface reflectance pattern for the both sensors signifies the consistency and utilities when integrated for combined use. For comparative assessment in index values derived from Landsat 8 and Sentinel-2, two widely used vegetation indices as NDVI and EVI was equated for known observations. The linear regression model demonstrates the potentiality of translation from Landsat data to equivalent Sentinel-2 data and vice-versa, where the relationship was observed strong for cropland and agriculture fallow compared to dense and open forest. This could be inferred that the similarity is high for homogeneous features as cropland and agriculture fallow compared to heterogeneous features as forest. High similarity (high correlation and lower RMSE) confirms the differences between MSI and OLI imagery are subtle and allows their complementary use. However, chances of errors are significant in heterogeneous class as forest, where the conversion of index values or integration between Landsat 8 and Sentinel-2 may lead to bias with a limited accuracy.

References

Anderson JH, Weber KT, Gokhale B, Chen F (2011) Intercalibration and evaluation of Resourcesat-1 and Landsat-5 NDVI. Can J Remote Sens 37:213–219

Astola H, Häme T, Sirro L, Molinier M, Kilpi J (2019) Comparison of Sentinel-2 and Landsat 8 imagery for forest variable prediction in boreal region. Remote Sens Environ 223:257–273

Banskota A, Kayastha N, Falkowski MJ, Wulder MA, Froese RE, White JC (2014) Forest monitoring using Landsat time series data: a review. Can J Remote Sens 40(5):362–384

Bawa K, Rose J, Ganeshaiah KN, Barve N, Kiran MC, Umashanker R (2002) Assessing biodiversity from space: an example of Western Ghats, India. Conserv Ecol 6(2)

Behera MD, Gupta AK, Barik SK, Das P, Panda RM (2018) Use of satellite remote sensing as a monitoring tool for land and water resources development activities in an Indian tropical site. Environ Monit Assess 190(7):401

Claverie M, Masek JG, Ju J, Dungan JL (2017) Harmonized Landsat-8 Sentinel-2 (HLS) product user's guide. National Aeronautics and Space Administration (NASA), Washington, DC, USA

Das P, Behera MD, Pal S, Chowdary VM, Behera PR, Singh TP (2019) Studying land use dynamics using decadal satellite images and Dyna-CLUE model in the Mahanadi River basin, India. Environ Monit Assess 191(3):804

Delegido J, Verrelst J, Alonso L, Moreno J (2011) Evaluation of Sentinel-2 red-edge bands for empirical estimation of green LAI and chlorophyll content. Sensors 11(7):7063–7081

Forkuor G, Dimobe K, Serme I, Tondoh JE (2018) Landsat-8 vs. Sentinel-2: examining the added value of sentinel-2's red-edge bands to land-use and land-cover mapping in Burkina Faso. GISci Remote Sens 55(3):331–354

Frampton WJ, Dash J, Watmough G, Milton EJ (2013) Evaluating the capabilities of Sentinel-2 for quantitative estimation of biophysical variables in vegetation. ISPRS J Photogram Remote Sens 82:83–92

French NH, Kasischke ES, Hall RJ, Murphy KA, Verbyla DL, Hoy EE, Allen JL (2008) Using Landsat data to assess fire and burn severity in the North American boreal forest region: an overview and summary of results. Int J Wildland Fire 17(4):443–462
FSI (2019) State of forest report forest survey of India, Dehradun
Ghosh SM, Saraf S, Behera MD, Biradar C (2017) Estimating agricultural crop types and fallow lands using multi temporal Sentinel-2A imageries. Proc Natl Acad Sci India, Sect A Phys Sci 87(4):769–779
Gómez C, White JC, Wulder MA (2016) Optical remotely sensed time series data for land cover classification: a review. ISPRS J Photogram Remote Sens 116:55–72
Hansen MC, Loveland TR (2012) A review of large area monitoring of land cover change using Landsat data. Remote Sens Environ 122:66–74
Hedley JD, Roelfsema C, Brando V, Giardino C, Kutser T, Phinn S, Mumbye PJ, Barrilerof O, Laporteg J, Koetz B (2018) Coral reef applications of Sentinel-2: coverage, characteristics, bathymetry and benthic mapping with comparison to Landsat 8. Remote Sens Environ 216:598–614
Hill MJ (2013) Vegetation index suites as indicators of vegetation state in grassland and savanna: an analysis with simulated SENTINEL-2 data for a North American transect. Remote Sens Environ 137:94–111
Huete A, Didan K, Miura T, Rodriguez EP, Gao X, Ferreira LG (2002) Overview of the radiometric and biophysical performance of the MODIS vegetation indices. Remote Sens Environ 83(1–2):195–213
Immitzer M, Vuolo F, Atzberger C (2016) First experience with Sentinel-2 data for crop and tree species classifications in central Europe. Remote Sens 8(3):166
Kobayashi H, Suzuki R, Kobayashi S (2007) Reflectance seasonality and its relation to the canopy leaf area index an eastern Siberian larch forest: multisatellite data and radiative transfer analyses. Remote Sens Environ 115:3615–3624
Krishnaswamy J, Bawa K, Ganeshaiah KN, Kiran MC (2009) Quantifying and mapping biodiversity and ecosystem services: utility of a multi-season NDVI based Mahalanobis distance surrogate. Remote Sens Environ 113:857–867
Lessio A, Fissore V, Borgogno-Mondino E (2017) Preliminary tests and results concerning integration of Sentinel-2 and Landsat-8 OLI for crop monitoring. J Imaging 3(4):49
Li P, Jiang L, Feng Z (2014) Cross-comparison of vegetation indices derived from Landsat-7 enhanced thematic mapper plus (ETM+) and Landsat-8 operational land imager (OLI) sensors. Remote Sens 6(1):310–329
Mallinis G, Mitsopoulos I, Chrysafi I (2017) Evaluating and comparing Sentinel-2A and Landsat 8 OLI spectral indices for estimating fire severity in a Mediterranean pine ecosystem of Greece. GISci Remote Sens 55(1):1–18
Mandanici E, Bitelli G (2016) Preliminary comparison of Sentinel-2 and Landsat 8 imagery for a combined use. Remote Sens 8(12):1014
Matushita B, Yang W, Chen J, Onda Y, Qiu G (2007) Sensitivity of the enhanced vegetation index (EVI) and normalised difference vegetation index (NDVI) to topographic effects: a case study in high-density cypress forest. Sensors 7:2636–2651
Meng J, Du X, Wu B (2013) Generation of high spatial and temporal resolution NDVI and its application in crop biomass estimation. Int J Digital Earth 6(3):203–218
Mkhabela MS, Bullock P, Raj S, Wang S, Yang Y (2011) Crop yield forecasting on the Canadian Prairies using MODIS NDVI data. Agric For Meteorol 151(3):385–393
Roy PS, Behera MD, Murthy MSR, Roy A, Singh S, Kushwaha SPS, Gupta S (2015) New vegetation type map of India prepared using satellite remote sensing: comparison with global vegetation maps and utilities. Int J Appl Earth Obs Geoinf 39:142–159
Sahoo T, Panda PC, Acharya L (2017) Structure, composition and diversity of tree species in tropical moist deciduous forests of Eastern India: a case study of Nayagarh forest division Odisha. J For Res 28(6):1219–1230

Son NT, Chen CF, Chen CR, Minh VQ, Trung NH (2014) A comparative analysis of multitemporal MODIS EVI and NDVI data for large-scale rice yield estimation. Agric For Meteorol 197:52–64

Steven MD, Malthus TJ, Baret F, Xu H, Chopping MJ (2003) Intercalibration of vegetation indices from different sensor systems. Remote Sens Environ 88:412–422

Teillet PM, Ren X (2008) Spectral band difference effects on vegetation indices derived from multiple satellite sensor data. Can J Remote Sens 34:159–173

Tucker CJ (1979) Red and photographic infrared linear combinations for monitoring vegetation. Remote Sens Environ 8:127–150

Van der Werff H, Van der Meer F (2016) Sentinel-2A MSI and Landsat 8 OLI provide data continuity for geological remote sensing. Remote Sens 8(11):883

Vuolo F, Żółtak M, Pipitone C, Zappa L, Wenng H, Immitzer M, Weiss M, Baret F, Atzberger C (2016) Data service platform for Sentinel-2 surface reflectance and value-added products: system use and examples. Remote Sens 8(11):938

Xie Y, Zhao X, Li LL, Wang H (2010) Calculating NDVI for Landsat 7-ETM data after atmospheric correction using 6S model: a case study in Zhangye City, China. In: 18th International conference on geoinformatics, Peking University, Beijing, China, June

Zhang D, Zhou G (2016) Estimation of soil moisture from optical and thermal remote sensing: a review. Sensors 16(8):1308

Chapter 8
Comparative Assessments of Forest Cover Change in Some Districts of West Bengal, India using Geospatial Techniques

Mitrajit Chatterjee and Atma Deep Dutta

Abstract Forests are one of the most important components on our planet as they regulate a number of natural systems viz. the food chain, the water cycle, the carbon cycle etc. In this study, we have focused on the forest cover of districts that has a very small percentage of area covered by forests. The study has been performed for a time period of 30 years i.e. from 1990 to 2020 and a time series analysis of the changes in forest cover has been done. The forested areas are divided into three type namely very dense forest, moderately dense forest and open forest. Five districts in the state of West Bengal have been selected namely Hooghly, Nadia, Purba and Paschim Bardhhaman (considered together as Burdwan) and Purulia. These districts particularly belong to the South Bengal region and out of them namely Hooghly, Nadia and Bardhhaman are also a part of deltaic region of the Lower Ganga which is known as the Bhagirathi-Hooghly River in West Bengal. These districts are chosen because they are one of the most populous districts both in the state and also in the country but they lack adequate amount of natural vegetation cover, the reasons for which can be cited are the availability of fertile land for agriculture which consequently makes these places one of the most suitable areas for human beings to survive and thrive. In this study a standardised and simple index namely the Normalized Difference Vegetation Index (NDVI) has been used to delineate various kinds of forest and a Land Use Land Cover (LULC) classification has been done to study the present land use condition of these districts. It has been found that the district of Purulia has the maximum forest cover while the district of Hooghly lacks areas with very dense forest. Studies have shown that forest cover conditions has improved for all of these districts since the last decade but most of these improvements has been observed in the open forest category which signifies that social forestry might have been taking place which proves an increasing concern among people and the government in saving the environment.

M. Chatterjee (✉)
Geography, Kabi Sukanta Mahavidyalaya, Bhadreswar, Hooghly, India
e-mail: chatterjeemitrajit56@gmail.com

A. D. Dutta
Geoinformatics, Birla Institute of Technology, Mesra, Ranchi, India
e-mail: atmadeepdutta@gmail.com

P. K. Shit et al. (eds.), *Spatial Modeling in Forest Resources Management*, Environmental Science and Engineering,
https://doi.org/10.1007/978-3-030-56542-8_8

Keywords NDVI · LULC · Landsat · Sentinel data

8.1 Introduction

Assessment of forest cover and estimating its changes along with Land Use and Land Cover (LULC) Mapping is very important in today's dynamically changing urban world as forests are at maximum threat and hence proper policies and strategies are required for their protection. Forests play a very significant role in shaping the earth's surface but in this era of human civilization, they are at an increased risk of extinction mainly due to the anthropogenic activities. Moreover with the gradual reduction of forest area and increasing urban sprawl, significant change in normal climatic conditions are also observed which will also be a threat to the entire planet in the coming future. Remote sensing and GIS play a vital role in mapping and monitoring forest cover along with other land use land cover classes in comparison to the conventional ground based techniques as it saves a lot of time and in much more economical. Conventional classification techniques have proved to be very beneficial in mapping of land use land cover classes but with the advent of research in this field various indices have been developed which have enhanced the accuracy in LULC classification and also helps in proper monitoring and modelling of the required entities (Bera and Prakash 2018).

Image classification is a very important process in the preparation of LULC maps and this process of image processing depends on a number of factors few of which are quality of satellite imageries available, resolutions of the satellite imageries depending on our requirement (spatial, spectral, radiometric and temporal), cloud cover in the imageries, expertise of the producer etc. (Rwanga and Ndambuki 2017).

Rawat et al. in their study have used remote sensing and GIS techniques to study the LULC of Ramnagar town and Hawalbagh district located in the state of Uttarakhand. Supervised classification has been done using the Maximum Likelihood Classifier to prepare the LULC maps of the study area and finally change detection has been done to analyse the changes in various land use and land cover classes of the study area (Rawat et al. 2013; Rawat and Kumar 2015).

The forest cover of India has rapidly declined due to increased urbanisation and industrialisation and it has declined to 21.3% of the total geographical area of the country. In a report of the Department of Economic and Social Affairs, Secretariat of the United Nations Forum on Forests "Review of the effectiveness of the International Agreement on Forests" published in the year 2005 it has been stated that "In India (another country where forest policy and planning are a concurrent responsibility of the central government and the state governments), there is a goal of increasing forest tree cover to 33% over 20 years, with forests being treated primarily as environmental and social resources and only secondarily as a commercial resource". According to a report of the PIB published by the Ministry of Environment, Forest and Climate Change, Government of India it has been stated that "Total Forest and Tree Cover rises to 24.56% of the total geographical area of the country". Although there has been a good progress in improving the forest cover of our country but still better efforts are to be made to improve the forest cover at a very ground level and a bottom up approach would be more effective in achieving the goal of 33% forest cover by 2025. In this scenario research should be done and strategy should be made and implemented at the district level so that it contributes largely to the increase in forest cover of our country. This chapter focuses on the forest cover conditions of four districts present in southern part of Bengal out of which three of the districts are poorly covered with dense forest area. The objective of this paper is to analyse the forest cover of four districts of southern part of West Bengal namely Burdwan, Hooghly, Nadia and Purulia and assess the rate of change of forest over the time period of 1990–2020. Another objective of this study is to understand the Land Use Land Cover pattern of these districts so that it can help in implementing joint forestry and social forestry schemes in the coming future which can also be a future scope of this study.

8.2 Study Area

Five districts in the Southern part of the state of West Bengal namely Hooghly, Nadia, Purba and Paschim Bardhhaman (considered together as Burdwan District) and Purulia has been selected for study in this chapter (Fig. 8.1). Hooghly is one of the districts in West Bengal named after the River Hooghly. It has a latitudinal extension of 22° 35′ N to 23° 15′ N and a longitudinal extension of 87° 25′ E to 88° 35′ E. The headquarters of the district are located at Chinsurah. The Hooghly district has four subdivisions namely Arambagh, Chinsurah Sadar, Chandannagar and Srirampur. Burdwan is the seventh most populous district in India which is situated in West Bengal. This district houses the two most industrialised cities of the state and the country namely Asansol and Durgapur. This district extends from 22° 55′ N latitude to 23° 50′ N latitude and 86° 40′ E longitude to 88° 20′ E longitude. It is also known as the 'rice bowl' of West Bengal. Nadia is a district in the state of West Bengal sharing its border in the East with Bangladesh. It also shares its borders with Hooghly district in the South and Burdwan district in the west. It has a latitudinal extent of 22° 45′ N to 24° 05′ N and a longitudinal extent of 87° 55′ E to 88° 55′ E. Purulia is also a district in the state of West Bengal which extends from 22° 40′ N latitude to 23° 35′ N latitude and from 85° 25′ E longitude to 87° 20′ E longitude. Some of the important towns of this district are Raghunathpur, Adra, Jhalda and Balarampur.

These districts have been selected as the study area because three of the districts namely Burdwan, Hooghly and Nadia have the least dense forest cover and efforts should be made to improve the forest cover of these regions through joint and social forestry. The district of Purulia although has the highest forest cover among the four of them but they are at a threat due to increasing rate of urbanisation and hence monitoring of the forest area is required.

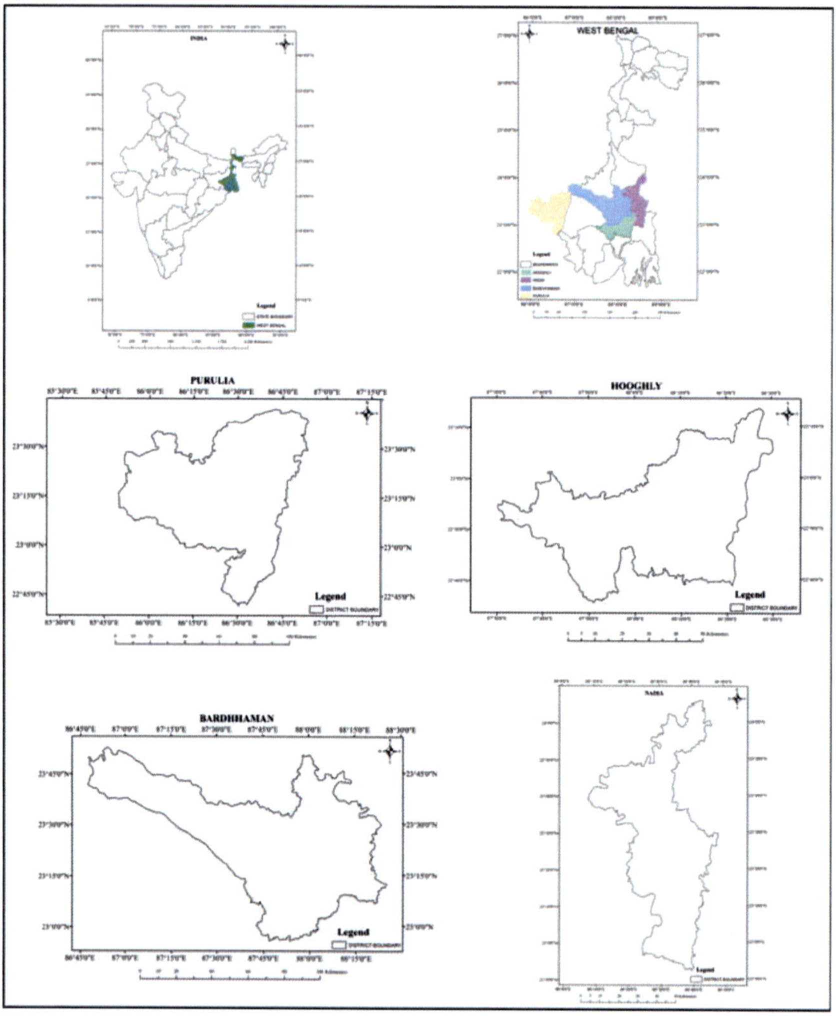

Fig. 8.1 Location map of the study area

8.3 Materials and Methods

8.3.1 Data Used

Satellite imageries of Landsat 5 Thematic Mapper (TM) and Landsat 8 Operational Land Imager (OLI) have been used for computing the NDVI and delineating the various forest covers in the above district. Higher resolution Sentinel 2 data has been used for executing the image classification process to produce the LULC maps. The details of the Landsat satellite and Sentinel imageries used along with their details are provided in Tables 8.1 and 8.2 respectively.

Table 8.1 Details of Landsat Satellite imageries

Satellite	District	Date of capture	Resolution (m)	Bands used	Wavelength (μm)
Landsat 5	Burdwan and Hooghly	05-01-1990	30		
		30-01-1990			
		17-01-2000			
		26-01-2000			
		13-02-2010			
		21-01-2010		Blue	0.45–0.52
	Nadia	30-01-1990		Green	0.52–0.60
				Red	0.63–0.69
		10-01-2000		NIR	0.77–0.90
		26-01-2000		SWIR 1	1.55–1.75
		21-01-2010		TIR	10.40–12.50
				SWIR 2	2.09–2.35
	Purulia	05-01-1990		Panchromatic	0.52–0.90
		12-01-1990			
		21-03-2000			
		24-01-2000			
		13-02-2010			
		19-01-2010			
Landsat 8	Burdwan and Hooghly	02-02-2020		Coastal aerosol	0.43–0.45
		24-01-2020		Blue	0.45–0.51
	Nadia	02-02-2020		Green	0.53–0.59
		17-01-2020		Red	0.64–0.67
	Purulia	24-01-2020		NIR	0.85–0.88
		27-01-2020		SWIR 1	1.57–1.65
				SWIR 2	2.11–2.29
				Panchromatic	0.50–0.68
				Cirrus	1.36–1.38
				TIRS 1	10.60–11.19
				TIRS 2	11.50–12.51

Table 8.2 Details of Sentinel Satellite imageries

Satellite	District	Date of capture	Resolution (m)	Bands used	Wavelength (nm)
Sentinel 2	Burdwan	20-03-2020	10		
		23-03-2020		Band 2—Blue	490
		25-03-2020		Band 3—Green	560
	Nadia	20-03-2020		Band 4—Red	665
		25-03-2020		Band 8—NIR	842
	Purulia	12-02-2020			

8.3.2 Methodology

The imageries of Landsat and Sentinel satellites have been downloaded from USGS Earth Explorer (https://earthexplorer.usgs.gov/). A flowchart of the methodology is being provided below in Fig. 8.2.

All the imageries used for the study have both land and scene cloud cover of less than 10%. LULC classification for the district of Hooghly has been carried out using Landsat OLI data captured in the year 2020 (specific dates mentioned in Table 8.1) as cloud free Sentinel imageries were not available for the year 2020. The Landsat 5 TM data downloaded for the year 2010 of all the districts were clipped with the help of scene shape file as the left and right edges of the imageries had finger-like discontinuation which was producing error in further investigations. The shape files of the Landsat scenes were downloaded from https://www.usgs.gov/media/files/landsat-wrs-2-descending-path-row-shapefile.

The Landsat satellite imagery bands that have been used to calculate the NDVI (marked in red) were mosaiced and then they were clipped using the district boundaries. After clipping, the NDVI were calculated based on the formula mentioned below.

$$\text{NDVI} = \frac{\text{NEAR INFRARED} - \text{RED}}{\text{NEAR INFRARED} + \text{RED}}$$

On calculation of NDVI, threshold values were selected to delineate the various forest type namely very dense forests, moderately dense forest and open forest (Sonawane and Bhagat 2017). The threshold values were completely selected on trial and error basis. From the accuracy assessment table of the NDVI, it can be seen that this method has produced significant results but better results needs to be produced for which indices that can remove the interference of agricultural lands needs to be developed (Table 8.6). Area of various forest types of the different districts has been calculated and their change has been analysed. LULC classification of different districts has been done using Sentinel 2 satellite imagery because of its very high resolution except for the district of Hooghly. Area of various LULC classes has been calculated and the differences with those of total forest area are analysed. The LULC classification has been done using the Maximum Likelihood Classifier (MLC) which

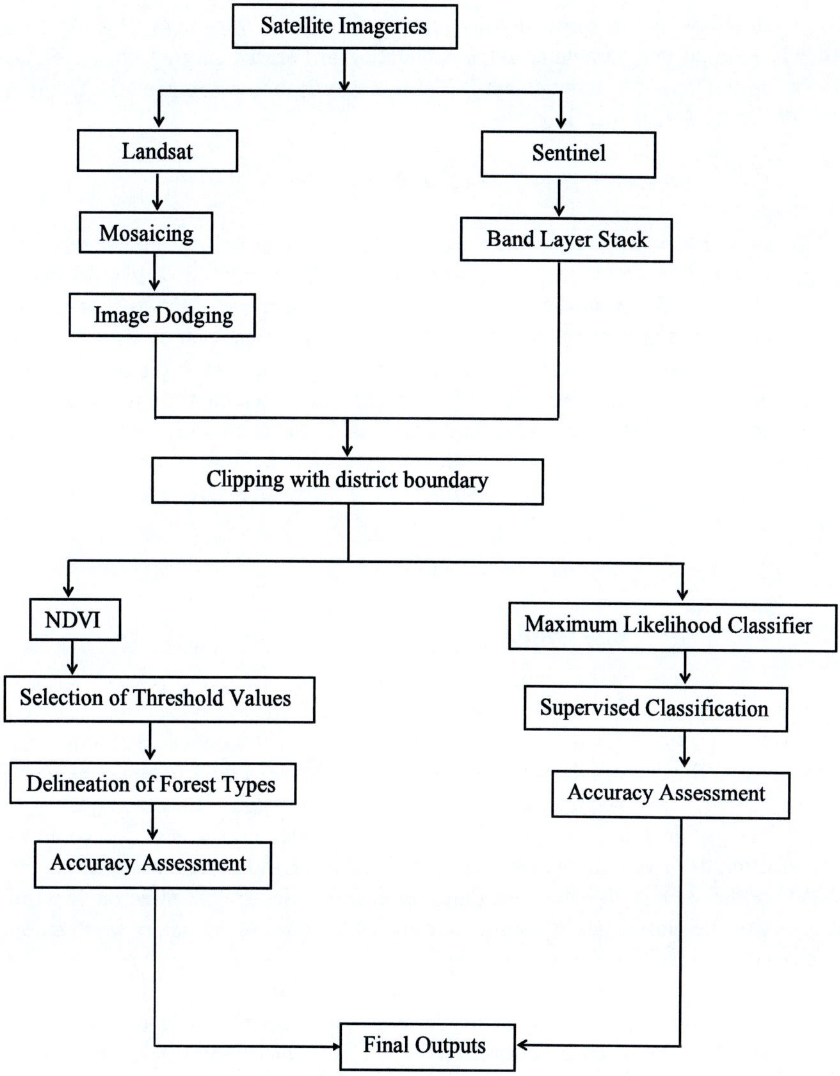

Fig. 8.2 Flowchart of methodology

is an algorithm of supervised classification technique (Rawat et al. 2013; Rawat and Kumar 2015).

An attempt has been made to calculate the Advanced Vegetation Index (AVI) for the selected study area. Vegetation and non-vegetated areas are generally differentiated with the help of NDVI. But being a ratio based index, it at times fails to highlight certain subtle differences due to canopy density in the infrared and red. By using power degree of infrared response, these subtle differences can be highlighted.

AVI is also sensitive to forest density and physiognomic vegetation classes. This index is an important parameter in the calculation of Forest Canopy Density (FCD) and hence an attempt to estimate the FCD for the study area was made. The formula for calculating AVI is mentioned below.

$$\text{AVI} = \{(\text{B4} + 1)\ (256-\text{B3})\ (\text{B4} - \text{B3})\}^{1/3}$$

The above formula is applicable for Landsat TM imageries only and B2 refers to the green band, B3 refers to the red band and B4 refers to the near-infrared (NIR) band (Sahana et al. 2015). In most of the literature it was found that AVI was specifically used for a particular forest and not for identifying the forest canopy density in an entire district, hence when it was applied on the Nadia district, it did not produce good results. Moreover, we selected the Nadia district for testing the applicability of this index as the forest cover of Nadia district highly matched with the forest cover report of Forest Survey of India (FSI). But since this index did not produce expected results, we did not use it for other districts of our study area and hence FCD could not be calculated.

8.3.3 Accuracy Assessment

Accuracy assessment of remote sensing product is a feedback system for checking and evaluating the objectives and results. Accuracy assessments determine the correctness of the classified image. Accuracy is a measure of the agreement between a standard that is assumed to be correct and a classified image of unknown quality. If the image classification corresponds closely with the standard, it is said to be accurate (Bhatta 2017). Accuracy assessment for Land use Land cover classification and NDVI for the year 2020 have been done for each district to determine the level of accuracy of the classification with the ground. A random set of points were generated and classification results were compared with the true information classes in the reference image (Google Earth image). In the evaluation of classification errors, a classification error matrix was typically formed by the software for each district and henceforth, the overall accuracy for each district has been calculated. Kappa Coefficients have also been taken into account to confirm the acceptability of the accuracy results. Kappa Coefficient (κ) is a statistic that is used to measure inter-rater reliability (and also Intra-rater reliability) for qualitative (categorical) items (McHugh 2012). Values <0 as indicating no agreement and 0–0.20 as slight, 0.21–0.40 as fair, 0.41–0.60 as moderate, 0.61–0.80 as substantial, and 0.81–1 as almost perfect agreement (Landis and Koch 1977; Rwanga and Ndambuki 2017). Table 8.3 is a glimpse of the generated error matrix along with the accuracy level and Kappa value for LULC classification of Burdwan District for the year 2020.

Similarly, error matrices for the remaining districts were also determined for LULC and NDVI classifications.

Table 8.3 Error matrix of LULC classification of Burdwan district for the year 2020

Sl No.	Classified	Water bodies	Agricultural land	Built up	Vegetation	Total	Users accuracy	Kappa
1	Water bodies	8	0	1	1	10	0.8	0
2	Agricultural land	0	8	2	0	10	0.8	0
3	Built up	0	3	7	0	10	0.7	0
4	Vegetation	0	1	0	9	10	0.9	0
	Total	8	12	10	10	40	0	0
	Producer accuracy	1	0.67	0.7	0.9	0	0.8	0
	Kappa	0	0	0	0	0	0	0.73

8.4 Results and Discussion

8.4.1 LULC Feature

The LULC features for the year 2020 of the districts have been classified namely vegetation, built up area and barren land, agriculture land and agricultural fallow and water bodies (Figs. 8.3, 8.4, 8.5 and 8.6). The area covered by the different LU/LC class is represented (Table 8.4). Maximum vegetation (841.17 km^2) in terms of area

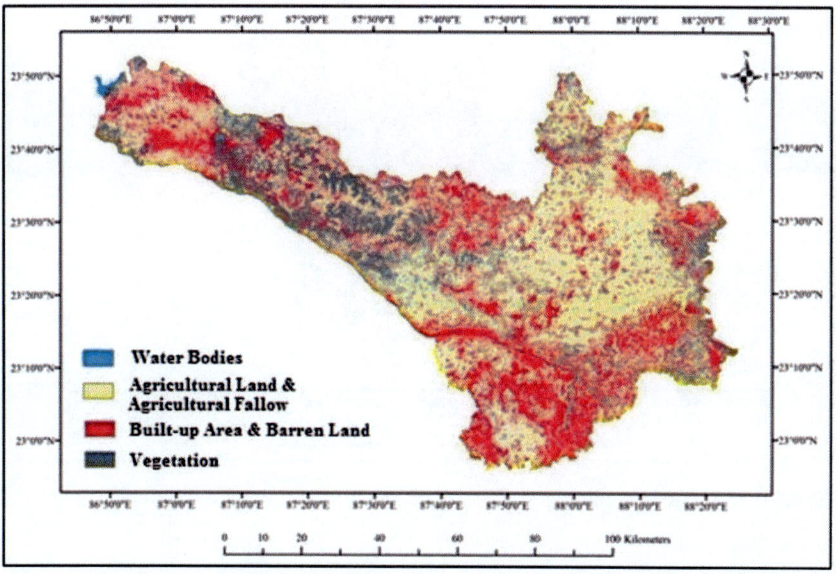

Fig. 8.3 LULC map of Burdwan

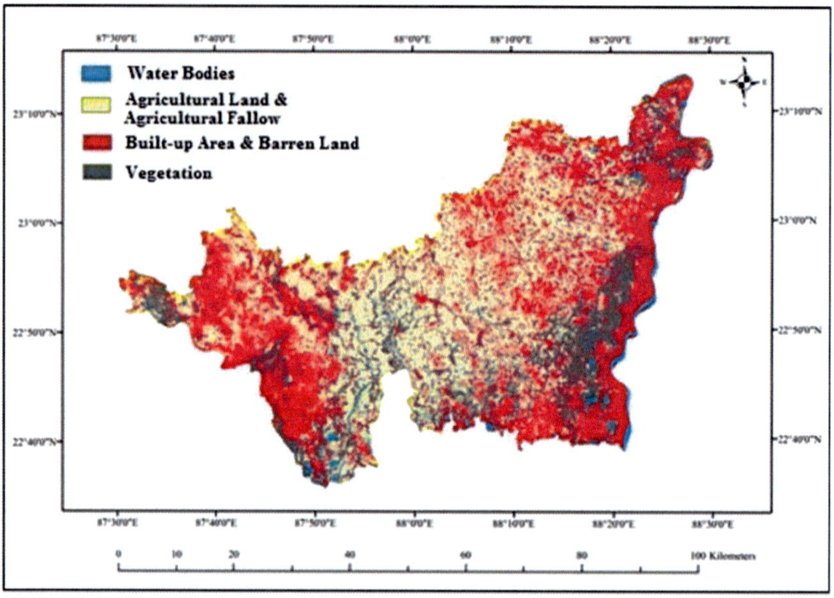

Fig. 8.4 LULC map of Hooghly

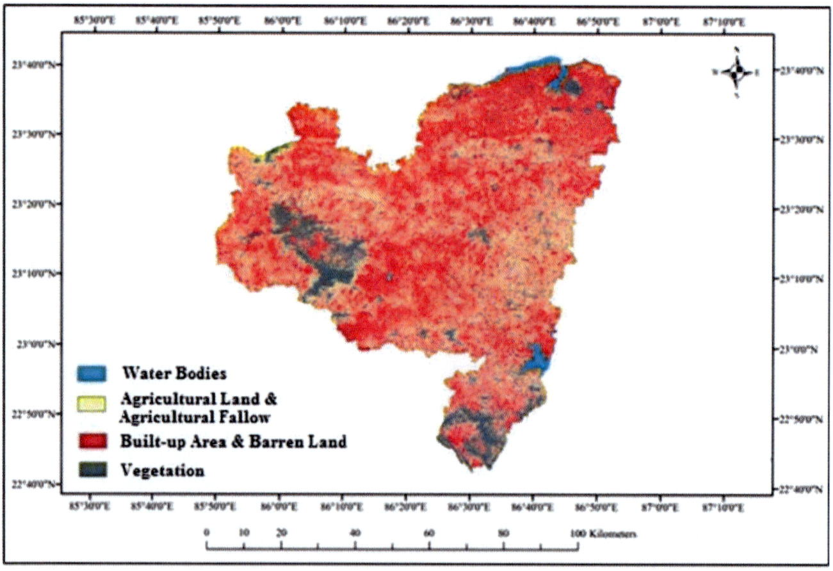

Fig. 8.5 LULC map of Purulia

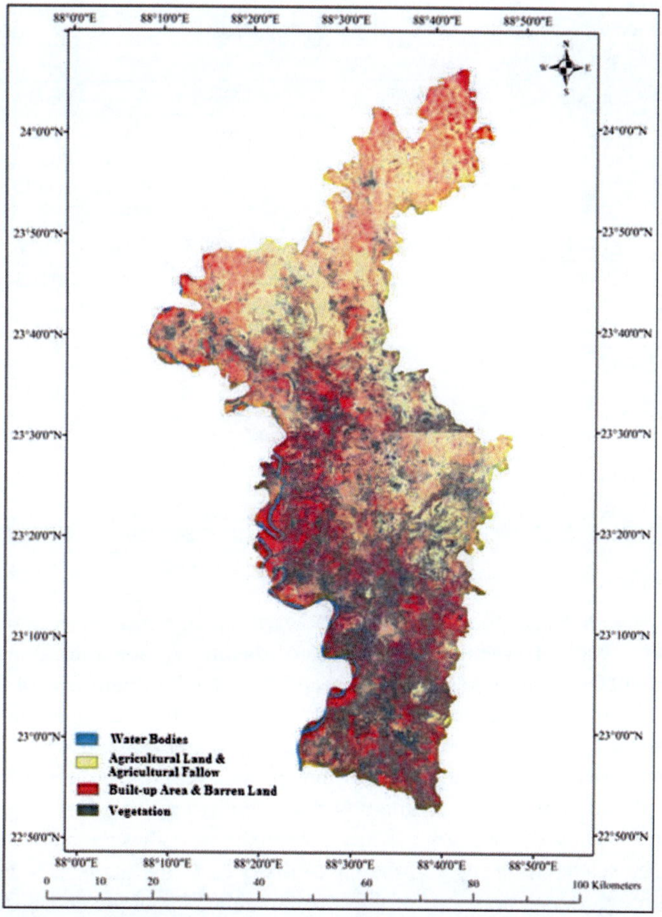

Fig. 8.6 LULC map of Nadia

is found in the Purulia district. The remaining districts have forest cover less than 500 km^2 with Hooghly district having the least vegetation cover (200.14 km^2). The forests are mostly scattered within the districts.

For the district of Burdwan, the forested areas are scattered throughout the entire district but a dense patch of forest can be seen in the western and the southern parts of the district. Majority of the district is covered with agricultural land (46.94%) which proves the title of "Rice Bowl of West Bengal". Settlements are mainly concentrated in the southern and south-eastern part and at the extreme western ends. Presence of a few water bodies are observed in a scattered manner along the entire district (3%).

The district of Hooghly has mostly agricultural land and built up area as the major land use types which are 38% and 49% respectively of the entire geographical area of the district. The major water body in the district namely Hooghly River runs

Table 8.4 Land use land cover (LULC) area for the year 2020

LULC classes	DISTRICT							
	Burdwan		Hooghly		Nadia		Purulia	
	Area (km^2)	% of area to total area	Area (km^2)	% of area to total area	Area (km^2)	% of area to total area	Area (km^2)	% of area to total area
Vegetation	418.51	5.946307	481.91	15.29985	406.25	12.6801	741.17	10.8248
Built-up and barren land	2897.83	41.17318	1285.41	40.80965	1050.9	32.8013	3087.64	45.0949
Agricultural land and agricultural fallow	3476.96	49.40162	1201.7	38.15199	1670.74	52.148	2277	33.2555
Water bodies	244.85	3.478897	180.75	5.738514	75.95	2.37059	741.17	10.8248
Total area	7038.15	100	3149.77	100	3203.84	100	6846.98	100

along the eastern end of the district and is clearly distinguishable from the LULC classification. Natural vegetation covers 6% of the entire geographical area of the district. A patch of natural vegetation is seen in the south-eastern part of the study area. The district of Nadia also has its major portion of its land use in built-up and agricultural land use sector which covers 31% and 32% area respectively of the entire geographical area of the district. Most of the northern part of the district is covered with agricultural land and the southern part of the district comprises mostly of the settlements. 8% area of the district is covered with water bodies which include small water bodies scattered throughout the entire district and the major Hooghly River running through the western part of the district. Only 12% of the entire area of the district is covered with vegetation which is scattered throughout the entire district. The district of Purulia has a dearth of water bodies which is evident from the LULC classification of the district. Only 1% of the total geographical area of the district is covered with water bodies and the major water body of this district are two dams in its northern and southern parts. The major part of the district is covered by built-up land use which covers 49% of the entire geographical of the district. The built-up areas are scattered throughout the entire district with its major concentration at northern and north western parts. Purulia having a rough terrain has less area under agricultural land in comparison to the other districts. The agricultural land use in Purulia covers 35% of the entire geographical area of the district. 13% of the geographical area of the district is covered with natural vegetation. Dense patches of natural vegetation are observed in the western and southern part of the district and smaller patches of vegetation are scattered throughout the entire district.

Table 8.5 Accuracy assessment of LULC for the year 2020

District	Accuracy (%)	Kappa coefficient
Burdwan	80	0.73
Hooghly	85.23	0.78
Nadia	80	0.73
Purulia	70	0.6

Accuracy assessment is done to analyse the effectiveness of the LULC classification. A classification accuracy assessment has been performed on the classified maps of 2020 using an error matrix algorithm. The LULC maps indicate an overall accuracy of 75–90% (Table 8.5) with Hooghly district showing the highest accuracy and Purulia the lowest. From the accuracy assessment table (Table 8.5), it is evident that all the LULC have acceptable accuracy and substantial Kappa coefficient values.

8.4.2 Vegetation Dynamics

The NDVI is the most commonly used index for forest vegetation biomass monitoring. The range of the NDVI lies between −1 to 1, but the absolute value of NDVI for vegetation change analysis is between 0 and 1. Healthy vegetation yields high positive NDVI values because they exhibit high reflectance in the NIR and low in visible wavelength and the negative values are mainly due to the barren lands, cloud covers etc.

NDVI for four different years 1990, 2000, 2010 and 2020 have been calculated and mapped for all the districts and there is a significant variation in the forest cover in a span of 30 years (Figs. 8.7, 8.8, 8.9 and 8.10). District wise threshold values have been ascertained for each year on the basis of the NDVI range obtained in order to categorize the vegetation into broadly three categories viz. very dense forest, moderately dense forest and open forest (Table 8.6). There is a significant decrease in the forest area for the districts of Hooghly and Nadia owing to the increase in the built-up areas. The maximum forest cover is in the district of Purulia while the forest cover in Burdwan has shown a stable figure over the time span of 30 years. Threshold values in NDVI are selected to differentiate the forested areas from the non-forested areas. These threshold values are selected completely on trial and error and hence it is different for the various districts and various time periods.

The differences in the threshold values can be attributed to the reasons which are different times (year) of capture of the imageries and the differences in the extent and distribution forested and non-forested areas of the various districts along with other land use and land cover classes. By comparing the NDVI values over the years, it indicates that the NDVI value range have decreased significantly. The decrease in positive value of NDVI indicates the change of healthy forest and vegetation. The accuracy assessment of NDVI based classification is much higher than the LULC

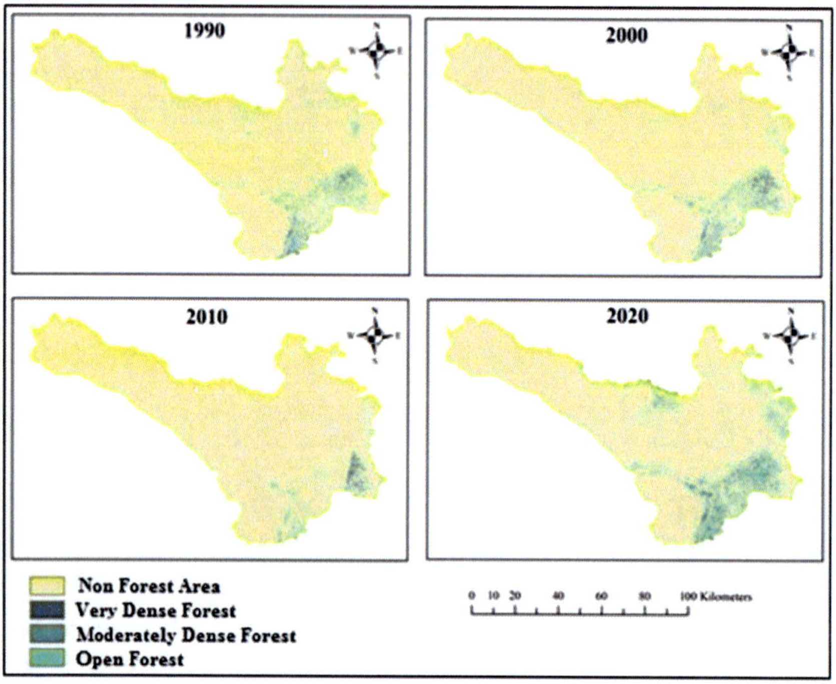

Fig. 8.7 NDVI based classification of Burdwan

and the Kappa Coefficients are also high (Table 8.7). So, the classification based on NDVI values can be taken into consideration.

From the results of NDVI for the district of Burdwan, it is seen that the district had an almost stable forest cover throughout the entire study period. A significant decrease in the forest cover has been observed for the year 2010 but it has shown an improvement for the year 2020. In fact the report of Forest Survey of India (FSI) also claims a similar increase in the forest area from 2011 to 2019 in reports of the respective years. Most of the forest cover is concentrated in the southern and south-eastern part of the district and the very dense forest category can also be traced in the south eastern part of the district. The presence of this very dense forest in the study area can be attributed to the presence of Ramnabagan Wildlife Sanctuary situated in the south eastern part of the district. The other two forest categories namely moderate dense forest and open forest are spread sporadically throughout the entire district. The Hooghly district is the only district in the study area among the four of the districts which has no recorded dense forest. Although the other two forest categories are spread throughout the entire district but a decrease in area has been observed in forest cover for the period of study. The district of Nadia shows similar trends with that of the district of Hooghly. Although the Nadia district has a protected forest area namely the Bethuadahari Wildlife Sanctuary but the area covered by dense forest cover as is seen from the study is relatively low. The forest cover in Nadia

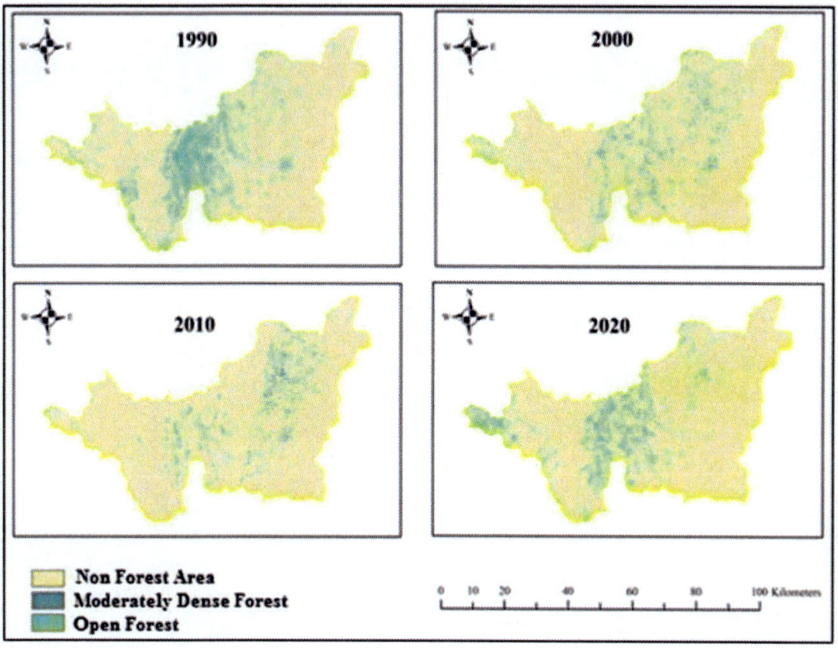

Fig. 8.8 NDVI based classification of Hooghly

district also shows a drastic reduction from 1990 to 2010 but the increase in forest cover for the year 2020 proves effective implementation of the various efforts by the government and NGOs in increasing forest cover through various joint forestry and social forestry schemes. The district of Purulia has shown a stable increase in the forest covered areas from 1990 to 2010 but an interesting change to observe in regards to the forest cover of this district that there has been a significant decrease in the dense forest cover category for the year 2020. Although forest cover under moderately dense forest and open forest has increased but efforts must be made to preserve the dense forests which contributes mostly to the biodiversity of an area. The reduction in dense forest cover of the district can be attributed to the cause of felling of trees by the tribal for their livelihood as Purulia houses a huge population of the various tribal communities of West Bengal.

8.5 Conclusion

This study gives an overview of the forest cover in four districts if West Bengal out of which the district of Hooghly lacks a proper forested area and most of the forests cover type in this district is moderately dense and open. Low accuracy for some cases has been observed which is mainly due to fact that in some areas the agricultural

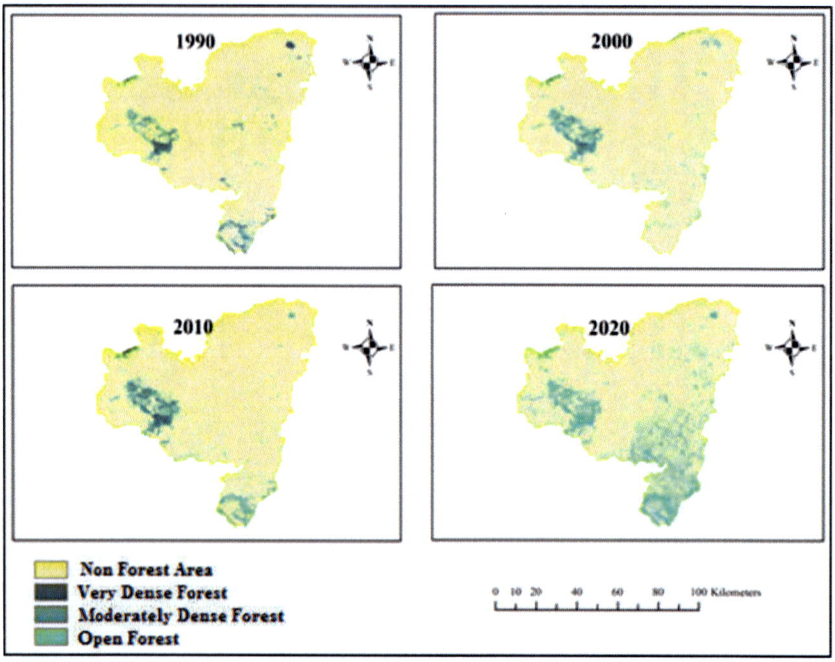

Fig. 8.9 NDVI based classification of Purulia

lands were included in the open forest category. Since these areas lack very dense forest and dense tree canopy, indices should be developed that can easily differentiate areas under natural vegetation from other LULC classes. Hooghly district completely lacks dense forest cover and hence areas must be found out where reforestation can be done. Purulia has a good forest cover in comparison to the other districts. This can be attributed to the moderately undulating topography in this district which has prevented rapid growth of human population as well as the climate and soil which are less favourable for agricultural practices.

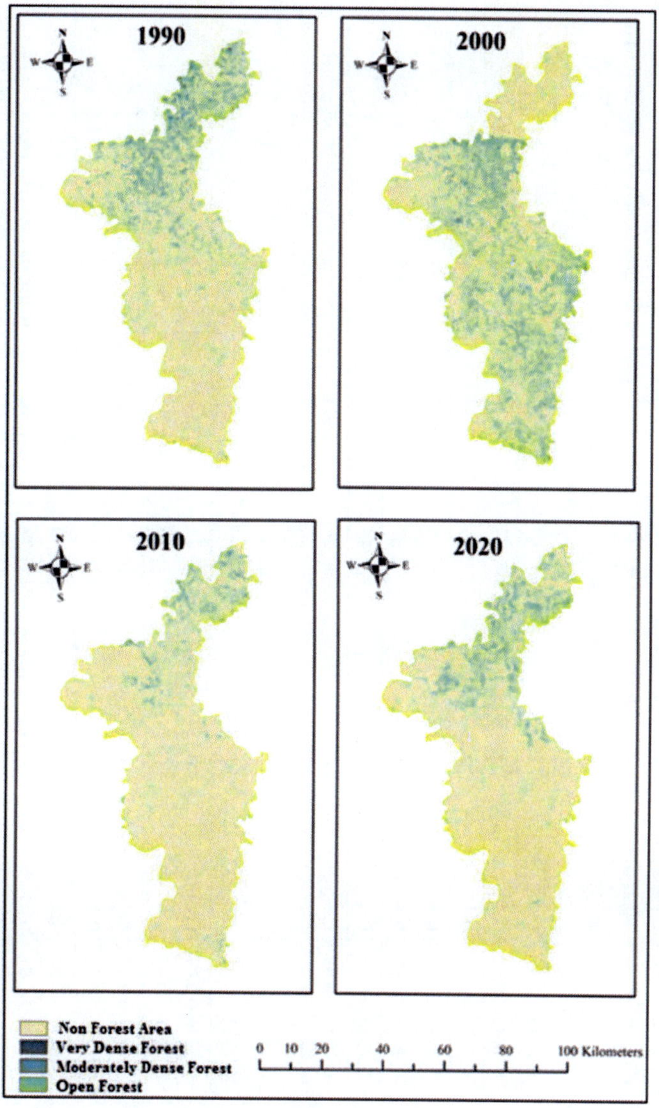

Fig. 8.10 NDVI based classification of Nadia

Table 8.6 Forest area estimation and categorization on the basis of NDVI values

District	Area (km^2)	Year	NDVI range	Threshold value	Area (km^2)			Total area (km^2)	% Forest area to total area
					Very dense forest	Moderately dense forest	Open forest		
Burdwan	7024	1990	−0.95 to +0.97	0.4	20.43	137.39	323.24	481.06	6.85
		2000	−0.68 to +0.95	0.4	35.78	156.84	215.21	407.83	5.81
		2010	−0.37 to +0.70	0.4	33.69	129.39	132.83	295.91	4.21
		2020	−0.23 to +0.52	0.4	62.03	154.88	257.86	474.77	6.76
Hooghly	3149	1990	−0.95 to +0.97	0.4	0	17.67	544.19	561.86	17.84
		2000	−0.73 to +0.92	0.4	0	25.48	328.72	354.20	11.25
		2010	−0.40 to +0.69	0.4	0	49.06	222.05	271.11	8.61
		2020	−0.15 to +0.52	0.3	0	16.00	198.92	214.92	6.82
Nadia	3927	1990	−0.32 to +0.96	0.3	10.19	197.44	537.24	744.88	18.97
		2000	−0.72 to +0.96	0.5	13.76	218.17	754.49	986.42	25.12
		2010	−0.39 to +0.64	0.4	0.19	29.48	363.05	392.72	10.00
		2020	−0.12 to +0.47	0.3	0.60	6.10	501.92	508.62	12.95
Purulia	6259	1990	−0.95 to +0.96	0.4	113.29	154.56	229.30	497.16	7.94
		2000	−0.93 to +0.82	0.2	24.74	146.87	400.18	571.79	9.14
		2010	−0.41 to +0.56	0.2	59.79	163.88	434.49	658.16	10.52
		2020	−0.24 to +0.53	0.2	39.02	454.78	536.49	1030.29	16.46

Table 8.7 Accuracy assessment of NDVI based classification

District	Accuracy (%)	Kappa coefficient
Burdwan	89	0.85
Hooghly	82.72	0.78
Nadia	90.9	0.86
Purulia	90	0.63

Acknowledgements The authors are thankful to National Aeronautics and Space Administration (NASA) and the United States Geological Survey (USGS) for providing Landsat satellite imageries for free and helping them in all possible during the processing of these imageries. They also express their gratitude to European Space Agency (ESA) for providing the high resolution Sentinel 2 data for open access use.

References

Bera S, Prakash A (2018) Forest cover change analysis based on remote sensing & GIS of West Singhbhum District, Jharkhand. Int J Res Appl Sci Eng Technol 6(5):1039–1050

Bhatta B (2017) Remote sensing and GIS. Oxford University Press, New Delhi, pp 375–376

Landis JR, Koch GG (1977) The measurement of observer agreement for categorical data. Biometrics 33(1):159–174

McHugh ML (2012) Interrater reliability: the kappa statistic. Biochem Med 22(3):276–282

Rawat JS, Kumar M (2015) Monitoring land use/land cover change using remote sensing and GIS techniques: a case study of Hawalbagh block, district Almora, Uttarakhand, India. Egypt J Remote Sens Space Sci 18:77–84 (National Authority of Remote Sensing and Space Sciences, Elsevier)

Rawat JS, Biswas V, Kumar M (2013) Monitoring land use/land cover change using remote sensing and GIS techniques: a case study of Ramnagar town area, district Nainital, Uttarakhand, India. Egypt J Remote Sens Space Sci 16:111–117 (National Authority of Remote Sensing and Space Sciences, Elsevier)

Rwanga S, Ndambuki JM (2017) Accuracy assessment of land use land cover classification using remote sensing and GIS. Int J Geosci 8:611–622 (Scientific Research Publishing)

Sahana M, Sajjad H, Ahmed R (2015) Assessing spatio-temporal health of forest cover using forest canopy density model and forest fragmentation approach in Sundarban reserve forest, India. Model Earth Syst Environ 1(49):1–10 (Springer)

Sonawane K, Bhagat V (2017) Improved change detection of forest using Landsat TM and ETM+ data. Remote Sens Land 18–40 (Gatha Cognition)

Chapter 9
Assessment of Forest Health using Remote Sensing—A Case Study of Simlipal National Park, Odisha (India)

Partha Sarathi Mahato, Kathakali Bandhopadhyay, and Gouri Sankar Bhunia

Abstract Forest ecosystems fulfill the entire ecosystem functions that are essential for life on our planet. However, an unprecedented level of anthropogenic influences is reducing the resilience and stability of our forest ecosystems as well as their ecosystem functions. In the present study we focused to determine forest health pattern of Simlipal National Park (Odisha, India) based on Remote Sensing and GIS techniques. Multitemporal Landsat 8 operational Land Imager (OLI) data are derived from USGS Earth Explorer Community. Normalized difference vegetation index (NDVI), SARVI (Soil and Atmospherically Resistant Vegetation Index), Modified Chlorophyll Absorption Ratio (MCARI), and Moisture Stress Index (MSI) have been used to create different vegetation indices to estimate forest health. Finally, Weighted overlay analysis is performed on GIS platform to identify the forest health pattern in the national park. NDVI index showed the maximum accuracy for identifying vegetation classes. Results showed in the eastern and central part of the study area having excellent vegetation cover. Good to moderate vegetation cover areas are observed in the south and small pockets in north of the study area. The excellent vegetation coverage area also increases day by day. To exclude the agricultural lands and cloud cover from forest area images from the month of January are selected.

Keywords Vegetation health · Landsat data · Vegetation indices · Weighted analysis · Simlipal national park

P. S. Mahato
Department of Geology, Government General Degree College, Gopiballavpur-II, Paschim Medinipur, West Bengal, India
e-mail: parthamahato1@gmail.com

K. Bandhopadhyay (✉)
Department of Geography, Vidyasagar University, Subarnarekha Mahavidyalaya, Jhargram, West Bengal, India
e-mail: titichatterjee5@gmail.com

G. S. Bhunia
Remote Sensing and GIS Expert, TPF Gentisa Euroestudio SL, Gurgaon, Haryana, India

P. K. Shit et al. (eds.), *Spatial Modeling in Forest Resources Management*, Environmental Science and Engineering,
https://doi.org/10.1007/978-3-030-56542-8_9

9.1 Introduction

Continuing biodiversity loss has prompted researchers and decision makers around the globe to reconsider current natural resource management policies and pursue alternate strategies that are successful in avoiding more habitat destruction and species extinction, while encouraging sustainable resource use (Karjalainen et al. 2010; Brockerhoff et al. 2017). India is one of the world's mega-biodiversity nations, with a deep dependency on its natural resources for economic development. For their livelihood, a significant part of India's rural and tribal population also relies directly on certain natural resources (Dash and Behera 2012). Many research studies have pointed out in recent years that the depletion of biodiversity (as regards species extinction) has risen significantly as a result of increasing human interference in the natural world (Das and Das 2008). Humans are predicted to vanish at a rate more than 1000 times faster than historically known (Dash and Behera 2013).

The Simlipal National Park in Odisha district of Mayurbhanj, which is mainly populated by indigenous peoples, was a source of tension between the tribal communities and the department of forest. The Field, spread out over 2750 sq. Km, a Tiger Reserve was proclaimed in 1973, and a sanctuary in 1979. The central government in 1986 numbered 845 sq. Km of reserve as a central area; the same area was later extended to 1,194.75 sq in 2007. Miles, encompassing 5 of the 65 villages that existed within park boundaries. Forest production accounts for over 50% of local household income and fuelwood is a 100 percent source of electricity in the Similipal Biosphere Reserve (SBR). Since 1995, nearly 20% of forest land in the biosphere reserve has been occupied by local people for farming activities are sources are primarily solely biotic by non-timber forest products (NTFP) miners, smugglers, poachers, and grazers. In the years 1991–2000, approximately 100 sq km of land was destroyed as a result of forest fire which is a significant cause of soil erosion which soil flora and fauna died. A few villages hold wells and pipes for the pipeline. For domestic duties and drinking needs, lakes and water sources are used mainly. Unhygienic a contaminated drinking water leads to multiple diseases including diarrhoea and jaundice (especially water-related vector-borne diseases).

Forest health is measured by the amalgamation of various factors like age, composition, function, vigour, presence of unusual levels of insects or diseases and reliance to disturbance (Martin and Aber 2006). Forest health monitoring has become increasingly important, especially in recent years as climate and other stressors have adversely affected forest conditions in many regions (Lindner 2014; Millar and Stephenson 2015). There has been a broad gap between the knowledge needed by forest managers and researchers and the information available for recognizing and evaluating forest health drivers 'nature and multidimensionality, stressors, threats, and consequences'. Decision- need high spatial and temporal precision forest health information, from local to global, for short and long-cycles that can be tracked at low cost. Such knowledge within regions should be analogous and based on standardized methods (Lausch et al. 2018).

Anthropogenic and environmental stressors and disruptions impact all stages of biotic organization, potentially impacting their resiliency in forest environments. Interactions between drivers, stressors, threats, and effects on sequential and spatial scales are dynamic, frequently non-linear, and multifaceted. A holistic approach is important in order to achieve a clearer understanding of the impact of various stressors and threats and in assisting land managers, decision-makers and policymakers in their forest management decisions. It involves data collection, reporting, interpretation and forest health evaluation at all stages of forest health monitoring (Trumbore et al. 2015) as well as biodiversity monitoring (Guerra et al. 2017) and forest habitat change through successful regional management of healthy biodiversity and environmental change findings, measures and scenarios (Krug et al. 2017).

Though forest health terrestrial assessment is time intensive and is typically carried out on monitoring plots, remote sensing (especially satellite-based) allows for cost-effective and time-efficient persistent analysis of forest conditions across vast areas (Frolking et al. 2009). Forest health reports are widely available from national forest inventories and surveillance systems (Traub et al. 2017), forest ecology laboratory experiments (Bruelheide et al. 2014), and reports from remote sensing (RS) and RS research items (White et al. 2016). Vegetation spectral properties derive from solar radiation contact with the composition of the cells, chlorophyll and other pigments. The amount of pigments is related to the degree of damage and the amount of chlorophyll decreases with increasing damage rates. Vegetation indices (Vis) are built from reflectance measurements over the optical spectrum of two or more wavelengths to detect different plant features, such as total leaf area and water content. Instead of measuring single-VIs and their classifications, the latest pattern in this area lies in time series applications and trajectory or feature recognition in datasets. Landsat satellite scenes have been used in forestry for decades (Coleman et al. 1990), but their use grew significantly after free data policy was implemented in 2008 (Wulder et al. 2012). All Landsat satellites routinely survey the Surface of the earth approximately once every 16 days in a spatial resolution of 30–60 m per pixel in visible, near infrared and medium infrared wavelengths in multiple spectral channels (Roy et al. 2014). Based on the above consideration, present study is undertaken to provide systematic description of remote sensing- based forest health monitoring and mapping of Simlipal National Forest of Odisha (India).

9.2 Study Area

Simlipal National Park (SNP) forest derives its name from the abundance of area-growing red silk cotton trees, and is India's 7th largest national park. The park has some stunning cascades such as Joranda and Barehipani Falls. It is home to Bengal tiger, Asian lion, chausingha and gaur. It is extended between 21° 10′ to 22° 12′ N latitude and 85° 58′ to 86° 42′ E longitude, located in Mayurbhanj district in Odisha within Mahanadian east-coastal bio-geographic region of tropical eastern India (Fig. 9.1). The study area is covered by 2750 sq km. The average height is

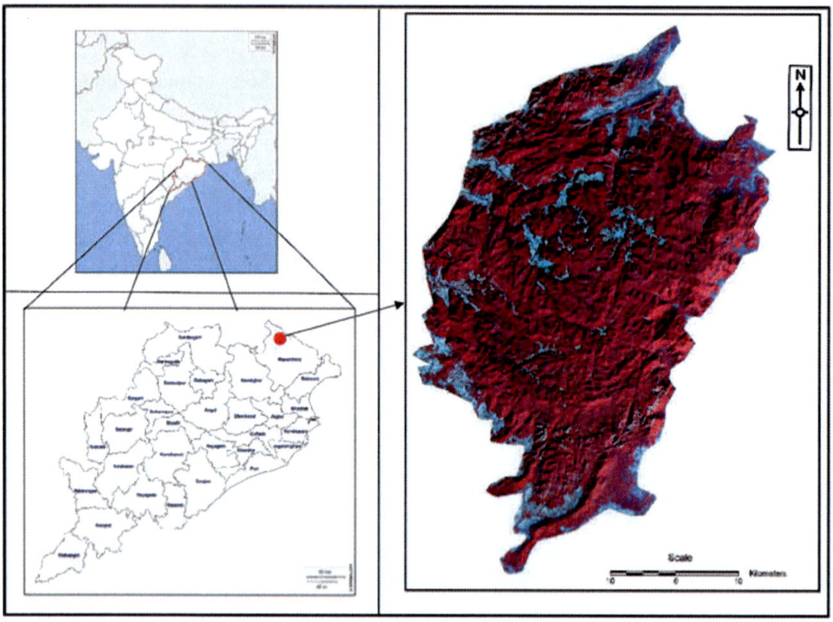

Fig. 9.1 Location map of Simplipal National Park, Mayurbhanj district (Odisha, India)

559.31 m. The Similipal region is therefore undulating, rising from 600 to 1500 m. The tropical monsoonal climate with three distinct seasons i.e. summer, monsoon and winter prevail over the Similipal. The southern flank of Similipal adjoining Devasthali, Upper Barhakamuda, Bhanjabasa and Nawana valley experience occasional frost during winter. The average maximum temperature during May is 33.5 °C and the average minimum temperature is 9.8 °C during December. High relative humidity prevails throughout the year which goes up to 90% during rainy season. Spring breeze is quite common throughout the park. The park is a treasure house containing 1076 plant species belonging to 102 families. Here 96 orchid species have been described (Jena 2005). It occurs in the moist deciduous forest ecoregion of the Eastern Highlands, with tropical moist broadleaf forest and tropical moist deciduous forests with dry deciduous hill forests and high-level Sal forests. Simlipal National Park has recorded a total of 42 species of mammals, 242 species of birds and 30 species of reptiles.

Table 9.1 Details of satellite data are used in this study

Satellite/sensor	Date of acquisition	Spectral resolution (micrometres)	Spatial resolution (m)	Radiometric resolution (bit)	Temporal resolution (day)
Landsat 8 operational land imager (OLI)	21/02/2019	Band 1: 0.43–0.45	30	16	16
		Band 2: 0.45–0.51	30	16	16
	15/01/2017	Band 3: 0.53–0.59	30	16	16
		Band 4: 0.64–0.67	30	16	16
	26/01/2015	Band 5: 0.85–0.88	30	16	16
		Band 6: 1.57–1.65	30	16	16
		Band 7: 2.11–2.29	30	16	16
		Band 8: 0.50–0.68	15	16	16
		Band 9: 1.36–1.38	30	16	16
		Band 10: 10.60–11.19	100	16	16
		Band 11: 11.50–12.51	100	16	16

9.3 Materials and Method

9.3.1 Data Sources and Pre-processing

The survey of India Toposheets number 73 J/4, 73 J/8, 73 J/12, 73 K/1, 73 K/2, 73 K/5, 73 K/6, 73 K/9, 73 K/10 with scale 1:50,000 covering the study were used for reference purpose. The latest medium-resolution satellite imagery Landsat 8 OLI satellite imagery (Path/Row–139/45) of 2019, 2017 and 2015 were acquired from USGS Earth Explorer Community (https://earthexplorer.usgs.gov/). The details of the satellite data are described in Table 9.1. The Landsat OLI image were geo-referenced to the Universal Transverse Mercator (UTM) projection zone 43 and World Geodetic System (WGS) 84 datum. All the images are radiometrically corrected to histogram equalization.

9.3.2 Estimation of Forest Health

To study the forest health of Simlipal National Park, multispectral spatial data from Landsat8 series were used. The Normalized difference vegetation index (NDVI), which distinguishes actively growing vegetation from background features, is the most widely used vegetation index in remote sensing studies (Hansen et al. 2000). Other indices for measuring cover commonly used include, the Soil Adjusted Vegetation Index (SAVI), Enhanced Vegetation Index (EVI), Modified Soil Adjusted

Vegetation Index (MSAVI) and the Transformed Soil Adjusted Vegetation Index (TSAVI) (Qi et al. 1994). In the present study NDVI, MSI (Moisture Stress Index), MCARI (Modified Chlorophyll Absorption Ratio Index), MSAVI (Modified Soil Adjusted Vegetation Index), and SARVI (Soil and Atmospherically Resistant Vegetation Index using) QGIS and ERDAS Imagine 2014. The calculated vegetation indices are reclassified using raster classify tool based on the pixel values into fourfold classification.

9.3.3 Estimation of Normalized Difference Vegetation Index (NDVI)

NDVI quantifies vegetation vigor by measuring the difference between near-infrared (NIR) and red wavelength of the electromagnetic spectrum. The pigment of plant leaves and chlorophyll strongly absorbs visible light (from 0.4 to 0.7 μm; $Band_4$) for the use of photosynthesis. The cell structure of the leaves, on the other hand, strongly reflects NIRwavelength (from 0.7 to 1.1 μm; $Band_5$). The NDVI is calculated from these individual measurements as follows:

$$NDVI = \frac{(NIR - RED)}{(NIR + RED)}$$

The value of NDVI ranges from -1 to $+1$. The NDVI amount of water is less than 0, the bare soils are between 0 to 0.1 and more than 0.1 vegetation. Increasing NDVI's optimistic outlook would make plants greener.

9.3.4 Estimation of Soil and Atmospherically Resistant Vegetation Index (SARVI)

By using the red-blue channel instead of the red channel, Kaufman and Tanre (1992) suggested the Soil Atmospherically Vegetative Index (SARVI). It uses a self-correction process for the atmospheric effect on the red channel, using the difference in the radiance between the blue and the red channels to correct red reflectance for atmospheric scattering. It is most useful in regions of high atmospheric aerosol content, including tropical regions contaminated by soot from slash-and-burn agriculture.

$$SARVI = \frac{(NIR - RB)(1 + L)}{(NIR + RB + L)}$$

where

RB denotes $Red - \gamma(Blue - Red)$,
L is the soil adjustment factor (1.0).

The value of this index ranges from −1 to 1, with higher positive pixel values corresponding to healthier and greener vegetation.

9.3.5 Modified Chlorophyll Absorption Ratio (MCARI)

MCARI indicates the relative abundance of chlorophyll (Daughtry et al. 2000). This provides a soil modification element while retaining LAI exposure and chlorophyll tolerance.

$$MCARI2 = \frac{1.5[2.5(\rho_{800} - \rho_{670}) - 1.3(\rho_{800} - \rho_{550})]}{\sqrt{(2 * \rho_{800} + 1)^2 - (6 * \rho_{800} - 5 * \sqrt{\rho_{670}}) - 0.5}}$$

Illumination patterns, surface substrate reflections and other non-photosynthetic factors are not influencing MCARI values.

9.3.6 Estimation of Moisture Stress Index (MSI)

MSI calculates the leaf water content. Moisture stress index detects the leaf water content using the near- and middle infrared reflection ratio. This index is used for canopy stress analysis. Higher values of the index indicate greater plant water stress and in inference, less soil moisture content. As the water content of leaves in vegetation canopies increases, the strength of the absorption around 1599 nm ($Band_6$) increases. Absorption at 819 nm ($Band_5$) is nearly unaffected by changing water content, so it is used as the reference.

$$MSI = \frac{\rho_{1599}}{\rho_{819}}$$

The value of this index ranges from 0 to more than 2. The higher value of MSI indicates greater water stress and less water content.

9.3.7 Accuracy Assessment

To check the accuracy, the ground truth data was compared with the graded picture. For check the accuracy of classification, the customer and the manufacturer were determined. The producer's accuracy can be achieved by dividing the number of correctly identified pixels by the actual precision of the class. Errors were identified in misclassified pixels and in another class of misclassified. The device's accuracy depends, however, on the form of pixels the same community has obtained (Congalton 1991). The overall precision for the exactness and reliability of the consumer and supplier was determined based on the uncertainty matrix. The overall accuracy definition is given below:

$$Overall\,accuracy = \frac{Total\,number\,of\,correct\,classified}{Total\,number\,of\,pixel} \times 100$$

9.3.8 *Weighted Sum Analysis*

To assess the forest health more precisely weighted sum analysis have been applied where the values of different indices added according to the given weightage (Kumari and Ashok 2017). In this analysis highest weightage given to the NDVI value followed by MCARI and SARVI, and MSI has been assigned as lowest value. In this analysis, the weightage value for NDVI is assigned as '4', for MCARI as '2', for SARVI as '3' and for MSI as '1'. The values are assigned based on the literature survey and expert knowledge. For the preparation of weighted sum of the vegetation indices the following formula is used using raster calculator of QGIS software version 3.0:

$$\mathbf{((''NDVI'' * 4) + (''MCARI'' * 2) + (''SARVI'' * 3) + (''MSI'' * 1))/10}$$

All the four indexes have similar input in weighted sum analysis dependent on the weighted values. The distinction depends on the pixels of all four indices falling within the full classification region. Finally, a forest health map was developed which scaled from low to excellent vegetation.

The Weighted Sum function provides the weighting capability and the combination of multiple inputs to create an integrated analysis. It is analogous to the Weighted Overlay feature in that multiple raster inputs can be conveniently combined to represent various variables, integrating weights or relative value. The Weighted Sum feature does not rescale the reclassified values back to an evaluation scale, it also requires floating-point and integer values to preserve their resolution.

9.4 Results and Discussion

On the global scale forests are threatened by population growth and human activities, including deforestation, air pollution and climate change (Anon. 2001; Kimmins 1997). Climate change is likely to cause increasing forest damage and tree mortality from direct and indirect causes (Martinez-Vilalta 2002; Tenow et al. 1999; Auclair 1993).

9.4.1 NDVI Analysis

The NDVI value of the study area ranges from −0.05 to 0.488 in 2015 with an average value of 0.293 ± 0.048 (Table 9.2). In 2017, the NDVI value of the study area ranges between −0.19 and 0.47 (mean ± SD–0.28 ± 0.056). In 2019, the average NDVI value is calculated as 0.293 ± 0.057. According to Gross (2005), very low NDVI values (0.1 and lower) correspond to barren areas of rock, sand and snow. Moderate values (0.2–0.3) indicate shrub and grassland, while temperate and tropical rainforests are represented by high NDVI values (0.6–s0.8). As per the value of NDVI the vegetation cover has been divided in 4 types based on NDVI values, where poor vegetation represent the less photosynthetic activity and has lower values of NDVI whereas Excellent Vegetation represent highest photosynthetic activity and has the highest NDVI values. The area covered by the three types of vegetation from NDVI values are summarized in Table 9.2. In the study area, the higher average value of NDVI is recorded in 2019 (0.293), followed by 2017 (Table 9.2).

Figure 9.2 shows the areal distribution of various type of forest cover in the study

Table 9.2 Descriptive characteristics of NDVI value from 2015–2019

Year	Maximum	Average	Standard deviation	Minimum
2019	0.488	0.293	0.057	−0.21
2017	0.47	0.28	0.056	−0.19
2015	0.427	0.245	0.048	−0.05

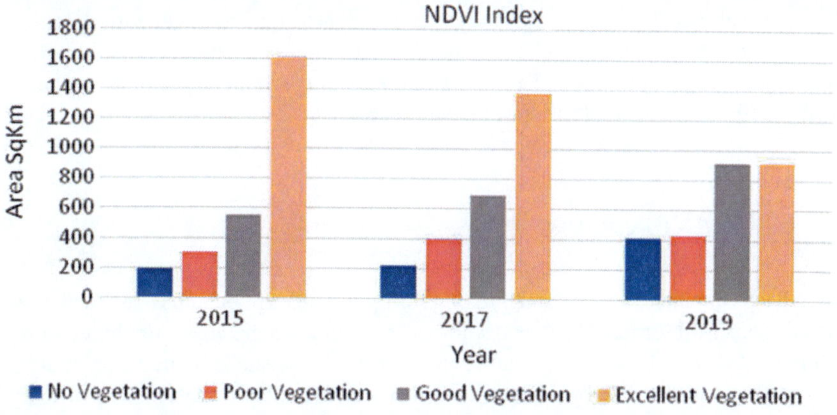

Fig. 9.2 Areal distribution of vegetation characteristics derived through NDVI index

area during the period between 2015 and 2019 on the basis of NDVI value. Results showed the areal distribution of excellent vegetation cover in the study area gradually decreases during the study period. The areal distribution of good vegetation and poor vegetation cove do shows a disruptive and dynamic variation. The changes are quite prominent indicating a positive response in terms of photosynthetic capacity of the plants, in the study area. However, an extreme exploitation as per spatial expansion is viewed mainly for poor vegetation and no vegetation. Most of the central and eastern part of the study area is characterized by excellent vegetation coverage. Good vegetation coverage is mostly found in the central part of the park and poor vegetation coverage are observed in the south-east and north of the park. A complete scarcity of vegetation coverage is found west and extreme north-west and some small pockets of the central part of the study area (Fig. 9.3).

9.4.2 SARVI Analysis

The highest value of SARVI indicate the high vegetation coverage and vice-versa. In 2019, the highest value of SARVI is calculated as 0.812 with an average value of 0.301 ± 0.291. The maximum SARVI value in 2017 is calculated as 0.833 with an average value of 0.307 ± 0.297. In 2019, the SARVI value ranges between -0.053 and 0.789 within an average value of 0.289 ± 0.277 (Table 9.3).

As per SARVI analysis, a similar trend has been noticed as being observed in NDVI index. Excellent Vegetation provides the highest coverage; however, the trend is not linear. However, a declining trend has been noticed for both moderate vegetation and poor vegetation cover. This may be due to the anthropogenic activities by increasing non-vegetative areas (Fig. 9.4). The area of no vegetation coverage is more or less similar.

Figure 9.5 displays the spatial distribution of vegetation coverage of Simlipal national forest area derived through SARVI index. Results of the analysis showed in the eastern and central part of the study area is characterized by excellent vegetation cover. In south and small pockets of north of the park are characterized by good to moderate vegetation cover. Less vegetation coverage is observed in the extreme north, north-west and south-east of the study area.

9.4.3 MCARI Analysis

MCARI gives a measure of the depth of chlorophyll absorption and is very sensitive to variations in chlorophyll concentrations. Chlorophyll is the green pigment present in the leaves and plays an important role in photosynthesis i.e. conversion of light energy to chemical energy. Hence, it is a direct indicator of the plant's primary production and photosynthetic potential. The MCARI value in 2015 ranges between

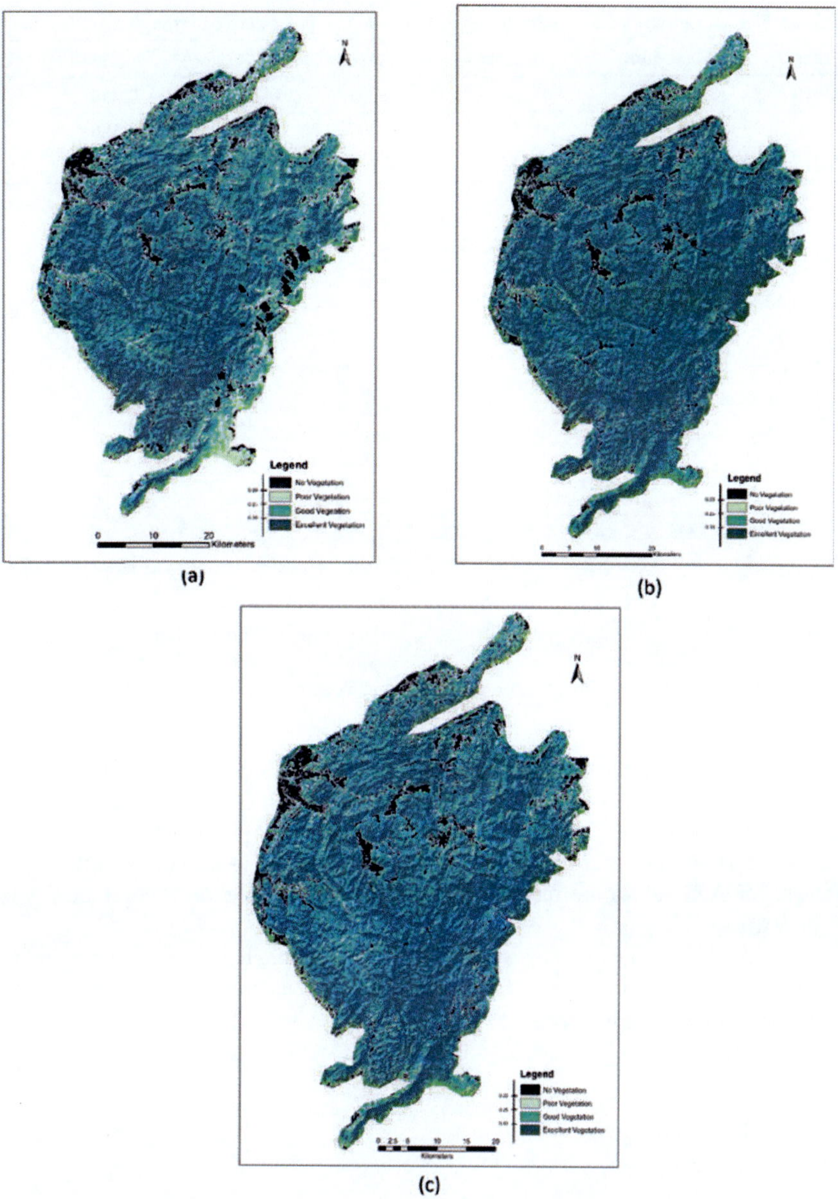

Fig. 9.3 NDVI classified of year 2019 (**a**), 2017 (**b**), 2015 (**c**) respectively

Table 9.3 Descriptive characteristics of vegetation characteristics derived through SARVI index

Year	Maximum	Average	Standard deviation	Minimum
2019	0.812	0.301	0.291	−0.289
2017	0.833	0.307	0.297	−0.386
2015	0.789	0.289	0.277	−0.053

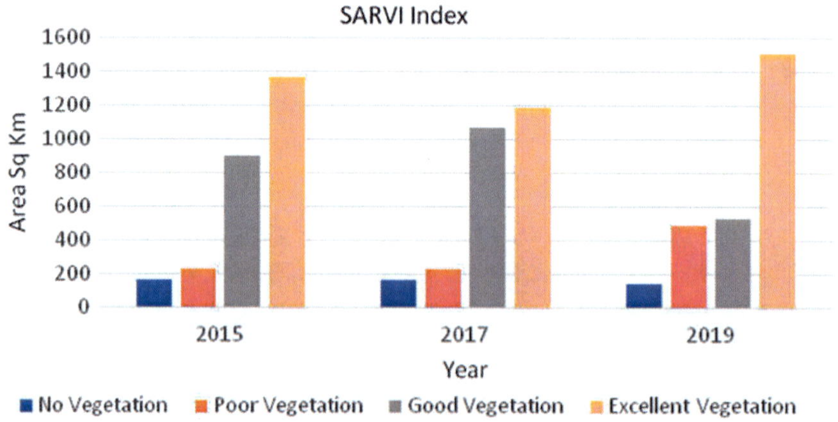

Fig. 9.4 Represent the areal distribution of vegetation characteristics using SARVI index

−0.045 to 0.592. The average value of MCARI is calculated as 0.219. The average MCARI value in 2017 is calculated as 0.238 (−0.484 to 0.631). In 2019, the MCARI of the study area varies from −0.389 to 0.633 (Table 9.4). The MCARI value for will increase with increasing vegetation quality. With reference points from NDVI image, MCARI values are recorded for good, moderate and poor vegetation type (Fig. 9.6).

The MCARI analysis shows a decline in chlorophyll concentration with in moderate vegetation, increasing the poor vegetation cover graph. The non-vegetation area also shows a positive trends to recent years (Fig. 9.7).

9.4.4 MSI Analysis

By reference point set over NDVI, MSI values are recorded for the three set of classes. The MSI values at high vegetation area is 0.2–0.6, in moderate vegetation area 0.6–0.8 and low vegetation area is 0.8–1. The rest of the MSI image is then classified according to this values. The area coverage of the three types of vegetation class as per MSI index are summarized in Table 9.5 and Fig. 9.8.

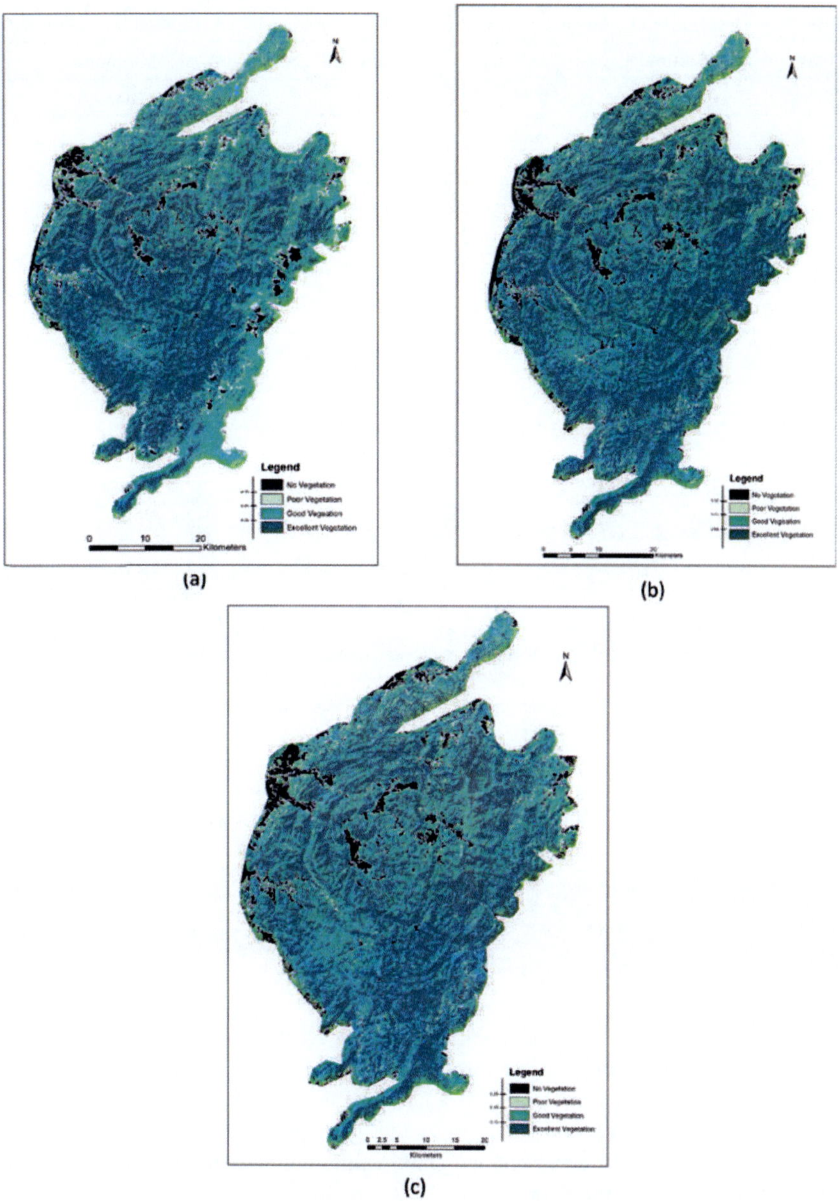

Fig. 9.5 **a, b, c** SARVI classified of year 2019, 2017, 2015 respectively

Table 9.4 Descriptive characteristics of vegetation characteristics derived through MCARI index

Year	Maximum	Average	Standard deviation	Minimum
2019	0.633	0.241	0.229	−0.389
2017	0.631	0.238	0.229	−0.484
2015	0.592	0.219	0.210	−0.045

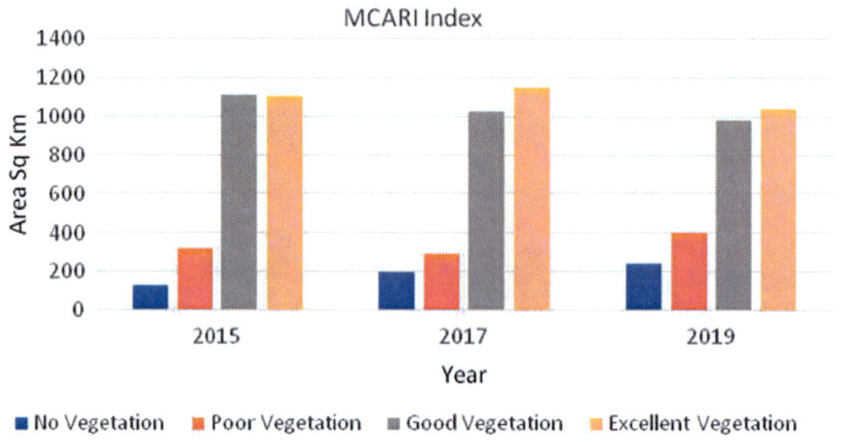

Fig. 9.6 Areal distribution of vegetation cover calculated from MCARI index

The MSI and MCARI indices also do reflects the similar trend that was predominant for SARVI and NDVI index (Fig. 9.9). It is here, thus a weighted sum analysis was prepared for better assessment of the forest health of Simlipal forest area.

9.4.5 Accuracy Assessment

The accuracy assessment has been done for all the vegetation indices. The maximum overall accuracy is calculated for the NDVI index (94.51%), followed by SARVI index (93.90%), MCARI index (92.31%), and MSI index (89.20%). The details of overall accuracy, user accuracy is represented in Table 9.6. However, based on the accuracy of the vegetation indices, weighted values are assigned for the vegetation classes to identify the forest health.

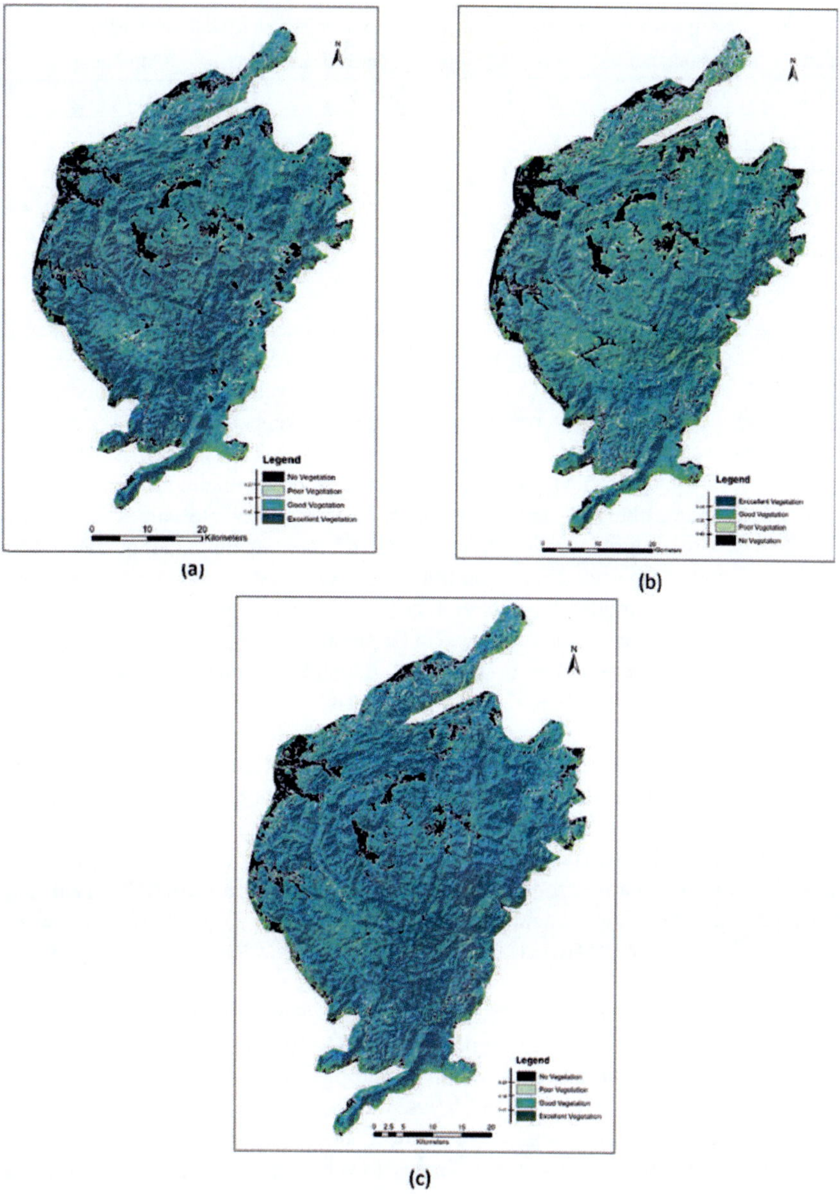

Fig. 9.7 MCARI classified output for the year **a** 2019, **b** 2017, **c** 2015 respectively

Table 9.5 Descriptive characteristics of vegetation characteristics derived through MSI

Year	Maximum	Average	Standard deviation	Minimum
2019	1.504	0.748	0.114	0.528
2017	1.451	0.754	0.110	0.543
2015	1.435	0.760	0.108	0.581

9.4.6 Weighted Sum Analysis

The weighted sum analysis of all three vegetation indices resulted in a forest health map and the values ranges from 0 to 4 (Fig. 9.10). As the healthy vegetation will show high value of NDVI, MCARI, SARVI and low value for MSI indices, area with suitable condition is referred to as most healthy vegetation of the regime. On the contrary the lowest value of weighted sum index which has low values of NDVI, MCARI, SARVI and high MSI determined as low photosynthetic activity and high water stress index likely to be no vegetation area. Intermediate values have been classified base on weightage values to moderate and poor vegetation vigour. Based on the derived output, the study area is classified four categories, such as 1-no vegetation, 2-poor vegetation, 3-moderate vegetation and 4-excellent vegetation (Table 9.7).

Non vegetation areas mostly observed in Gopinathpur and adjacent town/villages, drainage channels, roads, cultivation land etc. Excellent vegetation is the type show higher values for NDVI, SARVI and MCARI and lower values in MSI indices, indicating healthy vegetation cover. The high value of index value indicates excellent vegetation, which has a greater NIR reflectance over RED band, and hence indicates a greater amount of photosynthetic activity. The excellent vegetation area covered by 59.84% in 2015 and its gradually decreased upto 43.85% in 2019 (Fig. 9.11). The higher SARVI value for excellent vegetation does corrects any distortions or anomalies that can cause from soil or atmospheric scattering of RED band and hence supports the NDVI conclusion. The MCARI index represents the amount of chlorophyll content present in leaf. The higher value of MCARI covered by vegetation that must have higher chlorophyll content in healthy tree leaf. The lower MSI values indicate lower water stress for region with excellent vegetation i.e. water retention by leaf are higher or adequate to support a healthy vegetation cover.

Excellent vegetation designates dense and healthy vegetation vigour. The Moderate Vegetation covered area shows somewhat lower NDVI, SARVI and MCARI values with slightly higher MSI value, hence can be determined as moderate vegetation; whereas poor vegetation shows lower values of NDVI, SARVI and MCARI with higher in MSI index among the vegetation covers, and is considered as the less vegetation coverage and bare ground.

The non-vegetative areas are marked by negative NDVI, SARVI and MCARI values with highest value of MSI index that indicates high moisture stressed area with soft stem plant and scrub land. These areas are mostly covered by human habitat,

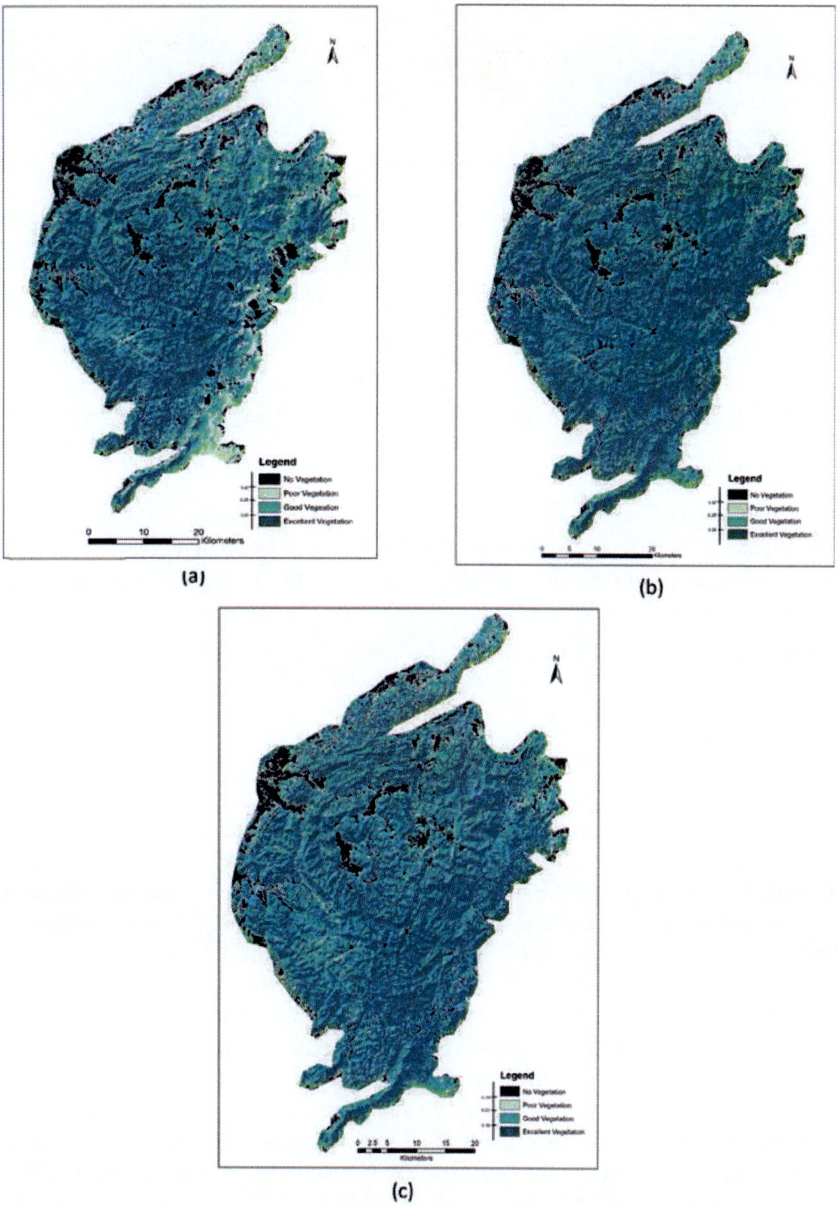

Fig. 9.8 MSI classified output for the year **a** 2019, **b** 2017, **c** 2015 respectively

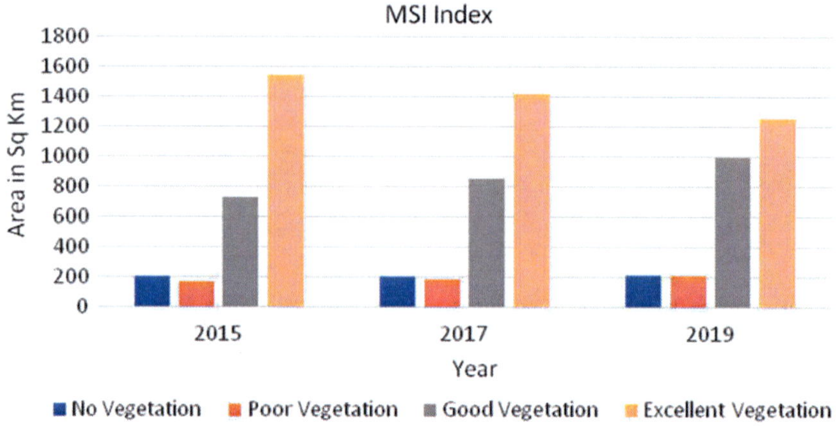

Fig. 9.9 Areal distribution of vegetation coverage derived through MSI index

river channels, roads and other man made infrastructure. To avoid repetitive indexing agricultural crops from forest, images from the month of January are selected so that agricultural land do not have growing crops to avoid misleading data on reflected data.

9.5 Conclusion

The Simlipal forest spread over 2600 sq km has 11.46% of land with poor vegetation, 32.84% land of moderate to good vegetation and 43.85% of land with excellent vegetation cover with 11.83% of land has no vegetation as of present study from Landsat OLI image. It shows a monotonous surface cover of the different vegetation types when compared to past years. Results of the analysis also showed, the excellent vegetation cover in the study area is gradually decreased; whereas, the poor vegetation coverage area is increased progressively. The agricultural lands are excluded from forest and are classified as the non-vegetation area that also includes land altered by human for construction, tourism and other actives and natural drainage channels. Identifying and quantifying forest quality and destruction of woodland has been a troubling problem that needs to be tackled as quickly as possible. This idea only became widespread when the population became aware of the intrinsic importance of forest land, which were unsustainably abused to satisfy their demands. Remote sensing and GIS have made a significant advance in this area. Its technology is user friendly, feasible usability. The monitoring should be continued to assessment variation and effect on forest health over time and to take necessary possible action in time to protect our forests.

Table 9.6 Accuracy assessment report of vegetation classes for MCARI

Class	Reference pixel	Classified pixel	Number correct	Producers accuracy (%)	User accuracy (%)
NDVI					
Excellent vegetation	56	54	53	94.64	98.15
Good vegetation	50	53	49	98.00	92.45
Poor vegetation	32	34	30	93.75	88.24
No vegetation	26	23	23	88.46	100.00
Totals	164	164	155		
Overall Classification Accuracy = 94.51%					
MSI					
Excellent vegetation	52	55	47	90.38	85.45
Good vegetation	48	52	44	91.67	84.62
Poor vegetation	33	24	23	69.70	95.83
No vegetation	31	33	30	96.77	90.91
Totals	164	164	154		
Overall Classification Accuracy = 89.20%					
MCARI					
Excellent vegetation	50	51	49	98.00	96.08
Good vegetation	53	56	50	94.34	89.29
Poor vegetation	35	34	30	85.71	88.24
No vegetation	26	23	22	84.62	95.65
Totals	164	164	154		
Overall Classification Accuracy = 92.31%					
SARVI					
Excellent vegetation	54	56	52	96.30	92.86
Good vegetation	56	50	50	89.29	100.00
Poor vegetation	32	33	30	93.75	90.91
No vegetation	22	25	22	100.00	88.00
Totals	164	164	155		
Overall Classification Accuracy = 93.90%					

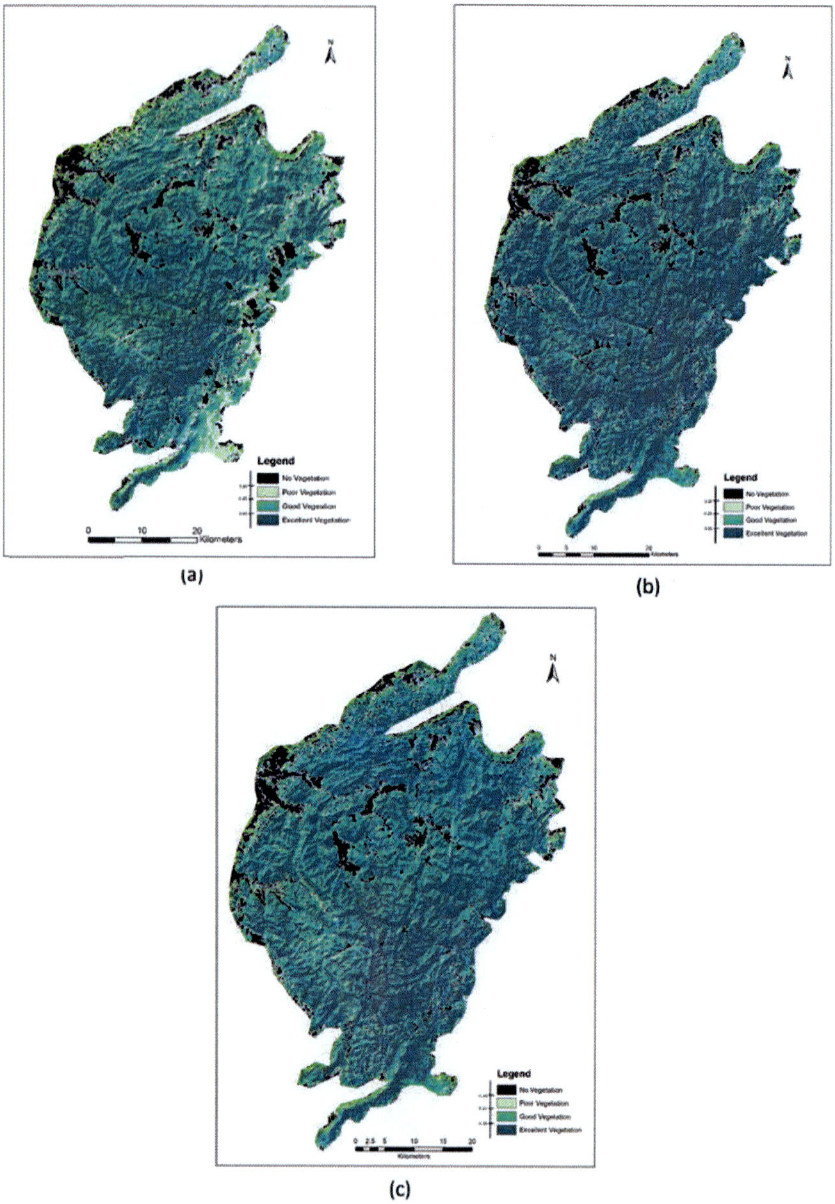

Fig. 9.10 Weighted sum classified of year 2019 (**a**), 2017 (**b**), 2015 (**c**)

Table 9.7 Areal distribution of vegetation derived from weighted

Index value	Interpretation	Area covered (sq km)			Area covered (%)		
		2015	2017	2019	2015	2017	2019
1	No vegetation	195	235	316	7.32	8.79	11.83
2	Poor vegetation	315	356	306	11.83	13.32	11.46
3	Good vegetation	559	717	877	20.99	26.84	32.84
4	Excellent vegetation	1593	1363	1171	59.84	51.02	43.85

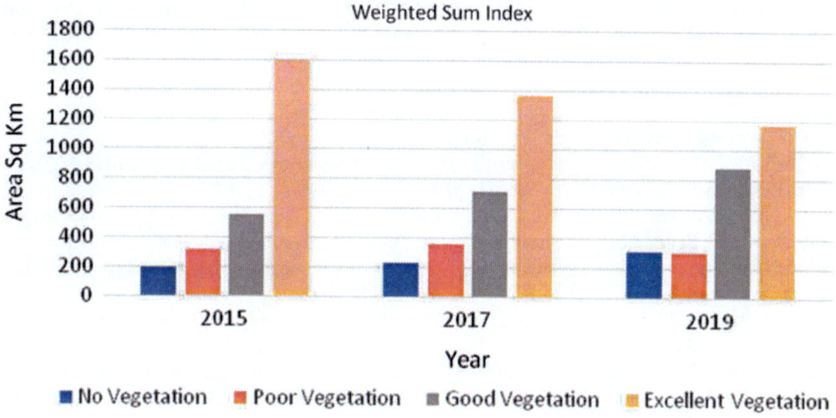

Fig. 9.11 Areal distribution of vegetation coverage calculated through weighted sum index

References

Brockerhoff EG, Barbaro L, Castagneyrol B et al (2017) Forest biodiversity, ecosystem functioning and the provision of ecosystem services. Biodivers Conserv 26:3005–3035. https://doi.org/10.1007/s10531-017-1453-2

Bruelheide H, Nadrowski K, Assmann T, Bauhus J, Both S, Buscot F, Chen XY, Ding B, Durka W, Erfmeier A et al (2014) Designing forest biodiversity experiments: general considerations illustrated by a new large experiment in subtropical China. Methods Ecol Evol 5:74–89

Coleman TL, Gudapati L, Derrington J (1990) Monitoring forest plantations using landsat thematic mapper data. Remote Sens Environ 33:211–221

Congalton RG (1991) A review of assessing the accuracy of classifications of remotely sensed data. Remote Sens Environ 37(1):35–46

Fredric County, Maryland (2009) Forest resource ordinance, pp 1–14. https://www.stormwatercenter.net/Model%20Odinances/misc-forest-conservation

Das S, Das BP (2008) Similipal biosphere: genesis of historicity. Orissa review

Dash M, Behera B (2012) Management of similipal biosphere reserve forest. Adv. Forest Lett (AFL) 1(1):7–15

Dash M, Behera B (2013) Biodiversity conservation and local livelihoods: a study on similipal biosphere reserve in India. J Rural Dev 32(4):409–426

Daughtry CST, Walthall CL, Kim MS, Colstoun EBD, Mcmurtrey JEI (2000) Estimating corn leaf chlorophyll concentration from leaf and canopy reflectance. Remote Sens Environ 74:229–239

Eden MJ (1996) Forest degradation in the tropics: environmental and management issues. In: Eden JM, John TP (eds) Land degradation in the tropics. Printer, A Cassell Imprint, New York, USA, pp 41–47

Frolking S, Palace MW, Clark DB, Chambers JQ, Shugart HH, Hurtt GC (2009) Forest disturbance and recovery: a general review in the context of spaceborne remote sensing of impacts on aboveground biomass and canopy structure. J Geophys Res 114:3–27

Goutam NC (1983) Satellite remote sensing technique for natural resource survey. In: Singh RL et al (eds) Environmental management. Allahabad Geographical Society, pp 177–181

Gross D (2005) Monitoring agricultural biomass using NDVI time series, food and agriculture organization of the United Nations (FAO). Italy, Rome

Guerra C, Navarro LM, Kissling WD, London MC, Turak E, Balvanera P, Costello MJ, Delavaud A, El Serafy GY, Ferrier S et al (2017) Monitoring biodiversity change through effective global coordination. Curr Opin Environ Sustain 29:158–169

Jena ML (2005) Similipal's scenic splendor. Women's Era 32(752):110–112

Karjalainen E, Sarjala T, Raitio H (2010) Promoting human health through forests: overview and major challenges. Environ Health Prev Med 15(1):1–8

Kaufman YJ, Tanré D (1992) Atmospherically resistant vegetation index (ARVI) for EOS-MODIS. IEEE Trans Geosci Remote Sens 30:261–270

Krug CB, Schaepman ME, Shannon LJ, Cavender-bares J, Cheung W, Mcintyre PB, Metzger JP, Obura DO, Schmid B, Strassburg BBN et al (2017) Observations, indicators and scenarios of biodiversity and ecosystem services change—a framework to support policy and decision-making. Curr Opin Environ Sustain 29:198–206

Kumari R, Asok S (2017) Remote sensing based forest health analysis using gis along fringe forests of Kollam district, Kerala. Int J Res Appl Sci Eng Technol (IJRASET) 5(x):740–751

Lausch A, Borg E, Bumberger J, Dietrich P, Heurich M, Huth A, Jung A, Klenke R, Knapp S, Mollenhauer H, Paasche H, Paulheim H, Pause M, Schweitzer C, Schmulius C, Settele I, Skidmore AK, Wegmann M, Zacharias S, Kirsten T, Schaepman ME (2018) Understanding Forest health with remote sensing, part III: requirements for a scalable multi-source forest health monitoring network based on data science approaches. Remote Sens 10(7):1120. https://doi.org/10.3390/rs10071120

Lindner M, Fitzgerald JB, Zimmermann NE, Reyer C, Delzon S, van der Maaten E et al (2014) Climate change and European forests: What do we know, what are the uncertainties, and what are the implications for forest management? J Environ Manage 146:69–83

Forest Management (2008) www.manage.gov.in

MEA (2005) Ecosystems and human wellbeing. Island Press, Covelo, CA, Millennium Ecosystem Assessment

Millar CI, Stephenson NL (2015) Temperate forest health in an era of emerging mega disturbance. Science 349:823–826

Philip H (1990) Forest, forest reserve and forest land in Thailand. Geograph J 156(2) Blackwell Publishing on behalf of The Royal Geographical Society (with the Institution of British Geographers), pp 166–174

Roy DP, Wulder MA, Loveland TR, Woodcock CE, Allen RG, Anderson MC et al (2014) Landsat-8: science and product vision for terrestrial global change research. Remote Sens Environ 145:154–172

Traub B, Meile R, Speich S, Rösler E (2017) The data storage and analysis system of the Swiss national forest inventory. Comput Electron Agric 132:97–107

Trumbore S, Brando P, Hartmann H (2015) Forest health and global change. Science 349:814–818

Turkmen, N, Duzenli A, Cakan H, Ozturk M (1996) Impact of environmental problems on the natural vegetation in the east mediterranean part of Turkey. In: Kapur AS, Eswaran HA, Kelling

G, Vita-Finzi C, Mermut AR, Ocal AD (eds) Proceedings of 1st international conference on land degradation, Cukurova University Press, pp 245–253, ISBN 975-4870519
White JC, Coops NC, Wulder MA, Vastaranta M, Hilker T, Tompalski P (2016) Remote sensing technologies for enhancing forest inventories: a review. Can J Remote Sens 8992:1–23
Wulder MA, Masek JG, Cohen WB, Loveland TR, Woodcock CE (2012) Opening the archive: how free data has enabled the science and monitoring promise of landsat. Remote Sens Environ 122:2–10

Part II
Modeling, Risk Assessment and Vulnerability

Chapter 10
Forest Health Monitoring using Hyperspectral Remote Sensing Techniques

Narayan Kayet

Abstract Hyperspectral Remote sensing is a handy tool for forest health monitoring. This study focuses on forest health monitoring using hyperspectral satellite data and validates it with tree spectral data. In the study area, increasing mining and anthropogenic activities within and near forest lands have caused threats to forest health. All these necessitate assessing the forest health in the areas surrounding mines. We have used two methods for the forest health assessment: one is VIs (vegetation indices) based model, and another is tree spectral analysis. The supervised classification (SAM) method was used for forest health classification based on spectral data. The results showed that a healthy forest portion was located in the hilly side of the study area while an unhealthy segment was situated alongside the mines. Hyperion data-based VIs model shows better accuracy than spectral based other methods. Also, it was found that the hyperspectral data based forest health classification gave a higher accuracy than multispectral data. Finally, forest health results were justified by ground tree spectral data. This work provides an effective guideline for forest planning and management.

Keywords Forest health · Hyperspectral data · Vegetation indices · GIS · Mining area

10.1 Introduction

Hyperspectral remote sensing is a very useful tool for forest health monitoring. Hyperspectral data are spectrally overdetermined, which means that they provide spectral information to identify and distinguish spectrally unique materials. Also this data has the potential for more accurate information extraction than possible from multispectral data (Apostolescu et al. 2016; Navinkumar and Parmar 2016). The hyperspectral sensor provides hundreds of narrow spectral bands of the Earth's

N. Kayet (✉)
Department of Mining Engineering, Indian Institute of Technology, Kharagpur, India
e-mail: narayankayet@gmail.com

P. K. Shit et al. (eds.), *Spatial Modeling in Forest Resources Management*, Environmental Science and Engineering,
https://doi.org/10.1007/978-3-030-56542-8_10

surface features. NASA EO-1 satellite, equipped with the Hyperion spectrometer, which has a sampling distance of 10 nm within the 7.7 km swath path, provides 242 spectral bands within the range of 400–2500 nm wavelengths of the EM spectrum (Hyperion user guide). This sensor easily detects forest biochemical and biophysical properties (Asner et al. 2015; Thenkabail et al. 2001; Chambers et al. 2007). Hyperspectral data can detect pest and blight conditions in a forest as well as be used for assessing areas for timber harvesting. Narrow-banded hyperspectral data has been used for forest health mapping (Tuominen et al. 2009; Kayet et al. 2019a, b), to depict the overall healthy and unhealthy portions of a forest. Recent developments in hyperspectral data acquisition from satellite-borne spectrometers have opened new areas of research that could bring revolutionary changes in the current approaches to forest management (Franklin 2001). Some researchers have worked on vegetation stress from the derivative chlorophyll index and leaf area index (Broge and Leblanc 2001; Brantley et al. 2011; Darvishzadeh et al. 2011). They have used the airborne hyperspectral image for this analysis. Detection of vegetation stress by hyperspectral remote sensing technique is based on the assumption that vegetation stress factors interfere with photosynthesis or the physical structure of the vegetation and affect the absorption of light energy and thus alter the reflectance spectrum of the vegetation (Zarco-Tejada et al. 2009; Jacquemoud et al. 2009; Bellvert et al. 2014). Airborne hyperspectral data is used for the estimate and distribution of different species in the forest areas. Hyperspectral data has used to develop the Photochemical Reflectance Index (PRI) for distinguishing the species-wise variations of leaves (Cho et al. 2008; Mashimbye et al. 2012; Darvishzadeh et al. 2008). The hyperspectral narrow banded data-based NDVI (normalized difference vegetation index) and LAI (leaf area index) has been used for plant health detection by Zarco-Tejada et al. (2005). They had used red, red edge, and NIR bands for this analysis. The International Institute for Geo-Information Science and Earth Observation (ITC) has studied vegetation health and tree species classification using hyperspectral data. They have used full pixel classification methods for vegetation health and tree species based on ground tree spectral data (Vauhkonen et al. 2011; Dalponte et al. 2014). The decision tree classifier tool of ENVI will be used to classify the pixels of a Hyperion image for necessary information acquisition for forest management purposes (vegetation indices tutorial ENVI). Ma et al. (2017) used airborne Hyperspectral data to develop photochemical reflectance index (PRI) for distinguishing the species wise variations of leafs. A number of researchers have assessed the vegetation stress from derivative chlorophyll index and leaf area index estimation, using compact airborne spectrographic image (Wu et al. 2010; Zarco-Tejada et al. 2002; Lee et al. 2004). The detection of vegetation stress by hyperspectral remote sensing techniques is based on the assumption that vegetation stress factors interfere with photosynthesis or the physical structure of the vegetation and affect the absorption of light energy and thus alter the reflectance spectrum of vegetation (Zarco-Tejada et al. 2009; Calderon et al. 2013; Li and Guo 2016).

The study area is located in the in an region that has many mines. Mining fields are under high-stress conditions showing signs of dry and dying plant species. In the study area, increasing mining and anthropogenic activities within and near forests pose threats to forest health. All these necessitate monitoring of the forest health in surrounding mining areas.

10.2 Materials and Methods

10.2.1 Study Area

The present study has been done in the Saranda Forest and its surrounding areas, which are located in the West Singhbhum district of the Indian state of Jharkhand (Fig. 10.1). It is famous for having Asia's largest Sal forests and is an important elephant habitat. Over the last few decades, in this region, many iron ore mining towns have emerged, e.g. Gua, Chiria, Megataburu and Kiriburu. The Saranda forest of Jharkhand is endowed with some rich iron ore deposits. The location of the forest is within latitude 22° 3′ 7.98″–22°14′ 0.67″ N and longitude 85° 21′ 31.52″–85° 25′ 53.18″ E with elevation of 850 m above the mean sea level (MSL). Saranda forest is fed by two major rivers, Karo and Koina. The catchment of these rivers comprises of a drainage system with stream order of up to six (Kayet et al. 2016).

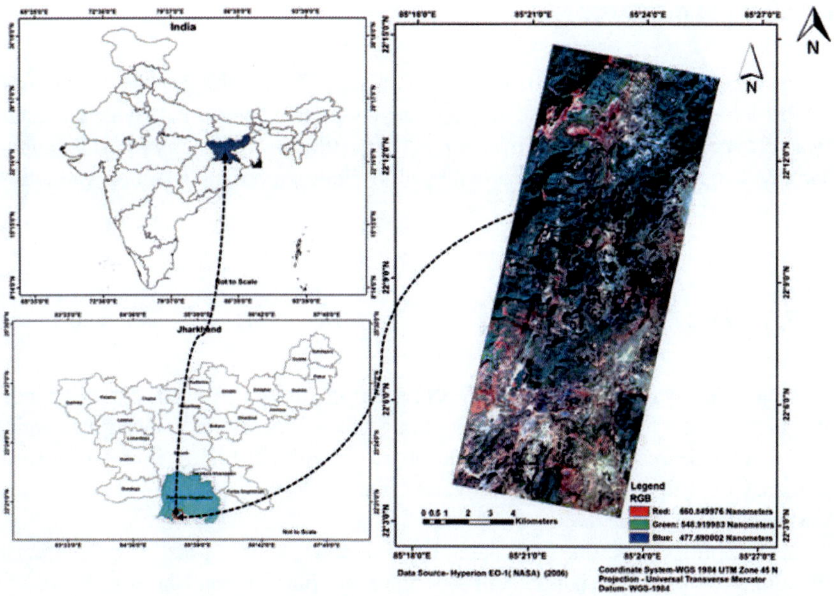

Fig. 10.1 Location map of the study area

Fig. 10.2 Trees spectral data collected from field

10.2.2 Data Source

Hyperspectral data (Hyperion) were downloaded from the USGS website and used for forest health monitoring. Healthy and unhealthy tree spectra data were collected by a field spectroradiometer instrument. Also, healthy and unhealthy forest locations (latitude and longitude) were recorded by GPS. Photographs are taken for validation purposes during the field survey (Fig. 10.2).

10.2.3 Data Pre-processing

Bad bands removal: The delivered USGS Hyperion product level LIR has 242 bands, of which 44 were not calibrated. The main reason for not calibrating the entire band was the decreased sensitivity of the detectors within the non-calibrated spectral regions. Out of the total collected 242 Hyperion bands, 44 (Table 10.1) bands do not work (Han et al. 2002).

De-stripping: Hyperion L1R data shows a severe striping effect by imprecise co-calibration of individual detectors on the focal plane array. The first 12 visible near-infrared bands and many short waves infrared bands are affected by striping and

Table: 10.1 List of unused bands of the hyperion sensor

Bands	Description
1–7	Not illuminated
58–78	Overlap region
120–132	Water vapour absorption Band
165–182	Water vapour absorption band
185–187	Identified by hyperion bad band list
221–224	Water vapour absorption Band
225–242	Not illuminated

bad columns. An uncorrected striping effect will lead to a faulty interpretation of the results. The vertical stripes error values were replaced by the average DN values of the adjacent columns. Hyperspectral data are affected by different noise sources which can be grouped into two main classes: random noise and fixed pattern noise. The photon and thermal noise are random noise; striping noise is a fixed pattern noise and created from push-broom sensors. Hyperspectral images are affected by those noises. Geometric correction: Geometric distortion often has to be removed from remotely sensed data. There are two main approaches to remove the geometric errors. One is the systematic approach and the other is the non-systematic approach. Some of these errors can be corrected by using the ephemeris of the platform and previously known internal sensor distortion characteristics. Other errors can be rectified by matching image coordinates of physical features recorded by the image to the geographic coordinates of the same feature collected from a map or by using a Global Positioning System (GPS).

Radiometric correction: Cross-track illumination- ENVI cross-track illumination tool was used to remove variation of illumination of the image. Cross-track illumination errors may be due to vegetating effects, instrument effects or scanning or other non-uniform illumination effects. The EFFORT algorithm calculates the mean values of an extended track polynomial function, and fit mean values remove this error. We have used this algorithm to remove the variation of illumination of the image.

Atmospheric correction (FLAASH): Atmospheric correction reduces the effects of the atmospheric components (water vapor, dust, gasses) on the electromagnetic radiation reflected or emitted from the surface. We have used the FLAASH model (Fast Line-of-sight Atmospheric Analysis of Spectral Hypercubes) in ENVI software for atmospheric correction and convert to surface reflectance. This model was developed by Spectral Sciences, Inc., under the sponsorship of the US Air force Research Lab. The calibration model based on the theory of atmospheric radiation is according to the physical process of radiation transfer, building by radiation transfer equation, using a theoretical formula (Cooley et al., 2002) to proceed with atmospheric correction (Eq 1).

$$L = \left(\frac{Ap}{1 - peS}\right) + \left(\frac{Bpe}{1 - pes}\right) + La$$

$$Le = \left(\frac{(A + B)pe}{1 - peS}\right) + La$$

FLAASH starts from a standard equation for spectral radiance at a sensor pixel, L; that applies to the solar wavelength range and flat, Lambert a material or their equivalents. FLAASH atmospheric correction removes this absorption feature. FLAASH MODTRAN has outperformed other radiative transfer codes especially in the water region 940 and 1130 nm and CO2 at 2055 nm (Pathak et al. 2016).

10.2.4 Methodology

10.2.4.1 Vegetation Indices (Vis)-Based Forest Health Mapping

The forest health analysis tool will generate a spatial map that shows the overall health and vigor of a forested area (Tuominen et al. 2008). It is good at detecting pest and blight conditions in a forest. The forest health tool uses the following vegetation index categories (ENVI Forest heath tutorial):

1. Narrow band Greenness, to show the distribution of green.
2. Leaf pigments, to show the concentration of arytenoids and anthocyanin pigments for stress levels.
3. Canopy water content, to show the concentration of water.
4. Light use efficiency, to show forest growth rate.

Greenness indices: Greenness vegetation indices generally measure the vigor and green vegetation (Kumar et al. 2015). They measures the various aspects such as chlorophyll concentration, canopy area, and canopy structure. Greenness vegetation indices are based on measuring the reflectance peak in the NIR region. Red wavelengths where the chlorophyll absorption is strongest are used as a reference (Lloret et al. 2004). Leaf pigments indices: Leaf pigment vegetation indices measure the amount of stress-related pigment in the vegetation (Jenkins et al. 2007). In stressed vegetation, there is a higher concentration of carotenoids and anthocyanins. Carotenoids are the leafs pigment that prevents vegetation light conditions. Anthocyanin pigment contents are high in new leaves (Gamon et al. 1999). Canopy water content indices: Water content vegetation indices are designed to estimate the canopy water content (Colombo et al. 2008). However, water content is an important vegetation property that controls vegetation growth and also correlates with vegetation health (Adam et al. 2010). The use of water content vegetation indices needs high spectral resolution data.

Table 10.2 Narrow banded VIs for forest health mapping

Narrow banded Indices	Indices	Algorithms	References
Greenness	(i) Modified red edge normalized difference vegetation index (ii) Vogelmann red edge index	mNDVI705 = (750nm 705nm)/(750nm + 705nm (2 $*$ 445nm)) $VREI1 = {}^{740\text{nm}}/_{720\text{nm}}$	Datt (1999), Zarco-Tejada et al. (2013)
Light use efficiency	Structure insensitive pigment index	SIPI = (800nm–445nm)/(800nm–680nm)	Penuelas et al. (1994)
Leaf pigments	(i) Carotenoid reflectance index (ii) Anthocyanin reflectance index	$ARI1 = \frac{1}{550\text{nm}} - \frac{1}{700\text{nm}}$ CRI2 = (1/510nm)(1/550nm)	Gitelson et al. (2002)
Canopy water content	Normalized difference infrared index	$NDII = 819\text{nm} - 1649\text{nm}/819\text{nm} - 1649\text{nm}$	Hardisky et al. (1983)

Light Use Efficiency indices: Light use efficiency vegetation indices provide the efficiency with which vegetation can use incident light for photosynthesis (Wilson et al. 1981). It is correlated with carbon uptake efficiency and growth rate. However, by the use of light use efficiency vegetation indices measure the growth rate and production of vegetation. We have used six narrow banded Vis for forest health mapping (Table 10.2).

10.2.4.2 Spectral Analysis Based Forest Health Mapping

The following procedures of Hyperspectral analysis were employed, including the Minimum Noise Fraction (MNF) transformation for reducing the spectral data, the Pixel Purity Index (PPI) for identifying those extreme or spectrally pure pixels, and the n-dimensional visualizer for determining the endmember directly from the image. Spectral Angle Mapper (SAM) was applied to estimate the abundances of each endmember to produce the final map.

Minimum Noise Fraction (MNF): MNF can reduce the inherent dimensionality of the dataset and reduce noise from the dataset. MNF also reduces the computational requirement for subsequent processing. The first step in MNF transforms the data in which the noise has a unit variance and there is also no band to band interrelationship (Denghui and Le 2011). Secondly, the MNF can compute/process the principal component analysis for noise-whitened data (Gamon et al. 2004). MNF inversion produces much smaller spectral angles than those derived in transformed space (Peddle et al. 2008). The first ten inverse MNF bands contain 95% of the total information.

Pixel purity index (PPI): Image transformation techniques typically use statistical analysis and reduce the dimensionality of the data. One such transformation is done through the principal component analysis or principal component transformation (Chang and Plaza 2006). The pixel purity index (PPI) is a means of finding the most spectrally pure, or extreme, pixels in multispectral and hyperspectral images (Chaudhry et al. 2006). The most spectrally pure pixels typically correspond to mixing endmembers (Plaza et al. 2006).

N-dimensional (N-D) Visualizer: N-D is applied after correcting the image. The distribution of bands in N space can be used for the estimation of some spectral endmembers and their pure spectral signature (Kruse et al. 1999). The N-dimensional visualization is applied after gathering the data through MNF or PPI algorithms. The pre-clustering result attempts to find in the corner pixel of N-dimensional using a spectral scatter algorithm.

Since the purest pixel is found in the neighbourhood of the data cloud (Wang et al. 2015). The n-dimensional visualizer allows for the interactive rotation of data in n-D space, selection of groups into different classes (Boardman et al. 1995).

Spectral classification techniques: Classification and feature extraction methods have been commonly used for many years for the mapping of forest health and vegetative cover from hyperspectral datasets. However, conventional classification algorithms cannot be applied to hyperspectral data due to the high dimensionality of the data. Spectral Angle Mapper (SAM) mapping techniques were used in the present study to map of forest health in the study area. Spectral Angle Mapper (SAM) is an algorithm, which is widely used for hyperspectral image correction (Petropoulos et al. 2013). It is a supervised image correction process. A pixel with a minimum spectral angle comparison with reference spectra is assigned to the pixel vector. This algorithm determines the spectral similarity between two spectra by calculating the angle between the spectra and treating them as vectors in space with dimensionality equal to the number of bands (Girouard et al. 2004).

10.2.4.3 Accuracy Assessment

Accuracy assessment is an important work in the classification validation system. Remote sensing technology is a great source of thematic map presentation although accuracy assessment assists how far the classification represents the real world. Accuracy assessment can produce user accuracy, producer accuracy, total accuracy, and also kappa coefficient value.

The producer accuracy of the classified pixels compared to the ground truth. The overall research methodology is shown in Fig. 10.3.

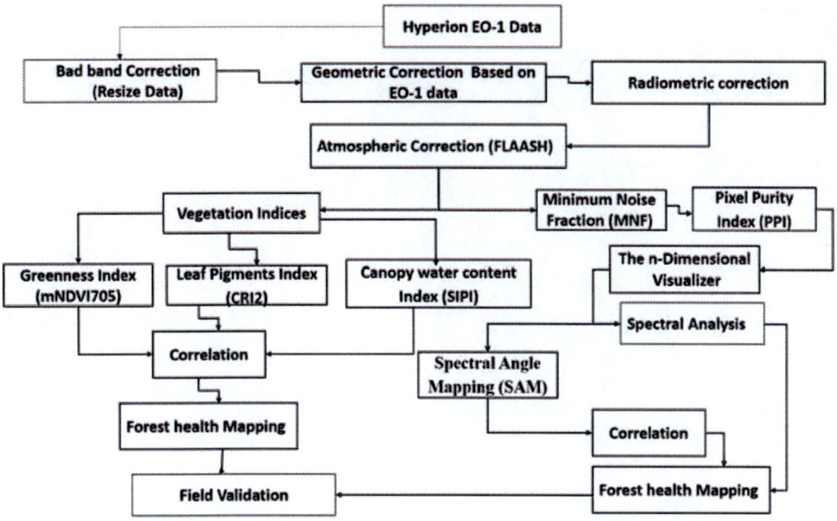

Fig. 10.3 Overall methodology for forest health monitoring

Table 10.3 Minimum, maximum, mean, and standard deviation values of before and after FLAASH correction

FLAASH correction	Min	Max	Mean	Stdev
Before correction	−77	3383	84,444.44	66,423.37
After correction	0	70.89	11.63	10.21

10.3 Results and Discussion

10.3.1 Result of FLAASH Atmospheric Correction

FLAASH is an effective atmospheric correction process where all bands are corrected to follow their proper geometric and radiometric characteristics. FLAASH can also remove the de-striping of the image, path radiation of the image, and various systematic and non-systemic effects. Pre and post FLAASH correction statistics value are shown in Table 10.3. Spectral variability of after FLAASH correction is shown in Fig. 10.4.

10.3.2 Vegetation Indices (Vis) Based Forest Health Mapping

Vegetation Indices (VIs) were calculated for 60 test sample pixels. Mean and standard deviation values for both healthy and unhealthy classes are shown in Table 10.4.

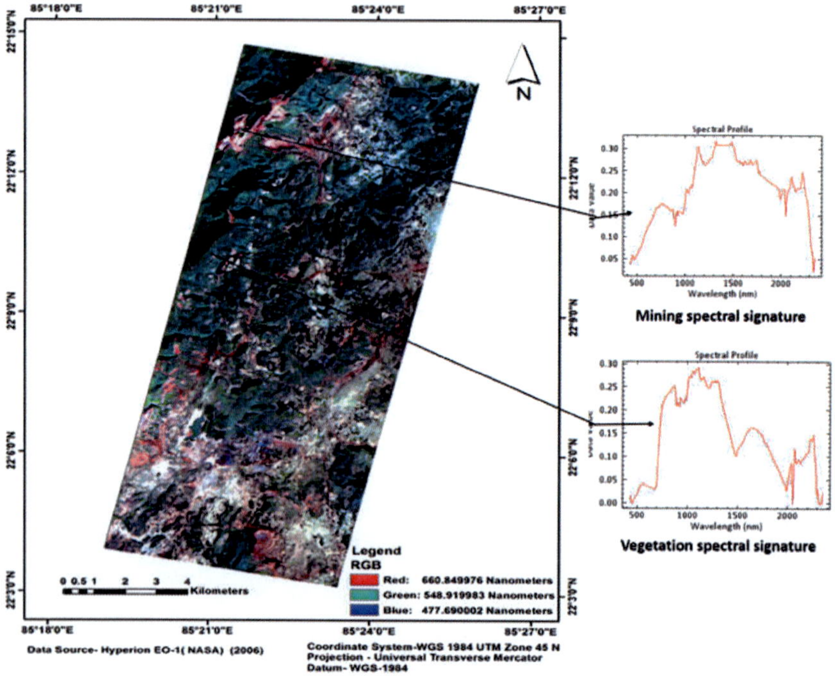

Fig. 10.4 Image spectra after atmospheric correction

Table 10.4 Mean, standard deviation and separability values for each index test

Vegetation indices	Healthy		Unhealthy		Separability
	Mean	Std.	Mean	Std.	S
MNDVI705	0.66	0.03	0.17	0.08	4.21
VREI1	1.44	1.13	0.06	0.05	2.60
CRI1	33.22	7.92	17.17	2.24	1.57
ARI1	9.65	4.78	7.71	2.29	0.27
NDII	−0.49	0.07	−1.23	0.27	2.11
SIPI	1.09	0.02	0.51	0.33	1.60

Generally, the separability values obtained for greenness and vegetation indices were relatively high. The highest separability values were obtained for the modified red edge normalized difference vegetation (MNDVI705) index due to good chlorophyll content. The MNDVI705 correlates well with good chlorophyll content, so its good performance could be expected as the result of forest health (Kayet et al. 2019a, b). ENVI software provides nine forest health classes (Fig. 10.5).

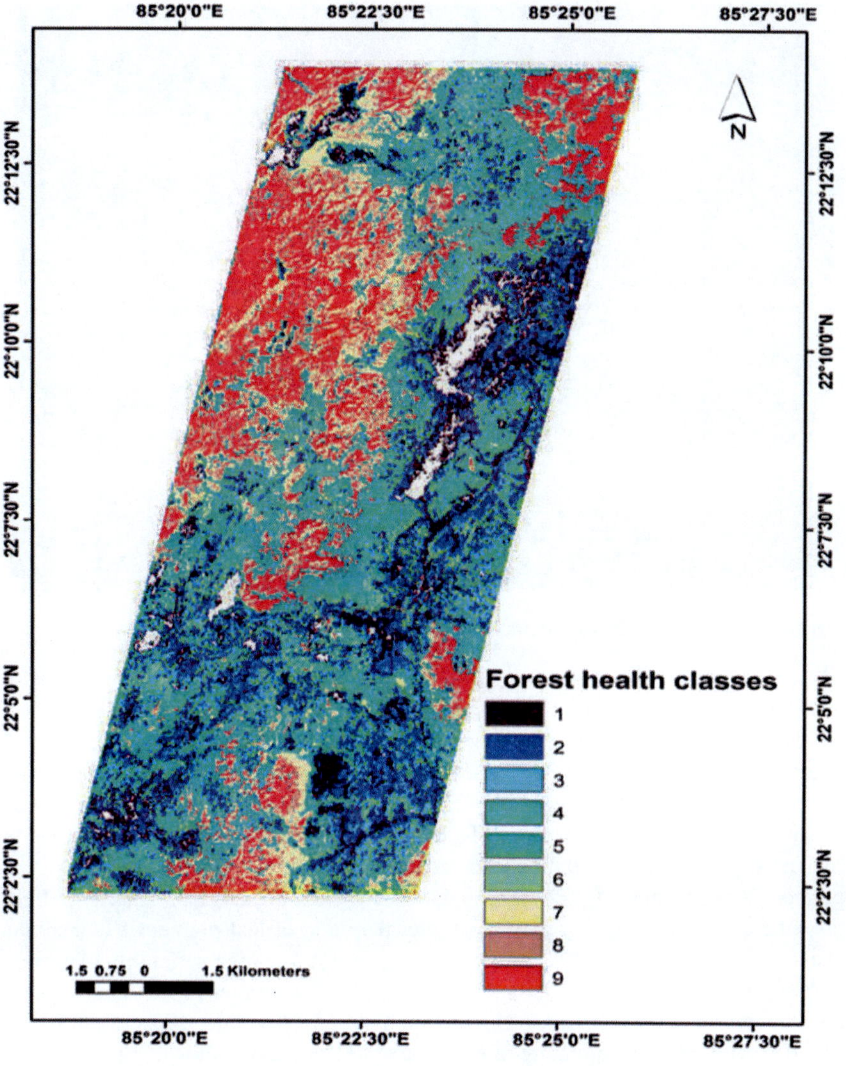

Fig. 10.5 Forest health map (Class 9 very healthy and class 1 unhealthy)

The classifications are relative to the particular input scene only and cannot be generalized to other areas or other scenes. The healthy and unhealthy value range is 0.5–0.8 and 0.1–0.3 (Kayet et al. 2019a, b). The classification map rates the scene from one, representing the least healthy forest (weakest) to the healthiest forest (strongest) which help to assess relative forest health conditions within the scene (ENVI forest health tool tutorial).

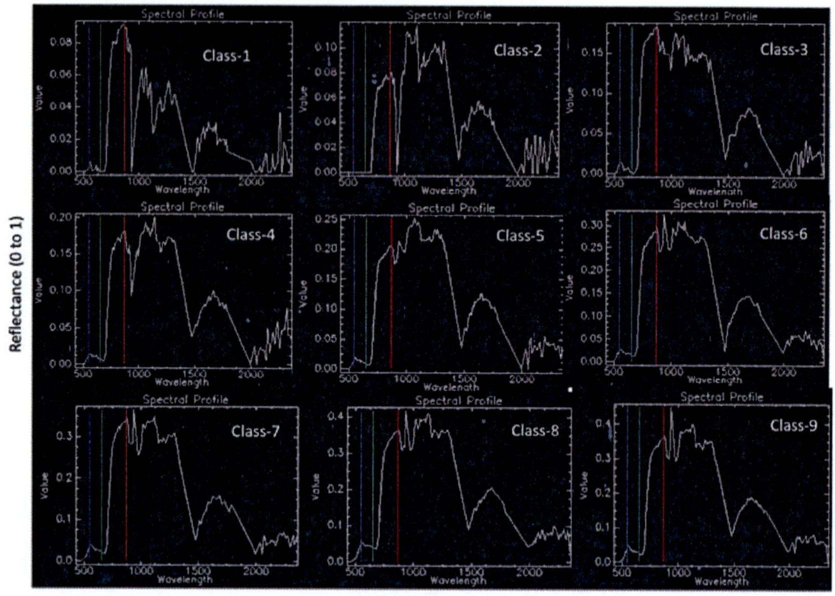

Fig. 10.6 Hyperion reflectance spectra of forest health classes (1–9)

Forest health classification was done by the use of three narrow-band vegetation indices which are MNDVI705 (Greenness or chlorophyll vegetation Index), CRI_1 (Leaf pigment Vegetation Index), and SIPI (Light efficiency vegetation Index). The MNDVI705 index works well with the lower chlorophyll content, so it is accepted for appreciable forest health result (Kumar et al. 2015). Test of forest health result for leaf pigments VIs is relatively lower than any other vegetation indices. In leaf pigment VIs, the value of ARI1 is high than other VIs (Serrano et al., 2002). For light use efficiency, Vis value is relatively higher than that of leaf pigment VIs (Jenkins et al. 2007). NDNI is a useful index where there exhibits a high variability in canopy or leaf pigment structure (Rodriguez et al. 2007). The test result for canopy water content index was low because of vegetation canopy structure (Sims et al. 2002). The Hyperion spectral signature of each forest health class is shown in Fig. 10.6.

10.3.3 Spectral Analysis Based Forest Health Mapping

Forest health is mainly dependent on various physiological parameters such as climate, temperature, geology, soil, slope, aspect, hill shade direction, and much more. The study area is mainly covered by three types of forest health (healthy, moderated healthy, and unhealthy). The healthy and unhealthy value range is 0.5–0.8 and 0.1–0.3 (Kayet et al. 2019a, b). The spectral signature of the forest health

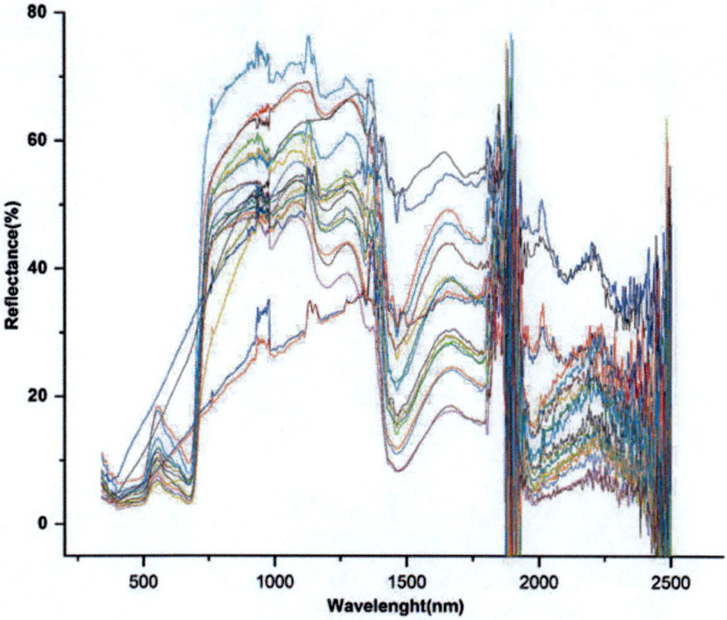

Fig. 10.7 Field collected healthy and unhealthy trees spectra

classes were collected from the image and these matche the spectral signature from field collected tree spectral library (Fig. 10.7) and field sample data in ENVI software.

Forest health mapping is done by the SAM classification technique. Classification of forest health into three classes (healthy, moderated healthy, and unhealthy) is shown in Fig. 10.8.

Healthy forest (0.5–0.8) cover comprises 49.87% of the study area, 14.15% area is covered by moderate healthy forest (0.3–0.5), and unhealthy forest (0.1–0.3) covers 38.03% area (Kayet et al. 2019a, b). Healthy forests are present mostly in the north and the north-east part of the study area and unhealthy is situated in mines surrounding area. George et al. (2014) shown the better forest classification through Hyperspectral remote sensing and compared the classification results obtained from Hyperion and Landsat TM sensors for the study of Western Himalaya and obtained collective accuracies of 81.52% and 69.62% respectively. Thenkabail et al. (2004) compared the classification results of different sensors viz., Hyperion, IKONOS, ALI, and ETM + sensors for the study of African rainforests and obtained collective accuracies of 93.2%, 87.46%, 81.53%, and 76.9% respectively.

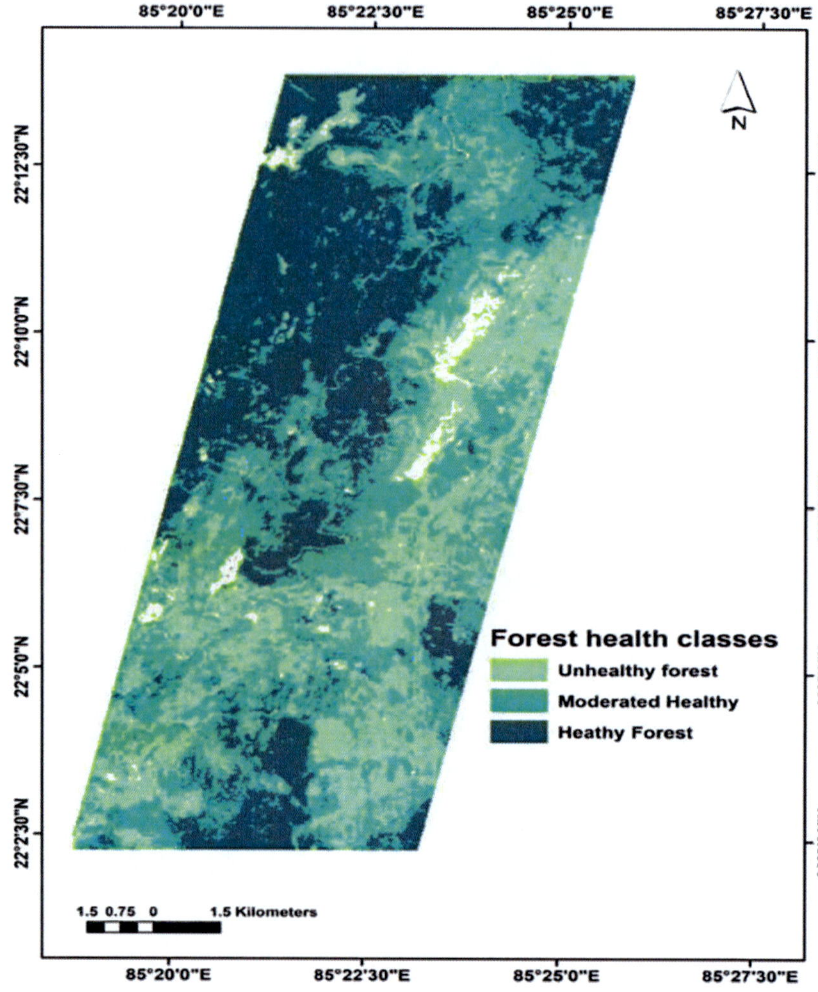

Fig. 10.8 Forest heath mapping by hyperion data

10.3.4 Accuracy Assessment

Classification accuracy was done by spectral angle mapper (SAM) classification technique in ENVI software. SAM was implemented to the collection of the spectral signature of healthy, moderated healthy and unhealthy forests. Forest health classification is accuracy by the USGS spectral library and sample field points. Thus, the overall accuracy of 76.53% and 0.71 kappa coefficient were determined (Table 10.5).

Table 10.5 Accuracy assessment results of SAM based on hyperion data

SVM based on hyperion	Healthy	Moderated healthy	Unhealthy	Total	UA
Healthy	11	0	1	11	81.52
Moderated healthy	0	13	0	13	74.54
Unhealthy	2	0	10	12	70.66
Total	13	13	11	36	
PA	91.83	85.53	80.11		
Overall accuracy: 76.53%, kappa statistics: 0.71					

10.3.5 Forest Health Validation

The healthy, moderately healthy, and unhealthy components constituting the forest of the study area were evaluated, both at ground level and pixel- level, having the highest reflectance data from the NIR wavebands region. The correlation determination (R^2) and RMS error values were evaluated from ground level and pixel-level spectral data (Fig. 10.9). A correlation ($R^2 = 0.84$) was observed between the ground level and

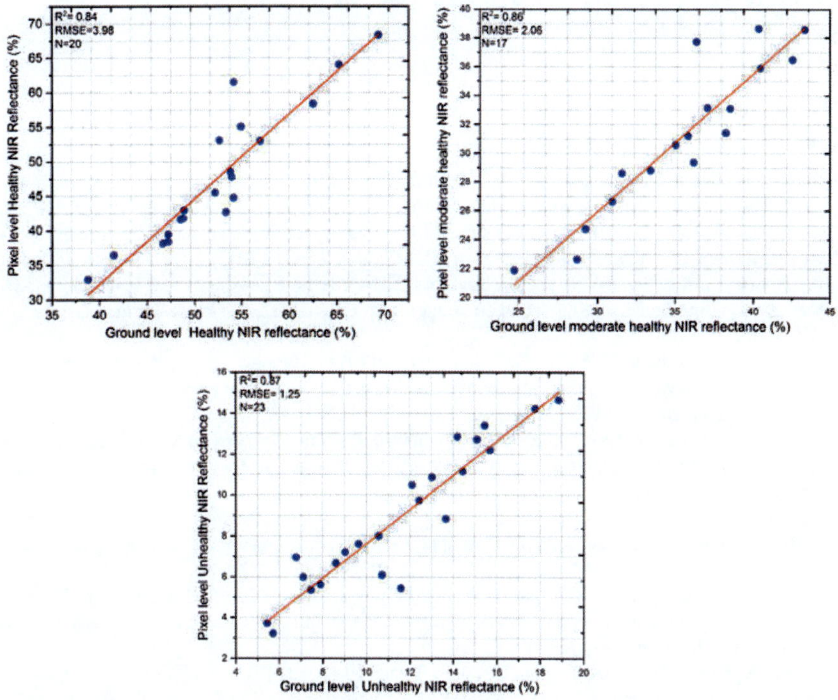

Fig. 10.9 Correlation between field reflected spectra and pixel reflected spectra of healthy, moderate healthy and unhealthy forest class

pixel-level for class healthy, and an RMS error of 3.98 was found. A correlation ($R^2 = 0.86$) was observed between the ground level and pixel-level for class moderately healthy, and an RMS error of 2.06 was found. And a correlation ($R^2 = 0.87$) was observed between the ground level and pixel-level for class unhealthy, and an RMS error of 1.25 was found.

10.4 Conclusion

The article has summarized forest health monitoring sounding mines areas. This case study demonstrated by hyperspectral data. Hyperspectral data has more capability than multispectral data. In this study, a good correlation was shown between forest health and distance from mines. It means that as the mining area increases forest as well as environment will also get affected. This methodology would be capable of monitoring various categories of forest region routinely irrespective of the different climate condition, forest structure, and soil condition. Hyperspectral remote sensing-based forest health monitoring is today's need so that the forest department, local self-government, and mining companies must adopt an adequate policy for reclamation and restoration of the forest ecosystem affected by mining activities.

References

Adam E, Mutanga O, Rugege D (2010) Multispectral and hyperspectral remote sensing for identification and mapping of wetland vegetation: a review. Wetlands Ecol Manage 18(3):281–296

Apostolescu N, Baran D (2016) Spectral similarity in hyperspectral image. Acta Electrotehnica 57

Asner GP, Ustin SL, Townsend PA, Martin RE, Chadwick KD (2015) Forest biophysical and biochemical properties from hyperspectral and LiDAR remote sensing. Land resources monitoring, modeling and mapping with remote sensing. CRC Press, Taylor & Francis Group, pp 429–448.

Bellvert J, Zarco-Tejada PJ, Girona J, Fereres EJPA (2014) Mapping crop water stress index in a 'Pinot-noir'vineyard: comparing ground measurements with thermal remote sensing imagery from an unmanned aerial vehicle. Precision Agric 15(4):361–376

Boardman JW (1995) Analysis, understanding, and visualization of hyperspectral data as convex sets in n space. In: Imaging spectrometry (vol 2480). International Society for Optics and Photonics, pp 14–22

Brantley ST, Zinnert JC, Young DR (2011) Application of hyperspectral vegetation indices to detect variations in high leaf area index temperate shrub thicket canopies. Remote Sens Environ 115(2):514–523

Broge NH, Leblanc E (2001) Comparing prediction power and stability of broadband and hyperspectral vegetation indices for estimation of green leaf area index and canopy chlorophyll density. Remote Sens Environ 76(2):156–172

Calderón R, Navas-Cortés JA, Lucena C, Zarco-Tejada PJ (2013) High-resolution airborne hyperspectral and thermal imagery for early detection of Verticillium wilt of olive using fluorescence, temperature and narrow-band spectral indices. Remote Sens Environ 139:231–245

Chambers JQ, Asner GP, Morton DC, Anderson L, O, Saatchi SS, Espírito-Santo FD, Souza, C Jr. (2007) Regional ecosystem structure and function: ecological insights from remote sensing of tropical forests. Trends Ecol Evolut 22(8):414–423

Chang CI, Plaza A (2006) A fast iterative algorithm for implementation of pixel purity index. IEEE Geosci Remote Sens Lett 3(1):63–67

Chaudhry F, Wu CC, Liu W, Chang CI, Plaza A (2006) Pixel purity index-based algorithms for endmember extraction from hyperspectral imagery. Recent Adv Hyperspect Signal Image Process 37(2):29–62

Cho MA, Sobhan I, Skidmore AK, De Leeuw J (2008) Discriminating species using hyperspectral indices at leaf and canopy scales. The International Archives of the Spatial Information Sciences, pp 369–376

Colombo R, Meroni M, Marchesi A, Busetto L, Rossini M, Giardino C, Panigada C (2008) Estimation of leaf and canopy water content in poplar plantations by means of hyperspectral indices and inverse modeling. Remote Sens Environ 112(4):1820–1834

Cooley T, Anderson GP, Felde GW, Hoke ML, Ratkowski AJ, Chetwynd JH, Bernstein LS (2002, June) FLAASH, a MODTRAN4-based atmospheric correction algorithm, its application and validation. In: IEEE International Geoscience and Remote Sensing Symposium, Vol 3. IEEE, pp 1414–1418

Dalponte M, Ørka HO, Ene LT, Gobakken T, Næsset E (2014) Tree crown delineation and tree species classification in boreal forests using hyperspectral and ALS data. Remote Sens Environ 140:306–317

Datt B (1999) Remote sensing of water content in Eucalyptus leaves. Aust J Bot 47(6):909–923

Darvishzadeh R, Atzberger C, Skidmore A, Schlerf M (2011) Mapping grassland leaf area index with airborne hyperspectral imagery: a comparison study of statistical approaches and inversion of radiative transfer models. ISPRS J Photogram Remote Sens 66(6):894–906

Darvishzadeh R, Skidmore A, Schlerf M, Atzberger C, Corsi F, Cho M (2008) LAI and chlorophyll estimation for a heterogeneous grassland using hyperspectral measurements. ISPRS J Photogram Remote Sens 63(4):409–426

Denghui Z, Le Y (2011) Support vector machine based classification for hyperspectral remote sensing images after minimum noise fraction rotation transformation. In: 2011 International conference on internet computing and information services. IEEE, pp 132–135

Franklin SE (2001) Remote sensing for sustainable forest management. CRC press

Gamon JA, Surfus JS (1999) Assessing leaf pigment content and activity with a reflectometer. The New Phytol 143(1):105–117

Gamon JA, Huemmrich KF, Peddle DR, Chen J, Fuentes D, Hall FG, Miller JR (2004) Remote sensing in BOREAS: Lessons learned. Remote Sens Environ 89(2):139–162

George R, Padalia H, Kushwaha SPS (2014) Forest tree species discrimination in western Himalaya using EO-1 Hyperion. Int J Appl Earth Obs Geoinf 28:140–149

Girouard G, Bannari A, El Harti A, Desrochers A (2004) Validated spectral angle mapper algorithm for geological mapping: comparative study between QuickBird and landsat-TM. In: XXth ISPRS congress, geo-imagery bridging continents, Istanbul, Turkey, pp 12–23

Gitelson AA, Stark R, Grits U, Rundquist D, Kaufman Y, Derry D (2002) Vegetation and soil lines in visible spectral space: a concept and technique for remote estimation of vegetation fraction. Int J Remote Sens 23(13):2537–2562

Han T, Goodenough DG, Dyk A, Love J (2002) Detection and correction of abnormal pixels in Hyperion images. In: IEEE international geoscience and remote sensing symposium, vol 3. IEEE, pp 1327–1330

Jacquemoud S, Verhoef W, Baret F, Bacour C, Zarco-Tejada PJ, Asner GP, Ustin SL (2009) PROSPECT+ SAIL models: a review of use for vegetation characterization. Remote Sens Environ 113:S56–S66

Jenkins JP, Richardson AD, Braswell BH, Ollinger SV, Hollinger DY, Smith ML (2007) Refining light-use efficiency calculations for a deciduous forest canopy using simultaneous tower-based carbon flux and radiometric measurements. Agric Meteorol 143(1–2):64–79

Kayet N, Pathak K, Chakrabarty A, Sahoo S (2016) Spatial impact of land use/land cover change on surface temperature distribution in Saranda Forest Jharkhand. Model Earth Syst Environ 2(3):127

Kayet N, Pathak K, Chakrabarty A, Singh CP, Chowdary VM, Kumar S, Sahoo S (2019) Forest health assessment for geo-environmental planning and management in hilltop mining areas using Hyperion and Landsat data. Ecol Ind 106:105471

Kayet N, Pathak K, Chakrabarty A, Kumar S, Chowdary VM, Singh CP, Basumatary S (2019) Assessment of foliar dust using hyperion and landsat satellite imagery for mine environmental monitoring in an open cast iron ore mining areas. J Clean Product 218:993–1006

Ke LIU, Qing-bo ZHOU, Wen-bin WU, Tian XIA, Hua-jun TANG (2016) Estimating the crop leaf area index using hyperspectral remote sensing. J integrative agriculture 15(2):475–491

Kruse FA, Boardman JW, Huntington JF (1999) Fifteen years of hyperspectral data: Northern grapevine mountains, Nevada. In: Proceedings of the 8th JPL Airborne earth science workshop: Jet Propulsion Laboratory Publication, JPL Publication, pp 99–117

Kumar K, Singh NM, Kerr JM (2015) Decentralisation and democratic forest reforms in India: moving to a rights-based approach. Forest Policy Econ 51:1–8

Lee KS, Cohen WB, Kennedy RE, Maiersperger TK, Gower ST (2004) Hyperspectral versus multispectral data for estimating leaf area index in four different biomes. Remote Sens Environ 91(3–4):508–520

Li Z, Guo X (2016) Remote sensing of terrestrial non-photosynthetic vegetation using hyperspectral, multispectral, SAR, and LiDAR data. Prog Phys Geogr 40(2):276–304

Lloret F, Peñuelas J, Ogaya R (2004) Establishment of co-existing Mediterranean tree species under a varying soil moisture regime. J Veg Sci 15(2):237–244

Ma J, Sun DW, Qu JH, Pu H (2017) Prediction of textural changes in grass carp fillets as affected by vacuum freeze drying using hyperspectral imaging based on integrated group wavelengths. LWT-Food Sci Technol 82:377–385

Mashimbye ZE, Cho MA, Nell JP, De Clercq WP, Van Niekerk A, Turner DP (2012) Model-based integrated methods for quantitative estimation of soil salinity from hyperspectral remote sensing data: a case study of selected South African soils. Pedosphere 22(5):640–649

Navinkumar TU, Parmar SK (2016) Unmixing of hyperpectral image using nonlinear parametric model. In: 2016 IEEE international conference on recent trends in electronics, information & communication technology (RTEICT), IEEE, pp 1206–1211

Pathak VN (2016) Development of an atmospheric correction method for retrieval of surface reflectance from satellite data. Doctoral dissertation, Sardar Patel University

Peddle DR, Boulton RB, Pilger N, Bergeron M, Hollinger A (2008) Hyperspectral detection of chemical vegetation stress: evaluation for the Canadian HERO satellite mission. Can J Remote Sens 34(sup1):S198–S216

Petropoulos GP, Vadrevu KP, Kalaitzidis C (2013) Spectral angle mapper and object-based classification combined with hyperspectral remote sensing imagery for obtaining land use/cover mapping in a Mediterranean region. Geocarto Int 28(2):114–129

Plaza A, Valencia D, Plaza J, Chang CI (2006) Parallel implementation of endmember extraction algorithms from hyperspectral data. IEEE Geosci Remote Sens Lett 3(3):334–338

Rodríguez-Pérez JR, Riaño D, Carlisle E, Ustin S, Smart DR (2007) Evaluation of hyperspectral reflectance indexes to detect grapevine water status in vineyards. Am J Enology and Viticulture 58(3):302–317

Serrano L, Penuelas J, Ustin SL (2002) Remote sensing of nitrogen and lignin in Mediterranean vegetation from AVIRIS data: decomposing biochemical from structural signals. Remote Sens Environ 81(2-3):355–364

Sims DA, Gamon JA (2002) Relationships between leaf pigment content and spectral reflectance across a wide range of species, leaf structures and developmental stages. Remote Sens Environ 81(2-3):337–354

Thenkabail PS (2001) Optimal hyperspectral narrow bands for discriminating agricultural crops. Remote Sens Rev 20(4):257–291

Thenkabail PS, Enclona EA, Ashton MS, Van Der Meer B (2004) Accuracy assessments of hyperspectral waveband performance for vegetation analysis applications. Remote Sens Environ 91(3-4):354–376

Tuominen J, Lipping T, Kuosmanen V (2008) Assesment of ENVI forest health tool in detection of dust and seepage contaminated forest areas. In: IGARSS 2008–2008 IEEE international geoscience and remote sensing symposium, Vol 3. IEEE, pp. III–1358

Tuominen J, Lipping T, Kuosmanen V, Haapanen R (2009) Remote sensing of forest health. In: Ho P-GP (ed) Geoscience and remote sensing. InTech, Shanghai, China, pp 29–52

Vauhkonen J, Mehtätalo L, Packalén P (2011) Combining tree height samples produced by airborne laser scanning and stand management records to estimate plot volume in Eucalyptus plantations. Can J Res 41(8):1649–1658

Wang W, Lawrence KC, Ni X, Yoon SC, Heitschmidt GW, Feldner P (2015) Near-infrared hyperspectral imaging for detecting Aflatoxin B1 of maize kernels. Food Control 51:347–355

Wilson JW (1981) Analysis of growth, photosynthesis and light interception for single plants and stands. Ann Bot 48(4):507–512

Zarco-Tejada PJ, Berni JA, Suárez L, Sepulcre-Cantó G, Morales F, Miller JR (2009) Imaging chlorophyll fluorescence with an airborne narrow-band multispectral camera for vegetation stress detection. Remote Sens Environ 113(6):1262–1275

Zarco-Tejada PJ, Guillén-Climent ML, Hernández-Clemente R, Catalina A, González MR, Martín P (2013) Estimating leaf carotenoid content in vineyards using high resolution hyperspectral imagery acquired from an unmanned aerial vehicle (UAV). Agric for Meteorol 171:281–294

Zarco-Tejada PJ, Miller JR, Mohammed GH, Noland TL, Sampson PH (2002) Vegetation stress detection through chlorophyll a+ b estimation and fluorescence effects on hyperspectral imagery. J Environ Qual 31(5):1433–1441

Zarco-Tejada PJ, Ustin SL, Whiting ML (2005) Temporal and spatial relationships between within-field yield variability in cotton and high-spatial hyperspectral remote sensing imagery. Agron J 97(3):641–653

Chapter 11
Estimating Above Ground Biomass (AGB) and Tree Density using Sentinel-1 Data

Sambhunath Roy, Sujoy Mudi, Pulakesh Das, Sujit Ghosh, Pravat Kumar Shit, Gouri Shankar Bhunia, and John Kim

Abstract Assessment of the forest above ground biomass (AGB) and tree density is essential for various studies related to forest structure, productivity, carbon cycle, atmospheric processes, climate change etc. including forest cover management activities and framing the conservation policies. The freely available C-band Sentinel-1 microwave data allows to estimate forest cover biomass at high spatial resolution; moreover, the Sentinel-2 optical data enables to integrate the biophysical attributes. In the current study, the AGB has been estimated in a *Shorea robusta* (sal) dominated forest cover in sub-tropical region employing the Sentinel-1 microwave data. The inventory of forest cover attributes has been collected in 40 sample plots, where the in-situ AGB was geo-statistically linked with the satellite observations. Employing the univariate linear regression, it has been observed that the microwave backscatter obtained in the VH band well explained ($R^2 = 0.63$) the variability of the in-situ AGB in comparison to the backscatter received in VV band ($R^2 = 0.44$) and optical data derived EVI image ($R^2 = 0.45$). The predicted biomass map verified with the test data points indicated an accuracy of $R^2 = 0.45$, with low RMSE (± 17 tonnes/ha) and slight underestimation (bias = -0.024). However, the accuracy in tree density estimation obtained from the AGB map was observed much higher ($R^2 = 0.87$). The field observed AGB varied between 88.56 tonnes/ha and 170.29 tonnes/ha, where the

S. Roy (✉) · P. K. Shit
PG Department of Geography, Raja N. L. Khan Women's College (Autonomous), Gope Palace, Midnapore 721102, West Bengal, India
e-mail: sambhuroynath95@gmail.com

S. Mudi · P. Das
Department of Remote Sensing & GIS, Vidyasagar University, Midnapore, India

S. Ghosh
Centre for Oceans, Rivers, Atmosphere and Land Science (CORAL), IIT Kharagpur, Kharagpur, India

G. S. Bhunia
Seacom Silk University, Bolpur, India

J. Kim
USDA Forest Service, Pacific Northwest Research Station, Corvallis, USA

P. K. Shit et al. (eds.), *Spatial Modeling in Forest Resources Management*, Environmental Science and Engineering,
https://doi.org/10.1007/978-3-030-56542-8_11

satellite data derived AGB estimated the range of 44.1 tonnes/ha and 249 tonnes/ha for the entire study area. Majority of the biomass was estimated in the range of 100–200 tonnes/ha, which was contributed by majority of the tree density region varying between 69/ha and 75/ha. However, few patches are observed to have much higher and lower AGB, which could be indicating the highly dense and less dense forest cover regions in the study area, respectively. The uniform AGB map indicates the selected region to be more homogenous forest cover area.

Keywords Above ground biomass (AGB) · Sentinel-1 · Sentinel-2 · Tree density · Regression

11.1 Introduction

Forest ecosystem is one of the most important carbon sinks of the terrestrial ecosystem and plays a very important role in the global carbon cycle by sequestering a substantial amount of CO_2 from the atmosphere (Houghton et al. 2012). Deforestation or natural death of plants followed by burning, decomposition or natural decaying leads release of the stored carbon back into the terrestrial ecosystem. The natural cycling of the carbon is preserved and controlled by a dynamic equilibrium between the biological and inorganic processes since the geological history of the earth. In the 19th century, with the advent of industrial revolution, humans are burning a huge amount of fossil fuels, and releasing the carbon stored in it back into the atmosphere. Since the industrial insurgency, the CO_2 level in the atmosphere has increased tremendously causing remarkable changes in the global carbon cycle. The CO_2 concentration in the atmosphere is continuously being measured and recorded accurately since 1957 (Keelinget al. 1976), indicates increasing trend from about 280 ppm during the pre-industrial period to approximately 390 ppm in the present day (Hofmann et al. 2009). Besides, the combustion of fossil fuels, other human activities such as land use change and deforestation have a considerable impact on the ability of the terrestrial biosphere to remove CO_2 from the atmosphere (Houghton 2012). About 60% of the observed global climate change is attributed to this increasing CO_2 concentration in the atmosphere. The main carbon pool in forest ecosystems consists of the living biomass of trees, under story vegetation, dead mass of litter, woody debris and soil organic matter. The tropical forest cover encompasses majority of the earth's terrestrial biodiversity and account for a large proportion of terrestrial carbon stored. The world's forest and forest soils currently store more than one trillion tonnes of carbon, which is twice the amount found floating free in the atmosphere. According to Global Forest Resource Assessment Report 2010 (FRA 2010), the total forest carbon stock of the world is 652 Giga tonnes (161.8 t/ha). Out of this, the forest biomass contains 289 Giga tonnes (71.6 t/ha), the dead organic matter contains 72 Giga tonnes (17.8 t/ha), and forest soil organic carbon contains 293 Giga tonnes (72.3 t/ha) of carbon (Fig. 11.1).

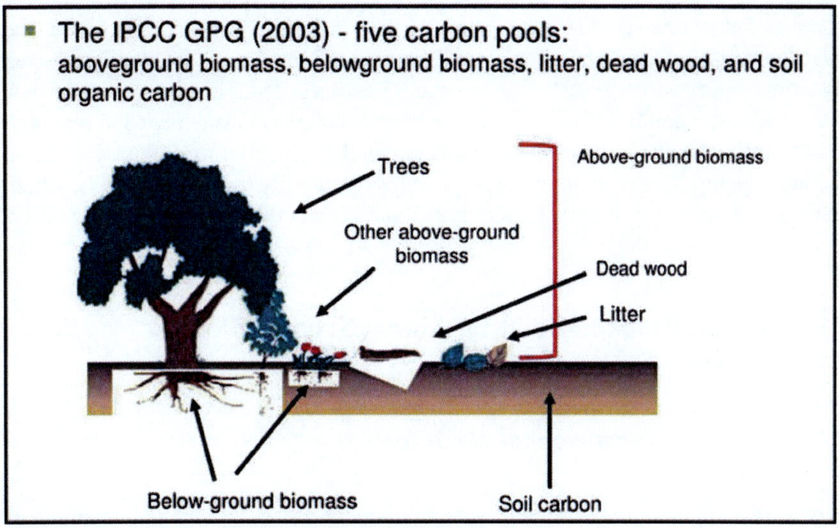

Fig. 11.1 Carbon pools (CIFOR, Forest Carbon Toll box 2008)

In forest ecosystem, the stored carbon is classified into five pools (Table 11.1). The living portion of biomass carbon is classified in two pools: the 'above ground biomass' (AGB) and 'below ground biomass' (BGB), which are stores of significant amount of carbon. The 'dead organic matter' (DOM) is also classified into two pools: 'dead wood' and 'litter'. The fifth pool is 'soil organic matter' (SOM), which

Table 11.1 Carbon pool classification according to IPCC (2003)

	Pools	Description
Living biomass	Above ground biomass (AGB)	All living biomass above the soil including stem, stump, branches, bark, seeds and foliage
	Below ground biomass (BGB)	All living biomass of live roots
Dead organic matter	Dead wood	Includes all non-living woody biomass not contained in the litter either standing or lying on the ground
	Litter	Includes all non-living biomass with a diameter less than a minimum diameter chosen by the country (for FSI 5 cm), lying dead, in various states of decompositions above the mineral or organic soil
Soil	Soil organic matter (SOM)	Includes organic carbon in mineral and organic soils (including peat) to a specific depth chosen by the country (for FSI 30 cm)

contains substantial amount of organic carbon. Among the five components, the AGB accounts the largest carbon stored in tropical region and it is directly affected by deforestation and degradation (Kumar and Sharma 2015). The change in the forest areas and corresponding changes in forest biomass due to management and regrowth greatly influence the transfer of carbon between the terrestrial forest ecosystem and the atmosphere. Assessment of the AGB helps to quantify the carbon stocks, which in turn enables to analyse the current status and project the near future changes.

11.2 Methods for Estimating Above Ground Biomass (AGB)

11.2.1 Field Measurement Methods

Primarily two field measurement methods are used—(i) destructive method or harvesting method—direct estimation of AGB and the carbon stocks stored in the trees. Although, this method determines the biomass accurately, limited to a small tree sample sizes being time consuming, strenuous, destructive and expensive. Usually, this method is used for developing biomass equation to be applied for assessing biomass on a larger area. The other method is (ii) non-destructive measurements, involves climbing the tree to measure the various parts or by simply measuring the DBH, tree height, volume of the tree and wood density to calculate the biomass using allometric equations (Sullivan et al. 2018; Lucas et al. 2015). The allometric equations are developed and applied to forest inventory data to assess the biomass and carbon stocks of forests by establishing a relationship between the various physical parameters of the tree such as the DBH, tree height, wood density, tree species, etc. (Vargas-Larreta et al. 2017). Haripriya (2000) used forest inventories to compute the species-wise forest biomass and carbon stock for various forest strata in different states. Numbers of allometric equation have been developed for estimating AGB for different species and in different climate zones, where the relationship between AGB and forest structural information have been statistically linked. The allometric equation developed by Brown et al. (1989), Chamber et al. (2001) and Chave et al. (2005) are given as follows:

$$AGB = \exp\left(-0.370 + 0.333 \ln D + 0.9333 \ln D^2 - 0.122 \ln D^3\right) \text{ (Brown et al. 1989)}$$

$$AGB = 13.2579 - (4.8945D) + 0.6713D^2 \text{ (Chamber et al. 2001)}$$

$$AGB = 0.0776\left(\rho D^2 H\right)^{0.940} \text{ (Chave et al. 2005)}$$

Chave et al. (2014) revised the parameters for the pantropical region employing significant numbers of ground data. Nath et al. (2019) tested four models for biomass estimation for the forest cover of North East India. Baishya et al. (2009) estimated AGB in the humid tropical forests in North-East India employing the ground-based forest inventory technique for the natural semi-evergreen and *Shorea robusta* (sal) plantation forest; and observed higher AGB in plantation regions (406.4 tonnes/ha) compared to natural forest (323.9 tonnes/ha). While assessing the AGB in seven districts of Madhya Pradesh state, India, Salunkhe et al. (2016) utilised the species-wise allometric equations and density provided by FSI and FRI; and observed a lower range of average AGB as 31.8 tonnes/ha and 20.7 tonnes/ha for the tropical dry and mixed deciduous forest, respectively.

11.2.2 Remote Sensing Approaches

The remote sensing-based approach is primarily a non-devastating method and uses the previously developed allometric equations. The field measured plot-level forest biomass is integrated with the remotely sensed data derived forest biophysical and structural parameters or proxies. The developed relation is then extrapolated to generate the continuous forest biomass map. In this process, the input data (field plots) is divided into training and testing data points. The training data points are used to train the model or develop the relation employing statistical or machine learning approaches; whereas, the test data is used to validate the model performance and derive the accuracy metrices.

The biomass estimation using optical remote sensing data is usually realized by revealing the correlation between biomass and spectral responses (vegetation indices derived from multispectral images). Several studies including Lu et al. (2012) and Karna et al. (2015) have used circular plots of 0.05 ha for field data collection while estimating AGB and carbon stock by integrating lidar and other optical remote sensing. Sader et al. (1989) studied the relationship between forest biomass and the Landsat data derived Normalized Difference Vegetation Index (NDVI); and observed better performance for low biomass region with gregarious formation, which was not suitable for high biomass and mixed forest cover regions. Lee and Nakane (1997) also employed the Landsat data and modelled the biomass employing a number of vegetation indices, where they observed acceptable accuracy for different forest types for different indices. Roy and Ravan (1996) studied the relationship between optical data derived parameters and forest biomass and suggested that the brightness and wetness have strong contribution in explaining the AGB variability. Madugundu et al. (2008) used the Indian Remote Sensing (IRS) LISS-IV data derived leaf area index (LAI) and NDVI to estimate the AGB in the deciduous forests in Western Ghats and observed an acceptable accuracy.

As, the optical remote sensing technologies have limited capability to predict forest biomass since the recorded spectral responses in optical images are mainly represent top canopy properties of forest stand. Thus, the satellite data derived optical properties can model the biomass up to a certain threshold and experiences a saturation at higher biomass regions. On the contrary, the microwave (radar) remote sensing operates at much higher wavelength (1 mm to 1 m), and have unique features compared to the optical remote sensing operates mostly in visual to middle infrared

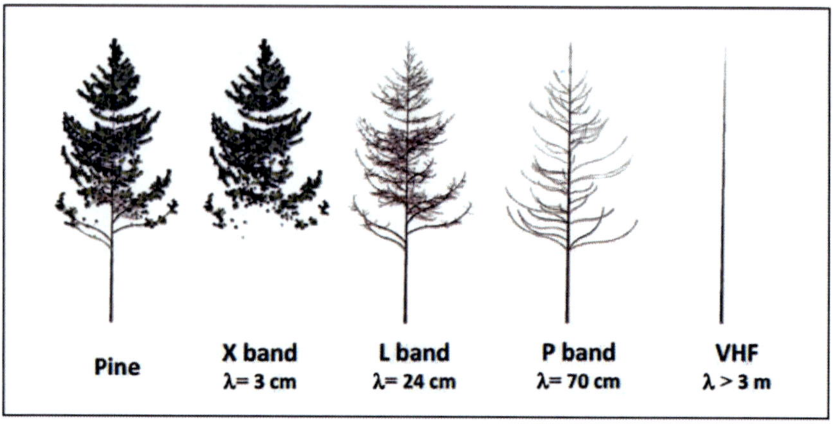

Fig. 11.2 Characteristics of microwave bands in relating to forest stand parameters

region of electromagnetic (EM) spectrum. In forestry, the microwave data are extensively used in forest cover and type mapping, discrimination of forest compartments, and estimation of forest stand parameters and biomass, etc. The most commonly used microwave data processing technique involve conversion of digital number (DN) values or pixel values to backscatter coefficient and generation of three-dimensional surface using Differential SAR Interferometry (DInSAR) technique (Otukei and Emanuel 2015). The accuracy of microwave remote sensing data for biomass and carbon assessment depends on the choice of the suitable wavelength bands and polarization. The synthetic aperture radar (SAR) data are acquired in several wavelength bands i.e., X, C, L and P. Each of these bands has different characteristics in relating to the forest stand parameters (Fig. 11.2). Due to its lower wavelength, the microwave signal in 'X-band' is highly scattered by leaves and canopy cover and enables to extract information about the surface layer of the trees. A microwave signal operating in a higher wavelength as 'C-band' penetrates through leaves and scatters by small branches and under layer elements. In comparison, the 'L-band' operates at much higher wavelength and penetrates through the surface layers and is scattered by the trunk and main branches. The 'P-band' with very high operating wavelength (around 70 cm) has very high penetration capacity through the canopy cover and the majority backscatter in 'P-band' is caused by trunk and the trunk ground reflectance. The next important parameter of microwave data is the polarization of the signals, i.e., the direction of electric field of the electromagnetic waves. The microwave signals are emitted and received in horizontal (H) or vertical (V) polarizations. Depending on the polarization of emitted and received (backscatter) signal, it has four combinations: (1) HH: horizontally emitted and horizontally received, (2) HV: horizontally emitted and vertically received, (3) VH: vertically emitted and horizontally received, and (4) VV: both the emitted and received signals have vertical polarization. ALOS-PALSAR is an example of quad-pol sensor commercial data operates in L band.

Radarsat-1, Radarsat-2 and the latest Sentinel-1 are examples C band microwave data, where the Sentinel-1 data is available freely.

Number of local, regional and global assessments have been carried out in mapping the AGB. Santos et al. (2002) used the L-band JERS-1 data to estimate forest and savanna biomass in two contact zones in the Brazilian Amazonia. Pandey et al. (2010) used the Envisat ASAR to generate the AGB map for the Dudhwa National Park of Lakhimpur-Kheri district in Uttar Pradesh state, India, and observed higher accuracy for the HV band than HH polarization band, where the model fitting was found better for the low biomass region compared to the high biomass regions. Otukei and Emanual (2015), Goh et al. (2014), Carreiras et al. (2013) used ALOS PALSAR microwave data while estimating the AGB, where the field data was collected with circular plot design different radius of 15 m, 20 m, and 25 m, respectively. Hamdan et al. (2015) carried out a similar assessment using L-band ALOS PALSAR data in a Peninsular forest and Mangrove forest in Malaysia. With the availability of the latest no-cost Sentinel-1 microwave data, number of studies have included this data for estimating forest biomass in different forest types and climate zones. Laurin et al. (2018) tested the accuracy of AGB prediction using the Sentinel-1 in a Mediterranean forested landscape and suggested to use multi-temporal data and differential analysis for various forest types, where the biomass can be reliably estimated up to 400 tonnes/ha. The advances in data processing approaches are leading to the application of both regression-based approach as well as the incorporation on machine learning models to develop the relation between predictor variables and field measurements. The regression-based approach allows the use of single or multiple predictor variables while generating the relationship and provides the statistical measures on the accuracy, and thereafter the developed relationship is used for extrapolation (Hamdan et al. 2015; Laurin et al. 2018). On the contrary, although, the machine learning models are mostly work as black box, are observed to better perform while linking the dependent and independent variables. Castillo et al. (2017) employed both the regression-based and machine learning models to estimate the AGB in Mangrove forest; and, observed good performance of the optical (Sentinel-2) derived LAI and best performance for the backscatter values from the microwave (Sentinel-1) data using machine learning model. Numbers of researchers have also tested the performance of integrated approach, where the optical data derived forest biophysical and microwave data derived structural and textural information are merged. Ghosh and Behera (2018) used the synergistic approach, where they integrated the Sentinel-2 data derived vegetation indices [Normalized Difference Vegetation Index (NDVI), Green Normalized Difference Vegetation Index (GNDVI), Renormalized Difference Vegetation Index (RDVI), and Soil-Adjusted Vegetation Index (SAVI)] and Sentinel-1 backscatter and textural information employing the decision tree-based machine learning models. Sinha et al. (2020) also adopted the integration approach, where they used the Landsat TM optical data and multiple microwave data as COSMO-Skymed (X-band), Radarsat-2 (C-band), and ALOS-PALSAR (L-band) data for multiple linear regression to model the AGB. Avitabile et al. (2016) combined two datasets and generated the AGB map for the pan-tropical region.

LiDAR (Light Detection and Ranging) is an active system offering tremendous potential for estimating forest biomass with the major advantage as the acquisition of three-dimensional data of the forest structure including canopy cover characteristics, leaf area index, crown cover and volume, etc. The ability of the laser altimeters to penetrate through forest canopies to the ground level better identifies the plant structure (Rosillo-Calle and Woods 2012). Previous research has indicated that use of LiDAR data is a promising approach for biophysical parameters estimation (Drake et al. 2002; Hyde et al. 2007), such as timber volume and stand height (Naesset 1997), tropical forest biomass (Drake et al. 2002), Douglas fir western hemlock biomass (Lefsky et al. 1999) etc. Long wavelength radar data have the advantage in AGB estimation for complex forest stand structures and LiDAR data have the potential to provide vertical structure information (Zimble et al. 2003).

The present study is undertaken to estimate above ground biomass (AGB) and carbon stock for *Shorea* robusta (sal) forest of Jhargram block, Paschim Medinipur district, West Bengal, India. Using the no-cost high resolution Sentinel-1 microwave data, the relationship between AGB and VH and VV polarization was derived using regression analysis. Moreover, using the AGB map, the tree density map of the study area was also produced.

11.3 Study Area

Jhargram is a community development block that forms an administrative division in Jhargram district in the Indian state of West Bengal (Fig. 11.3). As per the 2011 Census of India, the total population of Jhargram CD Block was 170,097. The terrain is regulated by the Chotanagpur Plateau, which gradually slopes down creating an undulating area having infertile laterite soils. About 75% of the cultivated area has lateritic soil and 20% has alluvial soil and 5% has red gravelly soil in this region. The topography of this area is relatively flat or gently undulating with an average elevation of 81 m above the mean sea level. The main streams in the area is Kangsabati River also known as Kasai and Cossye in the northern part of the study area. The diurnal temperature reaches as high as 46 °C in the hot and dry months of May and June and drops to 4 °C during winters (December and January). The average annual rainfall is about 1370 mm with a concentrated rainfall during monsoon (June to September) leading to comparatively dry pre and post-monsoon. Such a climate condition lead to dry deciduous tropical forest, where *S. robusta* (Sal) is dominated and partially mixed with various species like Mahua, Palash, Kusum, Kendu and Neem, etc.

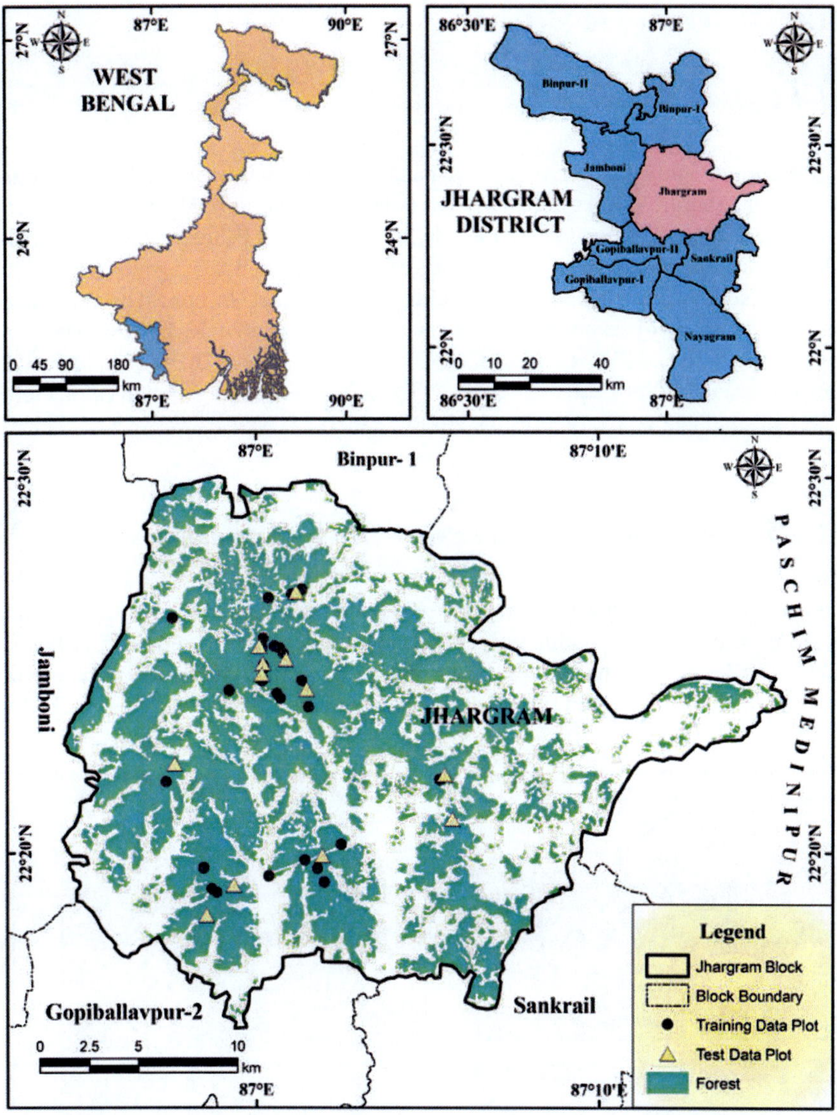

Fig. 11.3 Location map of the study area in West Bengal and field plots (training and testing) overlaid on forest cover area

11.4 Materials and Method

11.4.1 Data Sources

In the current study, the Sentinel-1A C-band SAR data has been used for biomass estimation. Sentinel-1A data contains two polarization bands as VV and VH. Based on the field data collection period, the satellite scene of 9th February 2018 was used. The satellite scene was acquired from the ESA (European Space agency) scientific data hub (scihub.copernicus.eu/dhus/#/home). The Sentinel-2A optical data was also used in this study to map the forest cover area. It consists of 13 spectral bands in the visible, NIR, and SWIR region of the EM spectrum. The SNAP tool and ArcGIS software was used for data processing, AGB and corresponding carbon estimation. The statistical analysis was performed in MS excel.

11.4.2 Field Data Collection and AGB Measurement

The plot level field data was collected in 40 locations having a rectangular plot size of 30 m × 30 m area. Within each plot, the number of trees, diameter at breast height (DBH) and height of trees were measured, and multiple geo-locations were recorded inside the plots and corner location using a handheld GPS device (Fig. 11.4).

Fig. 11.4 Field observations and plot level plant inventory data collection

The DBH was measured at height 1.3 m above the base of a tree. A revised and improved allometric equation (Eq. 11.1) was adapted from Chave et al. (2014) for AGB estimation; whereas the carbon fraction approach (Eq. 11.2) was used for carbon stock estimation (Hirata et al. 2012).

$$\mathbf{AGB = 0.0673(\rho \times D^{2\times}H)^{0.976}} \tag{11.1}$$

where, $\boldsymbol{\rho}$ = Wood density in gcm^3, **D** = DBH in cm, and **H** = Tree height in m

$$\mathbf{C = B \times CF} \tag{11.2}$$

where, **C** = Carbon stored in tonnes, **B** = AGB, and **CF** = Carbon fraction (0.5).

11.4.3 Methodology

The field collected data on tree count, DBH and height was used to compute the AGB using the suitable allometric equation, which was statistically linked with satellite data derived proxies. Based on the develop relation with the training data, a continuous AGB map was prepared and validated with test data points. The carbon stock map was prepared by multiplying the carbon fraction. The overall methodology flow diagram has been shown in Fig. 11.5.

11.4.3.1 Pre-processing of Optical Data and Generation Forest Cover Area Map

The atmospheric correction of Sentinel-2 scene was performed using the QGIS software. The atmospherically corrected data was then used to derive the land use forest cover map of the study area. The maximum likelihood classifier was used for image classification, which was validated with the field observed data points for accuracy assessment. The forested area in the study area was extracted and used to mask the microwave data before estimating the AGB.

Calculation of Enhanced Vegetation Index (EVI)

The Enhanced Vegetation Index (EVI) was computed using the Sentinel-2 data. The EVI is an optimized vegetation index designed to enhance the vegetation signal with improved sensitivity to high biomass and de-coupled the canopy background signal and influences of atmospheric scattering using suitable correction factors.

The atmospherically corrected Sentinel-2 data was employed to derive EVI map using the following equation:

$$\mathrm{EVI} = \mathrm{G} \times \frac{(\boldsymbol{NIR - RED})}{(\boldsymbol{NIR + C1 \times RED - C2 \times BLUE + L})} \tag{11.3}$$

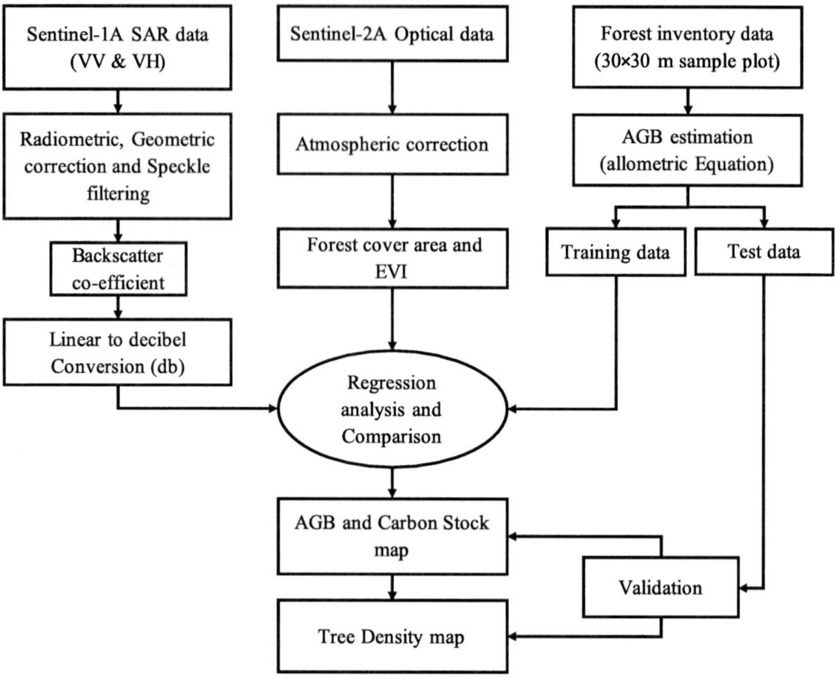

Fig. 11.5 Methodology of the study area

where, NIR, Red and Blue bands are atmospherically corrected bands, L is the canopy background adjustment, and C1 and C2 are the coefficients factors of aerosol resistance term, which uses the blue band to correct the influence of aerosol content. The coefficients values used in the MODIS-EVI were utilized: L = 1, C1 = 6, C2 = 7.5, and G (gain factor) = 2.5.

11.4.3.2 Pre-processing of Microwave Data

The sentinel-1 GRD data product (level 1) was used in this study. Using the SNAP tool, the radiometric (e.g., conversion of DN values to backscatter coefficient $\sigma°$) and geometric correction (e.g., conversion of slant range to ground range, multi look and azimuth compression) were performed prior to the data analysis. Thereafter, the Gamma filter (with 5 × 5 kernel size) was applied to remove the speckles in the image. Then, Doppler terrain correction was applied employing the shuttle radar topographic mission (SRTM) digital elevation model (DEM) data. The pre-processed image was then exported in GeoTIFF format for further processing in ArcGIS software.

11.4.3.3 Modelling and Estimation of AGB, Carbon Stock, and Tree Density

A total of 40 plot data was available, which was segregated into training and testing to develop the model and validation purpose, respectively. A random selection process was applied to divide the entire data into training (25 plots or 62.5% data) and testing (15 plots or 37.5% data). To corroborate with the pixel size and plot area, the backscatter coefficient from 9 neighbouring pixels centring the plot location was considered to develop the relation between backscatter values and the field measured in situ AGB. The univariate linear regression analysis was performed to develop the relationship between AGB and independent variables as radar backscatter bands (VV and VH), and EVI. The best performing predictor variable was chosen by comparing the coefficient of determination (R^2) value. The best performing variable was then employed to create the continuous biomass map, and the test data was utilized to validate it. The prediction accuracy was assessed via coefficient of determination, RMSE and bias values. The carbon stock map was prepared using the predicted AGB map by multiplying the carbon fraction factor of 0.5 (Watson 2009; Hirata et al. 2012). To estimate tree density map per pixel, the predicted AGB map was used, where the equation used for biomass estimation was reversed.

$$NP = \frac{Bmp \times 1000}{0.0673 \times (0.73 \times D \times H)0.976} \tag{11.4}$$

where, **Np** = No. of predicted tree, **Bmp** is predicted biomass and **0.0673** and **0.976** is model coefficient, and **D** is Diameter at breast height (DBH) in cm and **H** is tree height in meter.

11.5 Results and Discussion

It is essential to map the forest cover area for reliable biomass estimation. The Sentinel-2 multi-spectral data was used to create the land use forest cover map employing the four 10 m VNIR bands. The maximum likelihood classifier was used, where the forest covered areas were extracted to mask the microwave data before biomass estimation (Fig. 11.6). Field visits were carried out to collect plot level data, where the geo-location of *S. robusta* trees along with the DBH, tree density and height data were recorded in 40 sample plots (Fig. 11.6). At 30 m × 30 m plot area, the measured DBH varied between 34 and 67 cm, tree height varied between 12 and 15 m, and the numbers of trees varied between 44 and 157 per plot. A part of plot data was used to create signature file during image classification and rest of the data points were used to verify the classification accuracy indicated all confirm forest area occurrences. The total identified forest area is 55.36 km^2.

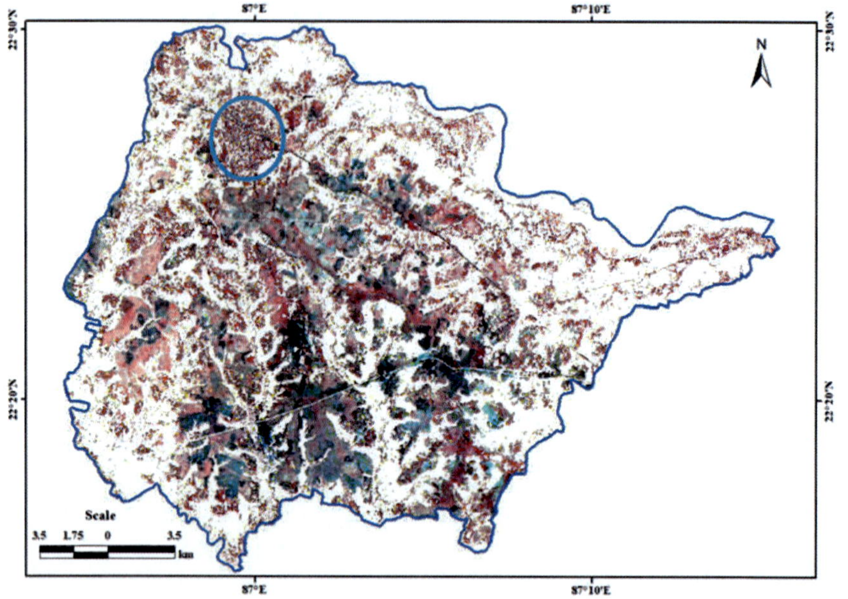

Fig. 11.6 The forest cover area map of the study area. The cyan coloured oval shaped area was not considered and masked from subsequent spatial maps being an urban settlement area (Jhargram town)

The allometric equation developed by Chave et al. (2005) was employed to estimate the AGB in the current study. The wood density value (0.73 gm/cm^3) for *S. robusta* given by the Forest Survey of India (FSI) was used in this case. The plot level in situ AGB varied between 88.56 tonnes/ha and 170.29 tonnes/ha, with an average biomass value of 122.93 tonnes/ha. The simple linear regression model was tested to assess the variability explained by each of the independent parameters as backscatter in VV and VH polarization bands, and EVI. During modelling, the obtained result indicated the highest coefficient determination value for VH polarized band (R^2 = 0.63) and a comparatively lower R^2 (0.44) for VV polarized band, which was 0.45 (R^2) for EVI (Fig. 11.7a–c).

As, the highest coefficient of determination value was obtained with the VH band, it was thus used for estimating AGB and carbon content map. The coefficients derived from the regression analysis was used to generate the continuous AGB map. Using the test data points, the coefficient of determination, RMSE and bias values were estimated for the predicted AGB map, which showed moderate correlation (R^2 = 0.45) with significantly low RMSE value (±17 tonnes/ha) and slight underestimation (bias value = −0.024) (Fig. 11.8).

The direct relationship between microwave backscatters values and plot-wise tree density was observed comparatively weak ($R^2 = 0.4$ for VH and 0.23 for VV), whereas, the relation with AGB was observed much stronger ($R^2 = 0.87$) (Fig. 11.9).

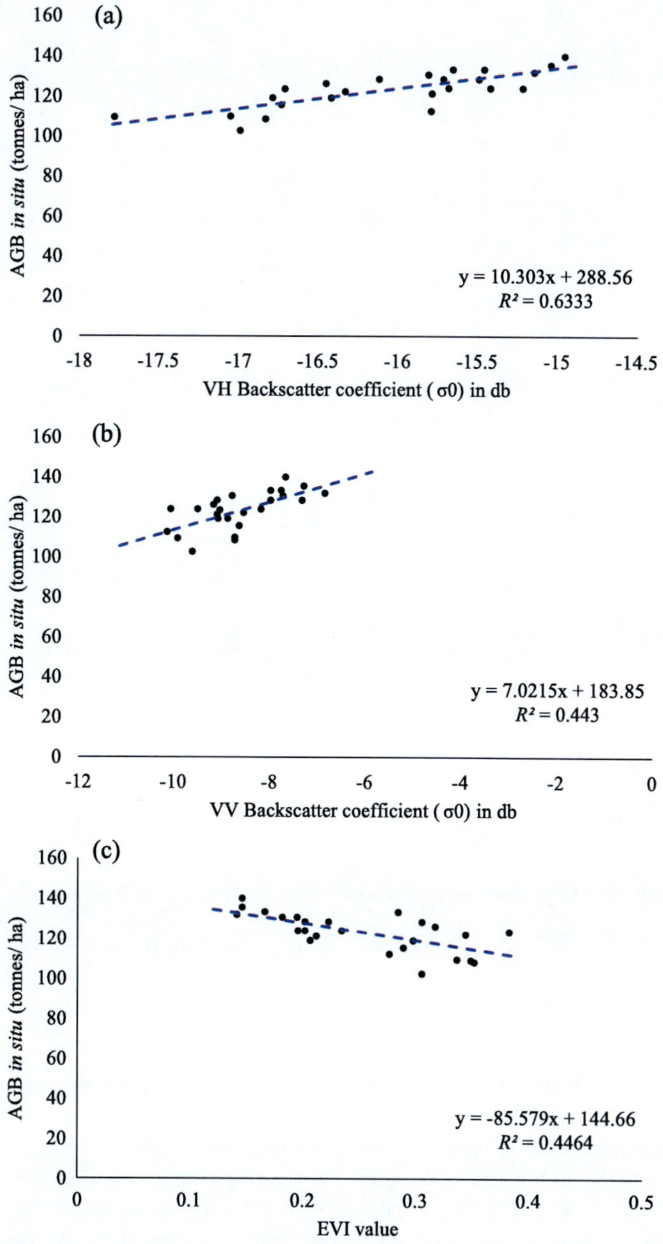

Fig. 11.7 Relationship between AGB and **a** backscatter coefficient (σ°) in db in VH band, **b** backscatter coefficient (σ°) in db in VV band, and **c** EVI

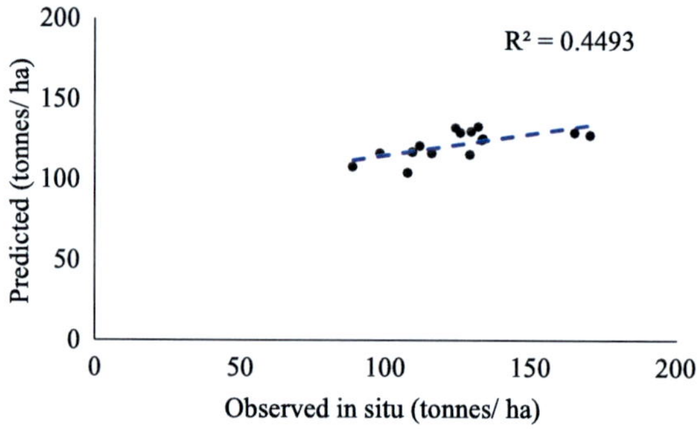

Fig. 11.8 Predicted AGB is plotted against the observed AGB for validation

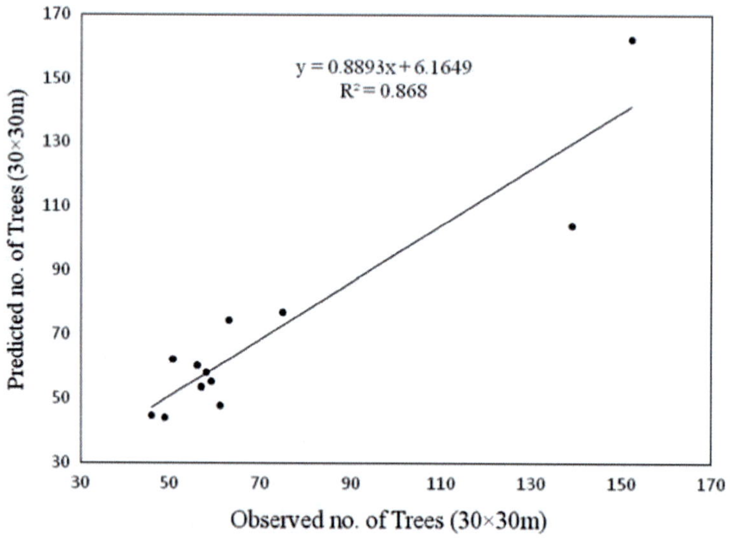

Fig. 11.9 Observed versus predicted tree density

Thus, the tree density map for the entire study area was produced using the AGB map.

The estimated AGB and carbon stock maps are shown in Fig. 11.10a, b. The result indicates that the amount of AGB varies between 44.1 tonnes/ha and 249 tonnes/ha with the total AGB as 33,102 tonnes estimated for the study area. Subsequently, the estimated carbon stock map follows the similar pattern of the AGB map. The AGB map indicates majority of the areas having the AGB ranging between 100–200 tonnes/ha. On the other hand, areas with biomass below 100 tonnes/ha and above 200 tonnes/ha are comparatively less. The areas below 100 tonnes/ha are mostly indicating the sparse and comparatively new plantation areas; whereas, the areas

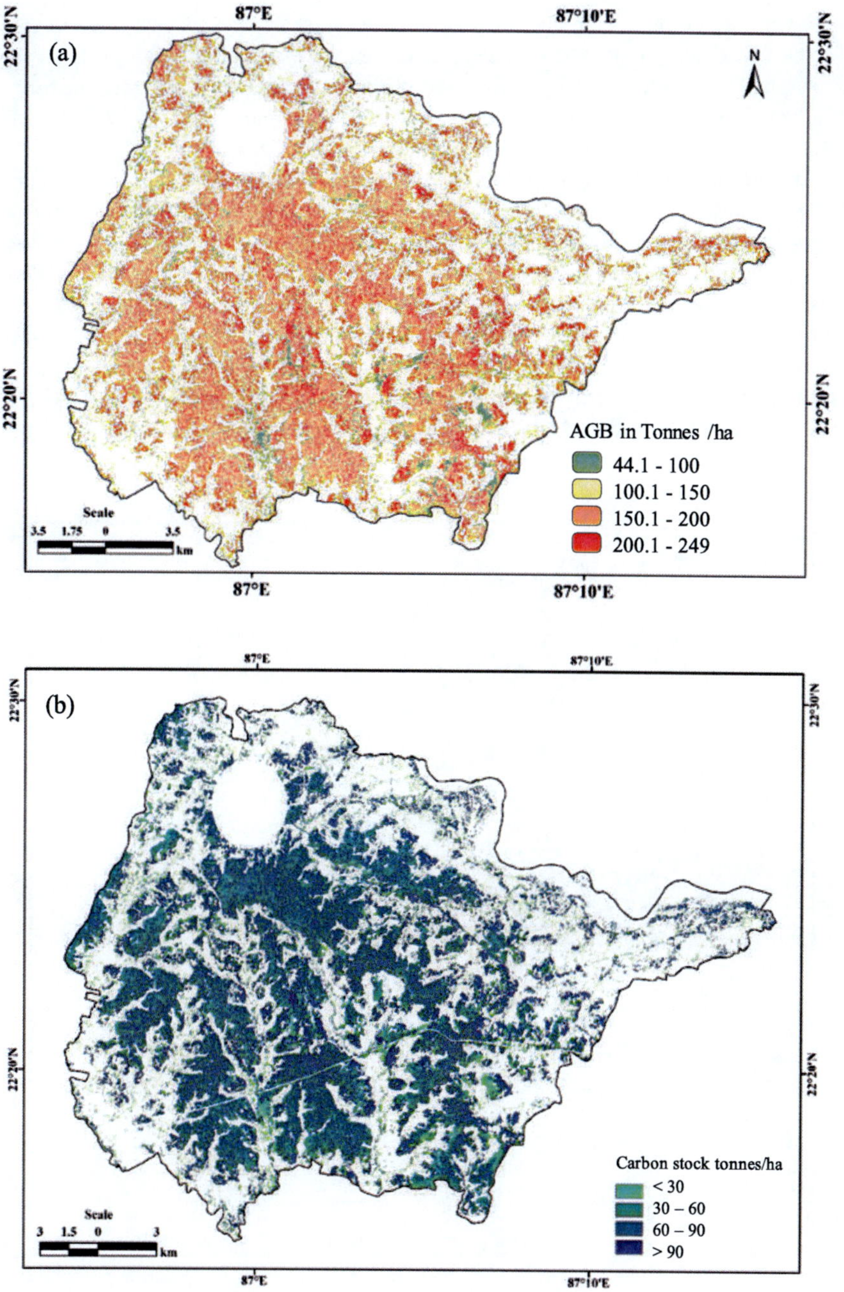

Fig. 11.10 Estimated **a** AGB and **b** Carbon stock for the study area

with very high biomass are the old grown dense forest areas. This was corroborated by the tree density map and statistics, which shows that the contribution from the lowest and highest tree density class is about 20% only, where the intermediate range (69/ha to 75/ha) contributes rest of the 80% AGB in this region (Fig. 11.11 and Table 11.2). The total numbers of trees estimated for the study region is 19,872,061.

The ground sampled data exhibits the variation in DBH as 34–67 cm, infers that's the incorporation of comparatively newer and old plants in the current analysis. However, the variations in average tree height was comparatively less. Moreover, the tree density mostly varies between 44/ha and 75/ha, where only two plots were observed to have exceptional high tree density. The supervised image classification well captured the vegetation cover area, from where the settlement region was masked out to avoid the differential scattering from the buildings/artificial constructions. The regression analysis indicted the higher performance by VH band ($R^2 = 0.63$) in capturing the plants structural information in comparison to VV band

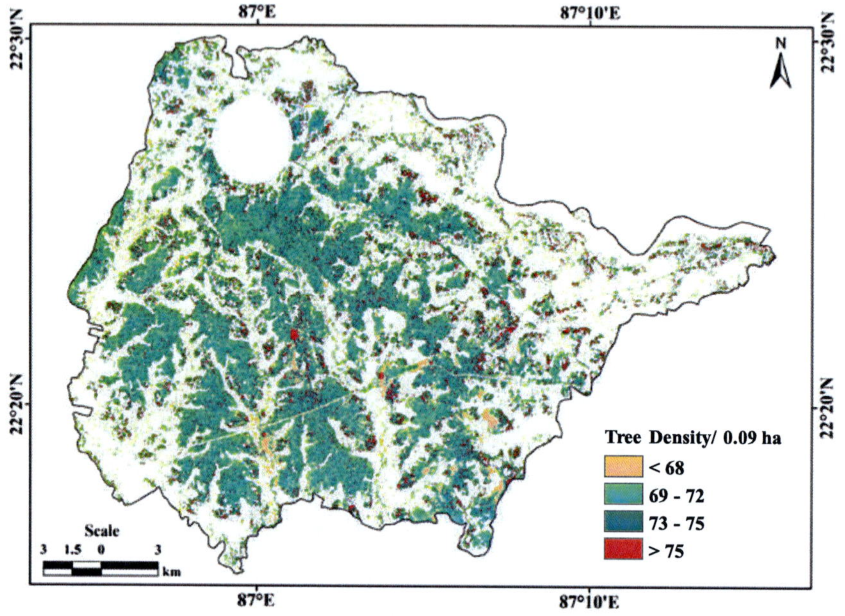

Fig. 11.11 Tree density map of Jhargram Block

Table 11.2 Estimated tree density statistics

Density range	Number of trees	Total number of trees
<68	2,194,751	19,872,061
69–72	8,684,798	
72–75	7,256,195	
>75	1,736,317	

($R^2 = 0.44$), where higher backscatter corresponds to higher AGB. Similarly, the biophysical characteristic as EVI, derived from the optical data could not perform well as explained by the structural information obtained from VH band. Moreover, the estimated AGB was found well accurate ($R^2 = 0.87$) while deriving the tree density. At plot level, the observed AGB varied between 88.56 tonnes/ha and 170.29 tonnes/ha, which was estimated between 44.1 tonnes/ha and 249 tonnes/ha for the entire study region and majority of the AGB was observed in the range of 100–200 tonnes/ha. This could be indicating a uniform forest cover regime with comparatively lower number patches having lower and higher AGB. Forest patches below the average AGB range could be the new plantation sites; whereas, patches with higher AGB could be the old and dense patches. Moreover, in case of tree density, the majority (>80%) of the estimated tree density varies between 69/ha and 75/ha, which supporting the uniformity in tree cover in this region. Behera et al. (2016) used the L-band ALOS-PALSAR data to estimate AGB in Katerniaghat Wildlife Sanctuary (KWS), a tropical forested region and observed better performance of VV band compared to VH band employing non-linear regression. However, their prediction accuracy ($R^2 = 0.47$) was similar as derived in the current study ($R^2 = 0.45$); and, observed a higher AGB ranges up to 625 tonnes/ha for *S. robusta* dominated forest covers. Thumaty et al. (2016) also estimated the AGB for the state of Madhya Pradesh mostly contains the moist and dry deciduous forest, *Tectona grandis* (teak), and *S. robusta* forests employing the L-band ALOS-PALSAR data. Considering all the forest type together, they have observed the HV band to better explain the variability of AGB ($R^2 = 0.51$) with a similar RMSE value (±19.32 tonnes/ha) as estimated in the currently study (±17 tonnes/ha). Behera et al. (2016) vividly studied the forest biomass for three distinct plant functional types (PFTs) employing filed measurements in KWS and estimated the AGB ranging 290.82– 455.99 tonnes/ha, and their principle component analysis (PCA) of microclimatic condition with forest structural information indicates higher contribution from humidity and air-temperature. While estimating the AGB integrating backscatter and textural information derived from Sentinel-1 data and vegetation index from Sentinel-2 data, Ghosh and Behera (2018) observed higher coefficient of determination value ($R^2 = 0.71$) with poor RMSE value (105.027 tonnes/ha) for *S. robusta* in KWS. The range of AGB observed in the current study also supported by the observation of Pandey et al. (2010) estimated the range as 119–520 tonnes/ha for sal and sal dominated forest cover in the tropical forest of Dudhwa National Park of Lakhimpur-Kheri district in Uttar Pradesh state, India.

11.6 Conclusion

The present study assessed the potential of Sentinel-1 microwave data for AGB and tree density estimation over a dry deciduous *S. robusta* (sal) dominated forest in the Jhargram CD block of West Bengal state, India. Moreover, the Sentinel-2 optical remote sensing data was used to demarcate the forest area and to derive the forest biophysical parameter as EVI and to assess its relationship with AGB. The sentinel-1 provided the backscatter values in VV and VH polarizations, which were linearly regressed with the field measured AGB data. 40 sample plot data were collected on the field, which indicates a better variation explained by the VH band ($R^2 = 0.63$) compared to the VV band ($R^2 = 0.44$), where the relationship was observed weaker for EVI also ($R^2 = 0.45$). Using the regression equation, the continuous forest AGB map was thus estimated employing the VH band. The validation results showed good accuracy ($R^2 = 0.45$) with low RMSE value (17 tonnes/ha), and slight underestimation (bias $= -0.024$). The regression analysis and validation result clearly indicates the applicability of Sentinel-1 microwave data for AGB estimation with a limited accuracy. However, the accuracy for the estimated tree density map was observed much higher ($R^2 = 0.87$). The variability in field measurement via DBH indicates incorporation of older and younger trees dominated plots in the analysis, and the estimated AGB varied between 44.1 tonnes/ha and 249 tonnes/ha, which corroborated the AGB observations from similar climate zones in India. Majority of the estimated AGB ranged between 100–200 tonnes/ha, which could be indicating the uniform forest cover in this region, which was also indicated by the tree height and tree density data. However, few regions were observed to have lower and higher AGB indicating the newer and densely forest cover regions, respectively. The spatial biomass and tree density maps are highly useful for forest cover management activities, especially to adopt the climate adaptive conservation measures. The study outcome confirms the use of the freely available Sentinel-1 C-band microwave data for reliable AGB estimation. It has further proposed to perform differential analysis for other gregarious forest types and mixed forest types in different climate zones and segregating the young and old forests.

Acknowledgments We are grateful to the PG Department of Geography, Raja N. L. Khan Women's College (Autonomous), affiliated to Vidyasagar University, Midnapore, West Bengal, India for supporting this research. The author (P. K. Shit) grateful acknowledges West Bengal DSTBT for financial support through R&D Research Project Memo no. 104(Sanc.)/ST/P/S&T/ 10G-5/2018).

References

Avitabile V, Herold M, Heuvelink GB, Lewis SL, Phillips OL, Asner GP, Armston J, Ashton PS, Banin L, Bayol N, Berry NJ (2016) An integrated pan-tropical biomass map using multiple reference datasets. Global Change Biol 22(4):1406–1420

Baishya R, Barik SK, Upadhaya K (2009) Distribution pattern of aboveground biomass in natural and plantation forests of humid tropics in northeast India. Tropical Ecol 50(2):295

Behera MD, Tripathi P, Mishra B, Kumar S, Chitale VS, Behera SK (2016) Above-ground biomass and carbon estimates of Shorea robusta and Tectona grandis forests using QuadPOL ALOS PALSAR data. Adv Space Res 57(2):552–561

Brown S, Gillespie AJ, Lugo AE (1989) Biomass estimation methods for tropical forests with applications to forest inventory data. For Sci 35(4):881–902

Carreiras J, Melo JB, Vasconcelos MJ (2013) Estimating the above-ground biomass in miombo savanna woodlands (Mozambique, East Africa) using L-band synthetic aperture radar data. Remote Sens 5(4):1524–1548

Castillo JAA, Apan AA, Maraseni TN, Salmo SG III (2017) Estimation and mapping of above-ground biomass of mangrove forests and their replacement land uses in the Philippines using Sentinel imagery. ISPRS J Photogram Remote Sens 134:70–85

Center for International Forestry Research (2008) Forest carbon toolbox. Forest and climate change toolbox. Retrieved from https://www.cifor.org/fctoolbox/

Chamber JQ, dos Santos J, Ribeiro RJ, Higuchi N (2001) Tree damage, allometric relationships, and above-ground net primary production in central Amazon forest. For Ecol Manage 152(1–3):73–84

Chave J, Andalo C, Brown S, Cairns MA, Chambers JQ, Eamus D... Lescure JP (2005) Tree allometry and improved estimation of carbon stocks and balance in tropical forests. Oecologia 145(1):87–99

Chave J, Réjou-Méchain M, Búrquez A, Chidumayo E, Colgan MS, Delitti WB, Duque A, Eid T, Fearnside PM, Goodman RC, Henry M (2014) Improved allometric models to estimate the aboveground biomass of tropical trees. Global Change Biol 20(10):3177–3190

Drake JB, Dubayah RO, Knox RG, Clark DB, Blair JB (2002) Sensitivity of large-footprint lidar to canopy structure and biomass in a neotropical rainforest. Remote Sens Environ 81(2–3):378–392

FAR (2010) Global forest Resources assessment main report, Rome, 2010

Ghosh SM, Behera MD (2018) Aboveground biomass estimation using multi-sensor data synergy and machine learning algorithms in a dense tropical forest. Appl Geogr 96:29–40

Goh J, Miettinen J, Chia AS, Chew PT, Liew SC (2014) Biomass estimation in humid tropical forest using a combination of ALOS PALSAR and SPOT 5 satellite imagery. Asian J Geoinf 13(4)

Hamdan O, Hasmadi IM, Aziz HK, Norizah K, Zulhaidi MH (2015) L-band saturation level for aboveground biomass of dipterocarp forests in peninsular Malaysia. J Tropical For Sci, 388–399

Haripriya GS (2000) Estimates of biomass in Indian forests. Biomass Bioenerg 19(4):245–258

Hirata Y, Takao G, Sato T, Toriyama J (2012) REDD-plus cookbook: How to measure and monitor forest carbon. REDD research and development center, forestry and forest products research institute, Tsukuba

Hofmann DJ, Butler JH, Tans PP (2009) A new look at atmospheric carbon dioxide. Atmos Environ 43(12):2084–2086

Houghton RA (2012) Carbon emissions and the drivers of deforestation and forest degradation in the tropics. Curr Opin Environ Sustain 4(6):597–603

Houghton RA, House JI, Pongratz J, Van Der Werf GR, DeFries RS, Hansen MC, Le Quéré C, Ramankutty N (2012) Carbon emissions from land use and land-cover change. Biogeosciences (12):5125–5142

Hyde P, Nelson R, Kimes D, Levine E (2007) Exploring LiDAR–RaDAR synergy—predicting aboveground biomass in a southwestern ponderosa pine forest using LiDAR, SAR and InSAR. Remote Sens Environ 106(1):28–38

IPCC (2003) IPCC good practice guidance for land use, land-use change and forestry (adopted by decisions 2/CP.17, 17/CP.18) IGES, Japan

Karna YK, Hussin YA, Gilani H, Bronsveld MC, Murthy MS, Qamer FM, Karky BS, Bhattarai T, Aigong X, Baniya CB (2015). Integration of WorldView-2 and airborne LiDAR data for tree species level carbon stock mapping in Kayar Khola watershed, Nepal. Int J Appl Earth Observ Geoinf 38:280–291

Keeling CD, Adams JA Jr, Ekdahl CA Jr, Guenther PR (1976) Atmospheric carbon dioxide variations at the South Pole. Tellus 28(6):552–564

Kumar A, Sharma MP (2015) Assessment of carbon stocks in forest and its implications on global climate changes. J Mater Environ Sci 6(12):3548–3564

Laurin GV, Balling J, Corona P, Mattioli W, Papale D, Puletti N, Rizzo M, Truckenbrodt J, Urban M (2018). Above-ground biomass prediction by Sentinel-1 multitemporal data in central Italy with integration of ALOS2 and Sentinel-2 data. J Appl Remote Sens 12(1):016008

Lee NJ, Nakane K (1997) Forest vegetation classification and biomass estimation based on Landsat TM data in a mountainous region of west Japan. In: The use of remote sensing in the modeling of forest productivity, pp 159–171. Springer, Dordrecht

Lefsky MA, Harding D, Cohen WB, Parker G, Shugart HH (1999) Surface lidar remote sensing of basal area and biomass in deciduous forests of eastern Maryland, USA. Remote Sens Environ 67(1):83–98

Lu D, Chen Q, Wang G, Moran E, Batistella M, Zhang M, Vaglio Laurin G, Saah D (2012). Aboveground forest biomass estimation with Landsat and LiDAR data and uncertainty analysis of the estimates. Int J For Res 2012

Lucas RM, Mitchell AL, Armston J (2015) Measurement of forest above-ground biomass using active and passive remote sensing at large (subnational to global) scales. Curr Forestry Rep 1:162–177. https://doi.org/10.1007/s40725-015-0021-9

Madugundu R, Nizalapur V, Jha CS (2008) Estimation of LAI and above-ground biomass in deciduous forests: Western Ghats of Karnataka, India. Int J Appl Earth Obs Geoinf 10(2):211–219

Naesset E (1997) Estimating timber volume of forest stands using airborne laser scanner data. Remote Sens Environ 61(2):246–253

Nath AJ, Tiwari BK, Sileshi GW, Sahoo UK, Brahma B, Deb S, Devi NB, Das AK, Reang D, Chaturvedi SS, Tripathi OP (2019) Allometric models for estimation of forest biomass in North East India. Forests 10(2):103

Otukei JR, Emanuel M (2015) Estimation and mapping of above ground biomass and carbon of Bwindi impenetrable National Park using ALOS PALSAR data. S Afr J Geomat 4(1):1–13

Pandey U, Kushwaha SPS, Kachhwaha TS, Kunwar P, Dadhwal VK (2010) Potential of Envisat ASAR data for woody biomass assessment. Tropical Ecol 51(1):117

Rosillo-Calle F, Woods J (2012) The biomass assessment handbook. Routledge

Roy PS, Ravan SA (1996) Biomass estimation using satellite remote sensing data—an investigation on possible approaches for natural forest. J Biosci 21(4):535–561

Sader SA, Waide RB, Lawrence WT, Joyce AT (1989) Tropical forest biomass and successional age class relationships to a vegetation index derived from Landsat TM data. Remote Sens Environ 28:143–198

Salunkhe O, Khare PK, Sahu TR, Singh S (2016) Estimation of tree biomass reserves in tropical deciduous forests of Central India by non-destructive approach. Tropical Ecol 57(2):153–161

Santos JR, Lacruz MP, Araujo LS, Keil M (2002) Savanna and tropical rainforest biomass estimation and spatialization using JERS-1 data. Int J Remote Sens 23(7):1217–1229

Sinha S, Mohan S, Das AK, Sharma LK, Jeganathan C, Santra A, Santra Mitra S, Nathawat MS (2020). Multi-sensor approach integrating optical and multi-frequency synthetic aperture radar for carbon stock estimation over a tropical deciduous forest in India. Carbon Manag 11(1):39–55

Sullivan MJP, Lewis SL, Hubau W, Qie L, Baker TR, Banin LF, Phillips OL (2018) Data from field methods for sampling tree height for tropical forest biomass estimation. ForestPlots.net. https://doi.org/10.5521/forestplots.net/2018_1

Thumaty KC, Fararoda R, Middinti S, Gopalakrishnan R, Jha CS, Dadhwal VK (2016) Estimation of above ground biomass for central Indian deciduous forests using ALOS PALSAR L-band data. J Indian Soc Remote Sens 44(1):31–39

Vargas-Larreta B, López-Sánchez CA, Corral-Rivas JJ, López-Martínez JO, Aguirre-Calderón CJ, Álvarez-González JG (2017) Allometric equations for estimating biomass and Carbon stocks in the temperate forests of north-western Mexico. Forests 8:269. https://doi.org/10.3390/f8080269.

Watson C (2009) Forest carbon accounting: overview and principles. Forest carbon accounting: overview and principles

Zimble DA, Evans DL, Carlson GC, Parker RC, Grado SC, Gerard PD (2003) Characterizing vertical forest structure using small-footprint airborne LiDAR. Remote Sens Environ 87(2–3):171–182

Chapter 12
Forest Fire Risk Assessment for Effective Geoenvironmental Planning and Management using Geospatial Techniques

Narayan Kayet

Abstract Forest are essential natural resources having the role of supporting economic activity, which plays a significant role in regulating the climate and the carbon cycle. Forest ecosystems increasingly threatened by fires caused by a range of natural and anthropogenic factors. Hence, spatial assessment of fire risk is critical to reducing the impacts of wildland fires. In the current research, the evaluation of forest fire risk (FFR) assessment performed by geospatial data of Melgaht Tiger Reserve Forest (MTRS), Maharashtra, India. We have used eleven natural and anthropogenic parameters (slope, altitude, topographic position index (TPI), aspect, rainfall, land surface temperature (LST), air temperature, wind speed, normalized differential vegetation index (NDVI), distance to road and distance to settlement) for FFR assessment based on the Analytic hierarchy process (AHP) and Frequency ratio (FR) models in a GIS framework. The results from AHP and FR models shown similar trends. The AHP model was significantly higher accuracy than the FR model. AHP and FR models based FHR maps were classified into five classes (very low, low, moderate, high, and very high). According to the generated FFR maps, the very high-risk class was found at some forest blocks (Mangtya, Kund, Gudfata, Katharmal, Amyar). The sensitivity analysis showed that some parameters (wind speed, air temperature, LST, slope, altitude, distance to settlement, and distance to the road) were more sensitive to forest fire risk. The FFR results were justified by the forest fire sample points (Forest Survey of India) and burn images (2010–2018). This work will provide a basic guideline for effective geo-environmental planning and management of Melgaht Tiger Reserve Forest.

Keywords Forest fire risk assessment · Analytical hierarchy process (AHP) · Frequency ratio (FR) · Remote sensing and GIS

N. Kayet (✉)
Department of Mining Engineering, Indian Institute of Technology Kharagpur, Kharagpur, India
e-mail: narayankayet@gmail.com

P. K. Shit et al. (eds.), *Spatial Modeling in Forest Resources Management*, Environmental Science and Engineering,
https://doi.org/10.1007/978-3-030-56542-8_12

12.1 Introduction

Forest is a natural resource, and it has a significant role in maintaining the balance of the environment. Fire is a rapid combination of fuel, oxygen, and heat. The population of India is increasing day by day, and it has become a problem, especially for the forest. Most of the forest fires in India are caused by human activity. A forest fire has loosely related to the socio-economic activity of the people who live in the forest areas near the forest area. Grazing, shifting cultivation, and deforestation of forest resources are the main reasons for a forest fire. According to the Forest Survey of India (FSI) state-wide study in the year 2000, about 1.45 million per year of forest area affected by the fire. So it is essential to monitor forest fire in Melgaht Tiger Reserve (MTR) Forest for effective geo-environmental planning and management.

Due to natural or anthropogenic reasons, forest ecosystems are threatened by increasing the rate of a forest fire. So, forest fire risk zonation or spatial assessment of fire risk is a very important task nowadays to reduce the impact of wildfire. The occurrence of forest fires is one of the major environmental concerns (called wildfires), which has an effect on forest preservation, ecological and economic damage as well as causes to human suffering (Cortez and Morais 2007). Many environmental and anthropogenic variables are related to fire, and recognizing them is required to understand fire occurrences and to determine fire risk (Krivtsov et al. 2009). But most of the forest fires are the consequences of anthropogenic activity. It may cause deforestation, timber harvesting, for the use of forest in another land use purpose such as agricultural land, pastureland, etc. Wildfire can be defined as the rapid combustion of fuel, heat, and oxygen. It is a chemical reaction of any substance that will ignite and burn to release a lot of energy in the form of heat and light (Rawat 2003).

Remote sensing has become a very useful and reliable technology for the monitoring as well as for management of forest fire. Many kinds of literature have been conducted to develop a forest fire risk zone mapping by the use of RS and GIS techniques. MODIS data commonly applied for the identification of forest fire all over the world; which can provide 12-bit data (Radiometric resolution) in 36 spectral bands at a range of 0.4–14.4 μm (Pourghasemi et al. 2016). Roy (2003) applied geospatial techniques that are very useful to assess the forest fire risk and degradation assessment. In this paper, he had described the present states of forest, forest fire risk, and degradation assessment in the context of India. Sharma et al. (2012) to find out the forest fire risk zone by the Fuzzy AHP (Analytical Hierarchy Process) approach in Shimla Forest Division (Himachal Pradesh) in 2012. The result has a strong agreement with actual fire occurrences in the past years.

In present study area is located in Amravati district of Maharashtra, India. Melghat forest famous for the Tiger Reserve forest. This area has dry deciduous forests and excellent tiger habitat. The local forest department said the Melghat forest area destroyed in the fire would be anything between 2000 and 2200 ha of forestland in the years. That why we chose this study area for forest fire risk assessment.

In this study, a proposed methodology is prepared for forest fire risk (FFR) assessment using multi-criteria models in a GIS framework. The objectives of the study

Table 12.1 Data used in the present study

Data layer	Source of data
Satellite Images (Landsat 8 OLI)	Earth Explorer/GLOVIS
Cartosat 1 DEM	NRSC Bhuvan
Quick Bird Image	Google Earth
Forest fire location points	Forest Survey of India
Climate data	SWAT

were: (1) Forest fire zoning mapping using the multi-criteria model in a GIS framework. (2) To identify very high forest fire risk zones in the Melghat Tiger Reserve forest. (3) Forest risk zoning results justified by FSI training points.

12.2 Materials and Methods

12.2.1 Data Source

For the study, the Landsat satellite image of Maharashtra state was acquired for the year 2016. The data was obtained from GLOVIS, an Earth Science Data Interface, while that of the Digital elevation model for Melghat forest was acquired from NRSC Bhuvan. It is also essential for the study because it provides topographical information regarding. Climate data obtained from SWAT global weather data. The details of the data used are given in Table 12.1.

12.2.2 Study Area

The Melghat Tiger Reserve situated in the Satpura hill ranges of Central India and it lies in Melghat forest of Amravati district in the Vidarbha region of Maharashtra, bordering Madhya Pradesh in North and East (Fig. 12.1). The MTR area of 1571.74 km^2 lies in the heart of the Melghat forest and was declared the Melghat Tiger Reserve on 22nd February 1974. The MTR is within latitude 2100 1500 to 2100 4500 N and longitude 7600 5700 to 7700 3000 E at elevations of 312 to 1178 m MSL. The yearly maximum temperature averages 42.7 °C, with annual rainfalls 2700 mm. Winds are generally light to moderate; the strongest winds are less than 22 km/h.

Fig. 12.1 Location map of Melghat Tiger Reserve Forest

12.2.3 Frequency Ratio Model (FR)

The frequency ratio model, a simple geospatial assessment tool for computing the probabilistic relationship between dependent and independent variables, including multi-classified maps, can be obtained by an FR model (Choi et al. 2012). Based on this assumption, the relationships between fires occurring in an area and the forest fire-related factors can be distinguished from the relationships between fires not occurring in an area and the forest fire-related factors. This thing can be represented as an FR, which represents the quantitative relationship between forest fires events and different causative parameters. The FR can be defined as the probability of occurrence of a certain attribute (Mohammady et al. 2012).

12.2.4 Analytical Hierarchy Process (AHP)

The analytical hierarchy process is a theory of measurement for considering tangible and intangible criteria that have been applied to numerous areas, such as decision theory and conflict resolution (Vargas 1990; Yalcin 2008). The AHP is an eigenvalue technique for the pair-wise comparisons approach. It is based on three principles: decomposition, comparative judgment, and synthesis of priorities (Chen and Xu 2010). The decomposition principle is applied to structure a complex problem into a hierarchy of clusters, sub-clusters. The AHP provides a fundamental numerical scale, which ranges from 1 to 9 to calibrate the quantitative and qualitative performances of priorities (Saaty 2008). This matrix ultimately enters expert choice (EC) software and will calculate the final weight for each conditioning factor with consistency ratio (CR). If CR is less than 10%, then the matrix can be considered as having an acceptable consistency (Saaty 2008).

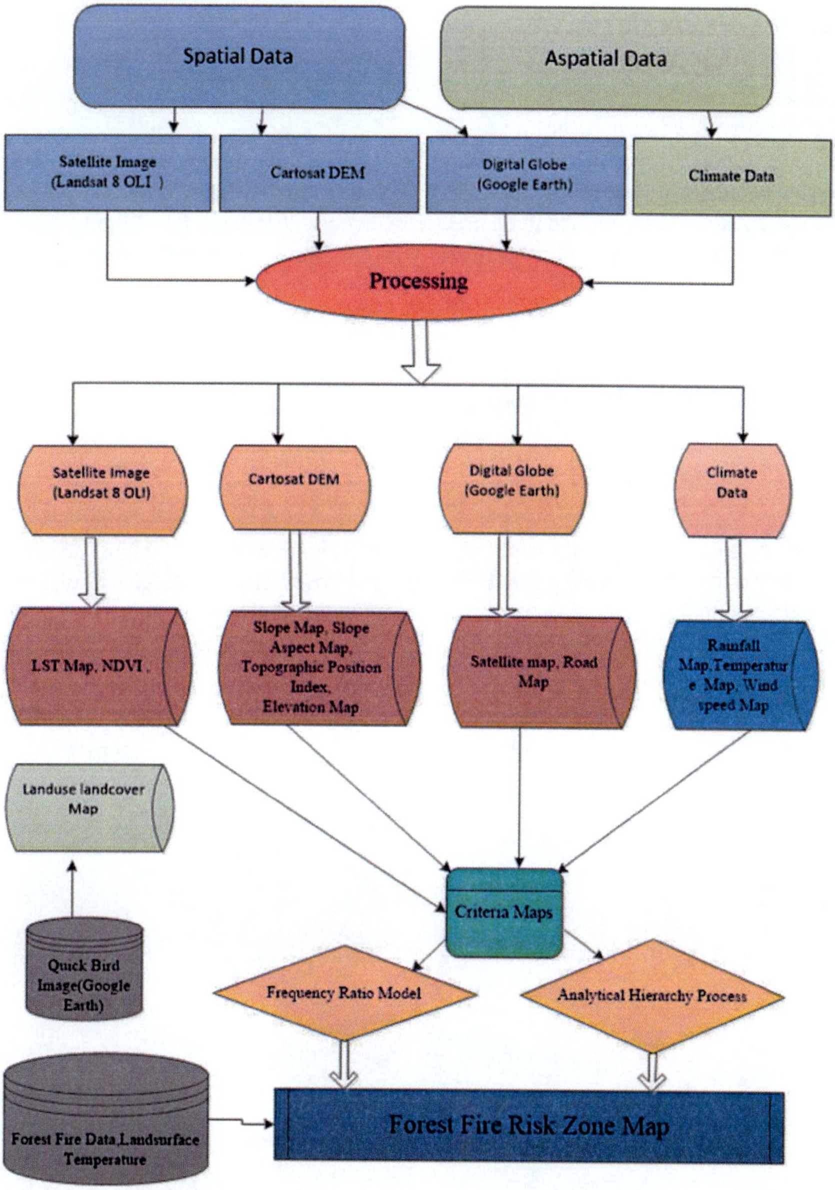

Fig. 12.2 Overall methodology of forest fire risk mapping

12.2.5 Meteorology

Figure 12.2 shows the used methodology in this study area with a flowchart. The figure shows the factors used in the analysis and the processes applied according to the methods. At the first stage, required data were collected, then those data will be processed with proper correction, and various factors are prepared to performed Frequency Ratio (FR) and Analytic hierarchy process (AHP) models. Finally, forest fire risk zone map prepared and validate with the previous record of forest fire in Melghat forest (2010–2016).

12.3 Results and Discussion

12.3.1 Land Use and Land Cover (LULC)

The study area is comprised mainly of forest areas, so most of the domain is belong to different forest categories. Land use and land cover map of Melghat forest are prepared using the Support Vector Machine (SVM) classifier technique. The LULC map of the study area is shown in Fig. 12.3. Area of different LULC classes are present in Table. 12.2.

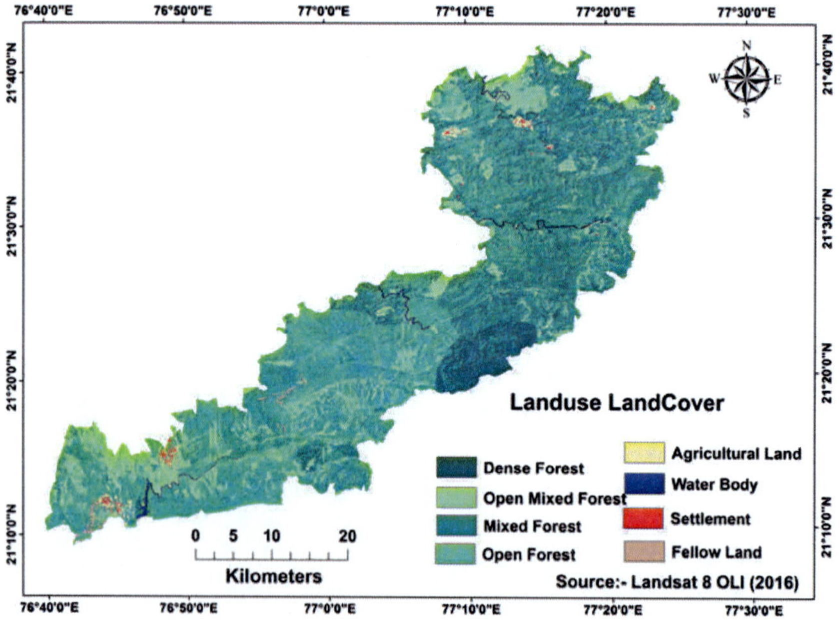

Fig. 12.3 Land use and land cover map of Melghat Tiger Reserve Forest

Table 12.2 Area of LULC classes

Classes name	Area in km^2	Area in percent
Dense Forest	257.91	17.47
Mixed Forest	379.81	25.74
Open Mixed Forest	395.89	26.83
Agricultural Land	9.24	0.62
Fellow Land	7.33	0.49
Settlement	4.63	0.31
Water Body	7.37	0.49
Open Forest	413.33	28.01

12.3.2 Land Surface Temperature (LST)

The land surface temperature of a particular date of an area can be measured from the satellite image by using a remote sensing technique. In this study, the land surface temperature of Melghat forest calculated for the date of 12th April 2016, the LST of the Melghat area is to be measured by LST model. Figure 12.4 shown in land surface temperature in Melghat Tiger Reserve Forest.

In this Fig. 12.4, the maximum temperature is to be found 39.34 °C, and the minimum is 180 °C. Forest fire points are plotted in this figure to find out in which

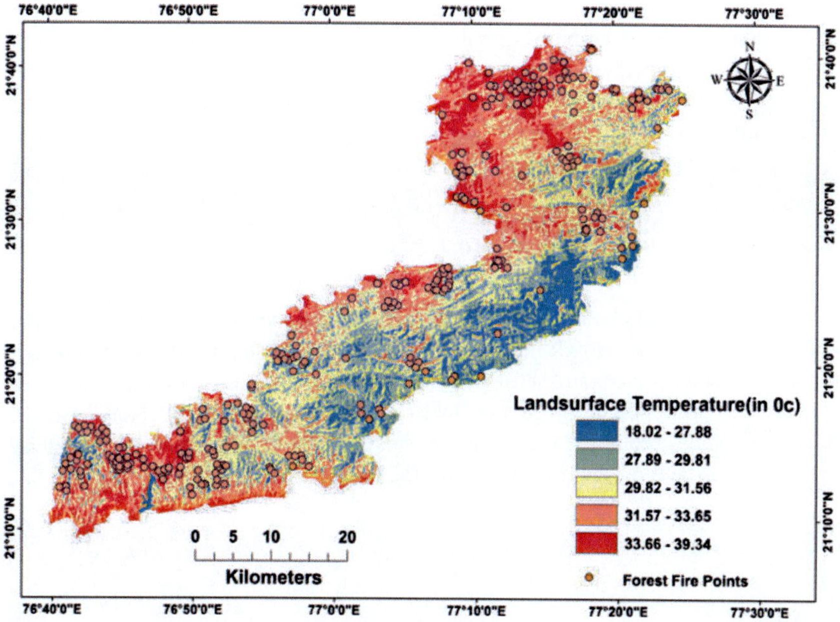

Fig. 12.4 Land surface temperature of Melghat Tiger Reserve Forest

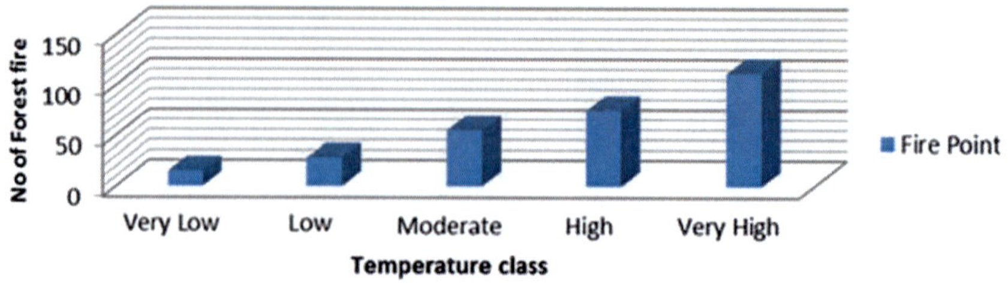

Fig. 12.5 Relation between forest fire occurrence and LST

temperature there is a maximum no of fire points that have occurred. The relation between forest fire occurrence and temperature is shown in the following figure by a simple bar graph (Fig. 12.5).

12.3.3 Criteria for Forest Fire Risk Zoning

Forest fire is one type of combustion of vegetation which mainly occurs in the forest area. The frequent occurrence of forest fires is a significant reason for depletion and extinction of our valuable plants and animal species. There are many natural and anthropogenic factors that are responsible for a forest fire, here natural factors such as elevation, slope, aspect, topographic position index, rainfall, wind speed, temperature, DR, DS, and NDVI are used as criteria. Again forest fire is a result of anthropogenic activity, so distance to Settlement and distance to the road are also used to prepare forest fire risk zone mapping as anthropogenic factors. Both natural and anthropogenic factors that are used in this study as criteria are shown in Fig. 12.6.

12.3.4 Frequency Ratio Based FFR

The results of the spatial relationship between forest fire and forest fire-related factors using the frequency ratio model are illustrated in Table 12.3. In the table we found that temperature, wind speed, and settlement are the highest frequency ratio value, then other factors and slope aspects get the lowest values. So, we can say that temperature, wind speed, and settlement factors are most responsible for a forest fire in the study area than other factors. The forest fire risk (FFR) zoning map shown in Fig. 12.7.

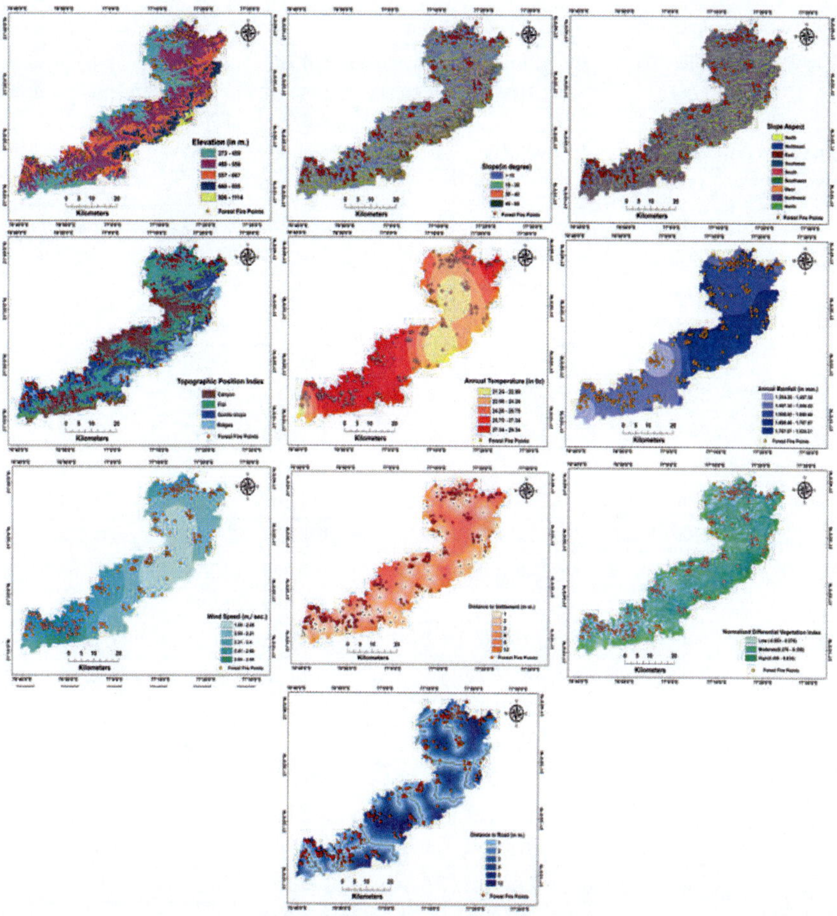

Fig. 12.6 All criteria maps for forest fire risk zoning mapping

12.3.5 Analytical Hierarchy Process Based FFR

The idea of multi-criteria techniques has been implemented with different uncertainty levels of AHP. Forest fire risk zonation map by weighted mean overlay analysis using AHP is thus obtained in GIS mode using MCDA. AHP model is actually based on Saaty's pairwise comparison method (Table 12.4). Here each and individual criteria are compared as a pair, and importance between these two criteria is measured and then put the weight as per Saaty's comparison scale. Figure 12.8 shown the forest fire susceptibility map by the AHP model.

Table 12.3 Frequency ratio value for each criteria

Factors	Class	No. of pixels	Percentage of pixels	No. of forest fire point	Percentage of forest fire points	Frequency ratio value
Elevation (m)	273–459	76,500	30.35	466,663	26.63	0.303
	459–556	120,600	47.85	713,390	43.51	0.478
	556–667	41,400	16.42	328,022	20.00	0.164
	667–805	12,600	5	134,911	8.22	0.050
	805–1114	900	0.35	26,456	1.61	0.003
Slope (°)	>15	103,500	41.07	571,217	34.84	0.410
	15–30	90,900	36.07	542,794	33.10	0.360
	30–45	40,500	16.07	370,035	22.57	0.160
	45–60	17,100	6.78	155,396	9.47	0.067
Slope aspect	Flat	20,700	8.21	184,908	11.27	0.082
	North	30,600	12.14	161,102	9.82	0.121
	North-east	19,800	7.85	154,762	9.43	0.078
	East	26,100	10.35	166,863	10.17	0.103
	South-east	31,500	12.50	193,884	11.82	0.125
	South	30,600	12.14	186,325	11.36	0.121
	South-west	40,500	16.07	182,038	11.10	0.160
	West	29,700	11.78	194,847	11.88	0.117
	North-west	22,500	8.92	214,713	13.09	0.089
Topographic position index (TPI)	Canyon	54,000	21.42	376,471	22.96	0.414
	Ridges	115,200	45.71	685,056	41.78	0.307
	Gentle Slope	61,200	24.28	404,390	24.66	0.242
	Flat	21,600	8.57	173,525	10.58	0.085
Normalized differential vegetation index (NDVI)	−0.563 to 0.276	153,000	60.71	486,777	29.69	0.287
	0.276–0.358	90,900	36.07	996,163	60.76	0.199
	0.358–0.830	8100	3.21	156,502	9.54	0.113
Rainfall (mm)	1354.26–1606.82	119,700	47.5	621,516	37.91	0.475
	1606.82–1787.87	93,600	37.14	650,652	39.68	0.371
	1787.87–1924.21	38,700	15.35	367,274	22.40	0.153
Temperature (°C)	21.24–24.26	52,200	20.71	557,364	33.99	0.207
	24.26–27.34	92,700	36.78	477,214	29.10	0.367

(continued)

Table 12.3 (continued)

Factors	Class	No. of pixels	Percentage of pixels	No. of forest fire point	Percentage of forest fire points	Frequency ratio value
	27.34–29.34	107,100	42.50	604,864	36.89	0.425
Wind speed (m/s)	1.85–2.21	80,100	31.78	792,860	48.36	0.317
	2.21–2.68	116,100	46.07	660,030	40.25	0.460
	2.68–2.98	55,800	22.14	186,552	11.37	0.482
Distance to settlement (km)	0–1	18,000	7.14	180,984	11.03	0.214
	1–2	54,000	21.42	381,139	23.24	0.067
	2–3	61,200	24.28	385,454	23.51	0.202
	3–4	55,800	22.14	291,708	17.79	0.221
	4–5	30,600	12.14	183,744	11.20	0.121
	5–12	32,400	12.85	216,413	13.20	0.128
Distance to road (km)	0–1	74,700	29.64	505,533	30.83	0.296
	1–2	55,800	22.14	374,338	22.83	0.221
	2–3	62,100	24.64	309,287	18.86	0.246
	3–4	30,600	12.14	216,808	13.22	0.121
	4–5	22,500	8.92	124,243	7.57	0.089
	5–12	6300	2.5	109,233	6.66	0.025

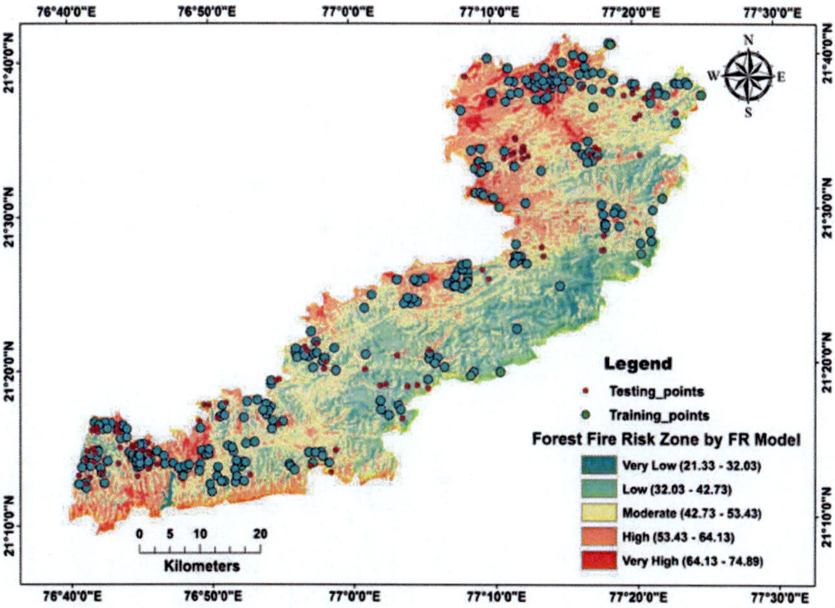

Fig. 12.7 Forest fire risk map using FR model

Table 12.4 Pairwise Comparision Matrix using AHP

	Elevation	Slope	Slope aspect	TPI	Temperature	Rainfall	Wind	NDVI	Road	Settlement
Elevation	1.0	4	5	6	3	4	7	2	6	5
Slope	0.25	1.0	4	5	3	2	5	4	5	6
Slope Aspect	0.2	0.25	1.0	2	3	2	5	3	5	4
TPI	0.16	0.2	0.5	1.0	2	3	4	2	5	6
Temperature	0.33	0.33	0.33	0.5	1.0	4	6	2	6	5
Rainfall	0.25	0.5	0.5	0.33	0.25	1.0	4	3	5	6
Wind	0.14	0.2	0.2	0.25	0.16	0.25	1.0	4	3	2
NDVI	0.5	0.25	0.33	0.5	0.5	0.33	0.25	1.0	6	7
Road	0.16	0.2	0.2	0.2	0.16	0.2	0.33	0.16	1.0	3
Settlement	0.2	0.16	0.25	0.16	0.2	0.16	0.5	0.14	0.33	1.0

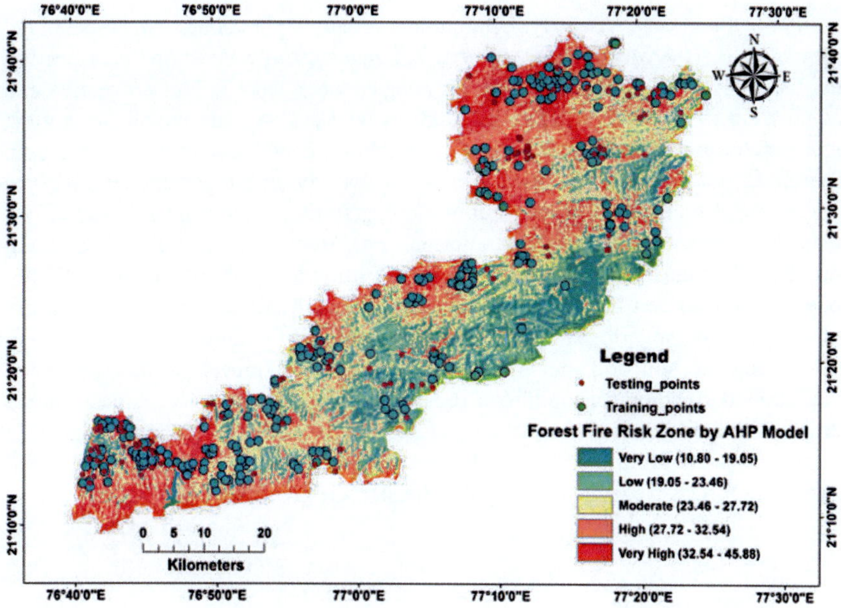

Fig. 12.8 Forest fire risk map using AHP model

12.3.6 Comparative Analysis Between FR and AHP Models for FFR

FR and AHP-based fire risk pixel values are used to comparative analysis in different forest risk classes (Fig. 12.9).

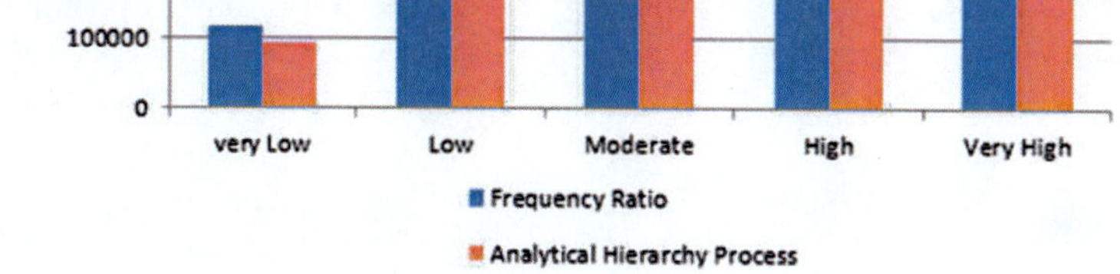

Fig. 12.9 Area of fire risk zones using FR and AHP models

Fires in the forest have always been studied as natural phenomena, and more of the efforts were on the way to suppress them. Not much effort was put on the inspection and the analysis of the causes. Policy strategies were according to preventive and suppression measures (Prasad et al. 2008) as well as, fire risk models are a great approach for precautionary measures for the environmental protection of the forests. Therefore, wildfires are the result of several underlying factors; model variables were slope degree, slope aspect, altitude topographic position index, wind effect, distance to roads, rivers and villages, normalized difference vegetation index, annual temperature, annual rain, and land use. In the study area, forest fire susceptibility maps were produced using the frequency ratio and Analytical Hierarchy Process models. The findings revealed that the most important conditioning factors were NDVI, land use, soil, and annual temperature. Therefore, preventive measures need to be applied to ecological conditions. Prasad et al. (2008) stated that the mean annual temperature could strongly influence on forest fires in the Deccan Plateau, India. In another research, Motazeh (2013) indicated that based on expert choice in hardwood Hyrcanian forests, vegetation coverage allocated the greatest weight.

12.4 Conclusion

In the current study, we find that the north, northeast, and southwest parts are in highly sensitive zones because that area contains high temperature, unhealthy vegetation, and the best part is that the human activity of those areas is very significant. So the tiger reserve of Melghat forest should be placed on deep forest areas, which is not fire porn also, human activity of those areas should be prohibited to protect the natural environment. Again to minimize the fire activity in the Melghat forest, it is essential to reduce the human activity and also should perform the plantation program because the vegetation health condition is not satisfactory. These models can be used in other cities by exchanging the variables and the weights but cannot use the same weights and variables in different regions because forest fire in each part of the earth has its characteristics and indices should be improved over areas with different environmental conditions. However, more simulation results need to compare other methods. In addition, remote-sensing data with its spatial information, when combined with GIS and statistical models, allow fire managers and personnel to predict 'where and when' forest fires will most likely. In general, the mentioned methods can be applied to early warning fire suppression resources planning and allocation works in the study area.

References

Chen C, Xu Z (2010) Forest ecosystem responses to environmental changes: the key regulatory role of biogeochemical cycling, pp 210–214

Choi J, Oh HJ, Lee HJ, Lee C, Lee S (2012) Combining landslide susceptibility maps obtained from frequency ratio, logistic regression, and artificial neural network models using ASTER images and GIS. Eng Geol 124:12–23

Cortez P, Morais ADJR (2007) A data mining approach to predict forest fires using meteorological data

Krivtsov V, Vigy O, Legg C, Curt T, Rigolot E, Lecomte I, Pezzatti GB (2009) Fuel modelling in terrestrial ecosystems: an overview in the context of the development of an object-orientated database for wild fire analysis. Ecol Model 220(21):2915–2926

Mohammady M, Pourghasemi HR, Pradhan B (2012) Landslide susceptibility mapping at Golestan Province, Iran: a comparison between frequency ratio, Dempster-Shafer, and weights-of-evidence models. J Asian Earth Sci 61:221–236

Motazeh AG, Ashtiani EF, Baniasadi R, Choobar FM (2013) Rating and mapping fire hazard in the hardwood Hyrcanian forests using GIS and expert choice software. Acknowledgement to reviewers of the manuscripts submitted to Forestry Ideas in 2013, p 141

Pourghasemi HR, Beheshtirad M, Pradhan B (2016) A comparative assessment of prediction capabilities of modified analytical hierarchy process (M-AHP) and Mamdani fuzzy logic models using Netcad-GIS for forest fire susceptibility mapping. Geomat Nat Hazards Risk 7(2):861–885

Prasad VK, Badarinath KVS, Eaturu A (2008) Biophysical and anthropogenic controls of forest fires in the Deccan Plateau, India. J Environ Manage 86(1):1–13

Rawat GS (2003) Fire risk assessment for forest fire control management in Chilla Forest Range of Rajaji National Park, Uttaranchal, India. ITC

Roy PS (2003) Forest fire and degradation assessment using satellite remote sensing and geographic information system. Satellite Remote sensing and GIS applications in agricultural meteorology, p 361

Saaty TL (2008) Decision making with the analytic hierarchy process. Int J Serv Sci 1(1):83–98

Sharma LK, Kanga S, Nathawat MS, Sinha S, Pandey PC (2012) Fuzzy AHP for forest fire risk modeling. Disast Prevent Manag Int J

Vargas LG (1990) An overview of the analytic hierarchy process and its applications. Eur J Oper Res 48(1):2–8

Yalcin A (2008) GIS-based landslide susceptibility mapping using analytical hierarchy process and bivariate statistics in Ardesen (Turkey): comparisons of results and confirmations. CATENA 72(1):1–12

Chapter 13
Forest Disturbance Analysis of Selected Blocks of Midnapore Subdivision using Digital Remote Sensing Technique

Ratnadeep Ray, Swarnali Biswas, and Ahona Bej

Abstract Change is ubiquitous in forest ecosystems. Forests undergo both seasonality ns well as enduring escalation cycles which may vary in long term. These long-term changes are punctuated by habitually interim disturbances from fire, insects, disease, and harvest which strongly alter the state and functioning of the forest. The spatio-temporal information on process-based forest loss as well as change is essential for a wide range of applications. The disturbances over forest cover and resulted changes alter the water and carbon cycles of forest stands as well as bang the habitat and biodiversity of these ecosystems. To effectively understand how forest disturbance impacts forest state and functioning, the disturbance and related effects on forest cover is needed to be quantified at spatial scale where human management and natural strife occur. In this present study, the spatial pattern of forest cover of the four blocks like Garhbeta 1, Garhbeta 2, Garhbeta 3 and Salboni under Midnapore Sadar Sub-division was analysed on temporal scale. The forested area of this region is region is lying under the Midnapore forest division, the total area of which is 50,267.49 ha. Forest is the one of the important natural resources of this area and the important source of rural livelihood as well as ecological sustainability. But it is changing temporally rather is fading and maximum stress is seen onto the dense forest and open forest. So for the restoration practice, forest disturbance indexing as well as identification of disturbed sites and an account on forest regeneration and degeneration are so needful. Therefore, as per the goal of the study a Multicriteria based Forest Disturbance Index and forest fragmentation analysis were deployed. Finally it is so pertinent to mention that the forest covers under the blocks like Garhbeta 2 and Salboni were in a miserable state.

R. Ray (✉)
Department of Geography and Environment Management, Vidyasagar University, Midnapore, India
e-mail: ratnadiprsgis@gmail.com

S. Biswas
Directorate of Science and Technology, Kolkata, India

A. Bej
Department of Geography, University of Calcutta, Kolkata, India

P. K. Shit et al. (eds.), *Spatial Modeling in Forest Resources Management*, Environmental Science and Engineering,
https://doi.org/10.1007/978-3-030-56542-8_13

Keywords Forest disturbance index · Fragmentation analysis · Spatio-temporal · Regeneration and degeneration

13.1 Introduction

Change is ubiquitous in forest ecosystems. Forests undergo both seasonality and long-term growth cycles, which can range in span of 50–500 years or more (Waring and Running 2007). Such long-term changes are punctuated by often short-term fire, pest, disease, and harvest disturbances, which strongly alter the forest state and function (He and Mlandenoff 1999). Both climate change and increasing global demand for wood and fiber products are likely to cause forest destruction levels to rise (Kurz et al. 2008; Nepstad et al. 2008). Forest status of a region is closely allied to the annual amount of precipitation. Continuous and increasing exhaustive climate change could have a explicit impact on the status of natural vegetation, particularly water availability (Malik et al. 2019a, b).Such disturbance changes will alter the forest stands' water and carbon cycle as well as affect the environment and biodiversity of those habitats (Lindenmayer et al. 2006; Gardner et al. 2009). Forest destruction is now recognized as a significant driver of non-fossil-fuel-related terrestrial fluxes into the atmosphere with respect to the carbon cycle (Running 2008; Amiro et al. 2010). Unplanned deforestation in a very precarious way hampers both the human being and the lives of the other animals (UN Report 2019). In addition, monitoring and assessing the vegetation status is the most important aspect of environmental health assessment and global climate change alleviation (Alberdi et al. 2019; Bjorkman et al. 2019; UN Environmental Report 2019).

In order to understand effectively how forest disturbance affects the condition and work of the environment, disturbance levels need to be quantified in the spatial grain where human activity and natural disturbances occur; usually less than 10 ha (Miller 1978; Cohen et al. 2002; Kuemmerle et al. 2007; Frolking et al. 2009). In addition, disturbances need to be quantified at a time and spatial scale appropriate to the affected processes (e.g. globally, annually), and over a time-limit appropriate to the development of baselines applicable to forest policy initiatives (historically, at least as far back as the 1990s) (Böttcher et al. 2008; Masek et al. 2008; Kennedy et al. 2012).

Forest is the one of the important natural resources of the area under study and the important source of rural livelihood as well as ecological sustainability. Besides these the majesticity of forest of this area has made it unique in the map of West Bengal. But it is changing day by day rather is deteriorating and maximum stress is seen onto the dense forest and open forest. West Bengal (WB) has lost 1612 km^2 of classified forest during last 100 years and within this, Paschim Medinipur stands third among all the districts of WB in terms of forest degradation with a loss of 110 km^2 of classified forest. Degradation of forest cover, on the marginal laterite regions with low potential agricultural productivity, is often linked to the soil characteristics and associated geomorphic processes on rarh lateritic regions of West

Bengal (Shit and Pati 2012; Bhunia et al. 2012). State Forest Report of 2011–12, shows an achievement in growing forest cover due to plantation activity in Paschim Medinipur District but it is well accompanied by destruction at the local level as 19 km^2 of forested track was degraded in 2017. The Tropical dry deciduous forest in WB co-exists with human settlement where anthropogenic interactions with the forest have aggravated the conflict between man and forest (Nagendra et al. 2009; Mahapatra and Tiwari 2005). In order to keep the forest sustainable, recognizing forest gaps within a specific area can be an important approach in order to provide adequate strategic management (Pal et al. 2018). Considering in this respect the gap as disturbed area forest disturbance indexing as well as identification of possible forest disturbance zone is so important.

13.2 About the Study Area

In the present study, the study area is consisted of 4 blocks like-Garhbeta 1, Garhbeta 2, Garhbeta 3 and Salboni of Midnapore Sadar Subdivision. It is situated within the latitude and longitudinal extensions of 22°28′ N to 22°52′ N and 87°5′ E to 87°31′ E. The total area under study is 1624.81 km^2 (Fig. 13.1).

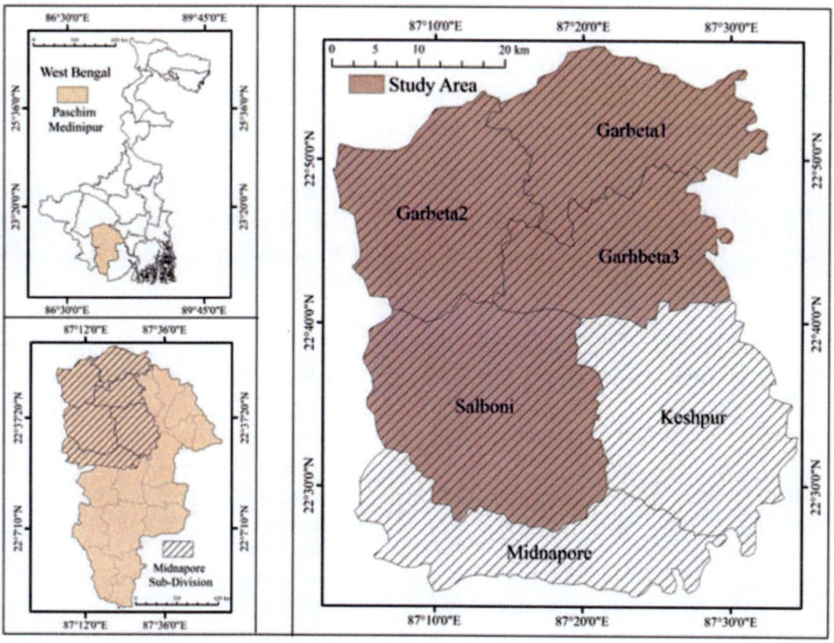

Fig. 13.1 Location of the study area

The general ground configuration is having gentle slope towards east and the slope of this region is ranging from 2° to 25°. Geomorphologically this area can be divided into 4 categories like Deep buried pediment, Flood plain deposits, Valley fill deposits, Deep to moderately buried pediments with laterite capping. Furthermore, it is the flank of Chotanagpur erosional plateau area. The relative relief of the region is varying from 5 to 28.92 m. Silaboti is the main river of this region and is the main river to geomorphologically modify this region through gully erosion and ample numbers of 1st order streamlets are seen as rill Chanel. The forested area of this region is region is lying under the Midnapore forest division, the total area of which is 50,267.49 ha. The area under reserved forest is the highest out of the other forest divisions of this district and that is 3814.05 ha. And the area under protected forest is 43,715.23 ha. Typically, the area occupied by sal, open scrub and plantation under this division are 19,677 ha, 9620 ha and 4955 ha respectively. Champion and Seth (1968) classified the forest types of Paschim.

Medinipur as Topical dry deciduous where dry peninsular Sal (Shorearobusta) is the dominant species. Undergrowth is mostly formed during post-monsoon season where Kurchi (Holarrhaenaantidesenterica), Kendu (Dispyrosmelanoxylan) and Satamul (Aparagusreasemossa) dominate.

13.3 Materials Used

Three cloud-free Landsat digital data of TM, ETM+ and OLI/TIRS sensors for the years of 1991, 2009, and 2018 covering the study area were downloaded freely from the U.S. Geological Survey's (USGS) Earth explorer website (https://earthexplorer.usgs.gov/).

The Landsat scenes were chosen for this study due to their affordability, availability, and medium to high spatial resolution. Details about the data are given in Table 13.1. These three Landsat digital data are level-one terrain-corrected (L1T)

Table 13.1 Details of satellite data used

Satellite and sensor	Date	Path/Row	Nos. of bands	Spatial resolution	Radiometric resolution
Landsat 4 (TM)	18.02.1991	139/44,45	7	Optical bands—30 m, Thermal band—120 m	8 bit
Landsat 7 (ETM+)	08.11.2000, 14.11.2009	139/44,45	9	Optical bands—30 m, 2 Thermal bands—60 m (each) Panchromatic—15 m	8 bit
Landsat 8 (OLI)	08.01.2018	139/44,45	11	Optical bands—30 m, 2 Thermal bands—100 m (each) Panchromatic—15 m	16 bit

in WGS84 geodetic datum, Universal Transverse Mercator map projection (UTM, Zone 45 N), and north-up image orientation. Besides SOI toposheets and SRTM DEM (30 m) have been used for the completion of the study.

13.4 Methodology

13.4.1 Atmospheric Correction

Remote Sensing satellite digital data acquired at temporal span provide a potentially ideal source for detecting change and analyzing trends. In view of the fact that multi-temporal images, acquired by different sensors under variable atmospheric conditions, solar illumination and view angles, atmospheric correction is obligatory to remove radiometric distortions and make the images analogous using the retrieved true reflectance values (Mahmoud et al. 2008).

The parameters used in different environmental condition detection algorithms, it require physical units, such as at-sensor radiance or top-of-atmosphere (TOA) reflectance, rather than the raw quantized calibrated pixel value (DN). TOA reflectance can be obtained from the quantized calibrated pixel value, as given by (Chander et al. 2009):

$$\rho\lambda = \left(\pi \cdot L\lambda \cdot d^2\right)/ESUN\lambda \text{Cos}\theta \tag{13.1}$$

where $\rho\lambda$is the TOA reflectance of wavelength λ (unit less), d is the earth-sun distance (astronomical units), $ESUN\lambda$ is mean exoatmospheric solar irradiance (W/(m^{2}μm)), θ is the solar zenith angle (degrees), and $L\lambda$ is the spectral radiance at wavelength λ at the sensor's aperture [W/(m^{2}srμm)]. $L\lambda$ can be obtained from the quantized calibrated pixel value also as given by (Chander et al. 2009):

$$L\lambda = (L_{\text{max}} - L_{\text{min}}/QCAL_{\text{max}} - QCAL_{\text{min}}) * (DN - QCAL_{\text{min}}) + L_{\text{min}} \tag{13.2}$$

where L_{max} is the spectral at-sensor radiance that is scaled to $QCAL_{\text{max}}$ [W/(m^2 sr μm)], L min is spectral at-sensor radiance that is scaled to $QCAL_{\text{min}}$ [W/(m^2 sr μm)], $QCAL_{\text{max}}$ is the maximum quantized calibrated pixel value corresponding to L_{max} (DN), $QCAL_{\text{min}}$ is the minimum quantized calibrated pixel value corresponding to L_{min} (DN), and $QCAL$ is the quantized calibrated pixel value (DN). The parameters in Eqs. (13.1) and (13.2) can be read from the header files of the ALI, TM, and ETM+ datasets or be retrieved from the USGS website.

13.4.2 Forest Cover Mapping

Satellite images have been an important basis for vegetation mapping and monitoring and understanding of eco-system functions, primarily through relationships between reflectance and vegetation structure and composition. Mapping is a method of portraying nature and the classification permits the mapper to approximate the true condition as clearly as possible. Although maps show objects with respect to attributes, their principal purpose is to depict object in terms of their relative location (Thakker et al. 1999). A good and useful mapping implementation requires large amount of information that comes from various sources like satellite images, ground truth etc. (Behera 2000). This study discusses a vegetation mapping methodology using Multi-criteria analysis technique. It agrees with the spectral limitation of remote sensing but argues that their predictive mapping power can be used more effectively using different floristic bio-physical parameters. This approach outlined aims to relate the reflectance information contained in multi-spectral imagery and spectral indices to traditionally accepted ecological classifications of the forests.

Since last two decades remote sensing is being used as an effective tool for forest vegetation assessment in the country. National Remote sensing agency (Anon. 1983) carried out nationwide forest cover mapping using satellite images from 1972–75 and 1980–82 on 1:1 million scale. National level forest cover assessment is now being done periodically by Forest Survey of India (FSI) using visual and digital interpretation techniques.

The advent of digital classification is that it incorporates spatial adjacency in addition to spectral signature (Argialas and Harlow 1990; Moller-Jensen 1990; Wickman and Norton 1994).The traditional digital classification implies a loss in the internal variability of the cover types considering the spatial mixtures of land covers (Ray and Mondal 2014).In this respect using the vegetation bio-physical variables as a criteria, calculated from different spectral indices an overlay analysis can classify the forest cover of region.

13.4.3 Shadow Index (SI)

The crown arrangement in the forest stand leads to shadow pattern affecting the spectral responses. The young and even aged stands have low Shadow Index (SI) compared to the mature natural forest stands. The later forest stands show flat and low spectral axis in comparison with open area.SI has been calculate using equation (Jennings et al. 1999).

$$S.I. = \sqrt{\{(QCAL_{\max} - GREEN) * (QCAL_{\max} - RED)\}} \tag{13.3}$$

13.4.4 Bare Soil Index (BI)

The bare soil areas, fallow lands, vegetation with marked background response and enhanced using this index. Similar to the concept of AVI the bare soil index (BI) is a normalized index of the difference sums of two separating the vegetation with different background viz. completely bare, sparse canopy and dense canopy etc. BI has been calculated using equation (Azizi et al. 2008).

$$B.I. = \{(NIR + GREEN) - RED\} / \{(NIR + GREEN) + RED\} \quad (13.4)$$

The value of this index is varying between 0 and 1, where the value towards 0 indicates the vegetation exploitation whereas value towards 1 indicates the vegetation abundance.

13.4.5 Modified Difference Vegetation Index (MAVI)

In this present study an object oriented enhancement algorithm has been designed using mathematical operators, which is supervised in nature and expressing the characteristics of plant chlorophyll 'a'. Two considerations have been taken in this respect like, pixel value having chlorophyll influence will be greater in near-infrared band than red band and pixel value having no chlorophyll influence will be greater in red band than near-infrared band. On the basis of these considerations following supervised enhancement algorithm: Modified Advance Vegetation Index (MAVI) has been applied (Ray et al. 2013). That can be calculated as

$$MAVI = [\{(\rho NIR + L) * (QCAL_{\max} - \rho RED)\} * (\rho NIR - RED)]^{1/3} \quad (13.5)$$

(where, ρRED = Reflectance value of Red band of respective Sensor, ρNIR = Reflectance value of Near-infrared band of respective Sensor, L = Coefficient, varies with the vegetation cover, Here L is the relational slope of *NIR* and *RED*).

It differs from NDVI in the perspective of physiognomic vegetation classes though there is almost positive relation-ship between these two. It is more sensitive than NDVI as it is using the power degree of Infrared response.

13.4.6 Vegetation Density (VD)

To calculate the Vegetation Density (VD), MAVI and BI are synthesized using PCA method as there is high correlation of negative between them. The PC 1 having high variance has been considered as Vegetation density (VD) criteria.

13.4.7 Scaled Shadow Index (SSI)

Alike VD, SSI has been developed by synthesizing MAVI and SI through PCA method as being of highly correlated of positive. The PC 1 having high variance has been considered as Scaled Shadow Index (SSI) criteria.

After studying the above vegetation indices, multi criteria based weightage analysis using Fuzzy AHP method using geometric mean (Buckley et al. 1985) has been applied to map the forest cover of the study area.

Fuzzy AHP (FAHP) is a artificial extension of classical AHP method, which considers the statistical fuzziness (uncertainty or insufficient information) of the decision makers. Fuzzy multiple attribute decision-making methods have been developed due to the elusiveness in computing the relative importance of attributes and the performance ratings of alternatives with respect to attributes (Chang 2011). There are varieties of reasons that may induce elusiveness: unquantifiable information, incomplete information, exclusive information. Conventional multiple attribute decision-making methods cannot overcome these problems successfully (Oguztimur 2011). Basically, Fuzzy-AHP method represents the embellishment of a standard AHP method into fuzzy domain by using fuzzy numbers for calculating instead of real numbers (Petkovic et al. 2012). Fuzzy AHP is of equal significance in the same hierarchy for each pair of factors (Das et al. 2019b).

In this process primarily conventional analytical Hierarchy Process (AHP) has been applied through creating pair wise comparison matrix (Table 13.2). This pair wise comparison matrix has been created with the help of scale of relative importance which are crisp numeric values (Table 13.2). Fuzzification and fuzzy membership functions are the two most important approaches in fuzzy system. In fuzzification linguistic terms are converted to fuzzy membership. Membership functions are of different types like triangulated, trapezoidal, bell shaped etc. In this study triangulated membership function has been conceived. The fuzzy values are generally represented by:

$$\mu\tilde{A}(x) = \tilde{A} = (l, m, u) \tag{13.6}$$

These (l, m, u) are known together as fuzzy members and these are lower, middle and upper ends of the triangular membership functions in the X axis.

Table 13.2 Pair wise comparison matrix with crisp values

	MAVI			BI			SSI			VD		
MAVI	1	1	1	4	5	6	4	5	6	6	7	8
BI	1/6	1/5	1/4	1	1	1	4	5	6	4	5	6
SSI	1/6	1/5	1/4	1/6	1/5	1/4	1	1	1	3	4	5
VD	1/8	1/7	1/6	1/6	1/5	1/4	1/5	1/4	1/3	1	1	1

Table 13.3 Fuzzy geometric mean

	Geometric mean		
MAVI	3.130169	3.637136	4.119534
BI	1.277886	1.495349	1.732051
SSI	0.539951	0.632456	2.364354
VD	0.255327	0.290715	0.344998
$\sum$	5.203334	6.055655	8.560937
Reciprocal	0.192184	0.165135	0.11681

In AHP matrix crisp numeric values are converted to the fuzzy numbers following Eq. (13.6). Whereas the reciprocal values are converted to fuzzy numbers (Table 13.2) using:

$$1/\tilde{A} = 1/(l, m, u) = (1/u, 1/m, 1/l) \tag{13.7}$$

Hence fuzzified pair wise comparison matrix can be calculated (Table 13.2).

After this fuzzy geometric mean ($\hat{r}i$) has been calculated (Buckley et al. 1985) to explore the criteria weight (Table 13.3). Fuzzy geometric mean ($\hat{r}i$) can be calculated as:

$$\tilde{A}_1 \otimes \tilde{A}_2 = (l_1, m_1, u_1) \otimes (l_2, m_2, u_2) \tag{13.8}$$

Finally the fuzzy weight can be calculated as:

$$\hat{w}_j = \hat{r}_i \otimes \left(\hat{r}_1 \oplus \hat{r}_2 \oplus \cdots \oplus \hat{r}_n\right) - 1 \tag{13.9}$$

After getting the fuzzy weights per criteria (Table 13.4), crisp numerical values can be calculated as weight ($\hat{w}j$) using defuzzification, which is known as Center of Area (COA) (Table 13.5) using,

$$w_j = (l + m + u)/3 \tag{13.10}$$

Table 13.4 Fuzzy weight

	Ŵ			W	W_{ij}
MAVI	0.365634	0.600618	0.791711	0.585987	0.540399
BI	0.149269	0.246934	0.332873	0.243026	0.224119
SSI	0.063072	0.10444	0.454392	0.207301	0.191174
VD	0.029825	0.048007	0.066303	0.048045	0.044307
			$\sum$	1.08436	1

Table 13.5 Criteria weight

Geometric mean				Ŵ		
MAVI	3.130169	3.637136	4.119534	0.365634	0.600618	0.791711
BI	1.277886	1.495349	1.732051	0.149269	0.246934	0.332873
SSI	0.539951	0.632456	2.364354	0.063072	0.10444	0.454392
VD	0.255327	0.290715	0.344998	0.029825	0.048007	0.066303
Reciprocal	0.192184	0.165135	0.11681			

All the indexed raster like MAVI, VD, SSI and BI have been assigned respective theme weights (w_j) and class score. The individual theme weights are multiplied by its respective class scores and then all the raster thematic layers are aggregated using a linear combination to get forest cover map (Rikimaru et al. 2002) as:

$$\sum(Criteria * w_j) \tag{13.11}$$

The final cumulative map has been reclassified into five categories of forest cover types viz; 'very dense to dense', 'dense to open', 'open to scrubs', 'scrubs to bare' and 'bare land'. This forest cover mapping as well as categorization has been done on temporal basis using digital satellite dataset of 1991, 2000, 2009 and 2018 so that a temporal dynamics can be visualized.

13.4.8 Identification of Forest Cover Dynamics

To explore the forest cover dynamics as well as the changing scenario all the temporal reclassified forest cover maps have been overlaid using Boolean OR operator on one to one basis i.e. 1991–2000, 2000–2009, 2009–2018 and 1991–2018. The class conversions have been registered as 'regeneration', 'degeneration' and 'unchanged'. From the attribute table of each pair of overlaid forest cover maps i.e. 1991–2000, 2000–2009, 2009–2018 and 1991–2018, combinations of 'very dense to dense' and other class category have been named as 'degeneration', combinations of 'bare lands' and other class category as have been named as 'regeneration' and lastly the combinations of identical class categories are named as 'unchanged'.

For the numerical depiction of the overall changing scenario, an index has been formulated as Disturbance Index (DI). DI is the function of the area under no forest cover degeneration and area with forest cover degeneration. DI can be calculated as:

$$DI = (Area\, under\ Regeneration + Area\, under\ Degeneration) \\ /Area\, under\ Regeneration \tag{13.12}$$

Higher the DI, lesser will be the distortions in the forest cover and vice versa.

13.4.9 Forest Fragmentation Analysis

Forest fragmentation is disintegration of contiguous forest cover to large numbers of forest patches on temporal basis as consequences of the edge effects (Rahman et al. 2016). In this present study forest fragmentation has been studied using one of the pattern metric tools like Area Weighted Mean Patch Fractal Dimension (AWMPFD) Index. It is analogous to the Mean Patch Fractal Dimension (MPFD) Index. Only difference is the use of area as an weighting parameter to each patch. Higher the value of AWMPFD lower will be the fragmentation and vice versa. The value ranges between 1 and 2. It can be calculated as:

$$P = \sum_{i=1}^{m}\sum_{i=1}^{n}\left[\left(\frac{2\ln(0.25P_{ij})}{\ln(a_{ij})}\right)\left(\frac{a_{ij}}{A}\right)\right] \quad (13.13)$$

where, P_{ij} is patch number, a_{ij} is patch area and A is total landscape area.

13.4.9.1 Multi-distance Spatial Cluster Analysis (Ripley's K Function)

Ripley's K function definition is a tool to portray the spatial structure of a point or polygon patterns by graph. It determines the values associated with features, exhibit statistically significant clustering or dispersion over a range of distances. It has a wide applicability in vegetation studies. The tool gives the output as a table with fields expected and observed K values, respectively. Mathematically, the Multi-Distance Spatial Cluster Analysis tool uses a common transformation of Ripley's k-function, which can also be said as L(d). With this L(d) transformed Distance values the Expected K values will always be matched. The L(d) transformation can be done as:

$$L(d) = \sqrt{A\sum_{i=1}^{n}\sum_{j=1,i\neq j}^{n} K_{i,j}/\pi n(n-1)} \quad (13.14)$$

Larger the observed K value than the expected K value indicates the clustered distribution than the random, whereas, smaller the observed K value than the expected K value, the distribution is more dispersed than random distribution at that distance. Furthermore, larger the observed K value is than the upper confidence envelope (HiConfEnv) value, the spatial clustering is statistically significant for that distance and when the observed K value is smaller than the lower confidence envelope (LwConfEnv) value; there will be a statistically significant spatial dispersion for that distance (Sayer et al. 2013).

13.4.9.2 Optimized Hot Spot (Getis-Ord Gi*) and LISA (Local Indicators of Spatial Association) Analysis

The spatio-temporal hotspot detection was based on the use of Getis-Ord Gi* statistic is a typical tool to detect spatio-temporal hot-spot of any spatial phenomenon. It statistically characterizes and captures significant spatial clusters as hotspots and cold spots using Gi-Bin values. In this present study, it has been devised to explore the spatial concentration of DI values in significant manner. It is considered as a useful tool for supporting sustainable management strategies (Zhu and Newsam 2016). Features with Gi-Bin values of ±3 were statistically significant at the 99% confidence level; the features with Gi-Bin value of 0 was not statistically significant, and the features with Gi-Bin values of ±2 were statistically significant at 95% confidence level and features with Gi-Bin values of ±1 were statistically significant at 90% confidence level (Wang 2010).

Alike the Optimized Hot Spot analysis, Local Indicators of Spatial Association or Autocorrelation (LISA) analysis was employed to characterize the spatial arrangement of Disturbance Index per spatial unit over time to generate cluster maps for identifying the existence of hot spots. It is such a spatial statistical technique which gives an indication of extent for significant existence of homogeneous clusters around a particular observation and the sum of local indicators of spatial association for all observations is proportional to a global indicator (Anselin 1995).

Local Moran *I* is a local indicator of spatial autocorrelation (LISA) is a decomposition of global indicators, such as Moran's *I* and shows the level of spatial autocorrelation at various individual locations within an area (Anselin 1995). LISA calculates an index value and Z-score for a feature. A high positive Z score indicates the association of a feature with similar value either above or below mean. On the contrary a high negative Z-score indicates the association of a feature with dissimilar values (Zhang et al. 2008). The LISA can be measured as:

$$I_i = \left(z_i \Big/ \left(\sum_i z_i^2/n \right) \sum_j w_{ij} z \right) \tag{13.15}$$

Spatial clusters consist of two categories: (i) high-high clusters indicate high values are surrounded by high values; (ii) low-low clusters indicate clustering of low values are surrounded by low values; (iii) high-low clusters indicate clustering of high values are surrounded by low values and (iv) low-high clusters indicate clustering of low values are surrounded by high values. Besides, "non-significance" is also a type of cluster, indicating no significant local spatial autocorrelation.

The distribution of spatial clusters and outliers can be shown in a Moran scatter plot which has four quadrants (Fig. 13.6b). The x-axis of the scatter plot represents the standardized values (area) of the geographical object and the y-axis measures the mean standardized values of neighboring objects. The upper-right quadrant of the scatter plot contains the cases where the geographical objects and their neighbors have high values. These so-called high-high (HH) situations are associated with clusters. The lower-left quadrant shows cases where the geographical objects and their neighbors are relatively small (i.e. a low-low or LL situation) which also signifies clustering. The lower-right quadrant contains the high-low (HL) cases while low-high (LH) cases are shown in the upper-left quadrant (Anselin 1995, 2003, 2005). The two quadrants having LH and HL represent outliers. HH cases are referred to as grouped hot spots and LL cases as grouped cold spots, whereas HL and LH are referred to as individual hot spots and cold spots, respectively (Anselin 2005). The HH, HL, LH and LL cases can be mapped to better understand the spatial distribution of hot and cold spots and to see where clustering occurs.

In this present study the calculations of local spatial autocorrelation has been performed in GeoDa platform (Anselin et al. 2006) at a significance filter of 0.01 and permutation level of 9999.

13.4.9.3 Identification of Forest Disturbance Potential Zone

In this present study, the word forest disturbance is mainly mean to say the spatial distortions in forest cover as the function of spatial effects of different parameters like settlement position, river, canals and road, physiographic nature in terms of relative relief, slope etc.

Irrespective of AHP and FAHP, to calculate the weights of each parameter for identifying the Forest Disturbance Potential Zones (FDPZ), 'Rank Sum method' under SMARTER (The Simple Multi Attribute Rating Technique exploiting rank) technique as suggested by Edwards and Barron. Using the principal of SMARTER technique the decision-makers arrange the criteria (C_i) into an importance order like C1 ≥ C2 ≥ C3... ≥ Cn. and assigns 'true' weights according to the Rank Order Distribution.

The criterion having mutual comparison pair wise can be synthesized by AHP or FAHP (Saaty 2008 and Kahraman et al. 2003). But in case of FDPZ the selected criterion cannot be compared pair wise mutually so that 'Rank Sum method' under SMARTER technique has been followed. The weight per criteria (Table 13.6) using this respective method has been calculated as:

$$w_j = (n - r_j + 1)\Big/\sum(n - r_k + 1) \quad (13.16)$$

where, w_j is the normalized weight for the jth criteria, n is the number of criteria under consideration ($k = 1, 2 \ldots n$) and r_j is the rank position of the criterion. Each criterion is weighted $(n - r_j + 1)$ and then normalized by the sum of all weights, i.e. $\sum(n - r_k + 1)$.

Table 13.6 Rank sum method to calculate FDPZ

R_k	P_i	R_j	$n - R_j + 1$	$n - R_k + 1$	W_{ij}
1	Can	2	5	6	0.238
2	River	6	1	5	0.047
3	Sett	5	2	4	0.095
4	Road	4	3	3	0.142
5	RR	3	4	2	0.190
6	Slp	1	6	1	0.285
$n = 6$			$\sum$	21	1

13.5 Result and Discussion

13.5.1 Forest Cover Dynamicity

The forest cover maps have been prepared from the Landsat digital data for the years of 1991, 2000, 2009 and 2018 using Multicriteria Fuzzy AHP technique (Fig. 13.2) and typical forest cover classes have been summerised in Table 13.7.

These data reveal that in 1991, about 9.33% (151.24 km^2) area of the study area was under bare surface area, 22.08% (357.88 km^2) under scrubs to bare surface area, 30.27% (490.46 km^2) under open to scrub area, 24.15% (391.39 km^2) under dense to open and 10.51% (170.32 km^2) under very dense to dense forest. During 2000 the area under these land categories was found about 6.57% (106.48 km^2) under bare surface area, 28.50% (461.82 km^2) under scrubs to bare surface area, 36.66% (594.07 km^2) under open to scrub area, 22.30% (361.41 km^2) under dense to open forest and 5.49% (89.01 km^2) under very dense to dense forest. During 2009 the area under these land categories was found about 16.85% (273.11 km^2) under bare surface area, 21.19% (344.38 km^2) under scrub to bare, 22.67% (367.34 km^2) under open to scrub land, 19.68% (319 km^2) under dense to open forest and 19.19% (2011.05 km^2) under very dense to dense forest area and lastly during 2019 the area under these land categories was found about 5.19% (84.24 km^2) under bare surface area, 9.78% (158.48 km^2) under scrub to bare, 20.28% (328.71 km^2) under open to scrub land, 48.49% (785.74 km^2) under dense to open forest and 16.23% (262.97 km^2) under very dense to dense forest area (Table 13.9).

Post-classification comparison method, which is the most common approach in change detection as well as the reconfiguration analysis (Araya 2009; Miller et al. 1998; Zhou et al. 2004), which has been applied in this study. The comparisons between the areal accounts of typical forest cover classes from the year of 1991–2018 reveals both the positive and negative changes in forest cover patterns. From Table 13.9, this has been revealed that within the 27 years span, the forest classes like open to scrub, scrubs to bare and bare land has decreased significantly by 9.98%, 12.30% and 4.13% respectively. On the other hand, the classes like very dense to

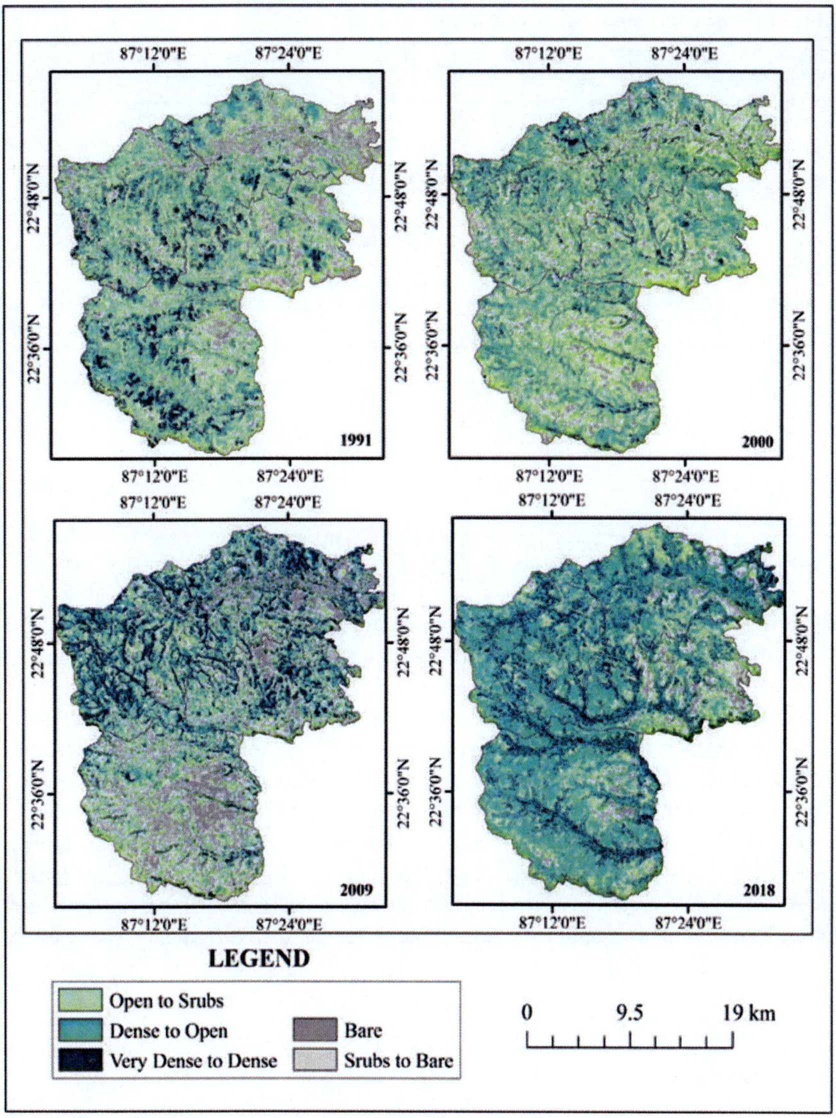

Fig. 13.2 Temporal forest cover maps of the study area

dense and dense to open forest area is seen to be increased by 5.71% and 4.34% respectively.

Table 13.7 Temporal forest cover area (km^2/%)

Classes	1991	%	2000	%	2009	%	2018	%	1991–2018 (%)
Very dense to Dense	170.33	10.51	89.02	5.49	311.05	19.20	262.97	16.23	5.72
Dense to open	391.39	24.16	361.41	22.31	319.01	19.69	785.75	48.50	24.34
Open to scrubs	490.46	30.27	594.07	36.67	367.34	22.67	328.71	20.29	−9.98
Scrubs to bare	357.89	22.09	461.82	28.50	343.39	21.19	158.49	9.78	−12.31
Bare	151.24	9.33	106.48	6.57	273.11	16.86	84.25	5.20	−4.13

In this present study the areal database registered in Table 13.8 the rate of change in area for each forest cover classes has been calculated using the following formula:

$$C_i = \{(Ae_i - Be_i)/Be_i\} * (1/t) * 100\% \quad (13.17)$$

where, C_i is the rate of change in the forest area, Ae_i and Be_i are the area under respective forest cover at the ending and beginning year of the span of study respective and t is the span of study.

Within the overall span of the study (1991–2018), the maximum areal gain is seen for 'dense to open forest' (24.34%) at the rate of 3.73% followed by very dense to dense forest (5.71%) at the rate of 2.05%. On the other hand, the maximum areal loss is seen for the scrubs to bare (12.30%) at the rate of 2.06% followed by open to scrubs (9.98%) and bare land (4.13%) at the rate of 1.22% and 1.64% respectively.

Forest cover dynamics can be best explained by forest cover fragmentation. Conversion from very dense to dense to the other category will obviously show the high fragmentation and vice versa. The database of metric analysis for every

Table 13.8 Rate of change in forest cover

Classes	1991	%	2000	%	2009	%	2018	%	1991–2018 (%)
Very dense to Dense	170.33	10.51	89.02	5.49	311.05	19.20	262.97	16.23	5.72
Dense to open	391.39	24.16	361.41	22.31	319.01	19.69	785.75	48.50	24.34
Open to scrubs	490.46	30.27	594.07	36.67	367.34	22.67	328.71	20.29	−9.98
Scrubs to bare	357.89	22.09	461.82	28.50	343.39	21.19	158.49	9.78	−12.31
Bare	151.24	9.33	106.48	6.57	273.11	16.86	84.25	5.20	−4.13

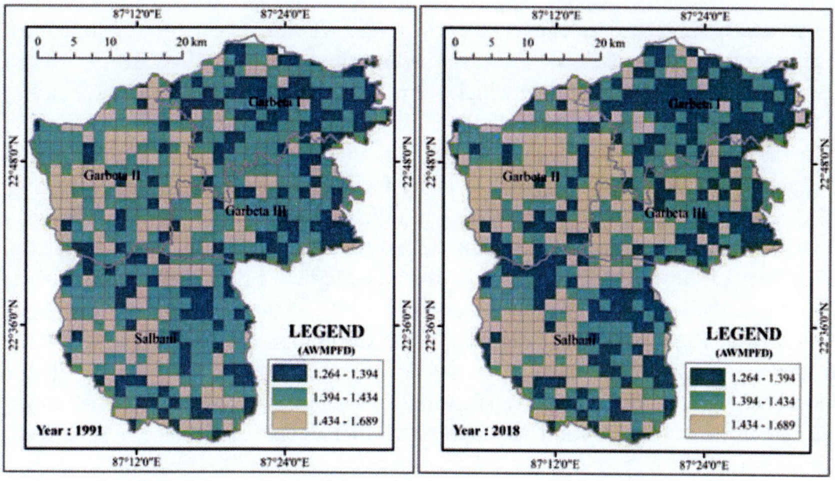

Fig. 13.3 Area weighted mean patch fractal dimension map of forest cover for the span of 27 years

square blocks (2.5 × 2.5 km^2) polygon covering the entire area under study. The thematic maps having the informations about AWMPFD depict the patterns of forest cover on temporal basis. From Fig. 13.3, an increasing trend in fragmentation can be demonstrated within the total span of the study. The happening of fragmentation is seen to be very distinct in blocks like Garbeta II and Salboni in 2018 rather than that of 1991. On the other hand in 2018, clustering of forest cover has been seen to be increased very distinctively in Garbeta I rather than in 1991.

13.5.2 Status of Forest Regeneration and Degeneration

In any region forest cover dynamics is the function of area under forest cover regeneration and degeneration. These two phenomenon are fundamentally the consequences of physical and anthropogenic impacts. Within the 27 years span of study i.e. from 1991 to 2019 the degenerated area was 276.46 km^2 (17.06%), regenerated area was 377.35 km^2 (23.29%) and the unchanged area was 82.52 km^2 (11.26%) (Fig. 13.6). So it is very pertinent that in this area under study, forest cover regeneration is overriding within the span of 27 years (Table 13.9).

Though forest cover regeneration was so prominent in the overall temporal span, there was a notable consistent declining state of regenerated area from 1991 to 2000 (565.48 km^2) towards 2009–2018 (372.86 km^2) followed by 2000–2009 (384.56 km^2) as per areal accounts have been registered in Table 13.10. For the three sets of spans (1991–2000, 2000–2009 and 2009–2018), in Garbeta 1 and Garbeta 3 there were a decreasing trend in regeneration coverage. Whereas in Garbeta 2 and Salboni regenerated area is seen to be increased within the span of 2009–2018 i.e. 108.27 km^2 and 92.04 km^2 respectively which were 70.10 km^2 and 71.60 km^2 respectively in the span of 2000–2009 (Table 13.9).

In the context of this overall forest cover dynamics within 27 years span Garbeta 1 is experiencing maximum areal extent under forest cover regeneration (197.55 km^2) followed by Salboni (69.70 km^2), Garbeta 3 (57.50 km^2), and Garbeta 2 (52.60 km^2). On the other hand maximum area under forest cover degeneration is seen in Salboni (91.80 km^2) followed by Garbeta 3 (69.50 km^2), Garbeta 1 (66.56 km^2) and Garbeta 2 (48.60 km^2).

However, Repley's K function or L(d) transformation can be a wise estimate to in interpret the nature of regeneration and degeneration in accordance to the relation between the clustered and dispersed factors. This function truly calculates the statistically significant clustering and dispersion at smaller and larger distances respectively. For the forest cover regeneration and degeneration area over the span of 27 years.

The resultant 'K' function for the regenerations showed that the observed spatial pattern started and finished with the statistically significant dispersion at largest distance in smaller space. On the other hand there was a statistically significant clustering at smaller distance in larger space (Fig. 13.4). In contrast for degeneration, statistically observed spatial patterns had statistically significant clustering at smaller distance between the clustered patterns and dispersed pattern (Fig. 13.5).

Forest disturbance index can be an ideal way to depict the forest cover status. DI indicates the nature of degeneration. Higher the DI value lower will be the disturbance and vice versa. In this respect within the overall span of the study Garbeta 3 is experiencing the maximum DI (1.10) followed by Salboni (1.35), Garbeta 2 (1.71) and Garbeta 1 (4.16) (Table 13.10). But as per the rendered z-score, spatial distribution forest disturbance (DI) is mostly prevalent in South and South-West part of Salboni block having the value ranges of $\geq$2.0 and in the Garbeta 1, there was a low to no disturbance regime can be identified having the value of ≤ -2.0 (Fig. 13.6a). Such clustering having those ranges of values is also persistent in nature over the other blocks under study. Gi-Bin values of Optimized Hot Spot analysis coupled with the Z-score rendering is also proving the existence of such concentration of forest disturbance as significant hot spot and cold spot clusters which have the spatial extent of 573.32 km^2 and 429.94 km^2 respectively (Fig. 13.6b).

Furthermore, a spatial auto correlative analysis of the spatial distribution of DI hot spot in local scale is very pertinent for the betterment in environmental management and sustainability. In this respect LISA or cluster and outlier analysis is improvising to describe the clustering nature of disturbance attributes. The Global Moran's I value of 0.578 indicates that the DI data is positively spatially auto-correlated regardless of neighborhood definition (Fig. 13.7b). The DI database calculated for the span of 27 years has been classified and mapped in terms of significant clusters of hot spots (HH and HL) and cold spots (LL and LH). The cluster association like HH and LL suggests clustering of similar values (positive spatial auto-correlation), whereas the cluster association like HL and LH indicate spatial outliers (negative spatial correlation). The LISA map (Fig. 13.7a) describes the situation of the prominent

Table 13.9 Areal accounts on forest cover regeneration and degeneration

Spans	1991–2000		2000–2009		2009–2018		1991–2018	
Blocks	Regeneration	Degeneration	Regeneration	Degeneration	Regeneration	Degeneration	Regeneration	Degeneration
Garbeta 1	159.44	61.68	134.54	86.10	107.94	101.02	197.55	66.56
Garbeta 2	132.81	110.92	70.10	69.50	108.27	103.23	52.60	48.60
Garbeta 3	125.01	68.00	108.32	111.31	64.61	77.23	57.50	69.50
Salboni	148.23	128.27	71.60	45.90	92.04	49.69	69.70	91.80
$\sum$	565.48	368.87	384.56	312.81	372.86	331.17	377.35	276.46
%	34.90	22.77	23.74	19.31	23.01	20.44	23.29	17.06

Table 13.10 Accounts on block wise Forest Disturbance Index

Block name	1991–2000	2000–2009	2009–2018	1991–2018
Garbeta 1	4.55	1.93	1.68	4.15
Garbeta 2	2.37	1.65	1.88	1.70
Garbeta 3	3.46	1.82	1.56	1.10
Salboni	2.46	2.73	3.70	1.34

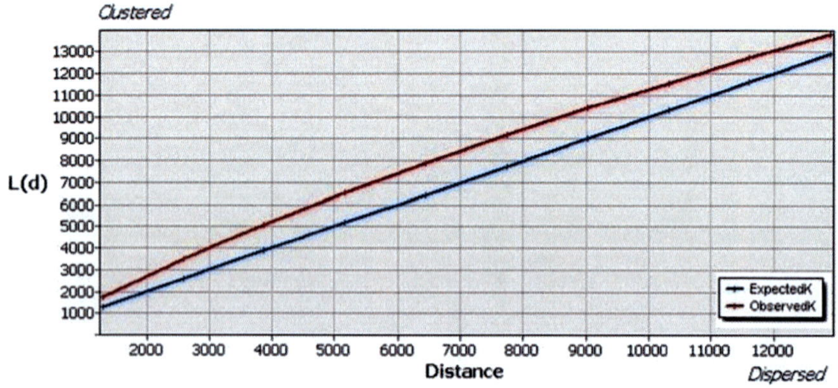

Fig. 13.4 'K' function plot of degenerated patches

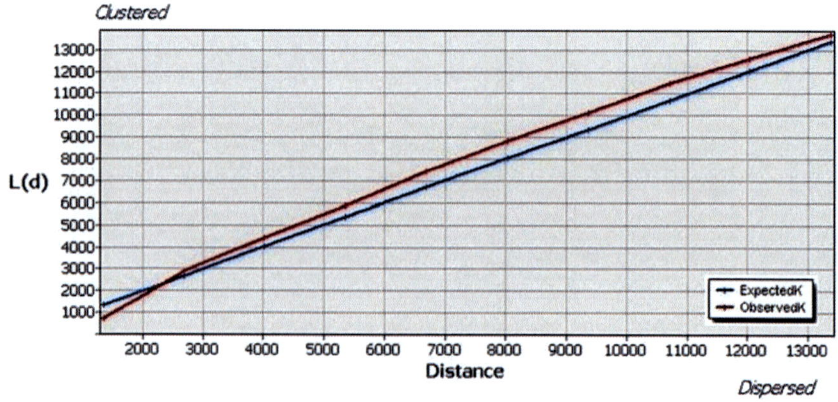

Fig. 13.5 'K' function plot of regenerated patches

patterns of HH and LL cluster having the concentration in Salboni and Garbeta 1 blocks respectively. These two blocks are have momentous contribution to the global Moran's *I* statistic.

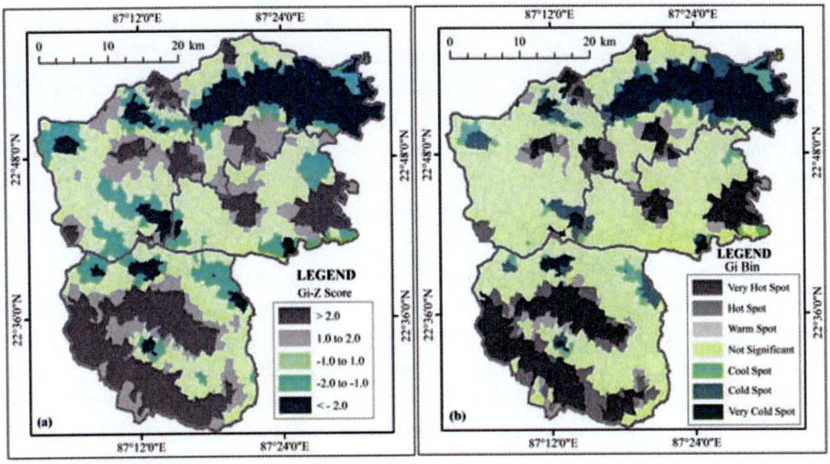

Fig. 13.6 **a** and **b** Distribution of DI and consequent Hot Spots

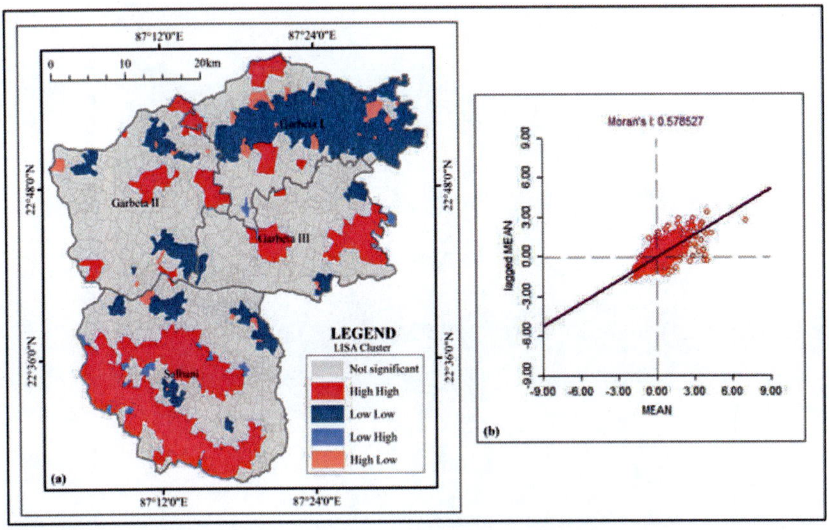

Fig. 13.7 **a** and **b** LISA and Moran's I plot of DI showing spatial clusters and outlier positions

13.5.3 Forest Disturbance Potential Zonation

Potential zonation of any functional environmental happening like ground water, soil loss, landslide etc.is an important function of regional planning and management. So that, a situational sustainability can be achieved. In this context forest disturbance potentiality zonation (FDPZ) is an important issue to sustainable management of a forested environment as well as socio-environmental ecology.

In the present study the total area has been classified in three FDP zones like 'High to Moderate' (413.89 km^2), 'Moderate to Low'(440.62 km^2), 'Low to Non Disturbed'(761.85 km^2) on the basis of six influencing criteria like relative relief, slope, distance from road, river, canal and settlement position (Fig. 13.8). Out of these criteria, distance from settlement, river, road and canal have their influences over the disturbance potentiality as 9.5%, 4.7%, 23% and 14% respectively (Table 13.6). On the other hand the two relief features like relative relief and slope have their influences over FDP have as 19% and 28% respectively (Table 13.6). Therefore on the basis of aforesaid accounts this can be said that canal alignment and slope factor have effective influences over the forest disturbance potentiality zonation.

To distinguish the FDPZ in ordinal manner, all the criteria are reclassified in three distinctive classes. For the distance based criteria like road, river, canal, settlement position three distinct distance buffers have been considered from each feature like

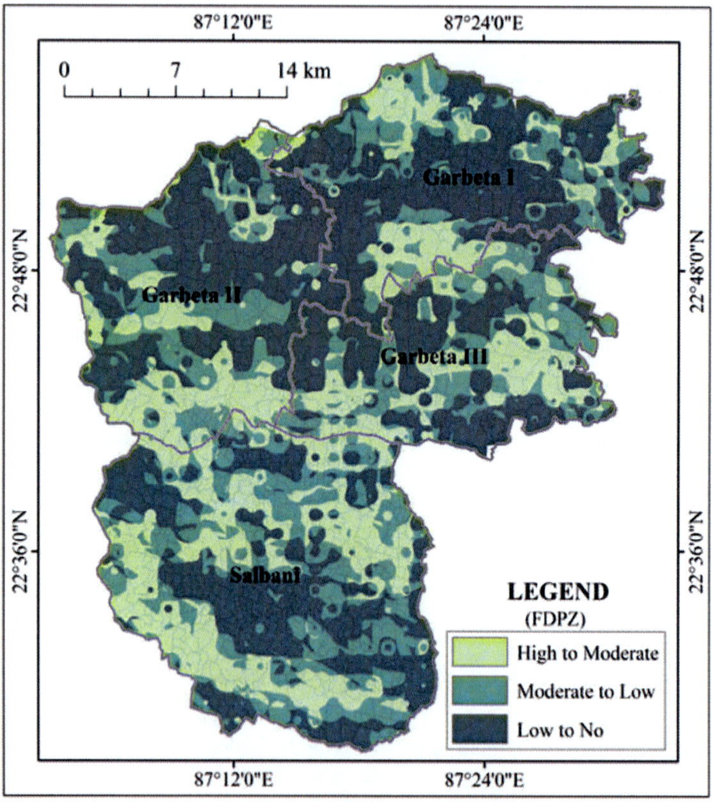

Fig. 13.8 Identified Forest Disturbance Potential Zones (FDPZ)

1 km, 3 km and 7 km. On the other hand the relief features like slope and relative relief also classified in three categories namely 0.5°–2.03° as 1; 2.03°–2.16° as 2; 2.16°–3.26° as 3 and 18–51 m as 1; 51–71 m as 2; 71–113 m as 3 respectively.

Furthermore each criterion has the categorical importance over each FDPZ. A data dimensionality reduction technique like Multiple Criteria Analysis (MCA) is an ideal approach to envisage that importance.

MCA is an exploratory data analysis technique as well as multivariate graphical technique used to analyze categorical data (Benzecri 1992; Sourial et al. 2010). A key feature of the analysis is the joint scaling of both row and column variables to provide information on the interrelationships between row and column variables. Correspondence analysis can be used on both of qualitative or quantitative data. A final step in the analysis involves rescaling of characteristic vectors into optimal scores. Normalising these optimal scores allows for assessment of relative importance of factors. Correspondence analysis can also be used to find the optimal ordering of variables for a given set of characteristics (Weller et al. 1990). Homogeneity between row element and column element specifies the correspondence between elements and this homogeneity can be addressed by the maximum between each of row elements and column elements (Table 13.11). In this present study the correspondence of FDPZ with the influencing criteria has been calculated.

However the corresponding clusters can clearly be identified from the perceptual maps (Fig. 13.9a–f). The perceptual maps are formed by plotting the row elements against column elements and vice versa based on the inertia or the mass values. As per the perceptual maps high FDPZ is associated to <1 km distance from canal and road, 1–3 km distance from river and settlement positions and lastly the surface having high elevation category along with medium slope amount is prone to forest disturbance. On the other hand low FDPZ is associated to >3 km distance from canal and road, <1 km distance from river and 1–3 km distance from the settlement positions and the surface having low elevation category along with high slope amount is prone to forest disturbance.

Contextually it is mentionworthy that positions of settlement clusters have a mixed impact over the forest disturbance potentiality. Within 1–3 km buffer distance from the settlement position both of high and low disturbance potentiality are seen to be associated. Besides <1 km distance from settlement position is merely effective to forest disturbance though moderate types of disturbance potentiality is closely associated to this range of the distance and beyond the distance of 3 km (i.e. 3–7 km) disturbance potentiality is getting low to no disturbance.

Therefore, it can be alleged that there is no such significant impact of settlement situation over the forest cover dynamicity. The symbiotic interaction between human livelihood and forest environment has a sustainable effect over the forest cover retention. Besides, an eco-tourism development initiative at Ganagani village under Garbeta 1, absence of urban development in large scale except some small Census town conversion in Garbeta 1and 3 and absence of local urban infrastructural development as well has kept the forest degeneration inhibited. However the development of roadways (SH or NH) and canal alignment have a significant effect over forest cover degeneration and there are an adjacency impact over the forest cover so

Table 13.11 Multiple correspondence analyses between FDPZ and driving criteria

		High FDPZ	Moderate FDPZ	Low FDPZ	Active margin
Elevation	Low	102.712	160.706	**308.057**	571.475
	Moderate	182.705	173.602	**289.851**	646.158
	High	128.408	106.172	**163.586**	398.166
	Active Margin	413.825	440.480	761.494	1615.799
Slope	Low	93.638	193.346	**242.548**	529.532
	Moderate	**308.955**	151.648	58.102	518.705
	High	11.228	95.473	**460.749**	567.450
	Active margin	413.821	440.467	761.399	1615.687
Distance from Settlement	1 km	141.835	159.881	**219.813**	521.529
	3 km	254.526	260.529	**477.919**	992.974
	7 km	17.469	20.101	**63.866**	101.437
	Active margin	413.830	440.511	761.598	1615.940
Distance from Canal	1 km	57.779	147.325	**330.448**	535.552
	3 km	**257.226**	147.875	134.532	539.633
	7 km	98.832	145.321	**296.611**	540.764
	Active margin	413.837	440.521	761.591	1615.949
Distance from River	1 km	193.825	237.040	**473.306**	904.171
	3 km	217.635	195.596	**271.607**	684.838
	7 km	2.370	7.875	**16.685**	26.930
	Active margin	413.830	440.511	761.598	1615.939
Distance from Road	1 km	209.030	156.303	**211.151**	576.484
	3 km	154.415	201.881	**333.136**	689.432
	7 km	50.382	82.317	**217.307**	350.006
	Active margin	413.827	440.501	761.594	1615.922

Bold numbers are indicating the maximum association between each of the Row and Column elements. As an example this can be said that there is the maximum association between Low elevation category and Low Forest Disturbance Potentiality which has been indicated by the value 308.057

that within 1 km buffer distance from road and canal high disturbance potentiality has found. In this respect, preferences over medium to gentle slope of terrain as well as high to medium relief height for canal alignment have defined the areas having elevation of 71–113 m and slope of 2.03°–2.16° as high disturbance potential zone.

Area under study composed of four CD Blocks is one of the most important areas under West Midnapore District for conservation of forest resources in West Bengal. The significance and use of forest land must be managed for their genetic diversity, productivity, regeneration capacity, vivacity and potential to fulfill the needs relevant to ecological and socio-economic functions at the local and regional levels (FAO 2000). FDPZ is worth of practicing to identify the suitable areas for forest

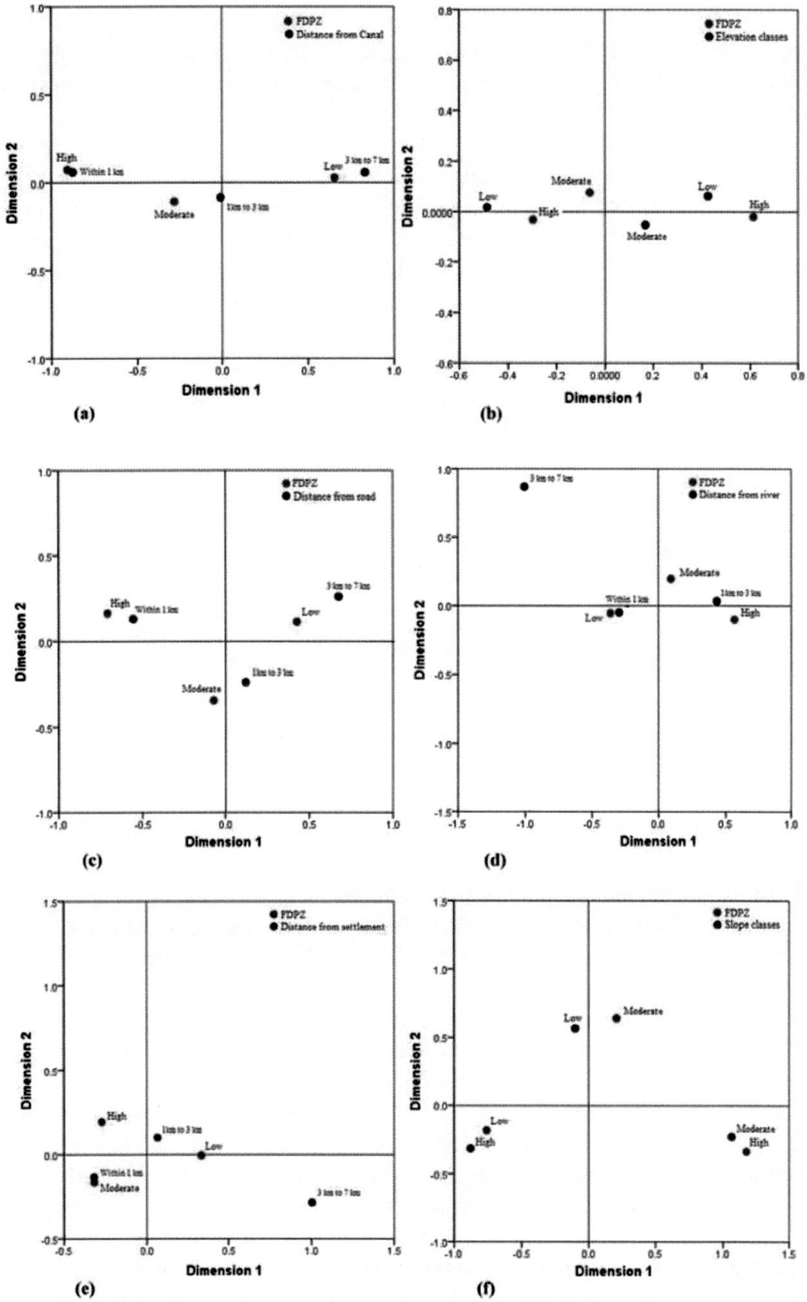

Fig. 13.9 a–f Perceptual maps showing the correspondence between FDPZ and driving factors

recovery so that environmental viability can be gained as well as forest based livelihood can be sustained. The conservation and restoration of forests will provide local communities with the best decision to get the profit and restore the forests' environmental and economic functions (Sanchirico and Siikamaki 2007). The challenges of achieving sustainable forest management and forest protection must be addressed jointly between stakeholders and communities (Sheil et al. 2002; Gunningham 2009). This can be said as joint forest management (JFM) and the increasing rate of regeneration in the forest cover is the consequence of effectiveness of it. JFM in West Midnapore district of West Bengal has a history that initiated during 1980s while a number of forest protection committees (FPC) and their members have been kept on escalating with time. During 1995, wasteland management scheme was implemented in several states of West Bengal. Plantation of *Acacia* and *Eucalyptus* species has increased the vegetation cover (SFR, 2011–12). Presently most of the forest cover is plantation forest where clearing of mature dry deciduous species was regulated properly along with planting new plants. Plantation activity has helped to gain increasing forest cover in West Medinipur district (SFR, 2011–12). All these factors were responsible for the dynamicity of vegetation cover in this selected part of West Midnapore district.

13.6 Conclusion

Forest disturbance analysis is very contemporary and realistic issue in forming sustainable environment. In global environmental forestry is an important ecosystem parameter as a source of environmental purifier fodder and shelter as well as for economic importance. But with the progress in civilization deforestation has come as regular practices though the act of join forest management has prevented the reckless of deforesting activities of human being.

This research has created a theoretical framework that can be implemented when preparing future forest management policies based on area. In the case of efforts made through participatory forest management schemes, the overall growth of forest cover was very convincing. Though area and intensity different vulnerable patches were established in this study where forest recovery schemes can be implemented on priority bases by the administration. Such a study concerning the innate dynamics of the certain attribute is needed in the field of planning and decision making. It can be embattled with creating sustainable forestry market opportunities for the betterment of forest based livelihood and to promote the use of forest based commodity. The disturbance potential areas can be set up for new forest generation to promote plantations that convey reimbursement to the environment and neighboring communities. Protecting and restoring forests could provide the best pronouncement with the local communities to get the benefit and restore the environmental and economic functions of the forests. The identified disturbance potential areas can be reused by afforestation to balance the forest degeneration. Therefore, the change detection analysis and periodic determination of potential forest disturbance areas could help to undertake new policy and planning for forest restoration and ecosystem viability.

References

Alberdi I, Nunes L, Kovac M, Bonheme I, Cañellas I, Rego FC, Gasparini P (2019) The conservation status assessment of Natura 2000 forest habitats in Europe: capabilities, potentials and challenges of national forest inventories data. Annals of Forest Science 76(2):34

Amiro BD, Barr AG, Barr JG, Black TA, Bracho R, Brown M, Chen J, Clark KL, Davis KJ, Desai AR, Dore S, Engel V, Fuentes JD, Goldstein AH, Goulden ML, Kolb TE, Lavigne MB, Law BE, Margolis HA, Martin T, McCaughey JH, Misson L, Montes-Helu M, Noormets A, Randerson JT, Starr G, Xiao J (2010) Ecosystem carbon dioxide fluxes after disturbance in forests of North America. J Geophys Res 115:G00K02

Anon. (1983) Mapping of forest cover in India from satellite imagery 1972–75 and 1980–82: Summery Report. Hyderabad, Nat Remote Sensing Agency, p 9

Anselin L (1995) Local indicators of spatial association-LISA. Geograph Anal 27(2):93–115

Anselin L (2003) Spatial externalities, spatial multipliers, and spatial econometrics. Int Reg Sci Rev 26:153–166

Anselin L (2005) Exloring spatial data with GeoDaTM: a workbook, p 198e218. https://geoda.uiuc.edu/pdf/geodaworkbook.pdf

Anselin L, Syabri I, Kho Y (2006) GeoDa: an introduction to spatial data analysis. Geographical Analysis 38(1):5–22

Araya Y (2009) Urban land use change analysis and modelling: a case study of Setubal-Sesimbra, Portugal

Benzecri, J.P.(1992). Correspondence analysis Handbook, ISBN. 058536303X, 9780585363035, CRC Press., 485–495.

Behera MD (2000) Biodiversity characterization at landscape level in Subansiri district of Arunachal Pradesh (Eastern Himalaya) using remote sensing and GIS. Ph.D. Thesis. Gurukul Kangri University, Hardwar, India

Bhunia GS, Samanta S, Pal DK, Pal B (2012) Assessment of Groundwater Potential Zone in Paschim Medinipur District, West Bengal—a Meso-scale study using GIS and remote sensing approach. Iiste 2:41–59

Bjorkman, A. D., Criado, M. G., Myers-Smith, I. H., Ravolainen, V., Jónsdóttir, I. S., Westergaard, K. B., Heiðmarsson, S. (2019). Status and trends in Arctic vegetation: Evidence from experimental warming and long-term monitoring (pp. 1–15). Ambio, Advance online publication. doi:10.1007/s13280-019-01161-6 UN report on environment https://wedocs.unep.org/bitstream/handle/20.500.11822/27539/GEO6_2019.pdf? sequence=1&isAllowed=y (Accesses on 08 June, 2019)

Böttcher H, Kurz WA, Freibauer A (2008) Accounting of forest carbon sinks and sources under a future climate protocol—factoring out past disturbance and management effects on age-class structure. Environ Sci Policy 11:669–686

Champion HG, Seth SK (1968) A revised survey of forest types of India. Government of India Press, New Delhi, p 404

Chander G, Markham BL, Helder DL (2009) Summary of current radiometric calibration coefficients for landsat MSS, TM, ETM+, and EO-1 ALI sensors. Remote Sensing of Environment 113(5):893–903

Cohen WB, Spies TA, Alig RJ, Oetter DR, Maiersperger TK, Fiorella M (2002) Characterizing 23 years (1972–1995) of stand replacement disturbance in western Oregon forests with landsat imagery. Ecosystems 5:122—137

FAO (2000) Asia-Pacific forestry commission: development of national-level criteria and indicators for the sustainable management of dry forests of Asia. Workshop Report, RAP Publication 2000/07, Food and Agriculture Organization. Available online: https://www.fao.org/3/x6896e/x6896e00.htm. Accessed on 5 Feb 2019

Frolking S, Palace MW, Clark DB, Chambers JQ, Shugart HH, Hurtt GC (2009) Forest disturbance and recovery: a general review in the context of spaceborne remote sensing of impacts on aboveground biomass and canopy structure. J Geophys Res 114:G00E02

Gardner TA, Barlow J, Chazdon R, Ewers RM, Harvey CA, Peres CA, Sodhi NS (2009) Prospects for tropical forest biodiversity in a human-modified world. Ecol Lett 12:561–582

Gunningham N (2009) The new collaborative environmental governance: the localization of regulation. J Law Soc 36:145–166

He HS, Mlandenoff DJ (1999) Spatially explicit and stochastic simulation of forest landscape fire disturbance and succession. Ecology 80:81–99

Kahraman C, Cebeci U, Ulukan Z (2003) Multi-criteria supplier selection using fuzzy AHP. Logistics Inf Manag 16(6):382–394

Kennedy RE, Yang Z, Cohen WB, Pfaff E, Braaten J, Nelson P (2012) Spatial and temporal patterns of forest disturbance and regrowth within the area of the Northwest Forest Plan. Remote Sens Environ 122:117–133

Kuemmerle T, Hostert P, Radeloff VC, Perzanowski K, Kruhlov I (2007) Post-socialist forest disturbance in the Carpathian border region of Poland, Slovakia, and Ukraine. Ecol Appl 17:1279–1295

Kurz WA, Dymond CC, Stinson G, Rampley GJ, Neilson ET, Carroll AL, Ebata T, Safranyik L (2008) Mountain pine beetle and forest carbon feedback to climate change. Nat Lett 452:987–990

Lindenmayer DB, Franklin JF, Fischer J (2006) General management principles and a checklist of strategies to guide forest biodiversity conservation. Biol Conserv 131:433–445

Mahapatra AK, Tewari DD (2005) Importance of non-timber forest products in the economic valuation of dry deciduous forests of India. For Policy Econ 7:455–467. https://doi.org/10.1016/j.forpol.2004.02.002

Mahmoud EH, Agnès B, Bruno L, Olivier H, Gérard D, Matthieu R (2008) Relative radiometric normalization and atmospheric correction of a SPOT 5 time series. Sensors 8:2774–2791

Malik S, Pal SC, Das B, Chakraborty R (2019a) Assessment of vegetation status of Sali River basin, a tributary of Damodar River in Bankura District West Bengal, using satellite data. Environ Dev Sustain. https://doi.org/10.1007/s10668-019-00444-y,SpringerNatureB.V.2019

Malik S, Pal SC, Das B, Chakraborty R (2019b) Intra-annual variations of vegetation status in a sub-tropical deciduous forest-dominated area using geospatial approach: a case study of Sali watershed Bankura, West Bengal, India. Geol Ecol Landscapes. https://doi.org/10.1080/24749508.2019.1633219

Masek JG, Huang C, Cohen WB, Kutler J, Hall FG, Wolfe R et al (2008) North American forest disturbance mapped from a decadal Landsat record: methodology and initial results. Remote Sensing of Environment 112:2914–2926

Miller DH (1978) The factor of scale: ecosystem, landscape mosaic and region. In: Hammond KA, Macinko G, Fairchild WB (eds) Sourcebook on the environment: a guide to the literature. The University of Chicago Press, Chicago. pp 63–88; natural disturbance rates could offset forest management impacts on ecosystem carbon stocks. Forest Ecology and Management, 308, 178–187. Nature 2000, 403, 84–86.

Miller, AB, Bryant ES, Birnie RW (1998) An analysis of land cover changes in the Northern forest of New England using multitemporal landsat MSS data. Int J Remote Sens 19:245–265

Moller-Jenson L (1990) Knowledge-based classification of an urban area using texture and context information in LANDSAT-TM imagery. Photogram Eng Remote Sens 56:899–904

Nagendra H, Paul S, Pareeth S, Dutt S (2009) Landscapes of protection: forest change and fragmentation in Northern West Bengal, India. Environ Manage 44:853–864. https://doi.org/10.1007/s00267-009-9374-9

Nepstad DC, Stickler CM, Soares-Filho B, Merry F (2008) Interactions among Amazon land use, forests and climate: prospects for a near-term forest tipping point. Philos Trans R Soc Sect B 363:1737–1746

Oguztimur S (2011) Why fuzzy analytic hierarchy process approach for transport problems? In: Proceedings of 51st congress of European Regional Science Association—ERSA, pp 1–10. Barcelona, Spain

Pal SC, Chakraborty R, Malik S, Das B (2018) Application of forest canopy density model for forest cover mapping using LISS-IV satellite data: a case study of Sali watershed, West Bengal. In:

Modeling earth systems and environment, Springer International Publishing AG, part of Springer Nature 2018. https://doi.org/10.1007/s40808-018-0445-x

Petkovic J, Sevarac Z, Jaksic ML, Marinkovic S (2012) Application of fuzzy AHP method for choosing a technology within service company. Tech Technol Educ Manag 7(1):332–341

Rahman MF, Jashimuddin M, Kamrul Islam, Kumar Nath T (2016) Land use change and forest fragmentation analysis: a geoinformatics approach on chunati wildlife sanctuary, Bangladesh. J Civil Eng Environ Sci 2(1):020–029. https://doi.org/10.17352/2455-488X.000010

Ray R, Mondal S (2014) Multi criteria based forest cover type mapping using Landsat ETM + digital data—a case study on Eastern part of Jungle Mahal Region, West Bengal. Indian J Regional Sci XLVI:77–84

Rikimaru A, Roy PS, Miyatek S (2002) Tropical forest cover density mapping. Trop Ecol 1:39–47

Running SW (2008) Ecosystem disturbance, carbon, and climate. Science 321:652–653

Saaty TL (2008) Decision making with the analytic hierarchy process. Int J Serv Sci 1(1):83–98

Sanchirico JN, Siikamaki J (2007) Natural resource economics and policy in the 21st century: conservation of ecosystem services. Resources 2007(165):8–10

Sayer D, Wienhold M (2013) A GIS-investigation of four early Anglo-Saxon cemeteries: Ripley's K function analysis of spatial groupings amongst graves. Soc Sci Comput Rev 31:71–89

Sheil D, Puri R, Basuki I, van Heist M, Wan M, Liswanti N, Rukmiyati Sardjono MA, Samsoedin I, Sidiyasa K, Chrisandini, Permana E et al (2002) Exploring biological diversity, environment and local peoples perspectives in forest landscapes, 2nd edn. CIFOR, Bogor, Indonesia

Shit PK, Pati CK (2012) Non-timber forest products for livelihood security of tribal communities : a case study in Paschim Medinipur District, West Bengal, 40, 149–156

Sourial N, Wolfson C, Zhu B, Quail J, Fletcher J, Karunananthan S, Bandeen-Roche K, Béland F, Bergman H (2010) Correspondence analysis is a useful tool to uncover the relationships among categorical variables. J Clin Epidemiol 63(6):638–46. https://doi.org/10.1016/j.jclinepi.2009.08.008

Thakker PS, Sastry KLN, Kandya AK, Kimithi MM, Jadav rN (1999) Forest vegetation mapping using remote sensing and GIS-a casestudy in Gir (Sasan) forest. In: Proceedings of ISRS National Symposium on Remote Sensing Application for Natural Resources: Retrospective and Perspective, Hyderabad, January, 18-20, 1998.

United Nation (2019) Global forest goals and targets of the UN strategic plan for forests 2030

Waring R, Running S (2007) Forest ecosystems: Analysis at multiple scales. 3rd ed. Academic Press

Wang Y (2010) Climate change and its ecological and socioeconomic impact: evidence from China's historical document for Qing Dynasty. Ph.D. Thesis, Rutgers University, New Brunswick, NJ, USA, 30 May, pp 1–174

Weller SC, Romney AK, Scaling M (1990) Correspondence analysis. Sage, Newbury Park, CA

Wickman JD, Norton DJ (1994) Mapping and analyzing landscape patterns. Landscape Ecol 1:7–23

Zhu Y, Newsam S (2016) Spatio-temporal sentiment hotspot detection using geo-tagged photos. In: Proceedings of the 24th ACM SIGSPATIAL international conference on advances in geographic information systems, Burlingame, CA, USA, 31 October–3 November, p 76

Zhang C et al (2008) Use of local moran's I and GIS to identify pollution hotspots of Pb in urban soils of Galway, Ireland. The Science of the total environment, 398 (1- 3), p. 212

Zhou Q, Li B, Zhou C (2004) Detecting and modelling dynamic landuse change using multitemporal and multi-sensor imagery. In: Paper presented at 20th ISPRS Congress, 12–23 July 2004, Istanbul, Turkey

Chapter 14
Comparison of AHP and Maxent Model for Assessing Habitat Suitability of Wild Dog (*Cuon alpinus*) in Pench Tiger Reserve, Madhya Pradesh

Dhruv Jain, G. Areendran, Krishna Raj, Varun Dutta Gupta, and Mehebub Sahana

Abstract Habitat suitability is an illustrious mission to conserve the ecosystems and its components. Human wildlife conflicts have increased tremendously due to habitat fragmentation; reason being human encroachment for their own use or for commercial purpose like forest products and unsustainable development programs leading to major decline in Biodiversity. Pench Tiger Reserve (PTR), Madhya Pradesh has great capacity to nurture young Wild dogs (*Cuon alpinus*) because it supports lots of biodiversity especially herbivores which will be great availability of food and have great tree cover on the landscape. The motive of this study is to find out the landscape which will suitable for Wild dog (*Cuon alpinus*) at PTR, MP using presence SDM (Species Distribution Model) at 900 m grain Size using Maxent, Ecological Niche Factor analysis and Bioclim and compare with AHP (Analytical hierarchical process). Maxent gives contributions of every factor in percentage also gives responses curve of every factor separately and also give permutation importance of every factor. Results show that the model with accuracy more than 0.90 AUC value. Bioclimatic variable 17 and LULC contributes the moderately high in the model. For AHP only need experts, who fill the Saaty table on the basis of environmental variables importance to wild dog (*Cuon alpinus*). Calculate or predict the habitat suitability of wild dog (*Cuon alpinus*) by experts rating to every environmental variable (comparison of every variables with respect to each other) from 1 to 9 or inverse like 1/9. Maxent shows that out of the about 750 km^2 areas, more than 150 km^2 found to be highly suitable at 900 m grain size SDM. This shows PTR, MP has the great potential to nurture or sustain Wild dog (*Cuon alpinus*) because wild dog (*Cuon alpinus*) need larger area to walk, they walk daily around 20–25 km to find their food and PTR, MP is immensely rich in large size herbivore animals like Chital (*Axis axis*). Strict implementation of laws is required this can be achieved by government agencies like forest department and others along with that awareness can be spread among the local residents of the

---

D. Jain · G. Areendran · K. Raj · V. D. Gupta · M. Sahana (✉)
Indira Gandhi Conservation Monitoring Centre (IGCMC), WWF-India, New Delhi, India
e-mail: mehebubsahana@gmail.com

P. K. Shit et al. (eds.), *Spatial Modeling in Forest Resources Management*, Environmental Science and Engineering,
https://doi.org/10.1007/978-3-030-56542-8_14

PTR, MP so the wildlife and their habitat can be conserve by the conservationists and wildlife experts.

Keywords Wild dog (*Cuon alpinus*) · Maxent species distribution model · AUC (area under curve) score · Pench tiger reserve (Madhya Pradesh) · AHP (analytical hierarchical process) tool box

14.1 Introduction

Wild dog (*Cuon alpinus*) commonly known as Dhole and some others local names in India like Jangli Kutta in Hindi, Balia Kukura in Oriya and many more local names in India also. Wild dogs (*Cuon alpinus*) are the only species in the genus *Cuon* (Hodgson 1838). Dole (*Cuon alpinus*) has been long placed in the subfamily Simocyoninae, along with bush dogs (*Speothos venaticus*) and African hunting dogs (*Lycaon pictus*). Recently it has become evident that the subfamily Simocyoninae is not correct entity, and the position if falling into incorrect use (Sheldon 1992). As long ago 1954, Thenius (1954, cited in Cohen 1977) pointed out those similarities in dentition within three species placed in the simocyoninae sub-family could be due to convergent or biological evolution and also pointed one more point i.e., other anatomical features are exhaustively different.

In ancient time Asiatic wild dog (*Cuon alpinus*) range are in big portion of East Asia and South-East Asian Countries like in Bangladesh, Cambodia, India, Kazakhstan, Kyrgyzstan, Lao Malaysia, Bhutan, Mongolia, Myanmar, Nepal, People's Democratic Republic, Russian Federation, Indonesia, Tajikistan, China, Thailand Viet Nam and many more South-East Asian small Countries. Wild dogs (*Cuon alpinus*) have two sub-Species those are Pocock (*Cuon alpinus adustus*) and other is Hardwick (*Cuon alpinus sumatrensis*). The sub species Pocock (*Cuon alpinus adustus*) is majorly found in the South-East Myanmar and South-west Thailand. and the other sub species i.e., Hardwick (*Cuon alpinus sumatrensis*) are generally found from the south of the Isthmus of Kra down to Sumatra and Java and the Hardwick (*Cuon alpinus sumatrensis*) are the smallest in size and their hairs are also smallest in the Genus *Cuon* (Fig. 14.1) (Kanchanasakha et al. 1998) Now Asiatic Wild dog (*Cuon alpinus*) are constrain to very limited areas like in central, south and some portion of north India also in some East Asian countries. In India wild dogs (*Cuon alpinus*) are generally found in central Indian landscape, peninsular region of India i.e., Southern India, North-East India and even they are sighted at most harsh conditions of north India i.e., in Ladakh also. Wild dog (*Cuon alpinus*) cubs are generally can only be confused with jackal cub, but can be recognized by bring longer and narrower. The shapes of the body, with small waist and deep chest, and the legs and their mode of attachment, are built for speed on the lines of greyhound as similar to the other big carnivores. The size and coloration of wild dogs (*Cuon alpinus*) varies regionally. The average full sized Asiatic wild dog (*Cuon alpinus*) stand 22 inches at the shoulder and weighs 19 kg on the other hand female Asiatic wild dogs

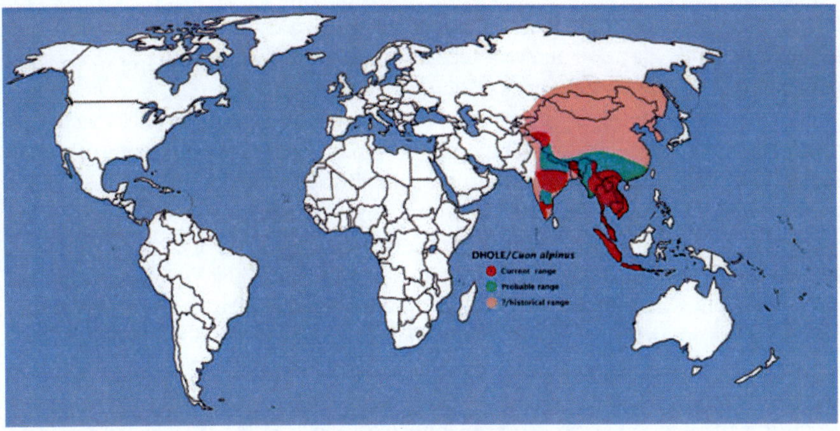

Fig. 14.1 Wild dogs ancient and present presence on global level

(*Cuon alpinus*) are 1.5 inches shorter and 2.5 kg lighter than males (Brandar 1982). Generally ground color of the coat is little reddish brown and the color on the belly is quite lighter than on the back as well as on the both sides. There is thick hair on the claws and between the toes. The dental formula of wild dogs (*Cuon alpinus*) differs from the usual formula in the Canidae. Usually there are three molars on the sides of each lower jaw but wild dogs (*Cuon alpinus*) have two molars on each side on the lower jaw. Thus, the dental formula for wild dog (*Cuon alpinus*) is 3–3 incisors on both sides in upper and lower jaw also similarly 4 pre-molars, 2 molars and 1 canines on both sides in the lower and upper jaw as well. Therefore, total 40 teeth [(4 + 3 + 2 + 1) * 4] (Novikov 1962). Another unique physical character of the wild dog (*Cuon alpinus*) is their projecting part of the face, including the nose and mouth i.e., muzzle is quiet shorter than that of most of canids with a larger projecting part of the face. They are always live and hunting in packs of 3 (minimum) or 8–10 (maximum) (Brander 1982) as unity is always big strength.

Most special characters of wild dog (*Cuon alpinus*) are they are extremely brave or daring even they fight with tiger (*Panthera tigris*) for their food. Gestation period of female wild dogs (*Cuon alpinus*) are 60–63 days (Cohen 1978; Sosnovskii 1967) and their most favorable season for denning is winter i.e., October–February. Wild Dog (*Cuon alpinus*) is a species which is considered to be highly social in nature and they have this social nature from the start of their life as in Pub life. Their rearing is considered to be a social activity. Breeding season of wild dog (*Cuon alpinus*) is confined to a narrow season as respect to other Canidae family members. In India it happens any time from end of the Monsoon season i.e., September and it goes up to February end. When a female wild dog (*Cuon alpinus*) give birth to his young ones, at a time she gives on an average 4–5 descendants: Maximum is 9 or 10 descendants (Sosnovskii 1967). In India young ones of wild dogs (*Cuon alpinus*) generally or maximum time observed that they are born in January or February (Prater 1965). Up to 12 young ones of wild dogs (*Cuon alpinus*) may be live in a single den, but these young ones are probably the descendants of two or female wild dogs (*Cuon alpinus*). Females may den and rear their descendants untidily. Prater (1965) stated that a number of female wild dogs (*Cuon alpinus*) may select a den site and form a breeding colony. A group of new young ones may have several different aged that create confusion to determine which one is more younger or offspring of which female wild dog (*Cuon alpinus*) then, Johnsingh (1982) clears that confusion and he stated that breeding is restricted to only a single female wild dog (*Cuon alpinus*) within a pack, as is also the case for the highly social, pack living African wild dogs (*Lycaon pictus*) and for wolves (*Canis lupus*). In a pack of Wild dogs (*Cuon alpinus*) every adult ones help to young ones to provide food at their den and protect all descendants from the predators like Tiger (*Panthera tigris*) and other big carnivores who can eat their descendants along with the lactating females. "Guards"—Wild dog (*Cuon alpinus*) that remain behind at the den site with the young ones while the rest are out for the hunting—are also fed (Johnsingh 1982; Fox 1984). Many times, in a group, members of a group of Wild dogs (*Cuon alpinus*) after hunting, it is not mandatory to all members of group to assemble again. In the big size group of wild dogs (*Cuon alpinus*) is very rarely to seem to do large hunting together otherwise the wild dogs (*Cuon alpinus*) are rapidly spilt into many small groups or say large size group breaks into many small groups for the hunting. Wild dogs (*Cuon alpinus*) use odor of the prey and find their prey then take turns pursue to catch or attempt for catch their prey. Most common pattern of hunting is group of wild dogs (*Cuon alpinus*) cover the prey in a circle like structure and attack on prey from all sides specially on those moments will extremely effective when prey is busy to defend from one side and another group member attack from back side or say when prey is unaware from back side attack. When sometime a pack of wild dogs (*Cuon alpinus*) were not find any prey for them to fulfill their food requirement then sometimes they eat their pups. According to Davidar (1975), he has been seen group of wild dogs (*Cuon alpinus*) whose strength is about 40. During the summer's i.e., from May to mid July and in Rainy or Monsoon Season (i.e., not in breeding Season) larger assemblage of wild dogs (*Cuon alpinus*) on a single place is quite common because all are in the resting stage of the year so that's why quite common to assemblage of very larger of wild dogs (*Cuon alpinus*) on a place. In the recent years sighting of large group of wild dogs (*Cuon alpinus*) is rare or frequency of sighting is frequently decreased because of anthropogenic pressure, encroachment of humans on their suitable habitat landscape, loss of suitable habitat and decline in prey population. During the end of Monsoon season i.e., end of the august or starting of the September and Start of breeding season or denning season sighting of small packs of wild dogs (*Cuon alpinus*) are frequent (Sheldon 1992). Wild dogs (*Cuon alpinus*) are generally amalgamate and disruptive way in a loosely or scattered pattern. There is no proof about the influence over the other gradable stratified class within their groups, even though in the free moving group of wild dogs (*Cuon alpinus*), one male wild dog (*Cuon alpinus*) was clearly show his influence over the rest members of this group. Belligerent interactions between the group members will seen extremely rare and group member showing willingness to each other for the killing purpose and in a group generally competition for the hunting at very low level. The relating places where wild dogs (*Cuon alpinus*) can live and their range of that kind of places in India is generally in a group is around 35–40 km^2, this home range is only when they are in no breeding season but when we talk about during their breeding season then their suitable habitat range will rapidly decrease from 35–40 km^2 to 10–12 km^2 (Johnsingh 1982).

Most preferable habitat is tropical deciduous forest, open forest and grasslands also (Brandar 1982). Sambar (*Rusa unicolor*), Chital (*Axis axis*), and Wild boar (*Sus scrofa*) are considered the principal prey of Wild dog (*Cuon alpinus*) within India (Davidar 1975; Johnsingh 1992; Venkataraman et al. 1995). From a study done in India, it was found out that the prey of the wild dogs (*Cuon alpinus*) are as follows 73% area spotted deer (*Axis axis*), 17% are sambar deer (*Rusa unicolor*), 5% are rodents [Order—Rodentia (Bowdich 1821)] and 2.5% are rabbit (*Oryctolagus cuniculus*). 7% of their excrement is the residue of grass (Johnsingh 1982) Main thing is they don't require lots of water like other organisms and they are generally walked a lot basically they eat and walk around 20–25 km per day to find their food. Generally, Wild dogs (*Cuon alpinus*) have a bimodal or say in other words crepuscular activity pattern of life style or daily routine. During heat of the day or commonly during summers high temperature day they are usually inactive or in lazy state or in other we can say that in rest mode, though in India during Monsoon i.e., in late June, July, August and September, they may be in active stage because when Monsoon season end their breeding season will about to start and they can search for their prey at any time of the day during monsoon season. Another reason for their hunting during Monsoon is because in Monsoon heavy rainfall help to grow small grasses on land or help to growing of herbs and shrubs also which the primary food for herbivores like chital (*Axis axis*), Sambar (*Rusa unicolor*), Indian bison (*Bos gaurus*) and other big or small herbivore which means probability of availability of food for wild dogs (*Cuon alpinus*) will be high. Hayward et al. (2014) found the prey preference index of the wild dogs (*Cuon alpinus*) is aviate slightly in different areas. During Monsoon season wild dogs (*Cuon alpinus*) hunt some time during night also it depends on their group size and their hungriness' (Johnsingh 1982; Krishnan 1972; Fox and Johnsingh 1975; Sosnovskii 1967; Cohen et al. 1978; Fox 1984; Prater 1965). Hayward et al. (2014) concluded that the prey of wild dogs (*Cuon alpinus*) from 24 different studies in the across 16 areas including habitat and distribution are Wild boar (*Sus scrofa*), Sambar (*Rusa unicolor*), Gaur (*Bos gaurus*), Spotted deer (*Axis axis*) and Muntjac (*Muntiacus*).

Wild dogs (*Cuon alpinus*) have threat from continuous loss of habitat, rapid decrease in the count of their prey, persecution and also due to increase in transfer of disease or several infected pathogens from domestic dogs (*Canis lupus familiaris*) and feral dogs (*Canis lupus familiaris*). The present range of wild dogs (*Cuon alpinus*) has been much reduced due to anthropogenic activities. Wild dogs (*Cuon alpinus*) have become rare and are depleted entirely from the parts of central Asia, large parts of India, and Eastern China (Müller-Using 1975). Across the globe any government or any organization doesn't take any attempt to census of the wild dogs (*Cuon alpinus*) population, either at country level or regional level wise, Even in places where prediction of the available, because of the chances of the same animal being counted more than once, or not at all, animal's wandering habits and the figures can be misleading.

The present study is important to identify most suitable landscape for wild dogs (*Cuon alpinus*) in the Pench tiger reserve, Madhya Pradesh with the help of geospatial techniques. The major research gap is most of scientists or researchers work on mostly on wild dog (*Cuon alpinus*) ecology, physical characteristics like in which pattern their body size varies during their upbringing, their distribution and habitat, taxonomy, diet preferences, daily activities, Reproductive behavior across all seasons, Social organization and the behavior in with their group members but present study break the tradition of study of the wild dog (*Cuon alpinus*) because now first time try to predict suitable landscape for the habitat of wild dog (*Cuon alpinus*) in Pench tiger reserve, Madhya Pradesh with the help of GIS (Geographic Information System) and remote sensing. The main objective of the study is to identify and assess the suitable habitat of wild dog (*Cuon alpinus*) and to compare the different model's prediction for the habitat suitability of wild dogs (*Cuon alpinus*) in the Pench tiger reserve, Madhya Pradesh. This study will contribute a new pillar to the house of literature of wild dogs (*Cuon alpinus*) because all the literatures are old and some new literatures also didn't predict the suitable habitat, they are also about their diet preferences and their percentage of food of different preys.

14.2 Study Area

This study is conducted on the Pench Tiger Reserve (PTR) Madhya Pradesh, India. It comprises the Pench Mowgli sanctuary, the Indira Priyadarshini Pench national park, and a buffer. The Pench Tiger Reserve, Madhya Pradesh occurs in the Karmajhiri, Gumtara, Kurai and Satpura range (southern slopes of the Satpura range of central India) with altitude between 300 and 671 m. The geographical extent 21° 63′ to 21° 95′ N latitude and 78° 94′ to 79° 53′ E longitude (Fig. 14.2). Satpura hills in Pench tiger reserve, MP have almost flat, gently, slopping top and steep sides. Pench Tiger Reserve, Madhya Pradesh is the one of the major protected areas of Satpura-maikal range. Pench Tiger Reserve, Madhya Pradesh is one of the highest herbivores densities in India. The Pench national Park, Madhya Pradesh is split into two parts by river Pench. The river Pench is the lifeline of the National Park, near the Pench River the area is flat and forming woodland and meadows. The Pench tiger reserve is the first interstate tiger project area of the country that is between Madhya Pradesh and Maharashtra. Pench Tiger Reserve, Madhya Pradesh is among the sites notified as important bird areas of India. Granites and amphibolites pegmatite and quartzite later intrude these rocks. The Deccan trap, otherwise prevalent in the other parts of both Seoni and Chhindwara districts are almost absent within the limits of the present boundary of the National park and sanctuary.

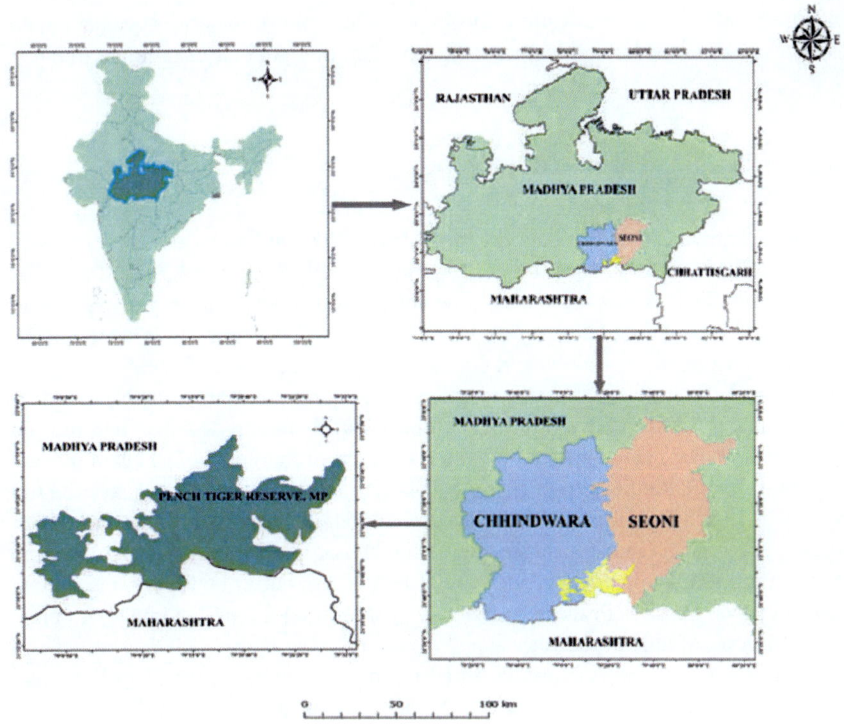

Fig. 14.2 The location of Pench Tiger Reserve (PTR) in Madhya Pradesh, India

As per the all India tiger estimation report 2018, highest number of tigers found in the Pench tiger reserve in Madhya Pradesh and according to the 4th round of the management effectiveness evaluation of tiger reserves (MEETR), which assess the India's all 50 tiger reserves, which says that Pench tiger reserve, Madhya Pradesh as best managed tiger reserve in the country with Periyar sanctuary of the Kerala. Pench national park which occurs under the Pench tiger reserve, Madhya Pradesh boundary, which has the maximum density of the herbivore animal's population like nilgai (*Boselaphus tragocamelus*), Sambar Deer (*Rusa unicolor*), Chital (*Axis axis*), Barking Deer (*Muntiacus muntjak*), Indian spotted chevrotain or Indian mouse deer (*Moschiola meminna*), Chousingha or Four horned antelope (*Tetracerus quadricornis*) and many more small and big size herbivore animals. All the above information simply depicts that Pench tiger reserve, Madhya Pradesh is highly rich in herbivore diversity which indicates availability of food for wild dog (*Cuon alpinus*) and other carnivores that's why it is the one of the highest tiger population tiger reserve in the India. Another major reason for Pench tiger reserve, Madhya Pradesh for wild dogs (*Cuon alpinus*) is one of the best managed tiger reserve in the India so that it shows there is less human interference for the wildlife as compare other Protected areas which means wild dogs (*Cuon alpinus*) can live in their natural habitat with their appropriate food availability. As per wild canids India project website wild dogs (*Cuon alpinus*) inhabit less than 50% of their potential habitats in the India.

14.3 Data Base and Methodology

We take Sentinel-2 satellite data of 14th May, 2019. My study area covers (with 10 km buffer boundary) four tiles of Sentinel-2 satellite i.e., T44QKK, T44QLK, T44QLJ and T44QKJ. After the satellite image download first we processed it in ERDAS IMAGINE 2014 software. First layer stack all the four tile in which every tile contain 12 bands in which band 8 have two sets that are band 8 band 8 A and all bands have different resolutions like 4 bands have 10 m resolution, 6 bands have 20 m resolution and 3 bands have 60 m resolution (Pokhriyal et al. 2020). Then mosaic all the tiles, mosaic is the process in which we joint or merge all tiles, again mosaic process is done with the help of ERDAS IMAGINE 2014 software and save the file in the *.img (Imagine) format. Then clip the Mosaic image of the all four tiles which become one image. Open the mosaic output in the Arc Gis 10.1 software and clip the Mosaic output image from the Study area (with 10 km buffer boundary) i.e., Pench Tiger Reserve, Madhya Pradesh and then we get sentinel-2 Satellite image of Pench Tiger Reserve, Madhya Pradesh.

We consider some Bioclimatic variables of change in every 30 s data like Annual Mean temperature (BIO 1), Mean diurnal Range [Mean of monthly (Maximum Temperature–Minimum Temperature)] (BIO 2), Isothermality (Mean Diurnal Range/Temperature Annual Range) (*100), Maximum Temperature of the Warmest Month, Minimum Temperature of driest Month, Annual Precipitation.

Sentinel-2 Satellite data for the year of 2018 were used to prepare the land use–land cover map of Pench Tiger Reserve, Madhya Pradesh (Table 14.1). The study area was extracted from the satellite imagery using 10 km buffer of Pench Tiger Reserve, MP boundary. As we mentioned above wild dog (*Cuon alpinus*) walk a lot and their movements across PTR is very fast so that we have taken 10 km buffer boundary (Fig. 14.3) to ease to estimate suitable habitat for wild dog (*Cuon alpinus*) in Pench Tiger Reserve, MP. As we mentioned above wild dog (*Cuon alpinus*) walk a lot and their movements across PTR is very fast so that we have taken 10 km buffer

Table 14.1 The details of the data sources used for this study

SI No.	Data	Data source	Resolution	Resample resolution (m)
1	Sentinel 2 (2019)	https://earthexplorer.usgs.gov/	10–60 m	900
2	ASTER DEM	https://earthexplorer.usgs.gov/	30 m	900
3	Bioclimatic data	https://worldclim.org/	900 m	900
4	Occurrence data	Steinmetz et al. (2012)	CSV table	900

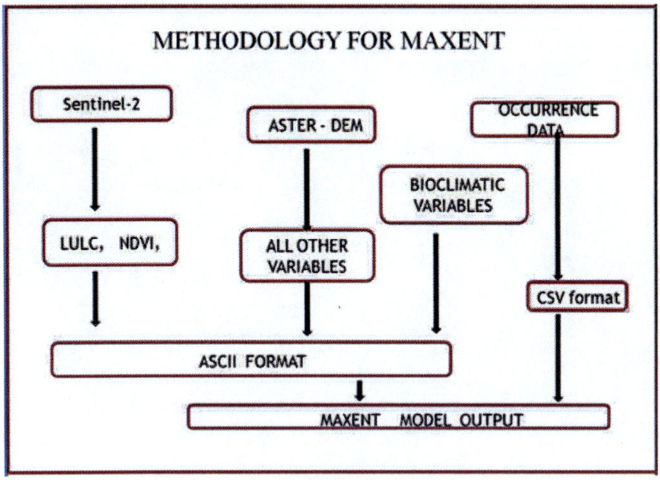

Fig. 14.3 Methodology for Maxent species distribution model

boundary to ease to estimate suitable habitat for wild dog (*Cuon alpinus*) in Pench Tiger Reserve, Madhya Pradesh (Figs. 14.4 14.5 and 14.6).

The statistics-based decision rules were applied for determining the land use-land cover identity of each pixel in the images. In the ERDAS IMAGINE 2014 software analysis process, the result is an assemblage of pixels with common features without the user giving sample classes i.e., Unsupervised Classification. For the preparation land use classification, the spectral classes were grouped into 60 classes. Classes like water bodies, Agricultural land, fallow land and barren land i.e., scrubland show similar kind of spectral reflectance. Recoding and cleaning process were followed by

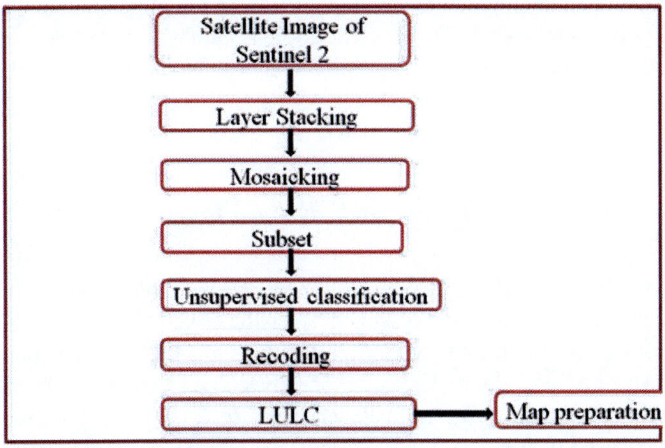

Fig. 14.4 Steps for making of LULC map

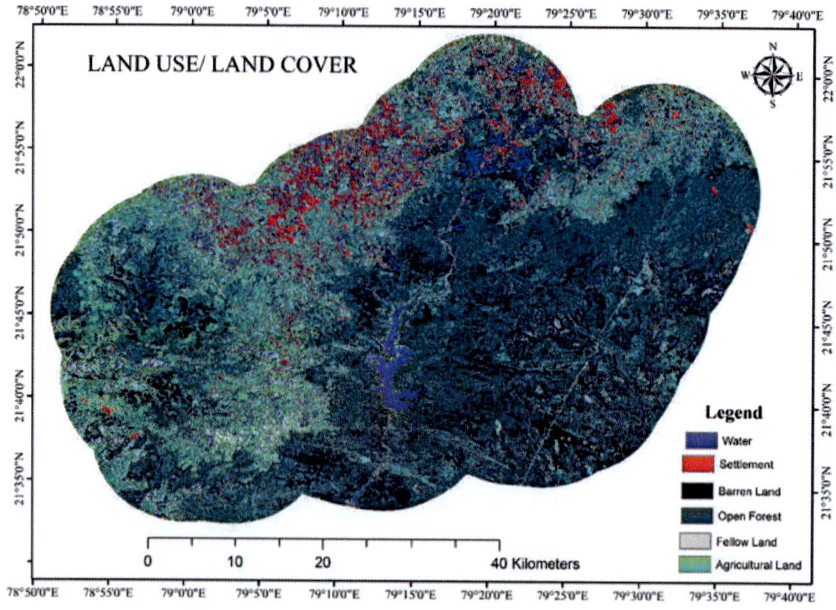

Fig. 14.5 Land use/ Land cover map (2019) of Pench Tiger Reserve, Madhya Pradesh

unsupervised classification in ERDAS IMAGINE 2014. Different aspects or factors which plays role in suitable habitat for wild dog (*Cuon alpinus*) like Pench river, Totladoh Reservoir, Settlements and roads were digitised in Google earth Pro. Shape file of Pench Tiger Reserve, MP was imported over the image under cleaning process and classes like Agricultural land, Fallow land, Barren land, Forest, Settlements and Water bodies or streams were updated on LULC using ground truth data and Google earth pro accuracy assessment was also done. Random sample points were laid down with a 100 points on the LULC to crosscheck the accuracy of LULC through which we obtained around 91 points were perfectly matches with google earth pro imagine that means 91% accuracy of LULC (Fig. 14.5).

14.3.1 Dactors derived form of Elevation layer

We used ASTER DEM data at 30 m resolution sample size. With the help of ASTER DEM we can calculate Elevation of Pench Tiger Reserve, MP with the help of Arc Gis 10.1 software in which we used Conversion tools in which first we convert DEM data into raster file to calculate Elevation of Pench Tiger Reserve, MP through we get Maximum elevation is 671 m in Pench tiger Reserve, MP which is nearly in chhindwara district region.

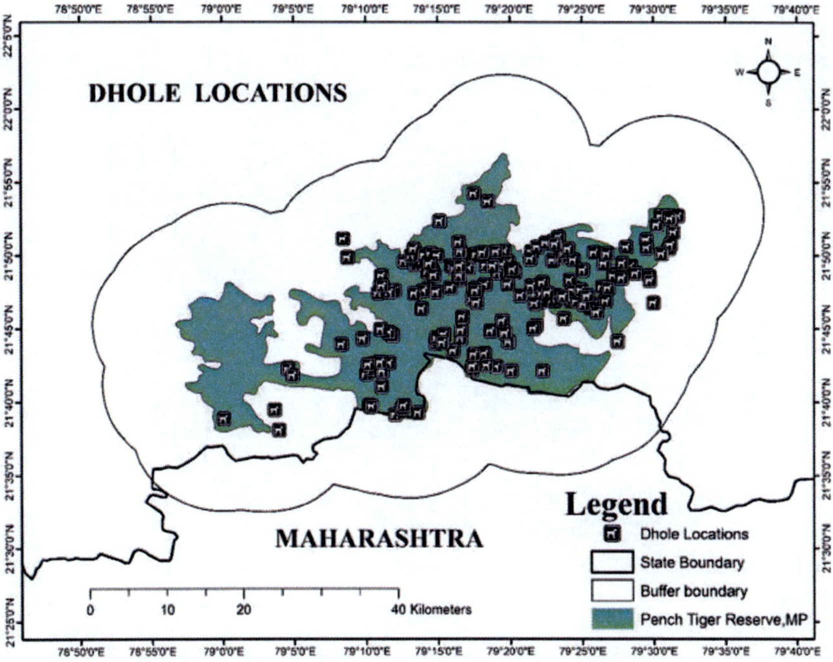

Fig. 14.6 Spatial distribution of Dhole in Pench Tiger Reserve, Madhya Pradesh

In Pench Tiger Reserve, MP as such no major elevation changes but in west Pench Tiger reserve, MP there is some portion of Satpura Hill range, but wild dogs (*Cuon alpinus*) are not prefer much height they like plain surface more. They prefer mostly plain area or slightly elevated areas not much elevated areas (Fig. 14.7).

With the help of Elevation we can calculate several other parameters which will affect the habitat of wild dog (*Cuon alpinus*) like Hill Shade, Slope, and Aspect in Arc GIs 10.1 software through surface analyst tools in which use surface option to generate layers of Hill shade, Aspect and Slope.

We observe when we calculate hill shade with the Z-factor value 0.1 then we get range of hill sahde is 0–183 similar way when try to calculate slope with z value is 1 and 50 interval class is maximum 83° angle in the study area Pench Tiger Reserve, MP and Aspect which value range is between −1 and 359.433. Wild dog (*Cuon alpinus*) preferred slight steep places, where slope is around 20–30° (Brander 1980). Apart from this Wild dog (*Cuon alpinus*) strongly prefer grassland terrain with low slope and low elevation and moderate slope like 10–25° (Fig. 14.8).

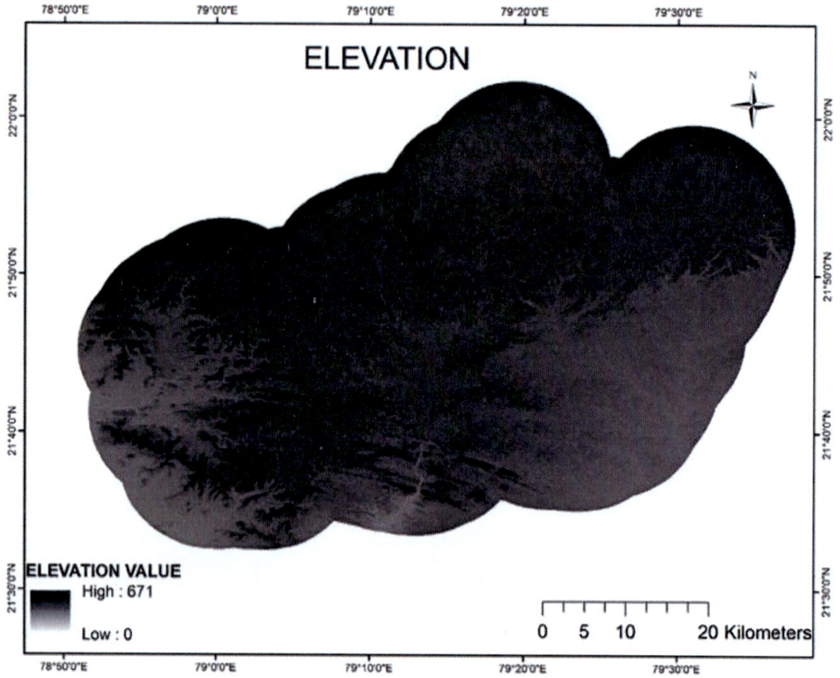

Fig. 14.7 Spatial variation of elevation (m)

14.3.2 Preparation of Other Factors

14.3.2.1 Fire

First we take fire data of India from FSI for last decade i.e., from 2008 to 2018 then clip that data to my Study Area. We got in Pench Tiger Reserve, MP have lots of fire incidence it indicates this Protected area is good for wild dog (*Cuon alpinus*) because when fire goes down then on the place of fire firstly grows small grasses which is the primary food of herbivores that means availability of food for wild dog (*Cuon alpinus*) (Sahana and Ganaie 2017). Major fire incidence sighted in chhindwara district region and more than 700 incidence are sighted in last decade (2008–2018) (Fig. 14.9).

14.3.2.2 NDVI (Normalised Difference Vegetation Index)

To calculate NDVI, I use Mosaic product of Sentinel-2 Satellite Image, using unsupervised classification Tool in ERDAS IMAGINE 2014. In Pench Tiger Reserve, MP, there no as such dense forest area and Wild dogs (*Cuon alpinus*) are generally

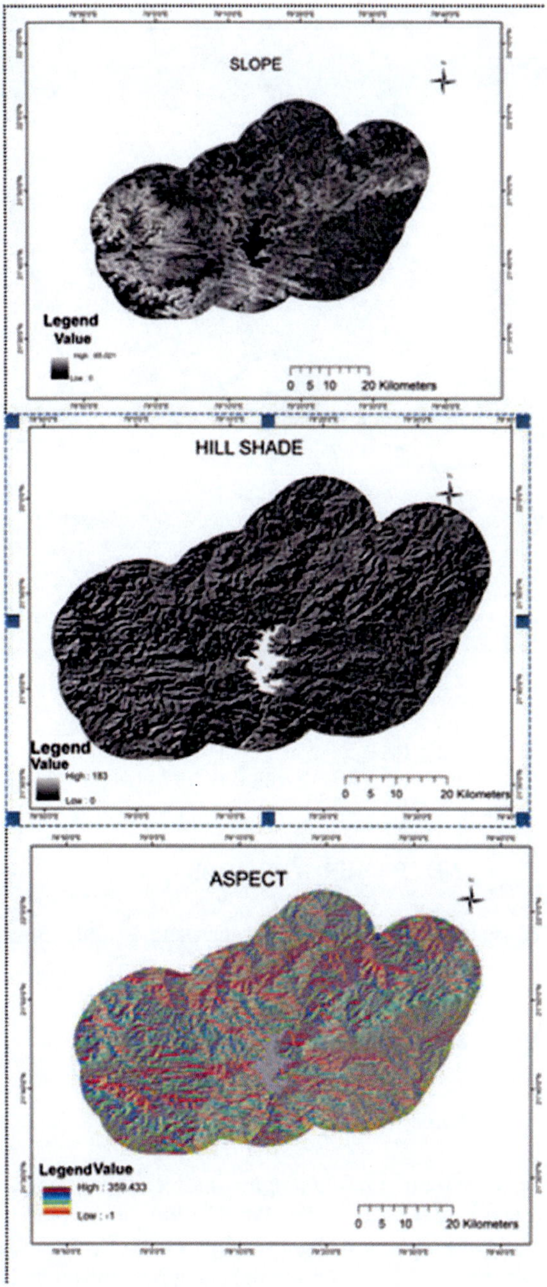

Fig. 14.8 Spatial distribution of thematic layers

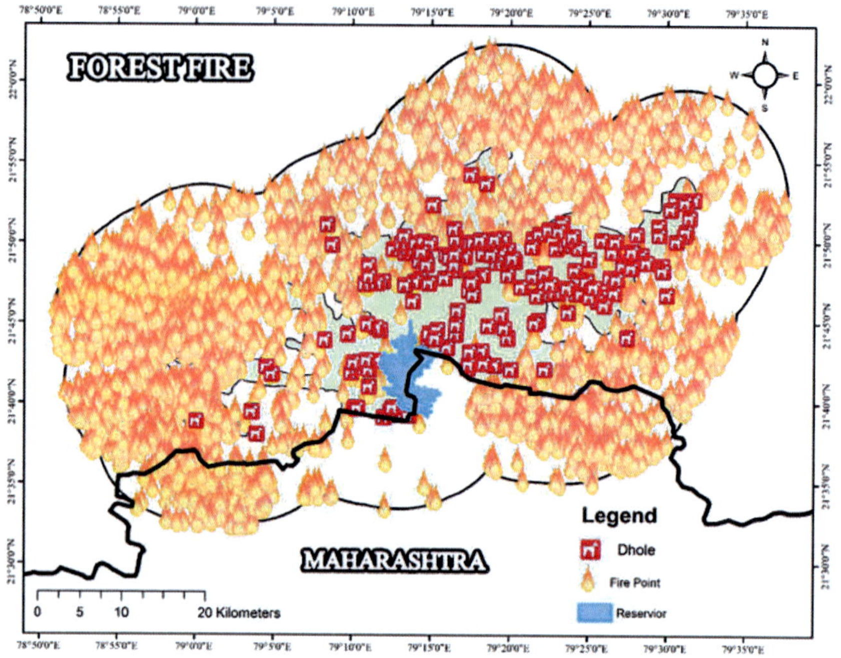

Fig. 14.9 Spatial distribution of forest fire risk

prefer open forest and grassland. NDVI value is lie between −1 and +1 in which zero indicates for water.

$$\text{NDVI} = \text{NIR} - \text{RED}/\text{NIR} + \text{RED}$$

For Sentinel-2, use band 8 (Near infrared) and band 4 (Red) (Fig. 14.10)

$$\text{NDVI} = \text{Band } 8 - \text{Band } 4/\text{Band } 8 + \text{Band } 4$$

14.3.2.3 Prey Occurrence

Major Prey species of Wild dog [(*Cuon alpinus*) are sambar (*Rusa unicolor*), Wild boar (*Sus scrofa*) and Chital (*Axis axis*)]. We take only those places ranges where maximum probability of sightings of all these major preys of wild dog (*Cuon alpinus*) from annual forest report of Madhya Pradesh state government annual forest report on their biodiversity 2018 (Fig 14.6; Fig. 14.11).

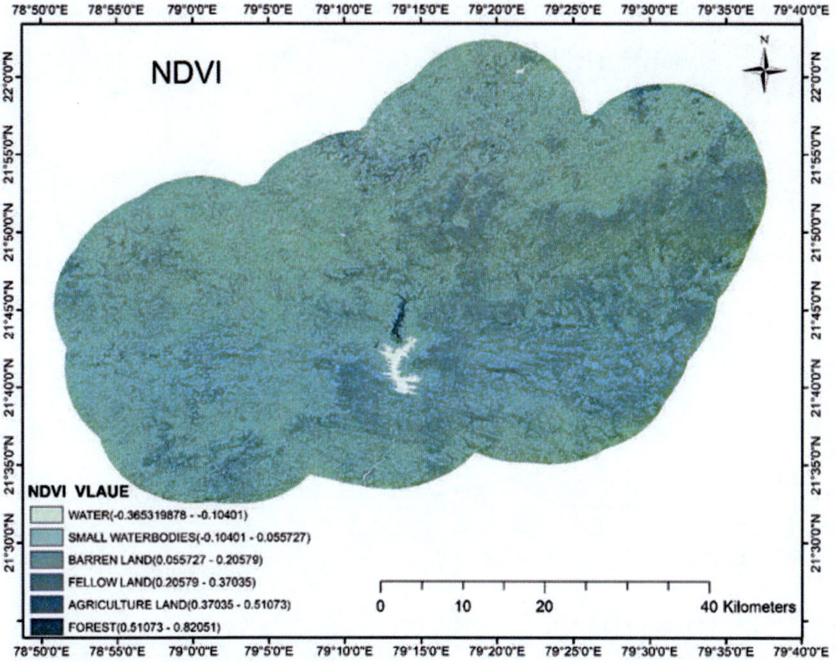

Fig. 14.10 Spatial distribution of NDVI

14.3.2.4 Tree Cover

Tree cover plays vital role for Wild dog (*Cuon alpinus*) because if tree cover will high or dense forest it will create some difficult to hunt prey for food but some times it is very good to hide and hunt preys. Tree cover data is taken by Hansen tree cover from tile which has data of 20–30 N and 70–80 E coordinates (Fig. 14.12).

14.3.3 Maxent Species Distribution Model

Wild dogs (*Cuon alpinus*) are poorly or little studied species in India except some protected area and biosphere reserve such as Central India specially in Madhya Pradesh and Southern India like in Tamil Nadu and nearby states such type of area covers less than 5% of potential suitable habitat of Wild dog (*Cuon alpinus*) in India (Johnsingh 1989; Yadav et al. 2020).

Maxent software are excessively use for assessing the habitat suitability of these elusive species.

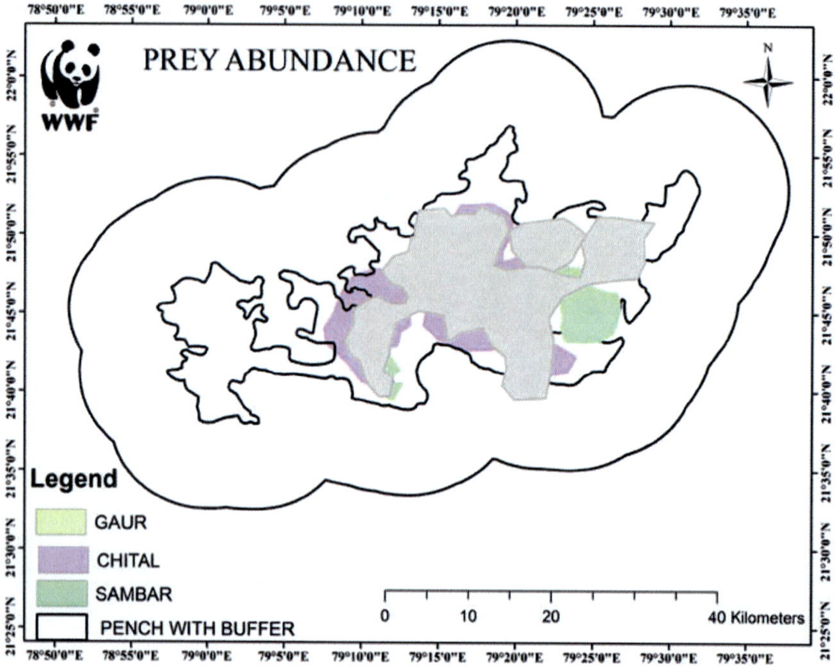

Fig. 14.11 Geographical area of prey abundance

Maxent is a sophisticated machine learning model, with very explicit statistical and mathematical formula (Jayne's 1957). It's widely used for predicting the habitat suitability and species distribution. Maxent provide the probability of species distribution and their suitable area which are based on concept of maximum deterioration (i.e., heavily diffuse out or close to uniformity). It uses the covariates from the species occupancy point for predicting the species distribution.

Maxent uses the occurrence or presence only data for species distribution. In this study occurrence data for wild dog (*Cuon alpinus*) were collected from Aniruddha Majumdhar et al. (2012). Recent updated version of Maxent 3.4.1. were used for predicting the habitat suitability of Wild dog (*Cuon alpinus*) in Pench Tiger Reserve, MP.

A total 150 plus occurrence data were used for wild dog (*Cuon alpinus*), 30% of occurrence data were used for verification of model, rest 70% were used for testing or build the model for wild dog (*Cuon alpinus*) species. Maxent accept the occurrence data only in environmental variables in ASCII format and *.CSV format with same size cell size, projection and extent.

All environmental variables which are used in habitat suitability modeling for (*Cuon alpinus*) were uploaded in Arc Map window then re-projected in to GCS (Geographic Coordinate System) WGS (World Geodetic System) by using project raster tool in Arc Map software. Re-projected variables were resembled into 30 m

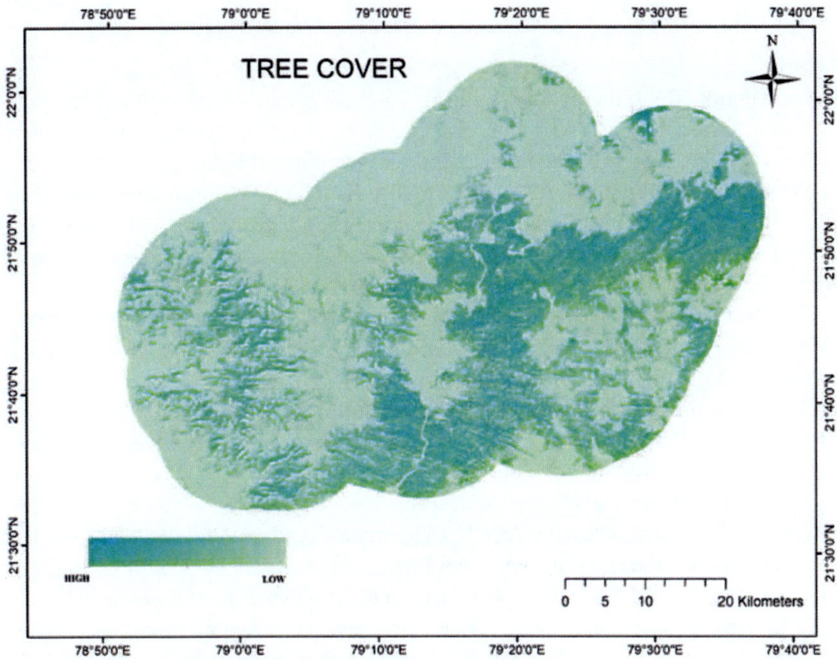

Fig. 14.12 Spatial variation of tree cover

meter resolution to 900 m. All variables were extracted from first layer that is BIO-I in this study, and then all variables were converted from raster to ASCII (American Standard Code for Information Interchange) format by using conversion from raster tool in Arc tool box.

Maxent model can run multiple of time, which enhance the model accuracy. In this study a total of 10 replicates with 5000 iteration were chosen, it gave the sufficient time to the model for convergence the result. If model does not have sufficient time for convergence the variables then there will be possibility of over and under prediction of model. Jackknife test are used for estimate the importance of each variable and their contribution in prediction of habitat suitability.

Sensitivity of each environmental and a biotic variable was done in logistic format. Accuracy of output were compared with ROC value, according to expert view ROC value 0.5–0.6 represent the unacceptable result, poor result represents by ROC value 0.6–0, Normal result with 0.7–0.8, good out represent by 0.8–0.9 and excellent output shows with ROC value 0.9–1.0 (Swets 1988).

Main thing is that when you try to run Maxent Species distribution model always remember your all layers have same cell size, projection and extent otherwise Maxent Species distribution model will not run and also not predict the suitable habitat for your which you want.

14.3.4 Methodology for Maxent Species Distribution Model

As we mentioned the methodology in Fig. 14.3 we make all the factors like Land Use/Land Cover, Hill shade, Slope, Aspect, Normalized Difference in vegetation index, Elevation, Fire, Prey Abundance, Tree cover, Annual Mean Temperature (Bio 1), Mean Diurnal Range (Bio 2), Isothermality (Bio 3), Maximum Temperature of warmest month (Bio 5), Minimum Temperature of the Coldest Month (Bio 6), Annual Precipitation (Bio 12), Precipitation of the wettest Month (Bio 13), Precipitation of Driest Month (Bio 14) and Precipitation of Driest Quarter (Bio 17).

First convert the entire layer Projection into GCS (Geographic Coordinate System) WGS (World geodetic system) 1984 with the help of Arc GIS 10.1 software. Then convert all the layer cell size into same cell size, we prefer all the layers cell size similar to bioclimatic variables cell size that is at 900 m cell size and then last to do extent of all the layers should be same that's why me make same extent of all layers with respect of bioclimatic variable layer. In Maxent Species Distribution Model, it requires all the layers are in same projection, same cell size and same extent. Then convert all the raster files into ASCII (American Standard Code for Information Interchange) with the help of raster tool box in the Arc GIS 10.1 with the help of batch mode option. Batch mode option help to do same process of more than 1 layer that's why we choose batch mode because we have 18 layers to which we convert into ASCII (American Standard Code for Information Interchange) format. ASCII (American Standard Code for Information Interchange) layer is the only format which Maxent Species Distribution Model can read. Then we georeferenced the map of wild dog (*Cuon alpinus*) sighted in Pench Tiger Reserve, Madhya Pradesh from Dr. Aniruddha Majumder Ph.D. Thesis (2011) and take coordinates and make it a shape file. Then in file manager we open shape file folder and open *.dbf (Data Base File) file and open that *.dbf (Data Base File) file in Microsoft Excel and save as it in *.CSV (Comma Separated Values) format file. When we get all the layers in the ASCII (American Standard Code for Information Interchange) with same cell size, Projection and Extent with all Coordinates of the wild dogs (*Cuon alpinus*) in the *.CSV (Comma Separated Values) format of Excel file and put in the Maxent Species Distribution Model (Fig. 14.13).

14.3.5 Overview of Factors that Affect Habitat of Wild Dog (Cuon alpinus)

See Fig. 14.14.

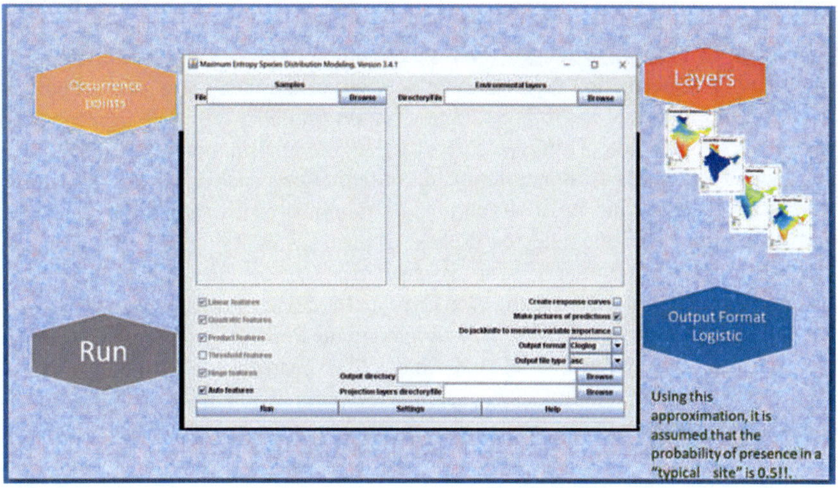

Fig. 14.13 Methodology frame work of habitat suitability assessment

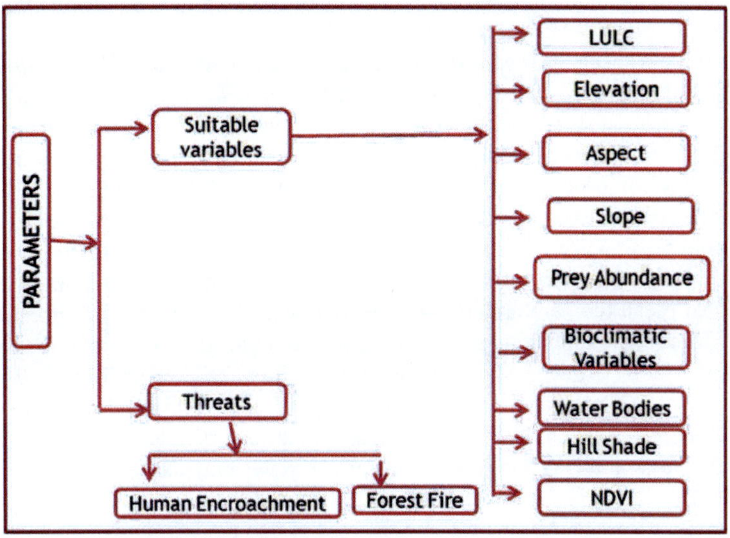

Fig. 14.14 Various aspects of habitat quality

14.3.6 Methodology for AHP (Analytical Hierarchical Process)

Analytical hierarchical process (AHP) was given by Thomas Saaty in 1997 in which he gave a table in which we compare two factors and rate both factors in 1–9 range if first factor is more contribute or more dominant to second factor then give rating in the range of 1–9 in 1–9 range 1 indicates to equal importance of both factors as when rate it 9 it means first factor is extremely important then second factor but on the other hand if second factor is extremely important then give range in reciprocal i.e., 1/9 or 1/8 like this. This is ensured that all the environmental Variables and other factors were used with a combination of techniques such as GIS (Geographic Information System) and AHP (Analytical hierarchical process) in this study. The relative importance of the selected criteria is revealed by using AHP (Analytical Hierarchical Process) and AHP also assisted in assigning the weightage to each criterion. The criterion was used in Index Overlay Model to arrive at the most suitable landscape sites for the wild dogs (*Cuon alpinus*) in the Pench Tiger Reserve, Madhya Pradesh. Local knowledge and field experience have been used to assign the weight for each factor in vulnerability analysis. Pair-wise comparison matrix is provided by AHP (Analytical Hierarchical Process), through which the criterions are structured according to their hierarchical order. A vector of weights can obtain by using 1–9 scale in each pair-wise comparison.

Saaty table is not restricting to two factors; Saaty table gives freedom to select factors up to nine factors. As per Saaty table compare every factor to each other and not repeat any combination and give rating between 1–9 and rate in reciprocal if second factor plays dominant role like 1/9 or 1/8 etc.

For the cross check the prediction of suitable habitat of wild dog (*Cuon alpinus*) in Pench Tiger Reserve, MP Maxent species distribution model we did AHP (Analytical hierarchical process) for the predication of suitable habitat of wild dog (*Cuon alpinus*) in Pench Tiger Reserve, MP. First we were doing AHP (Analytical hierarchical process) only for Bioclimatic Variables with the help of Saaty Table 14.2. Saaty table is filled by five experts who are expert in GIS (Geographic Information System) and remote sensing also has expertise in wildlife. Firstly I were gave empty Saaty table to my all experts to fill it and gave ratings to all bioclimatic variables which are Annual mean temperature (Bio-1), Mean annual Range (Bio-2), Isothermality (Bio-3), Maximum temperature of Warmest Month (Bio-5), Maximum temperature of coldest Month (Bio-6), Annual Precipitation (Bio-12), Precipitation of Wettest Month (Bio-13), Precipitation of Driest Month (Bio-14) and the Precipitation of Driest Quarter (Bio-17). Then we have two ways to further proceed Saaty table which filled by some GIS (Geographic Information System) and remote sensing and wildlife experts.

Firstly we renamed the all Factors name or say use the symbols like for Annual mean Temperature (Bio-1) denoted by A1, Mean annual range (Bio-2) denoted by A2, Isothermality (Bio-3) denoted by A3, Maximum temperature of Warmest Month (Bio-5) denoted by A4, Maximum temperature of coldest Month (Bio-6) denoted by A5, Annual Precipitation (Bio-12) denoted by A6, Precipitation of Wettest Month (Bio-13) denoted by A7, Precipitation of Driest Month (Bio-14) denoted by A8 and the Precipitation of Driest Quarter (Bio-17) denoted by A9 (Table 14.3).

Then we used Saaty table which filled by all experts for bioclimatic variables (Table 14.2) then we use AHP tool box in Arc GIS 10.1. In AHP (Analytical hierarchical process) tool box first fill the data of Saaty table which field by GIS (Geographic

Table 14.2 Pair-wise comparison matrix (after Saaty)

Intensity of importance	Definition	Explanation
1	Equal importance	Two activities contribute to the objective
2	Weak or slight	
3	Moderate importance	Experience and judgment slightly favor one activity over another
4	Moderate plus	
5	Strong importance	Experience and judgment strongly favor one activity over another
6	Strong plus	
7	Very strong or demonstrated importance	An activity is favored very strongly over another; its dominance demonstrated in practice
8	Very, very strong	
9	Extreme importance	The evidence favoring one activity over another is of the highest possible order of affirmation
Reciprocal of above	If activity *i* has one of the above non-zero numbers assigned to it when compared with activity *j*, then *j* has the reciprocal value when compared with *i*	A reasonable assumption
1.1–1.9	If the activities are very close	May be difficult to assign the best value but when compared with other contrasting activities the size of the small numbers would not be too noticeable, yet they can still indicate the relative importance of the activities

Information System) and remote sensing and Wildlife experts for bioclimatic variables then run the data and give output location to my result but there is a little bit problem because when we get result from AHP table in the form of a predicted map layer in *.GTF (Geographic Transformation File) file format but when you try again to open *.GTF (Geographic Transformation File) file layer of AHP (Analytical hierarchical process) bioclimatic map prediction which will not open again so to avoid this problem when you get first map prediction from AHP (Analytical hierarchical process) tool box then immediately export that layer in your desirable location in your suitable format.

Then again use AHP (Analytical hierarchical process), now this time we use last AHP (Analytical hierarchical process) Bioclimatic variables map projection (Fig. 14.15) used as a factor in next AHP (Analytical hierarchical process).

Table 14.3 Responds on AHP table for bioclimatic variables

Respondents	Expert A	Expert B	Expert C	Expert D	Expert E	Average
A1 versus A2	1	4	5	3	0.25	2.65
A1 versus A3	1	5	4	1	6	3.4
A1 versus A4	9	6	7	7	0.14	5.82
A1 versus A5	1	6	6	0.2	1	2.84
A1 versus A6	1	7	9	0.2	1	3.64
A1 versus A7	1	4	5	0.14	1	2.22
A2 versus A3	5	5	3	1	6	4
A2 versus A4	3	6	5	5	0.16	3.83
A2 versus A5	3	5	3	0.5	5	3.3
A2 versus A6	0.14	6	3	3	5	3.42
A2 versus A7	0.14	5	5	0.33	6	3.3
A3 versus A4	3	5	5	3	0.16	3.23
A3 versus A5	1	6	3	5	3	3.6
A3 versus A6	0.33	5	6	3	3	3.46
A3 versus A7	0.33	6	3	0.33	3	2.53
A4 versus A5	0.14	5	1	0.33	1	1.5
A4 versus A6	0.14	6	3	0.33	1	2.1
A4 versus A7	1	6	5	3	1	3.2
A5 versus A6	3	5	7	5	1	4.2
A5 versus A7	1	6	5	7	1	4
A6 versus A7	1	7	1	0.33	1	2.06

Now we take six factors which are Land Use/Land Cover, Prey occurrences, AHP (Analytical hierarchical process) predication layer of Bioclimatic variables (Fig. 14.15), Water, Tree cover or Forest canopy cover and Human Disturbances which include roads, settlements and highways. As in previous AHP (Analytical hierarchical process) process we again give codes to all factors which we've taken this time like A1 for LULC, A2 for Bioclimatic variable layer which we previously AHP (Analytical hierarchical process) process we got, A3 for Prey occurrences, A4 for Tree cover or Forest canopy cover, A5 for Water and A6 for Human disturbance (Sahana et al. 2018). Then again AHP (Analytical hierarchical process) table gave to experts which are this time again filled by GIS (Geographic Information System) and remote sensing and wildlife experts fill the table with respect to habitat suitability of wild dogs (*Cuon alpinus*) in Pench Tiger Reserve, Madhya Pradesh (Table 14.4).

Again, same process, put the values of experts rating Saaty table which we put into the AHP (Analytical hierarchical process) tool box in the Arc GIS 10.1 software and select the output file location in the desirable folder. When we precede the process we get our final product of Habitat suitability map of wild dogs (*Cuon alpinus*) in

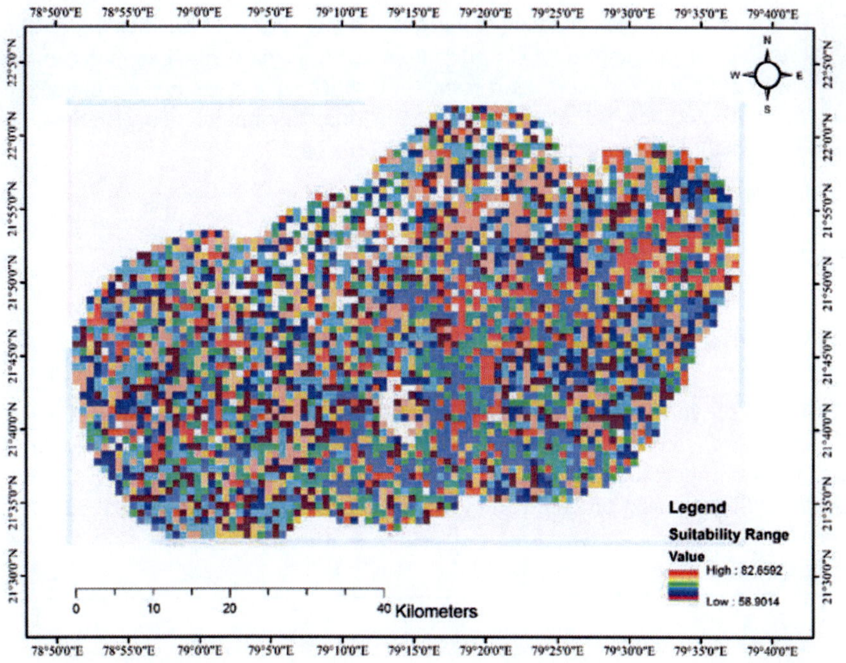

Fig. 14.15 AHP prediction for bio climatic variables

Table 14.4 Responds on AHP table for all variables

Respondents	Expert A	Expert B	Expert C	Expert D	Average
A1 versus A2	8	6	6	1	5.25
A1 versus A3	0.13	7	8	0.14	3.81
A1 versus A4	6	7	9	0.2	5.55
A1 versus A5	0.11	7	4	0.14	2.81
A1 versus A6	5	7	5	9	6.5
A2 versus A3	0.11	4	7	0.14	2.81
A2 versus A4	0.17	5	5	3	3.29
A2 versus A5	0.13	5	4	0.2	2.33
A2 versus A6	5	2	2	9	4.5
A3 versus A4	6	5	5	5	5.25
A3 versus A5	1	6	4	1	3
A3 versus A6	6	2	5	9	5.5
A4 versus A5	0.11	4	7	0.2	2.83
A4 versus A6	5	1	7	9	5.5
A5 versus A6	8	6	5	9	7

the Pench Tiger Reserve, Madhya Pradesh in the *.GTF format file then export the final product of the map in *.GTF file into suitable format which supports in the system like in *.img format or *.tiff format file which we can easily open in the any software. We get the result after export the final map layer prediction with the help of AHP (Analytical hierarchical process) in Sect. 14.4.2.

14.4 Results

14.4.1 Maxent Species Distribution Model Result

As we can see in Fig. 14.16 major Core area (No disturbance zone) of Pench Tiger Reserve, Madhya Pradesh light green color shows highly moderately suitable which covers most of core area of Seoni District of Madhya Pradesh, Pench Tiger Reserve and North-East region of Core region of Pench Tiger Reserve, Madhya Pradesh is

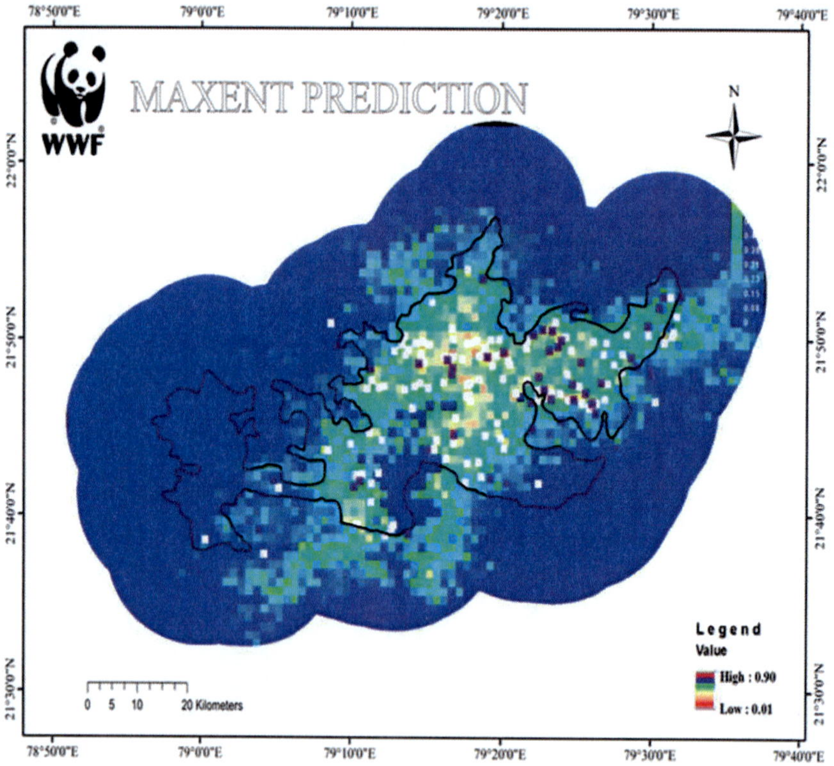

Fig. 14.16 Habitat suitability of Dhole in Pench Tiger Reserve, MP (Maxent prediction)

highly suitable zone which shows in yellow color and light blue color zone which covers the light green color (highly moderately suitable zone). The light blue color zone shows slightly suitable area. The last dark blue zone which indicates highly unsuitable zone for wild dogs (*Cuon alpinus*). We can observe some highly moderately area which shows by light green color which goes out from Pench Tiger Reserve, Madhya Pradesh and enters into Pench Tiger Reserve, Maharashtra and other extension we see in east region of Pench Tiger Reserve, Madhya Pradesh or say towards Kanha-Pench Corridor in Madhya Pradesh.

With the help of Jackknife of AUC for wild dog (*Cuon alpinus*) we can relate or predict the how much every variable importance with only variable shown by light blue color histogram, without variable denoted by dark blue color histogram and with all variables denoted by red color histogram, all these environmental variables conditions with respect to AUC value jackknifing create all histograms (Fig. 14.17).

We can see in the Upper Image of Fig. 14.18a which shows the curve between Sensitivity versus Specificity for wild dog (*Cuon alpinus*) in the Pench Tiger Reserve, Madhya Pradesh, AUC (Area Under Curve) value is 0.938 in training data, As per Maxent Species distribution model tutorial only those models are acceptable which AUC value for training data is more than 0.75 and AUC value for Test data also be

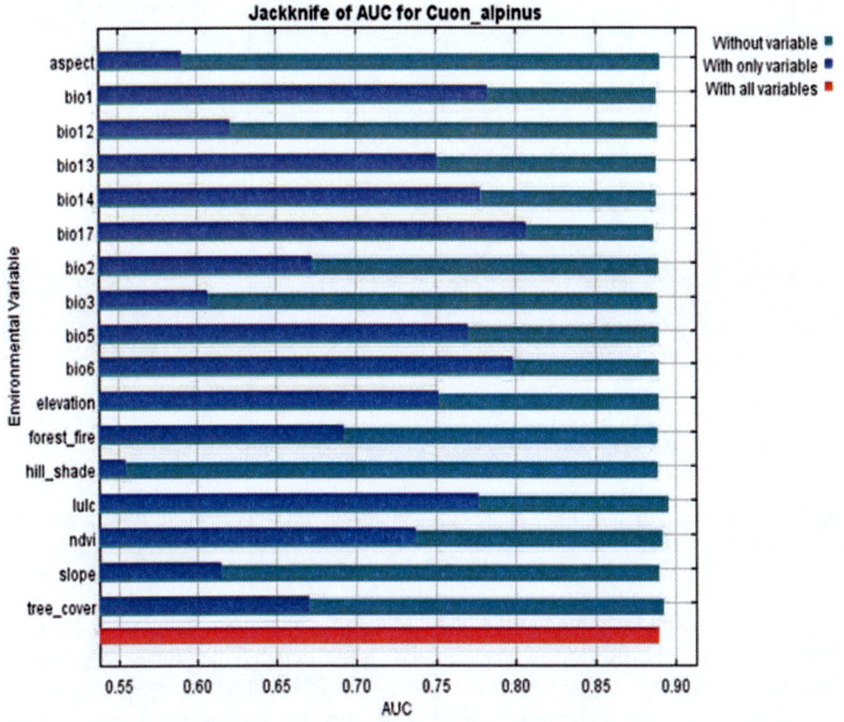

Fig. 14.17 Jackknife of AUC for *Cuon alpinus*

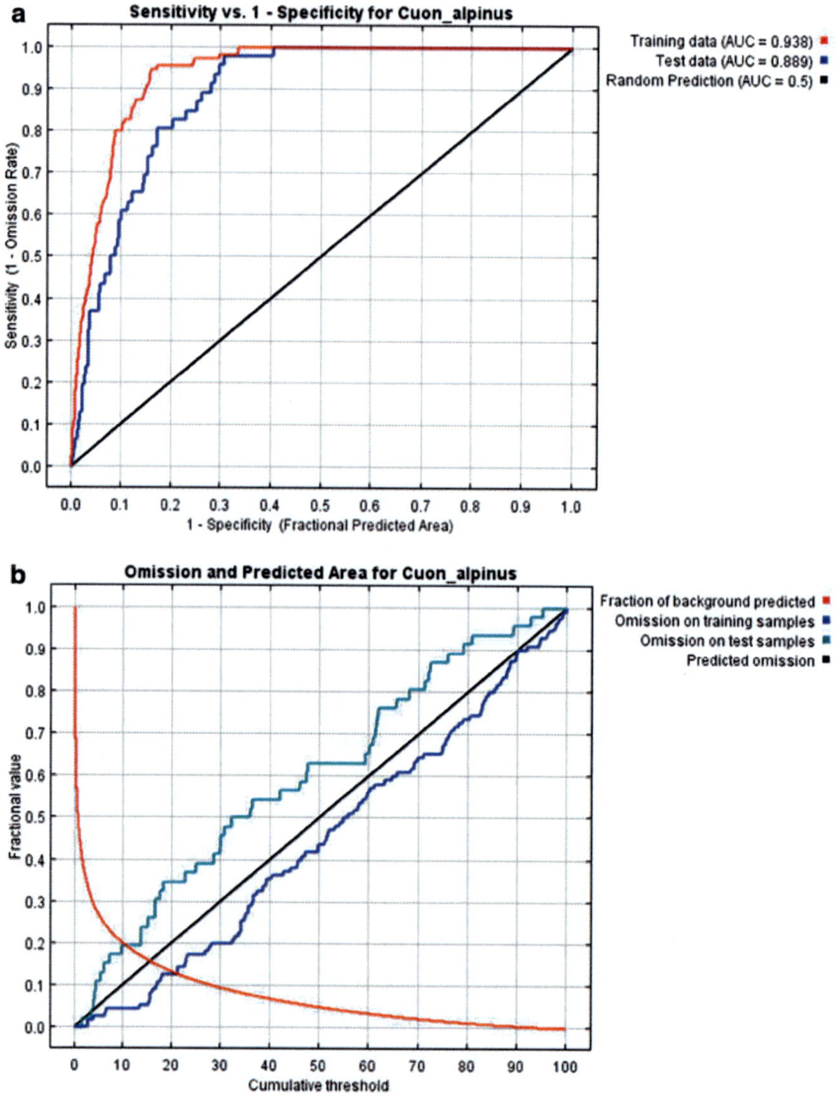

Fig. 14.18 **a** Sensitivity versus specificity for *Cuon alpinus*. **b** Omission and predicted areas for *Cuon alpinus*

more than 0.75 similarly we get 0.838 AUC value for test data that means Maxent Species Model for wild dog (*Cuon alpinus*) in the Pench Tiger Reserve, Madhya Pradesh is acceptable. In the Lower image of Fig. 14.18b which represents the curve between Omission and Predicted area for wild dogs (*Cuon alpinus*), we can see the red line which indicates the fraction of background prediction that is hyperbolic

(A hyperbola is formed by the intersection of a plane perpendicular to the bases of a double cone) curve which indicates the Fractional value is mirror image of Cumulative threshold value or in other words we can say that the set of all values of Cumulative threshold and fractional value of this curve have such ratio of the distance to a single focal point divided by the distance to the either Fractional value or Cumulative Threshold value is greater than one.

Final outputs for training and test data with respect to the regularized and un-regularized value. Regularized Training gain is 1.537, training AUC is 0.938. Un-regularized training gain is 1.772 and un-regularized training gain is 1.129.

Table 14.5 interpret the relative advantages of environmental variables to the Maxent species distribution model. Table 14.5 depicts the results which are useful to decide the utterance of training algorithm, i.e., an addition to the usual advantage beside the gain in the relative variables or if we deduct the gain in the relative advantages from the result, the value of lambda will be in the negative in numbers.

Likewise, Table 14.5 depicts the values of environmental variables which are training presence; the secondary data were randomly permuted. Further Table 14.5 also depicts that often extracting permuted data and consequently rapidly drop in training AUC (Area under Curve) data were shown in percentage in which observe that Bioclimatic variable 17 i.e., Precipitation of the driest quarter permuted 47.1%, which is highest but when we go on second highest which bioclimatic variable 13 i.e., precipitation of the wettest month which permuted 13.3%. As we can observe there

Table 14.5 Percent contribution and permutation importance (in %) of every factor

Variable	Percent contribution	Permutation importance
Bio 17	27	47.1
LU/LC	25.3	9.1
Tree cover	11.3	2.5
Bio 13	10.8	13.3
Forest fire	8.3	3.3
Bio 1	5.8	3.3
NDVI	5.1	11.3
Bio 14	1.7	0.6
Aspect	1.3	1.6
Slope	1.3	2.5
Hill shade	0.9	1.6
Elevation	0.5	1.7
Bio 2	0.4	0.1
Bio 3	0.3	0.7
Bio 6	0	1
Bio 12	0	0
Bio 5	0	0

is rapid drop in the Permutation importance of the all environmental variables. As with the variable jackknife (Fig. 14.17), Variable contributions should be interpreted when the predictor variables are correlated but with cautions when you interpret the data.

Maxent Species Distribution Model also predicts the how much percent every environmental variable in (Table 14.5) contribute in the habitat suitability of wild dog (*Cuon alpinus*) and how much every environmental variable permutation importance (in Table 14.5).

With the help of Table 14.5 we can conclude that Bioclimatic Variable 17 i.e., Precipitation of Driest quarter contribute 27%, Land Use/Land Cover contributes 25.3%, Tree cover contributes 11.3% are the major contributors to the suitable habitat of wild dogs (*Cuon alpinus*) in the Pench Tiger Reserve, Madhya Pradesh.

With the help of Table 14.5 we can say that Permutation Importance again Bioclimatic Variable 17 i.e., Precipitation of the Driest quarter is 47.1%, Bioclimatic Variable 13 i.e., Precipitation of the Wettest month is 13.3% and the Normalized Difference of vegetation Index (NDVI) which is 11.3% permutation of importance.

14.4.2 AHP (Analytical Hierarchical Process) Result

The motive of this study is to identify the suitable landscape of Pench Tiger Reserve zones in the two districts (Chhindwara and Seoni) of the Madhya Pradesh in the area protected for wild dog (*Cuon alpinus*) habitat. Weighted overlay model has been used for the declination by integrating RS-GIS and AHP techniques. The motive of using this integrated techniques to reveal the potentiality of the Pench Tiger Reserve, MP through the overlay analysis of various influencing factors. In suitability model, the result was generated on the basis of weightage given to each criteria based on its importance. With the help of AHP (Analytical hierarchical process) tool box technique we get various type of ranges of suitable habitat of wild dogs (*Cuon alpinus*) in the Pench Tiger Reserve, Madhya Pradesh like Pink color zone which is outside the protected area i.e., Pench tiger Reserve, Madhya Pradesh boundary. Main reason of take 10 km buffer from Pench Tiger Reserve, Madhya Pradesh boundary due to their large range of movement, now we can see in Fig. 14.19, Pink color zone which shows maximum suitable landscape for the habitat of wild dogs (*Cuon alpinus*) which covers East zone of buffer area (10 km from the east boundary of Pench tiger Reserve, Madhya Pradesh) another second highest suitable landscape for the wild dogs (*Cuon alpinus*) in the Pench tiger Reserve, Madhya Pradesh is represented by dark blue color, which doesn't cover as much area but AHP (Analytical hierarchical process) predicts the very little area just behind the most suitable area that is in pink color, we can say dark blue color zone is high suitable landscape for the Wild dogs (*Cuon alpinus*). Another zone which occupies South-East and West region of the study area with 10-km buffer boundary as per AHP (Analytical hierarchical process) prediction which represents from light blue color moderately suitable habitat landscape for the wild dogs (*Con alpinus*). Light blue color zone showing on three

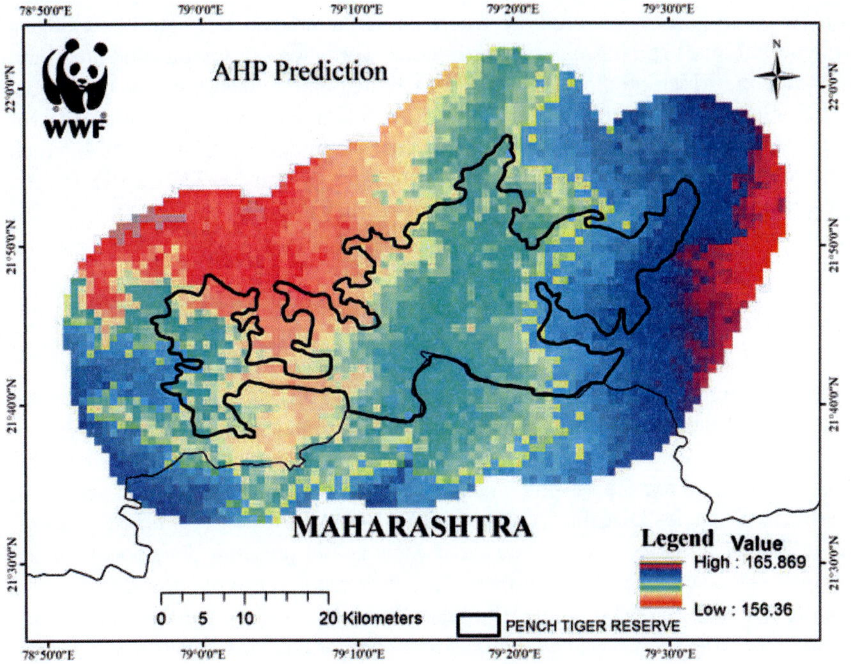

Fig. 14.19 Habitat suitability of *Cuon alpinus* in Pench Tiger Reserve, MP (AHP prediction)

directions that is in South and South-West zone but outside the protected area i.e., Pench Tiger Reserve, Madhya Pradesh boundary. Most of South zone, South-West and South-East part represents by light blue color and all these areas come under the boundary of Pench Tiger Reserve, Maharashtra which clearly indicates that not Pench Tiger Reserve, Madhya Pradesh region is suitable landscape for the habitat of wild dogs (*Cuon alpinus*) even Pench Tiger Reserve, Maharashtra have also capability to provide suitable landscape for the habitat of wild dogs (*Cuon alpinus*). Another region which is suitable landscape for the habitat of wild dogs (*Cuon alpinus*) which represented by green color, which covers the core zone of the Pench Tiger Reserve, Madhya Pradesh. Main reason of core region of the Pench Tiger Reserve, Madhya Pradesh is just suitable because that zone tree density is quite higher which create hurdle to see prey and also hunting of the prey due to barriers which are trees another reason is majority green color zone is near the Pench river and the huge Totladoh reservoir specially on southern boundary of Pench Tiger Reserve Madhya Pradesh and Pench Tiger Reserve, Maharashtra which share the Totladoh reservoir. We can see very little portion of green color zone on the western boundary i.e., in Chhindwara district region of the Pench Tiger Reserve, Madhya Pradesh. Another region which is not as such suitable landscape for the habitat of the wild dogs (*Cuon alpinus*) which represents from yellow color and occupies highest area in the among all zones of the suitable landscape for the habitat of the wild dogs (*Cuon alpinus*). Majorly

Yellow color zone occurs on those where human interference is high like villages, agricultural lands and barren lands or in other words we can say built-up areas and the last zone which predicted by AHP (Analytical hierarchical process) technique, which represents from red color which indicates highly unsuitable landscape for the habitat of the wild dog (*Cuon alpinus*) because in that zone lots of noise which created huge human disturbance and Major source of noise is Railway. In the red zone area Indian Railways spread their railway track network.

14.5 Discussion

The combine habitat suitability approach, as given in this study, is based on the integrated use of models which are purely based on surrounded in the GIS (Geographic Information System) and remote sensing environment and empirical evaluation models. GIS and remote sensing were help to produce those data which needed in the Maxent Species Distribution model and AHP (Analytical hierarchical process) technique, as a platform to execute the models and in presenting the results of the analysis. This study showed that many GIS (Geographic Information System) and remote sensing approaches and AHP (Analytical hierarchical process) technique are rapidly available for habitat suitability assessment for the wild dogs (*Cuon alpinus*). The one of the major advantages of these approaches are connected to the possibilities to consider suitable habitat factors on various scales, to combine suitability of habitat assessment for wild dogs (*Cuon alpinus*), and to empirical model and the knowledge of GIS (Geographic Information System) and remote sensing. It has been pointed out in many recent studies that habitat requirements of certain species are affected by factors measured on different scales (Areendran et al. 2020; Edenius and Elmberg 1996; Jokimäki and Huhta 1996; Saab 1999). Also, Wu and Smeins (2000) emphasized in their modeling approach the importance of consideration at various spatial temporal scales. Wu and Smeins (2000) used various levels of multiple scales in their study, but when the different scales were not to be incorporate; as an alternative, their models were constructed disassociated on various different parameters. Most of the recent suitable habitat models have been mainly constructed on a single parameter and therefore they are not as suitable as tools in large landscape management or conservation biology. In order to use the model's habitat suitability evaluation, the needed variables on appropriate parameters were calculated overall for the wild dogs (*Cuon alpinus*) in the Pench tiger reserve, Madhya Pradesh In our approach, all the environmental variables at different resolution scales.

We used the Maxent Species Distribution Model and AHP (Analytical hierarchical process) tool box for the prediction of the suitable landscape for the habitat of the wild dogs (*Cuon alpinus*) in the Pench Tiger Reserve, Madhya Pradesh. Indian landscapes have great capacity to nourish highly diverged biodiversity that why India consider as one of the mega diverse country in the world among all 12 mega diverse countries. Asiatic Wild dogs (*Cuon alpinus*) are found in huge range of South-East Asia and East Asian countries but now due to excess of anthropogenic pressure and climate change they found in very limited areas. In India they found majorly in central Indian landscapes. Southern Indian landscapes and some time in north India especially in Ladakh they sighted rarely. Wild dogs (*Cuon alpinus*) are play vital role for the balancing of the ecosystem because wild dogs (*Cuon alpinus*) are the top-level carnivore in the food web also wild dog (*Cuon alpinus*) are the only species in his genus *Cuon* in the family Canidae. In India ancient time wild dogs are found most of the part of the nation but now they are extremely confined to the protected regions of the Indian landscapes to understand what are the factors affects the habitat of wild dog (*Cuon alpinus*) in the Pench Tiger Reserve, Madhya Pradesh with the help of GIS (Geographic Information System) and Remote sensing through two different kind of models in which one is Maxent Species Distribution Model and Another one is AHP (Analytical hierarchical process) technique. We found that AHP (Analytical hierarchical process) technique give us various classes for the suitable landscape for the habitat of the wild dogs (*Cuon alpinus*) in the Pench Tiger Reserve, Madhya Pradesh but on the other hand doesn't give as much classes like AHP (Analytical hierarchical process) but Maxent species distribution model give how much Percentage contribution of each factor and give percentage permutation importance of the every factor and Major reason to accept Maxent Species distribution model is it gives Jackknifing of all environmental variables with respect to all other variable in three case. First case of jackknifing of test data gain for wild dog (*Cuon alpinus*), second case of jackknifing is jackknife of regularized training gain data for wild dog (*Cuon alpinus*) and last third case is jackknife of AUC (Area under Curve) for wild dogs (*Cuon alpinus*). Another major profit of Maxent species distribution model is gives curves show that how each environment variable affects the Maxent prediction. The curves (Fig. 14.20) shows how the predicted probability of presence changes as each environmental variable is varied, keeping all other environmental variables at their average sample value. If the variables are strongly correlated that it is hard to interpret the curviforms. Maxent species distribution model depend on correlation in evident ways in the curviforms. In the other words, curviforms shows marginally changing only and only one variable whereas variables change together in the Maxent species distribution model.

In contrary to the aforesaid marginal response curviforms, each of the following curviforms represents a unique model, a Maxent species distribution model was created using only corresponding environmental variables. The curviforms in the Maxent species distribution model depicts the dependency on predicted adaptability of the wild dogs (*Cuon alpinus*) in the Pench tiger reserve, Madhya Pradesh. The curviforms also reflect selected as well as dependent environmental variables which were weighted by correlations between the selected and other environmental variable, which will easier to interpret if there is strong correlation between the two or more. They may be easier to interpret if there are strong correlations between the variables (Fig. 14.21).

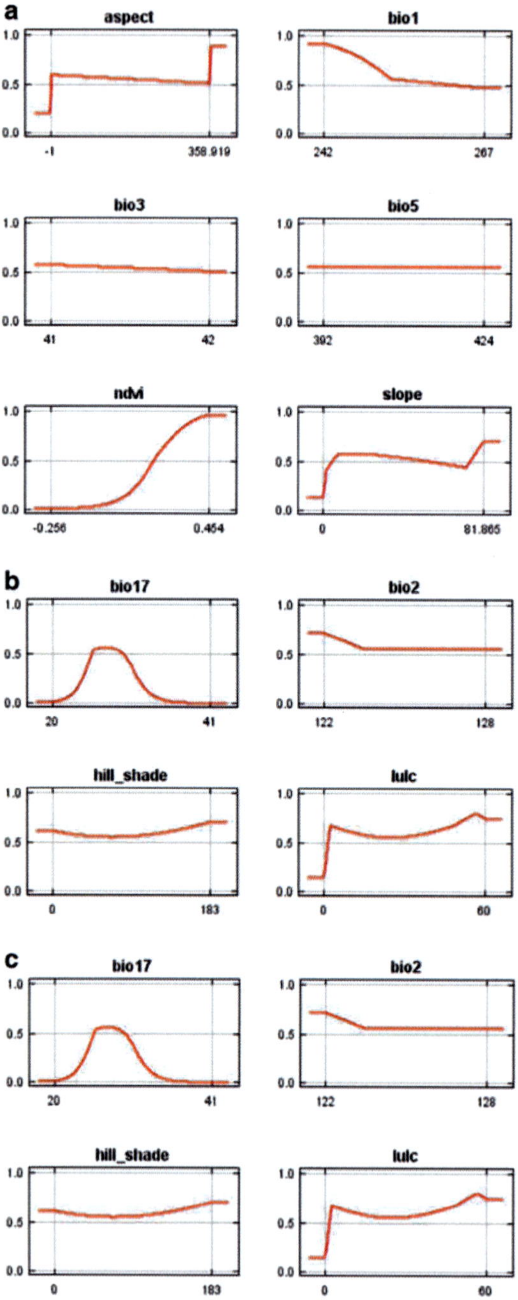

Fig. 14.20 **a** Predicted probability of presence changes as each environmental variable is varied. **b** Predicted probability of presence changes as each environmental variable is varied. **c** Predicted probability of presence changes as each environmental variable is varied

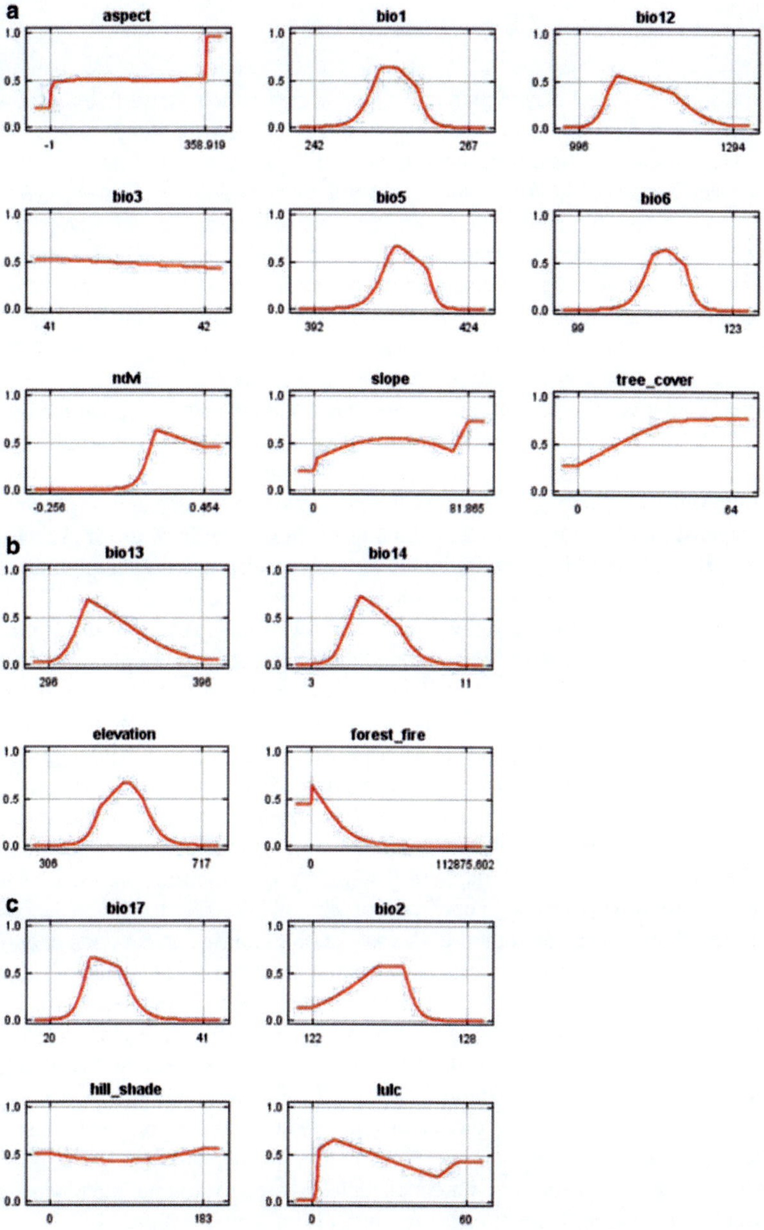

Fig. 14.21 **a** Plots reflect the dependence of predicted suitability both on the selected variable and on dependencies induced by correlations between the selected variable and other variable. **b** Plots reflect the dependence of predicted suitability both on the selected variable and on dependencies induced by correlations between the selected variable and other variable. **c** Plots reflect the dependence of predicted suitability both on the selected variable and on dependencies induced by correlations between the selected variable and other variable

14.6 Conclusion and Recommendations

This study gives a direct way to predict the landscape for the suitable habitat for wild dog (*Cuon alpinus*) with the help of geospatial techniques with comparison of AHP and Maxent model for assessing habitat suitability of wild dog (*Cuon alpinus*) in Pench Tiger Reserve, Madhya Pradesh. Several environmental variables have been used to predict suitable landscape for the suitable habitat for the wild dogs (*Cuon alpinus*). Maxent Species Distribution model and AHP technique both depicts the suitable habitat in the different classes but Maxent species distribution model gives less number of classes as compare to AHP also Maxent gives some statistical data like every environmental variable contribution for the suitable habitat for the Wild dog (*Cuon alpinus*) in the Pench tiger reserve, Madhya Pradesh and another good thing is Maxent also gives the permutation importance of every environmental factor like for wild dogs (*Cuon alpinus*) in the Pench tiger reserve, Madhya Pradesh precipitation of the driest quarter i.e., 17th Bioclimatic variable play most vital role or say most contributed and then Land use/Land cover respectively and when we talk about highest permutation importance then again precipitation of driest quarter top but now second highest is Precipitation of wettest month i.e., 13th Bioclimatic variable. As we know that monsoon season is very crucial season for the wild dogs (*Cuon alpinus*) because when rainy season starts wild dogs (*Cuon alpinus*) groups or packs become shorter in the strength and start hunting in small packs and when winter season is about to start, their denning season also start so they make small packs with females to reproduce offspring's. As per IUCN (International Union for Conservation of Nature) Canid specialist group and IUCN Red list Wild dogs (*Cuon alpinus*) are extinct from Uzbekistan, Singapore, Mongolia, Korea, Afghanistan, Tajikistan, Russian federation, republic of Kyrgyzstan, Kazakhstan and possibly extinct in the Vietnam also IUCN red list says that in some countries like Democratic People's republic of Pakistan and Korea, in those countries presence of wild dogs (*Cuon alpinus*) presence is uncertain or say not sure about their presence but on the other hand there are some South-east Asian countries and Asians countries those have Wild dogs in their forests or Protected areas like in Thailand, Myanmar, Lao People's Democratic republic, India, Cambodia, Bangladesh, Nepal, Malaysia, Indonesia, China and Bhutan. As per IUCN red list data across the globe about 950 to 2200 mature wild dogs (*Cuon alpinus*) are remained. To conserve wild dog (*Cuon alpinus*) countries government should take help from several wildlife organizations and wildlife experts to conserve them and some strict rules and policies in the favor of wild dogs (*Cuon alpinus*).

Another major work for government agencies is tried to stop people towards the deforestation or stop to destruct forest for agricultural land creation.

The results of this paper will help to overcome the finding of those landscapes which are more suitable for wild dogs (*Cuon alpinus*) in the Pench tiger reserve, Madhya Pradesh and it will to the conservation agencies or organizations as well as for the government also to especially conserve those areas which are most suitable for wild dogs (*Cuon alpinus*).

Maxent species Distribution model and AHP prediction will help to local people also because now they were aware about those places where wild dogs (*Cuon alpinus*) so that they were not going in those regions for collection of fire wood and other daily life using forest products and most important don't take their cattle's in that zone for grazing because it will very harmful for them also.

This research methodology will help in the future very well because it helps to understand to what are those factors which affect the habitat of the wild dogs (*Cuon alpinus*) in the Pench tiger reserve, Madhya Pradesh and help to overcome those factors so that in future wild dogs (*Cuon alpinus*) can conserve and another major reason to this study will help in future also because as we know only 950–2200 mature wild dogs (*Cuon alpinus*) are remain across the globe so India become prime country to conserve wild dogs(*Cuon alpinus*) in their landscape like India is very much successful to conserve tiger (*Panther tigris*) in the central Indian landscapes so it will very much helpful to government agencies which are working in the Pench tiger reserve, Madhya Pradesh.

Every research have certain limitations like this research is entirely based on present scenario but it helps many years because ecosystem and its constituents takes time to change their conditions but future prospect of this research is help to overcome or conserve those environmental factors which will affect currently habitat of wild dogs (*Cuon alpinus*) and enhance their population to stabilize ecosystem or food web in the Pench tiger reserve, Madhya Pradesh because as we already mention that according to all India tiger report 2018 Pench tiger reserve, Madhya Pradesh has the highest herbivore animal density so wild dogs (*Cuon alpinus*) plays vital role to stabilize their population.

Acknowledgements I am also obliged to Arun Kumar (Programme Officer), Akhil Sivadas (Programme Officer), Samrat Deb (Programme Officer), and Kumar Ranjan (Programme Officer) from, IGCMC, WWF-India for helping me throughout the project duration. I would also like to extend my thanks to Mr. Rajeev Kumar (Senior Programme Officer) and Mr. Sandeep Kumar (Programme Officer) from ENVIS, IGCMC, WWF-India for their constant encouragement. I would also like to thank Mr. Ravi Singh, CEO, WWF India and Dr. Sejal Wohra, Programme Director, WWF-India for allowing me to carry out my project in the esteemed organization. I would also like to extend my gratitude to the library staff of WWF-India for helping me with the resources for the literature review.

References

Areendran G, Sahana M, Raj K, Kumar R, Sivadas A, Kumar A, Gupta VD (2020) A systematic review on high conservation value assessment (HCVs): challenges and framework for future research on conservation strategy. Sci Total Environ 709:135425

Bowdich TE (1821) An analysis of the natural classification of mammalia for the use of students and travellers. Printed by J. Smith, Paris

Cohen JA (1977) A review of the biology of the Dhole or Asiatic wild dog (*Cuon alpinus*, Pallas). Anim Regul Stud 1:141–158

Cohen JA (1978) Cuon alpinus, Mammalian species, Issue 100, 21 September 1978, Pages 1–3 https://doi.org/10.2307/3503800

Cohen JA, Fox MW, Johnsingh AJT, Barnett BD (1978) Food habits of Dhole in South India. J Wildl Manage 42(4):933–936

Davidar ERC (1975) Ecology and behaviour of the dhole or Indian wild dog (Cuon alpinus Pallas). In: Fox MW (ed) The wild canids, Van Nostrand Reinhold, New York, pp 109–119

Dunbar Brander AA (1982) Wild animals in central India. Central Publication, India, Madhya Pradesh

Edenius L, Elmberg J (1996) Landscape level effects of modern forestry on bird communities in North Swedish boreal forests. Landscape Ecol 11:325–338

Fox MW (1984) The whistling hunters: field studies of the asiatic wild dog (Cuon alpinus). State University of New York Press, Albany, ISBN 0-87395-843-8.

Fox M, Johnsingh AJT (1975) Hunting and feeding in wild dogs. J. Bombay Nat Hist Soc 72(2):321–326. BIBL. 1 REF

Hayward MW, Lyngdoh S, Habib B (2014) Diet and prey preferences of wild dogs (*Cuon alpinus*): dietary competition within Asia's apex predator guild. J Zool 0952–8369. https://doi.org/10.1111/jzo.12171

Hodgson S (1838) Truths from the West Indies: including a sketch of maderia in 1833. Ball W

https://biodiversityinformatics.amnh.org/open_source/maxent/. Accessed on 2nd Aug 2019

https://currentaffairs.gktoday.in/tags/pench-tiger-reserve. Accessed on 5th Aug 2019

https://data.globalforestwatch.org/datasets/63f9425c45404c36a23495ed7bef1314. Accessed on 26th June 2019

https://earthenginepartners.appspot.com/science-2013-global-forest/download_v1.2.html. Accessed on 20th July 2019

https://earthenginepartners.appspot.com/science-2013-global-forest/download_v1.2.html. Accessed on 31st July 2019

https://earthexplorer.usgs.gov/. Accessed on 30th June 2019

https://vajiramias.com/current-affairs/pench-tiger-reserve/5c5003431d5def6d7fab5e89/. Accessed on 5th Aug 2019

https://wwf.panda.org/our_work/wildlife/profiles/mammals/dhole/. Accessed on 15th July 2019

https://www.penchnationalpark.com/. Accessed on 28th June 2019

https://www.penchnationalpark.com/about-pench.html. Accessed on 28th June 2019

https://www.penchtiger.co.in/. Accessed on 28th June 2019

https://www.penchtiger.co.in/conservation_history.html. Accessed on 28th June 2019

https://www.penchtiger.co.in/geology_rock_soil.html. Accessed on 28th June 2019

https://www.penchtiger.co.in/geomorphology.html. Accessed on 28th June 2019

https://www.penchtiger.co.in/history.html. Accessed on 28th June 2019

https://www.penchtiger.co.in/location.html. Accessed on 28th June 2019

https://www.wildcanids.net/dhole.html. Accessed on 6th Aug 2019

Jaynes ET (1957) Information theory and statistical mechanics. Phys Rev J Arch. American Physical Society. https://doi.org/10.1103/PhysRev.106.620

Johnsingh AJT (1989) The threatened gallery forest of the river Tambiraparani, Mundanthurai wildlife sanctuary, South India, Biol Conser 47(4):273–280. https://doi.org/10.1016/0006-3207(89)90070-0

Johnsingh AJT (1982) Reproductive and social behavior of the dhole, *Cuon alpinus* (Canidae). J Zool Lond 198:443–463

Jokimäki J, Huhta E (1996) Effects of landscape matrix and habitat structure on a bird community in Northern Finland: a multi-scale approach. Ornis Fennica 73:97–113

Kanchanasakha B, Simcharoen S, Than UT (1998) Carnivores of Mainland, South East Asia. Siam Tong Printing Co. Ltd., Thailand

Khajuria H (1963) The wild dog [*Cuon alpinus* (PALLAS)] and the tiger [*Panther tigris* (LINN)]. J Bombay Nat Hist Soc 60:448–449

Krishnan M (1972) An ecological survey of the larger mammals of peninsular India. J Bombay Nat Hist Soc 69(1):26–54
Majumdhar A et al (2011) Occupancy and abundance of dhole (Cuon alpinus) in Pench landscape of central India. J Bombay Nat His Soc 108(3)
Müller-Using D (1975) The red dog. In: Grzimek B (ed) Grzimek's animal life encyclopedia, vol 12. Van Nostrand Reinhold, New York, pp 264–266
Novikov GA (1962) Carnivorous mammals of the fauna of USSR. Israel Program Sci. Transl., Jerusalem
Pokhriyal P, Rehman S, Areendran G, Raj K, Pandey R, Kumar M, Sajjad H (2020) Assessing forest cover vulnerability in Uttarakhand, India using analytical hierarchy process. Model Earth Syst Environ 1–11
Prater SH (1965) The book of Indian animals, 2nd (rev) ed. Bombay Natural History Society and Prince of Wales Museum of Western India
Saab V (1999) Importance of spatial scale to habitat use by breeding birds in riparian forests: a hierarchical analysis. Ecol Appl 9:135–151
Sahana M, Ganaie TA (2017) GIS-based landscape vulnerability assessment to forest fire susceptibility of Rudraprayag district, Uttarakhand, India. Environ Earth Sci 76(20), 676
Sahana M, Hong H, Sajjad H, Liu J, Zhu AX (2018) Assessing deforestation susceptibility to forest ecosystem in Rudraprayag district. India using fragmentation approach and frequency ratio model. Sci Total Environ 627:1264–1275
Sheldon JW (1992) Wild dogs, the natural history of nondomestic Canidae. Academic Press Inc., San Diego, California, p 92101
Sosnovskii IP (1967) Breeding the red dog or Dhole (*Cuon alpinus*) at Moscow Zoo. Int Zoo Yearb 7:120–122
Swets JA (1988) Measuring the accuracy of diagnostic systems. Sci 240(4857):1285–1293
Steinmetz R, Seuaturien N, Chutipong W (2012) Tigers, leopards, and wild dogs (*Cuon alpinus*) in a half-empty forest: assessing species interactions in a guild of threatened carnivores. Biol Conserv xxx(2013):xxx–xxx
Thenius E (1954) ZurAbstammung der Rotwölfe (Gattung *Cuon* HODGSON). Osterr Zool Z 5:377–378
Wu B, Smeins F (2000) Multiple-scale habitat modeling approach for rare plant conservation. Landscape Urban Plan 51:11–28
Venkataraman AB, Arumugam R, Sukumar R (1995) The foraging ecology of dhole (Cuon alpinus) in Mudumalai sanctuary, Southern India. J Zool 237 (4):543–561
Yadav N, Areendran G, Sarma K, Raj K, Sahana M (2020) Susceptibility assessment of human–leopard conflict in Aravalli landscape of Haryana using geospatial techniques. Model Earth Syst Environ 1–15

Chapter 15
Assessment of Forest Cover Dynamics using Forest Canopy Density Model in Sali River Basin: A Spill Channel of Damodar River

Asish Saha, Manoranjan Ghosh, Subodh Chandra Pal, Indrajit Chowdhuri, Rabin Chakrabortty, Paramita Roy, Biswajit Das, and Sadhan Malik

Abstract In a spatio-temporal scale, changing conditions of forest land cover and its detection study is an important concern for sustainable forest management. Nowadays, the forest canopy density (FCD) model has been used for the analysis and management of forest resources through identifying the forest gap areas where afforestation should be started immediately. The present study applied FCD model to detect changes in forest land cover in Sali River basin between the years 2000 and 2018. Moreover, the vegetation indices like Bareness Index (BI), Greenness Vegetation Index (GVI), Normalized Difference Vegetation Index (NDVI), Perpendicular Vegetation Index (PVI), and Shadow Index (SI) along with weighted overlay analysis have been used to prepare FCD map of the Sali river basin. It has been noticed from FCD map that south and north-eastern part of the study area covered with high canopy density in comparison with north and north-western region in the year 2000. Whereas, in the year of 2018, high FCD has been found in the middle portion of the southern region and the rest of the area varies from low to medium FCD.

Keywords Forest canopy density · Sali River · Normalized difference vegetation index · Perpendicular vegetation index

15.1 Introduction

The natural vegetation or forest is essentially maintaining the rich terrestrial biodiversity, sustaining of human lives, proving resources and sustaining the contemporary environmental problems. In addition, forest covers are also regulating the impact of

A. Saha · S. C. Pal (✉) · I. Chowdhuri · R. Chakrabortty · P. Roy · B. Das · S. Malik
Department of Geography, The University of Burdwan, Burdwan, West Bengal, India
e-mail: geo.subodh@gmail.com

M. Ghosh
Rural Development Centre, Indian Institute of Technology Kharagpur, Kharagpur, India

P. K. Shit et al. (eds.), *Spatial Modeling in Forest Resources Management*, Environmental Science and Engineering,
https://doi.org/10.1007/978-3-030-56542-8_15

global warming, enriching the level of groundwater and playing a role of carbon sink area for maintaining the balance of atmospheric Carbon-dioxide (FAO 2014; Halliday et al. 2003). However, the area under forest cover is gradually decreasing mainly due to human interventions. It has been estimated that out of the earth's total land surface area, around 31% is covered with forest area; whereas, in case of India, it has been estimated that only 24% of country's area is under forest (FAO 2010). The global forest area has been losing its area at the rate of 0.08% per year (FAO 2015). Therefore, the conservation and management of forest areas have become very important to solve the above mentioned problems.

In a simple way, forest canopy or crown cover can be defined as the topmost layer of vegetation which covered the ground surface by the vertical projection of vegetation canopy or tree crowns (Howard 1991; Jennings et al. 1999). The status of forest canopy can be expressed in the physiognomic character of forest (Joshi et al. 2006; Nandy et al. 2003). The forest canopies have some environmental importance, such as providing cooling in the hot summer season, maintaining water and air quality and contributing a comfortable environment (Morrow et al. 2001). As mentioned earlier, the forest area is degrading due to natural and anthropogenic activities day by days and the large forest area becoming into smaller isolated patches which are known as forest fragmentation (Haila 1999). Therefore, canopy density monitoring is a prerequisite condition for sustainable forest management at various scales such as local, regional, national and global (Nandy et al. 2003; Sahana et al. 2015).

However, to assess the quality and quantity of forest cover is an important part of sustainable forest management; although it is a time taking and laborious work (Zhang et al. 2003). To overcome these problems, satellite remote sensing based forest canopy density (FCD) model helps us to monitor and assess the forest growth and forest health over the time (Mon et al. 2012; Azizi et al. 2008). Rikimaru et al. (2002) have mentioned that the FCD model could also use for identification the forest fragmentation areas. In addition, Pal et al. (2018) have argued that FCD can play a significant role for the planning and implementation of afforestation programmer.

In India, to monitor the afforestation and conservation practices of vegetation, every state government has set up a Remote Sensing-GIS cell under the 'Green India' mission (Raha 2014); although state government only monitor the quantitative area of forest cover. The present paper focuses on spatio-temporal assessment of forest cover based on advance index based methodology. However, there is various index based method for forest cover monitoring using satellite images (Cross et al. 1991; Boyd et al. 2002; Rikimaru et al. 2002; Joshi et al. 2006). The FCD model is widely used technique estimating the forest cover areas from remotely sensed satellite data (Baynes 2004; Lee and Lucas 2007; Mon et al. 2012; Rikimaru 1996). The FCD model is continuously used in the tropical forest areas to monitoring the forest health, forest coverage, afforestation, and deforestation (Azizi et al. 2008; Wang and Brenner 2009; Deka et al. 2013; Mensah et al. 2017). In India, some study has also been done using FCD in different parts of the country (Nandy et al. 2003; Prasad et al. 2009; Pal et al. 2018; Malik et al. 2019a, b); however, there are still significant lacks of studies in a micro-scale. Therefore, we choose Sali River basin as a case study area to identifying the changes of the vegetation cover during the period of 2000 to 2018 by using FCD model. The study used different types of vegetation indices, such as Bareness Index (BI), Greenness Vegetation Index (GVI), Normalized Difference Vegetation Index (NDVI), Perpendicular Vegetation Index (PVI), and Shadow Index (SI) using Landsat TM satellite imagery to understand the forest cover dynamics. Finally, the method of weighted overlay analysis has been used to prepare the FCD map of the present study area.

15.2 Materials and Methods

15.2.1 Study Area

Sali River basin, a spill channel of Damodar River in West Bengal has chosen as the study area. According to the Soil and Land Use Survey of India (SLUSI), Sali River basin has been codified by 2A2D4 alpha-numeric codification system (CWC and NRSC 2014). The study area has bounded from 23° 11′ 11″N to 23° 26′ 30″N latitudes and 87° 2′ 14″E to 87° 38′ 26″E longitudes with an area of 757.29 km^2 (Fig. 15.1). The entire study area belongs to the dry sub-humid tropical climatic with

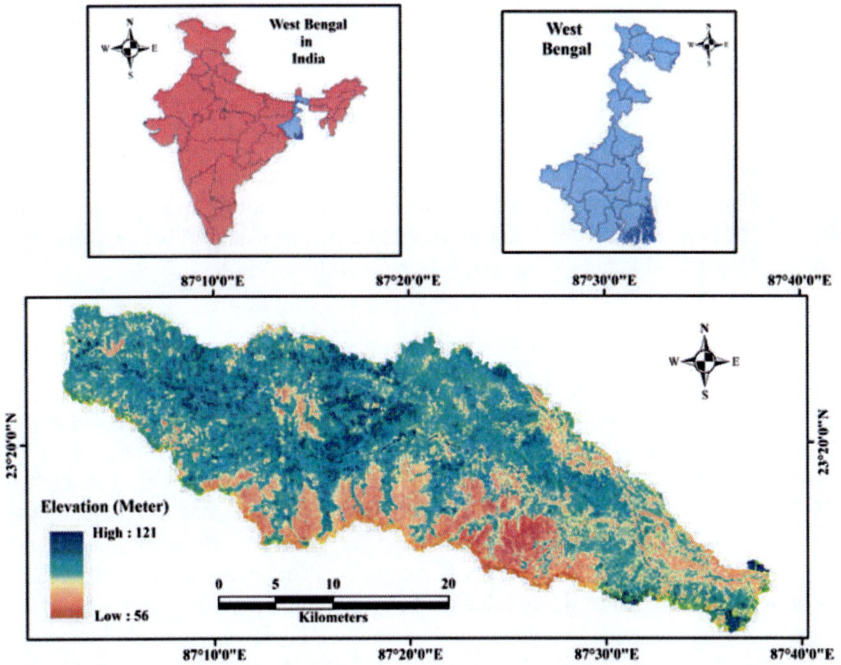

Fig. 15.1 Location map of the study area

a yearly rainfall of 1480.62 mm; the maximum and minimum temperature of the study area is 45 °C and 10 °C respectively (Malik et al. 2019a, b). Geologically, Sali River basin has classified into two geomorphic categories i.e. the upper part of the basin area represents Cenozoic laterite with Holocene sediment and the lower part of basin area represent middle Pleistocene to Holocene alluvium (Geological Survey of India 2003). Topographically, the upper part of the study area is undulating in nature covered by lateritic soil with the supremacy of Sal forest (O'Malley 1908).

15.2.2 Data Source

Satellite remote sensing provides the most efficient and foremost techniques for estimating the FCD than the traditional ground monitoring method (Blodgett et al. 2000). In this study, first Shuttle Radar Topographic Mission (SRTM)-30 m DEM data was used for extraction of the Sali River basin. Thereafter, DEM based prepared river basin was verified with topographical maps of 73 M/3, 73 M/7, 73 M/11, and 73 M/12 on 1:50,000 scale published by Survey of India (SOI). Finally, the Landsat Thematic Mapper (Landsat-TM) satellite images of 2000 and 2018 were used for monitoring and analysis the FCD.

15.2.3 Methods

Different types of vegetation indices like NDVI, BI, GVI, PVI, and SI have been considered for assessing the FCD of Sali River basin whereas, the NDVI is a reliable and widely used method to categorize the forest and non-forest areas (Telesca et al. 2008; Verbesselt et al. 2006; Avtar et al. 2014). Rouse et al. (1974) proposed the following equation to calculate the NDVI (Eq. 15.1).

$$NDVI = \frac{(NIR - RED)}{(NIR + RED)} \tag{15.1}$$

The BI is a normalized index, which helps indiscriminate between the background properties of soil such as completely bare, sparse bare and dense canopy with its spectral vegetation signature (Roy et al. 1996; Sahana et al. 2015; Pal et al. 2018). The main aim of the development of the Bareness index (BI) was to recognize the urban and open surface areas (Malik et al. 2019a). As the forest density decreases, BI value increases and simultaneously soil exposure also increases (Rikimaru et al. 2002; Baynes 2004). The BI has calculated by using the following Eq. 15.2.

$$BI = \frac{(Band5 + Band3) - (Band4 + Band1)}{(Band5 + Band3) + (Band4 + Band1)} * 100 + 100 \tag{15.2}$$

Further, GVI is one of the major vegetation indices to determine forest health status. The GVI index was developed by Kauth and Thomas (1976), using Landsat MMS data. The increasing value of GVI indicates very good greenery cover areas and the lower value of GVI indicates just the opposite condition (Malik et al. 2019a). The GVI has calculated by using the following Eq. 15.3.

$$GVI = [(0.2728 * Band2) - (0.2174 * Band3) - (0.5508 * Band4) + (0.7221 * Band5) + (0.0733 * Band6) - (0.1648 * Band7)] \quad (15.3)$$

The PVI is generally used to determine the quantity of vegetation cover. Richardson and Wiegand (1977) have applied the PVI is an indicator for viewing the development of plant density. The value of PVI depends on soil surface brightness and it is the significant parameters to differentiate the spatial vegetation cover (Pal et al. 2018). It is a well known fact that if PVI value is high, than it represents high brightness of the soil surface with a very good quality vegetation cover; on the other side, low PVI value indicates low brightness of soil surface and poor vegetation cover (Huete et al. 1997; Malik et al. 2019a). The PVI value of the present study area has calculated by using the following Eq. 15.4.

$$PVI = \sqrt{[\{(0.355 * Band5) - (0.149 * Band4)\}^2 - \{(0.355 * Band4) - (0.852 * Band5)\}^2]} \quad (15.4)$$

The SI generally developed for the detection of shadow and to determine the fluctuation of canopy level (Liu and Yamazaki 2012; Pal et al. 2018). One of the major advantages of this index is that the index can differentiate between the amount of vegetation canopy and ground level. SI is an essential vegetation index to understand the vegetation characteristics such as vegetation types, its growth, vegetation gap area, and its classification (Ono et al. 2010; Andersen et al. 2005). The SI has calculated by using the following Eq. 15.5.

$$SI = \sqrt{\{(256 - Band2) * (256 - Band3)\}} \quad (15.5)$$

Multi criteria decisions analysis method is very important method to assign weight for factors and sub factors and also to take decisions for many problems. This type of multi criteria decisions making methods have been used to mitigate the different problems in earth science like landslide susceptibility, groundwater potentiality, and flood susceptibility analysis with the help of Geographical information system (GIS) (Chakrabortty et al. 2018; Chowdhuri et al. 2020; Das et al. 2019; Pal and Chowdhuri 2019; Pal et al. 2019). Analytical hierarchical process (AHP) is the scientific decisions technique and it is applied in FCD in this study. Each and every factor has been compared through one to nine Saaty's scale based on the characteristics of various

Table 15.1 Pairwise comparison matrix, weights and consistency ratio of five thematic layers

Theme	NDVI	GVI	PVI	BI	SI	Weights
NDVI	1	1	2	3	4	0.28
GVI	1/1	1	2	5	2	0.22
PVI	1/2	1/2	1	4	2	0.22
BI	1/3	1/5	1/4	1	1	0.11
SI	1/4	1/2	1/2	1/1	1	0.17

Principal Eigen value (λmax) = 5.154
Consistency Ratio (CR) = 0.034
Consistency index (CI) = 0.0385
Random Consistency Index (RI) = 1.132

indices and their classes accordingly on the basis of literature review and field knowledge of the researcher (Saaty 1988). Thereafter, weights of different categories of vegetation indices have been calculated which is shown in Table 15.1. Consistency index (CI), Random Consistency index (RI), Eigen value ($\gamma_{\max}$) and Consistency ratio (CR) have showed the weights have fit for this FCD model (Eqs. 15.6 and 15.7). Rank and weight method have been applied for sub classes of every factor which is shown in Table 15.2. Finally, the following Eq. 15.8 (Pal et al. 2018) has been used to prepare the FCD map of the present study area of Sali River basin to analysis the FCD changes between the years 2000 and 2018.

$$CR = CI/RI \tag{15.6}$$

$$CI = \frac{(\gamma_{\max} - n)}{(n - 1)} \tag{15.7}$$

where, n is number of factors.

$$\begin{aligned} FCD = \{&(NDVI_W * NDVI_{wi}) + (BI_w * BI_{wi}) \\ &+(GVI_W * GVI_{wi}) + (PVI_w * PVI_{wi}) + (SI_w * SI_{wi})\} \end{aligned} \tag{15.8}$$

where, w = weight of a theme and wi = weight of individuals of a theme.

Table 15.2 Assigned weights of five thematic layers and their classes

Theme	Class	Weight (w)	Assigned weight (wi)	Normalised weight
NDVI	Very high	0.28	5	0.33
	High		4	0.27
	Medium		3	0.20
	Low		2	0.13
	Very low		1	0.07
GVI	Very high	0.22	5	0.33
	High		4	0.27
	Medium		3	0.20
	Low		2	0.13
	Very low		1	0.07
PVI	Very high	0.22	5	0.33
	High		4	0.27
	Medium		3	0.20
	Low		2	0.13
	Very low		1	0.07
BI	Very high	0.11	1	0.07
	High		2	0.13
	Medium		3	0.20
	Low		4	0.27
	Very low		5	0.33
SI	Very high	0.17	5	0.33
	High		4	0.27
	Medium		3	0.20
	Low		2	0.13
	Very low		1	0.07

15.3 Results and Discussion

15.3.1 Normalized Difference Vegetation Index

In the Sali River basin, NDVI values ranges from −0.463 to 0.649 for the year of 2000; which is classified into five categories such as very low (−0.463 to 0.001), low (0.001–0.085), medium (0.085–0.100), high (0.100–0.185) and very high (0.185–0.649) (Fig. 15.2). The very low NDVI values indicate surface without vegetation cover and present of water bodies, whereas the very high NDVI value indicates high-density forest as land cover. The high to very high NDVI values has found in the north-eastern, eastern and southern part of the study area which is mainly dominated by *Shorearobusta* trees. Medium to low NDVI values was dominated in the northern

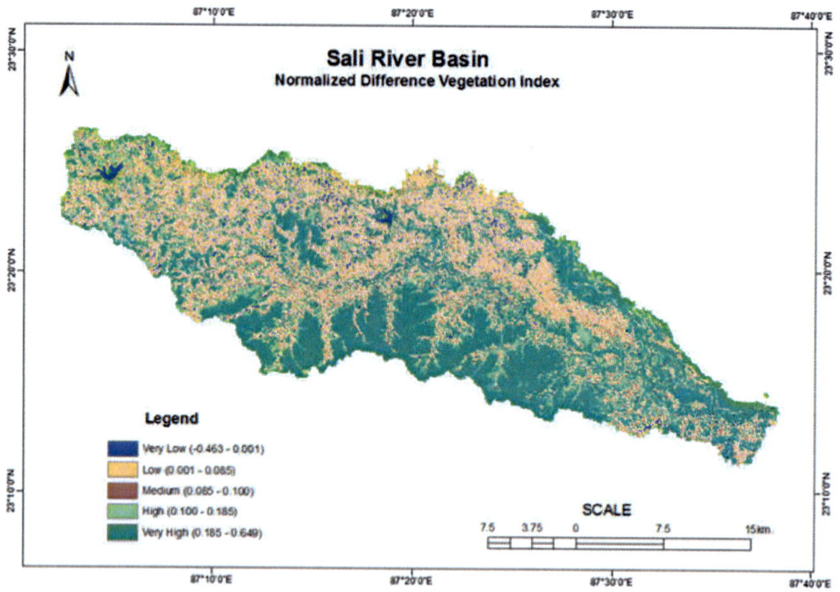

Fig. 15.2 NDVI map of Sali River basin for the year of 2000

and western part of the study area; this area is mainly dominated by agricultural land with grassland. The very low NDVI values were found in some isolated patches throughout the study area, this is the area of water bodies and bare soil surface.

On the other hand, NDVI map of the Sali River basin for the year 2018 has not as many changes compared to the year 2000. Here, also NDVI values classified into five categories as described for the year of 2000, and NDVI values range from −0.463 to 0.649 (Fig. 15.3).

15.3.2 Bareness Index

BI distinguishes between the surface covered with vegetation and bare soil area. Generally, very low BI values indicate forest cover area, moderate BI values indicates fallow agricultural land with shrubs, and high BI values indicate bare sandy surface with dry fallow land. In the Sali River basin, for the year of 2000, BI values range from 0 to 0.716 (Fig. 15.4) and the BI values have classified into five categories i.e. very low (0–0.259), low (0.259–0.344), medium (0.344–0.371), high (0.344–0.456) and very high (0.456–0.716). The high to very high BI was found in the southern, north-eastern and some isolated part of the north-eastern part of Sali River basin, which is mainly covered by dry fallow land and water bodies. The very low to low BI was found in the northern and throughout the north-western and south-western part of the basin area, which is mainly covered by *Shorearobusta* forest.

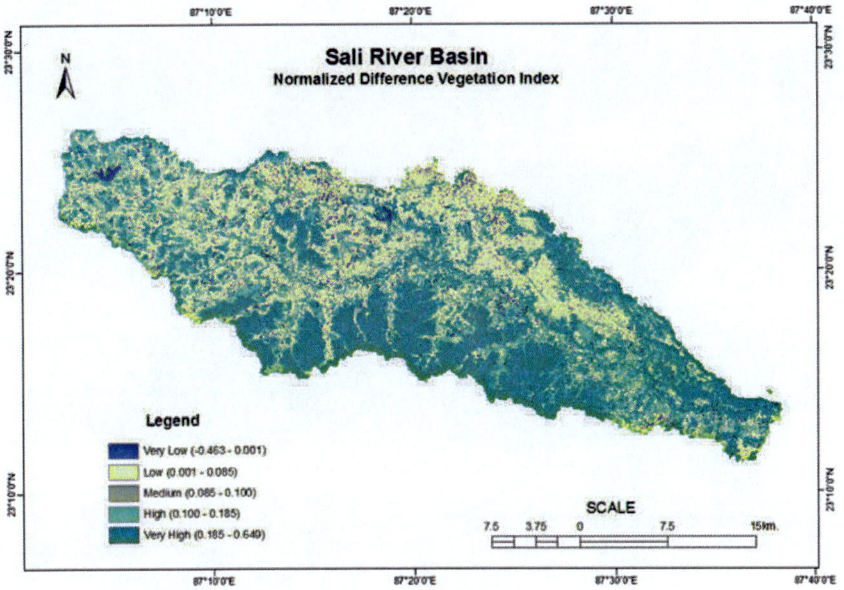

Fig. 15.3 NDVI map of Sali River basin for the year of 2018

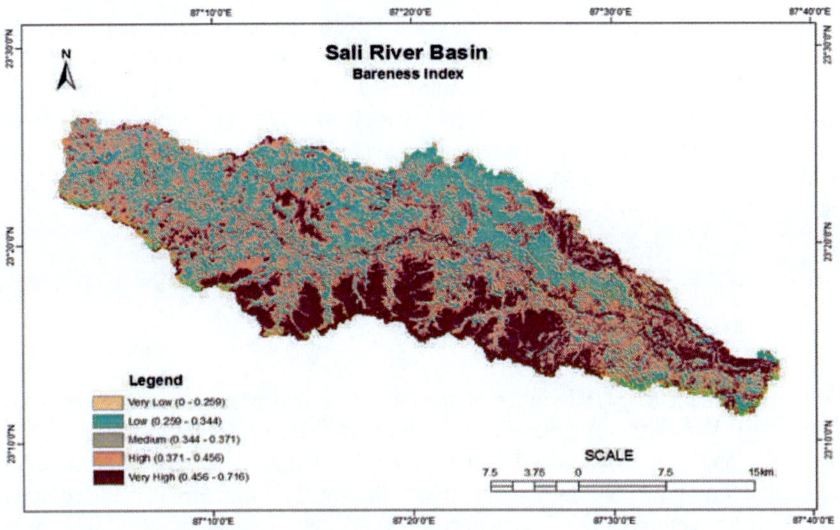

Fig. 15.4 BI map of Sali River basin for the year of 2000

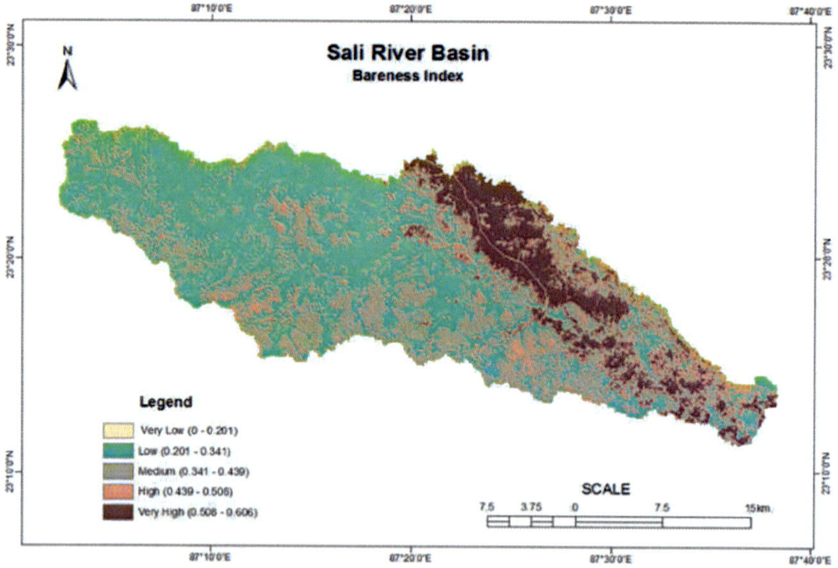

Fig. 15.5 BI map of Sali River basin for the year of 2018

However, on the other side, for the year of 2018, the BI values range from 0 to 0.606 (Fig. 15.5). Here also BI values have classified into five categories i.e. very low (0–0.201), low (0.201–0.341), medium (0.341–0.439), high (0.439–0.508) and very high (0.508–0.606). Here, high to very high BI was found in the patches from north to the south-eastern portion which is covered by bare soil surface with fallow land. The very low to low BI was found in the northern, western, and southern part of the Sali River basin covered by Sal dominated forest area.

15.3.3 Greenness Vegetation Index

For the year of 2000, GVI values of the Sali River basin ranges from 0 to 96.90 (Fig. 15.6). Very high GVI was found in the north-eastern, southern and along the river side zones which indicates the areas covered with high vegetation. Medium GVI was found throughout the study area with dominancy of the northern part and covered by mainly agricultural areas. The values of low GVI was found in isolated areas over the study area which represents mainly water bodies with very low chlorophyll and biomass content. On the other hand, for the year of 2018, GVI values range from 0 to 116.68 (Fig. 15.7). Very high GVI was found in the north-eastern and southern part with high vegetation-covered areas. Medium GVI was found in the northern, eastern and middle southern part with dominancy of agricultural areas. And low GVI was found in a small number of isolated areas with coverage of water bodies.

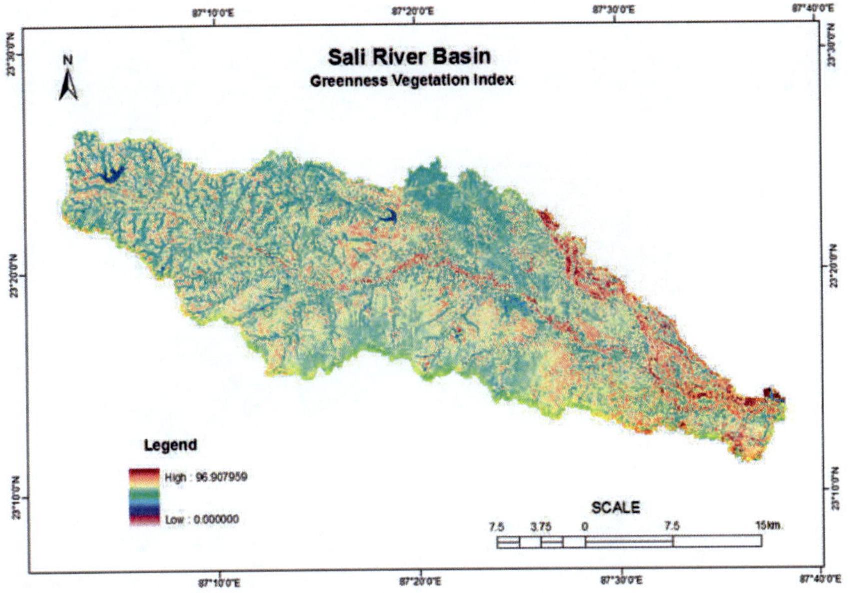

Fig. 15.6 GVI map of Sali River basin for the year of 2000

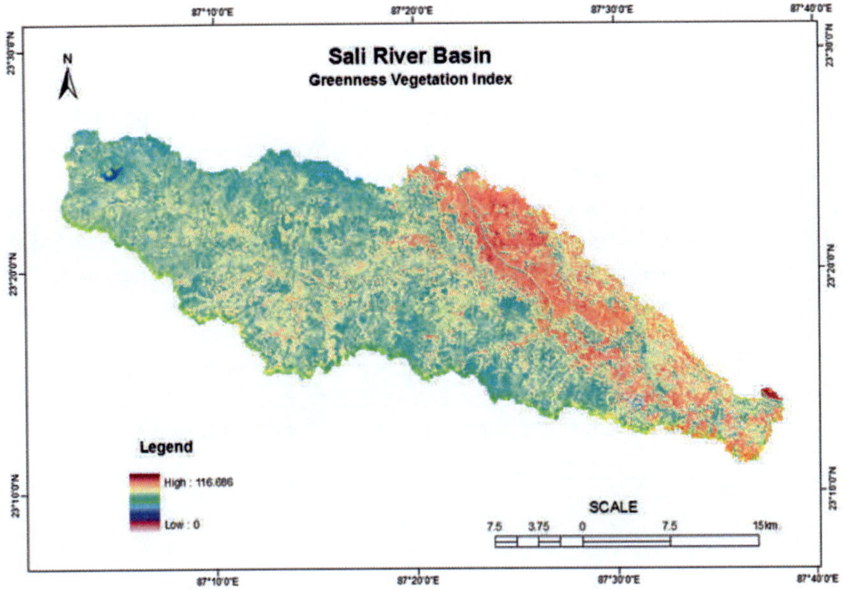

Fig. 15.7 GVI map of Sali River basin for the year of 2018

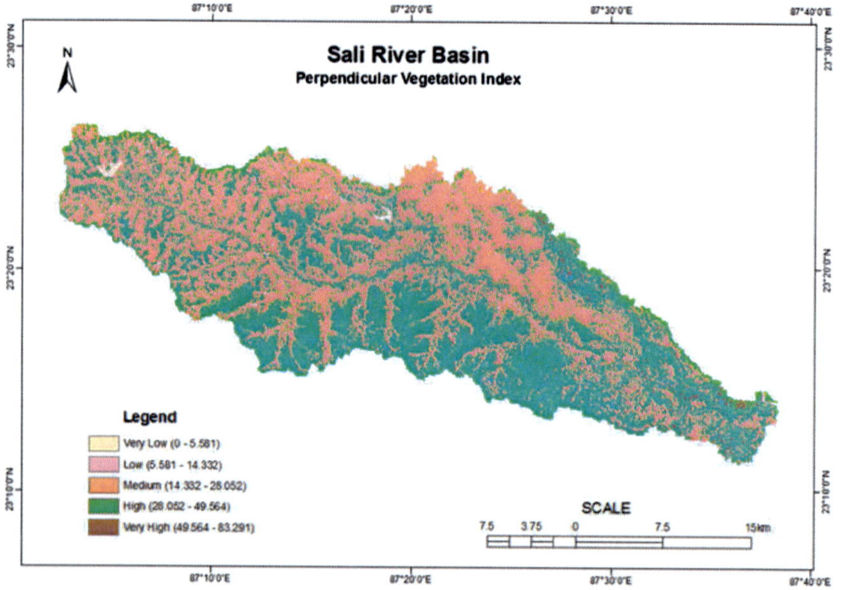

Fig. 15.8 PVI map of Sali River basin for the year of 2000

15.3.4 Perpendicular Vegetation Index

For the year of 2000, PVI values of the Sali River basin ranges from 0 to 83.29 (Fig. 15.8). Here, PVI values have classified into five categories i.e. very low (0–5.581), low (5.581–14.332), medium (14.332–28.052), high (28.052–49.564) and very high (49.564–83.291). Very low to low PVI values was found almost half of the basin area with dominancy in the northern and north-western portion, covered by mainly water bodies, settled areas with the dry bare surface. High to very high PVI values concentrated mainly some isolated patches in the northern, north-western, southern, and north-eastern portion, covered by Sal dominated forest areas. On the other hand, for the year of 2018, there are not as many changes in the PVI map in comparing to the year 2000. Here also PVI values range from 0 to 83.29 (Fig. 15.9) and classified them into five categories as described for the year of 2000.

15.3.5 Shadow Index

In the present study area of Sali River basin, for the year of 2000, SI values range from 173.65 to 256 (Fig. 15.10). SI values have classified into five categories i.e. very low (173.65–214.499), low (214.499–220.499), medium (220.499–226.499), high (226.499–236.486) and very high (236.486–256). Very low SI value was found

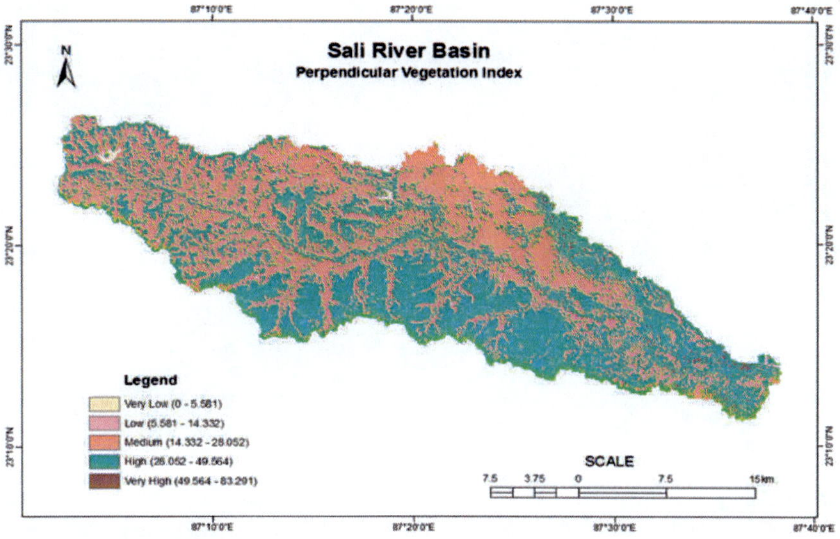

Fig. 15.9 PVI map of Sali River basin for the year of 2018

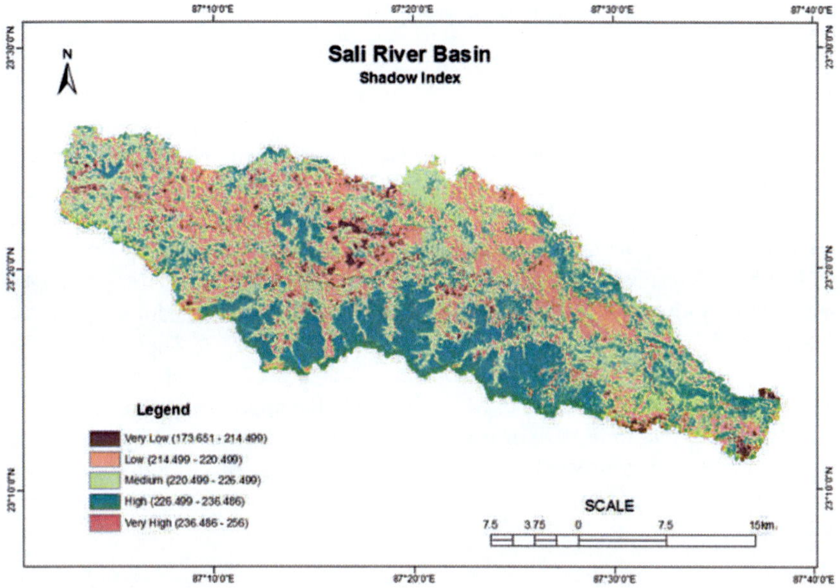

Fig. 15.10 SI map of Sali River basin for the year of 2000

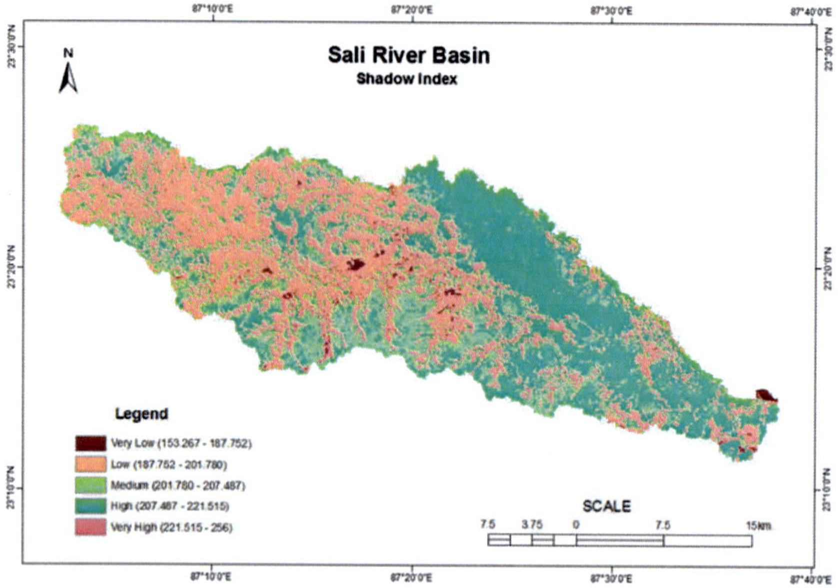

Fig. 15.11 SI map of Sali River basin for the year of 2018

within some isolated patches throughout the basin area and covered by mainly water bodies and sandy areas. A Low SI value was concentrated dominantly in the north-western, central and the direction from north-eastern to south-eastern part and this class is associated with the settled area and moist land. High to very high SI class was found in the southern and north-eastern portion of the basin area and this class is associated with forest areas. Similarly, for the year of 2018, SI values have classified into five categories and the values range from 153.267 to 256 (Fig. 15.11). These classes are very low (153.267–187.752), low (187.752–201.780), medium (201.780–207.487), high (207.487–221.515) and very high (221.515–256). The class of very low SI was found in the central part and some isolated pockets in the southern and south-eastern part. The low class of SI was found in the upper part of the basin, and some patches in the north-eastern and south-eastern part. High to very high class of SI was found dominantly in the north-eastern, south-eastern and some patches of south and north-western part. These classes are associated with forest covered with Sal trees.

15.3.6 *Forest Canopy Density*

FCD map of the Sali River basin was prepared based on weighted overlay analysis in GIS platform by using five vegetation indices i.e. Bareness Index (BI), Greenness

Vegetation Index (GVI), Normalized Difference Vegetation Index (NDVI), Perpendicular Vegetation Index (PVI), and Shadow Index (SI). The weights of different thematic layers and their classes have been assigned accordingly based on the field and basic knowledge (Table 15.1). Normalized weight and the geometric mean of each index was also calculated (Table 15.2).

For the year of 2000, FCD map (Fig. 15.12) of Sali River basin have been classified into five categories i.e. very low (0.088–0.127), low (0.127–0.155), medium (0.155–0.174), high (0.174–0.187) and very high (0.187–0.206) and covering the area of 3.64%, 30.67%, 12.43%, 20.14% and 33.12% respectively. On the other side, for the year of 2018, FCD map (Fig. 15.13) of Sali River basin have been classified into five categories as described for the year of 2000; but there was a significant change in the percentage of forest cover area. The covering area of very low, low, medium, high and very high zone is 5.72%, 32.67%, 14.98%, 17.23% and 29.41% respectively. Therefore, it is noticed that high and very high FCD areas decreased gradually from 2000 to 2018 and low to medium FCD zone areas increasing gradually due to the natural as well as human interference on forest degradation. The correlation matrix map for the year of 2000 (Fig. 15.14) and 2018 (Fig. 15.15) between the different vegetation indices show the understanding of the importance to prepare the FCD map of Sali River basin.

In the year of 2000, the correlation matrix map (Fig. 15.14) among the different vegetation indices shows that NDVI positively correlated with BI and PVI, and moderately correlated with GVI and SI. BI also positively correlated with NDVI,

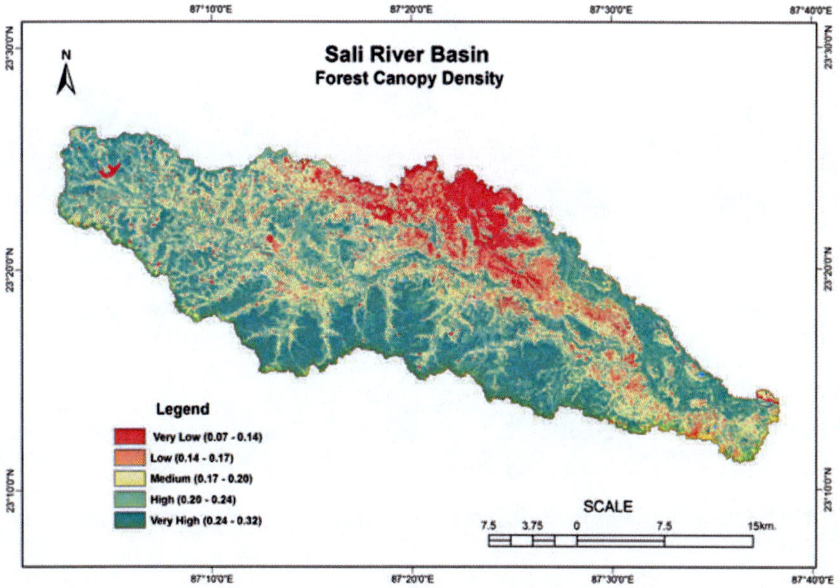

Fig. 15.12 FCD map of Sali River basin for the year of 2000

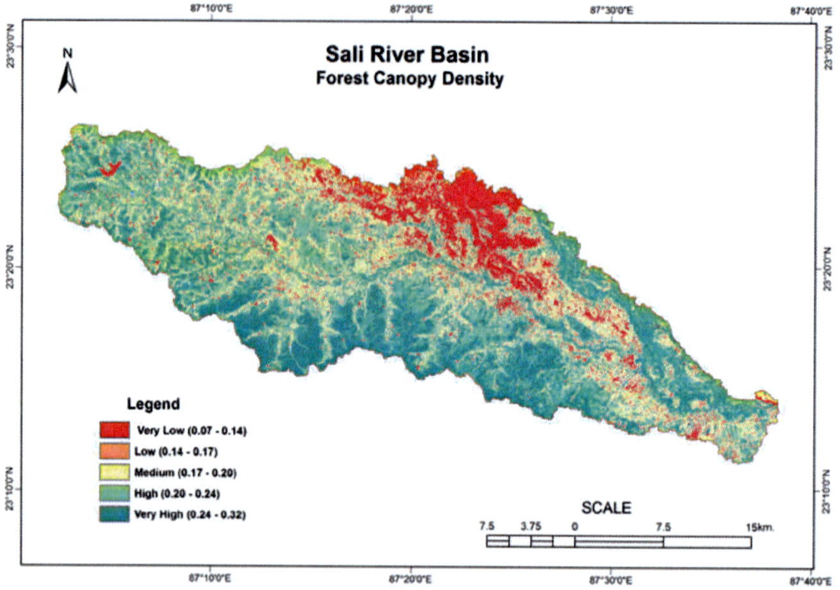

Fig. 15.13 FCD map of Sali River basin for the year of 2018

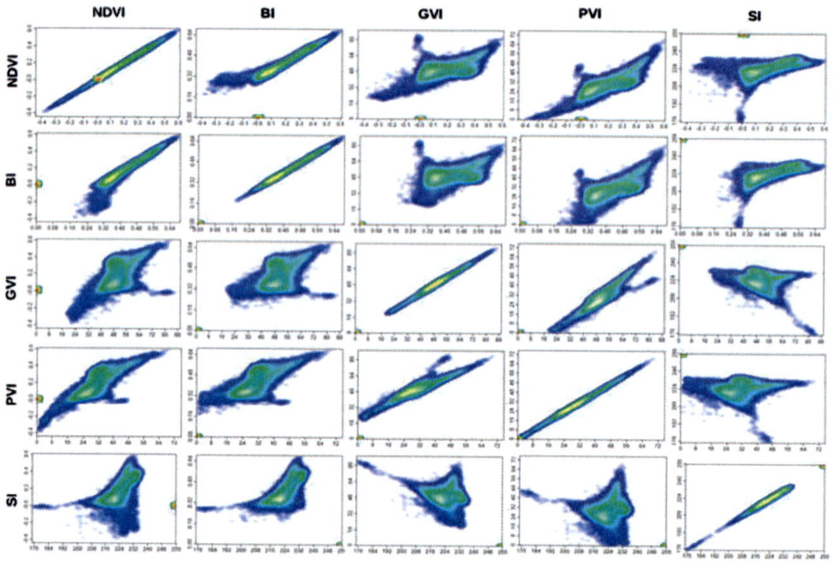

Fig. 15.14 Correlation matrix among the five selected vegetation indices using density plot for the year of 2000

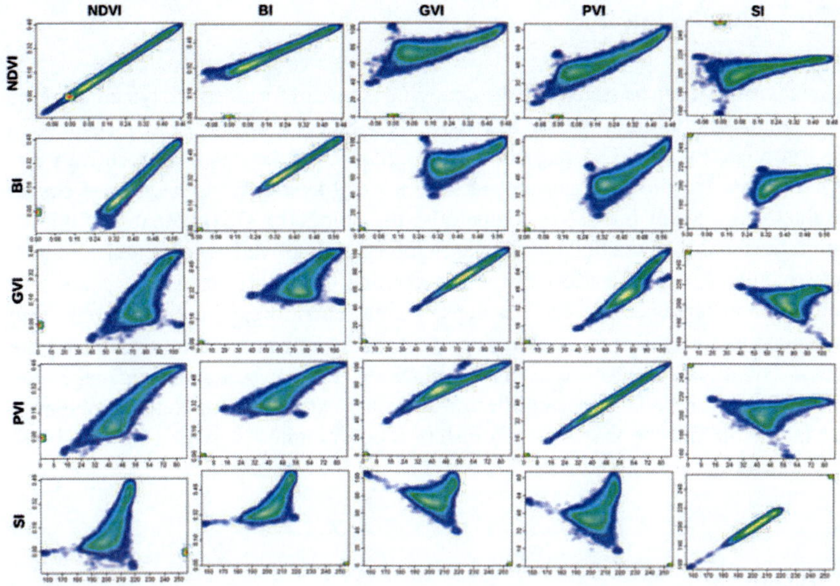

Fig. 15.15 Correlation matrix among the five selected vegetation indices using density plot for the year of 2018

GVI and PVI but moderately correlated with SI. This indicates that the entire study area covered with good forest and agricultural areas. GVI highly positively correlated with NDVI, BI, PVI and SI. PVI also positively correlated with NDVI, BI and GVI but moderately correlated with SI. SI strongly positively correlated with NDVI and BI, but moderately correlated with GVI and PVI. On the other side, for the year of 2018, the correlation matrix map (Fig. 15.15) among the different vegetation indices shows more or less similar output to the year of 2000. Though, it is found that there is a significance changes in the FCD covers areas between the years 2000 and 2018.

15.3.7 Validation of Results

Validation is the important part of the any model so to validate the FCD map of 2018 we used 92 random points over the river basin. The locations of each point on map have been compared with the actual field locations by Seed cell area index (SCAI). The result showed the FCD map is 87.26% accurate.

15.4 Conclusion

In conclusion, it can be stated that the changing nature of forest cover dynamics of the Sali river basin during the period of 2000 and 2018 has shown a significant change in the forest health. The FCD map of the study area during these two consecutive years revealed that there is a remarkable change in the forest cover areas. Forest covers gradually decreasing in 2018 as compared to the year 2000. This is because of human interferences in the case of increasing agricultural land as well as converted the forest area to build-up land. The forest area is becoming fragmented as well as some core densely forest area has become less densely area. Therefore, continuous monitoring is needed to restore forest health. The FCD model can be used to monitoring the forest dynamics and also be used as a planning tool for sustainable management of forest resources. The unscientific forest timber extraction is a major problem, it needs to stop and the scientific afforestation programmer is needed to restore the forest health.

References

Andersen HE, McGaughey RJ, Reutebuch SE (2005) Estimating forest canopy fuel parameters using LIDAR data. Remote Sens Environ 94(4):441–449

Avtar R, Herath S, Saito O, Gera W, Singh G, Mishra B (2014) Application of remote sensing techniques toward the role of traditional water bodies with respect to vegetation conditions. Environ Dev Sustain 16(5):995–1011

Azizi Z, Najafi A, Sohrabi H (2008) Forest canopy density estimating, using satellite images. In: The international archives of the photogrammetry, remote sensing and spatial information sciences, vol XXXVII. Part B8. Beijing

Baynes J (2004) Assessing forest canopy density in a highly variable landscape using Landsat data and FCD mapper software. Aust For 67:247–253

Blodgett C, Jakubauskas M, Price K, Martinko E (2000) Remote sensing-based geostatistical modelling of forest canopy structure. Annual Proceeding of ASPRS, Washington, DC, pp 22–26

Boyd DS, Foody GM, Ripple WJ (2002) Evaluation of approaches for forest cover estimation in the Pacific North West, USA using remote sensing. Appl Geogr 22:375–392

Chakrabortty R, Pal SC, Malik S, Das B (2018) Modeling and mapping of groundwater potentiality zones using AHP and GIS technique: a case study of Raniganj Block, Paschim Bardhaman, West Bengal. Model Earth Syst Environ 4(3):1085–1110

Chowdhuri I, Pal SC, Chakrabortty R (2020) Flood susceptibility mapping by ensemble evidential belief function and binomial logistic regression model on river basin of eastern India. Adv Space Res 65(5):1466–1489

Cross AM, Settle JJ, Drake NA, Paivinen RTM (1991) Sub-pixel measurement of tropical forest covers using AVHRR data. Int J Remote Sens 12(5):37–41

CWC and NRSC (2014) Watershed Atlas of India. Ministry of Water Resource, New Delhi

Das B, Pal SC, Malik S, Chakrabortty R (2019) Modeling groundwater potential zones of Puruliya district, West Bengal, India using remote sensing and GIS techniques. Geol Ecol Landsc 3(3):223–237

Deka J, Tripathi OP, Khan ML (2013) Implementation of forest canopy density model to monitor tropical deforestation. J Indian Soc Remote Sens 41:469–475

FAO (2010) Global forest resources assessment 2010: Food and Agriculture Organization of the United Nations, Rome

FAO (2014) Global forest land-use change 1990–2010. Food and Agricultural Organization of the United Nations with the E.U. Joint Research Centre (JRC). Rome

FAO (2015) Global forest resources assessment 2015. FAO Forestry Paper No. 1. The Food and Agricultural Organization of the United Nations, Rome

Geological Survey of India (2003) https://www.gsi.gov.in/webcenter/portal/OCBIS/page1478/page1872?_adf.ctrl-state=10i7wmzvd45&_afrLoop=1920464058426572#!

Haila Y (1999) Islands and fragments. In: Hunter ML (ed) Maintaining biodiversity in forest ecosystems. Cambridge University Press, UK, pp 234–264

Halliday J, Tate K, McMurtrie RE, Scott NC (2003) Mechanisms for changes in soil carbon storage with pasture to Pinusradiata land-use change. Glob Change Biol 4:1294–1308

Howard JA (1991) Remote sensing of forest resources, theory and application. Campbell and Hall, London

Huete AR, Liu H, van Leeuwen WJ (1997) The use of vegetation indices in forested regions: issues of linearity and saturation. In: Geoscience and remote sensing, 1997. IGARSS'97. Remote sensing—a scientific vision for sustainable development, 1997 IEEE international, vol 4, pp 1966–1968. IEEE

Jennings S, Brown N, Sheil D (1999) Assessing forest canopies and understory illumination: canopy closure, canopy cover and other measures. Forestry 72:59–73

Joshi PKK, Roy PS, Singh S, Agrawal S, Yadav D (2006) Vegetation cover mapping in India using multi-temporal IRS Wide Field Sensor (WiFS) data. Remote Sens Environ 103(2):190–202

Kauth RJ, Thomas GS (1976) The tasselled cap–a graphic description of the spectral-temporal development of agricultural crops as seen by Landsat. In: LARS symposia, p 159

Lee AC, Lucas RM (2007) A LiDAR-derived canopy density model for tree stem and crown mapping in Australian forests. Remote Sens Environ 111:493–518

Liu W, Yamazaki F (2012) Object-based shadow extraction and correction of high-resolution optical satellite images. IEEE J Sel Top Appl Earth Observ Remote Sens 5(4):1296–1302

Malik S, Pal SC, Das B, Chakrabortty R (2019a) Assessment of vegetation status of Sali River basin, a tributary of Damodar River in Bankura District, West Bengal, using satellite data. Environ Dev Sustain 1–35

Malik S, Pal SC, Das B, Chakrabortty R (2019b) Intra-annual variations of vegetation status in a sub-tropical deciduous forest-dominated area using geospatial approach: a case study of Sali watershed, Bankura, West Bengal, India. Geol Ecol Landsc 1–12

Mensah S, Veldtman R, Seifert T (2017) Allometric models for height and aboveground biomass of dominant tree species in South African Mistbelt forests. Southern For J For Sci 79(1):19–30

Mon MS, Mizoue N, Htun NZ, Kajisa T, Yoshida S (2012) Estimating forest canopy density of tropical mixed deciduous vegetation using Landsat data: a comparison of three classification approaches. Int J Remote Sens 33(4):1042–1057

Morrow V, Young J, Roberts C (2001) Mapping tree canopy in Broward County, Florida

Nandy S, Joshi PK, Das KK (2003) Forest canopy density stratification using biophysical modeling. J Indian Soc Remote Sens 31:291–297

O'Malley LS (1908) Bengal District Gazetteers, Bankura. Government of West Bengal, India

Ono A, Kajiwara K, Honda Y (2010) Development of new vegetation indexes, shadow index (SI) and water stress trend (WST). Int Arch Photogramm Remote Sens Spatial Inf Sci 38:710–714

Pal CS, Chakrabortty R, Malik S, Das B (2018) Application of forest canopy density model for forest cover mapping using LISS-IV satellite data: a case study of Saliwatershed, West Bengal. Model Earth Syst Environ 4:853–865

Pal SC, Chowdhuri I (2019) GIS-based spatial prediction of landslide susceptibility using frequency ratio model of Lachung River basin, North Sikkim, India. SN Appl Sci 1(5):416

Pal SC, Das B, Malik S (2019) Potential landslide vulnerability zonation using integrated analytic hierarchy process and GIS technique of Upper Rangit catchment area, West Sikkim, India. J Indian Soc Remote Sens 47(10):1643–1655

Prasad PRC, Nagabhatla N, Reddy CS, Gupta S, Rajan KS, Raza SH, Dutt CBS (2009) Assessing forest canopy closure in a geospatial medium to address management concerns for tropical islands-Southeast Asia. Environ Monit Assess 160:541

Raha AK (2014) Time series analysis of forest and tree cover of West Bengal from 1988 to 2010, using RS/GIS, for monitoring afforestation programmes. J Ecol Photon 108:255–265

Richardson AJ, Wiegand CL (1977) Distinguishing vegetation from soil background information. Photogramm Eng Remote Sens 43(12):1541–1552

Rikimaru A (1996) Landsat TM data processing guide for forest canopy density mapping and monitoring model. ITTO workshop on utilization of remote sensing in site assessment and planning for rehabilitation of logged-over forests, Bangkok, 30 July–1 Aug 1996, pp 1–8

Rikimaru A, Roy PS, Miyatake S (2002) Tropical forest cover density mapping. Trop Ecol 43(1):39–47

Rouse J Jr, Haas RH, Schell JA, Deering DW (1974) Monitoring vegetation systems in the Great Plains with ERTS, pp 309–317

Roy PS, Sharma KP, Jain A (1996) Stratification of density in the dry deciduous forest using satellite remote sensing digital data—an approach based on spectral indices. J Biosci 21(5):723–734

Saaty TL (1988) What is the analytic hierarchy process? In: Mathematical models for decision support. Springer, Berlin, pp 109–121

Sahana M, Sajjad H, Ahmed R (2015) Assessing spatio-temporal health of forest cover using forest canopy density model and forest fragmentation approach in Sundarban reserve forest, India. Model Earth Syst Environ 1–10

Telesca L, Lasaponara R, Lanorte A (2008) Intra-annual dynamical persistent mechanisms in Mediterranean ecosystems revealed SPOT-VEGETATION time series. Ecol Complex 5(2):151–156

Verbesselt J, Jonsson P, Lhermitte S, Van Aardt J, Coppin P (2006) Evaluating satellite and climate data-derived indices as fire risk indicators in savanna ecosystems. IEEE Trans Geosci Remote Sens 44(6):1622–1632

Wang Z, Brenner A (2009) An integrated method for forest canopy cover mapping using Landsat ETM imagery. In: ASPERS/MAPRS-2009 fall conference, 16–19 Nov, San Antonio, Texas

Zhang X, Friedl MA, Schaaf CB, Strahler AH, Hodges JCF, Gao F, Huete A (2003) Monitoring vegetation phenology using MODIS. Remote Sens Environ 84:471–475

Chapter 16
Estimation of Aboveground Stand Carbon using Landsat 8 OLI Satellite Image: A Case Study from Turkey

Alkan Günlü, Sedat Keleş, İlker Ercanli, and Muammer Şenyurt

Abstract Accurate and consistent measurement of carbon stocks and flows in forest ecosystems has recently gained global importance. This study aims to estimate the aboveground stand carbon (AGSC) using Landsat 8 OLI satellite image in pure Crimean pine stands and to compare the results of various modeling techniques. In this context, a total of 108 sample plots were firstly taken in a case study forest area. The AGSC of each sample area was calculated using a species-specific carbon equation developed for the case study area. The band values, vegetation indices, and texture values for each sample plot were also obtained from Landsat 8 OLI satellite image. The relationships between the AGSC and the band values, vegetation indices, and texture values were investigated by multivariate linear regression (MLR), support vector machine (SVM) and artificial neural networks (ANN) models. The results demonstrated that the ANN models with Bayesian regularization are better than the MLR and SVM models to estimate the AGSCin pure Crimean pine stands. Also, the band values showed better predictive performance in explaining the variation in AGSCthan vegetation indices and texture values.

Keywords Aboveground stand carbon · Landsat 8 OLI satellite image · Multiple linear regression · Artificial neural network · Support vector machine

16.1 Introduction

Forest ecosystems provide a lot of goods and services such as timber production, water conservation, soil protection, oxygen production, aesthetics, and recreation. Forest ecosystems are also an important part of the carbon pool. Changes to the size and efficiency of this pool can act as a carbon dioxide storage or source of forest ecosystems (Milne and Brown 1997; Karjalainen et al. 1999; Nowak and Crane 2002; Keleş and Başkent 2006). Forest trees behave like a CO_2 pool by holding

A. Günlü · S. Keleş (✉) · İ. Ercanli · M. Şenyurt
Faculty of Forestry, Çankırı Karatekin University, 18200 Çankırı, Turkey
e-mail: dr.sedatkeles@gmail.com

P. K. Shit et al. (eds.), *Spatial Modeling in Forest Resources Management*, Environmental Science and Engineering,
https://doi.org/10.1007/978-3-030-56542-8_16

CO_2 during photosynthesis and storing it in biomass. Forest ecosystems accumulate carbon mainly in biomass and soils. These are divided into sub-components as carbon stored in aboveground biomass (AGB), underground biomass, litter, dead cover, soil organic and inorganic matter. These are also the basis for carbon budget calculations (Lim et al. 1999; Ney et al. 2002; Kaipainen et al. 2004; Lal 2005). The total amount of carbon stored in a forest ecosystem at any given time is estimated as the sum of the carbon contents stored in live biomass, forest soil and wood products. With the calculation of these components, carbon stocks and flows in a forest can be calculated. On the other hand, the carbon held by the forest ecosystems from the atmosphere is released into the atmosphere with the decomposition of wood products, litter and organic matter in the soil. Also, some interventions such as production in forests, degradation as a result of excessive cuts and exploitation in forests, forest fires, insect attacks, tree species diseases, and the conversion of forests to different land uses such as agriculture and settlements are important factors causing CO_2 emission from forest ecosystems to the atmosphere. As a result, these processes cause forest ecosystems to be carbon sources (Karjalainen et al. 1999; Brown 2002; Keleş and Başkent 2006).

Accurate and consistent measurement of these carbon stocks and flows in forest ecosystems has recently gained global importance. Remote sensing techniques that are easier, cheaper and require less labor are an important alternative way to estimate the amount of carbon stored in forest ecosystems. In this context, different spectral variables such as band values (Rahman et al. 2008; Kelsey and Neff 2014; Safari et al. 2017; Li et al. 2018), vegetation indices (Myeong et al. 2006; Rahman et al. 2008; Kelsey and Neff 2014; Safari et al. 2017; Li et al. 2018) and texture values (Kelsey and Neff 2014; Li et al. 2018) have been used to estimate forest stand carbon using Landsat satellite data used extensively in forest resources management. Many statistical models based on the MLR analysis for predicting the AGSC values from the remote sensing data have been developed in forest studies. However, these MLR models require some statistical assumptions, normally distributed residuals and homoscedastic trends in the AGSC predictions, and when these assumptions are violated, these AGSC predictions can be biased and erroneously obtained in forest applications. To overcome this problem in tree predictions, Artificial Intelligence (AI) Techniques has been increasingly and successfully used as an alternative method in forest literature. Numerous prediction models based on AI, especially ANN, have been developed for modeling various individual tree and stand attributes such as tree volume (Diamantopoulou 2005a, b; Diamantopoulou et al. 2005; Özçelik et al. 2008; Görgens et al. 2009; Diamantopoulou and Milios 2010; Özçelik et al. 2010; Soares et al. 2011; Miguel et al. 2016; Sanquetta et al. 2017), tree taper (Diamantopoulou et al. 2005; Leite et al. 2011; Nunes and Görgens 2016), total tree height (Brandao 2007; Ranson et al. 2007; Diamantopoulou and Özçelik 2012; Özçelik et al. 2013; Ercanlı et al. 2015), tree mortality (Hasenauer et al. 2001), survival model (Guan and Gertner 1991), regeneration establishment and height growth (Hasenauer and Kinderman 2002), bark volume (Diamantopoulou 2005a, b), leaf area index (Ercanlı et al. 2018), diameter distributions (Ercanlı and Bolat 2017), biomass prediction

(Özçelik et al. 2017), basal area and volume increment growth model (Ashraf et al. 2013), the stand carbon (Ercanlı et al. 2016).

Besides many ANN studies, the other AI technique such as SVM stand out as another prominent AI technique. SVM may attractive potential usability to predict the stand carbon from the remote sensing data in forest inventory. The AI technique with SVM may be mainly regarded as a nonparametric technique with kernel functions, which have been proposed by Vladimir Vapnik and his co-workers in 1992 (Vapnik 1995). In the 1960s, the preliminary applications of SVM were introduced based on the nonlinear generalization of the Generalized Portrait algorithm (Vapnik and Lerner 1963; Vapnik and Chervonenkis 1964). As being artificial intelligence technique, the learning algorithm based on the SVM relies on simple ideas that originated in statistical learning theory (Vapnik 1995), thus the SVM may be utilized for both regression and classification tasks (Wang et al. 2005). In forest applications, SVM can be distinguished as a regression method, preserving completely the key structures that characterize the training algorithm. The other attractive feature of the SVM is the actual particular class training algorithm including kernel functions which considered a nonparametric technique. According to the knowledge of the forest biometric studies, no studies have been achieved to compare the SVM models and ANN models to predict the AGSC, especially based on the remote sensing data, and so the evaluation of the these AI techniques including SVM and ANN in predicting AGSC have been uncertain and needs to be clarified.

In this study, it is aimed to evaluate the capability of the usability of these AI techniques such as SVM and ANN models in predicting empirical relationships between the AGSC and the remote sensing data as a leading and innovative application.

16.2 Materials and Methods

16.2.1 Study Area

Yenice Forest Management Unit has been selected as the case study area since there is enough data about this forest area. It is located within the borders of Ankara Regional Directorate of Forestry, Ilgaz Forest Management Department in the north Turkey (Fig. 16.1). The average annual precipitation is 474 mm, and the average annual temperature is approximately 5 °C. Main tree species in the study area are Scotch pine (*Pinussylvestris* L.), Crimean pine [*Pinusnigra* Arnold. subsp. *pallasiana* (Lamb.) Holmboe], and fir (*Abiesbornmülleriana* Matttf.). The total forest area in the study area is 7144 ha and approximately 1700 ha of this area are composed of pure Crimean pine stands as the main object of the study.

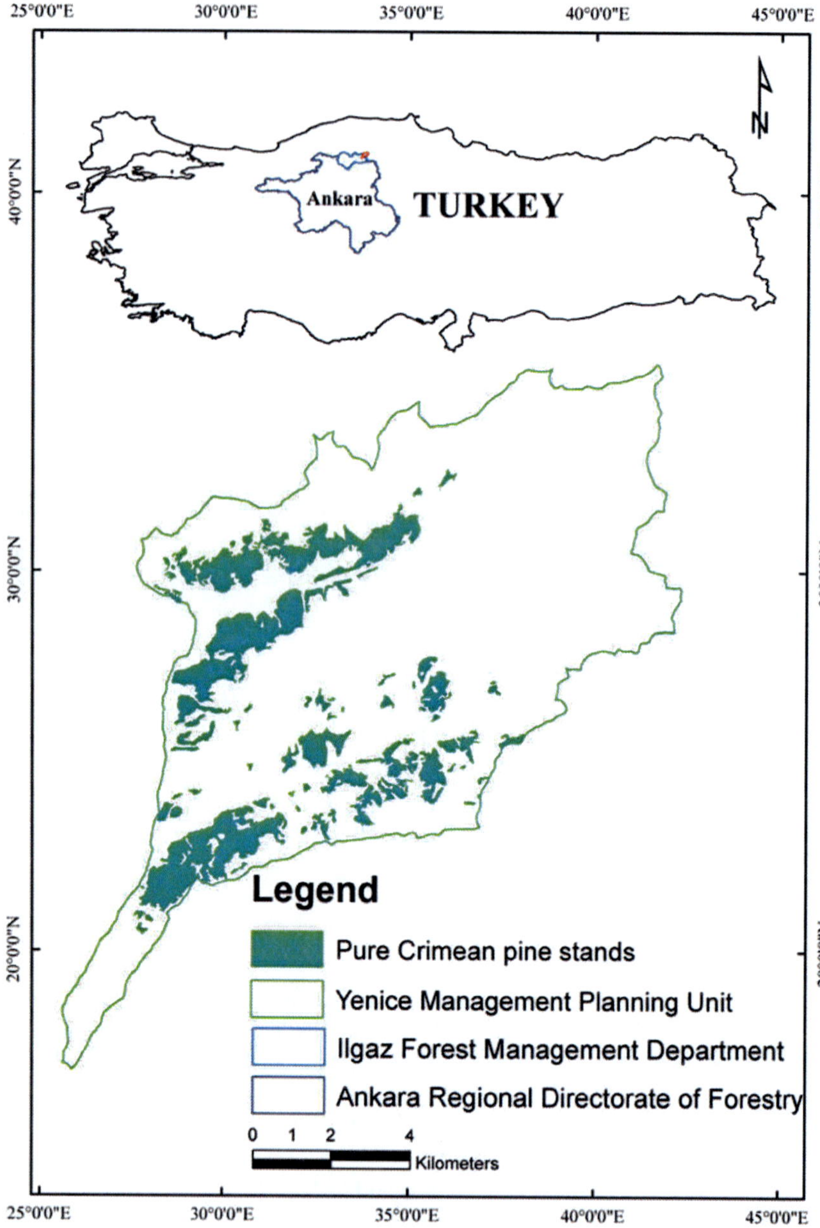

Fig. 16.1 Map of the study site and interaction area

16.2.2 Calculation of Aboveground Stand Carbon

A total of 108 sample areas were taken in this study. The size of the sample plots varies between 400 and 800 m^2 depending on stand crown closure. In each sample area, the diameter of all the trees with diameter 8 cm and over was measured at breast height level. The amount of AGSC by each tree in each sample area was calculated using the local equation for the Crimean pine forest stands developed by Sakıcı et al. (2018). Using this Eq. 16.1, the amount of AGSC by each tree in the sample plot was first calculated. Then, by collecting the amounts of AGSC by the trees in the sample plot, the AGSC amount was calculated for the sample plot. Finally, the AGSC amounts calculated for each sample area were converted to hectare using the conversion factor to the hectare.

$$C_{AG} = 0.054 \cdot (dbh)^{2.362} \tag{16.1}$$

where C_{AG}: Aboveground stand carbon (ton) and dbh is the tree diameter measured at breast height (cm).

16.2.3 Remote Sensing Data

A total of 108 sample areas based on site index, crown closure, and development stage and Landsat 8 OLI satellite image dated 14 August 2014 were used as materials. Six bands (band 2, band 3, band 4, band 5, band 6 and band 7) with a spatial resolution of 30 m were used in this study. Landsat 8 OLI satellite image was applied to some processing before being analyzed. These processing steps are listed below. The six bands used in the study were combined into a single satellite image and a single satellite image cut according to the outer boundary of the study area. Necessary atmospheric and geometric corrections were performed on the satellite image. 108 sample points were overlaid with satellite images and band brightness values were obtained for six bands and each sample plot. However, using the band brightness values obtained for each sample plot, some vegetation indices values in Table 16.1 are calculated. Also, the texture values were calculated for six bands and each sample plot using eight texture features (mean, correlation, homogeneity, dissimilarity, second moment, variance, entropy and contrast) and four different window sizes (3×3, 5×5, 7×7 and 9×9).

16.2.4 Multivariate Linear Regression

To model the empirical relationships between the AGSC values and, band brightness values, vegetation indices, and texture values obtained from Landsat 8 OLI satellite

Table 16.1 The vegetation indices used in this study

Vegetation indices	Description	Formula	References
NDVI	Normalized difference vegetation index	(Band4-Band5)/(Band4 + Band5)	Rouse et al. (1974)
SAVI	Soil adjusted vegetation index	(Band5-Band4) * (1 + L)/(Band5 + Band4 + L)	Huete (1988)
DVI	Difference vegetation index	Band5-Band4	Clevers (1988)
SR	Simple ratio	Band5/Band4	Jordan (1969)
TVI	Transformed vegetation index	Band5-Band4/Band5 + Band4 + 0.5	Deering et al. (1975)
ND64	Normalized difference	Band6-Band4/Band6 + Band4	Lu et al. (2005)
ND65	Normalized difference	Band6-Band5/Bnad6 + Band5	Lu et al. (2005)
ND67	Normalized difference	Band6-Band7/Band6 + Band7	Lu et al. (2005)
ND42	Normalized difference	Band4-Band2/Band4 + Band2	Lu et al. (2005)
ND74	Normalized difference	Band7-Band4/Band7 + Band4	Sivanpillai et al. (2006)
ARVI	Atmospherically resistant vegetation index	(Band5-2 * Band4-Band2)/(Band5 + 2 * Band4-Band2)	Kaufman and Tanré (1992)

L = 0.5

image, the MLR were used in this study. The Ordinary Least Squares technique was utilized to acquire the parameters and other statistics of the MLR model. To select predictive satellite data including the band brightness values, vegetation indices and texture values, the stepwise variable selection regression technique was used to obtain the AGSC predictions significantly ($p < 0.05$).

16.2.5 Artificial Neural Network Models

The ANN models based on the Feed Forward Backprop training algorithms were used to predict the AGSC from the satellite data in this study. This training algorithm has commonly been used to predict tree and forest attributes in forest literature. While estimating single tree and stand features with ANN, previous studies have frequently used a network training function, also called as *trainlm*, based on Levenberg-Marquardt optimization. Also, the Bayesian regularization including a network training function, called as *trainbr*, reduces a combination of squared errors and weights, and then regulates the correct combination of these values to improve predictions from the network. In this study, these network training function including *trainlm* and *trainbr* were used and compared to determine which training function gives the best predictive results.

In this ANN model, input variables were best predictive stand parameters, in which these independent variables were selected by the stepwise variable selection technique in regression analysis, and the target variable was the AGSC value obtained from the ground measurements. These ANN models include three layers such as the input layer, hidden layer, and output layer and these activation functions linking with these layers are another important network parameter in its structure. In our preliminary analyses, the activation function alternative based on the log-sig function between the input layer and hidden layer and tan-sig function between the hidden layer and output layer gave better predictions than those by other activation functions, thus this activation function alternative was selected to train the ANN models in this study. Further significant restriction of the network structure is the number of neurons in the hidden layer. Thus, some alternatives for the number of neurons that ranged from 1 to 100; 1, 2, 3, ..., 20, 30, 50, 70, 90, and 100 number of neurons were compared to select the best predictive neuron alternative in this study. Besides these values of parameters in ANN structure, the value of 200 for epochs, the value of 1×10^{-10} for performance goal, the value of 0.0001 for the learning rate and the value of 1×10^{-10} for minimum performance gradient gave the best predictive results to train these ANN models in the preliminary of this study and so these parameters were used to obtain the AGSC predictions.

A total of 200 ($100 \times 2 = 200$ alternatives) network alternatives including 100 number neurons and 2 network training function alternatives (*trainlm* and *trainbr*) based on the Multilayer Feed Forward Backprop training algorithms, were trained and used to obtain the AGSC predictions. These network trainings for 200 network alternatives for ANN models were carried out using newff syntax for the feed-forward back propagation network codded in MATLAB software (MATLAB 2014).

16.2.6 Support Vector Machine Models

Another alternative artificial intelligence model is the SVM technique, which is currently attracting attention and importance subject being an artificial intelligence technique. This SVM technique can be used to carry out general regression to predict forest and tree attributes and classification to obtain similar and dissimilar forest units and fractures. The concept of SVM has been proposed by Vladimir Vapnik and his co-workers for classification purposes, then the regression application of the SVM hasbeen developed based on the same principle with the SVM for classification.

It trains SVM models, the significant way of adding non-linearities in SVM is by the use of kernel functions, which this function defined by the input data into a high-dimensional feature space to improve the predictive ability of this network. Especially, Radial-based function (RBF) kernel can offer successful modeling results in nonlinear data. In applications involving regression estimation, including continuous number to predict, "eps-regression" type of the SVM can be trained to perform a regression task. In this SVM models, input variables were best predictive stand parameters as with ANN models, and target variable was the AGSC values. In this study, the SVM model was applied based on "eps-regression" type of the SVM and the RBF Kernel of the e1071 R package (R Development Core Team 2018).

16.2.7 Comparison Criteria

In this study, several comparison criterion values were used to compare and evaluate the predictions of AGSC that were obtained by the MLR, ANN and SVM models. These fitting criteria are (2) Sum of Squared Errors (SSE), (3) the root mean squared error (RMSE), (4) % root mean squared error (RMSE%), (5) the fit index (FI), (6) Akaike's information criterion (AIC) and (7) Bayesian information criterion (BIC). These criteria are calculated as follows:

$$SSE = \sum_{i=1}^{n} \left(SC_i - \widehat{SC}_i \right)^2 \tag{16.2}$$

$$RMSE = \sqrt{\sum_{i=1}^{n} \left(SC_i - \widehat{SC}_i \right)^2 / (n-k)} \tag{16.3}$$

$$RMSE\% = \left(\left[\sqrt{\sum_{i=1}^{n} \left(SC_i - \widehat{SC}_i \right)^2 / (n-k)} \right] / \overline{SC}_i \right) \cdot 100 \tag{16.4}$$

$$FI = \frac{\sum_{i=1}^{n} \left(SC_i - \widehat{SC}_i \right)}{\sum_{i=1}^{n} \left(SC_i - \overline{SC}_i \right)} \tag{16.5}$$

$$AIC = \ln(RMSE) + 2k \tag{16.6}$$

$$BIC = \ln(RMSE) + \ln(k) \tag{16.7}$$

where, SC_i is the measured AGSC value in the sample plot (observed values), $\overline{SC}_i$ is the average of observed AGSC values, $\widehat{SC}_i$ is the predicted AGSC value obtained by the MLR, ANN and SVM models, k is the number of inputs or independent variable in the prediction methods, ln is the natural logarithm with the base of the mathematical constant *e*. From these fitting criterion values, it is desired that the FI, which is between 0 and 1, should be as close to 1 as possible. Smaller values of other criterion values indicate that better predictive AGSC is obtained.

The flowchart of the method used in this study is given in Fig. 16.2.

16.3 Results and Discussion

The goodness-of-fit statistics of SSE, FI, RMSE, RMSE%, AIC and BIC for various prediction techniques and Landsat 8 OLI satellite data are given in Table 16.2. Seeing this Table 16.2, the ANN model with trainbr and # 85 neurons gave the best predictive fitting results including SSE value of 34,180.132, RMSE value of 17.957, the RMSE% value of 17.352, AIC value of 315.902, BIC value of 386.762, FI value of 0.887 for the band brightness values. For the vegetation indices, the ANN with trainbr and # 89 neurons resulted in the best predictive fitting statistics with SSE value of 48,670.930, RMSE value of 21.428, the RMSE% value of 21.461, AIC value of 334.987, BIC value of 405.847, FI value of 0.839. For the texture values, the ANN with trainbr and # 92 neurons presented the best predictive fitting results with SSE value of 54,862.021, RMSE value of 22.750, the RMSE% value of 21.644, AIC value of 341.453, BIC value of 412.313, FI value of 0.819. Considering completely these prediction techniques, the prediction model based on the ANN with trainbr and # 85neurons and the band brightness values showed better predictive performance in explaining the variation in AGSC than those by various prediction techniques and, vegetation indices and texture values.

In Table 16.3, the means of error and fitting values for SSE, FI, RMSE, RMSE%, AIC and BIC for various prediction techniques and, band brightness, vegetation indices and texture values were presented. From these mean values, the band brightness values and a network training function with the Bayesian regularization, trainbr, resulted in the best predictive AGSC compared the those for the vegetation indices and texture values, and prediction techniques. These results suggested that the band brightness values and the prediction technique based on the ANN with the Bayesian regularization outperformed by presenting better predictive results for the SSE, FI, RMSE, RMSE%, AIC and BIC than the vegetation indices and texture values, and prediction techniques.

The relationships between observed (x-axis) and predicted AGSC (y-axis) by (a) the ANN with trainbr and # 85 neurons and the band values, (b) ANN with trainlm and # 41 neurons and the band values, (c) SVM, (d) MLR were shown in Fig. 16.3. This graph presented the evidence of the best predictive ANN with trainbr and # 85 neurons and the band brightness values, which tend to more angle of 45 degrees with axes than those for other prediction methods and, vegetation indices and texture values.

Figure 16.4 presented the plot of residuals against the predicted AGSC obtained from by (a) the ANN with trainbr and # 85 neurons and the band brightness values, (b) ANN with trainlm and # 41 neurons and the band brightness values, (c) SVM,

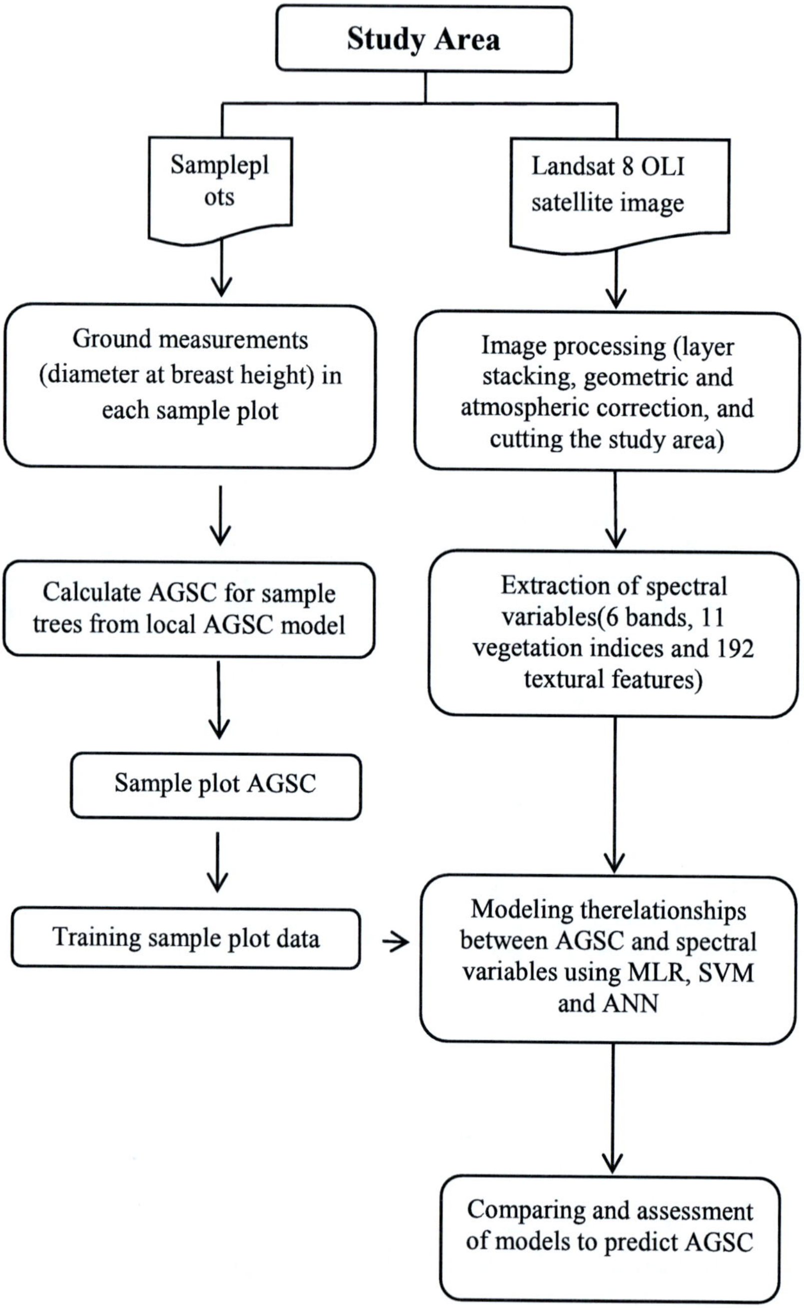

Fig. 16.2 The flowchart of the path followed in the study

Table 16.2 The goodness-of-fit statistics of SSE, FI, RMSE, RMSE%, AIC and BIC for various prediction techniques and satellite data

Satellite data	Technique	SSE	RMSE	RMSE%	AIC	BIC	FI
Band	MLR	79,203.562	27.335	26.414	361.282	432.142	0.738
	ANN with trainlm and # 41 neurons	55,333.613	22.848	22.078	341.916	412.776	0.817
	ANN with trainbr and # 85 neurons	34,180.132	17.957	17.352	315.902	386.762	0.887
	SVM	75,491.309	26.687	25.787	358.690	429.550	0.751
Vegetation indices	MLR	90,968.310	29.295	29.340	368.761	439.621	0.699
	ANN with trainlm and # 62 neurons	53,317.556	22.428	22.462	339.911	410.771	0.824
	ANN with trainbr and # 89 neurons	48,670.930	21.428	21.461	334.987	405.847	0.839
	SVM	83,208.216	28.018	28.061	363.946	434.806	0.725
Texture	MLR	161,319.467	39.011	39.072	399.696	470.556	0.467
	ANN with trainlm and # 86 neurons	86,497.352	28.566	27.177	366.039	436.899	0.714
	ANN with trainbr and # 92 neurons	54,862.021	22.750	21.644	341.453	412.313	0.819
	SVM	114,894.416	32.923	31.322	381.370	452.230	0.620

Table 16.3 The means of error and fitting values for SSE, FI, RMSE, RMSE%, AIC and BIC for various prediction techniques and satellite data

		SSE	RMSE	RMSE%	AIC	BIC	FI
Satellite data	Band values	61,052.154	23.707	22.908	344.447	415.307	0.798
	Vegetation indices	69,041.253	25.292	25.331	351.901	422.761	0.772
	Textures	104,393.314	30.813	29.804	372.140	442.999	0.655
The prediction techniques	MLR	110,497.113	31.880	31.609	376.580	447.440	0.635
	ANN with trainlm	65,049.507	24.614	23.906	349.289	420.149	0.785
	ANN with trainbr	45,904.361	20.712	20.152	330.781	401.641	0.848
	SVM	91,197.980	29.209	28.390	368.002	438.862	0.699

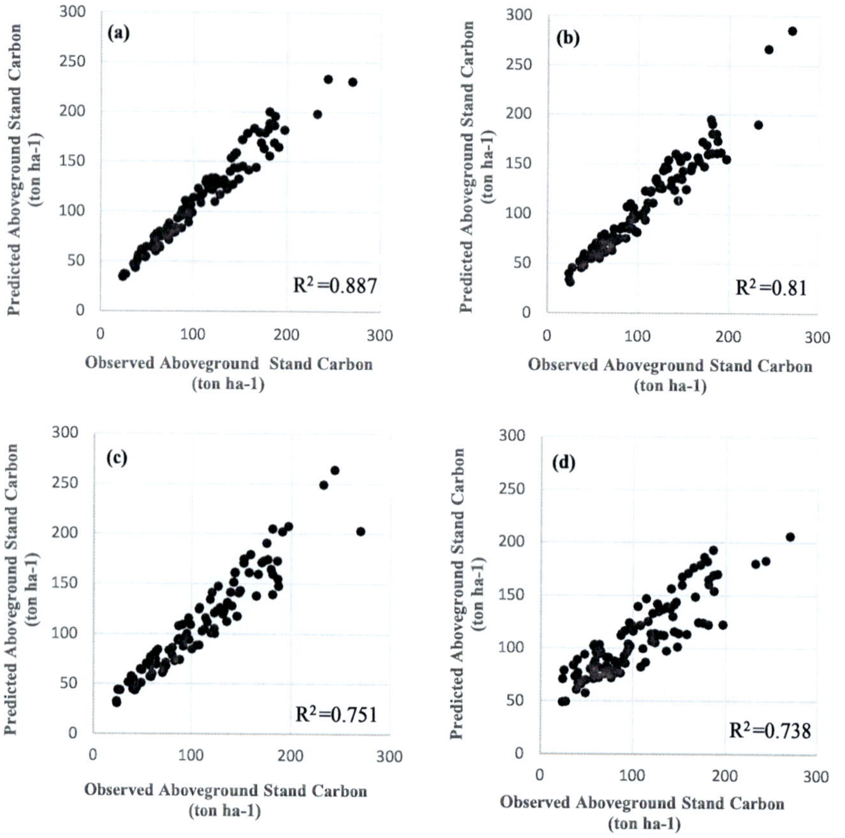

Fig. 16.3 The relationships between observed (x-axis) and predicted AGSC (y-axis) by **a** the ANN with trainbr and # 85 neurons and the band brightness values, **b** ANN with trainlm and # 41 neurons and the band brightness values, **c** SVM, **d** MLR

(d) MLR. This best predictive ANN with trainbr and # 85 neurons and the band brightness values resulted in significant improvement in these residuals with a lower range compared by other alternatives. These residual results in Fig. 16.2 indicate that predictive estimates of AGSC values have been achieved by this best predictive ANN with trainbr and # 85 neurons and the band brightness value.

The MLR, SVM and ANN models were applied for estimating the relationships between AGSC and the band brightness, vegetation indices and texture values obtained from a Landsat 8 OLI satellite image in this study. When the literature is analyzed, it is seen that there are many studies performed using different modeling techniques with different data obtained from different satellite images. The results obtained in this study were compared with other studies on this subject. Foody et al. (2003) estimated forest biomass using vegetation indices obtained from Landsat TM

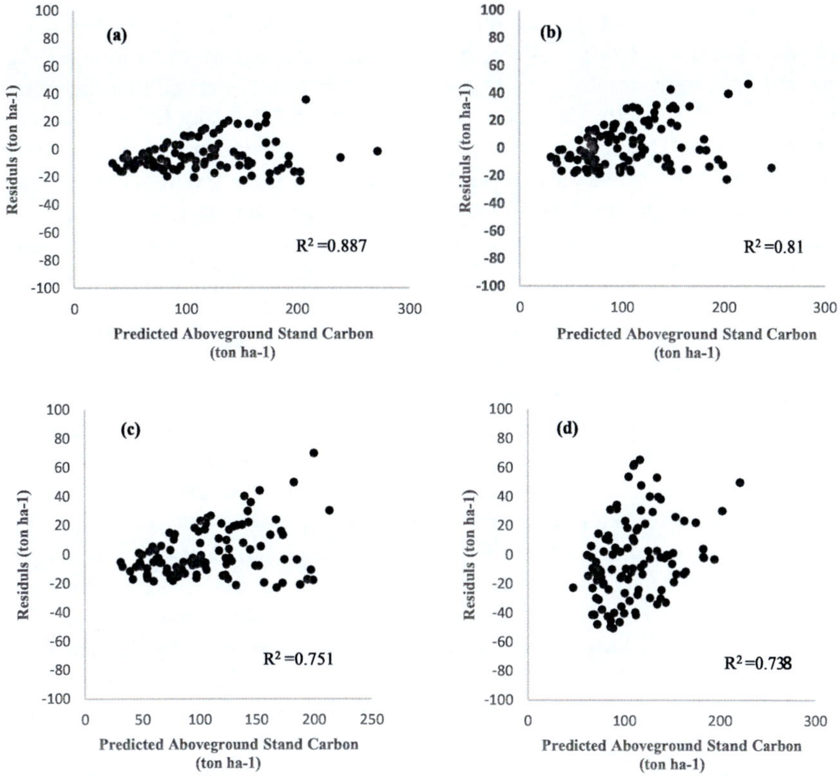

Fig. 16.4 The plot of residuals against the predicted AGSC obtained from by **a** the ANN with trainbr and # 85 neurons and the band brightness values, **b** ANN with trainlm and # 41 neurons and the band brightness values, **c** SVM, **d** MLR

satellite image. They predicted the relationships between forest biomass and vegetation indices by using MLR and ANN models. Their results showed that predictions derived from a neural network-based approach were strongly and significantly correlated with the forest biomass estimate derived from ground measurements in comparison to MLR model ($R^2 < 0.32$).In a study developed by Lu et al. (2005), it was tried to predict AGB with band, vegetation indices and texture values obtained from Landsat TM satellite image using MLR model. The results of the study showed that band values give better results in predicting AGB in simple forest stand structure. This is in line with the findings in our results. Conversely, it was seen that the texture characteristics gave better results in predicting AGB compared to the bands in complicated forest stand structure. In a study conducted by Min et al. (2009) the relationships between band and vegetation indices values obtained from Landsat TM satellite images in three different forest types (softwood forest, hardwood forest and mixed forest) and AGB were tried to be determined by MLR analysis. In the results obtained, the model produced with vegetation indices was found to be more successful Xu et al. (2011). Used linear regression, partial least-squares regression and ANN models to estimate AGSC based on the combined use of Landsat ETM+ data and field measurements. As in our study, their results showed that the ANN model ($R^2 = 0.912$) provided better results than the MLR model ($R^2 = 0.247$) in estimating AGSC Günlü et al. (2014). Developed MLR models to estimate the AGB using optical band brightness values and vegetation indices derived from Landsat TM data in pure Crimean pine stands. Contrary to our results, their results indicated that vegetation indices better estimation of AGB as compared to optical bands. Similar results were obtained by Günlü et al. (2016) in the same study area, the relationships between band brightness values and vegetation indices obtained from Landsat 8 OLI satellite image and the AGSC values were investigated by MLR. In the results obtained, it has been seen that vegetation indices give better results than band brightness values. The difference between this study (2016) and the present study is the method differences used in determining the AGSC of the sample areas. While the AGSC amounts of the sample areas were calculated using the coefficients developed by Tolunay (2014) in this study, in the current study, the AGSC amounts of the sample areas were calculated with the carbon equations developed by Sakıcı et al. (2018). However, in a study conducted by Turgut and Günlü (2020), different results were obtained. In their study, the relationships between the band brightness, vegetation indices and texture values obtained from the Landsat 8 OLI satellite image and the AGB values calculated from ground measurements were estimated by MLR method in pure Crimean pine stands. Contrary to our study, the model obtained with texture values gave better results in estimating the AGB. In this study, the success of the model obtained with texture values was higher from our study, whereas the success level of the model obtained with band and vegetation indices was lower than our study. In another study conducted by Günlü and Ercanlı (2020), the relationships between texture values obtained from Alos Palsar data and AGSC in pure beech stands were tried to be estimated by MLR and ANN modeling techniques. As in our study, in their study results demonstrated that the ANN was better than MLR models to estimate AGSC values Baloloy et al. (2018). Modeled that the relationships between band and vegetation indices values obtained from two different satellite images (Rapideyeand Sentınel-2) andAGB were estimated by MLR and Multivariate Adaptive Regression Spline (MARS) method. When the results obtained were examined, it was seen that MARS method gave better results than MLR method. However, models with vegetation indices (except for Rapideye satellite data) gave more successful results. The R^2 values between vegetation indices generated from Rapideye and Sentinel-2 satellite data and the AGB were found the 0.82 and 0.89, respectively. However, the R^2 values between band values and AGB were found the 0.92 and 0.62, respectively. Thapa et al. (2015) predicted the relationships between texture values obtained from Alos Palsar image and AGSC amounts by MLR method, the R^2 value was found to be 0.84. In this study, a certain amount of predictive ability in predicting AGSC can be obtained by with artificial intelligence models. Safari et al. (2017) estimated AGSC in coppice oak forests using Landsat 8 image and four machine learning algorithms. Their results showed that the AGSC estimation of SVM, boosted regression trees, random forest and multivariate adaptive regression splines algorithms (MARS) had

R^2 values of 0.64, 0.57, 0.64 and 0.58, respectively. They also found that the simple band ratios more accurate AGSC estimates in comparison to the use of Landsat 8 OLI satellite image derived raw bands and vegetation indices. In a study conducted by Dong et al. (2020) predicted the relationships between band, vegetation indices and texture values obtained from World View-2 image and AGB were estimated by using ANN and SVM methods. In the model with band values, the ANN method ($R^2 = 0.238$) gave better results than SVM ($R^2 = 0.046$) method. Similarly, the ANN ($R^2 = 0.248$) model with the band and vegetation values together gave better results than the SVM ($R^2 = 0.052$) method. On the other hand, the SVM ($R^2 = 0.970$) model with texture values better results than the ANN($R^2 = 0.831$) method. In addition to these predictive findings, another important finding is that, despite the intensive use of the *trainlm* training function based on Levenberg-Marquardt optimization in previous studies, better predictive results are found with the network training function based on the Bayes approach, *trainbr*, which may be an alternative to this *trainlm* training function. The network training function based on the Bayes approach, which can be used as an alternative to the standard training function based on Levenberg-Marquardt optimization, provides a slower training at an acceptable scale, but offers better predictive results than those by this standard training function. The reason for such a result may be fact that this network training function based on the Bayes approach can improve the ANN generalization, which the weights and biases of the network from this network training function are assumed to be random variables with specified distributions. Comparison of similar training functions was made by Kamble et al. (2015) and the best predictive results were obtained with the training function based on the Bayes approach.

16.4 Conclusion

The relationships between band brightness values, vegetation indices and texture values obtained from Landsat 8 OLI satellite image and AGSC values obtained by ground measurements were evaluated by using MLR, SVM and ANN (trainlm and trainbr) methods in pure Crimean pine stands. The ANN method outperformed than the SVM and MLC methods for AGSC prediction. Also, the best predictive accuracy was obtained for the ANN with trainbr model used band brightness values and followed it by ANN trainlm method with band brightness values. Using different modeling techniques such as deep learning, mixed effect modeling and MARS, and different satellite images such as passive, active and fused data (combined the active and passive satellite data) can improve the model achievement criteria in similar and different forest ecosystems.

Acknowledgements This work was supported by TÜBİTAK (Project No. 213 O 026). We would like to thank to for support to the TÜBİTAK.

References

Ashraf MI, Zhao Z, Bourque CP, MacLean DA, Meng FR (2013) Integrating biophysical controls in forest growth and yield predictions with artificial intelligence technology. Can J For Res 43(12):1162–1171

Baloloy AB, Blanco AC, Candido CG, Argamosa RJL, Dumalag JBLC, Dimapilis LLC, Paringit EC (2018) Estimation of mangrove forest aboveground biomass using multispectral bands, vegetation indices and biophysical variables derived from optical satellite imageries: rapideye, planetscope and Sentinel-2. In: ISPRS annals of the photogrammetry, remote sensing and spatial information sciences, vol IV-3, 2018 ISPRS TC III Mid-term Symposium "Developments, Technologies and Applications in Remote Sensing", 7–10 May, Beijing, China

Brandao FG (2007) Estimativa da altura total de eucalyptus sp. utiliandológica fuzzy e neuro fuzzy. Master's thesis dissertation, Universidade Federal de Lavras

Brown S (2002) Measuring carbon in forests: currebt status and future challanges. Environ Pollut 116:363–372

Clevers JGPW (1988) The derivation of a simplified reflectance model for the estimation of leaf area index. Remote Sens Environ 25:53–70

Deering DW, Rouse JJW, Haas RH, Schell JS (1975) Measuring forage production of grazing units home Landsat MSS data. In: Proceedings 10th international symposium remote sensing environment, Environmental Research Institute of Michigan, Ann Arbor, Michigan, pp 1169–1178

Diamantopoulou MJ (2005a) Artificial neural networks as an alternative tool in pine bark volume estimation. Comput Electron Agric 48:235–244

Diamantopoulou MJ (2005b) Predicting fir trees stem diameters using Artificial Neural Network models. Southern For: J For Sci 205:39–44

Diamantopoulou MJ (2005) Predicting fir trees stem diameters using artificial neural network models, southern forests. J S Afr For Assoc 20:39–44

Diamantopoulou MJ, Milios E (2010) Modelling total volume of dominant pine trees in reforestations via multivariate analysis and artificial neural network models. Biosys Eng 105:306–315

Diamantopoulou MJ, Özçelik R (2012) Evaluation of different modeling approaches for total tree-height estimation in Mediterranean Region of Turkey. For Syst 21(3):383–397

Diamantopoulou MJ, Özçelik R, Crecente-Campo F, Eler Ü (2015) Estimation of Weibull function parameters for modelling tree diameter distribution using least squares and artificial neural networks methods. Biosyst Eng 133:33–45

Dong L, Du H, Han N, Li X, Zhu D, Mao F, Zhang M, Zheng J, Liu H, Huang Z, He S (2020) Application of convolutional neural network on Lei Bamboo above-ground-biomass (AGB) estimation using Worldview-2. Remote Sens 12:958

Ercanlı İ, Kahriman A, Bolat F (2015) Applications of artificial neural network for predicting the relationships between height and age for oriental beech. In: The 10th international beech symposium, 1–6 Sept 2015, Kastamonu, Turkey

Ercanlı İ, Günlü A, Şenyurt M, Bolat F, Kahriman A (2016) Artificial neural network for predicting stand carbon stock from remote sensing data for even-aged scots pine (*pinussylvestris* L.) Stands in the Taşköprü-Çiftlik forests. In: 1st international symposium of forest engineering and technologies (FETEC 2016), Bursa Technical University, Faculty of Forestry, 2–4 June 2016, Bursa-Turkey

Ercanlı İ, Günlü A, Şenyurt M, Keleş S (2018) Artificial neural network models predicting the leaf area index: a case study in pure even-aged Crimean pine forests from Turkey. For Ecosyst 5–29

Ercanlı İ, Bolat F (2017) Diameter distribution Modeling based on Artificial Neural Networks for Kunduz Forests. In: International symposium on new horizons in forestry-ISFOR 2017. Isparta University, Faculty of Forestry, 18–20 Oct, Isparta, Turkey

Foody GM, Boyd DS, Cutler MEJ (2003) Predictive relations of tropical forest biomass from Landsat TM data and their transferability between regions. Remote Sens Environ 85:463–474

Görgens EB, Leite HG, Santos HN, Gleriani JM (2009) Estimação do volume de árvores utilizando redes neurais artificiais. Revista Árvore 33(6):1141–1147
Guan BT, Gertner G (1991) Modeling red pine tree survival with an artificial neural network. For Sci 37:1429–1440
Günlü A, Ercanlı İ (2020) Artificial neural network models by ALOS PALSAR data for aboveground stand carbon predictions of pure beech stands: a case study from northern of Turkey. Geocarto Int 35(1):17–28
Günlü A, Ercanlı İ, Başkent EZ, Çakır G (2014) Estimating aboveground biomass using Landsat TM imagery: a case study of Anatolian Crimean pine forests in Turkey. Ann For Res 57(2):289–298
Günlü A, Ercanlı İ, Şenyurt M, Keleş S (2016) Determining the amounts of aboveground carbon storage using landsat 8 satellite image in crimean pine stands (*Pinusnigra* Arnold. subsp. *Pallasiana* (Lamb.) Holmboe). In: International forestrysymposium, 07–10 Dec, Kastamonu, Turkey
Hasenauer H, Kinderman G (2002) Methods for assessing regeneration establishment and height growth in uneven-aged mixed species stands. Forestry 75(4):385–394
Hasenauer H, Merkl D, Weingartner M (2001) Estimating tree mortality of Norway spruce stands with neural networks. Adv Environ Res 5(4):405–414
Huete AR (1988) A soil adjusted vegetation index (SAVI). Remote Sens Environ 25:295–309
Jordan CF (1969) Derivation of leaf area index from quality of light on the forest floor. Ecology 50:663666
Kaipainen T, Liski J, Pussinen A, Karjalainen T (2004) Managing carbon sinks by changing rotation length in European forests. Environ Sci Policy 7:205–219
Kamble LV, Pangavhane DR, Singh TP (2015) Neural network optimization by comparing the performances of the training functions—prediction of heat transfer from horizontal tube immersed in gas–solid fluidized bed. Int J Heat Mass Transf 83:337–344
Karjalainen T, Pussinen A, Kellomaki S, Makipaa R (1999) Scenarios for the carbon balance of Finnish forests and wood products. Environ Sci Policy 2:165–175
Kaufman YJ, Tanre D (1992) Atmospherically resistant vegetation index (ARVI) for EOS-MODIS. IEEE Trans Geosci Remote Sens 30:261–270
Keleş S, Başkent EZ (2006) Orman Ekosistemlerindeki Karbon Değişiminin Orman Amenajman Planlarına Yansıtılması: Kavramsal Çerçeve ve Bir Örnek Uygulama (1. Bölüm). Orman ve Av Dergisi, Sayı 2, Cilt 83:36–41 (in Turkish)
Kelsey KC, Neff JC (2014) Estimates of aboveground biomass from texture analysis of Landsat imagery. Remote Sens 6:6407–6422
Lal R (2005) Forest soils and carbon sequestration. For Ecol Manage 220(1-3):242–258
Leite HG, Marques da Silva ML, Binoti DHB, Fardin L, Takizawa FH (2011) Estimation of inside-bark diameter and heartwood diameter for Tectonagrandis Linn. Trees using artificial neural networks. Eur J For Res 130:263–269
Li Y, Han N, Li X, Du H, Mao F, Cui L, Liu T, Xing L (2018) Spatiotemporal estimation of bamboo forest aboveground carbon storage based on Landsat data in Zhejiang, China. Remote Sens 10:898
Lim B, Brown S, Schlamadinger B (1999) Carbon accounting for forest harvesting and wood products: review and evaluation of different approaches. Environ Sci Policy 2:207–216
Lu D, Mausel P, Broondizio E, Moran E (2005) Relationships between forests stand parameters and Landsat TM spectral responses in the Brazilian Amazon Basin. For Ecol Manage 198:149–167
MATLAB (2014) MATLAB and statistics toolbox release 2014b, The Math Works, Inc., Natick, MA, USA
Miguel EP, Mota FCM, Menez IC, Téoeo SJ, Nascimento RGM, Leal F, Assis I, Pereira RS, Rezende AV (2016) Artificial intelligence tools in predicting the volume of trees within a forest stand. Afr J Agric Res 11:1914–1923
Milne R, Brown TA (1997) Carbon in the vegetation and soils of great Britain. J Environ Manage 49:413–433

Min L, Qu JJ, Xianjun H (2009) Estimating aboveground biomass for different forest types based on Landsat TM measurements. In: 17th international conference on geoinformatics. Proceeding of a meeting held on 12–14 Aug 2009, Fairfax. Virginia, pp 1–6

Myeong S, Nowak DJ, Duggin MJ (2006) A temporal analysis of urban forest carbon storage using remote sensing. Remote Sens Environ 101:277–282

Ney RA, Schnoor JL, Mancuso MA (2002) A methodology to estimate carbon storage and flux in forestland using existing forest and soils databases. Environ Monit Assess 78:291–307

Nowak DJ, Crane DE (2002) Carbon storage and sequestration by urban trees in the USA. Environ Pollut 116:381–389

Nunes, MH, Görgens EB (2016) Artificial intelligence procedures for tree taper estimation within a complex vegetation mosaic in Brazil. Plos One 11

Özçelik R, Diamantopoulou MJ, Wiant HR, Brooks JR (2008) Comparative study of standard and modern methods for estimating tree bole volume of three species in Turkey. For Products J 58(6):73–81

Özçelik R, Diamantopoulou MJ, Wiant HV, Brooks JR (2010) Estimating tree bole volume using artificial neural network models for four species in Turkey. J Environ Manage 91(3):742–753

Özçelik R, Diamantopoulou MJ, Crecente-Campo F, Eler U (2013) Estimating Crimean juniper tree height using nonlinear regression and artificial neural network models. For Ecol Manage 306:52–60

Özçelık R, Diamantopoulou MJ, Eker M, Gürlevık N (2017) Artificial neural network models: an alternative approach for reliable aboveground pine tree biomass prediction. For Sci 63(3):291–302

R Development Core Team (2018) R: A language and environment for statistical competta et al. 2017uting. R Foundation for Statistical Computing, Vienna, Austria

Rahman MM, Csaplovics E, Koch B (2008) Satellite estimation of forest carbon using regression models. Int J Remote Sens 29(23):6917–6936

Ranson KJ, Kimes D, Sun G, Nelson R, Kharuk V, Montesano P (2007) Using MODIS and GLAS data to develop timber volume estimates in central Siberia. In: IEEE International Geoscience and Remote Sensing Symposium, Barcelona, 2007, pp 2306–2309

Rouse JW, Haas RH, Deering DW, Schell JA, Harlan JC (1974) Monitoring the vernal advancement and retrogradation (green wave effect) of natural vegetation. NASA/GSFC Type III, Final Report, Greenbelt, MD

Safari A, Sohrabi H, Powell S, Shataee S (2017) Comparative assessment of multi-temporal Landsat 8 and machine learning algorithms for estimating aboveground carbon stock in coppice oak forests. Int J Remote Sens 38(22):6407–6432

Sakici OE, Seki M, Saglam F (2018) Above-ground biomass and carbon stock equations for Crimean pine stands in Kastamonu region of Turkey. Fresenius Environ Bull 27(10):7079–7089

Sanquetta CR, Dolci MC, Corte APD, Sanquetta MNI, Pelissari AL (2017) Form factors vs. regression models in volume estimation of Pinus taeda L. stem. Científica, Jaboticabal 45(2):175–181

Sivanpillai R, Smith CT, Srinivasan R, Messina MG, Benwu X (2006) Estimation of managed loblolly pine stands age and density with Landsat ETM data. For Ecol Manage 223:247–254

Soares FAA, Flôres EL, Cabacinha CD, Carrijo GA, Veiga ACP (2011) Recursive diameter prediction and volume calculation of eucalyptus trees using multilayer perceptron networks. Comput Electron Agric 78:19–27

Thapa B, Watanabe M, Motohka T, Shimada M (2015) Potential of high-resolution ALOS–PALSAR mosaic texture for aboveground forest carbon tracking in tropical region. Remote Sens Environ 160:122–133

Tolunay D (2014) Coefficients that can be used to calculate biomass and carbon amounts from increment and growing stock in Turkey. In: Proceedings of the International Symposium for the 50th Annıversary of the Forestry Sector Planning in Turkey, 26–28 November, pp 240–251, Antalya, Turkey

Turgut R, Günlü A (2020) Estimating aboveground biomass using Landsat 8 OLI satellite image in pure Crimean pine (*Pinusnigra* J.F. Arnold subsp. *Pallasiana* (Lamb.) Holmboe) stands: a casefromTurkey. Geocarto J. https://doi.org/10.1080/10106049.2020.1737971

Vapnik V (1995) The nature of statistical learning theory. Springer, New York, p 334

Vapnik V, Chervonenkis A (1964) A note on one class of perceptrons. Autom Remote Control 25

Vapnik V, Lerner A (1963) Pattern recognition using generalized portrait method. Autom Remote Control 24:774–780

Wang J, Neskovic P, Cooper LN (2005) Training data selectionforsupportvectormachines. In: International conference on natural computation. Springer, Berlin, pp 554–564

Xu X, Du H, Zhou G, Ge H, Shi Y, Zhou Y, Fan W, Fan W (2011) Estimation of aboveground carbon stock of Moso bamboo (Phyllostachysheterocycla var. pubescens) forest with a Landsat Thematic Mapper image. Int J Remote Sens 32(5):1431–1448

Chapter 17
Spatio-temporal Variation of Evapotranspiration Derived from Multi-temporal Landsat Datasets using FAO-56 Penman-Monteith Method

Tanushree Basu Roy, Dipanwita Dutta, and Abhisek Chakrabarty

Abstract Reference evapotranspiration (ET_0) is the representation of real-time crop-specific measurement of evapotranspiration and could be used for measuring the available water for agriculture. Accurate estimation of reference evapotranspiration (ET_0) is required for irrigation management and water resource allocation. Satellite remote sensing provides an opportunity to estimate its quantity and map the spatio-temporal distribution of evapotranspiration in an efficient way. There are several methods developed for estimating ET_0 but most of them are mainly based on daily meteorological data provided by weather station networks. This paper aims to estimate the monthly reference evapotranspiration (ET_0) by the FAO-56 Penman-Monteith method using the remote sensing data (LANDSAT 8-OLI and LANDSAT 7-ETM+) of 2014, 2015, 2016 and weather data (Maximum and minimum temperature, Dew point temperature, wind speed, relative humidity) over the Dwarakeswar river basin. Input parameters required for this model are emissivity, land surface temperature (LST), net radiation, soil heat flux (G), air temperature, actual and saturation vapor pressure and wind speed. This study indicates that evapotranspiration variation in this area is closely related to crop growth. Evapotranspiration values were found low (66–120 mm/month) when paddy fields were empty and the fields were covered by very sparse vegetation. Whereas, the estimated values were high (120–180 mm/month) in cropping season and in monsoon, when vegetation cover was dense. Furthermore, the evapotranspiration estimation results were analyzed and validated with MODIS data which shows a good agreement between them.

Keywords Reference evapotranspiration · Landsat · Penman-Monteith method

---

T. Basu Roy · D. Dutta (✉) · A. Chakrabarty
Department of Remote Sensing and GIS, Vidyasagar University, Midnapore, West Bengal, India
e-mail: duttarsgisvu@gmail.com

P. K. Shit et al. (eds.), *Spatial Modeling in Forest Resources Management*, Environmental Science and Engineering,
https://doi.org/10.1007/978-3-030-56542-8_17

17.1 Introduction

Estimation of evapotranspiration (ET) from vegetation has become increasingly important in agriculture, irrigation management and planning. Evapotranspiration by forest vegetation is a crucial component to understand the water budget of the land surface (Chen and Liu 2020). Estimation of evapotranspiration is important to know the amount of water required during growing season, improve crop water management and hydrological applications as well as to conserve soil and water resources. It is also useful to understand the water and energy cycles, environmental stress on natural and agricultural ecosystems, and monitoring drought (French et al. 2015). The redistribution of water on Earth's surface is mostly done by the evapotranspiration process, as the surface absorbs approximately 46–50% of the incoming solar radiation globally by the evapotranspiration process (Verstraeten et al. 2008). Evapotranspiration (ET), the fundamental process of the hydrological cycle is a combination of two separate processes, evaporation from soil-water surfaces and plant transpiration (Monteith and Unsworth 2013; Xiong et al. 2015). Evaporation is the process by which liquid water is converted into water vapor to the air from sources such as soils, vegetations, water bodies etc. Transpiration involves condensed water vaporization from plant tissues and atmospheric vapor elimination (Zotarelli et al. 2014). Its rate depends on crop characteristics, environmental aspects, and cultivation practices. Evaporation and transpiration occur simultaneously and both processes depend on solar radiation, air temperature, relative humidity and wind speed (Seenu et al. 2019). The evapotranspiration rate from an extensive surface, uniformly covered with actively growing vegetation having specified height, surface resistance, specific characteristics and with adequate water status in the soil, is called reference evapotranspiration (ET_0) (Da Rocha et al. 2009). It reflects the energy available for water evaporation and the wind used to move water vapors to the lower atmosphere from the ground. The ET_0 concept has been developed to investigate the evaporative demand of the atmosphere irrespective of crop size, crop growth and management practices. Soil factors do not affect ET_0 if adequate water is available at the reference surface.

Factors Affecting Evapotranspiration

1. **Weather Parameters**
   The major weather parameters controlling evapotranspiration are solar radiation, wind speed, air temperature and humidity.
2. **Crop Factors**
   Crop types, development stages or level of maturity are considered as important factors for estimating ET_0. Apart from these, crop height, crop surface roughness, reflection, ground cover and crop routing also affect its values.
3. **Management and environmental condition**

Evapotranspiration can be decreased by factors such as soil salinity, low soil fertility, minimal application of soil fertilizers, presence of rough or impenetrable horizons, poor soil management, and the absence of disease and pesticide control (Allen and Pruitt 1991). Ground cover, percentage of soil cover, plant density and soil water content can also affect the ET rate.

Weighing lysimeter are the most precise technique for the ET_0 assessment (Xu and Chen 2005). But, there are very limited lysimeter data is available and therefore the alternative methods have been developed (Cruz-Blanco et al. 2014). The ET_0 models from a remote sensing perspective can be divided into two groups, (1) temperature-based ET_0 and (2) conductance-based ET_0 models. Temperature based ET_0 models are appropriate for water stress conditions in sparsely vegetated areas and conductance-based ET models are well suited for both water-stressed and energy-limited conditions (Chen and Liu 2020). Temperature based models are based on surface temperature measurements; therefore, they are unable to estimate ET where heat flux is minimal. But conductance-based ET models avoid these issues of surface temperature, rather than using vegetation structural information which can be easily retrieved from satellite data and therefore, these conductance-based ET models are the best option for land surface modeling (Zeng 2003). Penman-Monteith method is a conductance-based ET model that depends on wind speed, vegetation height, surface roughness and atmospheric stability. Previous researches proved good agreement between PM-FAO56 and lysimeter measurements in semi-arid area with very precise ET_0 estimates (López-Urrea et al. 2006). This method is being considered as a standard ET_0 estimated method, which gives satisfactory results better than other methods under any region or climatic conditions (Chiew et al. 1995). High resolution satellite data is necessary for accurate ET estimations of different land-use types at the regional scale. As the LANDSAT series has a higher spatial resolution, LANDSAT OLI and LANDSAT ETM+ images were used in this study.

The aim of this study is to estimate monthly reference evapotranspiration (ET_0) for three years using remote sensing techniques which is important in agricultural and irrigation management. Irrigation is important to reduce rain dependency during the monsoon season and to help agriculture during non-monsoon season. The goal was achieved by fulfilling the following objectives:

- To estimate Net Radiation, Soil Heat Flux and Psychrometric Coefficient using remote sensing technique.
- To measure Gradient of Vapor Pressure curve, Actual Vapor pressure and Mean Saturated Vapor pressure using weather data.

17.2 Study Area and Dataset

17.2.1 Study Area

Dwarakeswar River basin has been selected as the study area (Fig. 17.1). The Dwarakeswar River is one of the major rivers of the western part of West Bengal originates from Tilaboni hill, and then flows through the hilly terrain of Purulia district

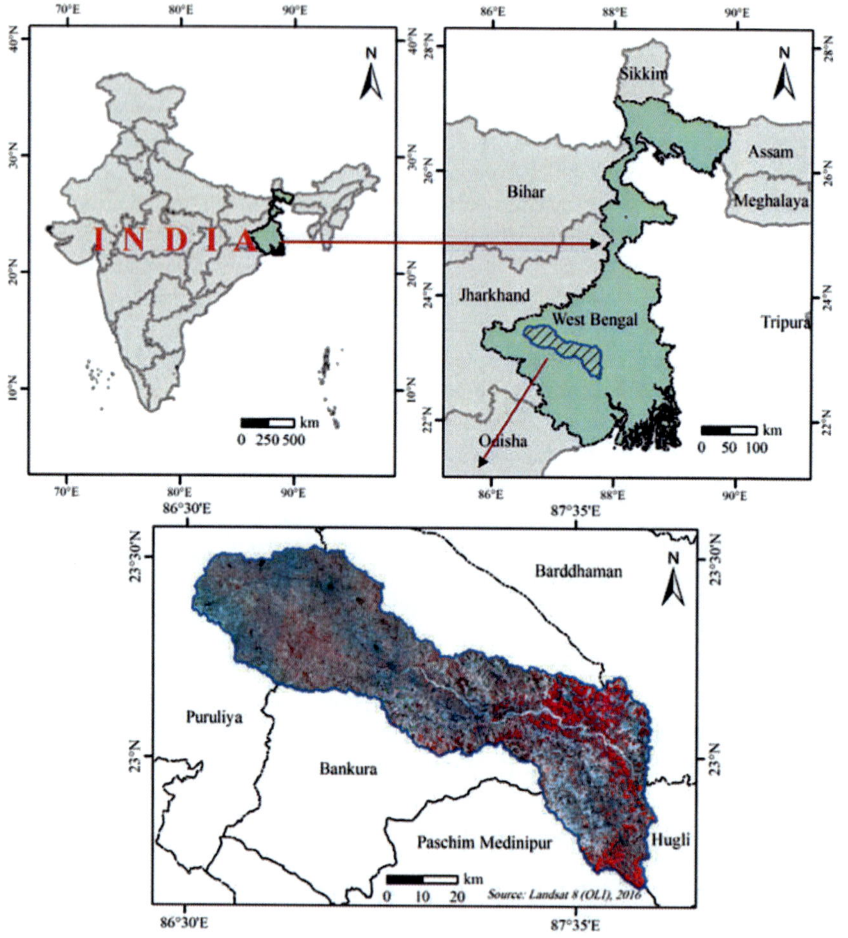

Fig. 17.1 Study area

from northwest to southeast direction. The river divides Bankura into two halves and then enters the south-eastern tip of Burdwan District. It flows through Hooghly District and ultimately joins River Silabati and forms Rupnarayan River.

The Dwarakeswar river basin, covering approximately 4341.765 km^2, is located at 23° 9′ 9″N and 87° 15′ 39″E on the southern part of the West Bengal (Fig. 17.2). The length of this river is approximately 214 km. It originates from an extended part of Chotanagpur plateau in Purulia District and ended in the alluvial tract of West Midnapore of West Bengal. The basin area is included in the survey of India, toposheet numbers 73-I/11; 73-I/12; 73-I/15; 73-I/16; 73-M/4 and 73-M/3.

Drought is a common phenomenon in the Bankura and Purulia districts of West Bengal where proper irrigation system is essential for crop growth. Estimating

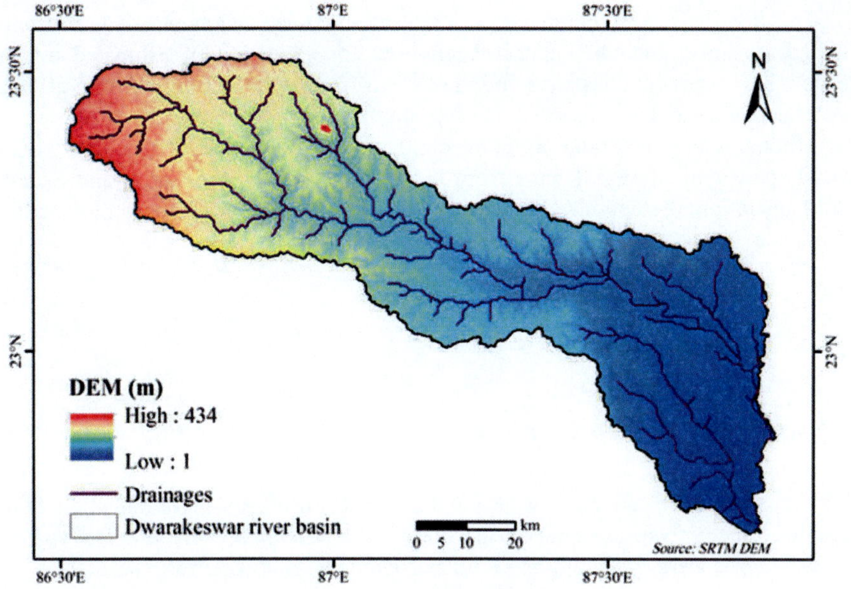

Fig. 17.2 Drainage and elevation map of the Dwarakeswar river basin

evapotranspiration rate in this region is crucial for improving the management, design and exploitation of the irrigation systems. It is important to note that the Dwarakeswar River originates from a region with having a history of chronic drought with insufficient precipitation.

The area falls under tropical dry sub-humid climate with varying annual precipitation from 1100 mm in the western part to 1400 mm in the eastern part. The mean daily temperature ranges from 12 °C (in winter) to a maximum of 46 °C (in summer). The diurnal range of temperature varies from 4 to 9 °C. Dry spells during drought years usually last for weeks which affects the crop growth and yields adversely.

17.2.2 Datasets

The Landsat 8 OLI TIRS, 7 ETM+ and SRTM DEM data for the years 2014, 2015 and 2016 were used for assessing the reference evapotranspiration. Twenty LANDSAT 8 OLI/TIRS and five LANDSAT 7 ETM+ images have been used for this work, which was obtained from the Earth Explorer, United States Geological Survey (USGS) site (https://glovis.usgs.gov/).

MODIS ET (February 2015) product downloaded from https://www.ntsg.umt.edu/data has been used for validation of estimated reference ET. This product has a resolution of 1-km^2 for the 109.03 million km^2 global vegetated areas and the algorithm is based on the Penman-Monteith method.

Minimum and maximum air temperature, dew point temperature, humidity and wind speed data were collected from NASA's (National Aeronautics and Space Administration) Prediction of Worldwide Energy Resources (https://power.larc.nasa.gov/) site.

17.3 Methodology

17.3.1 Image Pre-processing

As the images were collected for several dates under different solar illumination, Sun Angle correction was essential for the study. After Sun angle correction DN values of the images were converted into Radiance using the following formulas.

For LANDSAT 8

$$M_L \times band10 + A_L \tag{17.1}$$

$M_L = 0.0003342$, represents the band-specific multiplicative rescaling factor, $A_L = 0.1$, is the band-specific additive rescaling factor.

For LANDSAT 7

$$L_\lambda = \frac{(LMAX_\lambda - LMIN_\lambda)}{QCALMAX - QCALMIN} \times (DN - QCALMIN) + LMIN_\lambda) \tag{17.2}$$

LMIN and LMAX are the spectral radiances for each band at digital numbers 1 and 255, DN is the pixel DN value, λ is the wavelength.

The pre-processing of satellite dataset including FLAASH correction was carried out by image processing software. The detailed methodological flow chart of the study provided in Fig. 17.3.

17.3.2 Estimation of Land Surface Temperature (LST)

Land Surface Temperature is the radiative skin temperature of the ground. In most cases, it is a mixture of vegetation and bare soil temperatures. Most of the remotely

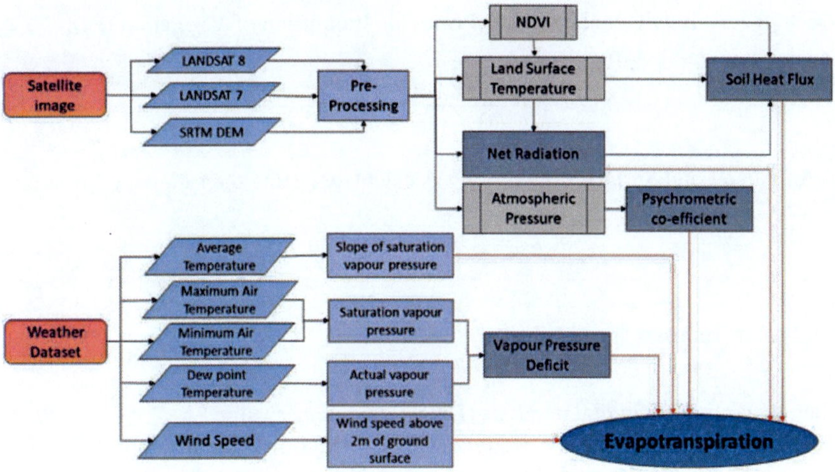

Fig. 17.3 Flow chart of the methodology

sensed ET models used land surface temperature as a primary boundary condition (Wang et al. 2016).

For the LANDSAT 7 Image

$$LST = \frac{K_2}{\ln[(K_1/L_\lambda) + 1]} \tag{17.3}$$

K_1 and K_2 stand for the band-specific thermal conversion constants from the metadata.

For the LANDSAT 8 Image

The input parameters, i.e. NDVI, Land Surface Emissivity, and Brightness Temperature were derived from the Landsat 8 images to estimate LST.

Normalized Difference Vegetation Index (NDVI)

The NDVI is a graphical indicator used extensively to monitor vegetation (Rouse et al. 1974). It represents the modulation ratio of the near-infrared and red bands. The NDVI values range from -1 to $+1$.

$$NDVI = \frac{NIR - R}{NIR + R} \tag{17.4}$$

Land Surface Emissivity (ε_λ)

The emissivity of a material is the ability of its surface to emit energy. It is the ratio of energy radiated by a particular material to energy radiated by a blackbody at the same temperature.

Surface Emissivity was calculated from the Proportion of Vegetation (P_v).

$$P_V = \left(\frac{NDVI - NDVI_{MIN}}{NDVI_{MAX} - NDVI_{MIN}}\right)^2 \tag{17.5}$$

And then Land Surface Emissivity was calculated using the following formula:

$$\varepsilon_\lambda = 0.004 P_V + 0.986 \tag{17.6}$$

Brightness Temperature (BT)
Conversion of Radiance to At-Sensor Temperature:

After converting the digital numbers (DNs) into radiance using Eq. (17.1), the TIRS band data was converted from spectral radiance to brightness temperature (BT) using the thermal constants provided in the metadata file.

$$BT = \frac{K_2}{\ln[(K_1/L_\lambda) + 1]} - 273.15 \tag{17.7}$$

where K_1 and K_2 are band-specific thermal conversion constants obtained from the metadata.

$$LST = \frac{BT}{\{1 + [(\lambda BT/\rho)\ln \varepsilon_\lambda]\}} \tag{17.8}$$

LST and BT in Celsius (°C), and λ is the wavelength of emitted radiance (λ = 10.895).

$$\rho = h\frac{c}{\sigma} = 1.438 \times 10^{-2}\,\text{mK} \tag{17.9}$$

where σ is the Boltzmann constant (1.38×10^{-23} J/K), h is Planck's constant (6.626×10^{-34} J s), and c is the velocity of light (2.998×10^8 m/s).

17.3.3 Estimation of Evapotranspiration (ET)

In last 50 years, a large number of empirical methods have been developed for estimating the evapotranspiration using different climatic variables. It was Penman, (1948) who was the first to establish an evaporation model which is based on an energy balance formula that supports the vaporization of latent heat and as well as wind speed and air-water vapor deficit. But the Penman model explains the influence of wind on evaporation using an empirical model and not directly taking into account the effects of surface conditions such as vegetation density, vegetation height, soil water content etc. Then, Monteith (1973) improved the model by using an aerodynamic conductance which depends not only on wind speed but also on surface ruggedness, atmospheric stability, vegetation height. This modified model is known as the Penman-Monteith model.

$$\lambda ET_0 = \frac{\Delta(R_n - G) + \left[86{,}400\frac{\rho_a C_p (e_s^0 - e_a)}{r_{av}}\right]}{\Delta + \gamma\left(1 + \frac{r_s}{r_{av}}\right)} \tag{17.10}$$

where ρ_a = air density (kg m^{-3}), C_p = specific heat of dry air, $e_s{}^o$ = mean saturated vapor pressure (KPa), r_{av} = bulk surface aerodynamic resistance for water vapor (s m^{-1}), e_a = mean daily ambient vapor pressure (KPa), and r_s = the canopy surface resistance (s m^{-1}).

The Penman-Monteith method was recommended as a standard procedure for estimating Reference Evapotranspiration (ET_0) in the meeting (1990) of the United Nations Food and Agriculture Organization (FAO) to review the "Irrigation and Drainage Paper 24" document aiming to determine new ET estimation procedures (FAO 2017). In 1990, FAO organized a consultation with the Internal Commission for Irrigation and Drainage and World Meteorological Organization, to update the previous methodologies and procedures on crop water requirements. The experts and researchers updated and simplified the Penman-Monteith Eq. (17.10) by using some assumed constant parameters such as a grass reference crop of height 0.12 m with surface resistance of 70 s m^{-1} and an albedo value of 0.23 (Smith et al. 1998).

The new equation was:

$$ET_0 = \frac{0.408\Delta(R_n - G) + \gamma\frac{900}{T+273}u_2(e_s - e_a)}{\Delta + \gamma(1 + 0.34u_2)} \tag{17.11}$$

where ET_o = reference evapotranspiration (mm day^{-1}); Rn = net radiation at the crop surface (MJ m^{-2} d^{-1}); G = soil heat flux density (MJ m^{-2} d^{-1}); T = mean daily air temperature at 2 m height (°C); u_2 = wind speed at 2 m height (m s^{-1}); e_s = saturation vapor pressure (kPa); e_a = actual vapor pressure (kPa); $e_s - e_a$ = saturation vapor pressure deficit (kPa); Δ = slope of the vapor pressure curve (kPa $°C^{-1}$); γ = psychrometric constant (kPa $°C^{-1}$). The parameters of the above equation described below.

17.3.3.1 Net Surface Radiance (R_n)

The radiation balance at the earth's surface is composed of four spectral radiation fluxes, incoming short wave radiation (0.14–4 μm) that arrives from the sun (Rs↓), the amount of the energy that is reflected from the surface (Rs↑), incoming long wave (>4 μm) radiation from the atmosphere (R_L↓) and the amount of long wave radiation emitted from the surface (R_L↑) (Krishna et al. 2014).

$$R_n = (1-\alpha)R_S\downarrow + (R_L\downarrow - R_L\uparrow) - (1-\varepsilon_\lambda)R_L\downarrow \quad (17.12)$$

where α is the surface albedo, ε_λ is surface emissivity, $R_S\downarrow$ = Incoming Short wave Radiation, $R_L\uparrow$ = Outgoing Long wave Radiation, $R_L\downarrow$ = Incoming Long wave Radiation.

Incoming Short Wave Radiation

$$R_S\downarrow = \tau S E_0 \cos\theta \quad (17.13)$$

where τ = atmospheric transmissivity,

$$\tau = 0.75 + (2\times 10^{-5}\times z) \quad (17.14)$$

z = elevation in m, S = solar constant = 1367 W/m^2,

E_0 = eccentricity correction factor of the Earth's orbit about the Sun and it was calculated by the following formula,

$$E_0 = 1 + 0.033\cos(2\pi d_n/365) \quad (17.15)$$

d_n = Julian day of the year (between 1 and 365/366), θ = solar zenith angle = (90°-sun elevation).

Outgoing Long wave Radiation

$$RL\uparrow = \varepsilon_a \sigma T_a^4 \quad (17.16)$$

where T_a = air temperature, σ = Stefan-Boltzman constant = 5.67×10^{-8} $W/m^2\,K^4$,

ε_a = atmospheric emissivity,

$$\varepsilon_a = 0.85\times(-\ln\tau)^{0.09} \quad (17.17)$$

Albedo (α)

Albedo, the measure for reflectance or optical brilliance is dimensionless and estimated on a scale from zero (relating to a dark body that ingests all incidental radiation) to one (comparing to a white body that mirrors all incidental radiation) (Liang et al. 2003; He et al. 2015). For the hypothetical crop reference surface, the albedo value was considered to be 0.23.

17.3.3.2 Soil Heat Flux Density (G)

The soil heat flux is the surface energy that is heating the soil. It is positive when soil is warming and negative when the soil is cooling and it varies according to the soil type and water content, vegetation type, and micro-climate (Hatfield et al. 2005). Different empirical equations were developed for calculating soil heat flux. Here, the following equation (17.18) developed by Bastiaanssen et al. (1998) was considered for this study:

$$G = R_n\left[\left(\frac{LST}{\alpha}\right)(0.0038\alpha + 0.0074\alpha^2)(1 - 0.98NDVI^4)\right] \quad (17.18)$$

17.3.3.3 Actual Vapor Pressure (e_a)

Actual Vapor Pressure is a measurement of the amount of water vapor in a volume of air and increases as the amount of water vapor increases.

$$e_a = 0.6108\exp\left[\frac{17.27T_{dew}}{T_{dew} + 237.3}\right] \quad (17.19)$$

where T_{dew} = dew point temperature (°C).

17.3.3.4 Mean Saturated Vapor Pressure (e_s)

Since, saturation vapor pressure is directly related with air temperature it can be calculated using the air temperature as input variable.

$$e_s = \frac{e_{(T\max)} + e_{(T\min)}}{2} \quad (17.20)$$

$e_{(T\max)}$ = Saturation vapor pressure for maximum air temperature, obtained from the following formula:

$$e_{(T\max)} = 0.6108\exp\left[\frac{17.27T_{\max}}{T_{\max} + 237.3}\right] \quad (17.21)$$

where $T_{\max}$ is the maximum air temperature in °C.

$e_{(T\min)}$ = Saturation vapor pressure for minimum air temperature, obtained from the following formula:

$$e_{(T\min)} = 0.6108\exp\left[\frac{17.27T_{\min}}{T_{\min} + 237.3}\right] \quad (17.22)$$

where T_{min} is the minimum air temperature in °C.

17.3.3.5 Slope of Saturation Vapor Pressure Curve (Δ)

Slope of the Saturation vapor pressure curve was obtained from the following formula.

$$\Delta = \frac{4098\left[0.6108\exp\left(\frac{17.27T_{mean}}{T_{mean}+237.3}\right)\right]}{(T_{mean}+237.3)^2} \tag{17.23}$$

where T_{mean} is the mean air temperature in °C.

17.3.3.6 Psychrometric Constant (γ)

It is the ratio of specific heat of humid air at constant pressure (C_p) to the vaporization of latent heat. The value specific heat depends on the air humidity. The C_p value of 101.013 x 10^{-3} MJ kg^{-1} °C^{-1} can be used for average atmospheric conditions. Since the average atmospheric pressure was considered for each location, the psychrometric constant for each location was kept constant depending upon the altitude.

$$\gamma = \frac{C_p P}{\varepsilon\lambda} = 0.000665P \tag{17.24}$$

where P = atmospheric pressure (kPa),

$$P = 101.3\left[\frac{293 - 0.0065z}{293}\right]^{5.26} \tag{17.25}$$

λ = latent heat of vaporization = 2.45 (MJ kg^{-1}), C_p = specific heat at constant pressure = 1.013 × 10^{-3} (MJ kg^{-1} °C^{-1}), ε = ratio molecular weight of water vapor/dry air = 0.622.

17.3.3.7 Wind Speed (u_2)

The average daily wind speed in meters per second (ms^{-1}) measured at 2 m above the ground level was required.

$$u_2 = u_h \frac{4.87}{\ln(67.8h - 5.42)} \tag{17.26}$$

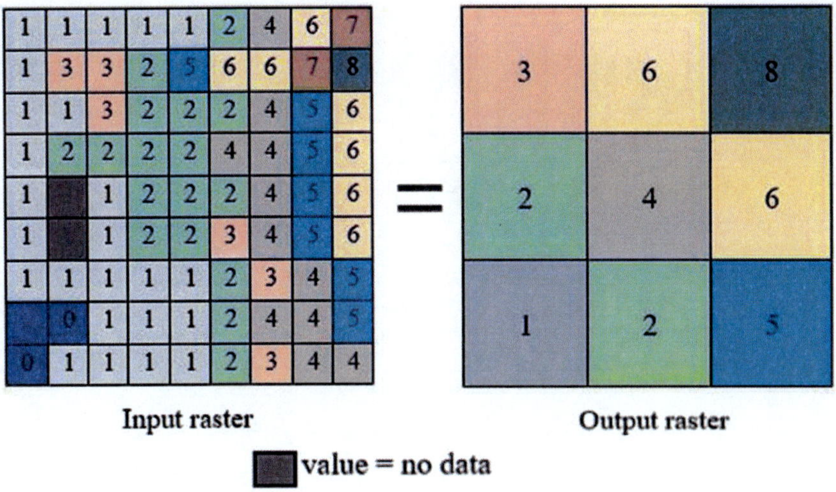

Fig. 17.4 The process of upscaling ET_0 image

where u_h = measured wind speed at height h (ms^{-1}) and h is the height of wind measurements above the ground surface (m).

17.3.4 Upscaling of Estimated ET_0

The estimated reference evapotranspiration (ET_0) of February 2015 was validated using MODIS ET products. The resolution of Landsat derived ET_0 was 30 m × 30 m whereas the image resolution of MODIS is 1 km × 1 km. The aggregation, upscaling method was used to generate a reduced-resolution version of the estimated ET_0 (Fig. 17.4).

17.4 Results and Discussion

17.4.1 Seasonal Variation of Reference Evapotranspiration (ET_0) Over Dwarakeswar River Basin

The spatio-temporal variation of estimated evapotranspiration rate was found coherent with the seasonal variation of temperature, precipitation, sowing time, harvesting time, crop height etc. In every year, the monthly ET_0 value followed an increasing trend starting from January–February to October and again decreased with the same rate in the winter months, November and December (Fig. 17.5). The varia-

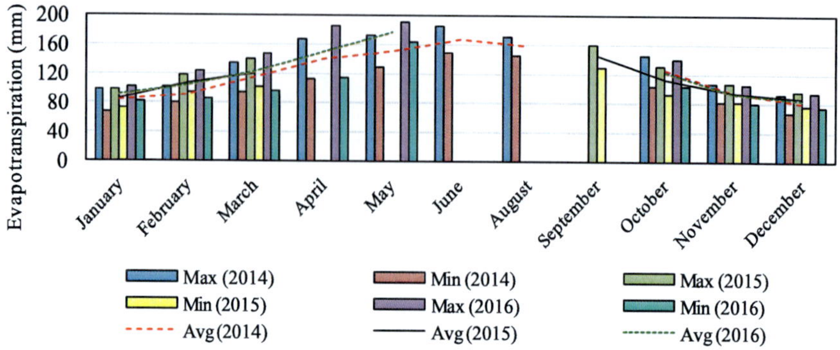

Fig. 17.5 Temporal variation of ET_0 of 2014, 2015 and 2016

tion of ET_0 value can be explained with the monthly variation of rainfall-temperature conditions. The low ET_0 value as observed in the months of January and February can be explained by cold temperature and limited rainfall. In March, April and May, ET_0 value was relatively higher than the previous two months because of high temperature causing significant increase in evaporation and transpiration. Since the areas receive monsoonal rainfall in the months of June and August ET_0 values were found high in these months. During post monsoonal period, especially in October the estimated ET_0 was high mainly due to high level of soil moisture. However, it followed an decreasing trend after October and the ET_0 value was low throughout in winter months.

The ET_0 value was comparatively higher in the year 2016 than the remaining years. In October 2015, the ET_0 rate was comparatively low, indicating a drought like condition. Likewise, in the pre-monsoon season of 2014, the values were remarkably less however, during post-monsoon (November–December) there was no significant variation in ET_0 values.

17.4.2 Variation of ET_0 Over Different LULC

While comparing the variation of evapotranspiration over different land use land cover of the area the study reveals high ET_0 in the cultivated areas (Figs. 17.6, 17.7 and 17.8). Though Purulia and Bankura are drought-prone areas, the productivity of crops especially in kharif season remains good in the years with moderate to high rainfall. Paddy, wheat, potato, mustard, til etc. are few major crops in these areas. The study reveals a good agreement between variation of ET_0 and cropping pattern of the area. It is worthy to mention that very low evapotranspiration in few areas of Purulia and Bankura districts is associated with sparse vegetation or bare soil. Very high ET_0 value as estimated during monsoon in the lower part covering portion of West Medinipur can be explained by rich kharif cultivation.

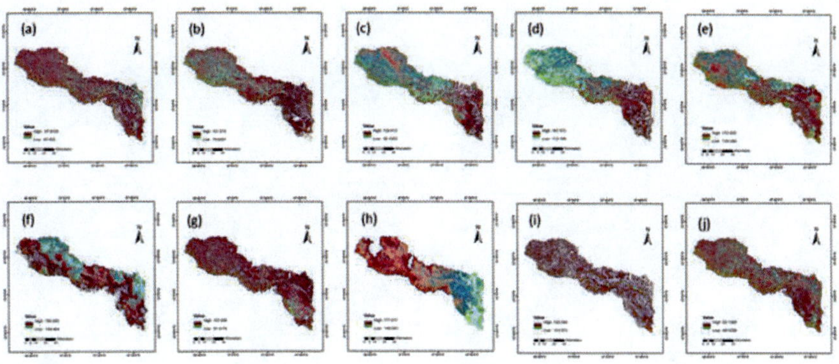

Fig. 17.6 Spatio-temporal variation of ET_0 in January–June (**a–f**), August (**g**), October–December (**h–j**), 2014

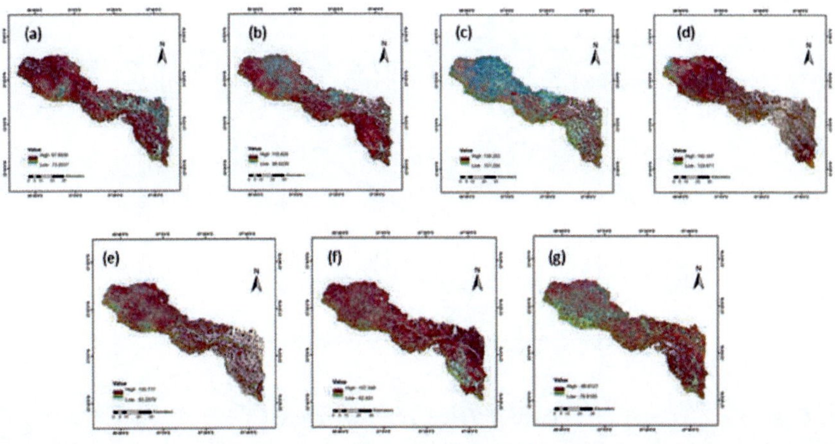

Fig. 17.7 Spatio-temporal variation of ET_0 in January–March (**a–c**), September–December (**d–g**), 2015

17.4.2.1 Correlation between Land Surface Temperature (LST) and Evapotranspiration (ET_0)

In order to identify the interrelationship between land surface temperature and ET_0 the correlation analysis was carried out which reveals a strong positive correlation between them (Fig. 17.9). As the temperature increases, the rate of evaporation and transpiration also increase. In tune with this fact, the Evapotranspiration rate was found higher in the areas with high land surface temperature.

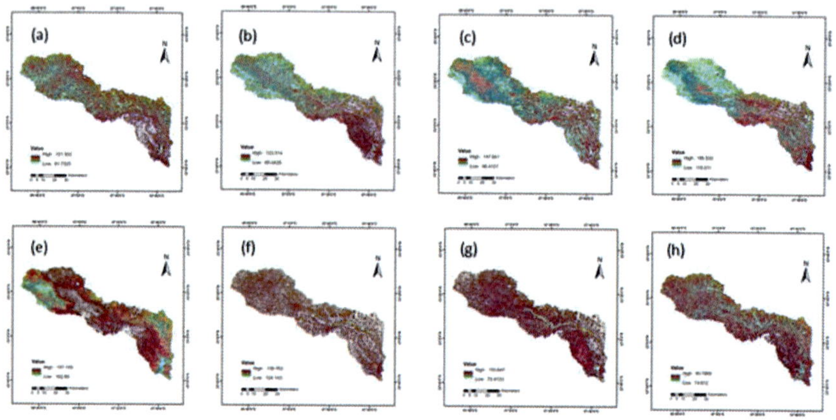

Fig. 17.8 Spatio-temporal variation of ET_0 in January–May (**a–e**), October–December (**f–h**), 2016

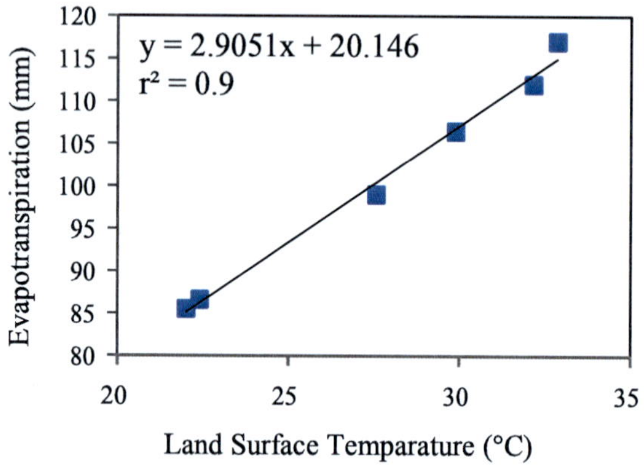

Fig. 17.9 Correlation between Land Surface Temperature (LST) and Evapotranspiration (ET_0)

17.4.2.2 Correlation Among Soil Heat Flux (G), Land Surface Temperature (LST) and Evapotranspiration (ET_0)

The thermal dynamics of soil are important for environmental processes as it plays an important role in controlling other variables. Soil heat is the conversion of incoming solar radiation into heat and the distribution of the corresponding heat flux which include long wave radiation, sensible heat flux caused by convection of air, latent heat flux resulting from evaporating water, heat flux into the ground, and strongly depends on temperature and soil water content. Therefore, Evapotranspiration is highly dependent on Soil Heat Flux and the present study also reveals a strong positive correlation between Soil Heat Flux and ET_0 (Fig. 17.10).

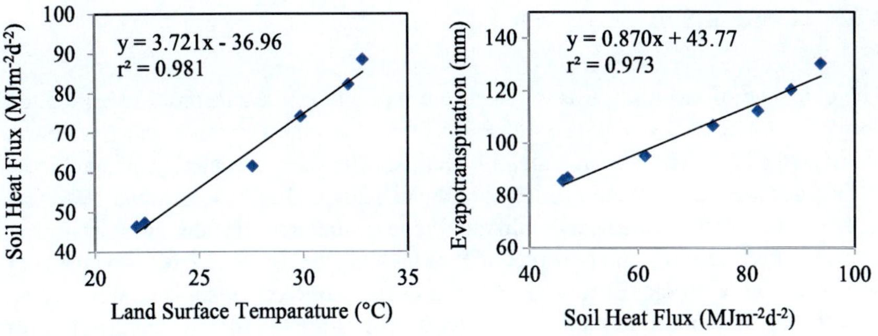

Fig. 17.10 Correlation among soil heat flux (G), land surface temperature (LST) and evapotranspiration (ET_0)

17.4.3 Validation of Estimated ET_0

The estimated ET_0 image of February 2015 was compared with the MODIS ET product of the same month and year (Table 17.1). The absolute error was found 6.12 (relative error 6.17%) for minimum ET_0 values, and it was 7.82 (relative error 6.85%) for maximum ET_0 value. Thus, the comparative analysis between Landsat derived and MODIS ET_0 reveals a good agreement between them (Fig. 17.11).

Table 17.1 Comparison between the estimated and the observed results of ET in 2015

	Minimum ET (°C)				Maximum ET (°C)			
Month	Estimated value	Observed value	Absolute error	Relative error (%)	Estimated value	Observed value	Absolute error	Relative error (%)
Feb	100.399	94.2	6.199	6.1743	114.12	106.3	7.82	6.8524

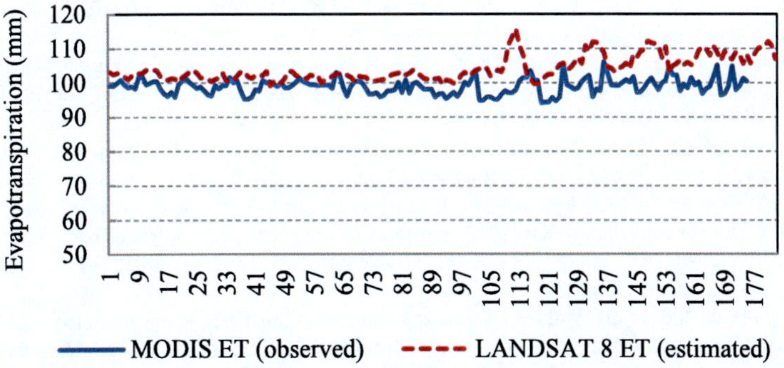

Fig. 17.11 Comparison between MODIS and LANDSAT 8 derived ET_0

17.5 Conclusion

The objective of the study was to assess the monthly evapotranspiration estimation derived from LANDSAT data and SRTM DEM over the Dwarakeswar river basin based on the FAO-56 Penman-Monteith method. The study revealed good agreement in the spatio-temporal pattern of Evapotranspiration and NDVI. Estimated ET was higher in the forested areas and cultivated lands, whereas very low evapotranspiration was observed in some portions of Purulia and Bankura districts with sparse or no vegetation coverage. Evapotranspiration change in River Basin is related to crop growth. The maximum monthly ET_0 values were estimated in irrigated cropland of the entire study area and especially, during the harvesting season. However, it was notably less in other months due lack of soil moisture in the croplands. The procedure for ET_0 assessment used for irrigation planning has shown itself as an essential factor in irrigation water management. In the summer season, a significant portion of agricultural lands of this regionis dependent upon irrigation. Spatial variation in yields of major crops are mainly caused by differences in irrigation scheduling, especially during the dry season when requirement of water is crucial for crop growth. Estimation of ET_0 is significant for determining the quality of the irrigation scheduling and as a whole, assessing the irrigation water management in an area.

References

Allen RG, Pruitt WO (1991) FAO-24 reference evapotranspiration factors. J Irrig Drain Eng ASCE 117(5):758–773

Bastiaanssen WGM et al (1998) A remote sensing surface energy balance algorithm for land (SEBAL): 2. Validation. J Hydrol 212–213(1–4):213–229. https://doi.org/10.1016/S0022-1694(98)00254-6

Chen JM, Liu J (2020) Evolution of evapotranspiration models using thermal and shortwave remote sensing data. Remote Sens Environ 237:111594. https://doi.org/10.1016/j.rse.2019.111594

Chiew FHS et al (1995) Penman-Monteith, FAO-24 reference crop evapotranspiration and class-A pan data in Australia. Agric Water Manage 28(1):9–21. https://doi.org/10.1016/0378-3774(95)01172-F

Cruz-Blanco M et al (2014) Assessment of reference evapotranspiration using remote sensing and forecasting tools under semi-arid conditions. Int J Appl Earth Observ Geoinf 33(1):280–289. https://doi.org/10.1016/j.jag.2014.06.008

Da Rocha HR, Manzi AO, Shuttleworth J (2009) Evapotranspiration. Geophys Monogr Ser 186:261–272. https://doi.org/10.4324/9781315081069-13

FAO (2017) Chapter 2—FAO Penman-Monteith equation, pp 1–12

French AN, Hunsaker DJ, Thorp KR (2015) Remote sensing of evapotranspiration over cotton using the TSEB and METRIC energy balance models. Remote Sens Environ 158:281–294. https://doi.org/10.1016/j.rse.2014.11.003

Hatfield JL et al (2005) Soil Heat Flux (January). https://doi.org/10.2134/agronmonogr47.c7

He T et al (2015) Land surface albedo estimation from Chinese HJ satellite data based on the direct estimation approach. Remote Sens 7(5):5495–5510. https://doi.org/10.3390/rs70505495

Krishna SVSS, Manavalan P, Rao PVN (2014) Estimation of net radiation using satellite based data inputs. Int Arch Photogramm Remote Sens Spatial Inf Sci-ISPRS Arch XL–8(1):307–313. https://doi.org/10.5194/isprsarchives-XL-8-307-2014

Liang S et al (2003) Narrowband to broadband conversions of land surface albedo: II. Validation. Remote Sens Environ 84(1):25–41. https://doi.org/10.1016/S0034-4257(02)00068-8

López-Urrea R et al (2006) Testing evapotranspiration equations using lysimeter observations in a semiarid climate. Agric Water Manage 85(1–2):15–26. https://doi.org/10.1016/j.agwat.2006.03.014

Monteith JL, Unsworth MH (2013) Principles of environmental physics: plants, animals, and the atmosphere. 4th Edition, pp 422

Monteith JL (1973) Principles of environmental physics, Edward Arnold, London, pp 241

Penman HL (1948) Natural evaporation from open water, hare soil and grass. Proc R Soc Lond Ser A Math Phys Sci 193(1032):120–145. https://doi.org/10.1098/rspa.1948.0037

Rouse JW, Haas RH, Schell JA, Deering DW (1974) Monitoring vegetation systems in the Great Plains with ERTS, In: Freden SC, Mercanti EP, Becker M (eds) Third Earth Resources Technology Satellite–1 Syposium. Volume I: Technical Presentations, NASA SP-351, NASA, Washington, DC, pp 309–317

Seenu N et al (2019) Reference evapotranspiration assessment techniques for estimating crop water requirement. Int J Recent Technol Eng 8(4):1094–1100. https://doi.org/10.35940/ijrte.d6738.118419

Smith M, Allen R, Pereira L (1998) Revised FAO methodology for crop-water requirements. International Atomic Energy Agency (IAEA), pp 51–58

Verstraeten WW, Veroustraete F, Feyen J (2008) Assessment of evapotranspiration and soil moisture content across different scales of observation. Sensors 8(1):70–117. https://doi.org/10.3390/s8010070

Wang YQ et al (2016) Is scale really a challenge in evapotranspiration estimation? A multi-scale study in the Heihe oasis using thermal remote sensing and the three-temperature model. Agric For Meteorol 230–231:128–141. https://doi.org/10.1016/j.agrformet.2016.03.012

Xiong YJ et al (2015) An evapotranspiration product for arid regions based on the three-temperature model and thermal remote sensing. J Hydrol 530:392–404. https://doi.org/10.1016/j.jhydrol.2015.09.050

Xu C-Y, Chen D (2005) Comparison of seven models for estimation of evapotranspiration and groundwater recharge using lysimeter measurement data in Germany. Hydrol Process 19(18):3717–3734. https://doi.org/10.1002/hyp.5853

Zeng X (2003) The common land model experience. Global Change Newsletter 55:19–20

Zotarelli L, Dukes MD, Romero CC, Migliaccio KW, Morgan KT (2014) Step by step calculation of the Penman-Monteith evapotranspiration. Institute of Food and …, pp 1–14. Available at: https://edis.ifas.ufl.edu/ae459

Chapter 18
Monitoring and Prediction of Dynamics in Sundarban Forest using CA–Markov Chain Model

Sarmistha Halder, Kaberi Samanta, and Sandipan Das

Abstract Mangrove ecosystems play an important functional role in providing coastal protection, carbon sequestration, coastal habitat, climate change resilience and socioeconomic services for coastal communities. The present research study investigates spatiotemporal variation, forest health status and predicts the land cover changes in Sundarban mangrove forests of India using multi-temporal satellite images and cellular automata and Markov Chain model. The results revealed that mangrove forest extent has decreased by 3.14% from 1994 and 2019. The image classification resulted in overall accuracy of 74% in 1994, 81% in 2004, 78% in 2014 and 84.5% in 2019 respectively. The satellite-based vegetation indices were analysed for assessing the health of the forests. The findings of present study indicate deteriorating health of the forest and observed significant vegetation stresses over the western to central part of the study region due to anthropogenic activities. The CA Markov model predicted that the extent of mangrove forests could possibly decline from 2011.60 km^2 to 1939.24 km^{2}by the period 2029. The results of the present study could foster better decisions, precise mitigation and sustainable development strategies for the region.

Keywords Mangrove forests · Remote sensing · CA–Markov · Forest health

---

S. Halder · S. Das (✉)
Symbiosis Institute of Geoinformatics (SIG), Symbiosis International (Deemed University) (SIU), Pune, Maharashtra, India
e-mail: sandipanraj2002@gmail.com

S. Halder
e-mail: sarmisthahalder14@gmail.com

K. Samanta
School of Oceanographic Studies, Jadavpur University, Kolkata 700032, India
e-mail: samanta.kaberi@gmail.com

P. K. Shit et al. (eds.), *Spatial Modeling in Forest Resources Management*, Environmental Science and Engineering,
https://doi.org/10.1007/978-3-030-56542-8_18

18.1 Introduction

Mangroves are enriched with diverse and highly productive ecological communities. Mangrove supports various ecosystems functions for human well-being as well as for other community and species. Mangroves forest cover protects coastal area communities from many natural and anthropogenic hazards such as tsunami, cyclone, maintain the coastal water quality, terrestrial sediments and rehabilitate nutrients (Sanders et al. 2010; Giri et al. 2011). It provide many important ecosystem services such as reduce the intensity of storm, flood damage, source of food, and aquatic habitat (Sandilyan and Kathiresan 2015; Tanner et al. 2019). Mangroves are found in one hundred and twenty countries and territories of the world. South Asia has the highest quantity of mangrove cover followed by West and Central Africa, South America and North Central America. The total area of mangrove cover in world is 150,000 sq.km. South Asia covers 10,344 sq.km which is the 6.8% of the world. The majority of mangroves are found alongside the tidal sea of Bangladesh, India, Pakistan and Sri Lanka countries of South Asia. India and Bangladesh contributes maximum amount of mangrove coverage in South Asia (Rahman et al. 2013; Ghosh et al. 2015). India has 45.8% of total mangroves in South Asia. The approximate area of mangroves in India is about 4921 sq.km which is 3.3% of world vegetation coverage. The major states of India where mangrove coverage is high comparatively are West Bengal, Andaman & Nicobar Islands, Gujarat, Andhra Pradesh, Odisha and Maharashtra etc. (FSI 2015).

Among all the mangrove forest in India, Sundarban is one of the beautiful and largest mangrove forest in South Asia. As per UNESCO, the Indian Sundarban forest is designated as the world heritage site in 1987 for its importance towards all the living beings and uniqueness. But many natural calamities, change of climate (sea level rise), land use patterns, deforestation encroachment of forested area, illegal cutting and extend of tidal creeks towards the land surface etc. are threats for mangroves (Gabler et al. 2017; Dasgupta et al. 2018). So there is need of assessing and monitoring the mangrove to design the proper conservation strategies for mangroves. The remote sensing and GIS technologies could play an important role in monitoring, change analysis and prediction of future forest cover dynamics (Heumann 2011; Yang et al. 2013). The satellite data provides valuable information of the past and the present through change detection method and that can act as an input for prediction studies. The integrated Cellular Automata-Markov chain model using remote sensing images have been commonly used for detecting forest cover changes and to simulate and predict the future forest cover scenarios (Adhikari and Southworth 2012; Mukhopadhyay et al. 2015; Feng and Liu 2016; Etemadi et al. 2018; Zubair 2019). The numerous previous studies have widely applied CA–Markov Chain model to predictive land use change models (Keshtkar and Voigt 2016; Ghosh et al. 2017; Perera et al. 2018; Mafi-Gholami et al. 2020). The present research study aimed to: (1) assess and monitor the forest cover dynamics using multi-temporal Landsat imageries pertaining to period 1994, 2004, 2014 and 2019, (2) analyse the spatial–temporal health of the forest and (3) predict the future change projections of forest cover (2019–2029) using CA Markov model in Sundarban region of India. The findings of this study can be useful in developing management strategies for sustainable land use planning.

18.2 Study Area

Sundarban, one of the largest mangrove forests is situated at the frontier of India and Bangladesh boundary.It is situated on the coastal tidal zone of lower deltaic plain and formed by the Ganges, Brahmaputra and Meghna rivers of the Bay of Bengal. The study area is the mangrove forest of Indian Sundarban which is found in southern part of the South 24 Parganas district in West Bengal state of India (Fig. 18.1). The location of Sundarban is between two districts of West Bengal viz., South 24 Parganas and North 24 Parganas (Sivakumar 2013; Ghosh et al. 2015). The major portions of mangroves are found in South 24 Parganas district. The approximate latitudinal and longitudinal extension of Sundarban Mangrove cover is about 21° 32′ to 22° 40′ north and 88° 05′ east to 89° 51′ among which 62% is in Bangladesh and other 38% is present in India.

Sundarban is known for its unique enriched biodiversity, productive ecosystem. Sundarban has various species, many endangered and extinct species such as Royal Bengal Tiger, estuarine crocodile, water buffalo. Sundarban consists of many islands and most of the southern islands namely Dalhousie, Jammudwip, Lothian, Bulchery, Bhangaduni, Chulkati, Dhanchi, Mousuni, Gosaba, Matla are covered by dense Mangrove forest. The dominant tree species in this region are Sundri (*Heritierafomes*)and Gewa (*Excoecariaagallocha*).

18.3 Materials and Methods

18.3.1 Data Collection

The multi-temporal satellite images pertaining to year 1994, 2004, 2014 and 2019 were downloaded from USGS Earth explorer (https://earthexplorer.usgs.gov/) and GloVis (https://glovis.usgs.gov/) portal. The satellite images from four-time epochs (1994, 2004, 2014 and 2019) were acquired to analyse temporal decadal changes of mangrove cover. The information of satellite data used in the study is shown in Table 18.1. The images were acquired during the nearby period to reduce the impact of phenology and temperatures variation. The cloud-free, orthorectified freely available Landsat images were used in the study. In addition, this research used related scientific, technical project reports, Google Earth and other maps for additional information in the study area.

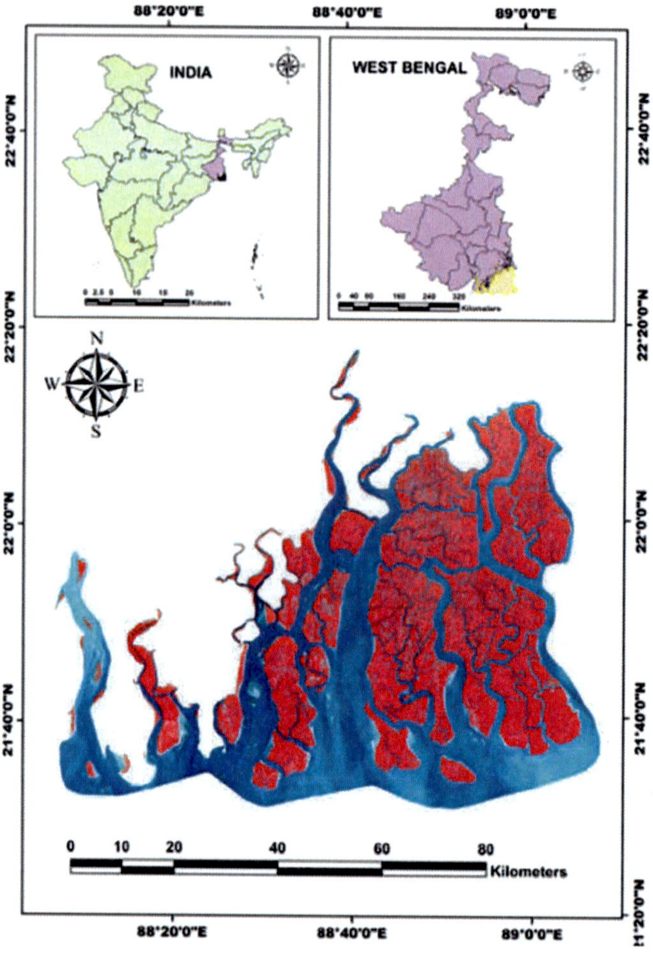

Fig. 18.1 Geographical location of study area

Table 18.1 Information about the Satellite Images

Year of acquisition	Satellite	Sensor	Spatial resolution (m)	Path/row
04 January, 1994	Landsat 5	TM	30	138/45
08 December, 2004	Landsat 5	TM	30	138/45
03 December, 2014	Landsat 8	OLI	30	138/45
14 January, 2019	Landsat 8	OLI	30	138/45

18.3.2 Image Classification and Accuracy Assessment

Image classification is an important remote sensing technique for the identification of different type of land covers. The year wise satellite images classification was carried out using the most commonly supervised classification algorithm in the study area. The seven major land cover categories were delineated in the study area viz. dense forest, degraded forest, creeks, sand (beaches/dunes), saline blanks, mudflats and river based on different reflecting spectral characteristics. The classification results were verified from field checks points for 2019 and Google Earth Prohistorical images of 01.01.1994 and 31.12.2019 available online. The error matrix evaluation method using Kappa coefficient (κ) and overall accuracy were computed to check the accuracy of classified satellite images (Congalton and Green 2009; Campbell and Wynne 2011). The year wise areal extent of mangrove forest cover has been calculated from the classified images.

18.3.3 Vegetation Health Indices

The assessment of forest health condition is important for sustainable management of natural ecosystem (Pan et al. 2011). Remote sensing provides accurate information about their health and functioning on a timely and consistent basis. Spectral vegetation indices (VIs) are important indicator of canopy greenness, foliage quality and stress levels of forest (Croft et al. 2015). The aim of the study was toassess the health of the Sundarban mangrove forest in India using the vegetation indices viz., NDVI, CVI, RDVI and RVI from the period 1994–2019.

18.3.3.1 Normalized Difference Vegetation Index (NDVI)

The NDVI is the most commonly used remote-sensed indicator to assess the photosynthetic process of vegetation (Piao et al. 2015). NDVI is calculated as the normalized ratio of the near-infrared band and red band from remote-sensed images. The near-infrared band is reflected from vegetation and red band is absorbed by vegetation. Generally, the high NDVI index value are associated with active photosynthetically vegetation condition growth, thus signifying good health conditions of the forest. In contrary, low NDVI index value indicate little photosynthetic activity of vegetation. This occurs when the vegetation is stressed due to diseases, drought and vegetation degradation, hence reflecting the poor health conditions of the forest. The NDVI is defined as follows

$$\text{NDVI} = (\text{NIR} - \text{RED}) / (\text{NIR} + \text{RED}) \tag{18.1}$$

18.3.3.2 Chlorophyll Vegetation Index (CVI)

Leaf chlorophyll content is an indicator of photosynthetic ability, nutritional stress, disease stress level, heavy metal pollution and growth status of the plants (Croft et al. 2015). Therefore, it has become important to quickly and accurately retrieve chlorophyll content information. Vegetation index-based methods using remote sensing technology have become the most popular methods to estimate chlorophyll content at spatial and temporal scales. Chlorophyll vegetation index (CVI) has been widely used to assesschlorophyll content at the leaf level. The high chlorophyll vegetation index value signifies higher proportion of chlorophyll density, hence representing healthy vegetation coverage. The CVI is defined as follows:

$$\mathrm{CVI} = \mathrm{NIR} * \left(\mathrm{RED}/\mathrm{GREEN}^2\right) \tag{18.2}$$

18.3.3.3 Renormalized Difference Vegetation Index (RDVI)

Spectral reflectance indices are the most common remote-sensing approach to determine biophysical variables such as foliage chlorophyll concentration, canopy structure, density, which is essential for understanding quantity and vigor of green vegetation. Renormalized Difference Vegetation Index (RDVI) is another important vegetation indices applied widely to classify forest growth status. RDVI is computed from the difference between near-infrared and red wavelengths, along with the NDVI (Roujean and Breon 1995). RDVI has been found to be increased sensitivity to canopy foliage activity and phytomas variation (Roy et al. 1996). The high RDVI value is usually related to healthy vegetation green cover. RDVI is calculated using the equation given below.

$$\mathrm{RDVI} = (\mathrm{R_{NIR}} - \mathrm{R_{Red}}) \Big/ \sqrt{\mathrm{R_{NIR}} + \mathrm{R_{Red}}} \tag{18.3}$$

18.3.3.4 Ratio Vegetation Index (RVI)

Theleaf area index information is essential in understanding the physical and biological processes associated with vegetation. The satellite based ratio vegetation index (RVI) has been a simple, effective and common approach to measure LAI for vegetation activity of status and succession. Ratio Vegetation Index is computed from the ratio of near infrared wavelength to red wavelength (Jordan 1969). RVI reduces the multiplicative effects of atmosphere and show increased sensitivity to change in LAI of plants. The high RVI index value is usually associated with dense vegetation foliage, thus indicating the healthy conditions of the vegetated surfaces. Ratio Vegetation Index was calculated using Eq. 18.4:

$$\text{RVI} = (\text{NIR}/\text{RED}) \quad (18.4)$$

18.3.4 Prediction of Mangrove Cover

In this study, the combined cellular automata (CA) and Markov chain model was used to predict the future land cover change over time (Fig. 18.2). The CA model is integrated to predict spatial dynamics of mangrove change, and Markov chain process is capable to simulate the probability change over time(Arsanjani et al. 2011; Mas et al. 2014). This research used IDRISI–TerrSet Geospatial Monitoring and Modeling System software to simulate the long-term landcover changes using CA_Markov approach. Thus, the CA–Markov model employed classified images from 2004 (past) and 2019 (present)as a baseline to produce the mangrove dynamic scenario. The output of CA–Markov algorithm was calculated based on following mathematical equation

$$L_{(t+1)} = p_{ij} \times L_{(t)} \quad (18.5)$$

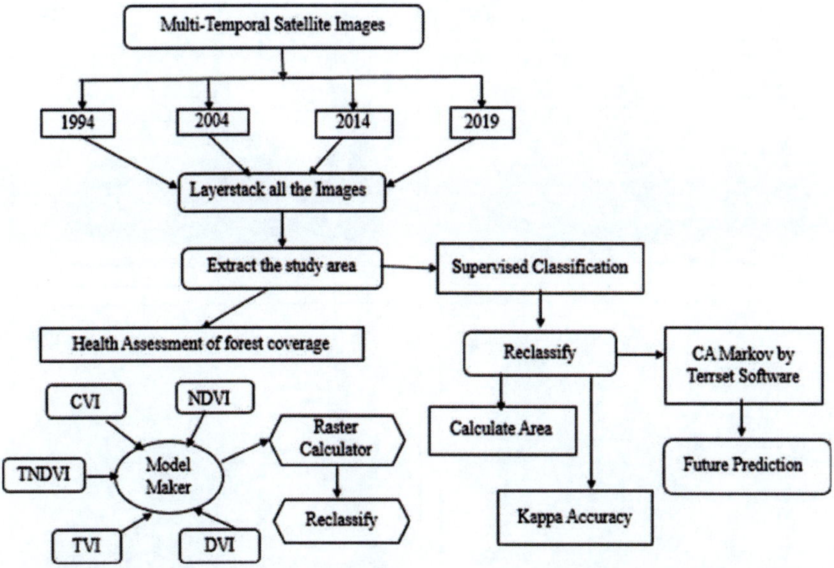

Fig. 18.2 Methodological framework of the study

$$p_{ij} = \begin{bmatrix} p_{11} & p_{12} & \cdots & p_{1m} \\ p_{21} & p_{22} & \cdots & p_{2m} \\ . & . & . & . \\ . & . & . & . \\ p_{m1} & p_{m2} & \cdots & p_{mm} \end{bmatrix} \tag{18.6}$$

where $L(t)$ is the LULC status at time t, $L(t+1)$ is the LULC status at time $t+1$; $0 \leqslant p_{ij} < 1$ and $\sum_{j+1}^{m} pij = 1, (i, j = 1, 2, \ldots, m)$ is the matrix of transition probability.

18.4 Results and Discussion

18.4.1 Changes in Land Cover Over Time

The landcover map generated for the years 1994, 2004, 2014 and 2019 are shown in Fig. 18.3. The images are classified into seven categories viz., dense forest, degraded forest, creeks, sand (beaches/dunes), saline blanks, mudflats and river. Forest covers include dense forest, degraded forest and saline blanks. The total area

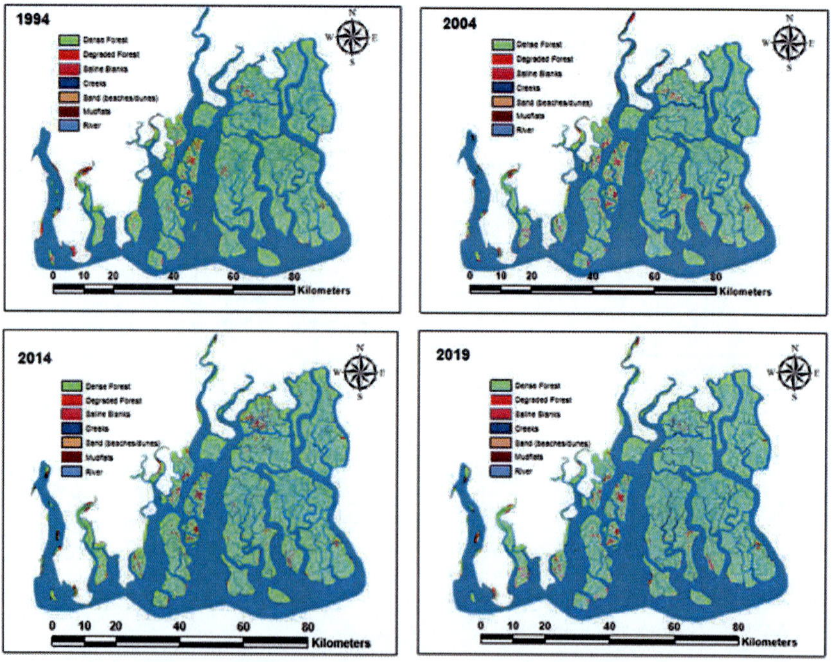

Fig. 18.3 Distribution of land cover classes in the study area (1990–2019)

Table 18.2 Areal estimates (in %) of land cover features over the years

LULC classes	1994	2004	2014	2019
Dense forest	45.31	44.69	43.65	43.22
Degraded forest	1.75	2.16	2.21	2.19
Saline blanks	1.16	0.95	1.05	1.20
Creeks	4.28	4.73	4.90	5.00
Sand (beaches/dunes)	0.21	0.20	0.35	0.37
Mudflats	0.45	0.34	0.33	0.27
River	46.84	46.93	47.51	47.75

Table 18.3 Overall accuracy and Kappa coefficient of classified images from 1994 to 2019

Year	Overall classification accuracy (%)	Kappa coefficient(κ)
1994	74	0.68
2004	81	0.73
2014	78	0.74
2019	84.5	0.81

of this Mangrove is 4305.685 sq.km. The mangrove forest cover density are primarily concentrated at the southern region of the study area. These areas need to be mainly preserved for biodiversity conservation. The fluctuation in mangrove area during the study periods are shown in Table 18.2. The mangrove forest covers area has significantly declined from 48.22%in 1994 to about 46.61% in 2019. The density of degraded forest is maximum at the western to central and near the shoreline portion due to anthropogenic and natural activities, respectively. The total areas under river, creeks and sand (beaches/dunes) have significantly increased during the analysis period.As per the classified images it can observed that saline blanks are associated with degraded forest for most of the year.

The creeks have been more prominently seen for the period 2014 and 2019. The classified map results were validated using ground reference data and high-resolution Google Earth satellite imagery. The results of the land use classification achieved overall accuracies of 74%, 81%, 78% and 84.5% with the Kappa coefficient of 0.68, 0.73, 0.74 and 0.81 for the years 1994, 2004, 2014 and 2019, respectively (Table 18.3).

18.4.2 Changes of Forest Health Conditions

The forest health condition is an important factor for the ecosystem as well as for natural resource. Each vegetation index shows the health of forest in different perspective. The vegetation indices including NDVI, CVI, RDVI and RVI were prepared for each year 1994, 2004, 2014 and 2019 using model maker tool in Erdas Imagine 2014 software. All vegetation indices output for each year were combined in a raster calculator' tool in ArcMap 10.3.1 software to obtain accurate health condition of the forest cover year wise. The vegetation health condition maps are grouped into three classes' viz., healthier mangrove, moderate mangrove and stressed mangrove for the period 1994, 2004, 2014 and 2019 as shown in Fig. 18.4.

The increasing overexploitations of natural resources and anthropogenic interference have caused severe damage to Indian Sundarbans mangrove forest ecosystem (FAO 2004). The research findings reveal that the forest health status was progressively deteriorating from 1994 to 2019. The amount of stressed vegetation is moderately constant from 1990 to 2004 and was observed only western region of the islands. However, the trend shows the increasing rate of stressed vegetation from 2014 to 2019 and was spread across the central portion islands of the mangrove forest. It has been witnessed that, that of about 3.14% mangrove forest were lost during 1994–2019 within the study area (Table 18.4).

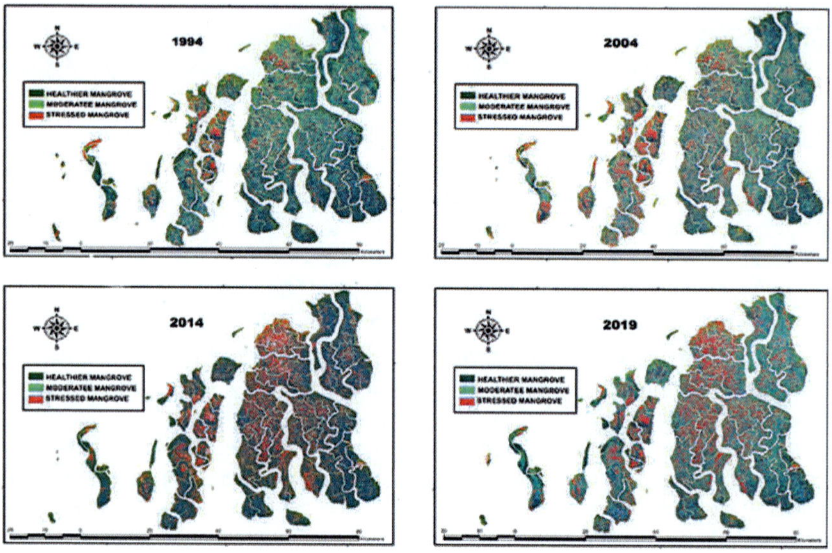

Fig. 18.4 Health condition of the vegetation cover in the study area (1990–2019)

Table 18.4 Changes in mangrove forests in the study area from 1994 to 2019

Year	Forest cover changing rate (in sq.km)	Forest cover changing rate (in %)
1994–2004	18.8	0.90
2004–2014	37.9	1.84
2014–2019	8.2	0.40
Total	64.9	3.14

The continuously growing population pressure; illegal cutting; encroachment of forest areas; increased settlement; construction activities; tourism; dam and road constructions; coastal pollution; exploitation of fishery resources have aggravate the environmental degradation process of Sunderbans mangrove ecosystem indicating the deterioration of forest health status (Rahman and Hossain 2015; Chowdhury and Maiti 2016). The natural stressors like sea level rise; global climate change; changes in hydrological regimes; sediment salinity; siltation; continuous erosion by high tidal forces and cyclonic storms have also equally contributed in declination of forest coverage of Indian Sundarban, thereby making the ecosystem most fragile (Dasgupta et al. 2018). The forest health status in the south east portion of the islands was observed comparatively healthier for almost every year. The forest coverage in the newly formed islands of the western region is observed to be healthier as shown in Fig. 18.4. This is mainly attributed to increase in natural uninterrupted growth of mangrove in perforated and edge forest. To conserve and protect this important ecological ecosystem, it is essential to develop and implement sound sustainable management strategies involving local people, government and other social organizations.

18.4.3 Mangrove Changed Prediction Results

In this study, CA–Markov algorithm was applied to predict dynamic spatial Sundarban land-cover changes in the near future for year 2024. The predicted spatial distribution of LULC in the year 2029 is shown in Fig. 18.5. The results

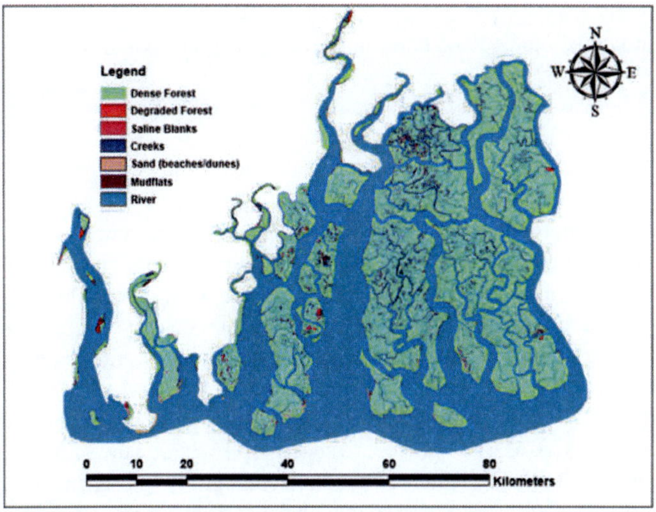

Fig.18.5 Predicted land cover map for the year 2029

Table 18.5 Expected change of future and land cover (2029) type of the study area

Land covers	2019		2029		Relative change 2019–2029	
	Area (in sq.km)	Area (in %)	Area (in sq.km)	Area (in %)	Area (in sq.km)	Area (in %)
Dense forest	1860.50	43.22	1765.02	40.99	105.48	2.23
Degraded forest	99.30	2.19	104.49	2.43	5.19	0.24
Saline blanks	51.80	1.20	62.35	1.45	10.55	0.25
Creeks	213.80	5.00	270.59	6.29	56.79	1.29
Sand (beaches/dunes)	17.50	0.37	23.54	0.55	6.04	0.18
Mudflats	11.50	0.27	9.30	0.21	2.20	0.06
River	2050.80	47.75	2069.71	48.08	18.91	0.33

of predicted LULC 2024 reveals an increase of river by 18.91 sq.km (0.33%), sand (beaches/dunes) 6.04 sq.km (0.18%), saline blanks 10.55 sq.km (0.25%) degraded forest 5.19 sq.km (0.24%) and creeks 56.79 sq.km (1.29%) from the 2019 LULC coverage respectively (Tables 18.5).

However, significant decreases are projected to occur in dense forest 105.48 sq.km (2.23%). The findings indicated that area of mangrove forests would decrease by 3.59% approximately in 2029 (Fig. 18.5). The Pearson correlation result indicated that both predicted LULC in 2029 and LULC of 2019 are significantly correlated ($r = 0.819$). The spatiotemporal changes in the extent of mangrove forests were driven by various factors including deforestation, increase in human populations, unprecedented LULC change and increase of water level (Yang et al. 2012). Thus, environmentally friendly policy approach is paramount significant to mitigate the trends of land-cover change and harmonize mangrove ecosystems in the region.

18.5 Conclusion

The Sundarban mangrove forests are the largest unique, diverse and highly productive ecological communities in the world. Remote sensing techniques have provided reliable and accurate results especially in mangroves areas where accessibility is difficult and greatly limits spatial coverage. The present research study aimed in analyzing multi-temporal change detection, prediction future changes and spatial–temporal health of the forest cover using geospatial technology and CA–Markov chain method. The historical LULC change data of 1994, 2004, 2014 and 2019 were further used as a baseline in the CA-Markov process to successfully predict the future mangroves change projection 2029. The results of predicted mangrove forests change using the Cellular Automata (CA) Markov model revealed that possible reduction of mangrove forest cover area 72.36 sq.km in area from 2019 (approximately 2011.60 sq.km) to

2029 (approximately 1939.24 sq.km). The findings of the study also specify that mangrove forests of Indian Sundarban have undergone remarkable degradation due to various natural and man-made pressures indicating deterioration of forest health. The results of this study indicated approximately 3.14% extent of mangrove forests were lost during the periods of 1994–2019. This can lead to serious impacts including loss of biodiversity, extinction, climate change and changes in the natural biogeochemistry. The damage could be minimized through restoration, restricting further anthropogenic disturbances and implementing sustainable strategies for management of forest cover. The findings of this study could provide invaluable quantitative information to environmentalists and policymakers for implementing sustainable management of mangrove ecosystems in the region.

References

Adhikari A, Southworth J (2012) Simulating forest cover changes of Bannerghatta National Park based on a CA–Markov model: a remote sensing approach. Remote Sens 4:3215–3243

Arsanjani JJ, Kainz W, Mousivand AJ (2011) Tracking dynamic land-use change using spatially explicit Markov Chain based on cellular automata: the case of Tehran. Int J Image Data Fusion 2:329–345

Campbell JB, Wynne RH (2011) Introduction to remote sensing. Guilford Press, New York, USA 670 pp. ISBN 978 160918 176 5

Chowdhury A, Maiti SK (2016) Identifying the source and accessing the spatial variations, contamination status, conservation threats of heavy metal pollution in the river waters of Sundarban biosphere reserve. J Coast Conserv India. https://doi.org/10.1007/s11852-016-0436-x

Congalton RG, Green K (2009) Assessing the accuracy of remotely sensed data: principles and practices, 2nd edn. CRC Press, Boca Raton, FL

Croft H, Chen JM, Zhang Y, Simic A, Noland TL, Nesbitt N, Arabian J (2015) Evaluating leaf chlorophyll content prediction from multispectral remote sensing data within a physically-based modelling framework. ISPRS J Photogr Remote Sens 102:85–95

Dasgupta N, Nandy P, Sengupta C, Das S (2018) Genetic variation in relation to adaptability of three mangrove species from the Indian Sundarbans assessed with RAPD and ISSR markers. J Forest Res 29(2):301–310

Etemadi H, Smoak JM, Karami J (2018) Land use change assessment in coastal mangrove forests of Iran utilizing satellite imagery and CA–Markov algorithms to monitor and predict future change. Environ Earth Sci 77(5):208

FAO (2004) Food and agriculture organization (FAO), state of world's forests 2003. FAO, Rome

Feng Y, Liu Y (2016) Scenario prediction of emerging coastal city using CA modeling under different environmental conditions: a case study of Lingang New City, China. Environ Monit Assess 188:1–15

FSI (2015) India state of forest report. Forest Survey of India. https://fsi.nic.in/isfr-2015/isfr-2015-mangrove-cover.pdf. Accessed date: 13 Mar 2017

Gabler CA, Osland MJ, Grace JB, Stagg CL, Day RH, Hartley SB, Enwright NM, From AS, McCoy ML, McLeod JL (2017) Macroclimatic change expected to transform coastal wetland ecosystems this century. Nat Clim Change 7(2):142

Ghosh A, Schmidt S, Fickert T, Nusser M (2015) The Indian Sundarban mangrove forests: history, utilization, conservation strategies and local perception. Diversity 7(2):149–169

Ghosh P, Mukhopadhyay A, Chanda A, Mondal P, Akhand A, Mukherjee S, Nayak SK, Ghosh S, Mitra D, Ghosh T, Hazra S (2017) Application of cellular automata and Markov-chain model in geospatial environmental modeling—a review. Remote Sens Appl Soc Environ 5:64–77

Giri C, Ochieng E, Tieszen LL, Zhu Z, Singh A, Loveland T, Masek J, Duke N (2011) Status and distribution of mangrove forests of the world using earth observation satellite data. Glob Ecol Biogeogr 20:154–159

Heumann BW (2011) Satellite remote sensing of mangrove forests: recent advances and future opportunities. Prog Phys Geog 35:87–108

Jordan CF (1969) Derivation of leaf area index from quality of light on the forest floor. Ecology 50:663–666

Keshtkar H, Voigt W (2016) A spatiotemporal analysis of landscape change using an integrated Markov chain and cellular automata models. Model Earth Syst Environ 2(1):10

Mafi-Gholami D, Zenner EK, Jaafari A (2020) Mangrove regional feedback to sea level rise and drought intensity at the end of the 21st century. Ecol Ind 110:105972

Mas JF, Kolb M, Paegelow M, Olmedo MTC, Houet T (2014) Inductive pattern-based land use/cover change models: a comparison of four software packages. Environ Modell Softw 51:94–111

Mukhopadhyay A, Mondal P, Barik J, Chowdhury SM, Ghosh T, Hazra S (2015) Changes in mangrove species assemblages and future prediction of the Bangladesh Sundarbans using Markov chain model and cellular automata. Environ Sci Process Impacts 17(6):1111–1117

Pan Y et al (2011) A large and persistent carbon sink in the world's forests. Science 333:988–993

Perera KARS, De Silva KHWL, Amarasinghe MD (2018) Potential impact of predicted sea level rise on carbon sink function of mangrove ecosystems with special reference to Negombo estuary, Sri Lanka. Glob Planet Change 161:162–171

Piao S et al (2015) Detection and attribution of vegetation greening trend in China over the last 30 years. Glob Change Biol 21(4):1601–1609

Rahman MR, Hossain MB (2015) Changes in land use pattern at Chakaria Sundarbans mangrove forest in Bangladesh. Bangladesh Res Public J 11(1):13–20

Rahman AF, Dragoni D, Didan K, Barreto-Munoz A, Hutabarat JA (2013) Detecting large scale conversion of mangroves to aquaculture with change point and mixed pixel analyses of high-fidelity MODIS data. Remote Sens Environ 130:96–107

Roujean JL, Breon FM (1995) Estimating PAR absorbed by vegetation from bidirectional reflectance measurements. Remote Sens Environ 51:375–384

Roy PS, Sharma KP, Jain A (1996) Stratification of density in dry deciduous forest using satellite remote sensing digital data—an approach based on spectral indices. J Biosci 21(5):723–734

Sanders CJ, Smoak JM, Naidu AS, Sanders LM, Patchineelam SR (2010) Organic carbon burial in a mangrove forest, margin and intertidal mud flat. Estuar Coast Shelf Sci 90(3):168–172

Sandilyan S, Kathiresan K (2015) Density of waterbirds in relation to habitats of Pichavaram mangroves. Southern India J Coast Conserv 19(2):131–139

Sivakumar K (2013) Coastal and marine biodiversity protected Areas in India. In: Venkataraman K, Sivaperuman C, Raghunathan C (eds) Ecology and conservation of tropical marine faunal communities. Springer Heidelberg, New York, pp. 463–476. ISBN 978-3-642-38200-0

Tanner MK, Moity N, Costa MT, Jarrin JRM, Aburto-Oropeza O, Salinas-de-Leon P (2019) Mangroves in the galapagos: ecosystem services and their valuation. Ecol Econ 160:12–24

Yang J, Gong P, Fu R, Zhang M, Chen J, Liang S et al (2013) The role of satellite remote sensing in climate change studies. Nat Clim Change 3:875–883

Yang X, Zheng X-Q, Lv LN (2012) A spatiotemporal model of land use change based on ant colony optimization, Markov chain and cellular automata. Ecol Model 233:11–19

Zubair H (2019) Prediction of land use and land cover (LULC) changes using CA Markov model in Mamuju Subdistrict. J Phys Conf Ser 1341(8):082033 (IOP Publishing)

Part III
Rural Livelihood and Sustainable Management

Chapter 19
Improving Potential Biodiversity and Human Footprint in *Nothofagus* Forests of Southern Patagonia through the Spatial Prioritization of their Conservation Values

Yamina Micaela Rosas, Pablo L. Peri, Josela Carrasco, María Vanessa Lencinas, Anna M. Pidgeon, Natalia Politi, Sebastián Martinuzzi, and Guillermo Martínez Pastur

Abstract The role of biodiversity in natural forests is critical as a regulator of ecosystem function, productivity, and provision of ecosystem services. The objective was to analyse the conservation value of *Nothofagus* forests in Southern Patagonia (Santa Cruz and Tierra del Fuego provinces), Argentina, through integration of maps of potential biodiversity (MPB) and human footprint (HFM), which can help to improve the natural reserve designs through the spatial prioritization of their

---

Y. M. Rosas (✉) · J. Carrasco · M. V. Lencinas · G. M. Pastur
Laboratorio de Recursos Agroforestales, Centro Austral de Investigaciones Científicas (CADIC), Consejo Nacional de Investigaciones Científicas y Técnicas (CONICET), Houssay 200 (9410) Ushuaia, Tierra del Fuego, Argentina
e-mail: micaela.rosas@cadic-conicet.gob.ar

J. Carrasco
e-mail: JoselaCarrasco@hotmail.com

M. V. Lencinas
e-mail: mvlencinas@conicet.gov.ar

G. M. Pastur
e-mail: gpastur@conicet.gov.ar

P. L. Peri
Instituto Nacional de Tecnología Agropecuaria (INTA), Universidad Nacional de La Patagonia Austral (UNPA), Consejo Nacional de Investigaciones Científicas y Técnicas (CONICET), CC 332 (9400) Río Gallegos, Santa Cruz, Argentina
e-mail: peri.pablo@inta.gob.ar

A. M. Pidgeon · S. Martinuzzi
SILVIS Lab, Department of Forest and Wildlife Ecology, University of Wisconsin-Madison, Linden Drive 1630 (53706), Madison, WI, USA
e-mail: apidgeon@wisc.edu

S. Martinuzzi
e-mail: martinuzzi@wisc.edu

P. K. Shit et al. (eds.), *Spatial Modeling in Forest Resources Management*, Environmental Science and Engineering,
https://doi.org/10.1007/978-3-030-56542-8_19

conservation values. To achieve the, first we determine that the different forest types presented different species assemblage with specific ecological niche requirements which justify differential conservation or management strategies. We used understory plants as proxy, and we selected indicator species of the understory plants for the following analyses. With these species we produce the MPB, and we found that the occurrence of MPB differ from the pattern of HFM according to the different forest types. After that, we identify woodland patches with special values of MPB and low HFM according to the different forest types, and analyzed if the distribution of MPB of the different forest types changed across the current natural protected reserve network, private and public lands. Finally, with these outputs, we propose new methodologies to enhance the current natural reserve network effectiveness. These outputs can be used as a tool to determine new strategies for management and conservation at landscape level in Southern Patagonia.

Keywords Habitat suitability · Biodiversity conservation · Human footprint · Forest landscapes · National parks · Provincial reserves · Spatial conservation prioritization

19.1 Introduction

The concept of ecosystem services (ES) refers to the goods and benefits that the society obtains from the natural ecosystems (Daily 1997). Society obtains material goods and services from these ES (e.g., timber products, food, clean water), as well as other economic (e.g., market value, salaries) and non-economic values for the well-being of individuals and communities. Forest ecosystems in particular provide critical ES to humanity (FAO 2010) and play a multifunctional role that balances human needs with the production of other goods and services, including habitat for native species (Lindenmayer and Franklin 2002). When ecosystem management goals include only a limited set of goods and services, such as timber forests, other ES may be overlooked and therefore undervalued (Perera et al. 2018).

We recognize four ES types coming from natural forests (MEA 2005): (i) provisioning ES, such as timber wood, fiber or firewood (Gealzquierdo et al. 2004), food (e.g., fruits, nuts, mushrooms, honey, or spices), pharmaceutical plants, and other non-woody industrial products (Quintas-Soriano et al. 2016); (ii) regulating ES, e.g. contributing to climate stability by removing greenhouse gases and other pollutants, increasing soil retention and water regulation, promoting natural pest control (e.g., wasps, owls and bats) and pollination services (De Groot et al. 2002; Panagos et al. 2015; González et al. 2015; Quintas-Soriano et al. 2016); (iii) aesthetic ES, related to

N. Politi
Facultad de Ciencias Agrarias, Universidad Nacional de Jujuy, Consejo Nacional de Investigaciones Científicas y Técnicas (CONICET), Juan Bautista Alberdi 47 (Y4600DTA), Jujuy, Argentina
e-mail: natalia.politi@fulbrightmail.org

beauty, recreation and ecotourism(De Groot et al. 2002) and provision of cultural and artistic inspiration and a sense of place for local communities (Martínez Pastur et al. 2016a); and finally, (iv) supporting ES, which are delivered by forests through soil formation (Lal and Lorenz 2012), nutrient and water cycling, (Prescott 2002), plant biomass (Perera et al. 2018), and provision of habitat for wild plants and animals, a necessary precondition for the provision of all the other ES, directly or indirectly (De Groot et al. 2002).

Biodiversity is assumed to be the critical factor for supplying ES (Mori et al. 2017). In fact, biodiversity itself is can be considered as one ES, e.g. when it is the basis of nature-based tourism or the regulation of diseases (Mace et al. 2012). The role of biodiversity in the natural forests is a critical regulator of ecosystem function, ecosystem productivity, and provision of ES. Biodiversity supports the delivery of ES in natural forests through various multi-faceted roles. Rosas et al. (2019a) describe the importance of these connections in Southern Patagonia forests, including: (i) Genetic diversity of wild species that are involved in quality and production of provisioning services (De Groot et al. 2010). (ii) Species that are of high significance with regard to the provision of economic values (Perera et al. 2018). (iii) Biotic interaction among species (predation, parasitism, competition and facilitation) that have important consequences for ES and include direct interactions, such as pollination, mycorrhizal fungi and nitrogen-fixing microorganisms that facilitate transfer of nutrients to tree roots (González et al. 2015; Quintas-Soriano et al. 2016). (iv) Biophysical structures that provide ecological conditions and habitat to support the biodiversity associated with ES provision (MEA 2005). (v) Functional traits that reflected adaptations to variation in the physical and biotic environment and trade-offs (eco-physiological and/or evolutionary) and assist to support biodiversity associated with ES provision (de Bello et al. 2010). And (vi) ecological processes of natural ecosystems that regulate the dynamics of ecosystems and the structure and dynamics of biological communities (Mace et al. 2012).

Recent methodological advancements have improved the assessment of species distributions, synergies and trade-offs among ES and biodiversity at different spatiotemporal scales (Raudsepp-Hearne et al., 2010; Martínez Pastur et al. 2017). Improved methods include: (i) species occurrence maps, which are useful when the link between species and services is well understood or when the species is the service itself (e.g., food) (Andrew et al. 2014), (ii) forest cover type maps, which use the spectral data of the dominant canopy or the heterogeneity of reflectance values within a set of pixels and relate them with the biodiversity, and (iii) modelling habitat species distributions and biodiversity which can be developed using remotely sensed environmental correlates, and which was applied around the world (Hirzel et al. 2002; Elith and Leathwick 2009) and Patagonia (Martínez Pastur et al. 2016b; Rosas et al. 2017, 2018, 2019b).

When considering ES and biodiversity in particular, it is important to consider the human influence, because humans are the major factor that affects life of all kinds in our world (Mace et al. 1998; Jacobson et al. 2019), and the cumulative effects generate the global phenomenon of its influence over nature. Many authors defined this new time as a new geological epoch, that some call as Anthropocene (Sanderson et al. 2002; Corlett 2015). The transformation generated by humans has resulted in declines of wilderness areas, loss of biodiversity, and degradation of the provision of many ES (Watson et al. 2016; Li et al. 2018). In this context, the understanding of the human influence on ecosystems and their services are crucial to achieve sustainable development (Costanza et al. 2017). A practical way to understand the influence and impact at global scale is to present it in a single map. Rapid advances in earth observation using satellite technology generate verifiable global maps of land use changes and cover (Loveland et al. 2000; Hansen et al. 2013). One proposal to quantify the impact of human activities in the nature is the human footprint index maps (HFM), which combine spatial data on human activities (roads, settlements, land use, etc.) to map the cumulative human pressure on the environment. The idea is that the footprint index is not only a single number, it must be understand as a continuum of human influence stretched across the land surface, revealing through its variation the major pattern of human influence on nature (Sanderson et al. 2002).

Southern Patagonia is composed of landscapes with extreme environmental conditions, from arid steppes to dense temperate forests (Peri et al. 2016), and with varying levels of natural and human impacts (e.g., livestock grazing, desertification) that can greatly affect local biodiversity (Rosas et al. 2017, 2018, 2019a, 2019b). Areas with little human influence often have outstanding global value for the preservation of endemic natural features, species and biodiversity (Inostroza et al. 2016). However, this fragile Patagonian ecosystem is very vulnerable to disturbance and it is widely acknowledged that environment and ecological services in this region changed significantly in the last century due to human activities and climate change (Peri et al. 2019a).

Sanderson et al. (2002) mapped the human footprint at the global scale and identified the last of the wild over the world (10% wildest areas in each biome) defined as continuous areas of large, intact tracts of relatively undisturbed ecosystems which can be particularly important for conserving biological diversity (from 5 to 100,000 km^2). There are many ways of using HFM to define areas of interest for conservation according to the objectives, e.g. last of the wilderness as was defined before but also as seeds of wilderness defined as the wildest areas in each biome, regardless of size. These authors also proposed that the HFM is roughly the inverse of the geography of natural processes and patterns in the region, expecting that where human influence is highest, ecosystems will be the most modified and species will stay under the most pressure from human activity. Similarly, where HFMis lower, the expectation is for more intact, as well as functional natural communities.

The design of natural reserve networks is mainly attributable to their location (e.g., remote areas or border areas between countries), or scenic values (e.g., by majestic landscapes or unique features like glaciers). However, there is today an increasing in the concern to add more areas with unique biodiversity values. Moreover, reserve networks can provide yet greater goods if they include consideration of, e.g. (i) different forest types, (ii) unique biodiversity values that these areas can host, (iii) provision of different ES, as well as the consideration of potential synergies and trade-offs among these services and biodiversity that the reserves host, and (iv) the human impacts or the economic activities that are carried out in the reserve networks.

Humans exert pressures on the planet in a great many ways that may lead directly or indirectly to changes in natural systems (McGowan 2016). For this, the rationale behind to consider the combination of maps of biodiversity and HFM is that more intense human footprint represent less value for conservation due to biodiversity loss, habitat degradation, exotic species invasion, contamination and different trade-offs between natural values and humans (Venter et al. 2016; Di Marco et al. 2018; Li et al. 2018).

The objective of this chapter was to analyse the forest landscapes and their conservation values in Southern Patagonia (Santa Cruz and Tierra del Fuego provinces) in Argentina, through maps of potential biodiversity (MPB) and HFM. Also, we want to define a spatial conservation prioritization for a large-scale conservation planning to improve the natural reserve network in both provinces, due to most of the reserves were selected for other values (e.g., closeness to the international borderlines) and not due to their biodiversity values. Additionally, we want to answer the following questions: (i) do forest types present different species assemblage with specific ecological niche requirements to justify different conservation or management strategies (we used understory plants as proxy)?, (ii) does the distribution of potential biodiversity differ from the pattern of human footprint according to forest types?, (iii) it can be possible to identify woodlands with special values of MPB and low HFM according to the different forest types?, (iv) what is the distribution of potential biodiversity of the forests in the current natural protected reserve network, private and public lands?, and (v) how can current natural reserve network effectiveness be enhanced?

19.2 Materials and Methods

19.2.1 Study Area

The study was carried out in Southern Patagonia (46°00′–55°03′ SL, 63°47′–73°32′ WL) including the provinces of Santa Cruz (243.9 thousand km^{-2}) and Tierra del Fuego (21.3 thousand km^{-2}), Argentina (Fig. 19.1). Ice fields and the Andes mountains (north to south direction) define relief and climate in Santa Cruz province, generating a rainfall and temperature gradients from west to east. Dominant vegetation is steppe and shrublands, and forests occupy a narrow fringe along the base of the mountains generating different normalized difference vegetation index (NDVI) (proxy of vegetation productivity) across the landscape. Santa Cruz province has a population density of 1.3 inhabitants km^2 mainly in small towns and cities ($n = 12$). National parks and provincial reserves mainly preserve forest at the foot of the Andes, however, some reserves were created to protect special heritage values (e.g., Bosque Petrificado national park), while few of them are designed to preserve unique biodiversity (e.g., Monte León national park and the Laguna de los Escarchados provincial reserve).In the province of Tierra del Fuego, the Andes Mountains runs from west to east and define the relief and climate of the region, but also here there is a great

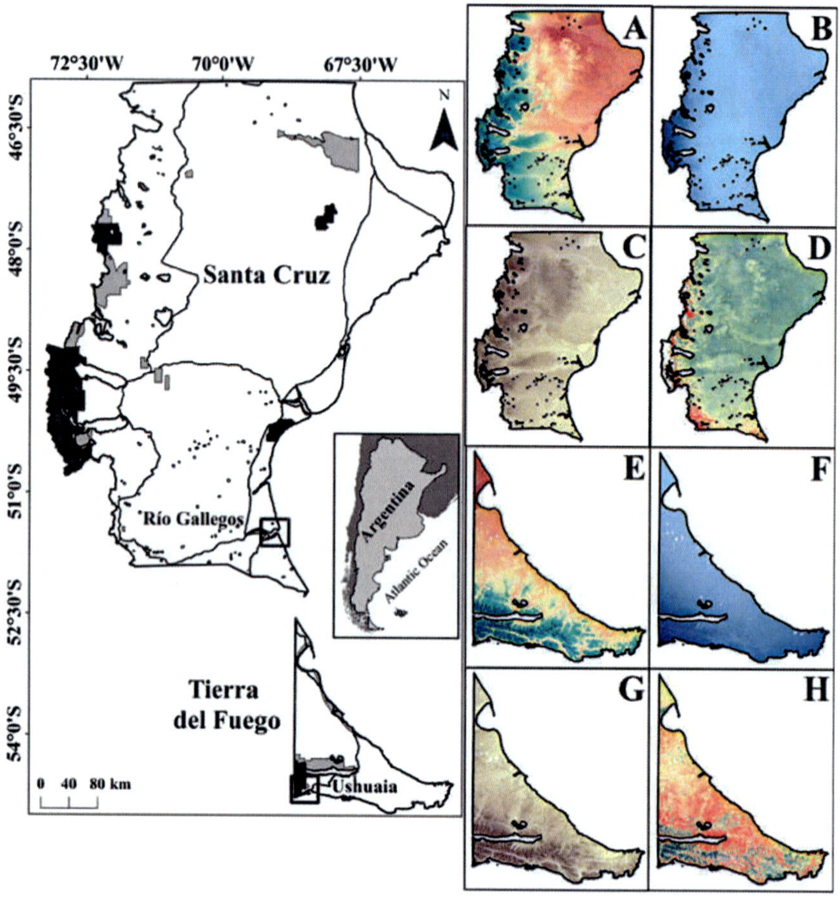

Fig. 19.1 Characterization of the study area indicating capital cities, national roads, and natural reserves (black = national parks, grey = provincial reserves). **a** and **e** show mean annual temperature for Santa Cruz (8.6 to 13.5 °C) and Tierra del Fuego (0.1–6.9 °C) (red is high and blue is low). **b** and **f** showed mean annual rainfall for Santa Cruz (136–1681 mm.yr^{-1}) and Tierra del Fuego (253–721 mm.yr^{-1}) (dark color means higher values). **c** and **g** showed elevation for Santa Cruz (0–3463 m.a.s.l.) and Tierra del Fuego (0–1447 m.a.s.l.) (dark color means higher values). **d** and **h** showed patterns of green vegetation using the normalized difference vegetation index (NDVI) (0–1) from satellite imagery for Santa Cruz and Tierra del Fuego (red is high and green is low)

influence of the Antarctica. A rainfall and temperature gradients from north to south defines the vegetation types in Tierra del Fuego, with grasslands in the north and forests in the south. This province has a population density of 6.0 inhabitants km^{-2} mainly located in two cities (97.5% of the total population), one close to ranching and oil extraction areas, and the other close to major tourism area. National parks and provincial reserves mainly preserved the forests, despite another ecological or heritage values (Fig. 19.1).

Two traditional provisioning ecosystem services prevailed in Southern Patagonian forests, livestock grazing and wood harvesting (Peri et al. 2016) supported by grasses, forbs, shrubs and trees found within the natural habitats and lands. Since the colonization, the use of forest resources have increased in terms of saw-timber for industry but decreased the extractions for firewood, since then the main source of energy, with the introduction of natural gas in the '70–80' decades. By the way, other alternative uses were not applied, as pulpwood export, because is forbidden by provincial laws. *N. pumilio* forests are economically and in extension the most important in Southern Chile and Argentina (Martínez Pastur et al. 2002, 2016a), while *N. antarctica* is one of the main deciduous native tree used for silvopastoral systems (Peri et al. 2017). They enable the diversification of farm products by sustaining sheep and cattle production, which provides income from meat, wool, and a range of wood products including poles, firewood and timber for rural construction purposes. In addition to livestock and timber production, the silvopastoral system in the region provide other ecosystem services such as water regulation, biodiversity conservation, soil and water quality, carbon sequestration, recreation, and cultural identity. These forests are growing in a mosaic of forest cover of varying ecological condition, with different structure and floristic composition as a result of livestock grazing and silvicultural management, in interaction with natural (e.g., drought) and anthropogenic (fires, introduction of species) factors (Peri and Ormaechea 2013). Beside this, forestlands were also used for tourism, and this activity increasing during the last decades (Martínez Pastur et al. 2016a). These forests presented invaluable cultural ecosystem services, e.g. (i) aesthetic values, which included unique natural landscapes (e.g., Perito Moreno glacier); (ii) existence values, which included unique species of flora and fauna; (iii) local identity, which included heritage, folklore, traditions, art and local workers (e.g., ranching, forestry, artisanal fishing, mining, and oil extraction); and (iv) many recreational activities, very appreciated for tourist of diverse parts of the World, including winter sports, hiking, trekking, climbing, riding, camping, kayaking and sport fishing.

Nothofagus forests are mainly associated to the mountain areas across both provinces and present greater variations in their over story dominant species along the latitude gradient. There are three main forest types in the region (Fig. 19.2). In Santa Cruz province, forests occurs mainly in the west, where north and central areas are dominated by *N. pumilio* and mixed evergreenforests, while *N. antarctica* prevails in the southern area (Peri and Ormaechea 2013; Peri et al. 2019b; Rosas et al. 2019a). In Tierra del Fuego province, forests occurs in central and southern areas, where the north are dominated by *N. antarctica*, central and southern areas are dominated by *N. pumilio*, andmixed evergreenforests occupied the middle hills in mountain valleys, and the shores of lakes and Beagle channel (Collado 2001; Mestre et al. 2017).

The proposed study included several modeling and combination of different analyses (Fig. 19.3) that will describe in the following subtitles. Most of the analyses maintaining similar methodologies for both provinces (Santa Cruz and Tierra del Fuego) trying to use the same variables and raster classifications. In the same way, the analyses were conducted for the different forest types, and the outputs were

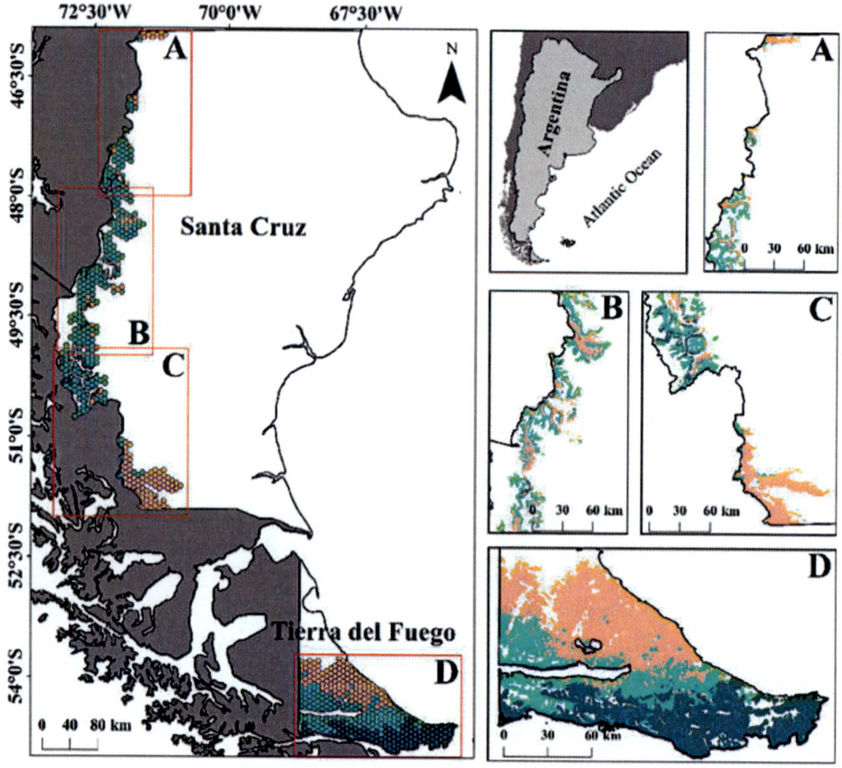

Fig. 19.2 Main forest types occurrence in the study area (*N. antarctica* forests = orange, *N. pumilio* forests = pale green, mixed evergreen forests = dark green). Based on Collado (2001), Peri and Ormaechea (2013) and Peri et al. (2019b). Each hexagon representing 5000 ha

combined to synthetize the outputs for the final outputs of the research at regional level.

19.2.2 Understory Assemblage Among the Different Forest Types

A main consideration to develop any conservation strategy at landscape level is to determine if biodiversity changes across the different ecosystems that we want to protect (Rosas et al. 2019a). For this, the first analyses that we conducted allowed to determine if the studied forest types presented different species assemblage with specific ecological niche requirements. For this, we used a database of vascular plants from 535 field plots sampled between 2000 and 2012 as part of a regional monitoring network in ecology and biodiversity (PEBANPA, Parcelas de Ecología y

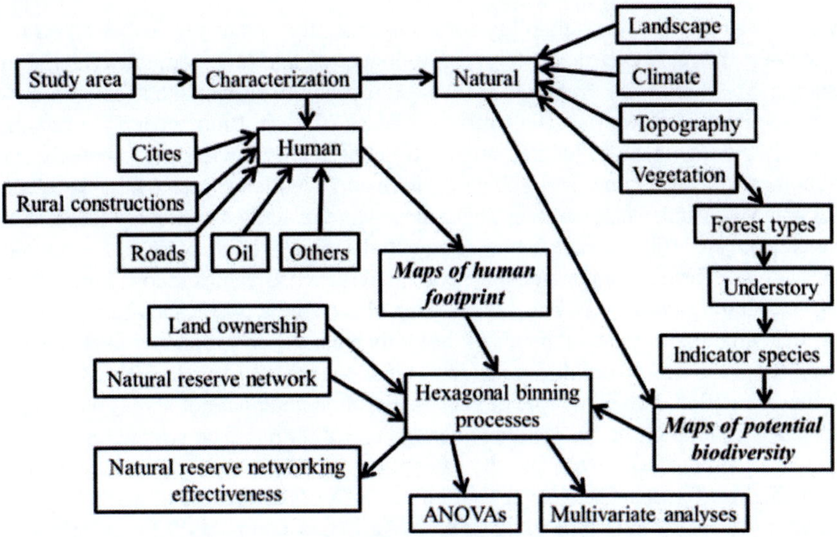

Fig. 19.3 Graphical framework of the material and methods of the study

Biodiversidad de AmbientesNaturales en Patagonia Austral) in which several federal institutions are involved (UNPA, INTA, CONICET) (Peri et al. 2016). This database includes data on plant species cover (%) and occurrence frequency (%) in the understory (D'Amato et al. 2009). A total of 35 vascular plant species indicators were selected by choosing the 20 most important species (cover × frequency of occurrence) for each of the three forest types. After that, a detrended correspondence analysis (DCA) using CANOCO5.0 (TerBraak and Šmilauer 2009) and an indicator value analyses using PC-ORD 5 (McCune and Mefford 1999) were used to determine the associated understory species cover for each forest type (NP = *N. pumilio* forests, NA = *N. antarctica* forests, MIX = mixed evergreen forests).

19.2.3 Maps of Potential Biodiversity for Nothofagus *Forests*

For these analyses, we employed the MPB developed by Martínez Pastur et al. (2016b) for Tierra del Fuego and Rosas et al. (2019b) for Santa Cruz. These maps used a large database of understory plants of *Nothofagus* forests (721 plots in Santa Cruz and 535 plots in Tierra del Fuego) from PEBANPA Network (Peri et al. 2016), and 635 plots belong to native forests provincial inventory and FAMA INTA laboratory (Forestal, Agricultura y Manejo del Agua). The database also was complemented with presence data of the selected species using the SistemaNacional de DatosBiológicos of Ministerio de Ciencia, Tecnología e InnovaciónProductiva (www.datosbiologicos.mincyt.gob.ar). Environmental Niche Factor Analysis (Hirzel et al. 2002) was used to map habitat suitability for each understory plant species based on 41 potential explanatory variables (climate, topography, and other variables related to landscape), which were rasterized at 90 × 90 m resolution using the nearest resampling technique on ArcMap 10.0 software (ESRI 2011) in Biomapper 4.0 software (Hirzel et al. 2004). The habitat suitability maps for the understory species were combined (average values for each pixel) to obtain a MPB for each forest type (NP, NA and MIX) and each province. These maps were rasterized to present scores that varied from 0 to 100 (average values of potential habitat suitability for all the studied species). For further comparisons, we classified the MPB values in low, medium or high depending on the analyses, where each class (forest type x province) contain an equal quantity of the total pixels of the study area. We analyzed the MPB considering the influence of the different forest types (based on Collado 2001; Peri and Ormaechea 2013; Peri et al. 2019b) and to determine potential hot-spots areas using a hexagonal binning processes (each hexagon = 5000 ha). The selected hexagons presented some forest cover, and the category assigned for each one corresponded to the highest forest species cover (NP, NA or MIX). The average of MPB for each hexagon of each forest type class and province was compared through one-way ANOVAs and Tukey post-hoc test.

19.2.4 Maps of Human Footprint for Provinces and the Different Nothofagus *Forest Types*

We created a HFM for Santa Cruz and Tierra del Fuego provinces based on the methodology proposed by Sanderson et al. (2002) but modified according to the available data (see Fig. 19.1 in example for main cities and roads). We used spatial data on (i) human settlements (n = 5 types) including the capital city, other cities, small towns, ranches, and other rural constructions; (ii) accessibility (n = 3 types) including national and provincial roads, secondary roads, and rough paths; and (iii) oil industry (n = 3 types) including oil exploration lines, ducts, and wells. This database was constructed using available data source of the terrestrial information system of Santa Cruz (SIT-Santa Cruz Argentina, www.sitsantacruz.gob.ar) and national energy inventory of the country (datos.minem.gob.ar). We defined the maximum impact distance (km) for each variable, which is an important step for this methodology (for more details see Sanderson et al. 2002). For human settlements we considered: 15–24.0 km for capital cities, 7 km for other cities, 2.2 km for small towns, 2 km for ranches, and 1 km for small rural constructions. Accessibility impact distance was defined according their importance and intensity uses: 2 km for national roads, 1 km for provincial roads, and 0.5 km for secondary roads and rough paths. The impact distance for oil impacts (oil exploration and oil wells) and ducts were defined as 0.5 km. Distances were defined according to the impacts that it can be possible to observe in the satellite images for each category. Then, we calculated the impact area of each variable using ArcMap 10.0 software (ESRI 2011) and Euclidian distances tool, and final grids were rasterized at 90 × 90 m resolution using the nearest resampling technique. For each variable we re-scale the grids by the logistic decreasing function (we used a Y factor between 80 and 96) through the defined impact distances. Human settlements, accessibility, ducts, oil wells presented values of impact from 1 (core) to 0 (the maximum impact distance), but oil exploration lines presented values of impact from 0.5 to 0 due to potential vegetation recovery (Dabros et al. 2017; Fuda et al. 2018). The final maps of the different variables were integrated into a single map, where the maximum values for each pixel was calculated using the cell statistics tool in ArcMap 10.0 software (ESRI 2011). The final map of human footprint index (HFM) had scores that varied from 0 (without impact) to 1 (maximum impact). As well as for MPB, we analyzed theHFM considering the influence of the different forest types (based on Collado 2001; Peri and Ormaechea 2013; Peri et al. 2019b) to determine potential hot-spots areas using a hexagonal binning processes (each hexagon = 5000 ha). The average of HFM for each hexagon of each forest type class and province was compared through one-way ANOVAs and Tukey post-hoc test.

19.2.5 Landscape Analyses of Potential Biodiversity and Human Footprint

Using the same hexagonal binning processes (hexagon = 5000 ha) we determined the status of potential biodiversity and human footprint index for each hexagon. The selected hexagons for the analyses are those that presented the same dominant forest type cover. We employed the following thresholds based on the rationale of dividing the number of hexagons in three equal proportions for the forested landscapes: (i) for Santa Cruz province the thresholds of potential biodiversity were: Low = <25%, Medium = 25–50%, and High = >50%, while for human footprint index were: Low = <0.3, and High = >0.3; and for Tierra del Fuego province the thresholds of potential biodiversity were: Low = <60%, Medium = 60–75%, and High = >75%, while for human foot print were: Low = <0.2, and High = >0.2. Beside this, each hexagon was classified according the different forest types described before, according to the most abundant forest type cover, and were used for further comparisons.

A second group of analyses was conducted considering the ownership and the protection status of the forest lands. First, using the same hexagonal binning processes (hexagon = 5000 ha) we determined the ownership status (public, private, National Parks and Provincial Reserves), and also define the forest type, the potential biodiversity and the human footprint index for each hexagon. The data of the land ownership was obtained from the official land registry of both provinces (year 2006 for Santa Cruz and 2015 for Tierra del Fuego), while natural reserves network was up to date (year 2019). For these analyses, we employed the following thresholds based on the rationale of dividing the number of hexagons in three equal proportions for the forested landscapes: (i) for Santa Cruz province the thresholds of potential biodiversity were: Low = <46%, Medium = 46–56%, and High = >56%, while for human footprint index were: Low = <0.1, and High = >0.1; and for Tierra del Fuego province the thresholds of potential biodiversity were: Low = <57%, Medium = 57–73%, and High = >73%, and the same values for human footprint. With these hexagon values, different analyses of variance (ANOVAs) were conducted for the different ownership status (public, private, National Parks, Provincial Reserves) and forest types (NP, NA, MIX) analyzing the MPB and HFM for both provinces (Santa Cruz and Tierra del Fuego). In all tests means were compared by Tukey test ($p < 0.05$).

19.2.6 *Natural Reserve Networking Effectiveness*

In this last section we want to determine the distribution of MPB of the forests in the current natural protected reserve network, private and public lands, and how current natural reserve network effectiveness can be enhanced. For this analysis, we employed HFM (Fig. 19.6), and classified the forested areas of both provinces according to the human impact (<0.1 and >0.1). We selecting the forest patches >1000 ha and HFM < 0.1, which were classified according to the different forest types (Fig. 19.2) and MPB (Fig. 19.5). When one patch presented more than one forest type, we assigned to the category with higher proportion of cover. The MPB was re-classified in three categories for both provinces according to the number of pixels with forest cover (e.g. each MPB category had the same quantity of pixels at each province and will be equally represented in the analysis). The thresholds for each category of Santa Cruz province were: (i) low: <57%, (ii) medium: 57–73%, and (iii) high >73%, and for Tierra del Fuego province were: (i) low: <46%, (ii) medium: 46–56%, and (iii) high >56% (these data were based on Martínez Pastur et al. 2016b and Rosas et al. 2019b) (Fig. 19.4).

19.3 Results and Discussion

19.3.1 *Understory Assemblage Among the Different Forest Types*

Results showed that each forest type presented a distinct species assemblage, however about half of the species were shared among the different types. The thirty three understory plant species that we included in the analyses were clearly associated to one or two forest types (Table 19.1 and Fig. 19.4). The twenty most important species of the three studied forest types included 2 ferns, 8 monocots and 25 dicots (Table 19.1). Eight species (2 monocots and 6 dicots) were exclusive for *N. antarctica*

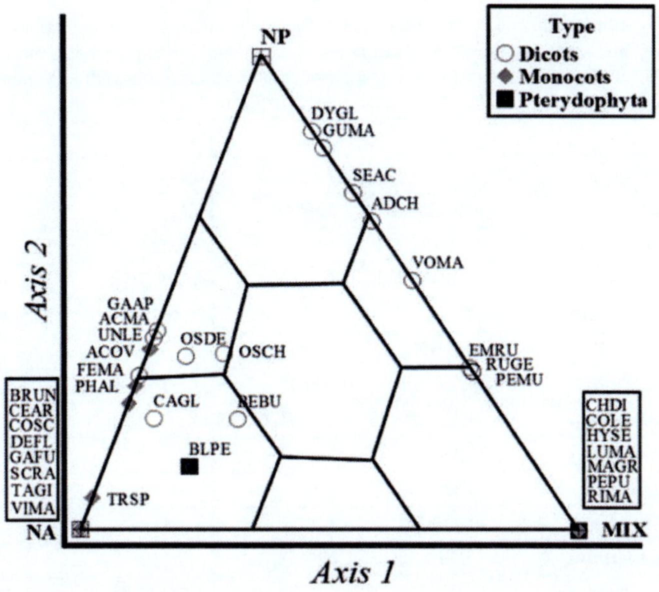

Fig. 19.4 Detrended Correspondence Analysis (DCA) for understory species cover for the 35 understory plant species (see Table 19.1 for species code) modelled for the three studied forest types (NA = *N. antarctica*, NP = *N. pumilio*, MIX = mixed evergreen) classified according to main taxonomic groups (DICO = dicots, MONO = monocots, PTERI = pteridophyta). Lines showed the areas where understory species are mostly related to one specific forest type or are shared among different forest types

forests, seven species (1 fern, 2 monocots and 4 dicots) were exclusive for the mixed evergreen forests, while no one species was exclusive for the *N. pumilio* forests. However, *N. antarctica* and *N. pumilio* forests shared seven understory species, *N. pumilio* and mixed forests shared eight species, while five species were shared by all three forest types. The selected species did not present high cover in these forests (a gradient from 0.06% for *Cardamine glacialis* in mixed forests to 6.32% for *Cotulascariosa* in *N. antarctica* forests), but most of these species had high frequencies of occurrence across the landscape (12% to 82%) (Fig. 19.4), and influenced over the understory species assemblage of the different forest type. This pattern can be graphically showed in the DCA, where no understory species were shared by the three forest types, indicating that most of selected species are not generalist and present some degree of specialization. This characteristic of the understory plant assemblage was previously indicated by Rosas et al. (2019b) through the analysis of the specialization and marginality in the habitat mapping modelling. Based on these results, we can conclude that all the forest types are important to be considered in conservation strategy at landscape level, and in order to protect the full biodiversity assemblage we need to protect all of them in the reserve networks.

Table 19.1 Species name and code, where values denote mean cover (%) (first value), frequency of occurrence (%) (second value between parentheses), and rank (1 to 20) for the 20 most important species (third value) for the three studied forest types (NA = *N. antarctica*, NP = *N. pumilio*, MIX = mixed evergreen) (based on Martínez Pastur et al. 2016b)

Species	Code	NA	NP	MIX
Acaenamagellanica	ACMA	0.72(52)14	0.49(24)17	–
Acaenaovalifolia	ACOV	1.31(47)13	0.63(25)12	–
Adenocaulonchilense	ADCH	–	0.53(21)20	0.28(12)19
Berberisbuxifolia	BEBU	1.19(70)08	0.52(23)18	0.51(18)15
Blechnumpenna-marina	BLPE	4.68(52)04	0.90(19)13	1.15(26)06
Bromusunioloides	BRUN	1.28(41)15	–	–
Cardamineglacialis	CAGL	0.69(61)11	0.23(41)09	0.06(12)20
Cerastiumarvense	CEAR	0.53(47)16	–	–
Chiliotrichumdiffusum	CHDI	–	–	0.80(12)17
Codonorchislessonii	COLE	–	–	0.62(42)04
Cotulascariosa	COSC	6.32(77)01	–	–
Deschampsiaflexuosa	DEFL	2.02(45)12	–	–
Dysopsisglechomoides	DYGL	–	1.65(41)04	0.31(14)18
Empetrumrubrum	EMRU	–	0.77(20)16	1.46(20)08
Festucamagellanica	FEMA	2.63(65)06	1.15(48)05	–
Galiumaparine	GAAP	3.08(82)03	2.24(37)03	–
Galiumfuegianum	GAFU	0.44(41)18	–	–
Gunneramagellanica	GUMA	–	2.23(29)06	0.53(18)14
Hymenophyllumsecundum	HYSE	–	–	1.09(18)11
Luzuriagamarginata	LUMA	–	–	0.46(20)12
Macrachaeniumgracile	MAGR	–	–	0.25(30)07
Osmorhizachilensis	OSCH	3.79(38)09	2.98(25)02	1.21(18)10
Osmorhizadepauperata	OSDE	5.01(76)02	3.31(45)01	0.70(18)13
Pernettyamucronata	PEMU	–	0.93(18)14	1.84(42)03
Pernettyapumila	PEPU	–	–	2.53(42)02
Phleumalpinum	PHAL	1.77(59)10	0.64(26)11	–
Ribesmagellanicum	RIMA	–	–	0.38(18)16
Rubusgeoides	RUGE	–	1.02(36)08	2.01(58)01
Schizeilema ranunculus	SCRA	2.28(64)07	–	–
Senecioacanthifolius	SEAC	–	1.52(38)07	0.61(34)05
Taraxacumgillesii	TAGI	0.60(36)19	–	–
Trisetumspicatum	TRSP	3.46(62)05	0.25(28)15	–
Uncinialechleriana	UNLE	0.89(39)17	0.55(21)19	–
Viciamagellanica	VIMA	0.33(27)20	–	–
Viola magellanica	VOMA	–	0.95(25)10	0.85(24)09

19.3.2 Potential Biodiversity Across the Different Nothofagus *Forest Types*

The combination of potential habitat suitability maps for understory plant species allowed us to develop the MPB for the different forest types in both provinces (Fig. 19.5). The lower biodiversity values occurred in Santa Cruz province: Lago Buenos Aires, Lago San Martín and Lago Argentino had medium and lower values, while the higher biodiversity values occurred in Lago Pueyrredón and Río Turbio. In general, MPB increased with latitude (north to south), presenting the highest values in *N. antarctica* forests, and decreased with longitude (east to west) with medium values in *N. pumilio* forests and the lowest values near glaciers in mixed evergreen forests (Rosas et al. 2019b). In Tierra del Fuego, *N. antarctica* forests with the highest potential biodiversity values were related to the ecotone areas with *N. pumilio* forests in the southern area of its distribution and with some large isolated patches surrounded

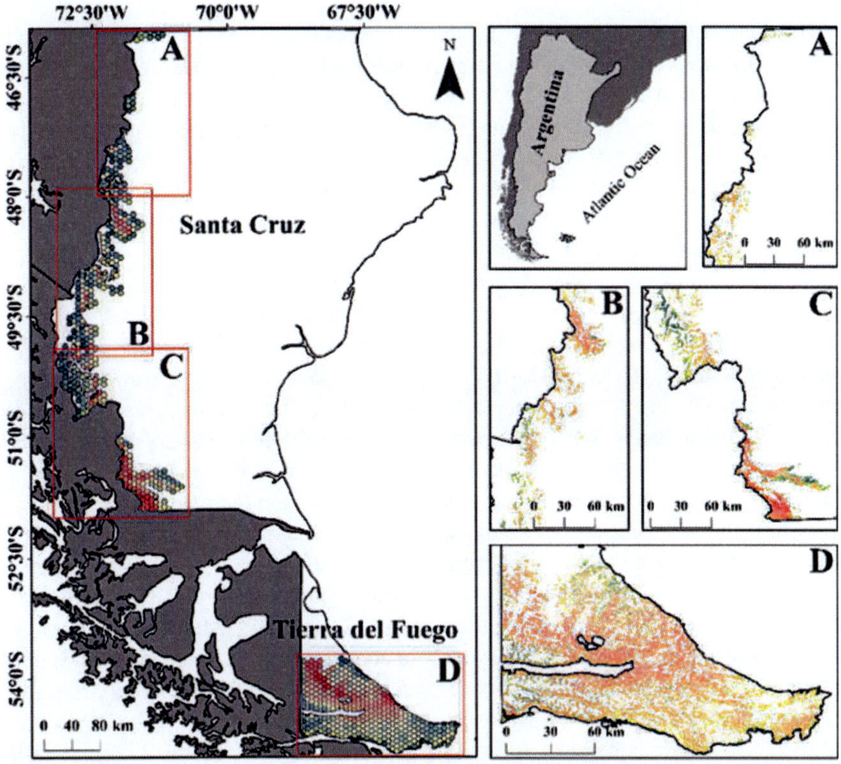

Fig. 19.5 Map of potential biodiversity of understory plants of *Nothofagus* forests in Southern Patagonia. Red represents the higher potential (value close to 1) and dark green the lower potential (value close to 0). **a** Lago Buenos Aires and Lago Pueyrredón, **b** Lago San Martín, **c** Lago Argentino and Río Turbio, and **d** Tierra del Fuego. Each hexagon representing 5000 ha

Table 19.2 Simple ANOVA analyses of the potential biodiversity values of understory plant species of the different forest types hexagons, where: NA = *N. antarctica*, NP = *N. pumilio*, and MIX = mixed evergreen at Santa Cruz (SC) and Tierra del Fuego(TDF) provinces. MPB = map of potential biodiversity values, HFM = human footprint map values

Province	Treatment	MPB	HFM
SC	NA	0.557 b	0.30 b
	NP	0.358 a	0.07 a
	MIX	0.433 ab	0.08 a
	F(p)	33.22(<0.001)	106.38(<0.001)
TDF	NA	0.749 c	0.27 c
	NP	0.702 b	0.13 b
	MIX	0.578 a	0.07 a
	F(p)	60.45(<0.001)	31.34(<0.001)

F = Fisher test, p = probability at 0.05

by grassland at the central area of the Island. The potential of these forests decreased with the closeness of the Atlantic Ocean. *N. pumilio* forests with the greater potential were also related to the ecotone areas with *N. antarctica* forests in the flat zones of its distribution at the central-east of the study area. MPB of this forest types generally increased from south-west to north-east. Beside this, the areas with greatest MPB in mixed evergreen forests were related to rainfall distribution, mainly in the lower elevations at the south-west of the study area and at the eastern forests of the island close to Mitre peninsula (Martínez Pastur et al. 2016b).

ANOVAs showed that MPB significantly changed across the forest types in both provinces (Table 19.2). The highest MPB values were found in *N. antarctica* forests(average values of 0.56 and 0.75) and are significant higher than in *N. pumilio* (0.36 and 0.70) and MIX (0.43 and 0.58 for Santa Cruz and Tierra del Fuego, respectively) (Table 19.2). In general, values for the different forests were higher at Tierra del Fuego than in Santa Cruz provinces. These differences were can be due to average values for the hexagons were in general higher in Tierra del Fuego, where forest quality are much better for biodiversity conservation (e.g., higher site quality, canopy closure and continuity, ecological functions integrity, and ecosystem health).

MPB is a powerful tool to understand the relationship among the different species assemblages and their environmental conditions around the world (e.g., Zaniewski et al. 2002; Hirzel and Le Lay 2008), as well as in Patagonia (Martínez Pastur et al. 2016b; Rosas et al. 2017, 2018, 2019a, b). In ecotone areas between grasslands and forests (e.g., grasslands and NA) and between different forest types (e.g., NA and NP), the MPB greatly increased due to multiple micro-environments that allowed the survival of a higher number of species (Lencinas et al. 2008; Antos 2009), as well as the existence of potential synergies among the species occurrence (Gargaglione et al. 2014). Beside this, multiple vegetation stratification promoted more species diversity, e.g. NA open forests support more shrubs and grasses species (Peri and Ormaechea 2013; Peri et al. 2016) than other close *Nothofagus* forests with one tree stratum supporting scarce understory biodiversity (Martínez Pastur et al. 2002; Lencinas et al. 2008, 2011).

19.3.3 Maps of Human Footprint for Provinces and the Different Nothofagus Forest Types

HFM revealed that most of the natural landscapes in the region, including grasslands, shrublands, peatlands and forests, all have some level of human influence (Fig. 19.6 where forests are highlighted and represented by hexagon average values). The highest human impact values (hexagon value up to 0.50) were located close to the capital cities and following the main national roads. In the northern areas of

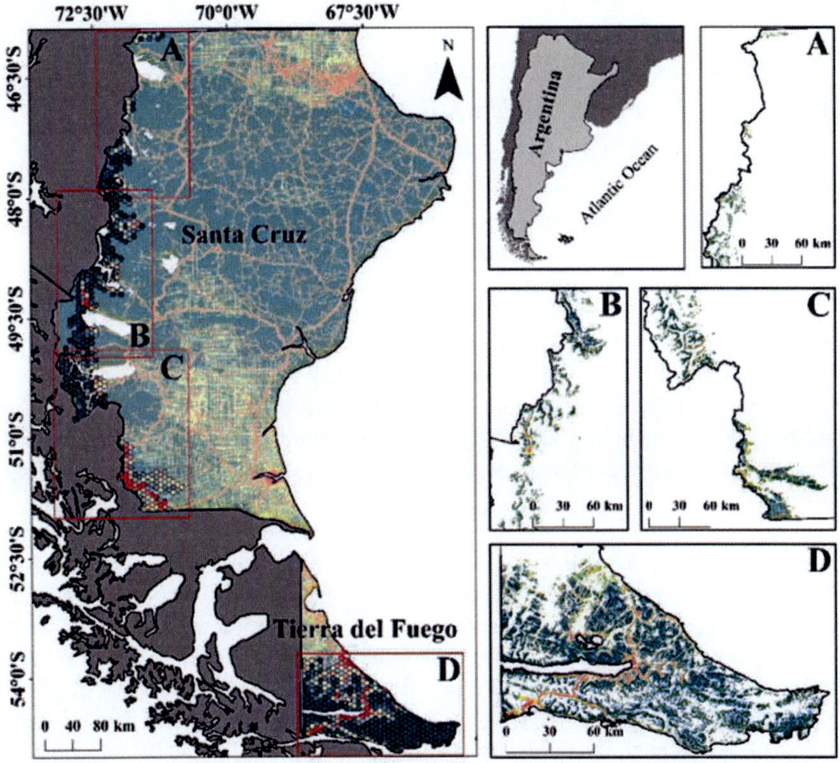

Fig. 19.6 Map of human footprint index for Southern Patagonia (left) and *Nothofagus* forests (right). Red represents the higher impact (value close to 1) and dark green, the lower impact (value close to 0). **a** Lago Buenos Aires and Lago Pueyrredón, **b** Lago San Martín, **c** Lago Argentino and Río Turbio, and **d** Tierra del Fuego. Each hexagon representing 5000 ha

Santa Cruz and Tierra del Fuego, oil industry generated a large impact over the landscape, while intermediate values were related to ranching activities located mainly in the south of Santa Cruz and center of Tierra del Fuego. Finally, the lowest HFM values were located in the lower productive lands (national parks and areas without accessibility like mountains and large peatlands).

ANOVAs showed that HFM significantly changed across the forest types in both provinces (Table 19.2). The highest HFM values were found in *N. antarctica* forests (average values of 0.30 and 0.27) compared to *N. pumilio* (0.07 and 0.13) and mixed evergreen forests (0.08 and 0.07 for Santa Cruz and Tierra del Fuego, respectively) which presented the lowest values (Table 19.2). In general, values for the different forests were quite similar in both provinces. These similarities were due to the economic activities that were implemented in the different forest types (e.g., silvopastoral systems in NA, timber harvesting in NP). According to these results, we can conclude that human footprint differentially occurred in forest types and landscapes across the studied provinces.

19.3.4 The Last of the Wild of the **Nothofagus** *Forests and Their Potential Biodiversity*

The different MPB and the cross-data with HFM allowed us to identify forests areas with low human impact and defined its potential conservation values (Fig. 19.7). The forests of Santa Cruz province presented a high dispersion in the landscape, in general with low potential biodiversity and human footprint. The best forests in Santa Cruz province (e.g., best forest patches for conservation) were located in the centre-south and close to Los Andes Mountains (e.g., Lago San Martín and Lago Argentino). Tierra del Fuego presented low HFM in the western areas, however those forests have low-medium MPB. The best forests were located in fragmented areas in the center of Tierra del Fuego, where most of the ranching and forest harvesting occurs.

The different forest types presented unequal values and received different human impact, mainly due to its timber potential or agroforestry uses (Peri and Ormaechea 2013; Martínez Pastur et al. 2017; Peri et al. 2019b), and their location in the landscape. To define conservation strategies it is necessary to know the status of each forest type, due to the different forests presented different understory assemblage (see Table 19.1 and Fig. 19.4), and other biodiversity values as were described for different authors for Southern Patagonia (Lencinas et al. 2008, 2011; Martínez Pastur et al. 2016b; Peri et al. 2016; Mestre et al. 2017; Rosas et al. 2017, 2019a, b).

The most abundant forest type in Southern Patagonia was *N. pumilio* (68% of Santa Cruz and 38% of Tierra del Fuego forests) (Table 19.3). The same trend can be observed in both provinces, where MPB increased as HFM also increased. This can be explained by the fact that higher site quality stands typically support greater biodiversity (e.g., Gallo et al. 2013), but also present higher timber values and are

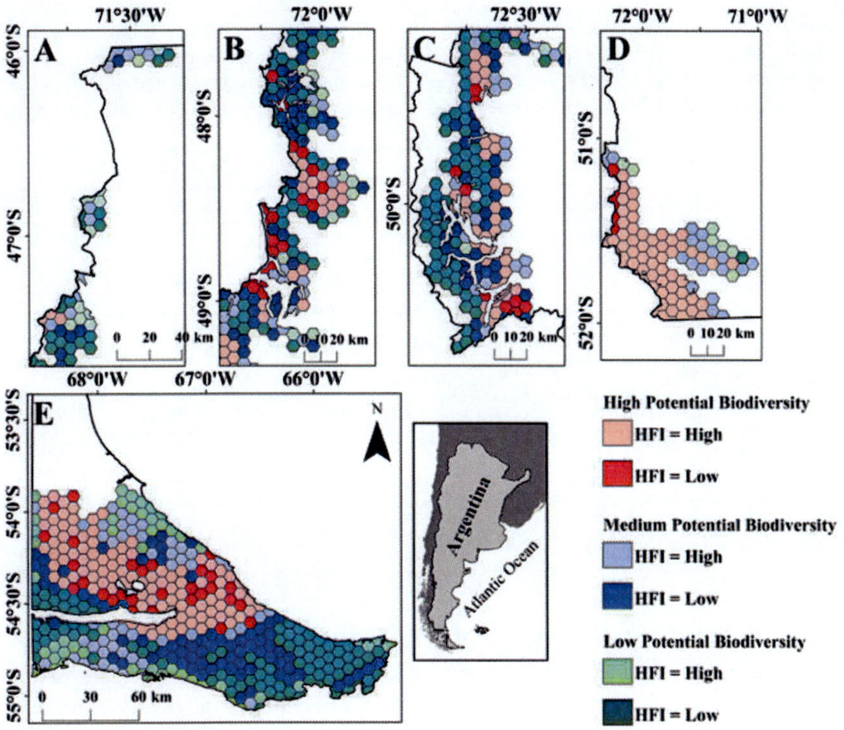

Fig. 19.7 Map of human footprint index for Southern Patagonia (left) and *Nothofagus* forests (right). Red represents the higher impact (value close to 1) and dark green the lower impact (value close to 0): **a** Lago Buenos Aires and Lago Pueyrredón, **b** Lago San Martín, **c** LagoArgentino, **d** Río Turbio, and **e** Tierra del Fuego. Each hexagon represents 5000 ha

therefore preferred for harvesting. Beside this, marginal forests usually were left aside of management due to other values (e.g., upper basin protection) or due to their inaccessibility (e.g., higher slope in mountain areas), and these forests usually presented lower MPB (Martínez Pastur et al. 2016b).

N. antarctica forests were the second most abundant type in the region (29% of the total forests) and the human impact followed different patterns in both provinces: (i) in Santa Cruz province the impact was higher than in *N. pumilio* but follow the same pattern (impact increased with MPB), and (ii) in Tierra del Fuego province the impact was higher at lower and higher than at medium MPB. This last trend was due to lower qualities occurred in ecotone areas with grasslands and close to sea shores (sheep production are greater at ecotone, and temperate lands close to sea shores are used as preferred winter refugee for cattle), while higher qualities occurred in ecotone areas with timber *N. pumilio* forests (harvesting intensity was higher in those areas due to relief presented lower slopes and greater accessibility) (Martínez Pastur et al. 2017). Finally, the mixed evergreen forests had a greater areal extent in Tierra del Fuegoprovince(33%) than in Santa Cruz province (4%). In Santa Cruz province,

Table 19.3 Forest types (NA = *N. antarctica*, NP = *N. pumilio*, and MIX = mixed evergreen) classified according the human footprint index (HFH = high, HFL = low impact) and the potential biodiversity (PBH = high, PBM = medium, PBL = low potential) for the studied provinces (SC = Santa Cruz, TDF = Tierra del Fuego). Values representing the hexagon (5000 ha) number or percentage for each category

SC	NA (29%)			NP (68%)			MIX (4%)		
	n	HFH	HFL	n	HFH	HFL	n	HFH	HFL
PBH	65	97%	3%	66	62%	38%	7	71%	29%
PBM	30	87%	13%	104	34%	66%	2	100%	0%
PBL	22	82%	18%	106	11%	89%	6	0%	100%
Total		**107**	**10**		**88**	**188**		**7**	**8**
TDF	NA (29%)			NP (38%)			MIX (33%)		
	n	HFH	HFL	n	HFH	HFL	n	HFH	HFL
PBH	63	84%	16%	64	70%	30%	0	0%	0%
PBM	29	72%	28%	46	33%	67%	49	27%	73%
PBL	21	95%	5%	37	27%	73%	78	23%	77%
Total		**94**	**19**		**70**	**77**		**31**	**96**

most of the greater values (PBH and PBM with HFH) were close to touristic places (Martínez Pastur et al. 2016a), while low quality forests (PBL) remains in inaccessible areas in the mountain areas (Peri et al. 2019b). In Tierra del Fuego, mixed evergreen forests presented lower and medium potential biodiversity values, and most of them presented lower human impact (>73%) due to mostly occurred in inaccessible areas close to Mitre peninsula (Mestre et al. 2017).

The proposal of Sanderson et al. (2002) to identify the last of the wild allowed us to quickly identify the areas with great conservation potential based on the intactness, and was successfully applied in the region (e.g., Magallanes in Chile) (Inostroza et al. 2016). On the other hand, not all the forests presented the same quality for conservation of understory plant species, due to some forests support greater biodiversity (Lindenmayer and Franklin 2002; Elith and Leathwick 2009) as was demonstrated for *Nothofagus* forests at southern Patagonia (Gallo et al. 2013; Martínez Pastur et al. 2016b; Rosas et al. 2019b). However, when we combine both approaches (e.g., potential biodiversity and human footprint), we obtained a powerful tool for decision-makers that allowed us to identify forests with high biodiversity and low human footprint that are much better targets for conservation purposes. For this, we arrive to the conclusion that it is possible to identify areas with special values of MPB and low HFM according to the different forest types combining both proposals.

The ownership and the legal status of the forest lands were unequally distributed between the two provinces (Fig. 19.8). In Santa Cruz province, the majority of the forest was in private lands, followed by National Parks, and with small areas under public lands. On the other hand, forest lands in Tierra del Fuego were equally distributed between private and public lands, and where the protected areas were equally distributed between national parks and provincial reserves.

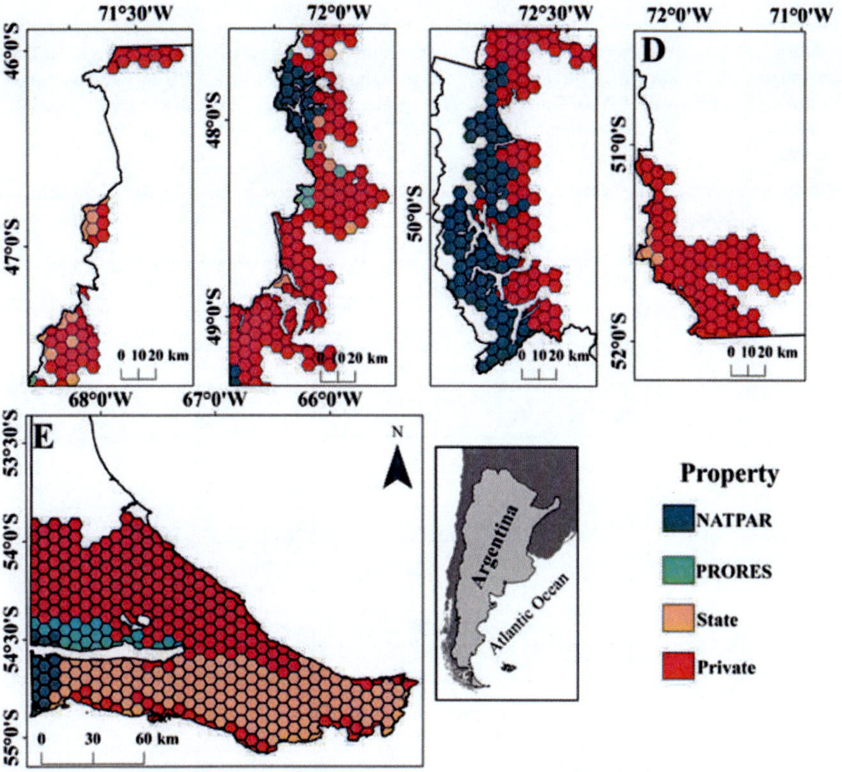

Fig. 19.8 Ownership status of the forest lands considering public, private and natural reserves (NATPAR = National Parks, PRORES = Provincial Reserves): **a** Lago Buenos Aires and LagoP-ueyrredón, **b** Lago San Martín, **c** Lago Argentino, **d** Río Turbio, and **e** Tierra del Fuego. Each hexagon represents 5000 ha

The forest types were not distributed similarly among private and public lands and protected areas (Table 19.4), where the protection status greatly varied among species. Also, the quality and naturalness also changed across the landscape and type of ownership. This is a very important consideration when the effectiveness of the natural reserves networks is analysed, due to private can change their management or conservation strategy in time, according their own interests. The outputs for forest types and province need special attention to design effective conservation strategies. If we considered the *N. pumilio* forests (Table 19.4), half of the area belongs to private ranchers (50% in Santa Cruz and 58% in Tierra del Fuego) and the other half belongs to public lands (6% in Santa Cruz and 32% in Tierra del Fuego) or are located in reserves (37% in Santa Cruz and 18% in Tierra del Fuego). Most of *N. pumilio*f orests presented low human impact (78% to 89%), and the greater impact are located in private lands in Santa Cruz (69%) and equally distributed between public and private lands in Tierra del Fuego. Beside this, most of the high quality *N.*

Table 19.4 Area occupied by the forest types (NP = *N. pumilio*, NA = *N. antarctica*, and MIX = mixed evergreen forests) classified according the human footprint index (HFH = high, HFL = low impact) and the potential biodiversity (PBH = high, PBM = medium, PBL = low potential) for the studied provinces (SC = Santa Cruz, TDF = Tierra del Fuego), and owner of the lands (%) considering public, private and natural reserves (NATPAR = National Parks, PRORES = Provincial Reserves)

Province	Forest type	Category	Area (thousand ha)	Public (%)	Private (%)	NATPAR (%)	PRORES (%)
SC	NP	Total	224.5	6	58	33	4
		HFH	23.4	18	69	13	1
		HFL	201.1	4	56	35	5
		PBH	76.9	8	74	13	5
		PBM	73.1	6	59	28	6
		PBL	74.5	2	36	60	1
	NA	Total	169.8	4	92	3	1
		HFH	78.6	5	91	3	0
		HFL	91.2	3	94	3	1
		PBH	84.2	5	92	2	1
		PBM	42.9	3	89	6	1
		PBL	42.7	2	96	1	0
	MIX	Total	18.0	2	42	53	3
		HFH	2.4	6	54	37	3
		HFL	15.6	1	40	56	3
		PBH	4.9	4	70	24	2
		PBM	4.9	0	49	46	4
		PBL	8.2	0	20	77	3
TDF	NP	Total	316.2	3	50	5	13
		HFH	69.5	40	51	1	8
		HFL	246.7	30	50	6	14
		PBH	127.6	26	59	2	13
		PBM	124.8	35	49	4	12
		PBL	63.8	38	32	14	16
	NA	Total	182.0	0%	99	0	1
		HFH	68.5	0	99	0	0
		HFL	113.5	0	99	0	1
		PBH	59.5	0	98	0	1
		PBM	61.8	0	99	0	0
		PBL	60.7	0	99	0	0
	MIX	Total	192.0	70	23	5	2

(continued)

Table 19.4 (continued)

Province	Forest type	Category	Area (thousand ha)	Public (%)	Private (%)	NATPAR (%)	PRORES (%)
		HFH	21.3	37	43	19	1
		HFL	170.7	74	21	3	2
		PBH	34.0	70	19	8	4
		PBM	95.8	76	20	3	1
		PBL	62.2	62	31	6	1

pumilio forests for conservation (PBH) are in private lands (74% in Santa Cruz and 59% in Tierra del Fuego) and only 15% to 18% are protected in reserves.

If we considered the *N. antarctica* forests (Table 19.4), most of the lands belong to private ranchers (92% in Santa Cruz and 99% in Tierra del Fuego) and just few belong to public lands (4% in Santa Cruz and very low represented in public lands of Tierra del Fuego) or are located in reserves (4% in Santa Cruz and 1% in Tierra del Fuego). *N. antarctica* forests presented higher human impacts (46% in Santa Cruz and 38% in Tierra del Fuego are considered HFH), and the best forests for conservation are in private lands without any consideration for their protection. Finally, the mixed evergreen forests presented a completely different status at both provinces (Table 19.4). In Santa Cruz 42% of mixed evergreen forests belong to privates, and 56% are under protection. Most of mixed evergreen forests were low impacted (87%), but the best conservation quality forests are in private lands (70%), due to only marginal forests were included in the natural reserves. In Tierra del Fuego most of the mixed evergreen forests are located in public lands (70%) compared to private lands (23%), and few is under protection (7%). Most of mixed evergreen forests were also low impacted (89%), and the best conservation quality forests are still in public lands (70%), and few are protected inside parks and natural reserves (12%).

We compared the MPB and HFM with different ownership status and in different provinces for the studied forest types (Table 19.5). In general, in Santa Cruz the best quality forests for conservation are located in private lands (47% MPB) and the worst are those located in the National Parks (29% MPB), and as was expected the forests located outside reserves presented more human impact (0.16–0.18 HFM compared to 0.00–0.02 HFM). In Tierra del Fuego, the best quality forests are also located in private lands (73 MPB) and the worst in National Parks (53 MPB), and presented greater impact in private lands (0.21 HFM) than the others types (0.04 to 0.09 HFM). As was described before, this is also can be showed for the different forest types (Table 19.5). *N. pumilio* forests presented significant differences for the comparisons, where the better conservation quality are in private lands (41 MPB in Santa Cruz and 77 MPB in Tierra del Fuego), but presented lower human impacts than in public lands (0.09 compared to 0.12 in Santa Cruz, and 0.15 compared to 0.17 in Tierra del Fuego). Beside this, the HFMI was lower inside the reserves as well as in the general trend. *N. antarctica* did not present significant differences,

Table 19.5 ANOVAs for the different ownership status (public, private, NATPAR = National Parks, PRORES = Provincial Reserves) and forest types (NP = *N. pumilio*, NA = *N. antarctica*, and MIX = mixed evergreen) analyzing the map of potential biodiversity (MPB) and the human footprint index (HFM) for the studied provinces (SC = Santa Cruz, TDF = Tierra del Fuego)

	Owner	SC		TDF	
		MPB	HFM	MPB	HFM
Public		38.2 ab	0.16 b	62.0 b	0.08 a
Private		47.3 b	0.18 b	72.6 c	0.21 b
NATPAR		28.6 a	0.02 a	52.6 a	0.09 ab
PRORES		29.8 ab	0.00 a	67.8 bc	0.04 a
F(p)		18.39(<0.001)	22.43(<0.001)	25.54(<0.001)	13.76(<0.001)
Forest type	Owner				
NP	Public	33.8 ab	0.12 b	66.8 b	0.17
	Private	40.8 b	0.09 b	76.5 c	0.15
	NATPAR	28.1 a	0.02 a	52.2 a	0.03
	PRORES	29.8 ab	0.00 a	67.8 b	0.04
	F(p)	8.90(<0.001)	8.35(<0.001)	26.70(<0.001)	3.05(0.030)
NA	Public	61.0	0.38		
	Private	55.6	0.30	75.0	0.27
	NATPAR	55.0	0.17		
	F(p)	0.06(0.939)	1.34(0.265)		
MIX	Public			59.5 b	0.04 a
	Private	62.7 b	0.14 b	52.5 a	0.12 b
	NATPAR	21.1 a	0.01 a	53.5 ab	0.18 b
	F(p)	32.63(<0.001)	8.36(0.012)	6.26(0.003)	4.46(0.013)

F = Fisher test, p = probability at 0.05

due to they are unequally represented in the different categories. Finally, the mixed evergreen forests presented significant differences for the comparisons, where the better conservation quality are in private lands (63 MPB) with higher human impact in Santa Cruz, and in public lands (60 MPB) with lower human impact in Tierra del Fuego. One particular observation is for National Parks in Tierra del Fuego, where HFM presented the higher values. This is due to in the Tierra del Fuego National Park the most valuable touristic places are located in *N. antarctica* forest type, despite the conservation values that these forests present.

It is possible to apply these methodologies to enhance the current natural reserve networking effectiveness? We combined the results for the study area, selecting the forest patches with low HFM (less than 0.1) and classified them according its average potential biodiversity (Table 19.6 and Fig. 19.9). Usually the proposals to improve the conservation strategies are based in only one of these methodologies, using MPB (e.g., Mace et al. 1998; Lencinas et al. 2008; Mori et al. 2017; Rosas et al. 2019a, b)

Table 19.6 Number of forest patches with HFM < 0.1 and area in hectares shown in parenthesis for the different forest types (NA = *N. antarctica*, NP = *N. pumilio*, and MIX = mixed evergreen) and classified according the potential biodiversity (H = high, M = medium, L = low) for the Santa Cruz and Tierra del Fuego provinces

Variable			Santa Cruz	Tierra del Fuego
Total			59 (3010)	77 (5096)
Forest type	MIX		3 (2176)	27 (4582)
	NP		41 (3439)	34 (6438)
	NA		15 (2004)	16 (3108)
Potential biodiversity	MIX	H	0	2 (7682)
		M	1 (1554)	20 (4662)
		L	2 (2587)	5 (3024)
	NP	H	15 (2373)	13 (8743)
		M	11 (4900)	18 (5004)
		L	15 (3434)	3 (5063)
	NA	H	7 (1728)	1 (2184)
		M	3 (1891)	5 (3054)
		L	5 (2459)	10 (3229)

or HFM (e.g., Sanderson et al. 2002; McGowan 2016; Inostroza et al. 2016; Venter et al. 2016; Di Marco et al. 2018; Li et al. 2018), however, the combination of both proposals generate a more powerful tool for decision-makers.

HFM allowed us to identify 59 forest patches in Santa Cruz (average area of 3010 ha) and 77 patches in Tierra del Fuego (average area of 5095 ha) (Table 19.6), which represented 43% and 57% of the total forest area respectively (see Table 19.4 for province total values). These patches are not equally distributed among the different forest types, and these inequity were mostly related to (i) forest type abundance, (ii) accessibility, and (iii) economical potential of the different forest types (Peri and Ormaechea 2013; Martínez Pastur et al. 2017; Peri et al. 2019b). In Santa Cruz (Table 19.6), *N. pumilio* was more abundant (69.5% of the patches), where 36.6% presented high potential biodiversity (n = 15 patches of 2373 ha in average). The other forest types presented less patches: (i) *N. antarctica* (25.4% of the patches) presented 46.7% of its patches with high MPB (n = 7 patches of 1728 ha in average), and (ii) mixed evergreen forests (5.1% of the patches) presented no one of its patches with high MPB and 33.3% of its patches with medium MPB (n = 1 patch of 1554 ha). In Tierra del Fuego (Table 19.6), *N. pumilio* was also the most abundant forest type (44.2% of the patches), where 38.2% presented high MPB (n = 13 patches of 8743 ha in average). The other forest types also presented less patches: (ii) mixed evergreen forests (35.1% of the patches), where 7.4% of its patches presented high MPB (n = 2 patches of 7682 ha in average), and (ii) *N. antarctica* (20.8% of the patches) presented 6.3% of its patches with high MPB (n = 1 patch of 2184 ha).

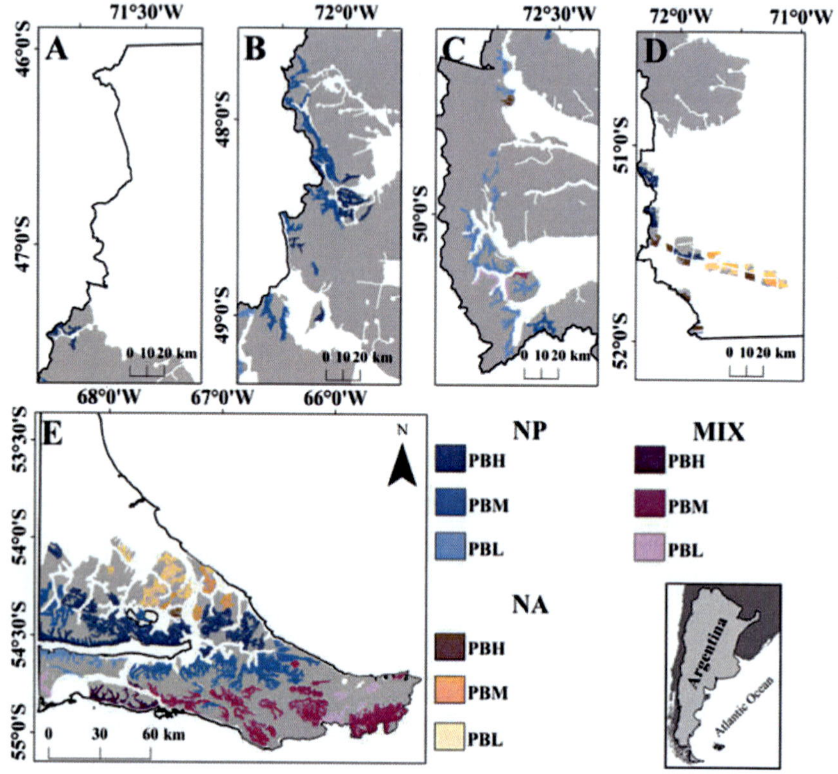

Fig. 19.9 Forest patches with HFM < 0.1 for the different forest types (NA = *N. antarctica*, NP = *N. pumilio*, and MIX = mixed evergreen forests) and classified according the potential biodiversity (PBH = high, PBM = medium, PBL = low) for Southern Patagonia: **a** Lago Buenos Aires and LagoPueyrredón, **b** Lago San Martín, **c** LagoArgentino, **d** Río Turbio, and **e** Tierra del Fuego

The number of patches with high MPB and low HFM identified outside of the current network of natural reserves (national parks and provincial reserves) can be important candidates to enhance the effectiveness of current natural reserve network. Conservation of biodiversity should be improved in forest types less represented in the protected areas network, such as *N. antarctica* (Rosas et al. 2019a). Most of the patches identified for *N. pumilio* in Santa Cruz are already included in national parks, while in Tierra del Fuego, those high quality patches are outside the reserves (Fig. 19.9). Most of the unprotected high-quality patches are in the most productive areas where timber harvesting is conducted, generating different trade-offs with biodiversity conservation. Also, the analyses for understory plant species showed that most of the protected areas do not adequately include the most valuables patches for conservation, which remains in private or public lands with low human impact.

High quality patches of mixed evergreen forests also were actually included in the formal reserves of Santa Cruz, but not in Tierra del Fuego (Fig. 19.9). Most of the high quality forest patches in Tierra del Fuego are in the mountain areas (both private and public lands), where low-impact tourism were conducted. Finally, *N. antarctica* forests were the most critical one, due to most of the high valuable patches at both provinces were outside the natural reserve network. These high quality patches are in private lands which were used for cattle or sheep breeding (Peri and Ormaechea 2013; MartínezPastur et al. 2017). This forest type was identified as one of the conservation priorities in Southern Patagonia due to its low representativeness in the national parks and provincial reserves of Santa Cruz (Rosas et al. 2019a) and Tierra del Fuego (Martínez Pastur et al. 2016b).

19.4 Conclusion and Recommendations

Biodiversity loss is occurring at global scale, especially in forest ecosystems (Mace et al. 1998; Gilliam 2007; Lindenmayer et al. 2012), and is related to different degradation processes (Peri et al. 2016; Gunn et al. 2019) generating trade-offs (e.g., cattle, harvesting, invasive species) with other ecosystem service provision (Martínez Pastur et al. 2017; Peri et al. 2017, 2019a). This degradation processes can be linked to the human footprint (Sanderson et al. 2002; McGowan 2016; Di Marco et al. 2018). Historically, the conservation strategy at Southern Patagonia was based on the conservation of intact large uninhabited areas with no consideration to biodiversity values. As Rosas et al. (2019b) suggested this opportunistic strategy presented two weaknesses: (i) land-sparing has been considered ineffective for biodiversity conservation (Coetzee 2017) due to many species need larger areas to survive; and (ii) not all the forest types were equally included in the natural reserves (also see Martínez Pastur et al. 2016b). Thus, most of the high biodiversity values were left outside the reserves as was detected in our analyses too.

Improving our understanding about biodiversity values and human footprint in natural forests is crucial to develop sustainable landscape management for an effective conservation planning (Gilliam 2007; Martínez Pastur et al. 2016b; Silva et al. 2014; Rosas et al. 2019a), and to predict the consequences of biodiversity loss due to anthropogenic activities (Lencinas et al. 2008, 2011; Corlett 2015). The methodology proposed here (crossing MPB and HFM) is a simple yet effective way to assist in the identification of new areas for conservation, or to detect potential hotspots of high biodiversity values and low human impact outside the natural reserves, e.g., promoting land-sparing strategies to improve the conservation values in the managed forests (Lindenmayer and Franklin 2002; Lindenmayer et al. 2012). However, the limitations due to the lack of other sound variables in the modeling (e.g., harvesting or livestock impacts, invasive species, poaching, etc.) can lead to biased results, so, the current proposals must be ground-truth checked before their implementation.

Acknowledgements This research is part of the doctoral thesis of YMR (Faculty of Ciencias Agrarias y Forestales in the Universidad Nacional de la Plata). We want to thank the Consejo Agrario Provincial of Santa Cruz Province, and the financial support of the project "Modeling endangered species forest habitats, and updating forest land use plans in Argentina in support of the UN sustainable development goals" (NRA ROSES 2018) USA (Project NASA 18-SLSCVC18-0019).

References

Andrew ME, Wulder MA, Nelson TA (2014) Potential contributions of remote sensing to ecosystem service assessments. Prog Phys Geogr 38(3):328–353

Antos J (2009) Understory plants in temperate forests. In: Owens JN, Gyde Lund H (eds) Forests and forest plants, vol I. Eolss Publishers Co. Ltd., Oxford, United Kingdom, pp 262–279

Coetzee B (2017) Evaluating the ecological performance of protected areas. Biodiv Conserv 26:231–236

Collado L (2001) Tierra del Fuego forest: analysis of their stratification through satellite images for the forest province inventory. Multequina 10:1–16

Corlett RT (2015) The anthropocene concept in ecology and conservation. Trends Ecol Evol 30(1):36–41

Costanza R, De Groot R, Braat L, Kubiszewski I, Fioramonti L, Sutton P, Farber S, Grasso M (2017) Twenty years of ecosystem services: how far have we come and how far do we still need to go? Ecosyst Serv 28(A):1–16

Dabros A, Hammon HEJ, Pinzon J, Pinno B, Langor D (2017) Edge influence of low-impact seismic lines for oil exploration on upland forest vegetation in northern Alberta (Canada). Forest Ecol Manag 400(15):278–288

Daily GC (1997) Nature's services. Island Press, Washington DC, USA

de Bello F, Lavorel S, Díaz S, Harrington R, Cornelissen JHC, Bardgett RD, Berg MP, Cipriotti P, Feld CK, Hering D, Martins da Silva P, Potts SG, Sandin L, Sousa JP, Storkey J, Wardle DA, Harrison PA (2010) Towards an assessment of multiple ecosystem processes and services via functional traits. Biodiv Conserv 19(10):2873–2893

De Groot RS, Wilson MA, Boumans RM (2002) A typology for the classification, description and valuation of ecosystem functions, goods and services. Ecol Econ 41(3):393–408

De Groot RS, Alkemade R, Braat L, Hein L, Willemen L (2010) Challenges in integrating the concept of ecosystem services and values in landscape planning, management and decision making. Ecol Complex 7(3):260–272

Di Marco M, Venter O, Possingham HP, Watson JE (2018) Changes in human footprint drive changes in species extinction risk. Nat Commun 9:e4621

Elith J, Leathwick J (2009) Species distribution models: ecological explanation and prediction across space and time. Annu Rev Ecol Evol Syst 40(1):677–697

ESRI (2011) ArcGIS desktop: release 10. Environmental Systems Research Institute Inc., Redlands, USA

Food and Agriculture Organization of the United Nations (FAO) (2010) Global forest resources assessment 2000. Technical Report, Rome, Italy

Fuda RK, Ryan SJ, Cohen JB, Hartter J, Frair JL (2018) Assessing the impacts of oil exploration and restoration on mammals in Murchison Falls conservation area Uganda. African J Ecol 56(4):804–817

Gallo E, Lencinas MV, MartínezPastur G (2013) Site quality influence over understory plant diversity in old-growth and harvested stands of *Nothofagus pumilio* forest. Forest Syst 22(1):25–38

Gargaglione V, Peri PL, Rubio G (2014) Tree–grass interactions for N in *Nothofagus antarctica* silvopastoral systems: evidence of facilitation from trees to underneath grasses. Agrofor Syst 88:779–790

GeaIzquierdo G, MartínezPastur G, Cellini JM, Lencinas MV (2004) Forty years of silvicultural management in southern *Nothofagus pumilio* primary forests. Forest Ecol Manage 201(2–3):335–347

Gilliam FS (2007) The ecological significance of the herbaceous layer in temperate forest ecosystems. Bio Sci 57(10):845–858

González E, Salvo A, Valladares G (2015) Sharing enemies: evidence of forest contribution to natural enemy communities in crops, at different spatial scales. Ins Conserv Div 8(4):359–366

Gunn JS, Ducey MJ, Belair E (2019) Evaluating degradation in a North American temperate forest. Forest Ecol Manag 432:415–426

Hansen MC, Potapov PV, Moore R, Hancher M, Turubanova SA, Tyukavina A, Thau D, Stehman SV, Goetz SJ, Loveland TR, Kommareddy A, Egorov A, Chini L, Justice CO, Townshend JR (2013) High-resolution global maps of 21st-century forest cover change. Science 342:850–853

Hirzel AH, Hausser J, Chessel D, Perrin N (2002) Ecological-niche factor analysis: how to compute habitat- suitability maps without absence data? Ecology 83(7):2027–2036

Hirzel AH, Le Lay G (2008) Habitat suitability modelling and niche theory. J Appl Ecol 45:1372–1381

Inostroza L, Zasada I, König HJ (2016) Last of the wild revisited: assessing spatial patterns of human impact on landscapes in Southern Patagonia Chile. Reg Environ Chan 16(7):2071–2085

Jacobson AP, Riggio J, Tait AM, Baillie JE (2019) Global areas of low human impact ('Low Impact Areas') and fragmentation of the natural world. Sci Rep 9:e14179

Lal R, Lorenz K (2012) Carbon sequestration in temperate forests. In: Lal R, Lorenz K, Hüttl R, Schneider B, von Braun J (eds) Recarbonization of the biosphere, Springer, Amsterdam, Holland

Lencinas MV, MartínezPastur G, Rivero P, Busso C (2008) Conservation value of timber quality versus associated non-timber quality stands for understory diversity in *Nothofagus* forests. Biodiv Conserv 17:2579–2597

Lencinas MV, MartínezPastur G, Gallo E, Cellini JM (2011) Alternative silvicultural practices with variable retention to improve understory plant diversity conservation in southern Patagonian forests. Forest Ecol Manag 262:1236–1250

Li S, Zhang Y, Wang Z, Li L (2018) Mapping human influence intensity in the Tibetan Plateau for conservation of ecological service functions. Ecosyst Ser 30:276–286

Lindenmayer D, Franklin J (2002) Conserving forest biodiversity: a comprehensive multiscaled approach. Island Press, Washington DC, USA

Lindenmayer D, Franklin JF, Lõhmus A, Baker S, Bauhus J, Beese W, Brodie A, Kiehl B, Kouki J, MartínezPastur G, Messier Ch, Neyland M, Palik B, Sverdrup-Thygeson A, Volney J, Wayne A, Gustafsson L (2012) A major shift to the retention approach for forestry can help resolve some global forest sustainability issues. Conserv Lett 5:421–431

Loveland TR, Reed BC, Brown JF, Ohlen DO, Zhu J, Yang L, Merchant JW (2000) Development of a global land cover characteristics database and IGBP DISCover from 1-km AVHRR data. Int J Rem Sen 21:1303–1330

Mace GM, Balmford A, Ginsberg JR (1998) Conservation in a changing world. Cambridge University Press, Cambridge, UK

Mace GM, Norris K, Fitter A (2012) Biodiversity and ecosystem services: a multilayered relationship. Trends Ecol Evol 27(1):19–26

MartínezPastur G, Peri PL, Fernández MC, Staffieri G, Lencinas MV (2002) Changes in understory species diversity during the *Nothofagus pumilio* forest management cycle. J Forest Res 7(3):165–174

MartínezPastur G, Peri PL, Lencinas MV, GarcíaLlorente M, Martín López B (2016) Spatial patterns of cultural ecosystem services provision in Southern Patagonia. Land Ecol 31(2):383–399

MartínezPastur G, Peri PL, Soler R, Schindler S, Lencinas MV (2016) Biodiversity potential of *Nothofagus* forests in Tierra del Fuego (Argentina): tool proposal for regional conservation planning. Biodiv Conserv 25(10):1843–1862

MartínezPastur G, Peri PL, Huertas Herrera A, Schindler S, Díaz Delgado R, Lencinas MV, Soler R (2017) Linking potential biodiversity and three ecosystem services in silvopastoral managed forests landscapes of Tierra del Fuego Argentina. Int J Biodiv Sci Ecosyst Ser Manage 13(2):1–11
McGowan PJ (2016) Mapping the terrestrial human footprint. Nature 537:172–173
Mestre L, Toro Manríquez M, Soler R, Huertas Herrera A, MartínezPastur G, Lencinas MV (2017) The influence of canopy-layer composition on understory plant diversity in southern temperate forests. Forest Ecosyst 4:6
Millennium Ecosystem Assessment (MEA) (2005) Ecosystems and human wellbeing: current state and trends. Island Press, Washington DC, USA
Mori AS, Lertzman KP, Gustafsson L (2017) Biodiversity and ecosystem services in forest ecosystems: a research agenda for applied forest ecology. J Appl Ecol 54(1):12–27
Panagos P, Borrelli P, Poesen J, Ballabio C, LugatoE MK, Montanarella L, Alewell C (2015) The new assessment of soil loss by water erosion in Europe. Environ Sci Policy 54:438–447
Perera A, Peterson U, MartínezPastur G, Iverson L (2018) Ecosystem services from forest landscapes: broadscale considerations. Springer, Cham, Switzerland
Peri PL, Ormaechea S (2013) Relevamiento de los bosques nativos de ñire (*Nothofagus antarctica*) en Santa Cruz: base para su conservación y, manejo. INTA, Buenos Aires, Argentina
Peri PL, Lencinas MV, Bousson J, Lasagno R, Soler R, Bahamonde H, Martinez Pastur G (2016) Biodiversity and ecological long-term plots in Southern Patagonia to support sustainable land management: the case of PEBANPA network. J Nat Conserv 34:51–64
Peri PL, López D, Rusch V, Rusch G, Rosas YM, MartínezPastur G (2017) State and transition model approach in native forests of Southern Patagonia (Argentina): linking ecosystemic services, thresholds and resilience. Int J Biodiv Sci Ecosyst Ser Manage 13(2):105–118
Peri PL, Rosas YM, Ladd B, Toledo S, Lasagno RG, MartínezPastur G (2019) Modelling soil nitrogen content in South Patagonia across a climate gradient, vegetation type, and grazing. Sustainability 11:e2707
Peri PL, Monelos L, Díaz B, Mattenet F, Huertas L, Bahamonde H, Rosas YM, Lencinas MV, Cellini JM, Martínez Pastur G (2019) Estado y usos de los bosques nativos de lenga, siempreverdes y mixtos en Santa Cruz: Base para su conservación y, manejo. INTA, Buenos Aires, Argentina
Prescott CE (2002) The influence of the forest canopy on nutrient cycling. Tree Physiol 22(15–16):1193–1200
Quintas-Soriano C, Martín-López B, Santos-Martín F, Loureiro M, Montes C, Benayas J, García-Llorente M (2016) Ecosystem services values in Spain: a meta-analysis. Environ Sci Policy 55(01):186–195
Raudsepp-Hearne C, Peterson GD, Bennett EM (2010) Ecosystem service bundles for analyzing tradeoffs in diverse landscapes. PNAS 107(11):5242–5247
Rosas YM, Peri PL, Huertas Herrera A, Pastore H, MartínezPastur G (2017) Modeling of potential habitat suitability of *Hippocamelusbisulcus*: effectiveness of a protected areas network in Southern Patagonia. Ecol Proc 6(1):28–42
Rosas YM, Peri PL, MartínezPastur G (2018) Potential biodiversity map of lizard species in Southern Patagonia: environmental characterization, desertification influence and analyses of protection areas. Amphib Reptil 39(3):289–301
Rosas YM, Peri PL, Bahamonde HA, Cellini JM, Barrera MD, Huertas Herrera A, Lencinas MV, Martínez Pastur G (2019a) Trade-offs between management and conservation for the provision of ecosystem services in the southern Patagonian forests. In: Stanturf J (ed) Achieving sustainable management of boreal and temperate forests. Chapter 6, 33 pp. Burleigh Dodds Science Publishing, Cambridge, UK
Rosas YM, Peri PL, Lencinas MV, MartínezPasturG (2019b) Potential biodiversity map of understory plants for *Nothofagus* forests in Southern Patagonia: Analyses of landscape, ecological niche and conservation values. Sci Total Environ 682:301–309
Sanderson EW, Jaiteh M, Levy MA, Redford KH, Wannebo AV, Woolmer G (2002) The human footprint and the last of the wild. Bioscience 52(10):891–904

Silva LD, Costa H, de Azevedo EB, Medeiros V, Alves M, Elias RB, Silva L (2014) Modelling native and invasive woody species: a comparison of ENFA and MaxEnt applied to the Azorean forest. International Conference on Dynamics, Games and Science. Springer, Cham, Switzerland, pp 415–444

Venter O, Sanderson EW, Magrach A, Allan JR, Beher J, Jones KR, Possingham HP, Laurance WF, Wood P, Fekete BM, Levy MA, Watson J (2016) Sixteen years of change in the global terrestrial human footprint and implications for biodiversity conservation. Nat Commun 7:e12558

Watson J, Shanahan DF, Di Marco M, Allan J, Laurance WF, Sanderson E, Mackey B, Venter O (2016) Catastrophic declines in wilderness areas undermine global environment targets. Curr Biol 26:2929–2934

Zaniewski AE, Lehmann A, Overton JM (2002) Predicting species spatial distributions using presence-only data: a case study of native New Zealand ferns. Ecol Modell 157:261–280

Chapter 20
The Role of Local Communities in Sustainable Land and Forest Management

Latif Haji, Naser Valizadeh, and Dariush Hayati

Abstract Conservation, protection, and proper utilization of forests play an important role in environmental sustainability of the globe. The ultimate goal of sustainable forest management is to create a balanced and appropriate solution for human well-being and preservation of forest ecosystems. However, one of the prominent obstacles to achieve this goal is the gap existing between governmental development aims and the perspectives of local people and communities. Forest conservation requires an integrated management that works in partnership with local communities. Local and community-based forest management is a multi-dimensional approach to sustainable forest management in which different stakeholders with different interests play a part in achieving a common goal. However, little research has been done in this area. In this regard, the main purpose of this chapter was to examine the role of participation of local community in sustainable land and forest management. This purpose fulfilled through six steps. In the first step, the role of community participation in sustainable forest management and its typology were explained. I the second and third steps, the barriers and drivers of local communities' participation were introduced, respectively. In the fourth step, techniques for participation of local communities in forest management were analyzed. In the fifth step, some practical experiences related to the participation of local communities in forest management were highlighted. In the sixtieth or final step, some social principles were introduced for agricultural system and interventions aiming at sustainable management of forests and lands.

Keywords Local communities · Participation · Forests and lands · Sustainable management · Community-based forest management (CFM)

L. Haji · N. Valizadeh (✉) · D. Hayati
Department of Agricultural Extension and Education, School of Agriculture, Shiraz University, Shiraz, Iran
e-mail: n.valizadeh@shirazu.ac.ir

L. Haji
e-mail: la.Haji@shirazu.ac.ir

D. Hayati
e-mail: hayati@shirazu.ac.ir

P. K. Shit et al. (eds.), *Spatial Modeling in Forest Resources Management*, Environmental Science and Engineering,
https://doi.org/10.1007/978-3-030-56542-8_20

20.1 Introduction

At the beginning of the twenty-first century, the issue of global sustainability was widely discussed by world leaders and consequently became a common theme among journalists, scientists, teachers, students, and citizens in many parts of the world. The idea of sustainability stems from a new order passed in 1969 by the International Union for Conservation of Nature (IUCN). This idea was the main theme of the United Nations Conference on the Human Environment (UNCHE), held in Stockholm, Sweden, in 1972. The concept of sustainability was explicitly stated to show that economic growth and industrialization cannot be achieved without environmental conservation. In the following decades, the mainstream of sustainable development thinking was gradually developed through the World Conservation Strategy (1980), Brutland Report (1987), the United Nations Conference on Environment and Development (1992), and Paris Agreement in 2015 (Hayati 2017).

Sustainable development is a complex concept. Since it has been constantly evolving, which has made it difficult to be defined in a narrow definition (Tibbs 2011). Brundtland Commission in 1987 defined sustainable development as a kind of development that meets the needs of present generation without compromising the ability of future generations to meet their needs (Robert et al. 2005). The concept of sustainable development developed based on three basic concepts namely "development" (socio-economic development in line with ecological constraints), "needs" (redistribution of resources to ensure quality of life for all), and "future generations" (long-term use of resources in order to ensure the quality of life for future generations). Although the term "sustainable development" initially referred to the ecological conservation, it immediately developed into a study of social and economic aspects (Klarin 2018). Therefore, sustainable development generally encompasses the three dimensions of the environment, society, and economy that are closely linked. In other words, sustainable development results from the balance of these three dimensions (Klarin 2018).

The issue of sustainability is not limited to one specific field and sector, but covers all areas. One of the most important areas is the agricultural sector, which plays an important role in the growth and development of countries (Hazran et al. 2017). Investigating the development process of different countries indicates that development of agricultural sector is one of the most important economic sectors and a necessary prerequisite for sustainable development. So that without removing the barriers to development in this sector, the other sectors including industry and service cannot be flourished and developed (Momeni et al. 2017). Sustainable agriculture is a system that while managing and utilizing resources to meet human nutrition needs, enhances the quality of environment and natural resource reserves (Ghanbari and Barghi 2009). In general, sustainable agriculture is an approach that seeks to maximize the benefits of agricultural capacity (both natural and human) and to minimize the adverse environmental impacts. According to the Food and Agriculture Organization (FAO), sustainable agricultural development must be technically sound, economically viable, socially acceptable, and environmentally supportive (Ghadiri-Masoum and Hajipour 2016). In other words, in order to institutionalize sustainable agricultural development, the emphasis on the issue of sustainability in all economic, social, and environmental dimensions is an inevitable necessity.

One of the sustainability aspects of agriculture is the sustainability of forests and forest lands. Sustainable development in forestry includes afforestation and harvesting of interconnected forests which should not damage the biological renewal of forests (Klarin 2018). Forests as the main body of the terrestrial ecosystem provide basic services to human beings (Feng et al. 2016). At the national and international scales, forests have several benefits such as biodiversity conservation (McKinley et al. 2011; Basso et al. 2018), soil and water conservation, carbon sequestration (Feng et al. 2016), pollution reduction (Escobedo and Nowak 2009), poverty alleviation and income inequality (Das 2010), and industrial wood production (Miller et al. 2009). In addition, in many developing countries, part of the income of people living in or around the forests comes from these resources (Zenteno et al. 2013). This refers to the benefits of forests at the local scale. More than 1.6 billion people worldwide depend on forest resources for their livelihoods, of which about 60 million are fully dependent and 350 million are partially dependent on forest revenue (Khosravi et al. 2014).

Today, the multifaceted importance of forests as one of the main pillars of development on the one hand, and the concern of their destruction on the other, have prompted global circles to develop frameworks and standards to assess the process of transformation, exploitation, restoration, development, and destruction of forests. These frameworks and standards can lead to sustainable exploitation of forests by integrating the activities of international communities (Gough et al. 2008). In recent decades, the continuous degradation of forest areas has led to much debate about the impact of forestry activities on diverse ecosystems around the world. These discussions have led to countless criticisms from the community, and have also forced the forestry sector to put forward mechanisms that demonstrate their commitment to environmental and social dimensions of forests. Their purpose in changing their point of view was to show that their activities were not illegal and predatory (Basso et al. 2018). Although some of damages may be due to geographical location and climatic conditions/changes, it cannot be denied that much of them have been caused by human factors (Gamoun et al. 2018). Combining these factors together and their synergetic effect lead to reduction of biological diversity, decreasing productivity, and high erosion rates in forest areas (Tarhouni et al. 2014). Several factors including turning a blind eye to the environmental values and biodiversity, increasing deforestation, degradation of forests and lands due to human behaviors and activities (to expand agricultural practices and produce fuel) (Young et al. 2005; Matsvange et al. 2016), illegal and low-yield exploitation (Randriamalala et al. 2011), extinction of forest species, and so on have profound effects on rural communities. This has resulted in more deforestation, rapid soil erosion, sedimentation of rivers and reservoirs, and accelerated flooding. Such problems pose a serious threat to the forests' resilience and revival capacity (Poffenberger 2000).

The persistence of deforestation indicates that current forest resource management system needs to be revised. Despite the actions of various public sector agencies, degradation of natural resources continues (Schmitz et al. 2010). The effort to formulate and enforce strict laws on the use of forests has not only failed to maintain and sustain them, but has often led to increased conflict between users and further invasion of the ecosystem (Poffenberger 2000). Therefore, reduction of conflicts in an optimal framework, appropriate decision-making on forestry, and sustainable management of forest require the support and involvement of a wide range of stakeholders (Kaya and Kahraman 2011).

Policies aimed at encouraging local participation, endogenous development, and bottom-up approaches have been expanding for the last three decades. In this regard, the need for greater involvement of local people in local environmental governance and natural resource management has been emphasized by experts, policy makers and even executives (Prager et al. 2015). These illustrate the need for a change in the relationships between environmental planners and local communities, as well as relationships between local communities and the nature. Because, these changes and developments are crucial factors in achieving sustainability (Noguera-Méndez et al. 2016). On the other hand, sustainability is not merely achieved through laws, regulations, and technical factors. Since social dimension is one of the main aspects of sustainability which is largely overlooked in programs. Attention to the social dimension mainly requires an understanding of social and individual concepts of nature, natural resources, future generations, and the relationships between these factors (Noguera-Méndez et al. 2016). Therefore, widespread public participation is one of the most well-known requirements of any sustainable management program including forest management projects (Prager et al. 2015) that seems to be a must to solve environmental problems and crises (Noguera-Méndez et al. 2016). Today, human beings have come to the conclusion that the survival of next generation is a function of the present generation's management. Therefore, the idea of sustainable development or management has been emphasized by the majority of international community as a way of preserving the survival of today's and future generations. The main outputs of these events have been designing, planning, and reviewing of forest policies and/or ways of efficient implementation, evaluation, monitoring and control of sustainable forest management (Khazaei et al. 2017). The goal of sustainable forest management is to create a win-win solution for human well-being and preservation of forest ecosystems. The gap between government goals and perspectives of local people is the major problem in this field (Boissière et al. 2009). Forest conservation requires a management style that is implemented in collaboration with local communities. Local community-based management (CFM) is a multi-sectoral approach to sustainable forest management in which different stakeholders play different roles (to achieve the common goal of sustainable forest conservation and exploitation) (Urech et al. 2013). CFM generally refers to local residents who develop institutions, norms, laws, and costs to conserve forest resources. Community-based management systems typically include one or more communities that protect and use a particular forest area (Poffenberger 2000). This collaborative approach empowers local communities through poverty alleviation, management and deforestation, and sustainable utilization of forest resources (Cronkleton et al. 2011; Chirenje et al. 2013). Sheppard (2005) developed a framework for participatory sustainable forest management. There were some important principles for participatory sustainable forest management in his framework. The principals included broad representation of stakeholders, open access to stakeholders, clearly structured decision-making process, engaging process, understandable and accurate information, appropriate scale and detail for participants and resource managers, focusing on assessing sustainability over time, credibility of the process, and mutual learning and capacity-building. Michon et al. (2007) also introduced a concept entitled "Domestic Forests" which highlights the importance of local community participation in forest management. These scholars stated that domestic forest constitutes a specific category of forest management which clearly differs from classic or more modern professional forestry as exposed in textbooks. The domestic forest is neither a production nor a conservation forest, and not even a forest management model that embodies multi-functionality. It is "a forest for living," a forest that integrates production and conservation with social, political, and spiritual dimensions. It clearly constitutes a "third way" in global forest management.

In a nutshell, it can be mentioned that the local and CFM is a multi-dimensional approach to sustainable forest management in which different stakeholders with different interests play a part in achieving a common goal. However, little research has been done in this area. In this regard, the main purpose of this chapter was to examine the role of participation of local community in sustainable forest management.

A review of the research literature on CFM or forest management with the participation of local communities shows that despite the adoption of various laws and extensive studies in this field, no significant attention has been paid in practice to the role of local communities. Many reasons (social, economic, political, and etc.) have been presented for the lack of serious focus on the real participation of local communities in forest management. This chapter highlights the lack of knowledge about the practical ways of local communities' participation as one of the most significant obstacles in this regard, which can be considered as a research gap. The most important original contribution of this chapter is that it emphasizes the potentials of local communities and the application of participatory methods in sustainable forest management. Filling this gap is important because it can help re-evaluate intervention activities to involve local communities in forest management. In this way, practitioners, decision makers, managers, and other stakeholders will focus on changing and improving participation techniques, rather than focusing on the economic and technical tools to increase participation in forest management. Also, one of the originalities of this study is that it introduces agricultural extension systems as potential and capable institutions in involving local communities in sustainable forest management. To achieve the main purpose of the chapter, six specific goals were defined:

- The role of community participation in sustainable forest management and its typology;
- Challenges and barriers for participation of local communities;

- Factors influencing local community involvement in sustainable forest management;
- Approaches and techniques for the participation of local communities;
- Practical experiences of local community participation in forest sustainability management; and
- Social principles for agricultural system and interventions aiming at sustainable management of forests and lands.

20.2 Participation of Local Communities as a Key Pillar of Sustainable Forest Management and Its Typology

Forest management has a long history and has traditionally been intended to provide timely timber supply (Puettmann et al. 2012). Concepts such as "the possibility of annual harvesting" and "sustainable management" were introduced in early stages of developing science of forestry. Sustainability of forest use in these stages was only measured by "the increase and stability of wood production". In the next stages of developing this science, the need to adopt community perceptions of forestry practices (such as the wider impacts of environmental movements and arguments on sustainable development) and the production of a diverse set of ecosystem goods and services led to a greater focus on sustainable forest management (Rist and Moen 2013). The concept of sustainability in forestry science was adopted as a principle from the early eighteenth century. But its interpretation has been developed over the time. It can therefore be mentioned that sustainable forest management is rooted in the science of forestry in the first decade of eighteenth century (Rist and Moen 2013). This thinking was then formulated in the form of "Agenda 21" in 1992, at the Rio Conference (Khazaei et al. 2017). Sustainable forest management is defined as "the stewardship and use of forests and forest lands in a way that preserves biodiversity, productivity, regeneration capacity, vitality, and the potential to achieve them for present and future". It also maintains the interdependence of environmental, economic, and social functions at the local, national, and global levels and does not harm other ecosystems (Rist and Moen 2013).

The multidimensional nature of sustainable development has given greater attention to the use of "social capital" which is designed and implemented in different countries for this purpose. Significant efforts have been made in recent years to improve social capital in the field of environmental protection. Given the expanding scope of environmental issues as well as the importance of using environmentally friendly technologies and methods, developing efforts in this area (attracting public participation) is inevitable (Schmitz et al. 2010). Given the importance of conservation of natural resources, the formulation of conservation strategies and sustainable exploitation of these resources are considered as a fundamental necessity. In this situation, the villagers have a major role to play as the main beneficiaries of these resources in local communities. People's participation in any project ensures that the project is implemented and sustained, and this is more important in protecting the forest (Urech et al. 2013). It is because of the fact that environmental and forest issues are intertwined with the lives of local communities and success of any program will require the participation of these communities (Matsvange et al. 2016). The role of people in decision making, planning, implementation, monitoring, and evaluation of any conservation program is crucial. Therefore, an appropriate management system for conservation of resources such as forests should be based on local community-based management (Chirenje et al. 2013). Proponents of community engagement argue that where local communities actively participate in decision making, achieving sustainability is facilitated. In other words, the main strategy for implementing sustainable development is to emphasize the involvement of local communities (Chirenje et al. 2013).

Local communities have long been involved in the exploitation, conservation, and manipulation of forests (McDermott 2009). Local communities base their forest management system on traditional knowledge and specific laws to maintain the biophysical status of forests resulting in long-term ecological sustainability consistent with local priorities (Rutt et al. 2015). However, government ownership of forests disrupts their use and management by local people, resulting in gradual deforestation and poverty of forest dwellers. Deprivation of local people from forest management is not a good policy when people depend on forest resources due to high poverty and high population density (Nath et al. 2016). Because it does not guarantee protection of forests from deforestation or avoiding more environmental problems. Following conservation programs by limiting people's access to forest resources reduces livelihood sources and increases migration and social conflict between rural poor. The failure of centralized or top-down supervisory approaches in forest resource management have led to the creation of sustainable policies and development of multidimensional resource management strategies that allow for a wider involvement of stakeholders (including the local population) in the management process. This has resulted in decentralization of forest management/governance in many countries (Nath et al. 2016).

The shift in international perspectives towards sustainable forest management practices has led policymakers in both developed and developing countries to distance themselves from purely technical interventions and consider "decentralization" as a strategy for improving public sector performance, achieving development goals, delivering services, and protecting environment (Ambrose-Oji et al. 2015; Nath et al. 2016). In this regard, in recent decades, different participatory approaches or decentralized policy frameworks for forest resource management through the transfer of authority to local communities have been put forward for better forest management and democratic participation (Kumar et al. 2015). Murphree (2009) states that CFM is one of the most important manifestations of decentralization. Gilmour (2016) identifies various events that have influenced the emergence of CFM. Details of these events have been presented in Table 20.1.

Decentralization allows stakeholders to participate in the joint management of forest resources. In other words, delegating forest management responsibilities to local communities facilitates collective decision-making processes in a fair, transparent, and rapid manner. By implementing policies and programs that reflect the

Table 20.1 Key global forest-related developments that have influenced the evolution of CBF

Period	Events	Response
1970s	Fuelwood crisis	Initiation of forestry for local community development
	Failure of forest industry development model to sustain forests and meet community needs	Establishment of fuelwood plantations (generally top-down); many failed
1980s	Large-scale deforestation; environmental degradation	Pilot projects tested CFM modalities in different settings to address environmental concerns
	Forest sector reforms: decentralization and devolution policies	Emergence of "people participation" and bottom-up development
1990s	Sustainable development paradigm	Focus on sustainable forest management and livelihoods as CBF objectives
	Recognition of indigenous people rights	Establishment of CFM regimes that formalize indigenous people rights to manage forests Expansion of CBF across all regions
2000s	Globalization, trade liberalization	Growing interest in commercialization of wood and non-wood goods and services produced under CFM
2010s	Global policy focus on climate change, illegal timber, and payment for environmental services	Additions to CFM objectives to address global policy interests

real needs and preferences of the people, decentralization leads to increased management efficiency and better oversight of decision makers (Andersson 2006). Decentralization of forest governance or management structure is seen as a key strategy to ensure efficiency, equity, and democracy in the forest management system and transfers forestry management responsibilities and functions from central governments to local communities (Mohapatra 2013). Decentralization of forest conservation programs has also the benefits of effective management, better adoption of management measures, improved stakeholder confidence, reduced law enforcement costs, increased public awareness, poverty alleviation, and improved economic well-being of rural people, especially in developing countries (Murphree 2009; Inoue and Shivakoti 2015; Kumar et al. 2015).

Since 1980s, many developed and developing countries have experienced various forms of decentralized forest management or CFM. The decentralized management plans of these countries have varied greatly in terms of level of success and failure. CFM programs which are known as community forestry, social forestry, participatory forestry, joint forest management, and etc. are regarded as innovative approaches to improve forest management and conservation strategies. These practices are famous for integrating environmental, economic, and social goals (Chomba et al. 2015; Ambrose-Oji et al. 2015; Nath et al. 2016). CFM is used as an effective strategy to achieve the multiple goals of sustainable forest resources management and poverty

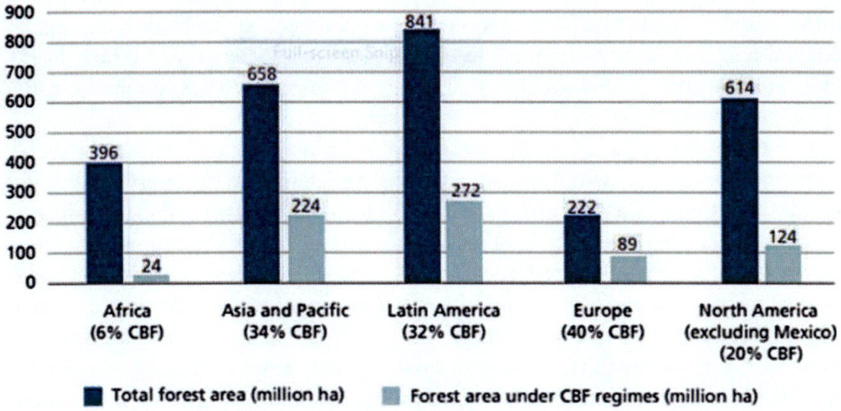

Fig. 20.1 Area of forest under CBF regimes, by region. *Source* Gilmour (2016)

reduction (Agbogidi et al. 2010). Integrating foresters' professional knowledge and skills with the knowledge and resources of the local community offers an alternative approach to achieve forest sustainability. This integration of knowledge results in empowerment of local communities and promotion their well-being (Nath et al. 2016). It is estimated that more than one-tenth of the world's forests are governed by participatory and community-based models (Casse and Milhøj 2011). Gilmour (2016) reported that about 18% of the world's forested areas are managed using a community-based forest (CBF) management approach (Fig. 20.1).

It should be noted that the literature on the management of natural resource by local communities does not claim that these communities always manage natural resources effectively and efficiently. Like governments and governmental approaches, these management practices sometimes fail to achieve this goal. What the research literature suggests is that local communities can do manage forests better than the central government. This does not mean that the government should have no role in forest resource management (Hutton et al. 2005). One of the benefits of local communities over central government in resource management is their ability to quickly access to accurate information on resources status and thus faster adaptability. In addition, the cost of law enforcement and monitoring is lower due to the high social capital available to local communities. The local community has a symbiotic relationship with the forests and can participate in forest development. The partnership is the cornerstone of CFM. Community-based forestry supports the empowerment of communities and the inclusion of all groups of society (minorities, youth, women, etc.) in decision-making. This participatory approach provides a framework for sustainable rural development (Agbogidi et al. 2010).

The concept of public participation has been widely understood so far. The UN considers public participation as one of the key elements of development process. Shuman (1994) sees participation as the first prerequisite for development. According to the World Bank, public participation empowers people and enhances the organizational skills and ability of them to be able to manage a society (World Bank 1995). Irvin and Stansbury (2004) believe that public participation leads to making better and more acceptable decisions on natural resources.

In the process of participation, the role of each member and partner in social, economic, and environmental partnerships must be determined by observing the rules and taking responsibility (Chirenje et al. 2013). Reed (2008) argues that participation means the voluntary involvement of individuals in decision making, implementation, operation, and evaluation of activities. Of course, this does not mean that in any partnership, people can make decisions and participate in all stages. Rather, it is important that people believe in the main issue and participate in the planning according to their level of understanding. Widespread use of the concept of development makes the participation more practical. With reviewing the concepts of economic growth, economic development, and sustainable development, it can be concluded that participation has even more fundamental implications. As such, participation in sustainable development has been called the goal and means of achieving development (Bessette 2012). Table 20.2 shows a classification of different types of participation by Chirenje et al. (2013).

There are another typology for participation of local communities in forest management and exploitation. Phongkaranyaphat et al. (2017) introduce a classification based on the different stages of participation in forest management and exploitation:

1. *Participation in decision making*
    - Participation in village meetings for the development of social forestry;
    - Participation in determining the forest area;
    - Participation in setting up the social forestry management committee;
    - Participation in the regulation of forest use laws; and
    - Participation in issuing warrants for community forest offenders.
2. *Participation in the implementation process*
    - Participation in the allocation of forest resources;
    - Participation in tree planting and tree care;
    - Participation in preventing offenders from deforestation;
    - Participation in forest conservation training;
    - Participation in forest fire fighting;
    - Participation in marking the forest area boundaries; and
    - Participation in youth awareness of forest values.
3. *Participation in the acquisition of benefits*
    - Use of non-wood forest products;
    - Use of forest area for livestock breeding and feeding;
    - Make more money and reduce household costs; and
    - Encouraging people to conserve forest resources.

Table 20.2 Typology of participation

Typology	Components of each type
Passive participation	People participate by being told what is going to happen or has already happened. It is a unilateral announcement by an administration or project management without listening to people responses. The information being shared belongs only to external professionals
Participation in information giving	People participate by giving answers to questions posed by external researchers and project managers using questionnaire surveys or similar approaches. People do not have the opportunity to influence proceedings, as the findings of the research or project design are neither shared nor checked for accuracy
Participation by consultation	People participate through consultation and external agents listen to their views. The external agents define both problems and solutions and may modify these in light of people's responses. Such a consultative process does not concede any share in decision-making and professionals are under no obligation to take on board people's views
Participation for material incentives	People participate by providing resources, for example labor, in return for food, cash or other material incentives. It is very common to see this so-called participation, yet people have no stake in prolonging activities when the incentives end
Functional participation	People participate by forming groups to meet pre-determined objectives to the project, which can involve the development or promotion of an externally initiated social organization. Such involvement does not tend to be at early stages of projects, but rather after major decisions have been made
Interactive participation	People participate in joint analysis, which leads to action plans. It tends to involve interdisciplinary methods that seek multiple perspectives and makes use of systematic and structured learning processes
Self-mobilization/active participation	People participate by taking initiatives independent of external institutions to change systems. They develop contacts with external institutions for resources and technical advice they need, but retain control over how resources are used. Such self-initiated mobilization and collective action may or may not challenge existing distributions of wealth and power

4. *Participation in the evaluation of results*

- Participation in tracking forest changes;
- Participation in outcome evaluation;
- Participation in discussions about forest benefits; and
- Participation in applying for a forest use approval.

20.3 Challenges and Barriers of Community Participation in Sustainable Forest Management

Decision makings on forest resource management are often associated with complexity and uncertainty due to the multi-functionality of these resources, difficulty of monetary valuation of ecological services, and involvement of large number of stakeholders (Sherry et al. 2005). However, this fact has been proven that the achievement or success of sustainable forest management depends on the support and input of a wide range of stakeholders. So that the benefits such as acquiring local knowledge, raising public awareness, and supporting forest management are achieved through public participation (Sheppard and Meitner 2005). Experiences have shown that successful forest management requires shared decision making between communities and other stakeholders. This requirement in turn depends on their (stakeholders) interconnection and understanding. For decades, national and international projects have been trying to stop the destruction of natural resources and find ways to create a sustainable forest management system. However, most of these CFM projects have rarely been successful (Pollini 2010). One reason for the failure of these projects is that they do not provide sufficient financial incentives to local people or are delayed due to socio-cultural problems (Urech et al. 2013).

Despite the importance of local people's participation in the conservation and development of natural resources and the emphasis of planners and experts on the need to use the people's power in this area, their participation has always been confronted with many problems and obstacles (Adam and Kneeshaw 2008; Chirenje et al. 2013). The problems and barriers to participation in natural resource management such as forests can be examined from two perspectives. The first perspective points to barriers and limitations that relate to the "internal conditions" of local stakeholders. These barriers include lack of awareness of the goals of programs, social heterogeneity, lack of trust, incompatibility with new conditions, short-term attitudes, and expectations of the government. But the most important problems can be found in the "external conditions" of target community which have been addressed in the second perspective. This perspective includes obstacles such as the overwhelming expansion of organizational structures, lack of rules for executive agencies to employ participatory approaches, centralized decision-making system, ineffective use of mass media, insufficient understanding of policymakers and planners about the importance of participation, the emphasis of academic and research centers on the technical aspects of research, inconsistency of technologies with the needs of individuals, lack of continued governmental support for organizations, lack of sufficient attention to the socio-economic dimensions of projects, lack of experienced experts and inappropriate policies. This study presents six of the most recognized major barriers and challenges for local community participation:

Economic barriers: Participations are costly in some cases and require investment from the participants. However, many people, especially in rural areas, lack the financial capacity to participate (Chifamba 2013; Kilewo and Frumence 2015). The economic benefits of local people have always been a challenge to their participation in forest management programs. The majority of locals, especially in less developed and developing societies, are poor and have little income (Colfer et al. 2011). Most of their livelihoods depend on agriculture and forestry. Therefore, CFM alone cannot improve the income of a growing population in these communities (Urech et al. 2013).

Social barriers: Lack of awareness about the goals of projects and their characteristics, illiteracy of the target group, competition and jealousy (Kilewo and Frumence 2015), lack of constructive relationship between individuals, lack of information sharing, lack of access to information (Chifamba 2013), lack of attention to indigenous knowledge in programs, and lack of community-oriented organizations (Khedrizadeh et al. 2017) are among the most important social constraints. Also, another societal challenge to local community participation can be the overriding of customary rights in forest use that cause conflicts between people (Andriamalala and Gardner 2010).

Traditionalism and its appreciation: Local people's distrust of government officials (Chifamba 2013), self-interests (Lahijanian and Shiehbeiki 2016), local power relations, and hierarchies between social groups and generations may lead to conflict of interests. So that the traditional social/structural order (in which the oldest member is the chairman) is disturbed. These traditional social orders may negatively increase the power of the elite minority (Urech et al. 2013). It should be noted that the extent of local people's participation in law-making and policy-making is not the same. For example, these processes may be affected by gender inequalities. Thus, women are less likely to participate (Fedele et al. 2011).

Practical barriers: Centralized planning, inadequate supply mechanisms, lack of local coordination, inadequate project content, and lack of local structures are considered as the practical obstacles. Implementing CFM in itself is a challenge. Continuous monitoring and control of CFM processes are important and worrying issues that need to be done by the communities themselves. Since doing so ensures its sustainable management. Long-term support, training, and guidance by foreign agencies will ultimately lead to failure of programs (Urech et al. 2013).

Structural barriers: Structural barriers to participation are fundamental problems in any forest management program that need be taken into account. The structure extends to the regional and local levels and encompasses all formal institutions and relationships (Lahijanian and Shiehbeiki 2016). The lack of institutionalization of community participation ethics (weak public belief in participation), weak institutional partnerships (Shaeri and Saadi 2003), inadequate institutional support, political instability, and poor governance are structural barriers to implementation of a program. Also, centralized planning, relations of production and power, and the structure of government organizations leave no room for participation. In the absence of government capacity to devise, implement, and finance environmental policies, external donors will dictate ways to implement conservation programs. In other words, most environmental activities will be carried out by foreign nongovernmental organizations. As a result, due to the lack of attention to the proved facts, interests, and concerns of the local context, they will lose credibility (Urech et al. 2013).

Perceptual barriers: Refer to the obstacles that can be overcome by the personal efforts of stakeholders or by changing the "cultural" status of a society. Personal values can be obstacles to people participation in forest conservation and management. Some stakeholders may value community participation, while others may not. Some stakeholders try to create a cynical atmosphere in the process of participation that can make other participants reluctant to participate. Social values can also be considered as the obstacles to people participation in forest conservation and management. Unless the social values in a society encouraging the civic engagement and open dialogue about important issues that affect stakeholders, participation may not take place (Offenbacker 2004).

20.4 Factors Influencing Local Community Involvement in Sustainable Forest Management

Partnership first of all needs to be facilitated. People need to know their role and duty well in the system of participation. The following factors play an important role in encouraging people to participate.

20.4.1 Partnership Belief

Prior to participation, democratic ethics must be institutionalized in society and become reality as a dynamic process in society. Lack of belief in participation is one of the most prominent problem of participation, especially in third world countries. In these countries, power rests with the political, economic, and local elites. Then, it is not possible for people to participate and create democracy in society. In this regard, the first step is the division of political, economic, and social power among the people. Following steps should be taken to build public trust:

- Equitable distribution of power between people: Without equitable distribution of political and social power, it is not possible for the weak/low-income groups to participate in society. Emphasizing the role of rural communities in all stages of the implementation of natural resource conservation projects and the transfer of project management to local people are effective in attracting people to participate (Rahmani et al. 2018).
- Building a sense of self-reliance in people: Understanding participation can only be achieved by encouraging self-reliance and increasing self-esteem. In this regard, changing people's attitudes is an indispensable necessity for participation and development (Khedrizadeh et al. 2017).

- Taking into account the social and cultural characteristics of the people: "adaptation of natural resource conservation schemes to the specific conditions of different regions", "empathy and unity between the villagers and those involved in the project", "preservation and development of indigenous skills", and "the attention to cultural conditions of local people" can lead to greater participation (Yaghoubi-Farani et al. 2017; Rahmani et al. 2018).
- Group actions and community development: It is necessary to help create spontaneous and popular organizations for the sustainable management of natural resources. Weak community groups are not able to participate in projects individually. But when they act in groups, they are more willing to participate (Rahmani et al. 2018).
- Voluntary participation: participations in natural resource (forests) management should be voluntary. People must participate consciously and voluntarily.
- Increasing awareness of natural resource users and expanding cultural and literacy centers (Khedrizadeh et al. 2017).
- Teaching the culture of participation to planners of participatory programs (Shaeri and Saadi 2003).
- People satisfaction with past programs and their attitude toward programs (Bagherian et al. 2009)

20.4.2 Existence of Participatory Institutions

The first step after establishing a public belief in participation is to establish participatory institutions. These institutions have a major purpose, which is to moderate the power at a high level of the government structure and deliver it to the lower levels of society. Such decentralization movement is usually faced with opposition through the upstream management structures. Also, the development of a law on investment in natural resources is one of the factors driving people participation (Shaeri and Saadi 2003).

20.4.3 Providing Sufficient Information and Establishing an Information Network

Without technical information, partnerships will not happen even if there are participatory institutions. In a participatory action, organizational communication, the level of knowledge and information of the people in the field of forestry, development of people's forest-related management skills, suitability of projects, and the real needs of the people should be considered (Yaghoubi-Farani et al. 2017). Also, awareness of rights, information on successful projects and experiences, commercialization of commodities, and CFM approaches can help attract more people (Bagherian et al. 2009). People in the process of participation may require two types of information:

- Technical Information: This information contains various skills and issues that people need to know in order to execute a specific project. This information help people identify their environment and their inner capacity to solve problems. The effectiveness of this information depends on the level of literacy of the community. Thus, literacy seems to be a significant factor in public participation.
- Management Information: This information include how and when people should participate in and execute projects. Management information include information that help monitor and execute a project. Most projects fail because of the lack of operational and management information (Shaeri and Saadi 2003).

20.4.4 Access to Financial and Technical Resources

In addition to the issues mentioned above, staffing and funding are needed to participation. In many cases, people have to reach a level of financial ability to participate in an activity. Thus, the extent of people's economic dependence on forests and income (Yaghoubi-Farani et al. 2017), awareness of the benefits and applicability of programs, and the benefits of participation are prerequisites for increasing participation (Khedrizadeh et al. 2017).

20.5 Participatory Approaches and Techniques in Forest and Natural Resource Management

Participation techniques and approaches are widely applied in subject areas such as agro ecosystems, natural resources, forestry, animal husbandry, fisheries, irrigation, health and nutrition, agricultural research systems, agricultural extension, marketing, organizational evaluation, and technology evaluation (Shabanali-Fami et al. 2004). Participatory approaches are useful in forest management and forest projects' implementation (that are supported by government or foreign organizations). They can encourage local people to become seriously involved in forest management projects and act as a decision maker and share the benefits thereof. Local community participation leads to learning, and learning is often a prerequisite for changing behaviors and practices (Apipoonyanon et al. 2020). In analyzing the methods of participation, a greater understanding about the entity (especially the culture of a territory and indigenous knowledge) is of the utmost importance. There are many applied studies that support participatory approaches as a tool for achieving sustainable development. These approaches easily make a lot of information available and enrich the body of sustainable development knowledge (Matteoli 2017). Here are four participatory approaches to CFM. These approaches include: rapid rural appraisal (RRA), participatory rural appraisal (PRA), participatory poverty appraisal (PPA), and participatory learning and action (PLA).

20.5.1 Rapid Rural Appraisal (RRA)

RRA was developed in the 1970s and was widely used in the 1980s. The main idea behind RRA is to collect data such as the physical characteristics of the area, available resources, suitable cropping patterns, and so on, in rural settings. These information are collected and produced in close cooperation with local people in the rural area (Falsafi et al. 2015). RRA is considered by researchers as an extensive system that can be used at both rural and public levels. RRA implies a process that is time and resource efficient. By comparing RRA with more classical perspectives, one can achieve more accurate and explicit results on learning from outsiders. In this way, multidisciplinary teams usually use different techniques. For example, to collect secondary resources for mapping local areas, the ranking matrix and Venn diagram or chart are created. Other tools used include interviewing key informants, questionnaires, subject workshops, case studies on local events and cultures, timelines, and other forms of direct observation (Matteoli 2017).

RRA leads to direct communication with local people and actors and facilitates explicit information gathering. These activities are done by a team of experts to learn more in a particular context. In most cases, RRA is conducted by a small research team or trained professionals in a one-to-three day workshop. The role of local people in RRA is to provide relevant local knowledge for research and development planning purposes. In this way, the initiative to use this information or decision-making power over projects is left to external agents/outsiders (Falsafi et al. 2015). In this participatory approach, information is channeled from a top-down system and society responds to external stimuli. The general feeling is that the end product of participation belongs to the researchers/outsiders not to the society (Matteoli 2017).

20.5.2 Participatory Rural Appraisal (PRA)

In the late 1980s, another approach was tested and used to build empowerment and legitimize local knowledge, building on the experiences gained from implementing the RRA method. This was the PRA that first appeared in Kenya in 1988. PRA is especially applicable to natural resource management as well as food security and social programs (Chambers 1997; Matteoli 2017). PRA refers to a set of methods and techniques aimed at stimulating people's participation in various socio-developmental processes and applying their indigenous ideas, theories, and knowledge in project planning/management (Hosein-Bar and Rastakhiz 2018). According to Chambers (1997), who is the most famous promoter of this approach, "PRA is a growing field of approach that strives to empower locals, increase their analytical power in life, and improve rural conditions. It is carried out "by" people and "with" their involvement in planning and implementation". Accordingly, PRA can be used by people in the city or village to assess their own situation. It also enables them to analyze, execute, monitor, and evaluate programs. This method also increases the empowerment of

Table 20.3 Comparison of RRA and PRA

Comparison criteria	RRA	PRA
Development history	Late 1970s–1980s	Late 1980s–1990s
Main initiators	Universities	NGO_s
Main users	Funding organizations and universities	NGO_s, local organizations, public organizations
Main source	Local people knowledge	Abilities of local people
Main innovation	Methods and techniques	Behaviors
The main method of outsiders	Extraction	Facilitation
Goals	Data collection	Empowerment
Main agents	Outsiders	Local people/insiders
Long-term results	Programs, projects, and publications	Sustainable local actions and institutions

women and the poor and ultimately leads them to have more control over their lives (Leurs 1996). This method consists of five basic principles including empowerment, respect, localization, enjoyment, and inclusiveness. External knowledge (outsiders' knowledge) must be absorbed and integrated with local knowledge through participation. Knowledge empowers societies and must therefore be understood as a value. In this way, researchers should try to understand and learn about the characteristics, cultures, and relationships of the community (Matteoli 2017).

Although RRA and PRA have similarities, the two approaches are not the same. RRA is a tool to outsiders through which they gather information and transmit it outside the local context. This information is then used as needed in planning and decision making. But PRA is not just a tool for gathering relevant, timely, and accurate information, but rather it is a process of empowering local people that represent the local knowledge capacities and its applicability in personal and social environments (Shabanali-Fami et al. 2004). Chambers (1997) has examined the differences between these two approaches appropriately in different respects (Table 20.3).

20.5.3 Participatory Poverty Assessment (PPA)

PPA mainly refers to those field surveys that were designed to assess poor countries. It was firstly developed by the World Bank in 1992. PPA focuses on the involvement of various stakeholders, especially the disadvantaged and vulnerable groups, and seeks to analyze poverty and identify specific solutions to this problem. It is very common for rural and poor people to feel that they are incapable of implementing development policies. Because they believe that decisions are made by outsiders and that their development plans are fully controlled. They feel they cannot influence the decision-making process. Therefore, the purpose of PPA is to overcome these social barriers. In fact, the poor are not only involved in the planning process, but also are considered as one of the most significant decision-makers. So that they can commit and address especially national policies (Matteoli 2017). PPA takes into account the fact that poor people have the capacity to analyze solutions and articulate their priorities, and their definition/understanding of poverty and its priorities is fundamentally different and often more realistic than policymakers' assessments. However, its main purpose is to improve the effectiveness of government measures to reduce poverty (Shabanali-Fami et al. 2004).

PPA is a process that seeks to understand poverty from the perspective of a wider range of stakeholders and engage them directly in the planning, implementation, and follow-up of programs. The most important stakeholders involved in the PPA process are poor men and women. PPA also includes decision makers from all levels of government, civil societies, and local elites. This helps identify different interests and perspectives and enhances local capacity and commitment to action. PPA seeks to understand poverty in its social, institutional, political, and local context. Regarding that PPA addresses national policy, micro-data is collected from a large number of communities to identify the characteristics of social groups and geographical areas. Identification of characteristics of social groups and geographical areas will then lead to the discovery of "patterns" in social groups. The most common tools used in PPA are micro level indices/data and Venn charts (Matteoli 2017).

20.5.4 Participatory Learning and Action (PLA)

PLA is an approach to learn and interact with communities. This is a growing set of collaborative and intuitive tools with interview techniques. PLA is used to facilitate the process of collective analysis and learning. This technique has traditionally been used in rural communities in developing countries (Thomas 2004). It is a practical and adaptive research strategy that enables different groups and individuals to learn and collaborate together, focus on common problems, identify challenges, and produce collaborative and democratic responses to the problems. This approach can be used to "identify needs" and "plan, execute, monitor, evaluate and follow up the projects" (Chambers 1997).

PLA is a powerful consulting tool and provides an opportunity for communities to actively participate in issues and interventions. Traditional research has focused more on "consulting" and "finding results" for the analysis, without providing any advice for action. In contrast, PLA analyzes the problems by sharing perspectives. PLA allows the poor people to be involved in providing realistic solutions. This allows local people to share their understanding and identify, prioritize, and evaluate issues using their knowledge of local conditions. PLA is increasingly used in a range of community-based poverty and regeneration projects where active community participation is a priority. PLA enables community members to participate in problem solving regardless of age, ethnicity, and literacy (Thomas 2004). In this way, key stakeholder groups are encouraged to learn from each other and share their ideas (Chambers 1997).

20.6 Practical Experiences of Local Community Participation in Forest Sustainability Management

Many countries around the world have successfully employed participatory techniques in management of forests and lands. Two examples were shortly explained here in this section.

20.6.1 Honduras

The relationship between local communities and the commercial exploitation of forests in Honduras goes back to the early colonial period (Tucker 2004). However, prior to the 1970s, local communities were not allowed to use forest resources for commercial purposes. In 1974, the law of Social Forestry System (SFS) was adopted to involve the rural population in the use and conservation of forest resources. Since then, Honduras has widely promoted agricultural forest cooperatives and other forms of community-based forestry enterprises (CFEs). Despite many problems, SFS is one of the lasting and successful examples of social forestry policy in Latin America (Trends 2013).

Over the past ten years, Honduras has experienced numerous efforts and initiatives to improve forest management. In the first half of the last decade, a pro-environmental grassroots movement has been formed to prevent deforestation. In 2005, the Honduran National Commission for Human Rights launched an independent forest monitoring project that led to the disclosure of illegal logging (Trends 2013).

In 2010, the National Institute for Forest and Wildlife Conservation and Development (ICF) launched a national strategy to control illegal forest exploitation and trade in Honduras (Trends 2013). Another important step for the Honduran government was negotiating with the European Union on a Voluntary Partnership Agreement (VPA). VPA in Honduras offers a new opportunity to recognize the rights of community-based forest, move away from top-down programs, and promote a more active regulatory framework. The common goal of all these initiatives is to improve forest management by reducing forest loss and destruction, enhancing forest-dependent livelihoods, and encouraging sustainable forest management (Trends 2013). Most importantly, the Honduran experience shows that these can be achieved through community-based forest enterprises (CFEs) which are involved in the production, processing, and trade of forest products (Trends 2013). Fortín et al. (2010) point out that while CFEs may not always be successful, there are many examples where CFEs have reduced deforestation/illegal exploitation and created employment for poorer communities. In other words, such successful examples of institutionalized forestry have created sustainable forms of forest use (Fortín et al. 2010).

A comparative study between 1997 and 1995 showed that there was a significant difference between CFE-controlled and uncontrolled areas in terms of deforestation. Deforestation rate was 0.03% in CFE-controlled areas and 1.5% in uncontrolled areas. Also, statistics show that over a six-year period (2011–2011), the rate of deforestation in CFE-controlled areas has decreased by 50% (Trends 2013). Overall, studies and evaluations show that social forestry cooperatives have prevented illegal forest use and deforestation in Honduras (Fortín et al. 2010).

20.6.2 Bangladesh

The role of forests in Bangladesh GDP is about 5% (FAO 2000). Forests in Bangladesh are being destroyed due to various socio-economic threats and competition for land use. Major problems affecting natural resource management in Bangladesh include poverty, high population growth rates, scarce financial resources, inappropriate use of technologies, institutional weakness, human resource weakness, poor quality of data on resources, and reduced sustainability of resources and forests (Biswas and Choudhury 2007). Although Bangladesh has more than 100 years of scientific forest management experience, CFM is a new intervention approach in this country. This approach is designed to support forest resources conservation with the help of people and forest dwellers. "Forest Policy Law" and "Social Forestry Rules" are among the most important Bangladesh legal acts to implement CFM. CFM programs have been implemented in Bangladesh for more than three decades (Nath et al. 2016). For many villagers, especially landless and small-scale farmers, such programs are highly attractive and legitimate. Involvement of local people in reforestation activities takes into account various aspects of sustainability including environmental, economic, and social dimensions. CFM has succeeded in reducing mistrust and conflict between forestry officials and local people, forest encroachment, and deforestation. CFM has also increased local community participation and awareness and reduced poverty (Nath et al. 2016).

The main components of CFM projects being implemented in Bangladesh include woodlot plantations, agroforestry plantations, strip plantations, empowering landless farmers in Chittagong, reforestation of villages, institutional planting and seedling distribution, the establishment of nurseries and training centers, the establishment of a plantation center, and training of various stakeholders. The main goals of these projects include increasing timber production, reducing poverty, and increasing the institutional capacity of the forestry sector. Forest-dependent and indigenous communities are the major beneficiaries of these programs (Nath et al. 2016).

20.6.3 Zimbabwe

Forest depletion and degradation are important problems in rural Zimbabwe and strategies to increase sustainable forest management are continually approached. Matsvange et al. (2016) carried out a research work to assess the impact of forests on communities from Nyanga, Guruve, and Zvimba districts of Zimbabwe. It was based on a Big Lottery Fund project implemented by Progressio-UK and Environment Africa. The main topic of this project was 'Conserving our Land and Producing Food' in Guruve, Zvimba and Nyanga districts of Zimbabwe. The anticipation was that at the end of the project, there will be an increase in family income and food security level of poor and marginalized communities through agro-ecology and sustainable, equitable farming approaches, and access to markets. An increase in sustainable management of forest, land, and water resources expected for the benefit of the most disadvantaged households. Local communities expected to be able to engage with local and national governments to ensure better management and use of natural resources. The project made use of community-based groupings, namely the Environmental Action Groups (EAGs) and Farmer Field Schools (FFS), which had been gone through an intensive capacity-building in forest management. The results of these researchers revealed that with a business-as-usual scenario, the level of deforestation could reach alarming rates because of rising trends such as increase in the veldt and forest fires, increase in demand for firewood for curing tobacco in addition to the conventional causes of deforestation. The FFS approach was not only a means to enhance the livelihoods of rural households, but also a vital tool for addressing environmental challenges such as deforestation, through empowering communities to take full custodianship of their natural resources. The results also demonstrate that giving communities control over resources within their local areas, which could otherwise be regarded as a common property, gives them the responsibility to manage the use of these resources and rehabilitate the environmentally degraded areas. In other words, communities motivated to come up with local initiatives to improve their environment.

20.7 Social Principles for Agricultural Extension and Interventions Aiming at Sustainable Management of Forests and Lands

The ongoing deforestation process and destruction of forests which have been resulted from human activities and climate changes are challenging problems for sustainable development around the world (United Nation General Assembly 2015). Investigations demonstrate that deforestation is declining, forest biomass is stable per hectare, and voluntary forest certification has increased (United Nation Economic and Social Council 2017). Despite these promising improvements, there are still many problems in the field of conservation, protection, and sustainable forest management in different areas (Liang et al. 2016). The need for government and social intervention at all levels has been acknowledged (United Nation General Assembly 2015; United Nation Economic and Social Council 2017) and this fact pushes researchers, decision makers, and policymakers to understand that "we are breaching planetary boundaries". These needs and issues require serious steps to be taken to establish a new paradigm for the continuous development of human societies and coordination between humans and forest ecosystems (Steffen et al. 2015; Ripple et al. 2017). Given that agricultural extension system is one of the institutions responsible for social interventions in the field of exploitation of forests and lands, inspired by Vanclay (2004), five social principles for this system in order to promote land and forest sustainability management are presented:

1. Taking socio-cultural dimensions of interventions into the account
   Instead of promoting and intervening as an interconnection process between science as the sole producer of ideas and farmers/forest dwellers (local communities) as passive recipients, interventions must accept that the process of adoption is a social process. Adoption of sustainable forest and land management measures is not an unthinking response. On the contrary, it is a deliberate decision that the respondent chooses from a wide range of options. In addition, it should be borne in mind that the adoption of sustainable forest and land management measures is not a singular act and the decision-maker is not in an isolated environment. Acceptance of these measures occurs in a social context. Social interventionists and agricultural change agents need to consider these issues in their plans.
2. Non-adoption of local communities is not the only reason for the unsustainability of land and forest management
   It is a misconception that the adoption of sustainable forest and land management measures by local communities leads to "definitive improvement" of the state of forests and lands. In other words, such a perspective not only simplifies the notion of "sustainability of forests and lands", but also introduces local communities as "the only major" contributors to this problematic situation. However, sustainability in forest and land management is a multidimensional problem that has social, economic, political, physical, and institutional roots. In addition, in many cases, inappropriate interventions and innovations have been the cause of current unsustainability of forests and lands, not the rejection of innovations by forest dwellers. Accordingly, agricultural extension and other social interventionists should pay special attention to this issue in planning for future interventions.
3. Local knowledge as an evaluation criterion for intervention programs of sustainable forest and land management
   Many interventions in the past and now are done with a "top-down" approach. Following such an approach shows that many forest and land management policymakers still believe that local people and forest dwellers do not have the knowledge and ability to make the necessary decisions. In other words, professionals and outsiders are more aware of local conditions and needs than locals. However, it should not be forgotten that in forest and land interventions, scientists are not the only source of knowledge. Accordingly, agricultural change agents should not

diffuse only the knowledge produced by scientists. All social and local groups create "knowledge" about their personal experiences. Accordingly, the information disseminated by agricultural extension evaluated by the personal information and knowledge of the individuals. Although their knowledge may have been gained through experimentation and trial, it is of special importance in local decision-making. The knowledge and information produced by scientists on sustainable forest and land management practices will be accepted by local people if it is consistent with their local knowledge. Although attention to local knowledge should be increased, it should not be overused.

4. Top-down interventioning is an inappropriate approach for sustainable forest and land management
   Top-down social interventions have many drawbacks. One of the main problems with these interventions is that they are biased against the innovations and ideas produced by scientists. In other words, innovations and ideas produced by scientists and policymakers are considered "useful innovations for the local community." However, in many cases, such judgment is inconsistent with the realities of the society and local context. In other words, the need for these innovations and ideas among locals is less than the need for them among scientists and policymakers. Another problem with top-down interventions in forest and land management is that they ignore social and ecological impacts. In addition, most of these interventions have used psychological and economic models to analyze the decision-making of local people. At the same time, they have ignored the social, political, cultural, and historical contexts of forest and land management. Accordingly, interventions in forest and land management need to avoid pursuing merely a top-down approach. Therefore, they should use participatory, bottom-up, and integrated approaches. Of course, this does not mean romanticizing a particular approach and recommending a one-size-fits-all approach. Because, holding a biased idea of the application of participatory approaches is again trapped in dogmatism.
5. Facilitating decision-making as the main aim of interventions
   Many forest and land management intervention programs seek to introduce ideas and innovations for "adoption". Such perspective represents a dogmatic top-down approach. Because, it has a predetermined goal: "adoption of an idea or innovation". However, there are scientific pieces of evidence that this approach is obsolete. The purpose of interventions in forest and land management programs should be "to facilitate decision-making". In such cases, local communities are not considered passive recipients, but "active decision-makers".

20.8 Summary and Conclusion

The main purpose of this chapter was to examine the role of local community participation in sustainable land and forest management. The results of first step revealed that the social dimensions are usually overlooked in land and forest management programs. Turning a blind eye to the social dimensions means ignoring the role of local communities in sustainable land and forest management. That is while, local communities are of great importance in achieving sustainability objectives of these programs. Also, the results related to the first specific goal showed that the participation of local communities in sustainable forest and land management has different types including passive participation, participation in information giving, participation by consultation, participation for material incentives, functional participation, interactive participation, and self-mobilization/active participation. In this regard, it is recommended that decision makers, managers, practitioners, policymakers, and other users increase their knowledge about the application of various types of local communities' participation in forest management. They should not seek to generalize one method of participation to all situations. Because it can reduce the level of dissatisfaction and non-participation of local communities.

The results of the second and third steps of this study showed that economic, social, cultural, practical, structural, and perceptual barriers are the most important obstacles to the local community participation in sustainable forest management. In addition, partnership belief, existence of participatory institutions, providing sufficient information and establishing an information network, and access to financial and technical resources are the most significant drivers of local community participation in sustainable forest management. Given that most participatory programs are carried out by the government or with the support of governments, and structural barriers in the body of the government are of the most important obstacles to the participation of local communities in forest management, it is suggested that decision makers and policymakers first remove the structural barriers to participation. A first important step in implementing this suggestion is to review top-down structures and replace them with bottom-top structures. Since top-down structures cannot be expected to involve local communities in forest management. This is a fact that has been proven over the past few decades. A review of the experience of Honduras, Bangladesh, and Zimbabwe in the fifth step also supports this conclusion. Because the participation of local communities in forest management in these countries has been achieved through participatory decisions and revising top-down intervention systems.

The results of the fourth step showed that Rapid Rural Appraisal (RRA), Participatory Rural Appraisal (PRA), Participatory Poverty Assessment (PPA), and Participatory Learning and Action (PLA) were among the most important practical techniques for participation of local communities in land and forest management. However, unfamiliarity with the practical techniques of local community participation is considered as a prominent barrier to their use in social interventions. In this regard, it is suggested that managers and practitioners of interventions related to sustainable forest and land management increase their knowledge in the field of participation methods and techniques. Holding courses and workshops in the field of participation techniques can be the first operational step in implementing this proposal.

The results of step six showed that the agricultural extension system is one of the entities responsible for social interventions in the field of exploitation of forests and lands. In this regard, it must follow certain guidelines and principals to be more effective in this context.

This chapter emphasizes that local communities construct their own knowledge and solutions for managing lands and forests in a sustainable manner. Therefore, interventions aiming at promoting management of forests should use local people knowledge, expertise, skills, and recommendations in different phases of program planning. But in order to do so, a paradigm shift from top-down approaches to bottom-up approaches should be occurred in the perspectives and practices of policy-makers and program planner. It means that interventions in forest and land management need to avoid following only top-down approaches. They need to use participatory, bottom-up, and integrated approaches. Of course, as it was previously mentioned, this does not mean romanticizing and singularizing a particular approach and thus suggesting a one-size-fits-all approach for all interventions. Because, holding a biased understanding of a particular approach is again trapped in dogmatism. However, the experiences reveal that participatory and bottom-up approaches are the only ones that leave a room for local people voices and priorities.

At the end it is worth mentioning that the results of this study by focusing decision makers, policymakers, and managers on the constructive role of local communities in the sustainable management of forests, can play a significant role in their sustainable management. Since delegating control of forests to local communities in local areas increases their sense of responsibility for proper forest management. In addition, the management of lands and forests by local people means that they do not have egoistic attitudes towards the exploitation of resources. This could also increase their willingness to rehabilitate destroyed forest areas. In addition, delegating management authority to local communities can help increase their cooperation with government agencies.

References

Adam MC, Kneeshaw D (2008) Local level criteria and indicator frameworks: a tool used to assess aboriginal forest ecosystem values. For Ecol Manage 255(7):2024–2037. https://doi.org/10.1016/j.foreco.2007.12.051

Agbogidi OM, Ofuoku AU, Dolor DE (2010) Role of community Forestry in sustainable forest Management and Development: a review. ASSET: Int J 7(1):44–54

Ambrose-Oji B, Lawrence A, Stewart A (2015) Community based forest enterprises in Britain: two organising typologies. For Policy Econ 58:65–74. https://doi.org/10.1016/j.forpol.2014.11.005

Andersson K (2006) Understanding decentralized forest governance: an application of the institutional analysis and development framework. Sustain Sci Pract Policy 2(1):25–35. https://doi.org/10.1080/15487733.2006.11907975

Andriamalala G, Gardner CJ (2010) L'utilisation du dinacommeoutil de gouvernance des ressourcesnaturelles: leçonstirés de Velondriake, sud-ouest de Madagascar. Trop Conserv Sci 3(4):447–472. https://doi.org/10.1177/194008291000300409

Apipoonyanon C, Kuwornu JK, Szabo S, Shrestha RP (2020) Factors influencing household participation in community forest management: evidence from UdonThani Province, Thailand. J Sustain For 39(2):184–206. https://doi.org/10.1080/10549811.2019.1632211

Bagherian R, Samah B, Samah AA, Ahmad S (2009) Factors influencing local people's participation in watershed management programs in Iran. Am-Eur J Agric Environ Sci 6(5):532–538

Basso VM, Jacovine LAG, Nardelli AMB, Alves RR, Silva EV, Silva ML, Andrade BG (2018) FSC forest management certification in the Americas. Int For Rev 20(1):31-42. https://doi.org/10.1505/146554818822824219

Bessette G (2012) People, land and water: participatory development communication for natural resource management. Routledge

Biswas SR, Choudhury JK (2007) Forests and forest management practices in Bangladesh: the question of sustainability. Int For Rev 9(2):627–640. https://doi.org/10.1505/ifor.9.2.627

Boissière M, Sheil D, Basuki I, Wan M, Le H (2009) Can engaging local people's interests reduce forest degradation in Central Vietnam? Biodivers Conserv 18(10):2743–2757. https://doi.org/10.1007/s10531-009-9627-1

Casse T, Milhøj A (2011) Community forestry and forest conservation: friends or strangers? Environ Policy Gov 21(2):83–98. https://doi.org/10.1002/eet.566

Chambers R (1997) Whose reality counts, vol 25. Intermediate Technology Publications, London

Chifamba E (2013) Confronting the challenges and barriers to community participation in rural development initiatives in Duhera district, ward 12 Zimbabwe. Int J Curr Res Acad Rev 1(2):01–19

Chirenje LI, Giliba RA, Musamba EB (2013) Local communities' participation in decision-making processes through planning and budgeting in African countries. Chin J Popul Resour Environ 11(1):10–16. https://doi.org/10.1080/10042857.2013.777198

Chomba S, Treue T, Sinclair F (2015) The political economy of forest entitlements: can community based forest management reduce vulnerability at the forest margin? For Policy Econ 58:37–46. https://doi.org/10.1016/j.forpol.2014.11.011

Colfer CJP, Pfund JL, Boucard A, Boissière M, Feintrenie L, Ingram V, Sunderland T (2011) An introduction to five tropical landscapes, their people and their governance. Collab Gov Trop Landsc 1

Cronkleton P, Bray DB, Medina G (2011) Community forest management and the emergence of multi-scale governance institutions: lessons for REDD+ development from Mexico, Brazil and Bolivia. Forests 2(2):451–473. https://doi.org/10.3390/f2020451

Das N (2010) Incidence of forest income on reduction of inequality: evidence from forest dependent households in milieu of joint forest management. Ecol Econ 69(8):1617–1625. https://doi.org/10.1016/j.ecolecon.2010.03.003

Escobedo FJ, Nowak DJ (2009) Spatial heterogeneity and air pollution removal by an urban forest. Landsc Urban Plan 90(3-4):102–110. https://doi.org/10.1016/j.landurbplan.2008.10.021

Falsafi P, Jangchi Kashani S, Mesbah A (2015) Application of rapid rural appraisal in empowering local people through collective action: case study of Hesar-Sorkh Village Javad Abad Dehestan, Varamin County. In: 1st international conference on new research in agricultural and environmental sciences, 15 Dec, pp 1–12

FAO (2000) Forest resources of Bangladesh: country report, working paper 15, The Forest Resources Assessment Programme, Forestry Department, Food and Agriculture Organization of the United Nations, Rome

Fedele G, Urech ZL, Rehnus M, Sorg JP (2011) Impact of women's harvest practices on Pandanus-guillaumetii in Madagascar's lowland rainforests. Econ Bot 65(2):158–168. https://doi.org/10.1007/s12231-011-9157-0

Feng J, Wang J, Yao S, Ding L (2016) Dynamic assessment of forest resources quality at the provincial level using AHP and cluster analysis. Comput Electron Agric 124:184–193. https://doi.org/10.1016/j.compag.2016.04.007

Fortín R, Butterfield R, Hodgdon BD (2010) The impacts of training, technical assistance and new market access for community forest enterprises in the Rio Platano Biosphere Reserve, Honduras. Rainforest Alliance

Gamoun M, Belgacem AO, Louhaichi M (2018) Diversity of desert rangelands of Tunisia. Plant Divers 40(5):217–225

Ghadiri-Masoum M, Hajipour M (2016) Ecological planning of place in order to promote sustainable agricultural development case: pistachio planting in Nehbandan County. Geogr Plan Space 6(22):109–124

Ghanbari Y, Barghi H (2009) Key challenges of sustainable agriculture and rural development in Iran. Strategy 16:2018–2234

Gilmour D (2016) Forty years of community-based forestry: a review of its extent and effectiveness. FAO

Gough AD, Innes JL, Allen SD (2008) Development of common indicators of sustainable forest management. Ecol Indic 8(5):425–430. https://doi.org/10.1016/j.ecolind.2007.03.001

Hayati D (2017) Literature review: a literature review on frameworks and methods for measuring and monitoring sustainable agriculture (No. 22). Technical report

Hazran Z, Norsida M, Nitty Hirawaty K, Nolila MN (2017) The post-flood impacts on farmers, agricultural sector and food security in Kelantan. Int J Acad Res Bus Soc Sci 7(9):175–184

Hosein-Bar MO, Rastakhiz B (2018) Participatory assessment approach: challenges and criticisms. Dev Strategy 53:138–159

Hutton J, Adams WM, Murombedzi JC (2005) Back to the barriers? Changing narratives in biodiversity conservation. Forum Dev Stud 32(2):341–370. https://doi.org/10.1080/08039410.2005.9666319

Inoue M, Shivakoti GP (eds) (2015) Multi-level forest governance in Asia: concepts, challenges and the way forward. SAGE Publications India

Irvin RA, Stansbury J (2004) Citizen participation in decision making: is it worth the effort? Public Adm Rev 64(1):55–65. https://doi.org/10.1111/j.1540-6210.2004.00346.x

Kaya T, Kahraman C (2011) Fuzzy multiple criteria forestry decision making based on an integrated VIKOR and AHP approach. Expert Syst Appl 38(6):7326–7333. https://doi.org/10.1016/j.eswa.2010.12.003

Khazaei H, Maleknia R, Ghazi-Nouri SS (2017) Evaluating the advantages and gaps of sustainable forest management technology. Wood For Sci Technol 24(4):147–161

Khedrizadeh M, Maleknia R, Adeli K, HanarehKhalyani J (2017) Survey of barriers and potential field to involve local people in the forest management process (case study: local communities in Nameshir, Baneh). Wood For Sci Technol 24(3):35–47

Khosravi S, Maleknia R, Khedrizadeh M (2014) The economic role of forest in rural householdS' livelihood in northern Zagros. Sustain Dev For 1(3):251–268

Kilewo EG, Frumence G (2015) Factors that hinder community participation in developing and implementing comprehensive council health plans in Manyoni District, Tanzania. Glob Health Action 8(1):26461. https://doi.org/10.3402/gha.v8.26461

Klarin T (2018) The concept of sustainable development: from its beginning to the contemporary issues. Zagreb Int Rev Econ Bus 21(1):67–94. https://doi.org/10.2478/zireb-2018-0005

Kumar K, Singh NM, Kerr JM (2015) Decentralization and democratic forest reforms in India: moving to a rights-based approach. For Policy Econ 51:1–8. https://doi.org/10.1016/j.forpol.2014.09.018

Lahijanian A, Shiehbeiki S (2016) Barriers of social participation of citizens in urban environment using the SWOT method and approaches for improving environmental sustainability. J Environ Sci Technol 17(4):147–166

Leurs R (1996) Current challenges facing participatory rural appraisal. Public Adm Dev 16(1):57–72. https://doi.org/10.1002/(SICI)1099-162X(199602)16:1<57::AID-PAD853>3.0.CO;2-Z

Liang J, Crowther TW, Picard N, Wiser S, Zhou M, Alberti G, De-Miguel S (2016) Positive biodiversity-productivity relationship predominant in global forests. Science 354(6309):aaf8957. https://doi.org/10.1126/science.aaf8957

Matsvange D, Sagonda R, Kaundikiza M (2016) The role of communities in sustainable land and forest management: the case of Nyanga, Zvimba and Guruve districts of Zimbabwe. Jamba: J Disaster Risk Stud 8(3):1421. https://doi.org/10.4102/jamba.v8i3.281

Matteoli F (2017) Role of participatory approaches and communication in national adaptation programmes of action: lessons learnt for national adaptation plans. Doctoral dissertation submitted in December 2017 to the Department of Economics, CaFoscari University of Venice, Italy

McDermott MH (2009) Equity first or later? How US community-based forestry distributes benefits. Int For Rev 11(2):207–220. https://doi.org/10.1505/ifor.11.2.207

McKinley DC, Ryan MG, Birdsey RA, Giardina CP, Harmon ME, Heath LS, Pataki DE (2011) A synthesis of current knowledge on forests and carbon storage in the United States. Ecol Appl 21(6):1902–1924. https://doi.org/10.1890/10-0697.1

Michon G, De Foresta H, Levang P, Verdeaux F (2007) Domestic forests: a new paradigm for integrating local communities' forestry into tropical forest science. Ecol Soc 12(2):1–24

Miller DA, Wigley TB, Miller KV (2009) Managed forests and conservation of terrestrial biodiversity in the southern United States. J For 107(4):197–203. https://doi.org/10.1093/jof/107.4.197

Mohapatra BP (2013) Decentralised Forest Governance, Institutions and Livelihoods in Odisha—a study of evolution of policy process and politics. Institutions and Livelihoods in Odisha-A Study of Evolution of Policy Process and Politics (January 23, 2013). https://doi.org/10.2139/ssrn.2205581

Momeni F, Dashtabani S, Bannouei AA (2017) The importance of agriculture in maintaining the socio-economic balance of Iranian urban and rural structure. Space Econ Rural Dev 4:17–46

Murphree MW (2009) The strategic pillars of communal natural resource management: benefit, empowerment and conservation. Biodivers Conserv 18(10):2551–2562. https://doi.org/10.1007/s10531-009-9644-0

Nath TK, Jashimuddin M, Inoue M (2016) Community-based forest management (CBFM) in Bangladesh, vol 22. Springer, Switzerland. https://doi.org/10.1007/978-3-319-42387-6

Noguera-Méndez P, Molera L, Semitiel-García M (2016) The role of social learning in fostering farmers' pro-environmental values and intentions. J Rural Stud 46:81–92. https://doi.org/10.1016/j.jrurstud.2016.06.003

Offenbacker BS (2004) Overcoming barriers to effective public participation. WIT Trans Ecol Environ 70:281–289. https://doi.org/10.2495/BF040281

Phongkaranyaphat K, Maiman S, Asanok L (2017) Factors influencing people participation in community forest management in Phrae Province, North Thailand. Int J Agric Technol 13(7.2):1707–1713

Poffenberger M (2000) Communities and forest management in South Asia. IUCN

Pollini J (2010) Environmental degradation narratives in Madagascar: from colonial hegemonies to humanist revisionism. Geoforum 41(5):711–722. https://doi.org/10.1016/j.geoforum.2010.04.001

Prager K, Nienaber B, Neumann B, Phillips A (2015) How should rural policy be evaluated if it aims to foster community involvement in environmental management? J Rural Stud 37:120–131. https://doi.org/10.1016/j.jrurstud.2014.12.006

Puettmann KJ, Coates KD, Messier CC (2012) A critique of silviculture: managing for complexity. Island Press, London

Rahmani MR, Mohammadi A, Heshmatpour A, Abbasi M (2018) Identification and analysis of the factors affecting the participation of rural people in ecosystem conservation projects (Caspian Hyrcanian Forests Project). Plant Ecosyst Conserv 6(12):77–92

Randriamalala H, Rasarely E, Ratsimbazafy J, Brizzi A, Ballet J, Razakamanarina N, Schuurman D (2011) Stocks de bois précieux de Madagascar–quellevoieemprunter? Madag Conserv Dev 6(2):88–96

Reed MS (2008) Stakeholder participation for environmental management: a literature review. Biol Conserv 141(10):2417–2431. https://doi.org/10.1016/j.biocon.2008.07.014

Ripple WJ, Wolf C, Newsome TM, Galetti M, Alamgir M, Crist E, Mahmoud MI, Laurance WF (2017) World scientists' warning to humanity: a second notice. BioScience 67(12):1026–1028. https://doi.org/10.1093/biosci/bix125

Rist L, Moen J (2013) Sustainability in forest management and a new role for resilience thinking. For Ecol Manage 310:416–427. https://doi.org/10.1016/j.foreco.2013.08.033
Robert KW, Parris TM, Leiserowitz AA (2005) What is sustainable development? Goals, indicators, values, and practice. Environ Sci Policy Sustain Dev 47(3):8–21. https://doi.org/10.1080/00139157.2005.10524444
Rutt RL, Chhetri BBK, Pokharel R, Rayamajhi S, Tiwari K, Treue T (2015) The scientific framing of forestry decentralization in Nepal. For Policy Econ 60:50–61. https://doi.org/10.1016/j.forpol.2014.06.005
Schmitz CL, Stinson CH, James CD (2010) Community and environmental sustainability. Crit Soc Work 11(3):83–100. https://doi.org/10.22329/csw.v11i3.5834
Shabanali-Fami H, Ali-Beigi AH, Sharifzadeh A (2004) Approaches and techniques of participation in agricultural EXTENSION and rural development. Institution for Rural Development in Iran, Tehran
Shaeri AM, Saadi HA (2003) A practical guide to participation and extension of natural resources. Pooneh Publication, Tehran
Sheppard SR (2005) Participatory decision support for sustainable forest management: a framework for planning with local communities at the landscape level in Canada. Can J For Res 35(7):1515–1526. https://doi.org/10.1139/x05-084
Sheppard SR, Meitner M (2005) Using multi-criteria analysis and visualisation for sustainable forest management planning with stakeholder groups. For Ecol Manage 207(1–2):171–187. https://doi.org/10.1016/j.foreco.2004.10.032
Sherry E, Halseth R, Fondahl G, Karjala M, Leon B (2005) Local-level criteria and indicators: an Aboriginal perspective on sustainable forest management. Forestry 78(5):513–539. https://doi.org/10.1093/forestry/cpi048
Shuman M (1994) Towards a global village: international community development initiatives. Pluto Press (UK)
Steffen W, Richardson K, Rockström J, Cornell SE, Fetzer I, Bennett EM, Folke C (2015) Planetary boundaries: Guiding human development on a changing planet. Science 347(6223):1259855. https://doi.org/10.1126/science.1259855
Tarhouni M, Ben Salem F, Ouled Belgacem A, Neffati M (2014) Impact of livestock exclusion on SidiToui national park vegetation communities, Tunisia. Int J Biodivers 2014. https://doi.org/10.1155/2014/620405
Thomas S (2004) What is participatory learning and action (PLA): an introduction. Centre for International Development and Training (CIDT), London
Tibbs H (2011) Changing cultural values and the transition to sustainability. J Futures Stud 15(3):13–32
Trends F (2013) Community forestry in Honduras: a path towards better governance. Forest Trends Association, 1203. www.forest-trends.org
Tucker C (2004) Aiming for sustainable community forest management. Working forests in the neotropics: conservation through sustainable management? p 178
United Nation Economic and Social Council (2017) Progress towards the sustainable development goals. E/2017/66, 11 May 2017
United Nation General Assembly (2015) Transforming Our World: The 2030 Agenda for Sustainable Development. A/RES/70/1, 21 Oct 2015
Urech ZL, Sorg JP, Felber HR (2013) Challenges for community-based forest management in the KoloAla site Manompana. Environ Manage 51(3):602–615. https://doi.org/10.1007/s00267-012-0011-7
Vanclay F (2004) Social principles for agricultural extension to assist in the promotion of natural resource management. Aust J Exp Agric 44(3):213–222
World Bank (1995) Social indicators of development; the socio-economic data division of the World Bank's
Yaghoubi-Farani A, Karimi S, Pormouzeh F (2017) Factors affecting people's participation in forest conservation plans in GuilanGharb County. Wood For Sci Technol 24(2):33–45

Young J, Watt A, Nowicki P, Alard D, Clitherow J, Henle K, Johnson R, Laczko E, McCracken D, Matouch S, Niemela J (2005) Towards sustainable land use: identifying and managing the conflicts between human activities and biodiversity conservation in Europe. Biodivers Conserv 14(7):1641–1661. https://doi.org/10.1007/s10531-004-0536-z

Zenteno M, Zuidema PA, de Jong W, Boot RG (2013) Livelihood strategies and forest dependence: new insights from Bolivian forest communities. For Policy Econ 26:12–21. https://doi.org/10.1016/j.forpol.2012.09.011

Chapter 21
Non-Timber Forest Products Based Household Industries and Rural Economy—A Case Study of Jaypur Block in Bankura District, West Bengal (India)

Debmita Nandi and Sumana Sarkar

Abstract Among the 22 blocks of the Bankura district, the Jaypur block occupies first position (5.81%, 2011) in the household industry sector which is also higher than district average (4.18%, 2011) and as well as the national average (3.81%, 2011). In this block, *Sal* leaf, *Khat Bel and Churung Khati* are the most economically important products among the available Non-Timber Forest Products (NTFPs) and these are primarily used as raw materials in women dominated cottage industries for making of *Sal* plate and beads chain. The rural people of the Jaypur block have collected NTFPs such as *Sal* leaf and seed, *Khat Bel*, *Churung Khati*, Mushroom, fruits of *Kendu* and *Mahul* etc. from the nearest reserve forest area for income generation and fulfilling the demand of their household food. The main objective of this study is to explore the role of NTFPs based household industries in rural economy at micro-level. To fulfil this objective, primary data are mainly used and these are collected through purposive sampling method with the help of questionnaire survey, participatory observation and focused group discussion. The study reveals that women from the Scheduled Caste and Muslim communities are dominantly engaged within these industries with traditional tools, primitive technology and middle man based marketing system and contributed in household maintenance by sharing their income to remove the poverty and sustain their livelihood. Hence, implementation of micro-level planning with the co-ordination of local people and the administration would be effective to increase the economic return to the villagers in general and rural women in particular.

Keywords NTFPs · Household industry · Rural economy · Marketing system · Micro-level planning · Economic independency

D. Nandi (✉) · S. Sarkar
Department of Geography, The University of Burdwan, Burdwan, West Bengal, India
e-mail: debmitanandi2012@gmail.com

P. K. Shit et al. (eds.), *Spatial Modeling in Forest Resources Management*, Environmental Science and Engineering,
https://doi.org/10.1007/978-3-030-56542-8_21

21.1 Introduction

Non-Timber Forest Products are those which are collected from the forest except timber (Ahenkan and Boon 2011). Fruits and nuts, vegetables, medicinal plants, resins, essences and a range of barks, fibres such as bamboo, rattans, and a host of other palms and grasses fish and game are mainly included here (CIFOR 2008). People of forest fringe area primarily depend upon common property resources such as forest and collected non-timber forest products for their livelihood support as there is either limited access to cultivated land or abundance of barren and low productive land (Saxena 2003). These products are mainly used by the poor people to meet the need of household food and income in rural forest fringe areas. Among these forest based products, some are very much useful for making crafts, construction and tools (Leakey 2012). It is further observed that rural people of Asia and Latin America mainly depend upon small-scale farming, raising livestock, fishing and non-farm activities type of common livelihood pattern for their survival and income generation (Mphande 2016). Poor landless and marginal farmers are mainly involved in non-timber forest-based household industry sector during the lean season of agriculture to support their economic base (Tambunan 1995).

Household Industry is the unregistered and unorganised sector under the Indian Factories Act and conducted by one or more members of the household at home or within the village in rural areas (Census Metadata 2001). In a country like India where 21.9% (2011–2012) people lived under below poverty line (Planning Commission Report 2014), small scale household industries become an attractive package of alternative source of income. Handicraft industry of India is the second-largest industrial sector for a rural and underemployed agricultural labourer who mainly reside in inaccessible area bounded by the traditional technology and they have not enjoyed the modern world benefits (Akilandeeswari and Pitchai 2016). According to the Census of India (2011), 3.8% of the total population is engaged in a household industry sector with a male and female share of 2.9% and 5.7% respectively.

In India, at about 50 million people reside near the forest area and rely on the non-timber forest product for their sustenance and cash income (Tewari 1992). At about 40% forest revenue and 55% forest-based employment of India came from the NTFPs and it is very much important for the source of the raw materials of household-based enterprises and the employment generation (Tewari and Campbell 1995). Forest fringe areas people usually reside in remote areas with poor living conditions and their business or forest-based enterprises condition does not improve for the existing system of poor infrastructure, service and market (Belcher et al. 2015). People usually collect forest-based raw materials for its easy accessibility and this will, in turn, help them to generate several NTFPs based activities like small industries or enterprise for poor and particularly its home-based character attracts women in rural India (Hasalkar and Jadhav 2004; Mahto and Chaterjee 2011). Forest dependent home-based small scale industries of rural area are mainly seasonal basis and it needed lower inputs of capital and technology with women and disadvantage group of people dominance (Hasalkar and Jadhav 2004).

Dry deciduous forest area of Orissa and Jharkhand also provided NTFPs to the villagers at the time of the agricultural slack season which reduces the number of seasonal migration to the town by making a value-added product like *Sal* (*Shorea robusta*) leaf and *Bidi* for continuing their livelihood (Mahapatra et al. 2005). Tribal people of the forest fringe area of the Western Ghats region of Kerala also collected NTFPs for livelihood maintenance and at about 56% of their household income come from these products (Krishnakumar et al. 2012). Tribal people of Udaipur district of Rajasthan prepared wine from the *Mahua* flower and this product sold to the village traders with lower price than the regulated market (Kumar and Meena 2018). In Malnaad region of Karnataka, collection of low volume and high-value product such as food products, medicinal product, cosmetics and crafts product can raise the standard of living of the poor rural people which also compatible with conservation of forest (Sills et al. 2003).

According to State Forest Report (2010–2011), among the total geographical area 13.38% of the West Bengal and 21.53% of the Bankura district covered with forest. Jaypur, Bishnupur, Ranibandh and Taldangra are the main forest enriched block in Bankura district where *Sal* dominated deciduous forest is mainly found. People of forest fringe areas in West Bengal specifically Bankura and West Medinipur district are belonging to a tribal community or other lower caste section (Goshal 2011; Shit and Pati 2012). They collect NTFPs viz. *Sal* leaf and seeds, mushroom and other products both for household purpose and commercial purpose with value addition and they also generate some home-based small industry for their economic earning (Goshal 2011; Shit and Pati 2012). *Belmala* making cottage industry is one of the important small scale industry in Bishnupur block and its surrounding area of Bankura district (Dasgupta et al. 2009). Environmental income from the forest resource is a good supportive component than non-forest based subsistence income in the situation of any shocks (crop failure or lost wage employment, illness or death of a productive adult, loss of land or livestock) in rural livelihoods in 12 months of a year (Angelsen et al. 2014; Shackleton and Shackleton 2004; Krishnakumar et al. 2012; Pandey et al. 2016).

Sal plate making and *Belmala* making are the important potential industries for local village people who reside in forest fringe area of Bankura district (Micro, Medium and Small Enterprises Report-MSME 2014–2015). Santal people of the forest fringe areas of Jaypur block are symbiotically related with the forest and they also earn money from the NTFPs (*Sal* leaf and seed, fuel wood, mushroom, *Mahul* flower, *Kendu* leaf and fruit etc.) where women dominated *Sal* plate making is the main income earning product (Saha and Sengupta 2014). Other than the Sal plate making, people mainly women section of forest fringe areas in the Jaypur block are also engaged in the non-timber forest-based *Belmala* and *Churung khati mala* making household industries to maintain their household economy. Katul, Madhabpur and Kunda Pushkarini are the three mouza where *Sal* plate, *Belmala* and *Churung Khati Mala* making household industry are dominant. In this context, objective of the present paper is framed to explore the nature of NTFPs based household industries and its impact in the rural economy at the micro-level.

21.2 Materials and Method

21.2.1 The Study Area

The rural people of the Jaypur block have collected Sal leaf and seed, *Khat bel, Churung khati,* Mushroom, fruits of *Kendu* and *Mahul* etc. from the nearest reserve forest area for income generation and fulfilling the demand of their household food. Besides, Jaypur block stands foremost place (5.81%, 2011) among the 22 blocks of the Bankura district in the household industry sector which is higher than district average (4.18%, 2011) and as well as the national average (3.81%, 2011). Thereby, we have selected this block as our area of interest. Geographically, the study area is located between 22° 53′ 00″ N to 23° 08′ 00″ N and 87° 20′ 00″ E to 87° 35′ 00″ E and administratively situated under the Bishnupur subdivision of Bankura district in West Bengal (Fig. 21.1). Nearly 21.88% geographical area of this block is covered with Reserve Sal forest of Jaypur Forest Range (DVD 2011), concentrated mainly in the western part of the block. A major portion (69.19%) of the block is characterized by agricultural land (DVD 2011) (Fig. 21.1). The District Human Development Report (DHDR 2007) found that the people of the forest range areas are landless or marginal farmers. Thereby, the families over here are extensively depending on the forest-based products to sustain their livelihood. They make a living by collecting some non-timber forest products for their own needs and for commercial purposes with some value addition. Such types of situation force the women to get involved in a variety of household industries such as small cottage industry like *Sal* plate making, beads chain making from *Churung Khati* and *Belmala* making from wood apple (*Aegle marmelos*) cover using the locally available NTFPs. As a result, Jaypur block has the highest percentage of female household workers (14.02%, 2011) among other blocks of the Bankura district. Three mouzas have been selected based on the dominance of the three different highly dependent household industries (Fig. 21.2) located in this block. These are viz. Kunda Pushkarini mouza (place of *Belmala* making), Katul mouza (place of *Sal* plate making), and Madhabpur mouza (place of *Churung Khati Mala* making) (Table 21.1). A brief outline of selected mouzas is given in Table 21.1.

21.2.2 Database and Methodology

The current work is inherently empirical and field-based in nature. An intensive field survey was conducted through personal interviews and focused group discussions. The Katul (making *Sal* plate), Madhabpur mouza (making *Churung Khati Mala*), and Kunda Pushkarini (making *Belmala*) have been selected purposefully following the H.J. Nelson's method of Influential Economic Activities (1955). Semi-structured questionnaires were applied to collect the primary data using the purposive sampling method to fulfil the objective of the study. During the field survey, 20 (25%), 35 (15%)

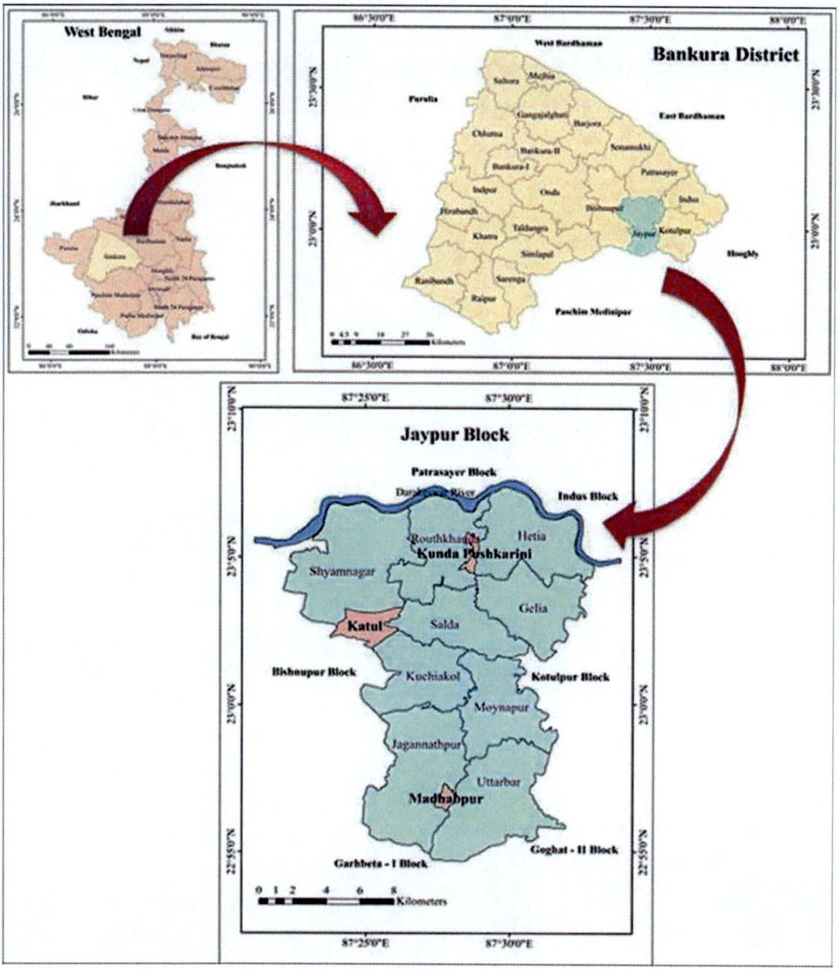

Fig. 21.1 Location map of the study area

and 25 (25%) numbers of female respondents have been selected from the female household industry workers of the selected Kunda Pushkarini, Katul and Madhabpur mouza. Therefore, to meet our objectives, a total of 80 female respondents have been selected from three different categories of female based household industries. Besides this, numerous secondary information were also collected which has been given below in Table 21.2 and the detailed method of work is shown in the flow chart (Fig. 21.3).

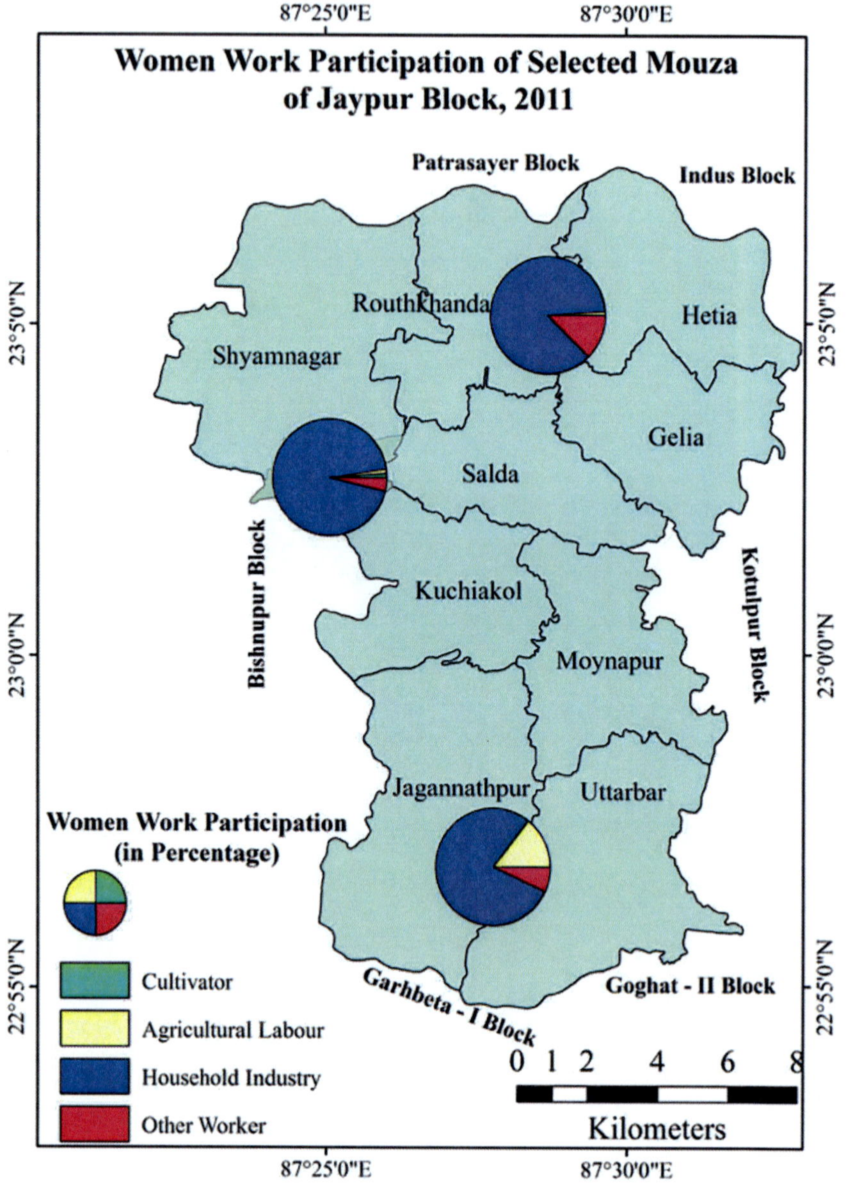

Fig. 21.2 Women work participation in different sector in selected mouzas where household industry sector is the dominant. *Source* District Primary Census Abstract (DPCA), Bankura (2011)

Table 21.1 Detail description of the selected Mouza of the study area

Name of the Mouza	Socio-economic characteristics of study area								
	Dominant household industry	Geographical area km^2	Gram panchayet	Number of households	Population	Forest cover	Total women workers	Engagement of women workers in NTFPs based household industry	Remarks
Kunda Pushkarini	*Belmala*	1.344	Routkhanda	187	885	–	82	77 (93.90%)	Mainly Muslim populations who are a marginal farmer or landless people
Katul	*Sal* plate	5.948	Shyamnagar	232	1073	3.85 km^2 (64.66%)	259	223 (86.49%)	Scheduled caste (SC) population share 97.58% share of the total population and have limited access to land

(continued)

Table 21.1 (continued)

Name of the Mouza	Socio-economic characteristics of study area								
	Dominant household industry	Geographical area km^2	Gram panchayet	Number of households	Population	Forest cover	Total women workers	Engagement of women workers in NTFPs based household industry	Remarks
Madhabpur	*Churung Khati Mala*	1.277	Jagannathpur	219	1062	0.40 km^2 (31.3%)	118	92 (77.97%)	Scheduled caste population have a share of about 55.93% of the total population and they are mainly marginal farmers or landless labourers

Source Based on Bankura District Village Directory (2011), DPCA (2011) and Field Survey (March, 2018)

Table 21.2 Details of data used in this study

Types of data and software	Nature of data	Source	Year
Primary data	Socio-economic	Field survey	2018
Secondary information's		District census handbook-DCH	1991
		District human development report-DHDR	2007
	Industry	Micro, medium and small enterprises Report-MSME	2014–2015
	Demographic	District primary census abstract-DPCA	2011
	Socio-economic	District village directory-DVD	2011
	Agricultural information	Comprehensive district agricultural plan report-CADP	2007–2012
	Land use and land cover	Land and land reform department, government of West Bengal and Google earth image	2013
Software	Arc GIS 10.2	Used for the map making	–

21.3 Results and Discussion

21.3.1 Non-timer Forest Products (NTFPs)

People of forest fringe areas of Jaypur block collected a number of NTFPs for maintaining their households throughout the year. Fuel wood, mushroom, *Sal* leaf, date leaf, *Churung Khati*, *Kurchi Khati*, wood apple, *Sal* gum etc. are the main non-timber forest product in this area (Development and planning department, West Bengal 2007). These forest products are most important natural capital for the landless or marginal people of weaker section who reside adjacent to the forest area. Male people of forest fringe areas have option to work as labours in different sector for their greater mobility but women of forest fringe areas are extremely dependent upon existing forest resources as there is little scope to absorb in other sectors. In forest fringe areas, women mainly collect NTFPs for giving economic support to their family and sustain their livelihood.

Among above NTFPs, some are used for household purpose and others are used as value added commercial products. Deciduous nature of forest gives opportunity to collect these products for 6–7 months yearly. Women are the main collector of this product because they have limited job opportunity in other sectors and with this work they finished both work i.e. household duties and industry work. In Katul mouza,

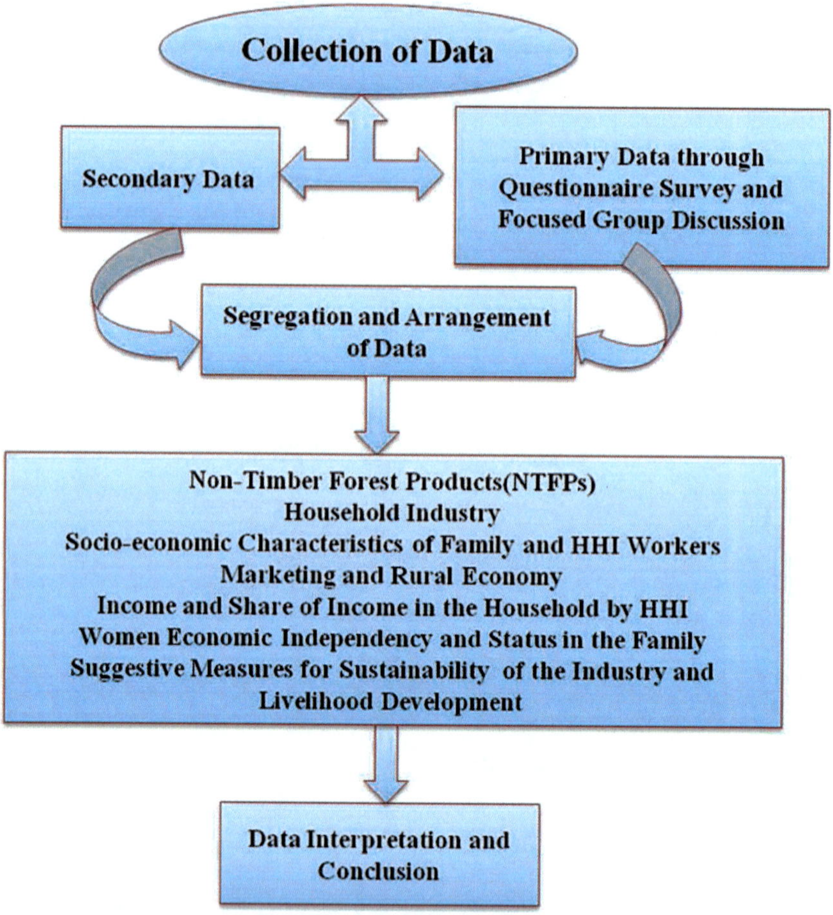

Fig. 21.3 Detailed methodological flow chart of the work

women have the facility to collect easily the required product from the nearest reserve forest area while in case of Madhabpur mouza it is relatively difficult as they reside too near of the reserve forest area. Women artisans of the Kunda Pushkarini mouza did not enjoyed these facilities for their far location from the forest area (Fig. 21.4). Among the above NTFPs, *Bel* cover (Kunda Pushkarini), *Sal* leaf (Katul), *Churung Khati* (Madhabpur) are widely used for making value added goods. *Belmala* making of Kunda Pushkarini village, *Sal* plate making of Katul village and *Churung Khati Mala* making of Madhabpur village are the three female based dominant household industries in this block.

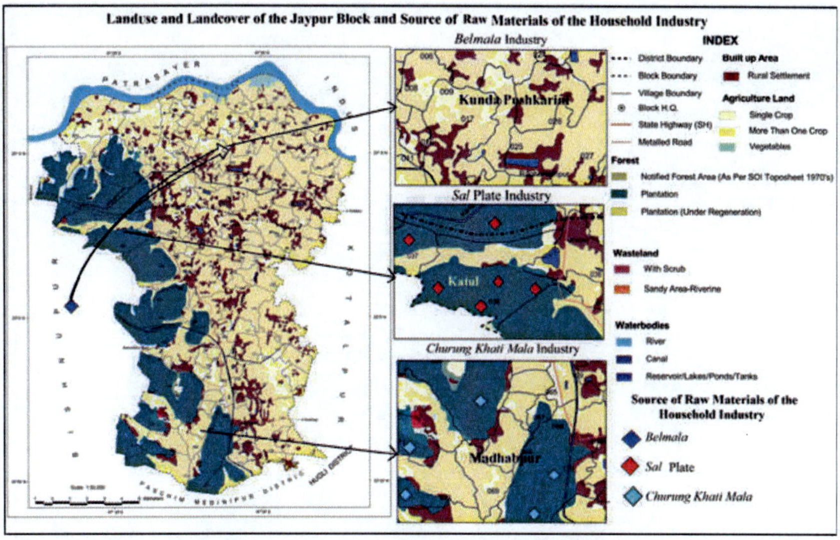

Fig. 21.4 LULC map of Jaypur block modified after Land and Land Reform Department, Government of West Bengal (2013) and source of raw materials of the household industry

21.3.2 Household Industry (HHI)

Three household industries viz. *Sal* plate, *Belmala* and *Churung Khati Mala* making have their history to grown up in the particular mouza (Fig. 21.5). Among the three industries, *Sal* plate is the oldest industry where *Belmala* making industry is newer one. *Belmala* makers have learned this technique from their relatives at about 50 years ago from *Darika* and *Bankhati* mouza of Bishnupur block. Age of *Churung Khati Mala* making industry is at about 100 years old and before this they had made *Mala* from stem of pigeon pea (*Cajanus cajan*).

According to CADP report (2007–2012), *Belmala* making industry is a potential cottage industry in Bankura district. Kunda Pushkarini village of Jaypur block is known for its female based *Belmala* making industry. Here, male people collect *Bel* fruits from nearest Jaypur forest range area but this raw material is limited in this reserve forest area and hence, they have to buy *Bel* cover from nearest Bihar and Jharkhand states (Dasgupta et al. 2009). Both male and female collect this *Bel* cover from Routkhanda Gram panchayet area where *Bel* covers have been unloaded. Female make *Belmala* from *Bel* cover with the help of traditional tools (*Chhank, Dhanuk* and *Fitchkari*) and techniques. Only Muslim female are engaged in this industry because at the time of preparation of beads from *Bel* cover, the feet of the makers required to touch the *Bel* cover that is why Hindu women avoided this work. They engaged in this industry for 6–7 (160–200 man days) months and finished *Belmala* are used for many sacred purposes by Hindus (*Marriage, Sradh, Annaprason* etc.) (Table 21.3).

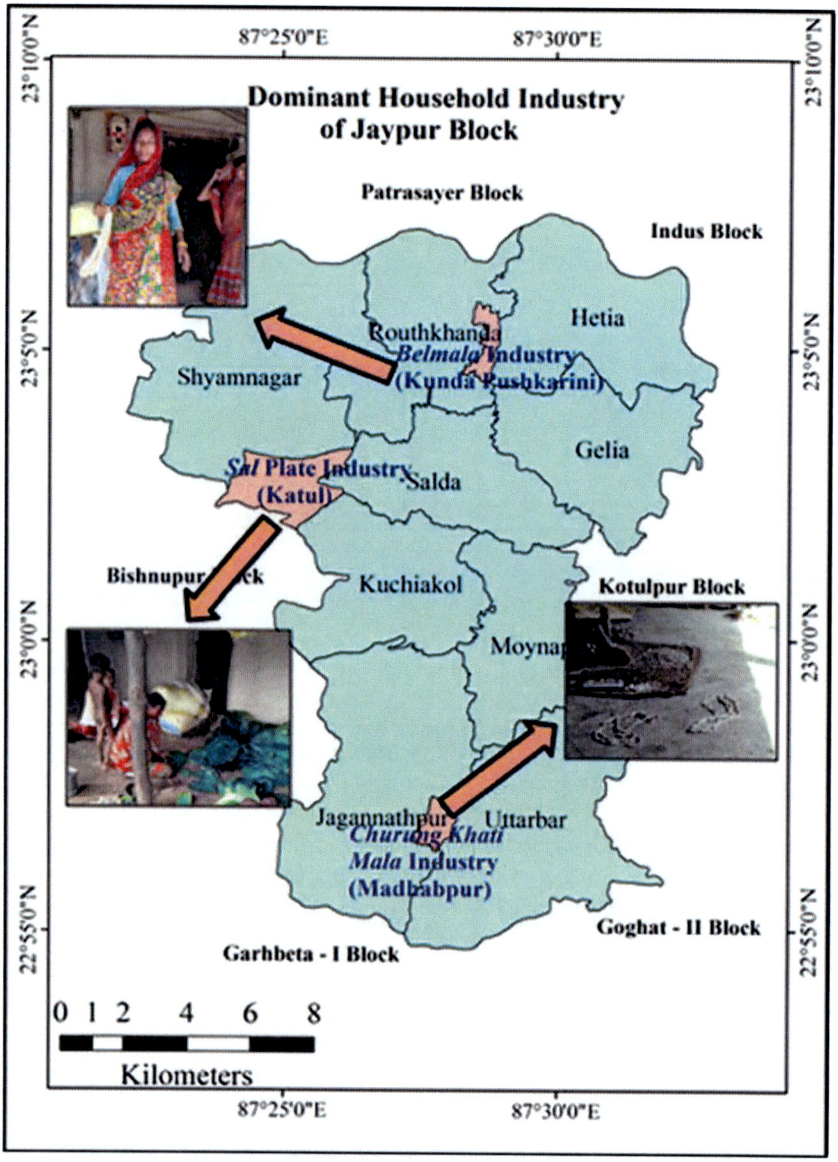

Fig. 21.5 Spatial pattern of dominant household industries with snapshots in Jaypur block. *Source* Field Survey, March, 2018

Table 21.3 History of the Household Industries

S. No.	J.L. number	Name of Mouza	Total area (km^2)	Forest area (km^2)	Name of dominant household industry	Generations	Age (approx)
1	025	Kunda Pushkarini	1.344	–	*Belmala*	2nd generation (learned from relative who resides in *Darika* and *Bankhati* mouza of Bishnupur block)	50 years
2	038	Katul	5.948	3.846	*Sal* plate	Very old (may be just after the beginning of the settlement)	Very old
3	068	Madhabpur	1.277	0.40	*Churung Khati Mala*	3rd generation (previously women made mala from stem of pigeon pea)	100 years

Source Based on Bankura District Village Directory (2011), DCH (1991) and Field Survey (March, 2018)

Traditional *Sal* plate making in Katul mouza is a dominant economy for the female to sustain their household. Laminated *Sal* plate making Industry is a potential cottage industry in Bankura district (MSME 2014–2015). Mainly female flock are collect *Sal* leaf from the nearest forest area (DHDR 2007). They go to the forest for collecting *Sal* leaf in the morning and return back their home at about 1–2 o'clock. They are also using traditional tools (*Neem Khati*) and technology for preparing *Sal* plates. They collect and store *Neem Khati* from Kotulpur, Moynapur, Jadavnagar and Bishnupur area at the time of lean season for stitching *Sal* leaf. They engaged in this industry 7–8 month (170–220 man days) in a year and rest of the period they remain jobless. *Sal* plate used in various purposes but mainly it is used in food serving container (Table 21.4).

Churung khati mala making is the dominant economy for the female of Madhabpur mouza. In earlier time female usually collected *Churung Khati* from the nearest forest area but now a days this product is unavailable in surrounding area as nearest forest area became degraded by the human interference. Hence, male people collect this product from 10 km away, mainly from *Kharkaisuli* area by bicycle. Female make *Mala* from the *Churung Khati* with the help of traditional tools (*Pati*, *Duol* and *Kund*) and technology. They engaged in this industry for 8–9 months (approx. 190–220 man days) in a year. Female engaged more days than other two household industries because the raw material of this industry is stem which more available than other two. In this village, female are also engaged in agricultural labour

Table 21.4 Details about the three dominant household industries

Mouza	Household industry	Main raw material			Availability (peak season)	Uses
		Name	Source	Collection		
Kunda Pushkarini	*Belmala* making	*Khat Bel* cover	Bishnupur, Bihar and Jharkhand	Both male and female collected it from nearest market area i.e. Routkhanda	6–7 month (July–December)	In *Sradh*, marriage, *Bujna* or *Annaprasan* and also in sacred thread wearing ceremony
Katul	*Sal* plate making	*Sal* leaf, *Neem Khati*	*Sal* leaf: nearest forest area *Neem Khati*: from Kotulpur, Jadavnagar, Bishnupur, Moynapur	Mainly by female	7–8 month (June–January)	Multiple use, mainly in food serving purpose
Madhabpur	*Churung Khati Mala* making	*Churung Khati*	Nearest forest area is at about 10 km (*Kharkaisuli*)	Mainly by male members	8–9 month (June–January)	Mainly use by *Sadhu and Sanysi* (Hindu Monks) as *Mala*

Source Field Survey, March, 2018

work but this opportunity is in very limited in nature. *Churung Khati Mala* is used mainly by the *Sanyasi* and *Sadhu* in various religious places of India (Table 21.4).

21.3.3 Socio-economic Characters of Family and HHI Workers

Landless and poor people of forest fringe areas are mainly depending upon NTFPs for their livelihood maintenance (DHDR 2007). Some people have marginal agricultural land which is not enough to meet their household needs. Hence, female also engaged in work for giving economic support to the family. They mainly involved these types of informal activities because of their little opportunity in other sector for illiteracy or low level of education, inefficiency and absence of job opportunity in agricultural labour sector. Therefore, they primarily rely on NTFPs for maintaining household and get economic return.

Among the above three household industries, *Belmala* making industry is confined within female belong to Muslim community and making of *Sal* plate and *Churung Mala* is confined within Scheduled Castes (SCs) women section in Jaypur block. It is very surprising fact to note that *Belmala* is made by Muslim women but it is used by the Hindu people to perform many rituals in various sacred ceremonies. The fact behind the monopoly of Muslim women in *Belmala* making activities is that feet of the *Mala* makers touched the cover of *Bel* during making of the beads from *Bel* covers. Therefore, Hindu female avoided it as they used the *Bel* fruit as a sacred worship material. Age-group of female belmala makers is mainly 20–50 years and they have low level of education (up to secondary level) and mobility. Female members of these Muslim households prefer to engage in this household industry because it is fully home based and marketing of *Belmala* is done from the home through middleman. The making process is tough and not suitable for aged women workers (More than 50 years of age group).

Female *Sal* plate makers of Katul mouza also have low level of education (up to upper primary level) and they are totally belonging to poor scheduled castes families. Some villagers have little amount of land (10–15% household) but a large section of households are landless. Hence, female members of these families play the role of bread earner to sustain their livelihood. Female mainly aged between 20–65 years, all are engaged in this industry because of easy processing method. Their household income is very less and they are all came from BPL category.

Churung Khati Mala makers are also belonging to Scheduled Caste family. They are very poor and come from BPL families. Their education level is also very low (up to secondary level). Female workers within the age group of 20–65 years old are mainly engaged in this industry with their traditional knowledge and techniques.

21.3.4 Marketing and Rural Economy

Marketing of goods is a challenging factor for rural economy to get proper price of goods. Maximum products of these three household industries are marketed from the home through middlemen. So, middlemen play a big role and often enjoy the economic profit after bypassing the producers. Low level of education of the female workers has limited their exposure to market system and modern technology to produce more value added products and hence, middleman making a hefty profit and condition of artisan is very pathetic (Development and Planning Department, WB 2007).

Belmala products of Kunda Pushkarini mouza are sold from the home to the local middleman. They only produced simple strands of *Belmala* which are used in scared ceremony purposes but they did not know about the other uses of *Bel* beads. Demand and price of *Belmala* have been increased at a slower pace from past few years (5 years ago, Rs. 1/piece and now Rs. 2/piece). *Bel* beads can also be used to make colourful *Mala*, designer *Kurti*, bag, shoes etc. This belmala has immense demand both in local (Jaypur, Bishnupur, Kotulpur and Kolkata) as well as in national (Delhi) markets with value addition in various places of India (Fig. 21.6).

Sal plates of Katul mouza is mainly handmade product and it has demand for local area only. But value added laminated *Sal* plates made by middleman or business man has more demand in internal and external market. Marketing of the *Sal* plate is

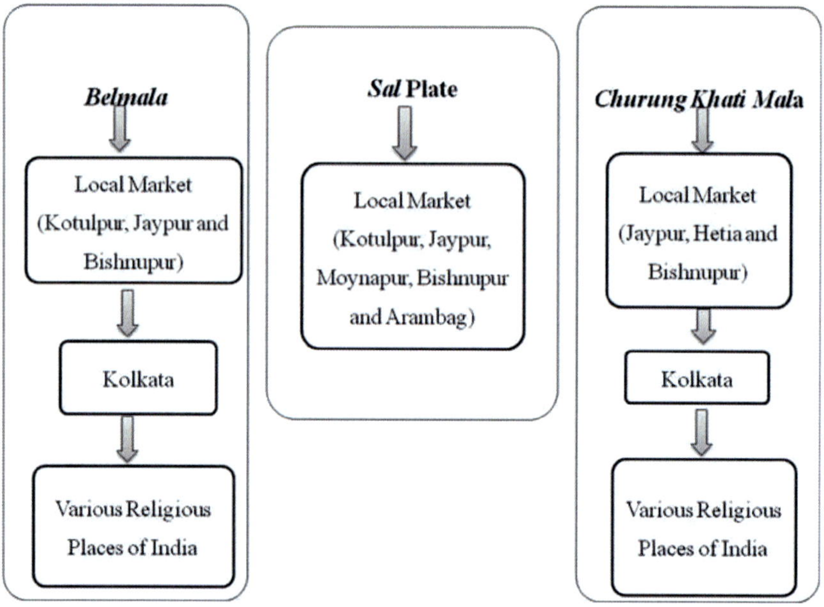

Fig. 21.6 Marketing of the household industry products by middleman. *Source* Field Survey, March, 2018

conducted mainly from the home but sometimes they sold their finished products in nearby markets viz. Jaypur, Kotulpur and Moynapur *bazaar* to have relatively better price. Transport facilities helped this industry to grown up than past and both the demand as well as price of *Sal* plates also increased in comparison to past years (5 years ago). Among three household industries, *Sal* plates have the higher demand in the local market because this product is frequently used by the villagers in any kinds of occasion. Laminated *Sal* plates are marketed by rich businessman in local market areas in Bishnupur, Kotulpur and Jaypur and others (Fig. 21.6).

Churung Khati Mala of Madhabpur mouza also a traditional technique based industry with local and national level demand (Religious places). Female of this industry also did not know about the modern value addition of *Churung Khati Mala*. Two types of mala they made with the help of small beads (3 Rs./piece) and large beads (2 Rs./piece) with different price. Local middlemen buy this raw product at a cheap rate and did not give any information to the female artisan related to further value addition. Bishnupur is the main market place of *Churung Khati Mala* and the exact area of marketing in this town is vague to the villagers. From Bishnupur this product has reached to Kolkata (main hub) and then distributed in various religious places of India (Fig. 21.6).

21.3.5 Income Generation and Its Share in Total Income of the Household

Income generation from the HHI and its share in the total family income is an important indicator of the economic significance of these sectors and for the survival of the engaged people in these HHIs. Women workers in this sector play an important role after making their contribution in household income to fight against poverty. The result showed that the earning of the *Belmala* makers maximum (60%) varies from below Rs. 8000/- per annum to Rs. 14,000/- per annum. It has been observed that 40% of *Belmala* maker women earned 8000–14,000 Rs./annum (Table 21.5). Only 10% of Belmala maker women earned the highest amount of more than 20,000 Rs./annum. Besides, 20% and 30% of Belmala maker women earned below 8000 Rs./annum and 14,000–20,000 Rs./annum respectively (Fig. 21.7). In case of the share of income in the household by HHI, 50% of *Belmala* maker women share more than 50% of their household income and 25% women share 50–70%, 20% women share 70–90% and 5% women share >90% of their household income (Table 21.5) (Fig. 21.8). The entire income from the *Belmala* making is used for household expenditure purpose. The male member of their family controls their earning. In the case of women, those who contribute more than 90% of their family income from HHI are either single women or widow or the male persons are incapable to contribute. Therefore, they are significantly dependent on this HHI for their survival and any alteration in this sector can hamper their economic conditions significantly. On the other hand, women who contribute less than 50% of household income from HHI are at less risk comparatively

Table 21.5 Details of family income, income from household industry and share of income

HHI Sector	Family income		Income from HHI		Share of income	
	Household income (Rs./annum)	% of household	Income (Rs./annum)	% of women	Share of income in %	% of women
Belmala	<20,000	10	<8000	20	<50	50
	20,000–35,000	25	8000–14,000	40	50–70	25
	35,000–50,000	45	14,000–20,000	30	70–90	20
	>50,000	20	>20,000	10	>90	5
Sal plate	<20,000	11.44	<8000	22.88	<50	34.29
	20,000–35,000	60	8000–14,000	45.71	50–70	42.86
	35,000–50,000	14.28	14,000–20,000	22.88	70–90	11.42
	>50,000	14.28	>20,000	8.53	>90	11.42
Churung Khati Mala	<20,000	20	<8000	32	<50	24
	20,000–35,000	60	8000–14,000	32	50–70	44
	35,000–50,000	16	14,000–20,000	24	70–90	20
	>50,000	4	>20,000	12	>90	12

Source Field Survey, March, 2018

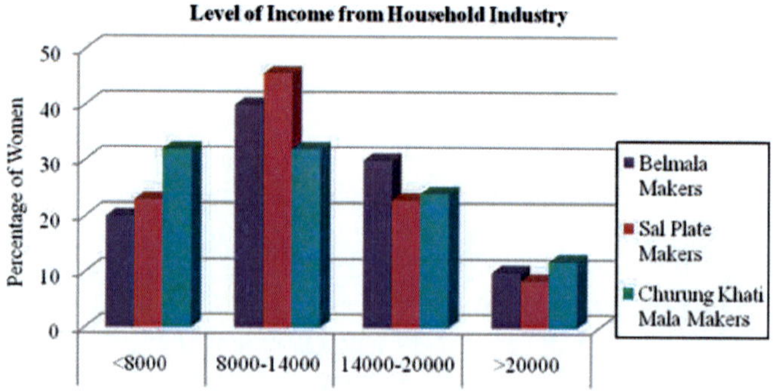

Fig. 21.7 Level of income of the women workers from household industry. *Source* Field Survey, March, 2018

as their male members are engaged in other economic activities. It is found that male members from the *Belmala* making households of the Kunda Pushkarini mouza are mainly engaged in informal activities as a truck driver, small trader, vegetable vendor and labourer etc. So, they have diversified their economic base to sustain their family. Besides, those who contribute 50–90% of their household income from HHI are landless or marginal landowners. Therefore, these families are moderately vulnerable compare to other families and if they adopt a diversified economic base, then only they will able to overcome this economic vulnerability. However, few

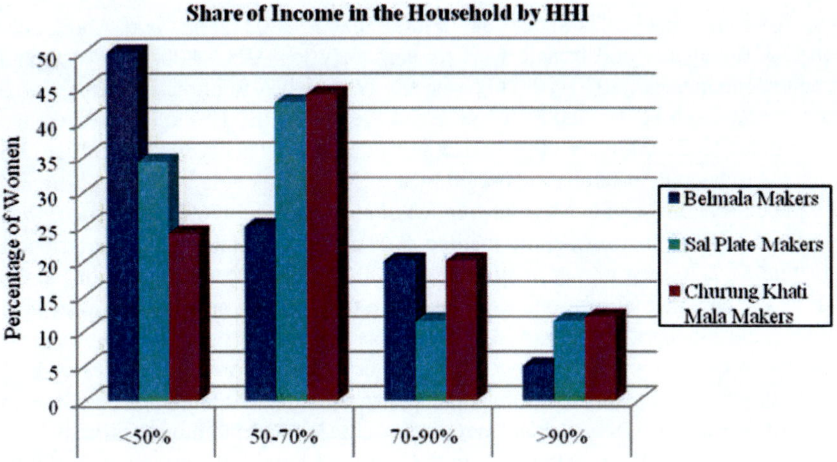

Fig. 21.8 Share of income in the household by household industry. *Source* Field Survey, March, 2018

women have started to buy some products from local hawkers for personal uses which is a good indication of women's economic empowerment in Muslim society.

The income of *Sal* plate makers is more important than ever. Because the families of female *Sal* plate makers are poorer than *Belmala* maker's. Therefore, income generation from these *Sal* plates has become very much vital for the survival of their families. The earnings from the *Sal* plates vary based on variation in the skill and the time invested to the making of *Sal* plate. Among the 35 women *Sal* plate makers, 22.88%, 45.71%, 22.88% and 8.53% of the women earned less than 8000, 8000–14,000, 14,000–20,000 and more than 20,000 Rs./annum respectively (Table 21.5) (Fig. 21.7). Among the total women *Sal* plate makers 34.29%, 42.86%, 11.42% and 11.42% of women share about <50%, 50–70%, 70–90% and >90% of their household income (Fig. 21.8). So, 65.71% of women share more than or equals to 50% income to their household budget. This indicates the dependency of these households on *Sal* plate making sector due to easy availability of raw materials near to the household associated with simple process of making of products. The higher percentage (>90%) share to the household income represents the widow or single income earner *Sal* plate makers. They spent their entire earning for family purposes but for their better mobility than *Belmala* makers, they get an opportunity in choosing their basic essential products. On the other hand, women who contribute less than 50% of their family income from the HHI, the male family members are mainly engaged as an agricultural labourer and marginal farmer. In addition, landless people and marginalized landholder families are economically 50–90% dependent on the plate making industry. The male members of these families are employed as labourers in sand mining and to some extent in agriculture to run their families.

A significant amount of income is also generated from *Churung Khati Mala* making. Among the 25 women engaged in *Churung Khati Mala* making, the earnings vary from less than Rs. 8000 per year to more than Rs. 20,000 per year. About one-third (32%) of the total female workers earn very low (Rs. <8000 per year) and medium-income (Rs. 8000–14,000 per year). Whereas, 24% of the women are able to generate Rs. 14,000–20,000 per year and the rest of the 12% of women earned more than Rs. 20,000/year (Table 21.5 and Fig. 21.7). This sector of HHI is also valuable to keep their family supported by a substantial income. A major portion of the population engaged in the *Churung Khati Mala* making (76%) share more than equals to 50% of their household income. Rest of the population (24%) contributes less than 50% to their family income (Fig. 21.8), where the male populations in the family are engaged in informal activities like a truck driver, small trader, vegetable vendor and labourer in the various informal sectors. Where single income earner women or widows are the main earners or male members of the family contribute very little financially, they contribute more than 90% of their household income. In addition, the landless or marginal landowners of the Madhabpur mouza earn 50–90% of their family income from the making of *Churung Khati Mala* because the earnings of their male members of the family are very little from agriculture. Among the 24% of total women share less than 50% of total household income as their families have others source of income due to possession of good productive agricultural land. Their income is fully utilised in household expenditure purpose and they can buy some basic necessary goods with their own choice.

21.3.6 Women Economic Independency and Status in the Family

Women are the main artisans of these three household industries where male engagement are very limited and confined only in collection process of the raw materials for making *Churung Khati Mala* and *Belmala.* Hence, in this rural economic process contribution of women is relatively high than other sectors. Household industry products are mainly sold from the home and most of the time women take the money from the middleman as advance. A large share of their earning is spent to meet the basic needs of the family. Women have the money in hand but they have no power to spend for their own purposes or household purposes without permission of the male members of the family. Among the three household industry workers *Belmala* makers are the most deprived one in terms of economic independency and decision making power in the family. Only widow or the main economic earner women (10%) of the family took decision about the household expenditure. Some families (20%) of the *Sal* plate makers mutually took decision about household expenditure and those women also sometimes consume little money for own purpose. But in case of *Churung Khati Mala* makers the picture is quite different. Here women (40%) took decision independently or mutually in family expenditure of their earning money. Their decision making power in family also better than the other two groups and they are more mobile than the other two to perform the various types of household responsibilities in the outside (banking and marketing). Therefore, it is clear that engagement of women in household industries did not so much improve their status in the family but enable them to share the burden of the cost of maintaining their families so as to reduce the economic stress in the household.

21.3.7 Suggestive Measure for Sustainability of the Industry and Livelihood Development

NTFPs based household industry adopted by landless, marginal farmers and forest dwellers is an important part of income generation, survival strategies and livelihoods that rely on conventional tools and techniques. Similarly, in this study it has been observed that since the beginning of the industry, they have produced similar products without any up-gradation or modification. Therefore, the producers of the traditional products do not get its proper economic value due to presence of middleman. Thus, they are being deprived from the benefits by the current system with existing chains of intermediaries as well as lack of information about modern techniques. In additions, backward location, poverty, low level of education, lower income, lack of awareness about the products and its value, inefficiency in modern techniques and less motivation plays a crucial role in backwardness of these household industries. They are very much comfortable with present system and less aspires to experiment with their skills and products. Thus, the next generation also started to follow the same kinds of steps and cannot get the chances to break the chain of this poverty cycle so their livelihoods remain in the similar condition. Therefore, the following steps have been suggested to uplift their livelihood as well as for the sustainability of household industries:

1. First of all awareness level about the value added goods, demand of goods and profit from industry and inspiration to the female artisans about modern technology must be increased. This can be done by encouraging formation of Self Help Groups (SHGs) in respective mouzas and making arrangements to train the group members.
2. After proving information, provision of hands-on training is an important part to be followed. Therefore, with the help of local government (Gram Panchayet and Block Development Office) training can be conducted about the uses of modern techniques and its uses in processing of crafted items and products. Development of both soft and hard skill should be another area of emphasis.
3. As the people engaged in these sectors belong to the economically poor section, therefore, Government should manage the availability of modern instrument through building a cooperative system in the villages under an appointed supervisor.
4. A micro-credit scheme also needed to motivate the artisans.
5. Variation of the product is another important part where one has to focus on making it more attractive by applying value-added products by machines

and modern techniques. Such as from *Bel Beads* (Kunda Pushkarini Mouza), colourful *Mala* with locket, bracelet, decorated bags and designable *Kurti* etc. can be produced; from *Sal* plate (Katul Mouza), Laminated *Sal* plate, cup, bowl (various sizes) etc. can be prepared and from *Churung Khati Mala* (Madhabpur Mouza), colourful *Mala* with locket, bracelet, decorated bags and designable *Kurti* etc. should be produced.

6. Initiatives should be taken from the Government side for the overall improvement of the existing infrastructure of the study area which will help to widen the market area for these local made products. More fair can be arranged and local craft based tourism can be promoted for better livelihood of rural people in these mouzas.

And finally, the government should provide adequate wages and financial assistance (loan from banks) for their sustainable livelihood.

21.4 Conclusion

NTFPs are important for the survival of economically poor people, landless people and marginal farmers in rural areas. Numerous poor people make some products from NTFPs based on traditional technologies and engaged in household industries to meet their basic needs thus, household industries became an important part of their source of income and livelihood. Here, in this study, it is observed that female-dominated HHIs are directly linked with the nearest forest for collection of raw materials and the produced products are mainly used for both household purpose as well as commercial purpose. Locally available forest resources facilitate the local people to engage in the HHI based on NTFPs. People engaged in making of *Sal* plate in the Katul mouza, *Belmala* in the Kunda Pushkarini and *Churung Khati Mala* in the Madhabpur mouza's belong to economically weaker sections. Hence the earning from these HHIs is very much crucial for their survival. The women from the Minority section mainly engaged in the *Belmala* making industry whereas the women from Scheduled castes population are associated with the making of *Sal* plates and *Churung Khati Mala*. In the case of *Sal* plate makers and *Churung Khati Mala* making families, they are mostly landless but some section of the *Belmala* maker families have marginal agricultural land for farming or the male people engaged within profitable informal activities. Therefore, their dependency on the HHI in the Katul mouza and Madhabpur mouza are comparatively higher than the Kunda Pushkarini mouza. Lower level education of the female has forced them to work in these sectors and they have no choice to do work in other fields. Several opportunities are available to the male population to work as truck driver, small trader, vegetable vendor and labourer in various informal sectors. The main reasons behind low income generation from these three HHIs are identified as poor capital of the households, lack of awareness among women workers about modern techniques and value added goods, less knowledge about prospective market areas and proper price of products. Thus, a large part of the profit directly goes to the middleman or intermediaries associated with these industries. Besides, their livelihoods are in jeopardy as deforestation continues which limited the supply of raw materials and again the local products have a tough competition with more value added and alternative products. Consequently a big question is arises whether these industries will survive or not. Local Government (Gram Panchayet and Block Development Office) should take initiatives to encourage these cottage based industries and prepare a systematic cluster framework from supply of raw materials to marketing of finished goods. In this context, provision of incentives to the skilled people, improvement of basic infrastructure, running vocational training for skill development, imparting information on market system, approval of easy and fast loan to women workers engaged in this sector will help them to maintain a sustainable livelihood.

References

Ahenkan A, Boon E (2011) Non-timber forest products (NTFPs): clearing the confusion in semantics. J Hum Ecol 33(1):1–9

Akilandeeswari SV, Pitchai C (2016) Pottery industry and its development by effective marketing through information and communication technologies (ICT). Int J Res 4(4):6–13

Angelsen A, Jagger P, Babigumira R, Belcher B, Hogarth NJ, Bauch S et al (2014) Environmental income and rural livelihoods: a global-comparative analysis. World Dev 64:S12–S28

Belcher B, Achdiawan R, Dewi S (2015) Forest-based livelihoods strategies conditioned by market remoteness and forest proximity in Jharkhand, India. World Dev 66:269–279

CIFOR (2008) Forests and non-timber forest products. Center Int For Res. https://www.cifor.org/Publications/Corporate/FactSheet/ntfp.htm. Accessed on 11/08/2019

Comprehensive district agricultural plan report (2007–2012). District Planning Committee, Bankura.

Dasgupta S, Biswas R, Mallik GK (2009) Heritage tourism: an anthropological journey to Bishnupur. Mittal Publications, New Delhi, p 96

District census handbook (1991). Census of India, Bankura

District human development report, Bankura (2007). Development and Planning Department, Government of West Bengal, Kolkata

District primary census abstract (2001, 2011). Census of India, Bankura. Retrieve from https://censusindia.gov.in. Accessed on 20/9/2019

District village directory (2011). Census of India, Bankura. Retrieve from https://censusindia.gov.in. Accessed on 20/9/2019.

Ghosal S (2011) Importance of non-timber forest products in native household economy. J Geogr Reg Plann 4(3):159–168

Hasalkar S, Jadhav V (2004) Role of women in the use of non-timber forest produce: a review. J Soc Sci 8(3):203–206

https://www.cifor.org/publications/corporate/factSheet/NTFP.htm

Krishnakumar J, Fox J, Anitha V (2012) Non-timber forest products. Econ Polit Wkly 47(52):132–139

Kumar S, Meena GL (2018) Procurement and marketing of non-timber forest products. Indian Forester 144(3):252–259

Land and Land Reform Department (2013). Government of West Bengal

Leakey RRB (2012) Non-timber forest products-a misnomer? J Trop For Sci 24(2):145

Mahapatra AK, Albers HJ, Robinson EJ (2005) The impact of NTFP sales on rural households' cash income in India's dry deciduous forest. Environ Manage 35(3):258–265
Mahto R, Chaterjee D (2011) Non-timber forest product based enterprise: leaf plate making in Angara, Jharkhand. J South Asian Anthropologist 11(1):59–65
Metadata (2001). Census of India. https://censusindia.gov.in/Metadata/Metada.htm. Accessed on 01/07/ 2019
Micro, medium and small enterprises report (2014–2015). Development Institute Kolkata Ministry of MSME, Government of India
Mphande FA (2016) Rural livelihood. In: Infectious diseases and rural livelihood in developing countries. Springer, Singapore, pp 17–34
Pandey AK, Tripathi YC, Kumar A (2016) Non timber forest products (NTFPs) for sustained livelihood: challenges and strategies. Res J For 10(1):1–7
Report of the expert group to review the methodology for measurement of poverty (2014). Planning Commission, Government of India, New Delhi
Saha M, Sengupta S (2014) Symbiotic relationship between forest and tribe: a case study of Santal tribe of Jaypur forest, Bankura District, West Bengal, India. Transactions 36(2):235–246
Saxena NC (2003) Livelihood diversification and non-timber forest products in Orissa: wider lessons on the scope for policy change? Overseas Development Institute, London
Shackleton C, Shackleton S (2004) The importance of non-timber forest products in rural livelihood security and as safety nets: a review of evidence from South Africa. S Afr J Sci 100(11):658–664
Shit PK, Pati CK (2012) Non-timber forest products for livelihood security of tribal communities: a case study in Paschim Medinipur district, West Bengal. J Hum Ecol 40(2):149–156
Sills EO, Lele S, Holmes TP, Pattanayak SK (2003) Nontimber forest products in the rural household economy. In: Forests in a market economy. Springer, Dordrecht, pp 259–281
State forest report (2010–2011). Government of West Bengal, Directorate of Forests Office of the Principal Chief Conservator of Forests and Head of Forests Force, Aranya Bhaban, Salt Lake, Kolkata
Tambunan T (1995) Forces behind the growth of rural industries in developing countries. A survey of literature and a case study from Indonesia. J Rural Stud 11(2):203–215
Tewari DN (1992) Tropical forestry in India. International Book Distributors, p 387
Tewari DD, Campbell JY (1995) Developing and sustaining non-timber forest products: some policy issues and concerns, India. J Sustain Forest 3(1):53–79

Chapter 22
Forest Ecosystem Services and Biodiversity

Afaq Majid Wani and Gyanaranjan Sahoo

Abstract Ecosystems, and the biodiversity and services they support, are intrinsically dependent on climate. During the twentieth century, climate change has had documented impacts on ecological systems, and impacts are expected to increase as climate change continues and perhaps even accelerates. This technical input to the National Climate Assessment synthesizes our scientific understanding of the way climate change is affecting biodiversity, ecosystems, ecosystem services, and what strategies might be employed to decrease current and future risks. In developing countries, the landscape encompassing agricultural land is vital for maintaining biodiversity and providing ecosystem services. The consequences of biomass shot on woody species richness and composition were analysed in forests underneath communal and government management. Interviews on forest use and perception of forest condition and system service delivery were conducted in farmer households bordering the forests. At the same time, the importance of forest ecosystem services has been progressively recognized. Though some initiatives aimed toward protective each biodiversity and ecosystem services are emerging, information gaps still exist regarding their relationships and potential trade-offs in forests. Considerably additional woody species were found within the community-managed forests. Species richness was negatively correlative with walking distance from the closest village and increasing levels of anthropogenic disturbance. There is increasing proof that diversity contributes to forest ecosystem functioning and also the provision of ecosystem services. Here, we are including some of the ecosystem services such as biomass production, production of atmospheric oxygen, soil formation and retention, nutrient cycling, water cycling, and provisioning of habitat. Community forests were

A. M. Wani (✉)
Department of Forest Biology and Tree Improvement, College of Forestry, SHUATS, Prayagraj, India
e-mail: afaqtree@gmail.com

G. Sahoo
Department of Forestry, Krishi Vigyan Kendra, Odisha University of Agriculture and Technology, Angul, Odisha, India
e-mail: gyanaranjan.sahoo3@gmail.com

P. K. Shit et al. (eds.), *Spatial Modeling in Forest Resources Management*, Environmental Science and Engineering,
https://doi.org/10.1007/978-3-030-56542-8_22

typically less degraded than government managed forests, giving support to common pool resource management. Woody vegetation depicted an important supply of fuel-wood, timber, fodder, edibles, aromatic and healthful plants. Employing a multidisciplinary framework to analyse system integrity and system service delivery enabled a finer understanding of those complicated agroecological systems, giving support to evidence-based management and conservation coming up with for the long run. Planting mixed-species forests ought to tend additional thought as they're probably to supply a wider range of ecosystem services at intervals the forest and for adjacent land uses.

Keywords Biodiversity · Ecosystem · Forestry · Conservation · Restoration

22.1 Introduction

Forest is that the largest terrestrial ecosystem within the Common Market covering around 40% of the territory and is home an excessive amount of the continent's diversity. Additionally, forests offer a large number of advantages to humans in terms of climate regulation, facility and regulation, timber, energy, habitats for biodiversity, clean air, erosion management and many others. For instance, European forests face multiple natural and anthropogenic pressures. For example, a dynamical climate affects tree species composition and assemblage. Climate-driven forest pressures foreseen foretold to extend and competitive socio-economic demands for forest service's end in multiple drivers of forest amendment. During this context, our activity on forest ecosystem services and diversity aims at mapping and assessing forest condition and forest ecosystem services, contributively of the MAES initiative in support to the diversity Strategy to 2020.

A wide range of mechanisms is projected to elucidate the relationships between biodiversity and ecosystem services. Niche complementarity in time and house, and complementarity of useful result traits and functional response traits are all seemed to be concerned (Isbell et al. 2013). Facilitation between plant species growing along has typically been found to guide to increased growth of sure tree mixtures (Thompson et al. 2011). For example, at nitrogen-limited sites, tree species that are nitrogen-fixers could enhance the expansion of different tree species in mixed stands (Resistance to disturbance is expedited by forest and tree diversity, resulting in a discount or dilution of resources (e.g., for herbivores), diversion or disruption, and multi-trophic interactions (e.g., increased abundance and action of natural enemies). Finally, the supposed 'sampling effect' will enhance the supply of ecosystem services, just because the presence of additional species will increase the chance that Associate in the Nursing ecosystem can contain a species that grows quicker, is additional immune to a specific disturbance, or has another an advantageous attribute that results in increased ecosystem functioning or provision of services, compared to communities with fewer species (Wardle 2001; Lefcheck et al. 2015).

Compared with 'natural forests' or mixed-species forests, planted forests sometimes have a lower level of the diverseness of cover trees and alternative species, and it's possible that their ability to supply bound ecosystem services is reduced. For instance, mixed forests tend to be more practical in delivering a spread of provisioning services (e.g., Gamfeldt et al. 2013; Forrester and Bauhus 2016), and are a lot of proof against varied disturbances than single-species planted forests. These relationships between forest sort, diverseness and ecosystem services are extremely relevant for informing forest policy and management. However, given the multitude of ecosystem services, it's tough to generalise concerning the role of forest diversity. Moreover, there are trade-offs between completely different ecosystem services looking on the tree mixture and stand sort concerned. Some tree mixtures are superior at providing bound services however alternative tree mixtures or maybe single-species forests are more practical for alternative services. Given the role of diverseness within the provision of ecosystem services, the widespread degradation of forests is probably going to own comprehensive effects, like reduced resistance (or raised susceptibility) to natural or anthropogenic disturbance.

There has been a lot of progress recently during this terribly active field of analysis, and therefore the International Union of Forest Research Organisations (IUFRO) established a task force to facilitate multi-disciplinary analysis collaboration and understanding the nature of the biodiversity-ecosystem services relationship and the possible effects of biodiversity loss on the delivery of ecosystem services (Harrison et al. 2014), in addition to literature reviews on the results of forest diversity on single and multiple ecosystem services. Pollination of crop plants and wild plants is a crucial ecosystem service worldwide. Taki et al. explore the employment of stable atom analysis to research however land use and climate have an effect on wild bee populations. This provides a helpful approach for the study of relationships between diversity, land use and therefore the provision of fertilisation services. The results of invasions of tree pests and pathogens on forest diversity and ecosystem services are assessed by, light the imperative ought to mitigate the danger of future invasions and to extend our ability to manage people who have already occurred. Cultural ecosystem services are extremely relevant in native communities.

Currently, there's a speedy advancement in our understanding of the relationships among biodiversity, ecosystem functions and services in forest ecosystems. A crucial example is an analysis of biodiversity-ecosystem functioning in forests. Biodiversity effects on ecosystem functions are historically evaluated in different systems like grasslands and aquatic systems, and therefore the relative lag in forest ecology domain is of great concern (Nadrowski et al. 2010). Early syntheses of this subject in forests (CBD 2009; Aerts and Honnay 2011) therefore for the most part relied on data from systems aside from forests. For the last several years, theoretical and experimental initiatives on practical biodiversity analysis are apace rising in forests (e.g. Hector et al. 2011; Morin et al. 2011; Baeten et al. 2013; Bruelheide et al. 2014; Verheyen et al. 2015). Though an outsized knowledge gap still exists in forests compared to raised studied systems, a growing analysis agenda on forest biodiversity system functioning is predicted to play a considerable role in mitigating future environmental change (e.g. temperature change mitigation; Hulvey et al. 2013; Poorter et al. 2015).

Another purpose of interest that exemplifies the necessity for additional studies on forest biodiversity and ecosystem services may be inferred from the marked distinction in framing socio-economic views on biodiversity conservation between agriculture and forestry sciences. For agriculture, detailed strategic plans have usually been planned for sustaining food production whereas reducing the environmental impacts of land use (e.g. Foley et al. 2011; Tilman et al. 2011). The rapid expansion of sustainable forest management (SFM) has resulted in the adoption of various forest management frameworks intended to safeguard biodiversity. Concurrently, the importance of forest ecosystem services has been increasingly recognized. Although some initiatives aimed at conserving both biodiversity and ecosystem services are emerging, knowledge gaps still exist about their relationships and potential trade-offs in forests. Given recent advancements, increasing opportunities and some lags in forest ecology, further research on biodiversity, ecosystem functions and services will play substantial roles in the development of SFM practices (Mori et al. 2017).

Forests are critical habitats for biodiversity and they are also essential for the provision of a wide range of ecosystem services that are important to human well-being. There is increasing evidence that biodiversity contributes to forest ecosystem functioning and the provision of ecosystem services. The biodiversity-ecosystem operation provides a useful framework to evaluate forest restoration in an ecosystem functioning context, but it also highlights that much remains to be understood, especially regarding the relation between forest functioning on the one side and genetic diversity and above-ground-below-ground species associations on the other. The strong emphasis of the biodiversity-ecosystem operation on functional rather than taxonomic diversity may also be the beginning of a paradigm shift in restoration ecology, increasing the tolerance towards allochthonous species (Aerts and Honnay 2011).

In view of recent advancements, knowledge gaps and also the future significance of analysis on forest biodiversity and ecosystem services, we tend to define many key analysis priorities addressing the necessity to make capability and resilience for social and ecological systems to face uncertainty within the ever-changing environment. The new opportunities that ecosystem services approaches provide for biodiversity conservation include: the development of broader constituencies for conservation and expanded possibilities to influence decision-making; opportunities to add or create new value to protected areas; and the opportunities to manage ecosystems sustainably outside of protected areas. Based on an in-depth literature review, we tend to envision a research agenda that may be fruitful for students similarly as practitioners.

Biodiversity may support ecosystem services directly and indirectly through ecosystem functions. (a) A study framework of biodiversity ecosystem functioning (BEF) is rapidly expanding in forests, with some notable examples of tree diversity effects on above-ground productivity (Morin et al. 2011). Tree BEF experiments have been recently launched in some regions (Bruelheide et al. 2014). (b) To obtain benefits from natural systems, society needs to consider the relationships between biodiversity and ecosystem multi-functionality (Gamfeldt et al. 2013). (c) To ensure the roles of biodiversity to deliver ecosystem services, the stability of ecosystem functioning is also important. Some studies have started to elucidate underlying mechanisms of the biodiversity–stability relationships using simulation (Morin et al. 2014) and using a retrospective approach (Jucker et al. 2014). (d) In addition to the local-scale evaluation, the assessment of ecosystem service provisions at the large scale (i.e. regional scale) is important to help resource management and decision-making (Eigenbrod et al. 2010). There are some key issues that can determine the direction of societal impacts on biodiversity and on ecosystem services in turn (e) Terrestrial systems may undergo regime shifts in response to precipitation changes, determining the degree of ecological resilience in a changing climate (Hirota et al. 2011). (f) Social-ecologically sustainable forestry can potentially provide a win-win solution to reconcile the trade-offs and conflicts between conservation and commodity production (Mori and Kitagawa 2014). (g) In theory, the order of species reintroduction affects restoration outcomes measured with functional complementarity and redundancy, determining the restoration outcomes for ecosystem functions (Devoto et al. 2012). (h) While local communities can buffer the impact of the random loss of species to maintain the fundamental functionality, non-random (realistic) loss of species can substantially alter functional characteristics of communities, leading to the emergence of novel communities (Mori et al. 2015b).

In many ecosystem classifications (including those which have been expanded to conceptualize ecosystems as assets), there appears to be no explicit place for the value of biodiversity. Indeed, a significant anxiety about recent ecosystem assessments is that the emphasis upon ecosystem services might ironically lead to the omission of the vital role which biodiversity plays in both the delivery of those services and as a source of value in itself. Mace et al. (2012) provide clarification of the issue, noting that biodiversity appears at three distinct points within the ecosystem service framework.

22.2 Forest Biodiversity

Forests are biologically diverse ecosystems that give an environment for a multiplicity of plants, animals and micro organism. In line with the Convention of Biological Diversity forest diversity may be a term that refers to any or all life forms found within forested areas and their ecological roles. Forest diversity is often thought of at totally different levels, as well as in the ecosystem, landscapes, species, populations and genetic science. In high biodiversity forests this quality permits organisms to adapt to continually ever-changing environmental conditions and to maintain ecosystem functions. Forest ecosystems and biodiversity are strongly interlinked. On the one hand, diversity levels rely to an outsized extent on the integrity, health and vitality of forests. On the other hand, losses of forest biodiversity cause attenuated forest productivity and property. From this perspective, analysing forest condition, the multiple services and trade-offs, and therefore the relationship with forest biodiversity provides decision instruments for targeted provisions within the context of the eco industry and bio economy in an exceeding transition to a greener environment (Cimon-Morin et al. 2013).

22.3 Ecosystem Processes, Functions, Services Are Distinct Concepts

It is widely known that biodiversity is a major driving force in ecosystem function. Many studies have self-addressed the consequences of tree species diversity on several forest ecosystem functions, together with primary production. During this terribly active field of analysis, the statement that tree diversity will improve "forest ecosystem function and (associated) services" has become quite common. However, the ideas of ecosystem operate and ecosystem services are typically confused, albeit they're totally different in terms of their definition and connectedness to scientists and managers. Whereas "function" is associate ecosystem-centred construct, "ecosystem service" is humanist that specialize in operate permits scientists to know however changes in forest biodiversity will modify the key ecological processes that are driving the functioning, integrity or maintenance of forest ecosystems. Given the linkages and relationships between ecosystem functions and services, forest managers or policy manufacturers might use such info to predict however biodiversity management or sweetening will have an effect on the delivery of products and services useful to the economy and to human well-being.

Forests typically are well placed to deliver most of the ecosystem services (ESs) listed in current frameworks like the, due to their wide distribution, made biodiversity and long history of human use. However, empirical studies that establish quantitative and causative relationships between forest diversity and ecosystem services are lacking for several vital ESs (Mori et al. 2017).

Ecosystem services are defined as the benefits that humans obtain from ecosystems. Employing the ecosystem service concept is intended to support the development of policies and instruments that integrate social, economic and ecological perspectives. In recent years, this concept has become the paradigm of ecosystem management. 1. The prolific use of the term 'ecosystem services' in scientific studies has given rise to concerns about its arbitrary application. A quantitative review of recent literature shows the diversity of approaches and uncovers a lack of consistent methodology. 2. From this analysis, we have derived four facets that characterise the holistic ideal of ecosystem services research: (i) biophysical realism of ecosystem data and models; (ii) consideration of local trade-offs; (iii) recognition of off-site effects; and (iv) Comprehensive but critical involvement of stakeholders within assessment studies. 3. These four facets should be taken as a methodological blueprint for further development. They should critically reveal and elucidate what may often appear to be ad-hoc approaches to ecosystem service assessments. Increased anthropogenic activities, especially within the river corridor, have progressively disrupted natural flow regimes and segmented channel–flood plain connectivity. Consequent alterations in flow dynamics have caused geomorphic and hydrological changes in channel morphology and behaviour, decreasing their natural replenishment capacity, thereby causing their degradation. Our dependence on biodiversity and ecosystem services (ES) is increasing, due to population expansion and economic growth. Consequently, maintaining biodiversity and sustaining ES supply should consistently be incorporated into conservation project objectives. Systematic conservation planning procedures based on site complementarity would increase the efficiency of both biodiversity and ES conservation. Economic valuation of ES, such as through cost–benefit analysis, could help to justify conservation actions by showing that the financial benefits of nature conservation greatly exceed the cost. Moreover, payments for ecosystem services could create new incentives and funding sources for the conservation of biodiversity.

A common approach to ecosystem service assessment is to use proxy variables, particularly land cover, to represent ecosystem processes and provide maps of ecosystem service. The measurement, modelling and monitoring of ecosystem functions are the foundation for ecosystem service assessments and are thus the basis for the sustainable use of biodiversity, ecosystems and natural resources in general (Carpenter et al. 2009). This requires relating ecosystem functioning to ecosystem service indicators. A variety of methodological approaches are available to describe these non-monotonous, non-linear and time-variant relationships that all require data, maps, monitoring, fieldwork or experiments.

22.4 Biodiversity and Ecosystem Services

Social ecological models evaluating the cost-effectiveness of conserving biodiversity and several ecosystem services (e.g. commodity production and carbon sequestration), such as seen in issues surrounding REDD+ (e.g. Koh and Ghazoul 2010), have been recently developed to reconsider land use; however, relevant case studies are still in short supply. Land-use models are often useful for considering tradeoffs between biodiversity conservation and ecosystem service conservation. Scenarios and policy options gained from these models would be fundamentally beneficial for decision making. However, models considering the synergies between biodiversity and ecosystem services are relatively scarce; that is, both tend to be considered as distinct and unlinked response variables affected by human activities. The functional consequences of biodiversity conservation on ecosystem services have not been well integrated into these models so far (Isbell et al. 2015b). In this regard, ecosystem services models that can also account for biodiversity conservation, such as the In-VEST (Integrated Valuation of Environmental Services and Tradeoffs) model (Kareiva et al. 2011), have great potential to inform SFM. To date, while numerous tools are available, the conceptual flows from biodiversity to the functionality and services of ecosystems have consequently been rarely integrated into models and analyses of forest ecosystems. In many parts of the world forest disturbance regimes have intensified recently, and future climatic changes are expected to amplify this development further in the coming decades. These changes are increasingly challenging the main objectives of forest ecosystem management, which are to provide ecosystem services sustainably to society and maintain the biological diversity of forests.

With the aim to provide ecosystem services to society while fostering biodiversity, ecosystem management requires a comprehensive understanding of the impacts of natural disturbances. Notwithstanding this high relevance, natural disturbances have hitherto been discussed inconclusively in the context of ecosystem management, with views and recommendations ranging from strict avoidance of disturbance (due to negative effects on selected ecosystem services) to emulating disturbance in management (to utilize their beneficial effects on biodiversity) (Thom and Seidl 2016). On the one hand, substantial efforts are undertaken in research and management to quantify disturbance risk, with the aim to minimize their negative impacts through increasing the resistance of forests to disturbances.

22.5 Biodiversity and Ecosystem Functions

In addition to the potential importance of higher productivity in mixed stands, there are other reasons why tree diversity may be important (Scherer-Lorenzen 2014). Tree diversity is often linked with major properties in forests, including the possible enhancement of diversity of other forest assemblages (Schuldt et al. 2014) and potential contribution to other functions, such as litter decomposition (Handa et al. 2014). Notably, these kinds of connections can have significant uncertainty. In sum, different databases and platforms that are currently available are expected to jointly play important roles in disentangling the underlying processes of diversity functioning relationships and thus further advancing functional biodiversity research. So far, we have described the present context of the forest BEF studies, with a primary focus on tree assemblages. Apart from trees, knowledge gaps exist between forest science and other domains. For example, while the importance of forests as a habitat for conserving pollinator communities a critical component of sustained crop production has been well studied (e.g. Garibaldi et al. 2011; Mitchell et al. 2014), the reverse relationship (the functional roles of pollinators for forest ecosystems) has been relatively little covered. Furthermore, while the functional consequences of non-random, realistic loss of diversity on trophic the structure has been recently demonstrated for grassland and aquatic systems (e.g. Zavaleta and Hulbey 2004; Bracken et al. 2008; Karp et al. 2013a, b), this issue has been rarely visited in forest ecosystems (but see Barnes et al. 2014). Thus, in understanding the contributions of forest biodiversity to ecosystem functions and services, a large amount of uncertainty still exists not only for trees but also for other groups of organisms. These issues undoubtedly need to be continually tested, in addition to the focus on trees, to gain a full picture of the flow from biodiversity and ecosystem functions to ecosystem services. However, considering the fundamental roles that trees play both directly and indirectly (via other facets of diversity in a given system) to support the overall functionality of forest ecosystems, the high likelihood that trees determine the assemblage structures of other taxa, and the fact that different silvicultural practices and land-use modifications directly alter tree diversity, tree diversity function studies represent a most promising frontier for the improvement of SFM practices.

Forest resilience for facing uncertainty High levels of biodiversity is effective and often essential for ecosystems to endure environmental changes and retain their fundamental functionality, largely contributing to the maintenance of resilience in ecosystems (Elmqvist et al. 2003; Mori et al. 2013a, b). Here, eco-logical resilience is defined as the capacity of a system for absorbing changes to maintain fundamental controls on function and structure (Chapin et al. 2009; Gunderson et al. 2009). Ecological resilience is the modern concept of facing uncertainty, unpredictability, nonlinearity and changeability in a system to be managed (Standish et al. 2014).

22.6 Forest Biodiversity, Multi-functionality and Trade-Offs Among Ecosystem Services

Forests area unit valued for multiple ecosystem services, together with timber production, climate regulation and recreation, and for biodiversity in its own right (Mace et al. 2012). A significant challenge for forest managers is to maximise as several of those services as possible, thereby increasing 'ecosystem multi-functionality'. Once totally different ecosystem services and biodiversity area positively related all completely associated with one another, meeting this goal is, a minimum of in theory, comparatively easy. However, in recent years, the biodiversity of studies have investigated relationships between forest ecosystem services and located that whereas some ecosystem services correlate completely, others show sturdy negative relationships at local scales (Gamfeldt et al. 2013) or large spatial scales. Because of these trade-offs, increasing all desired forest ecosystem services is challenging.

Some trade-offs between ecosystem services occur as a result of different tree species give the different ecosystem functions and services (Gamfeldt et al. 2013), whereas others are driven by forest management, which regularly maximises certain ecosystem services at the cost of others (Chhatre and Agrawal 2009). Hence, at local scales, promoting sure tree communities might maximise some, however not all, ecosystem services of interest. As a result, forest ecosystem multi-functionality usually will increase with each tree (Gamfeldt et al. 2013) and fungal (Mori et al. 2016) species diversity, though it's virtually not possible to maximise all desired ecosystem services and functions underpinning them at local scales. Therefore, recent studies have investigated whether or not larger scale biodiversity, caused by a high spatial turnover in species composition (i.e. high beta diversity) will promote system multi-functionality at the landscape scale. This has clothed to be the case, as high beta diversity ensures that totally different localities complement one another within the system functions and services they supply (Mori et al.2016).

Because of the large amount of information that's needed for analysis on biodiversity and ecosystem multi-functionality, this field has solely started out comparatively recently. Hence, despite several recent advances, there are still several unresolved queries concerning however biodiversity and ecosystem multi-functionality may be at the same time maximised in natural forests. For example, it's unknown whether or not the positive effects of local-scale tree species richness on ecosystem multi-functionality are even stronger once co-occurring species differ significantly in their traits or evolutionary origins, though such data may be crucial for planting multi-functional forests. Additionally, it's known that forests will give multiple ecosystem services to near landscape units, like agricultural fields (Mitchell et al. 2014). However, whether or not the advantages of diverse forests for neighbouring fields are bigger than those of species-poor forests remains an associate open question. With the increasing interest in understanding what drives multifunctional landscapes, it's possible that these and different connected queries are investigated within the future.

22.7 Forest Biodiversity, Ecosystem Functions and Services

A series of Biodiversity Ecosystem Functioning (BEF) studies (Cardinale et al. 2012) have revealed that biodiversity (including taxonomic, functional and phylogenetic diversity) promotes the functionality of ecosystems (e.g. primary production, decomposition, nutrient cycling, trophic interactions and so on) and consequently supports a broad range of ecosystem services (e.g. food production, climate regulation, pest control, pollination and numerous others). The development of BEF theory has mainly arisen from experimental and theoretical work, with a central contribution from experimental manipulation of plant assemblages in grassland ecosystems over the last several decades (e.g. Hautier et al. 2014; Isbell et al. 2015a). The knowledge coming from these biodiversity experiments has been widely used to make inferences about other systems, as it is often difficult to set up an equivalent experiment elsewhere. This is especially true for forests, which are characterized by higher structural complexity, longer life cycles of the dominant taxa and larger scale spatiotemporal dynamics than grassland communities (Scherer-Lorenzen 2014). Knowledge gaps in the discipline of forest ecology have been therefore largely complemented by studies in other systems such as grass-lands, aquatic systems and bacterial microcosms. Now, global meta-analyses and syntheses are also available for diversity functioning relationships in forests (e.g. Zhang et al. 2012; Chisholm et al. 2013), indicating some signs of progress in this research field in forest ecology.

22.8 Diversity Responses Under Environmental Fluctuations

An important issue in considering resilience is the way to secure the basic practicality of a focal system. In this regard, the insurance hypothesis (Yachi and Loreau 1999) deserves further attention. The insurance hypothesis predicts that ecosystem function is stabilized in species-rich communities wherever the redundancy of species contributes to a similar function and therefore reduces fluctuations in this function over space or time. That is, high diversity could make sure the high resiliency of a system. The conception of 'response diversity' adds more inference to the insurance effects of biodiversity (see Mori et al. 2013a, b). Briefly, additionally to the amount of functionally redundant species, the intraspecific variation in responses to environmental fluctuations is additionally critical; if this variation is reduced, a basic management on ecosystem function can be lost from local communities within the face of environmental change (Elmqvist et al. 2003; Mori et al. 2013a, b). To date, empirical proof of response diversity continues to be scarce, particularly for forest ecosystems. Notably, Karp et al. (2011) expressly demonstrated the importance of response diversity in forest communities (bird assemblages) for sustaining ecosystem services (pest management, seed dispersal and pollination). Probably vital mechanisms for enhancing response diversity embody interspecies desynchronizing of population dynamics and temporal niche differentiation among species, each of that cause useful compensation under environmental fluctuations (Mori et al. 2013a, b). However, the issue exists in assessing these processes, particularly in naturally assembled communities in which long-term monitoring data are usually required to quantify intraspecific variation. At this juncture, in progressive accumulation of long-run monitoring data for forest communities as well as plants and animals, that are progressively archived within the property right (open data), can provide opportunities to demonstrate whether or not diversity in responses among species exists and the way this ensures the practicality of ecosystems.

22.9 Diversity Effects to Stabilize Ecosystem Functions

Experimental and theoretical studies have prompt that biodiversity has the potential not solely to come up with however additionally to stabilize ecosystem functions (Tilman et al. 2006; Hautier et al. 2014; Morin et al. 2014; Isbell et al. 2015a). Though the diversity of theoretical explanations is given to resolve the mechanisms of diverseness stability relationships, massive uncertainty still sure enough exists concerning however and once diverseness contributes to stabilising and (consequently) making certain the very important practicality of ecosystems (Hautier et al. 2015). However, given the prominence of environmental stability for determinative the event, prosperity and security of human society (Hsiang et al. 2013), the potential of biodiversity is of practical importance (Isbell et al. 2015a). As we have a tendency to stressed earlier, primary productivity is one amongst the best issues in forests. Each study have prompt that whereas interspecies segregation of functional traits (e.g. shade tolerance and leaf display) will drive complementary effects in tree mixtures (implying temporal niche differentiation) as usually discovered in grasslands, species desynchronization remains vital however is weaker in tree communities than herbaceous communities. This evidence suggests that whereas the usually discovered relationship between diversity and stability in alternative ecosystems is applicable to forests, the underlying mechanisms don't seem to be essentially similar in different ecosystems. This evidence once more emphasizes the importance of getting specific frameworks for forest ecosystems. Given the importance of dominant tree species in supporting various ecosystem functions and services in forests (e.g. foundation species; Ellison et al. 2005), tree diversity could also be of restricted use for predicting productivity. Jucker et al. (2014) prompt a limitation of desynchronization among tree species because of the long-lived nature of trees. These findings imply that tree diversity might have limitations once buffering the impacts of huge and unpredictable environmental changes (e.g. climate extremes and bug outbreaks; Grossiord et al. 2014); that's, there could also be no direct contribution of tree diversity to forest ecosystem resistance and resilience. Forest ecologists so ought to additional investigate the difficulty of dominance versus diversity. Such debates might doubtless verify the long-run direction of SFM, as there's no current agreement over that forms of planting regimes (e.g. monocultures vs. mixed cultures) best maintain each forest commodity production and therefore the delivery of alternative forest ecosystem service.

22.10 Ecosystem Services Perspective for Multifunctional Forestry

Biodiversity conservation has historically cared-for target a set of diverseness that has iconic and/or endangered species, and rather neglected the functional roles of diverseness as a driver and supply of ecosystem functions and services (Mace et al. 2012). A series of approaches for forestry geared toward protective diverseness has had a bent to adopt that traditional perspective, and little explicit consideration has been given to multiple ecosystems functions and services so far. However there's nice potential in protective forest taxa from the perspective of ecosystem services, as conserved taxa in forest patches might contribute to providing and sustaining the functionality of forested ecosystems (see Karp et al. 2013a, b for associate example in agriculture). The potential edges embody biogeochemical processes supported by soil biodiversity maintained within the stand, pest control management and pollinations as a result of conserving trophic interactions, and water retention and erosion management by under-storey plant communities. These services are likely maintained in forests with high diversity in their assemblages, together with those of microbes, invertebrates, vertebrates and plants (Thompson et al. 2011), though expectations of such potential haven't been strictly tested in production forests managed with conservation approaches. Successive generation of multifunctional forestry studies ought to take under consideration ecosystem services provided as a result of conservation actions in forestry and therefore the tradeoffs created by this observe.

22.11 Management in a Changing Climate

In addition to focussing on the present standing of forests, completely different views also are necessary during this era of an ever-changing environment. Climate change is one in every of the best issues that may incur extra costs on forest management (Hanewinkel et al. 2013). Modelling approaches have known some key aspects of future forest management approaches that are aimed towards mitigating or adapting to an ever-changing climate. Duveneck and Scheller (2015) planned a climate-suitable planting regime within which species from outside the landscape are planted to anticipate a northward shift of the best thermal ranges of tree species. However, the study concluded that, as a result of climate change effects usually outweigh management actions, this management various incorporates a restricted ability to reinforce the (engineering) resilience. Furthermore, the results of climate change on forest landscapes are often amplified if more thought is given to the ever-changing disturbance regime. Seidl et al. (2014) estimated that there's a high probability of accelerating injury from wild fires, insect outbreak and wind throw within the returning decades, and this can possible offset management ways that are aimed toward increasing the forest carbon sink. The study concluded that exacerbating forest disturbances won't solely have an effect on the carbon sink, however, will have detrimental influences on different types of ecosystem services. For instance, the authors argue that management prices could increase due to fire suppression, pest control, salvage logging and so on, which will very likely have an effect on timber markets. These and alternative model primarily based studies have provided a crucial image of the longer term, though these results are for the most part enthusiastic about numerous situations. To further implement these and alternative projections into practice, novel approaches like coupled human natural system models ought to be thought of within which future changes in ecological and social systems and synergies and feedbacks between these subsystems are often integrated. Such a completely unique approach can facilitate policy and management by providing multiple choices that higher affect future uncertainty. An increasing number of alarms are raised regarding anthropogenic impacts on forest biodiversity (e.g. Gibson et al. 2011; Wilcove et al. 2013); land use change related to deforestation and forest degradation have vulnerable varied forest-dependent taxa across many regions. In response to the present biodiversity crisis, a large number of studies have delineated however, environment alteration and destruction have affected and will change forest biodiversity at completely different spatial, temporal and biological scales (e.g. Newbold et al. 2014). Note that human-induced species extinction doesn't occur haphazardly, however, there are instead multiple deterministic factors inflicting their loss. In this regard, some studies have expressly incontestable the underlying mechanisms of non-random loss of forest biodiversity (e.g. Mori et al. 2015a). Moreover, recent studies have conjointly quantified the implications of land-use change on ecosystem services in forest landscapes (e.g. Lavorel et al. 2011). Compared to such information regarding forest degradation, knowledge of the responses of biodiversity and, especially, of ecosystem services to forest restoration are comparatively restricted. However, knowledge of purposeful consequences of forest restoration have been step by step accumulating (e.g. Lamb 2005; Chazdon 2008; Bullock et al. 2011).

22.12 Restoring Biodiversity and Ecosystem Services

Under Sustainable Forest Management frameworks, it is highly encouraged to convert monoculture plantations into stands with multiple tree species in many regions because trees in mixed-species stands are likely to harbour higher biodiversity of other organism groups and provide more multiple ecosystem goods and services then those in monocultures (Knoke et al. 2008). However, it is still unclear whether and especially how biodiversity and ecosystem functions respond to forest restoration activities based on re-establishment of multiple tree species. Although the functional roles of tree diversity in generating these services have not been directly evaluated, some empirical evidence on ecosystem properties in mixed-species stands (Rothe and Binkley 2001; Knoke et al. 2008) implies that this potential of ecosystem services restoration (including economic benefits; e.g. Piotto et al. 2010) is likely. Furthermore, due to the reduced environmental variation (environmental homogenization) in plantations of a single or a small number of species, such as through the decreased the structural complexity of forest canopy and the reduced diversity of plant litter, plantations have the potential to homogenize communities of forest dwelling taxa (Chazdon 2008; Mori et al. 2015a). Note that this biotic homogenization occurs not only taxonomically (taxonomic homogenization) but also functionally (functional homogenization), importantly suggesting that the vital functionality of forest stands supported by forest-dwelling communities can be threatened in mono species plantations (Mori et al. 2015a). Responses of forest communities to the forest reconversion into mixed stands are largely variable among different taxa (Knoke et al. 2008). To date, studies of the potential of re-differentiating forest dwelling communities in terms of both taxonomic and functional characteristics (i.e. recovery from biotic homogenization) have been limited (Mori et al. 2015a), and more studies are necessary to inform society on how to restore and conserve the forest ecosystem services.

22.13 Theory for Restoration

A series of theories in functional ecology may guide applied ecologists and practitioners (Laughlin 2014; Ostertag et al. 2015), but large uncertainties remain. For instance, there are an increasing number of observational studies of biodiversity in stands managed and conserved under retention or reduced-impact logging forestry; however, the underlying processes of community assembly, which determine the actual responses of biodiversity to environmental alteration in logged stands, have been given little attention so far (but see Bassler et al. 2014). Such approaches founded in community ecology, which can give mechanistic understandings of biodiversity responses to human influences (Mouillot et al. 2013), should be further employed in future studies of applied ecology (Laughlin 2014). Notably, recent theoretical and experimental studies in community ecology have proved the importance of history (the order of species immigration) in determining community composition (Fukami and Nakajima 2011) and ecosystem function (Fukami et al. 2010). Similar findings have also been reported in woodlands in which the importance of plant reintroduction sequences for restoring ecosystem functions supported by plant pollinator networks was shown (Devoto et al. 2012). To integrate these latest findings into practice, the order of species reintroduction into restoration sites (e.g. re-vegetation, enrichment planting and assisted relocation) should not be random, and restoring biodiversity and biodiversity-based ecosystem functions and services should ideally, be theory-driven.

22.14 Challenges

The queries that rapid global climate change poses for reconsidering conservation goals don't seem to be straightforward to fathom, coupled with the answer. Even though the principles and observe of conservation have evolved within the past, there are formidable institutional and psychological barriers to shifting from current conservation paradigms and realigning goals (Jantarasami et al. 2010). Institutional barriers to reconsidering and recalibrating goals embody those involving: (1) legislation and regulations; (2) management policies and procedures; (3) human and monetary capital; and (4) information and science (Julius and West 2008). Legal mandates and policies, specifically, might gift troublesome barriers. Several existing laws, like the Endangered Species Act, mandate specific approaches that constrain managers' ability to switch goals and management objectives. Many recent legal reviews (for example, Fischman 2007; Glicksman 2009; Craig 2010; Ruhl 2010), supply some insights on the nuances of Federal conservation laws as they relate to global climate change adaptation. Psychological barriers, however, will be equally difficult to beat. Hagerman et al. (2010) underscore the actual fact that a lot of conservationists realize it troublesome to move on the far side the acquainted goals of restoring and protective existing patterns of diversity and a priori selected conservation targets due to a robust resistance to creating trade-offs a thought delineate within the scientific discipline literature as "protected values" (Gregory et al. 2006). Poiani et al. (2011) equally discovered that a general reluctance of conservation practitioners to "give up on anything" might make a case for the comparatively low range of adaptation methods focused on transformation compared to maintaining established order conditions.

Many conservationists believe in the High Conservation Value (HCV) approach as an effective mechanism to understand and negate the ill impacts of unregulated natural resource extraction. However, along with the widening and diversification of its scope, its suitability for biodiversity conservation has been subject to speculation, scrutiny and critique from HCV users and assessors (e.g., Paoli and Harjanthi 2011). The HCV concept is transitioning from a heuristic model of learning by experience to a holistic management tool that enables better prioritization of components like ecosystem services, biodiversity and socio economic-cultural values.

In this direction, some of the key challenges for research are:

1. Comprehensive spatio-temporal, contextual and specific coverage of primary and secondary research
2. Formulation of National/Regional/Local Interpretation of HCV components, categories and definitions
3. Challenges in Implementation of Management Plan at Landscape Level
4. Stakeholder engagement and practice of FPIC
5. Prioritization of HCVs to be conserved based on Threats and Human Utility.

The HCV approach has been promoted by social and environmental NGOs and private parties. This network of collaborators shows the significance of the HCV approach to a wide spectrum of stakeholders and forms a balance between environmental, social and economic considerations (Meijaard and Sheil 2012). However, it has often been misapplied in the context of agriculture, leading to HCV assessment poor in quality (Paoli and Harjanthi 2011). This further magnifies the challenge of isolating the approach's drawbacks from those due to misapplication.

22.15 Limitations

- In addition to the potential importance of higher productivity in mixed stands, there are other reasons why tree diversity may be important (Scherer-Lorenzen 2014). Tree diversity is often linked with major properties in forests, including the possible enhancement of diversity of other forest assemblages (Schuldt et al. 2014) and potential contribution to other functions, such as litter decomposition (Handa et al. 2014). Notably, these kinds of connections can have significant uncertainty. For example, the former issue needs caution as tree richness does not necessarily promote the diversity of other organism groups (Donoso et al. 2010). The latter issue also has limitations as diversity–decomposition relationships have been primarily tested for evaluating the effects of litter diversity (Gessner et al. 2010), and the effects of tree diversity have been inferred indirectly.
- As observed in other systems, examining biomass production is probably one of the greatest opportunities to disentangle the underlying mechanisms of diversity–function linkages in forests (e.g. Morin et al. 2011; Jucker et al. 2014, 2015, 2016; Lasky et al. 2014). Although higher productivity cannot necessarily be translated directly into higher levels of provision services (i.e. timber production, bioenergy and so on) (Chisholm et al. 2013), it is likely to contribute to some services, such as carbon sequestration and storage (Hector et al. 2011; Hulvey et al. 2013; Poorter et al. 2015). Analogous to biodiversity experiments in grasslands, tree experiments have been established world-wide (Verheyen et al. 2015). Furthermore, it is feasible to control the number of individuals in tree plantations, providing a greater experimental control over 'true diversity indices' (*sensu* Jost

2007) compared to grassland experiments. This approach allows one to rigorously quantify how contributions to different ecosystem functions and services scale across ecological levels (i.e. the individual, species and community). Knowledge of scaling effects of biodiversity on ecosystem function is still largely limited (Mori et al. 2016), and thus, an important frontier in BEF studies is tree diversity experiments.

- Another issue that needs attention is that tree diversity–function relationships may be confounded by variables other than tree diversity, such as tree density, biomass, age, edaphic conditions and other environmental factors (Toïgo et al. 2015; Jucker et al. 2016). Furthermore, compared to grasslands, there are generally high levels of difficulties to manipulate environmental factors (e.g. water, nutrients and so on). Given these limitations, it has been necessary and continues to be important to rely on data from inventory plots that have not been primarily designed for BEF studies. To use data from these non-experimental tree communities, methodological advancements such as statistical approaches that can control for the confounding effects of the other explanatory variables on diversity–productivity relationships are an important research priority (e.g. Healy et al. 2008). To date, the research field of tree diversity–ecosystem function relationships has a primary focus on biomass production, but these new platforms and initiatives will and should play further important functions and services beyond the focus on productivity (Gamfeldt et al. 2013).
- As such, tree diversity studies are still at the early stage of development, and thus, a large amount of knowledge can be gleaned from the comparatively mature field of biodiversity studies in grasslands. For example, a recent grassland experiment of Zuppinger-Dingley et al. (2014) may have application not only to the agriculture sector but also to the forestry sector. They suggested that using varieties that have been selected in diverse planting regimes may increase (timber) productivity. Given the long history of genetic engineering in forestry (Harfouche et al. 2011), their implication is intriguing. Given the increasing pressure to allocate more lands for nature reserves as proposed under the CBD framework, the use of multiple species provenances may provide an alternative solution in the context of SFM. However, testing this potential is far from easy. Theory suggests a high likelihood of natural enemy attacks when conspecific cohorts are grown densely (negative density dependence, or the so-called Janzen–Connell hypothesis), which is one of the rationales for encouraging mixtures rather than monocultures to enhance productivity (Schnitzer et al. 2011).
- Apart from trees, knowledge gaps exist between forest science and other domains. For example, while the importance of forests as a habitat for conserving pollinator communities—a critical component of sustained crop production—has been well studied (e.g. Garibaldi et al. 2011; Mitchell et al. 2014), the reverse relationship (the functional roles of pollinators for forest ecosystems) has been relatively little covered. Furthermore, while the functional consequences of non-random, realistic loss of diversity on trophic structure have been recently demonstrated for grassland and aquatic systems (e.g. Zavaleta and Hulbey 2004; Bracken et al. 2008; Karp et al. 2013a, b), this issue has been rarely visited in forest ecosystems (but see

Barnes et al. 2014). Thus, in understanding the contributions of forest biodiversity to ecosystem functions and services, a large amount of uncertainty still exists not only for trees but also for other groups of organisms. These issues undoubtedly need to be continually tested, in addition to the focus on trees, to gain a full picture of the flow from biodiversity and ecosystem functions to ecosystem services.

22.16 Conclusion

Previous studies have provided valuable information on the role of biodiversity in ecosystem service delivery from a theoretical prospective (Mace et al. 2012) explored the links between functional traits and ecosystem services, all examined how biodiversity influences the functioning of ecosystems and thus, their ability to provide ecosystem services (Cardinale et al. 2012).

Our review confirms that forest type and tree species richness have an effect on forest biodiversity which forest diversity will be a crucial consider system perform and also the provision of system services. In this review, it is illustrated the potential of applied ecology to assist conserve and restore forest ecosystems. However, there are clear mechanisms that tree diversity will improve ecosystem function and also the delivery of system services, for several ecosystem services, there's still uncertainty regarding the extent of a 'functional relationship' between biodiversity and also the provision of these services. We have a tendency to conjointly have to be compelled to evaluate the result of various levels of tree diversity; not solely species however conjointly genetic and functional diversity. And whereas cover trees are clearly a dominant feature of forests, the range of under-storey plants, vertebrates, invertebrates, fungi and microbes are additionally possible to be necessary for ecosystem services.

Furthermore, several ecosystem services stay relatively poorly studied in forests in relevance to biodiversity; this is applicable notably to cultural services however conjointly to some provisioning services. Under the substantial human influence, forests can seemingly be grossly altered, doubtless resulting in the emergence of novel ecosystems or change to different stable states. Management thus must be additional versatile and use novel measures to face such large uncertainties. Resilience based mostly approaches are key to foreseeing future changes and managing surprises, though the problems that we've addressed don't seem to be an entire list for a future analysis agenda in applied forest ecology.

This review emphasizes the interactions and therefore the interdependency between these problems, a number of that have attended be mentioned rather superficially, whereas others are well studied in disciplines outside the domain of forest ecology. For natural forests this discussion could seem somewhat educational, because it is unlikely that tree species composition and diversity would be altered considerably within the interest of system services. However, it's necessary to boost awareness regarding the role of natural forests and forest diversity within the provision of system services to spotlight their worth on the far side the availability of timber and recreation. However, for planted forests there's ample chance for optimising their composition and diversity as a result of replanting once harvest home may be a continual method. If it will be shown that there are opportunities for adding worth and for increasing the resistance or resilience of planted forests, this ought to be smart incentives for forest homeowners and managers to contemplate alternatives to the monoculture paradigm of most planted forests. We have a tendency to so endorse the plea of Paquette and Messier (2010) "for the implementation of well-conceived, diverse, multi-purpose (forest) plantations as some way to conserve forest diversity and system functions".

Functional group or community level attributes (e.g. functional richness, community/habitat area) can be important for ESPs that work on one or more species' population levels. The connection of forest system services doesn't stop at the forest edge. There's a lot of scope for synergies between forests and farming land uses; as an example, even tiny patches of forest will profit crop production by enhancing insect and natural enemy populations, though they'll also provide disservices (Decocq et al. 2016).

Further research to expand the search to cover more services and to explore how the linkages identified differ by ecosystems or biogeographical region would also be useful for targeting the management of biodiversity and service provision. However, to support management and protection strategies, future research should also take account of effects of socio-economic factors and land use decisions on different components of biodiversity and, as a result, ecosystem service delivery. Incorporating traditional conservation strategies for species and habitat protection within the broader context of social-ecological systems and ecosystem service delivery can lead to added benefits for biodiversity through closer integration of conservation policy with policies in other sectors (Haslett et al. 2010).

Biodiversity is not (just) a good to be conserved for its intrinsic value, but has a critical role in ecosystem processes (Mace et al. 2012) that provide essential services to humans (Cardinale et al. 2012). Finally, any planted forest plan ought to measure choices for mixed species forests as these are possible to produce a wider range of ecosystem services.

Possible future approaches for Biodiversity Ecosystem Services and Sustainability:

- Tools and frameworks for decision-support
- Methods for data collection
- Statistical approaches
- Modelling
- Linking science to policy
- Enhancing interdisciplinary research
- Challenges for researchers and interlinkages across the themes
- Commonalities and differences with other horizon scanning exercises.

References

Aerts R, Honnay O (2011) Forest restoration, biodiversity and ecosystem functioning. BMC Ecol 11:29

Baeten L, Verheyen K, Wirth C, Bruelheide H, Bussotti F, Finer L et al (2013) A novel comparative research platform designed to determine the functional significance of tree species diversity in European forests. Perspect Plant Ecol Evol Syst 15:281–291

Barnes AD, Jochum M, Mumme S, Haneda NF, Farajallah A, Widarto TH, Brose U (2014) Consequences of tropical land use for multitrophic biodiversity and ecosystem functioning. Nat Commun 5:5351

Bassler C, Ernst R, Cadotte M, Heibl C, Muller J, Barlow J (2014) Near-to-nature logging influences fungal community assembly processes in a temperate forest. J Appl Ecol 51:939–948

Bracken ME, Friberg SE, Gonzalez-Dorantes CA, Williams SL (2008) Functional consequences of realistic biodiversity changes in a marine ecosystem. Proc Natl Acad Sci U S A 105:924–928

Bruelheide H, Nadrowski K, Assmann T, Bauhus J, Both S, Buscot F et al (2014) Designing forest biodiversity experiments: general considerations illustrated by a new large experiment in subtropical China. Methods Ecol Evol 5:74–89

Bullock JM, Aronson J, Newton AC, Pywell RF, Rey Benayas JM (2011) Restoration of ecosystem services and biodiversity: conflicts and opportunities. Trends Ecol Evol 26:541–549

Cardinale BJ, Duffy JE, Gonzalez A, Hooper DU, Perrings C, Venail P et al (2012) Biodiversity loss and its impact on humanity. Nature 486:59–67

Carpenter SR, Mooney HA, Agard J, Capistrano D, DeFries RS, Diaz S, Dietz T, Duraiappah AK, Oteng-Yeboah A, Miguel Pereira H, Perrings C, Reid WV, Sarukhan J, Scholes RJ, Whyte A (2009) Science for managing ecosystem services: beyond the millennium ecosystem assessment. Proc Natl Acad Sci 106:1305–1312

CBD (2009) Forest resilience, biodiversity, and climate change—a synthesis of the biodiversity/resilience/stability relationship in forest ecosystems. Technical series, p 67. Convention on Biological Diversity, Montreal, Canada

Chapin FSI, Kofinas GP, Folke C (2009) Principles of ecosystem stewardship. Resilience-based natural resource management in a changing world. Springer, New York City, New York, USA

Chazdon RL (2008) Beyond deforestation: restoring forests and ecosystem services on degraded lands. Science 320:1458–1460

Chisholm RA, Muller-Landau HC, Abdul Rahman K, Bebber DP, Bin Y, Bohlman SA et al (2013) Scale-dependent relationships between tree species richness and ecosystem function in forests. J Ecol 101:1214–1224

Cimon-Morin J, Darveau M, Poulin M (2013) Fostering synergies between ecosystem services and biodiversity in conservation planning: a review. Biol Cons 166:144–154

Craig RK (2010) "Stationarity is dead"—long live transformation: five principles for climate change adaptation law. Harvard Environ Law Rev 34:9–75

Decocq G, Andrieu E, Brunet J, Chabrerie O, De Frenne P, De Smedt P, Marc Deconchat M, Diekmann M, Ehrmann S, Giffard B et al (2016) Ecosystem services from small forest patches in agricultural landscapes. Curr For Rep 2:30–44

Devoto M, Bailey S, Craze P, Memmott J (2012) Understanding and planning ecological restoration of plant-pollinator networks. Ecol Lett 15:319–328

Donoso DA, Johnston MK, Kaspari M (2010) Trees as templates for tropical litter arthropod diversity. Oecologia 164:201–211

Duveneck MJ, Scheller RM (2015) Measuring and managing resistance and resilience under climate change in northern Great Lake forests (USA). Landscape Ecol 31:669–686

Eigenbrod F, Armsworth PR, Anderson BJ, Heinemeyer A, Gillings S, Roy DB, Thomas CD, Gaston KJ (2010) The impact of proxy-based methods on mapping the distribution of ecosystem services. J Appl Ecol 47:377–385

Ellison AM, Bank MS, Clinton BD, Colburn EA, Elliott K, Ford CR et al (2005) Loss of foundation species: consequences for the structure and dynamics of forested ecosystems. Front Ecol Environ 3:479–486

Elmqvist T, Folke C, Nystrom M, Peterson G, Bengtsson J, Walker B, Norberg J (2003) Response diversity, ecosystem change, and resilience. Front Ecol Environ 1:488–494

Fischman RL (2007) From words to action: the impact and legal status of the 2006 national wildlife refuge system management policies. Stanford Environ Law J 77

Foley JA, Ramankutty N, Brauman KA, Cassidy ES, Gerber JS, Johnston M et al (2011) Solutions for a cultivated planet. Nature 478:337–342

Fukami T, Nakajima M (2011) Community assembly: alternative stable states or alternative transient states? Ecol Lett 14:973–984

Fukami T, Dickie IA, Paula Wilkie J, Paulus BC, Park D, Roberts A, Buchanan PK, Allen RB (2010) Assembly history dictates ecosystem functioning: evidence from wood decomposer communities. Ecol Lett 13:675–684

Gamfeldt L, Snall T, Bagchi R, Jonsson M, Gustafsson L, Kjellander P et al (2013) Higher levels of multiple ecosystem services are found in forests with more tree species. Nature Commun 4:1340

Garibaldi LA, Steffan Dewenter I, Kremen C, Morales JM, Bom-marco R, Cunningham SA et al (2011) Stability of pollination services decreases with isolation from natural areas despite honey bee visits. Ecol Lett 14:1062–1072

Gessner MO, Swan CM, Dang CK, McKie BG, Bardgett RD, Wall DH, Hattenschwiler S (2010) Diversity meets decomposition. Trends Ecol Evol 25:372–380

Gibson L, Lee TM, Koh LP, Brook BW, Gardner TA, Barlow J et al (2011) Primary forests are irreplaceable for sustaining tropical bio-diversity. Nature 478:378–381

Glicksman RL (2009) Ecosystem resilience to disruptions linked to global climate change: an adaptive approach to federal land management. Nebr Law Rev 87:833

Gregory R, Ohlson D, Arvai J (2006) Deconstructing adaptive management: criteria for applications to environmental management. Ecol Appl 16:2411–2425

Grossiord C, Granier A, Ratcliffe S, Bouriaud O, Bruelheide H, Checko E et al (2014) Tree diversity does not always improve resistance of forest ecosystems to drought. Proc Natl Acad Sci U S A 111:14812–14815

Gunderson LH, Allen CR, Holling CS (2009) Foundation of ecological resilience. Island Press, Washington, District of Columbia, USA

Hagerman S, Sowlatabadi H, Satterfield T, McDaniels T (2010) Expert views on biodiversity conservation in an era of climate change. Glob Environ Change 20:192–207

Handa IT, Aerts R, Berendse F, Berg MP, Bruder A, Butenschoen O et al (2014) Consequences of biodiversity loss for litter decomposition across biomes. Nature 509:218–221

Hanewinkel M, Cullmann DA, Schelhaas MJ, Nabuurs GJ, Zimmermann NE (2013) Climate change may cause severe loss in the economic value of European forest land. Nat Clim Change 3:203–207

Harfouche A, Meilan R, Altman A (2011) Tree genetic engineering and applications to sustainable forestry and biomass production. Trends Biotechnol 29:9–17

Harrison PA, Berry PM, Simpson G, Haslett JR, Blicharska M, Bucur M, Dunford R, Egoh B, Garcia-Llorente M, Geamănă N, Geertsema W, Lommelen E, Meiresonne L, Turkelboom F (2014) Linkages between biodiversity attributes and ecosystem services: a systematic review. Ecosyst Serv 9:191–203

Haslett JR, Berry PA, Bela G, Jongman RHG, Pataki G, Samways MJ, Zobel M (2010) Changing conservation strategies in Europe: a framework integrating ecosystem services and dynamics. Biodivers Conserv 19:2963–2977

Hautier Y, Seabloom EW, Borer ET, Adler PB, Harpole WS, Hillebrand H et al (2014) Eutrophication weakens stabilizing effects of diversity in natural grasslands. Nature 508:521–525

Hautier Y, Tilman D, Isbell F, Seabloom EW, Borer ET, Reich PB (2015) Anthropogenic environmental changes affect ecosystem stability via biodiversity. Science 348:336–340

Healy C, Gotelli NJ, Potvin C (2008) Partitioning the effects of biodiversity and environmental heterogeneity for productivity and mortality in a tropical tree plantation. J Ecol 96:903–913

Hector A, Philipson C, Saner P, Chamagne J, Dzulkifli D, O'Brien M et al (2011) The Sabah biodiversity experiment: a long-term test of the role of tree diversity in restoring tropical forest structure and functioning. Philos Trans Roy Soc B Biol Sci 366:3303–3315
Hirota M, Holmgren M, Van Nes EH, Scheffer M (2011) Global resilience of tropical forest and savanna to critical transitions. Science 334:232–235
Hobbs RJ, Higgs E, Harris JA (2009) Novel ecosystems: implications for conservation and restoration. Trends Ecol Evol 24:599–605
Hsiang SM, Burke M, Miguel E (2013) Quantifying the influence of climate on human conflict. Science 341:1235367
Hulme PE (2014) Bridging the knowing-doing gap: know-who, know-what, know-why, know-how and know-when. J Appl Ecol 51:1131–1136
Hulvey KB, Hobbs RJ, Standish RJ, Lindenmayer DB, Lach L, Perring MP (2013) Benefits of tree mixes in carbon plantings. Nat Clim Change 3:869–874
Isbell F, Tilman D, Polasky S, Binder S, Hawthorne P (2013) Low biodiversity state persists two decades after cessation of nutrient enrichment. Ecol Lett 16:454–460
Isbell F, Craven D, Connolly J, Loreau M, Schmid B, Beierkuhnlein C (2015a) Biodiversity increases the resistance of ecosystem productivity to climate extremes. Nature 526:574–577
Isbell F, Tilman D, Polasky S, Loreau M (2015b) The biodiversity-dependent ecosystem service debt. Ecol Lett 18:119–134
Jantarasami LC, Lawler JJ, Thomas CW (2010) Institutional barriers to climate change adaptation in U.S. national parks and forests. Ecol Soc 15:33
Jost L (2007) Partitioning diversity into independent alpha and beta components. Ecology 88:2427–2439
Jucker T, Bouriaud O, Avacaritei D, Coomes DA (2014) Stabilizing effects of diversity on aboveground wood production in forest ecosystems: linking patterns and processes. Ecol Lett 17:1560–1569
Jucker T, Bouriaud O, Coomes DA (2015) Crown plasticity enables trees to optimize canopy packing in mixed-species forests. Funct Ecol 29:1078–1086
Jucker T, Avacariței D, Barnoaiea I, Duduman G, Bouriaud O, Coomes DA, Gilliam F (2016) Climate modulates the effects of tree diversity on forest productivity. J Ecol 104:388–398
Julius SH, West JM (eds) (2008) Preliminary review of adaptation options for climate sensitive ecosystems and resources. A report by the U.S. Climate Change Science Program and the Subcommittee on Global Change Research. U.S. Environmental Protection Agency, Washington, DC
Kareiva P, Tallis H, Ricketts T, Daily G, Polasky S (2011) Natural capital: theory & practice of mapping ecosystem services. Oxford University Press, Oxford, UK, p 392
Karp DS, Ziv G, Zook J, Ehrlich PR, Daily GC (2011) Resilience and stability in bird guilds across tropical country side. Proc Natl Acad Sci U S A 108:21134–21139
Karp DS, Moeller HV, Frishkoff LO (2013a) Nonrandom extinction patterns can modulate pest control service decline. Ecol Appl 23:840–849
Karp DS, Mendenhall CD, Sandi RF, Chaumont N, Ehrlich PR, Hadly EA, Daily GC (2013b) Forest bolsters bird abundance, pest control and coffee yield. Ecol Lett 16:1339–1347
Knoke T, Ammer C, Stimm B, Mosandl R (2008) Admixing broad leaved to coniferous tree species: a review on yield, ecological stability and economics. Eur J Forest Res 127:89–101
Koh LP, Ghazoul J (2010) Spatially explicit scenario analysis for reconciling agricultural expansion, forest protection, and carbon conservation in Indonesia. Proc Natl Acad Sci U S A 107:11140–11144
Lamb D (2005) Restoration of degraded tropical forest landscapes. Science 310:1628–1632
Lasky JR, Uriarte M, Boukili VK, Erickson DL, John Kress W, Chazdon RL (2014) The relationship between tree biodiversity and biomass dynamics changes with tropical forest succession. Ecol Lett 17:1158–1167
Laughlin DC (2014) Applying trait-based models to achieve functional targets for theory-driven ecological restoration. Ecol Lett 17:771–784

Lavorel S, Grigulis K, Lamarque P, Colace M-P, Garden D, Girel J, Pellet G, Douzet R (2011) Using plant functional traits to under-stand the landscape distribution of multiple ecosystem services. J Ecol 99:135–147
Lefcheck JS, Byrnes JE, Isbell F, Gamfeldt L, Griffin JN, Eisenhauer N et al (2015) Biodiversity enhances ecosystem multi-functionality across trophic levels and habitats. Nat Commun 6:6936
Mace GM, Norris K, Fitter AH (2012) Biodiversity and ecosystem services: a multilayered relationship. Trends Ecol Evol 27:19–26
Meijaard E, Sheil D (2012) The dilemma of green business in tropical forests: how to protect what it cannot identify. Conserv Lett 5:342–348
Mitchell MGE, Bennett EM, Gonzalez A, Banks-Leite C (2014) Forest fragments modulate the provision of multiple ecosystem services. J Appl Ecol 51:909–918
Mori AS (2011) Ecosystem management based on natural disturbances: hierarchical context and non-equilibrium paradigm. J Appl Ecol 48:280–292
Mori AS, Kitagawa R (2014) Retention forestry as a major paradigm for safeguarding forest biodiversity in productive landscapes: a global meta analysis. Biol Cons 175:65–73
Mori AS, Furukawa T, Sasaki T (2013a) Response diversity determines the resilience of ecosystems to environmental change. Biol Rev 88:349–364
Mori AS, Spies TA, Sudmeier Rieux K, Andrade A (2013b) Reframing ecosystem management in the era of climate change: issues and knowledge from forests. Biol Cons 165:115–127
Mori AS, Ota AT, Fujii S, Seino T, Kabeya D, Okamoto T, Ito MT, Kaneko N, Hasegawa M (2015a) Biotic homogenization and differentiation of soil faunal communities in the production forest landscape: taxonomic and functional perspectives. Oecologia 177:533–544
Mori AS, Shiono T, Haraguchi TF, Ota AT, Koide D, Ohgue T et al (2015b) Functional redundancy of multiple forest taxa along an elevational gradient: predicting the consequences of non-random species loss. J Biogeogr 42:1383–1396
Mori AS, Isbell F, Fujii S, Makoto K, Matsuoka S, Osono T (2016) Low multifunctional redundancy of soil fungal diversity at multiple scales. Ecol Lett 19:249–259
Mori AS, Kenneth P, Lertzman, Gustafsson L (2017) Biodiversity and ecosystem services in forest ecosystems: a research agenda for applied forest ecology. J Appl Ecol 54:12–27
Morin X, Fahse L, Scherer-Lorenzen M, Bugmann H (2011) Tree species richness promotes productivity in temperate forests through strong complementarity between species. Ecol Lett 14:1211–1219
Morin X, Fahse L, de Mazancourt C, Scherer-Lorenzen M, Bug-mann H (2014) Temporal stability in forest productivity increases with tree diversity due to asynchrony in species dynamics. Ecol Lett 17:1526–1535
Mouillot D, Graham NA, Villeger S, Mason NW, Bellwood DR (2013) A functional approach reveals community responses to disturbances. Trends Ecol Evol 28:167–177
Nadrowski K, Wirth C, Scherer-Lorenzen M (2010) Is forest diversity driving ecosystem function and service? Curr Opin Environ Sustain 2:75–79
Nagendra H, Reyers B, Lavorel S (2013) Impacts of land change on biodiversity: making the link to ecosystem services. Curr Opin Environ Sustain 5:503–508
Newbold T, Scharlemann JPW, Butchart SHM, Sekercioglu CH, Joppa L, Alkemade R, Purves DW (2014) Functional traits, land-use change and the structure of present and future bird communities in tropical forests. Glob Ecol Biogeogr 23:1073–1084
Ostertag R, Warman L, Cordell S, Vitousek PM, Lewis O (2015) Using plant functional traits to restore Hawaiian rain forest. J Appl Ecol 52:805–809
Paoli GD, Harjanthi R (2011) Overcoming barriers to effective implementation of HCV in RSPO. Daemeter Consulting, Bogor, Indonesia
Piotto D, Craven D, Montagnini F, Alice F (2010) Silvicultural and economic aspects of pure and mixed native tree species plantations on degraded pasturelands in humid Costa Rica. New Forest 39:369–385
Poiani KA, Goldman RL, Hobson J, Hoekstra JM, Nelson KS (2011) Redesigning biodiversity conservation projects for climate change: examples from the field. Biol Cons 20:185–201

Poorter L, Van der Sande MT, Thompson J, Arets EJMM, Alarcon A, Alvarez-Sanchez J et al (2015) Diversity enhances carbon storage in tropical forests. Glob Ecol Biogeogr 24:1314–1328
Rothe A, Binkley D (2001) Nutritional interactions in mixed species forests: a synthesis. Can J For Res 31:1855–1870
Ruhl JB (2010) Climate change adaptation and the structural transformation of environmental law. Environ Law 40:363–431
Scherer-Lorenzen M (2014) The functional role of biodiversity in the context of global change. In: Burslem D, Coomes D, Simonson W (eds) Forests and global change. Cambridge University Press, Cambridge, UK, pp 195–238
Schnitzer SA, Klironomos JN, Hillerislambers J, Kinkel LL, Reich PB, Xiao K et al (2011) Soil microbes drive the classic plant diversity-productivity pattern. Ecology 92:296–303
Schuldt A, Bruelheide H, Durka W, Michalski SG, Purschke O, Assmann T (2014) Tree diversity promotes functional dissimilarity and maintains functional richness despite species loss in predator assemblages. Oecologia 174:533–543
Seidl R, Rammer W, Spies TA (2014) Disturbance legacies increase the resilience of forest ecosystem structure, composition, and functioning. Ecol Appl 24:2063–2077
Standish RJ, Hobbs RJ, Mayfield MM, Bestelmeyer BT, Suding KN, Battaglia LL et al (2014) Resilience in ecology: abstraction, distraction, or where the action is? Biol Cons 177:43–51
Thom D, Seidl R (2016) Natural disturbance impacts on ecosystem services and biodiversity in temperate and boreal forests. Biol Rev 91(3):760–781
Thompson ID, Okabe K, Tylianakis JM, Kumar P, Brockerhoff EG, Schellhorn NA, Parrotta JA, Nasi R (2011) Forest biodiversity and the delivery of ecosystem goods and services: translating science into policy. Bioscience 61:972–981
Tilman D, Reich PB, Knops JM (2006) Biodiversity and ecosystem stability in a decade long grassland experiment. Nature 441:629–632
Tilman D, Balzer C, Hill J, Befort BL (2011) Global food demand and the sustainable intensification of agriculture. Proc Natl Acad Sci U S A 108:20260–20264
Toigo M, Vallet P, Perot T, Bontemps JD, Piedallu C, Courbaud B, Canham C (2015) Overyielding in mixed forests decreases with site productivity. J Ecol 103:502–512
Verheyen K, Vanhellemont M, Auge H, Baeten L, Baraloto C, Barsoum N et al (2015) Contributions of a global network of tree diversity experiments to sustainable forest plantations. Ambio 45:29–41
Wilcove DS, Giam X, Edwards DP, Fisher B, Koh LP (2013) Navjot's nightmare revisited: logging, agriculture, and biodiversity in Southeast Asia. Trends Ecol Evol 28:531–540
Yachi S, Loreau M (1999) Biodiversity and ecosystem productivity in a fluctuating environment: the insurance hypothesis. Proc Natl Acad Sci U S A 96:1463–1468
Zavaleta ES, Hulbey KB (2004) Realistic species losses disproportionately reduce grassland resistance to biological invaders. Science 306:1175–1177
Zhang Y, Chen HYH, Reich PB (2012) Forest productivity increases with evenness, species richness and trait variation: a global meta-analysis. J Ecol 100:742–749
Zuppinger-Dingley D, Schmid B, Petermann JS, Yadav V, De Deyn GB, Flynn DF (2014) Selection for niche differentiation in plant communities increases biodiversity effects. Nature 515:108–111

Chapter 23
Transformation of Forested Landscape in Bengal Duars: A Geospatial Approach

Koyel Sam and Namita Chakma

Abstract The Bengal Duars, a landscape of foothill ecology in Eastern Himalaya asherb of rich biodiversity with unique physiography and climate. This landscape is now tremendously under threat disrupting by natural as well as anthropogenic activities. The recent phase of transformation of forest cover caused by illegal felling, encroachment, mining, quarrying activities, further enhancing flood and its associated vulnerability in such landscape. We assess the level of transformation of an area under deforestation, reforestation within forest boundary by using geospatial technology. Landsat imageries of two different periods has been used to find out zonal transformation of different land cover. The study also reveal that the rate of deforestation is more than rate of reforestation and major transformation has observed from dense forest to open forest within 20 years (1990–2010). The recent conversion and disturbances are highlighted through high resolution overview and field observation.

Keywords Bengal duars · Eastern Himalaya · Transformation · Deforestation

23.1 Introduction

Forest as the lungs of our mother earth, purifying air, water, soil and providing a life of billions of people. Transformation of the forested landscape and deterioration of the habitat causes loss of biodiversity and decrease primary productivity (Laurance et al. 1997; Debinski and Holt 2000; Li et al. 2009). Moreover, it also has a significant impact on the local and global environment and climate change (Xiao et al. 2004). Measuring, monitoring and mapping of spatio-temporal dynamics of forest cover using geospatial technology plays a vibrant role for the management and restoration of forest ecosystem (Rikimaru 1999). As traditional monitoring methods are time consuming, labour intensive, and uneconomical. Therefore, satellite data are often considered by several conservation agencies, governmental and non-government

K. Sam (✉) · N. Chakma
Department of Geography, The University of Burdwan, Bardhaman, West Bengal 713104, India
e-mail: koyelsam04@gmail.com

P. K. Shit et al. (eds.), *Spatial Modeling in Forest Resources Management*, Environmental Science and Engineering,
https://doi.org/10.1007/978-3-030-56542-8_23

organizations for monitoring, mapping of vegetation as it rapid and cost effective by using free source imagery (Kumar et al. 2010; Pringle 2010; Madin et al. 2011).

Remote sensing technique has widely used by several researchers for monitoring deforestation and periodical change of land use and land cover. Specially to study the up to date dynamics of reserve and protected areas, satellite data are the important source to solve complex and current environmental issues (Andrieu and Mering 2008; Baldyga et al. 2007). Park and Lee (2016) identified the deforested area by comparing high and medium satellite imageries. The periodical change of forest cover from 1988 to 2010 in Khadimnagar National Park, Bangladesh has studied by Redowan et al. (2014). Bonilla-Bedoya et al. (2014) systematically quantified transition of forested landscape in Ecuadorian Amazon Region (EAR). Since 1987, Forest Survey of India (FSI) has been periodically observing the changes of forest cover with the help of remote sensing techniques. It was reported that forest cover of India from 1987 to 2003 has nearly stabilized and increased marginally over the years (FSI 1987, 1989, 1991, 1993, 1995, 1997, 1999, 2001, 2003). FAO estimates a yearly loss of 3.37 million hectares in India and claims that today most of the Indian forests are 'degraded forest' (Agarwal 1996). According to state forest report of India (2019), forest and tree cover of India has increased around 5188 km^2 across the country. However north-east region witnesses continuous decrease of forest cover since 2009. Saikia (2014) has studied a brief analysis of "Over-Exploitation of Forest: A case study from North-East India". This study not only highlighting the status of forest but also give emphasis on underlying factors and processes behind fragmentation under the pressure of Anthropocene.

Other than North-East India, forested landscape in Himalayan region and its surrounding foothills has experiencing disruption and different level of transformation explored by several researchers as Rawat and Kumar (2015), stated anthropogenic causes led to loss forest density in several districts of Uttarakhand. Dynamics of forest landscape of two protected areas in Himalaya Foothills was analyzed by Joshi et al. in (2014). They have explored two important protected areas in Himalayan foothills as Rajaji and Jim Corbett National Park. The used geospatial modeling tool to observe the land use dynamics from 1990 to 2005. They estimated that dense forests have high level of probability to convert into open forest. Quantitative analysis of forest cover changes in the Himalayan foothill of Dehradun forest division was attempted by Munsi et al. (2012). On the basis the previous data sets and anthropogenic disturbance Land Change Modeller (LCM) was used to predict status of forest cover in 2010 and 2015. They also demonstrated the potentiality of geospatial tool to understand the present and future scenario. Prokop and Sarkar (2012) were analysed land use transformation in the piedmont zone of Sikkim-Bhutan Himalaya over last 150 years (1930–2010). Stable and dynamic areas has identified through visual interpretation of toposheets and satellite images. The study reveal that the shifting from natural to human landscape causes enlargement of areas with monoculture cultivation of tea and paddy in between 1930–2010. Furthermore, human induced deforestation intensified the fluvial activities in piedmont. Deb et al. (2018) was addressed anthropogenic impacts on forest cover and future probability of changes in Himalayan Terai (foothills) of West Bengal. They had chosen Jaldapara

National Park and its surrounding area as an interest of study. They evaluated agricultural proliferation, human disturbances are the major cause behind the transition of LULC. The foothill landscape with rich forest resources is under threat due to population pressure, unplanned development, Illegal activities, climatic variability, socio-economic stresses and so on. Exploitation of natural resources increases level of physio-social vulnerability of the landscape. Ignoring the ecological importance, profit from commercial logging of forestry is a major issue in Duars region. This article assesses the spatio-temporal scenario of disruption took place in the forested landscape of Bengal Duars, by using satellite based remote sensing techniques and field observation. The goal of the study is to understand the changing trajectories in forest pattern and contribute relevant information to decision maker for biodiversity conservation.

23.2 Study Area: The Bengal Duars

The study area, Bengal Duars extends from 26°30′N to 27°11′20″N and 88°25′18″E to 89°52′40″E. At present, it covers a large portion of Jalpaiguri, Alipurduar, Kalimpong district and small part of Koch Bihar district (Fig. 23.1). The study area located in the foothill of Himalaya, the narrow strip of land with a width of 32.2–48.3 km and about 289.7 km in length aggregation of forest cover stretching between the river Sankosh in the east and the river Tista in the west, and Cooch Behar on the south forms the Bengal Duars or Western Duars and other parts in Assam is popularized as Assam Duars or Eastern Duars (Gruning 1911). The river Sankosh acts like a physical divider of Bengal and Assam Duars region. Bengal Duars region,

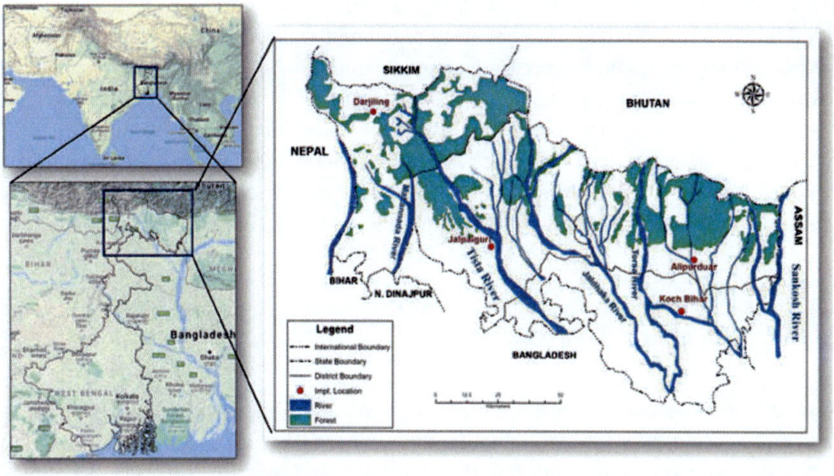

Fig. 23.1 Spatial distribution of Bengal Duars

Table 23.1 Descriptions of Landsat images used in the study

Year	Sensor	Resolution (m)	Path/Row	Date of acquisition
1990	TM5	30	138/042	14-11-1990
			139/041	23-12-1990
2010	TM5	30	138/042	05-11-2010
			139/041	30-12-2010

generally popularized for its rich forest resource and biodiversity. It covers around 36% of the total area of Bengal Duars. Physiographical, it is located in the foothill of Bhutan Himalaya which is fragile and erosion prone in nature. So, physically and socially forest plays an important role not only to maintain the sustainability of the ecosystem and climatic stability but also for the development of the local people per se.

23.3 Materials and Methods

23.3.1 Data Collection and Processing

Satellite images (Landsat TM) with 30 m resolution were downloaded from the USGS (https://earthexplorer.usgs.gov/) for the year of 1990 and 2010. The details of datasets in Table 23.1 were used in this study to detect spatiotemporal changes of forest cover from 1990 to 2010 in Bengal Duars. As spectral reflectance of vegetation varies from season to season, satellite data with similar sensors and same season are required in order to reduced spectral deviation. In the present study, the area covered by the cloud is treated as 'no-data'. Images were registered and geometrically rectified with the help of topographical sheet by Survey of India (SOI). The boundary of forest cover is obtained from toposheet and also verified with state forest map. The first order polynomial transformation has used in georectification of satellite images. After subset the boundary of forest cover on the images, a hybrid approach is applied in which NDVI transformation was performed in order to discriminate forest cover and non-forest cover area, using Erdas Imagine software (Fig. 23.2).

23.3.2 Accuracy Assessment

Accuracy is a measure of harmony between a standard one which is assumed to be a correct and classified image of unknown quality (Bhatta 2011). It is a feedback system for checking and evaluating the result. Ground information (GPS points) as well as high-resolution map were used to compare with classified maps. To evaluate

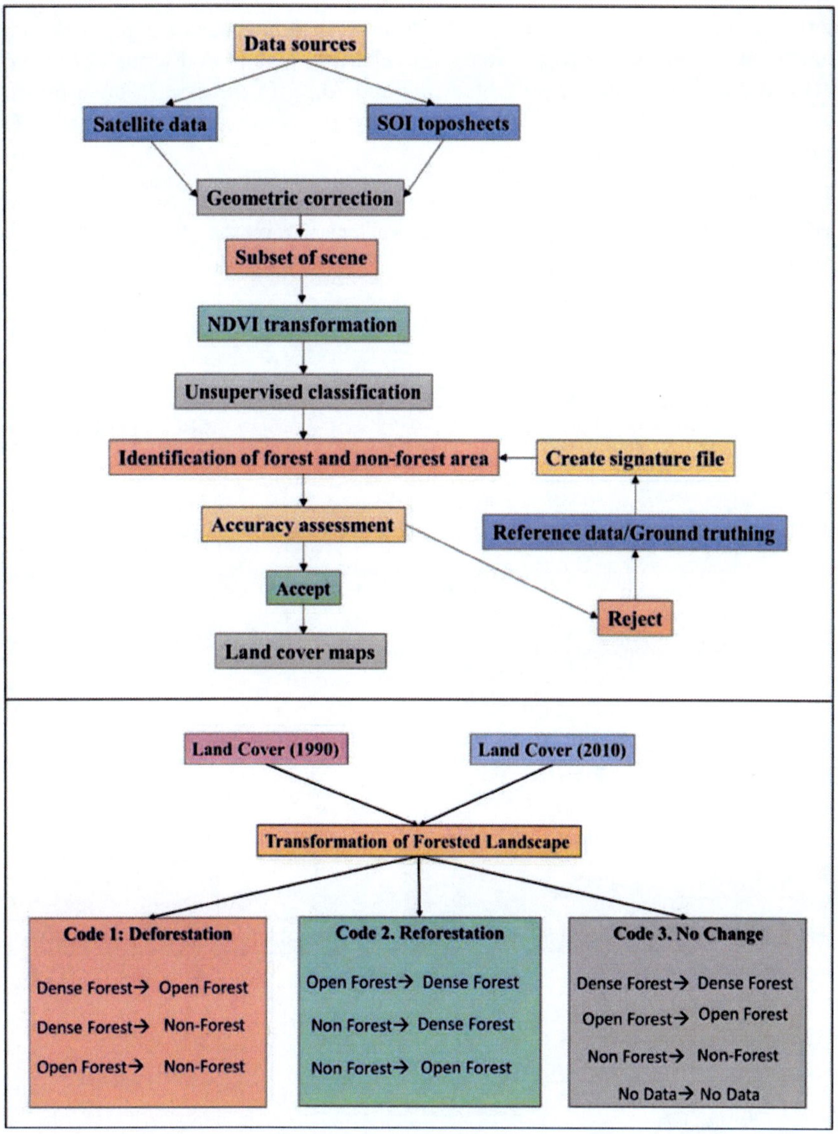

Fig. 23.2 Methodology followed in this study

the classification errors, confusion (contingency) matrix was generated. Individual land cover map was prepared for the year of 1990 and 2010 then change matrix was executed to work out the area under reforestation, deforestation and no change. The accuracy of the classified image is checked using equalized random sampling. Overall user's and producer accuracies are derived from the confusion matrix. The user's and producer accuracies are ranges from 62 to almost 92 percentages. User's accuracy of dense forest and non-forest are relatively high in both years (1990 and 2010) ranges from 84 to 92% (Tables 23.2 and 23.3). In the case of open-forest, user's accuracy is relatively less because open-forest area sometime intermixed with dense forest and non-forest cover.

Table 23.2 Contingency Matrix: Accuracy assessment for land cover classification (1990)

Year 1990		Reference image					
		Dense forest	Open forest	Non-forest	No data	Row total	User's accuracy
Classified image	Dense forest	146	24	3	0	173	84.39306358
	Open forest	13	44	7	0	64	83.01886792
	Non-forest	0	2	26	0	28	92.85714286
	No data	5	0	0	0	5	0
	Column total	164	70	36	0	256	
	Producer's accuracy	89.02	62.85	72.22	0		
	Kappa coefficient 0.81			Overall accuracy 84.37			

Table 23.3 Contingency Matrix: Accuracy assessment for land cover classification (2010)

Year 2010		Reference image					
		Dense forest	Open forest	Non-forest	No data	Row total	User's accuracy
Classified image	Dense forest	126	17	3	0	146	85.13513514
	Open forest	22	54	5	0	81	68.35443038
	Non-forest	2	0	25	0	27	92.59259259
	No data	2	0	0	0	2	0
	Column total	152	71	33	0	256	
	Producer's accuracy	82.89	76.05	75.75	0		
	Kappa coefficient 0.74			Overall accuracy 80			

23.4 Results and Discussion

Eastern Himalaya itself popularized as a biodiversity hotspot (Brooks et al. 2006). The foothill landscape of Himalaya had gone through abounded changes from pre-British to post British period and still continuing. Satellite-based observation and monitoring plays a pivotal role to provide real-time information about forest cover.

23.4.1 Spectral Characteristics of Forest Cover

The reflectance of vegetation covers in the near infrared region (0.74–1.3 μm) is high and strong absorption is found in 0.35–0.5 μm and 0.6–0.7 μm. Hence, the ratio in near-infrared and red is a good indicator of health of the vegetation. Systematic and extensive signature measurement has carried out in two different years in order to investigate the nature of change in forest cover. However, a place where forest cover is degraded, the range between NIR and RED bands also decreases. e.g. in case of Dumchi forest from 1990 to 2010, variation in spectral signature of reflects the same (Fig. 23.3). Hence remote sensing is a powerful tool for better analyzing the scope and scale of deforestation. Course to high-resolution data can provide a detail view of clear cut depletion and degradation of forest.

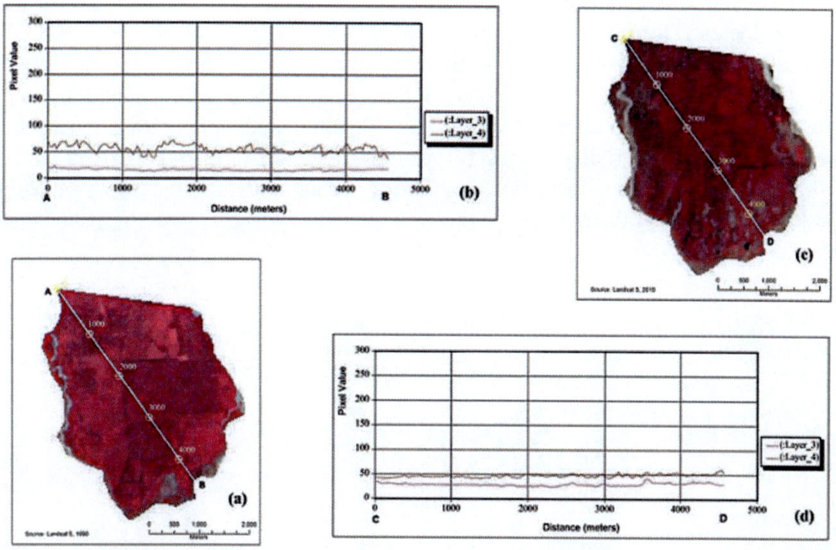

Fig. 23.3 Changing spectral characteristics in Dumchi Forest (1990–2010) **a** landsat TM satellite image 1990. **b** Spectral cross section along AB. **c** Landsat TM satellite image 2010. **d** Spectral cross-section along CD

23.4.2 Change Matrix

The NDVI is a widely used transformation for the enhancement of vegetation information (Nogi et al. 1993; Huete et al. 1997; Runnstrom 2000). The temporal dynamics of NDVI is very useful to detect changes in land cover characteristics, but season plays a vital role for spectral details. Figure 23.4a, b shows spatiotemporal variation of NDVI in 1990 and 2010. In 1990 NDVI value ranges from 0.96 to (−0.89) but in 2010 NDVI value decreases from 0.70 to (−0.80). Ground truth and high-resolution image further help in the identification of dense forest, open forest and non-forest areas. Whereas scarp lands, grasslands are included in open forest class and water bodies, bare lands, croplands in forest villages are considered as a non-forest area. After identification of individual land cover classes, the land cover map of 1990 and 2010 has prepared. Figure 23.5a, b shows the classification results of multi-temporal images for 1990 and 2010.

The change matrix represents the area in transition between two periods. In a simple way, diagonal box in the matrix represents that area remained the same class in both periods of time. On the other hand, non-diagonal values represent changes from one class to another class. Pixels that undergoes a change from non-forest to forest cover and open forest to dense forest are treated as 'reforestation'. Pixels which are changed from dense forest to open forest, dense forest to non-forest and open forest to non-forest are combined as 'deforestation'. Pixels that remained open forest and dense forest in both years are considered 'no change'. Following the above principles, a land cover change trajectories map 20 years has been prepared (Fig. 23.5c) in order to identify the spatial vulnerability of this landscape. However, the rate of deforestation is more than afforestation (Table 23.4) that's serious concerning issue in Bengal Duars landscape. Within 20 years' deforestation took place around 26% and afforestation has occurred around 15% (Table 23.5). Thus, the major transformation has taken place between dense forest to open forest cover from 1990 to 2010 (Table 23.4). The maximum transformation of deforestation has noticed in those forest cover located in Alipurduar district.

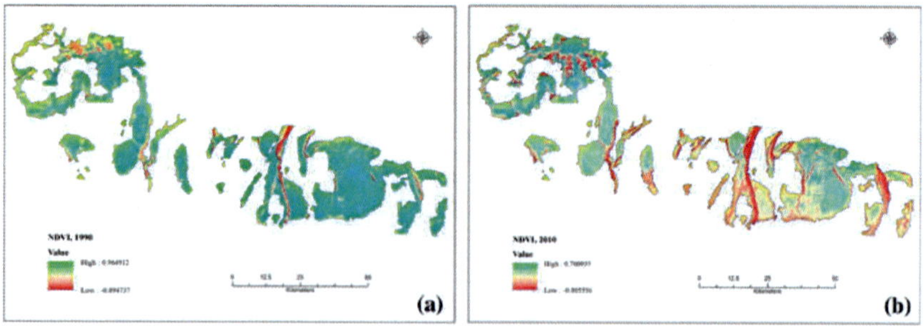

Fig. 23.4 NDVI detection of 1990 (**a**) and 2010 (**b**)

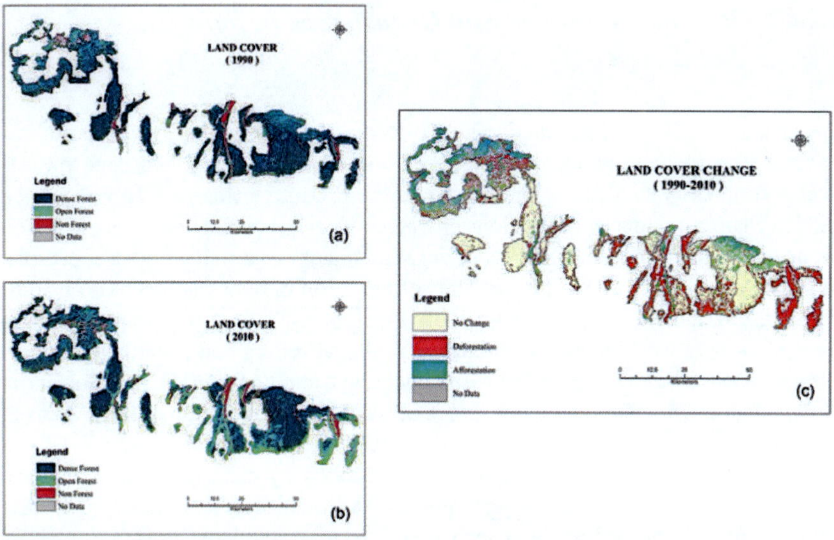

Fig. 23.5 Land cover map in 1990 (**a**), 2010 (**b**) and Change of Land Cover in between 1990–2010 (**c**)

Table 23.4 Change matrix of land Cover (1990–2010)

2010	Area (ha)				Total
1990	Dense forest	Open forest	Non-forest	No data	
Dense forest	94,368	47,754	3724	848	146,694
Open forest	20,213	16,111	3095	434	39,853
Non-forest	1930	8044	5477	21	15,472
No-data	1908	1208	618	44	3778
Total	118,419	73,117	12,914	1347	205,797

Table 23.5 Nature and rate of change of land cover (1990–2010)

Landcover status	Area (ha)	%	Rate of change
Deforestation	54,573	26.51	1.325893963
Reforestation	30,187	14.66	0.733416911
No change	116,000	56.36	2.818311248
No data	5037	2.44	

23.4.3 *Recent Conversion and Disturbance in Forested Landscape*

The year after 2010, forested landscape is disturbed by both natural as well as anthropogenic activities continuing to degrade the whole biodiversity and landscape towards losing its identity.As this region located in the foothills of eastern Himalaya, the sudden change of slope and notorious character of river in pediment make the region prone to flood. Mostly, each and every year due to heavy rainfall within a short span of time causes bank failure of river. The deforestation in river catchment is an additional disturbance to increase the level of vulnerability. This region always came in front of us because of human encroachment, illegal felling and poaching activities (Prokop and Sarkar 2012; Das 2012; Bhattacharyya and Padhy 2013; The Telegraph 2006, 2008, 2018). The clear cut fragmentation and transformation of some areas in Buxa Tiger Reserve has still recognised in 2019 (Fig. 23.7).

Recent disturbance in forest cover is accelerated by clear-cut felling of trees, poaching of animals, encroachment, mining and quarrying activities that further enhancing flood vulnerability and decreases vegetation stabilized gravel bars. Riverside bolder mining and deforestation caused shifting of river course from straight to meandering and braided (Prokop and Sarkar 2012) (Figs. 23.6 and 23.8a). Frequently

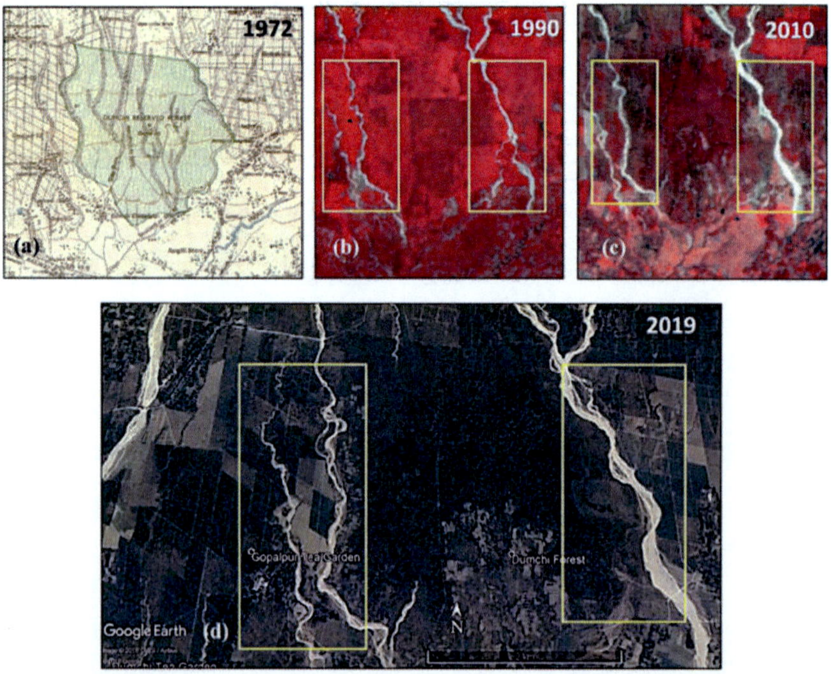

Fig. 23.6 Shifting course of river channel directed to cause by deforestation over time

reported illegal timber extraction, poaching of animal in local media and press reflects relatively poor enforcement and protection level that rises threat to the biodiversity hotspot. In north Bengal JFM is still evolving because the presence of tea garden in closeness to forest, presence of large protected area accompanist with forest villagers make things more complicated (Gupta 2005). Forest villagers are now also preferring to plant cash crops (betel nuts) around their surroundings rather than planning native trees (Fig. 23.8b).

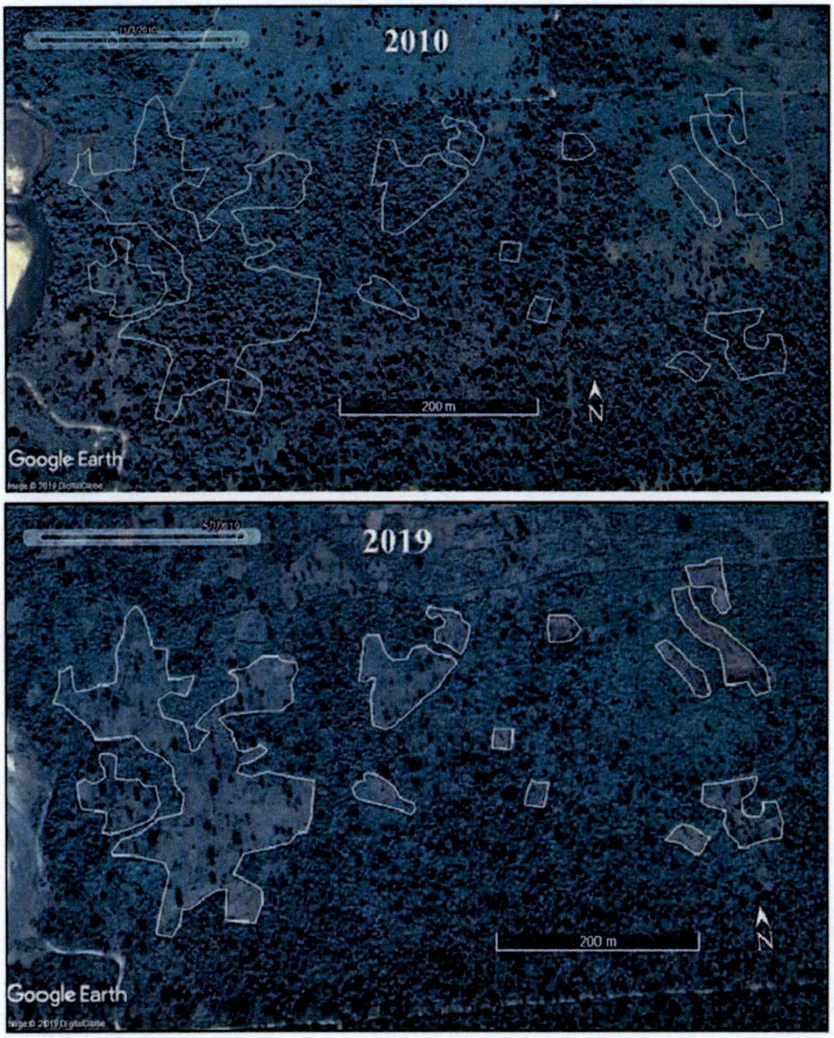

Fig. 23.7 New patches of forest fragmentation open up in 2019

Fig. 23.8 **a** River bed quarrying of boulder; **b** Betel Nut plantation at a forest village in Bengal Duars

23.5 Conclusion

The Eastern Himalaya is acknowledged as a 'biodiversity hotspot' and well identified as 'eco-crisis' zone. The major transformation has taken place during the British period, through the establishment of tea garden and expansion of settlement. The Recent phase of transformation carried out beyond shifting cultivation through illegal felling, encroachment, mining and quarrying activities that further enhancing flood vulnerability and decrease vegetation stabilized gravel bars. As well as riverside bolder mining, deforestation caused shifting of river course from straight to meandering and Braided. For conservation plans and policies, understanding of spatio-temporal disturbance is very crucial. The present study demonstrates the utilization of remote sensing data for the identification and monitoring of deforestation with proper methodology. This research can also help forest department to identify the vulnerable zones specially within inaccessible areas to restore the forest ecosystem. Further, quantification of carbon stock and estimation land surface temperature (LST) in Bengal Duars region can extend the scope of future research.

References

Agarwal A (1996) Pay-offs to progress. Down to Earth

Andrieu J, Mering C (2008) Vegetation land cover change mapping of West African coastal zone from remotely sensed images: example of "Rivières-du-Sud" from Salum Delta (Senegal) to Rio Geba (Guinea-Bissau). Revue Télédétection 8(2):93–118

Baldyga TJ, Miller SN, Driese KL, Mainagichaba C (2007) Land cover change assessment within Kenya's Mau forest region using remotely sensed data. Afr J Ecol 46:46–54

Bhatta B (2011) Remote sensing and GIS. Oxford University Press, New Delhi

Bhattacharyya MK, Padhy PK (2013) Forest and wildlife scenarios of Northern West Bengal, India: a review. Int Res J Biol Sci 2(7):70–79

Bonilla-Bedoya S, Molina JR, Macedo-Pezzopane JE, Herrera-Machuca MA (2014) Fragmentation patterns and systematic transitions of the forested landscape in the upper Amazon region, Ecuador 1990-2008. J For Res 25(2):301–309.

Brooks TM, Mittermeier RA, da Fonseca GA, Gerlach J, Hoffmann M, Lamoreux JF, Mittermeier CG, Pilgrim JD, Rodrigues AS (2006) Global biodiversity conservation priorities. Science 313:58–61

Das BK (2012) Losing biodiversity, Impoverishing Forest Villagers: analysing forest policies in the context of flood disaster in a national park of Sub Himalaya Bengal, India. Occasional Paper 35, Institute of Development Studies, Kolkata

Deb S, Debnath MK, Chakraborty S, Weindorf DC, Kumar D, Deb D, Choudhurya A (2018) Anthropogenic impacts on forest land use and land cover change: modelling future possibilities in the Himalayan Terai. Anthropocene 21:32–41

Debinski DM, Holt RD (2000) A survey and overview of habitat fragmentation experiment's. Conserv Biol 14:342–355

Forest Survey of India (1987) State of forest report. Ministry of Environment and Forests, Dehradun

Forest Survey of India (1989) State of forest report. Ministry of Environment and Forests, Dehradun

Forest Survey of India (1991) State of forest report. Ministry of Environment and Forests, Dehradun

Forest Survey of India (1993) State of forest report. Ministry of Environment and Forests, Dehradun

Forest Survey of India (1995) State of forest report. Ministry of Environment and Forests, Dehradun

Forest Survey of India (1997) State of forest report. Ministry of Environment and Forests, Dehradun

Forest Survey of India (1999) State of forest report. Ministry of Environment and Forests, Dehradun

Forest Survey of India (2001) State of forest report. Ministry of Environment and Forests, Dehradun

Forest Survey of India (2003) State of forest report. Ministry of Environment and Forests, Dehradun

Forest Survey of India (2019) India state of forest report, vol I & II. Ministry of Environment Forest and Climate Change, Dehradun

Gruning JF (1911) Jalpaiguri, Eastern Bengal and Assam District Gazetteers. WBDG, Govt. of West Bengal

Gupta K (2005) The political economy of forest management; importance of institution and social capital. Allied Publishers Pvt, New Delhi

Huete AR, Liu HQ, Batchily K, Leeuwen WV (1997) A comparison of vegetation indices over a global set of TM images for EOS-MODIS. Remote Sens Environ 59:440-451. https://doi.org/10.1016/S0034-4257(96)00112-5

Joshi AK, Joshi PK, Chauhan T, Bairwa B (2014) Integrated approach for understanding spatio-temporal changes inforest resource distribution in the central Himalaya. J For Res 25(2):281–290. https://doi.org/10.1007/s11676-014-0459-9

Kumar P, Meenu R, Pandey PC, Majumdar A, Nathawat MS (2010) Monitoring of deforestation and forest degradation using remote sensing and GIS: a case study of Ranchi in Jharkhand (India). Rep Opin 2(4):55–67

Laurance WF, Laurance SG, Ferreira L, Rankin-de Merona J, Gascon C, Lovejoy TE (1997) Biotas collapse in Amazonians forest fragment's. Science 278:1117–1118

Li HM, Ma YX, Liu WJ, Liu WJ (2009) Clearance and fragmentation of tropical rain forest in Xishuangbanna, SW, China. Biodivers Conserv 18:3421–3440

Madin EM, Madin JS, Booth DJ (2011) Landscape of fear visible from space. Nat Sci Rep 1(14). https://doi.org/10.1038/srep00014

Munsi M, Areendran G, Joshi PK (2012) Modeling spatio-temporal change patterns of forest cover: a case study from the Himalayan foothills (India). Reg Environ Change 12:619–632

Nogi A, Sun W, Takagi M (1993) An alternative correction of atmospheric effects for NDVI estimation. Int Geosci Remote Sens 3:1137–1139

Park J, Lee J (2016) Detection of landuse/landcover changes using remotely-sensed data. J For Res 27(6):1343–1350

Pringle H (2010) Google Earth shows clandestine worlds. Science 329:1008–1009

Prokop P, Sarkar S (2012) Natural and human impact on land use change of the Sikkimese-Bhutanese Himalayan piedmont, India. Quaestiones Geographicae Bogucki WydawnictwoNaukowe, Poznań 31(3):63–75. https://doi.org/10.2478/v10117-012-0010-z,ISSN0137-477X
Rawat JS, Kumar M (2015) Monitoring land use/cover change using remote sensing and GIS techniques: a case study of Hawalbaghblock, district Almora, Uttarakhand, India. Egyp J Remote Sens Space Sci 18:77–84
Redowan M, Akter S, Islam N (2014) Analysis of forest cover change at Khadimnagar National Park, Sylhet, Bangladesh, using Landsat TM and GIS data. J For Res 25(2):393–400
Rikimaru A (1999) Concept of FCD mapping model and semi expert system. FCD mapper's user's guide. International Tropical Timber Organization and Japan Overseas forestry Consultants Association, p 90
Runnstrom M (2000) Is northern China winning the battle against desertification? Satellite remote sensing as a tool to study biomass trends in the ordos plateau in semiarid China. Ambio 29:468–476. https://doi.org/10.1579/0044-7447-29.8.468
Saikia A (2014) Over-exploitation of forest: a case study from North-East India. Springer. ISBN: 978-3-319-01407-4
The Telegraph (2008) Timber smuggling on train thrives—Forest minister alleges GRP mafia nexus. West Bengal, August 6
The Telegraph (2006) Two courts to save forest. West Bengal, February 15
The Telegraph (2018) More cameras for forest. West Bengal, February 16
Xiao XM, Zhang Q, Braswell B, Urbanskib S, Boles S, Wofsy S, Berrien Moore III B, Ojimac D (2004) Modeling gross primary production of temperate deciduous broadleaf forest using satellite images and climate data. Remote Sens Environ 91:256–270

Chapter 24
Forest-Based Climate Change Social Interventions: Towards a Theoretical Framework

Naser Valizadeh, Sahra Mohammadi-Mehr, and Dariush Hayati

Abstract The main purpose of this chapter was to develop a framework for forest-based climate change social interventions which was fulfilled using a multi-stage process. There is no doubt that forests are important to humans, plants, animals, and the planet as a whole. In other words, the ongoing deforestation process and land degradation caused by human activities and climate changes are considered as major challenges for sustainable development around the world. Despite the improvements achieved, there are still many problems with the sustainable protection, conservation, and management of forests in different areas. This necessity has been acknowledged by the need for government interventions at all levels. In the first step of this study, the importance of forest was highlighted in terms of Sustainable Development Goals (SDGs). In the second step, the planned/preventive forest-based climate change social intervention introduced as an effective way to reduce deforestation under climate change. In the third step, some enabling and constraining factors were proposed for successful implementation of forest-based climate change social interventions. In the fourth step, different types of the uses of forest-based climate change social interventions were critically analyzed. In the fifth stage, a typology were introduced for forest-based climate change social interventions. Finally, in light of the results of previous steps, a practical framework for forest-based social interventions under climate change was developed. In general the results of this chapter showed that in all types of forest-based climate change social interventions, the most important constraining factors include structural, political, organizational, economic, executive, collaborative, network building, and follow-up barriers. In addition, enabling

N. Valizadeh (✉) · S. Mohammadi-Mehr · D. Hayati
Department of Agricultural Extension and Education, School of Agriculture, Shiraz University, Shiraz, Iran
e-mail: n.valizadeh@shirazu.ac.ir

S. Mohammadi-Mehr
e-mail: smohammadimehr@gmail.com

D. Hayati
e-mail: hayati@shirazu.ac.ir

P. K. Shit et al. (eds.), *Spatial Modeling in Forest Resources Management*, Environmental Science and Engineering,
https://doi.org/10.1007/978-3-030-56542-8_24

factors of these interventions consist flexible designing, institutional analysis, long-term intervention, risk assessment, prioritizing local knowledge, site-specific intervention, socio-cultural forestry, non-profit incentives, social learning, and participation. The framework presented in this study can provide useful insights for forest ecosystem managers, policy-makers, decision-makers, and practitioners who are directly involved in the process of designing and implementing social interventions.

Keywords Climate change · Social interventions · Forest management · Collaboration of local communities

24.1 Introduction

Forests are an important component of the global carbon cycle (Dixon et al. 1994; Mader 2019; Ray et al. 2019). Around the world, approximately 60% of ecosystem services in the fields of freshwater, fishing, air and water purification, and regional-local climate regulations are being degraded. Since trees are the oldest and largest organisms on the earth, they are therefore an important criterion for environmental health and quality (USDA 2008, 2013; Walsh et al. 2019). Forests are used as key elements of landscapes to provide many environmental and ecosystem services such as flood regulation, urban climate change, reducing air pollution, and preserving biodiversity (Calder et al. 2008; Torkar et al. 2014; Humphries et al. 2020). Forests have a significant role to play in mitigating the negative impacts of climate change, which means adapting to climate change (Maraseni et al. 2019). They affect local and regional climate by modifying temperature and humidity and affect photosynthesis and growth of trees through carbon capture. Forests comprise a significant portion of the world's biodiversity and help societies resist climate changes and pursue different livelihood strategies. Understanding the importance of forests under climate change circumstances leads to formulation of forestry and land use policies to counter the negative effects of climate change. Therefore, the formulated policies should take into account many of the effects of forests, in particular their relative impacts on climate change and their long-term sustainability. In this regard, it is important to assess and coordinate the long-term effects of climate change on forests and to analyze and determine future policies and measures to respond to these threats (Paudel et al. 2019). For example, when using wood in furniture or packaging materials, the carbon is stored in wood products. In addition, using wood in two ways can replace fossil fuels such as gas, oil, and coal. On the one hand, wood can be burnt directly as a fuel (alternative energy) and on the other hand, production and access to wood products usually require less energy than competing products made of materials such as plastic, metal, or concrete. While the potential of forests to reduce net greenhouse gas emissions is widely recognized, quantifying it on a national scale is a very challenging task. However, some studies have recently been conducted in this context. For example, Bösch et al. (2019) in their study assessed the effects of CO_2 and the cost of forest management alternative measures in Germany.

Forest ecosystem services contribute to human well-being in various ways. On the global scale, all people benefit from reducing negative effects of climate change on forest trees and crops. Most of the world's population use forest-based products such as wood furniture or timber for housing. In addition, it is estimated that 3 million rural residents are heavily dependent on forests for food security, livelihood, and energy. At the same time, 5 million indigenous people depend entirely on forests to sustain their lives (World Bank 2008).

24.2 Forests and Sustainable Development Goals

Forestry is very important from the sustainable development perspective. The original definition of sustainable development, published in the eighteenth century, was first used for afforestation. Today, this sector can play an important role in transition to a sustainable society. International debates on sustainable development have increased significantly with the introduction of sustainable development goals. On the other hand, exacerbating the problems of climate change required the actions of political activists at all levels (Baumgartner 2019). Given that sustainability issues are not new to the forest sector, they simply cannot be reduced to the initial definition of sustainability. Three parts of the forest sector including wood production, forest products, and tourism, are considered in its sustainability issues (Baumgartner 2019). These sections are examined in forest sustainability studies in the form of economic, environmental, and social effects. Economic sustainability refers to the competitiveness of companies or the economic viability of non-profit organizations. This means that topics such as innovation and technology management, collaboration, knowledge management, and organizational processes are used to measure the effects of economic sustainability. Environmental sustainability refers to "the use of renewable and non-renewable resources", "greenhouse gas emissions in air, water, and soil", "hazardous waste", "usage of ecosystems", and "impacts on biodiversity". In terms of social responsibility, ISO 26000 defines seven key issues that must be considered by any interventionist organization to improve sustainable forest performance. These include organizational governance, human rights, fair working practices, consumer issues, and community participation. Furthermore, it should be noted that the role of indigenous people in forestry is also closely related to the social dimension of forest sustainability (Hengeveld et al. 2015; Baumgartner 2019).

In general, human well-being ultimately is based on global natural resources and biodiversity. Sustainable use of these resources is the foundation of sustainable development (Holden et al. 2017). The world's forest areas are made up of 3695 million hectares of natural forest and 291 million hectares of artificial forest. However, the area of natural forests is declining and the area under cultivation of artificial forests is increasing. From 2010 to 2015, the world's natural forest areas declined by 6.5 million hectares per year. While artificial forest areas increased by 3.3 million hectares per year (FAO 2016).

Much of the discussion at international level is focused on the Sustainable Development Goals (SDGs). Yale's of the International Society of Tropical Foresters (Yale's ISTF) in 2015 decided to put its annual conference on the specific role of forests in achieving SDGs, due to the great importance of tropical forests for human societies. The conference was called "Tropical Forests for Sustainable Development: Shaping our Post-2015 Future with Knowledge from the Field" to provide an opportunity to discuss the role of forests with a wide range of disciplines, methods, and perspectives. The tropics provided a sustainable development plan between researchers and policymakers (Baumgartner 2019). Forests play a significant role in a wide range of issues consisting poverty reduction, food security, human welfare, water conservation, gender empowerment, access to energy, sustainable economic growth, production and consumption, coping with climate change, and promoting development and sustainable resource. The impact of forests on reducing the poverty of human societies can be impressive. Therefore, the effects of deforestation on forest-based livelihoods can be very complex and negative (Bukoski et al. 2018).

Forests have different functions that are aligned with the United Nations SDGs. Not only do they provide livelihoods of forest residents (Silva 1994), they can provide plant and animal products including food and medicine, and directly play a major role in eradicating hunger (SDGs 2) and ensuring health and welfare (SDGs 3). Forests increase employment (SDGs 8) and forest revenues can be used to buy food and thus reduce food insecurity (SDGs 1). Nhem et al. (2018) examined the relationship between declining forest resources and income inequality and poverty. They argue that forest revenue sources should be integrated with ecosystem service payments to diversify households' income and protect them from shocks and other stressors. Forests also affect "hydrological cycles and water supply in low-lying areas" and contribute to health (SDGs 6). Forest biomass can play a major role in reducing global dependence on fossil fuels for energy supply (SDGs 7). Forests can replace non-renewable materials by providing renewable materials and contribute to consumption and production (SDGs 12). Forest areas support industrial development and innovation (SDGs 9). Forests are essential for carbon storage and climate regulation (SDGs 13). They provide supportive services such as food cycles and pollination of crops, which play a key role in sustainable agricultural production. In addition, forests help protect coastlines and stability of coastal areas against weather hazards. Forest-based cultural ecosystem services have recreational, spiritual, and religious benefits. These benefits are important for rural and urban communities and contribute to physical-mental well-being (SDGs 3 and SDGs 4), flexibility, and sustainability of communities (SDGs 11). In addition, forests have the highest biodiversity in the world (SDGs 15) (Bukoski et al. 2018; Katila et al. 2019).

24.3 Forest-Based Climate Change Social Intervention as a Way to Prevent Deforestation Under Climate Change

Many researchers (Jindal et al. 2012) believe that climate changes can have far-reaching negative impacts on forests. In addition, climate changes could affect livelihoods of local people and rural development in forest communities (Ofoegbu et al. 2017). Climate change is directly affecting forest ecosystem services and exacerbating the impacts of natural anthropogenic stressor factors. Wildfires, native and non-native invasive species, and extreme climatic events are among the most important stressors that have negative impacts on forests. In general, the most significant impacts of climate change on forests can be summarized in the following Table 24.1 (Bernier and Schoene 2009).

Although forests are vulnerable to the impacts of climate changes, they do play an important role in adapting to climate change. For example, they help provide

Table 24.1 Impacts of climate change on the forests and forest management goals

1	Longer, warmer growing seasons
2	Altered fire regimes
3	Shifts in seasonality of hydrological processes
4	Intense droughts
5	Climate change interacting with current stress factors such as insect pests and disease, wildfire, legacy of past management and air pollution
6	Shifts in forest species composition
7	Increased erosion events impairing watershed condition
8	Large-scale forest dieback or vegetation type conversions as a result of more frequent extreme events
9	Altered landscape and successional dynamics
10	Increasing fragmentation of forest ecosystems and wildlife habitat
11	Increased air and stream temperatures
12	Reduced snowpack
13	Altered in-stream flows
14	Poor accessibility or lack of current information on climate change projections, ecosystem impacts and socio-economic impacts on local communities
15	Uncertainty associated with that information
16	Exacerbation of the stress that urban environments place on ecosystems, as a result of warming temperatures
17	Increased wildfire and drought risks in surrounding landscapes, which may compromise ability to maintain water quality and availability
18	Need for management tools that incorporate climate change considerations
19	Need to revise current management practices that are based on assumptions about ecosystems and climate that may be invalid in the future

livelihoods for forest dwellers and increase their livelihood resilience to climate change (Dlamini 2014). Many rural communities are dependent on forests. Therefore, sustainable use and management of forests are essential to reduce the impacts of climate change and increase their resilience (FAO 2015). Forest-based climate change social interventions are one of the most effective and efficient ways to reduce the impacts of climate change and adaptation of forest dwellers to climate change (Rennaudet al. 2013; Chia et al. 2013; FAO 2015). These interventions are aimed at achieving goals such as forest ecosystem sustainability, social equity, and livelihood sustainability (Hajost and Zerbock 2013; Ofoegbu et al. 2017). Forest-based climate change social interventions can be addressed on a variety of topics, such as "reducing emissions from deforestation and forest degradation" and "participatory forest management" (FAO 2015; Davis et al. 2020; Duong and De Groot 2020; Purnomo et al. 2020). Although forest-based climate change social interventions can have many benefits, such as employment, income opportunities, and forest protection, they may fail because they do not meet the real needs of forest communities (Shiba and Michael 2012; Ofoegbu et al. 2017).

There are different approaches to forest-based climate change social interventions. In general, this interventions can be classified into three groups: no intervention, reactive intervention, and planned/preventive intervention. No-intervention approach means that continuation of the current situation and implementing present management measures are satisfying and it should be hoped that forests will become more adaptable to climate change. In other words, it means "business as usual" (Bernier and Schoene 2009). Reactive intervention approach refers to taking action after occurring a change. In other words, in this approach, intervenors seek to treat the complication (Madani 2014). This complication of forests can be climate change as well as its negative impacts on forests and forest communities. In such cases, the government or other institutions, as intervenors, use specific measures to reduce the damage and increase resilience and adaptability to the situation. On the other hand, the planned/proactive intervention approach includes redefining forestry goals and practices in advance. These goals and activities are designed to reduce the risks of uncertainties associated with climate change in a preventive and planned manner. Planned/preventive intervention approach includes deliberate and anticipatory interventions at different levels and across sectors (Bernier and Schoene 2009).

At the community level, preventive intervention may include diversification of forest-based and non-forest-based income sources, better local governance of forest resources, and capacity building to monitor and deal with possible calamities (Blate et al. 2009). Within the industrial forest sector, planned/preventive interventions may include introduction of bioenergy as a product or development of wood products for low carbon footprint. At the national and international levels, preventive/planned interventions include timely monitoring and reporting systems and development of tools for assessing vulnerability and adaptability under climate change (Bernier and Schoene 2009). It should also be noted that forest managers may be required to take into account the global consequences of local interventions. Regarding that forests are part of biophysical and biogeochemical cycles, they must be recognized by global programs and policies. At the stand level, preventive intervention may include planting a greater diversity of species or planting trees resistant to projected stresses (such as drought, salinity, and etc.). At the landscapes level, preventive/planned interventions may include the use of methods to reduce the negative effects of fire, insects, and diseases, and construction of biodiversity corridors, and restoration of destroyed forests. At the forest management unit level, intervention options include vulnerability assessments and maximum preparedness for disasters (Blate et al. 2009). Finally, it should be noted that planning for the management of forests and lands cannot depend solely on growth and yield trajectories over time. Forest and land management programs should also pay attention to the uncertainties and probabilities of extreme events. Intensive monitoring of forests is a key factor in preventive interventions and requires more technical and human resources. Monitoring can provide early warning of various stresses such as pests and diseases. In addition, a proper monitoring system in such interventions will help reduce uncertainty in planning and decision-making and minimize losses. After an extreme event, rapid risk assessment and timely planning for timber salvage, forest protection and forecasting effects on timber supply, markets, and socio-economic conditions are very useful (Bernier and Schoene 2009; Blate et al. 2009). Many developed countries, which have many forests, follow the planting of trees and improve the condition of forests. It should also be noted that such activities have sometimes been further developed in developed countries and included dimensions such as carbon monitoring and forest health. In developing countries, a lack of funding and expertise for monitoring and evaluation may hinder the recognition of the negative effects of climate change and timely responses. In such situations, preventive interventions should begin with capacity building for periodic forest evaluations. Integrating forest monitoring and existing knowledge about the potential effects of climate change into the vulnerability and risk assessment is the first major step in developing an intervention. Such assessments can be used to identify hotspots of risk management and draw response options. Developing, testing, and improving risk assessment methods can also be a new focus for monitoring (Bernier and Schoene 2009).

Both reactive and preventive intervention approaches may be used to counter the negative effects of climate change on forests. A reactive intervention approach may be employed when uncertainties or costs appear to be much higher than the expected impacts and risks. Also, this intervention approach may be used in situations where after a climate change-based shock, cost savings, and benefits from implementation are significant. For instance, replanting an area with resistant trees to the climate change and severe droughts can be considered as one of these interventions (Bernier and Schoene 2009). However, it should not be forgotten that in many cases, planned/preventive intervention approaches may be less costly and more effective in achieving goals such as forest sustainability under climate change circumstances. In such an approach, adaptation methods and tools are used quickly and before a crisis occurs. In other words, these methods and options are done before climate change and emergence of its impacts. The main elements of a planned/preventive intervention approach to adapt to climate change are (Blate et al. 2009):

- Reviewing, identifying, and improving forest management goals;
- Assessing the challenges that climate change poses to forest management goals and planned/preventive activities' implementation;
- Monitoring ecosystems and forest management responses to provide the necessary information for vulnerability and risk basic assessments;
- Paying attention to the uncertainties related to the impacts of climate change on forest management approaches; and
- Developing a toolbox and guide to forest management strategies.

Such an approach requires extensive institutional and inputs' coordination. Preventive/planned intervention approaches should be evaluated continuously and simultaneously with climate change and ecological system responses. Because, in some cases, such assessments may indicate that the goals of forest intervention and management need to be changed.

24.4 Enabling Factors in Forest-Based Climate Change Social Interventions

In any social intervention program, there are factors that contribute to the sustainability of social interventions and their positive effects, called "enabling factors". The development and institutionalization of enabling factors can play a significant role in the effectiveness of various approaches to social intervention (non-intervention, reactive intervention, and preventive intervention). Given the speed and extent of deforestation and the urgent need for a variety of interventions at various levels, it is important to identify and address these factors. In this regard, in this section, some of the most important enabling factors that should be considered in forest-based climate change social interventions are discussed. These enabling factors borrowed from IFAD technical report on "the sustainability of rural development projects" (Bhandari 2009) and outstanding article of Franck Vanclay on "social principals for agricultural extension to assist in the promotion of natural resources management" (Vanclay 2004).

- *Creating an effective link between intervention components*

One of the key elements in any intervention and sustainable management program such as sustainable forest and land management, is its design based on a holistic view of livelihood systems, needs, and opportunities. Narrow and partial interventions can pose many risks to sustainability. For example, the improved economic status of forest dwellers can easily be affected by the negative impacts of climate change. In this regard, establishing an effective link between the various elements of the intervention project is a necessity in any forest-based climate change social intervention. To strengthen effective link in this area, a few points need to be considered. First, intervenors should not seek sector-focused interventions. Because these interventions do not take into account the various dimensions (social, economic, environmental, and institutional) of forest issues due to narrowness. However, carelessness to these dimensions may exacerbate the problem. The second thing to keep in mind is that planners and executives must use broad mixed interventions. This is because these interventions, due to the diversity of goals, address a wide range of forest community needs (Bhandari 2009).

- *Community participation in forest-based climate change social interventions*

Although many forest-based climate change social interventions employ participatory methods at different stages, only interventions succeed in this context that are seriously committed to the participation of different stakeholders. In other words, these interventions put the participation into the practice. Dedication, close monitoring, and the use of various executive tools are among the most important prerequisites for participation of stakeholders in a forest-based climate change social intervention. If interventions want to be successful, they must try to use bottom-up approaches to determine the real priorities and needs of different stakeholders. Many interventions are currently being done with a "top-down" approach. Following such an approach shows that many forest and land management policymakers believe that farmers do not have the knowledge and ability to make the necessary decisions. In other words, professionals and outsiders are more aware of the local conditions and needs than local people. However, it should not be forgotten that in forest and land focused interventions, scientists are not the only source of knowledge. Accordingly, intervenors should not diffuse only the knowledge produced by scientists. All social and local groups construct their own knowledge based on personal experiences (Vanclay 2004).

- *Flexibility in designing interventions' programs*

Flexibility in the design of forest-based climate change social intervention programs increases their ability to be demand-driven. Because, these interventions employ methods that take the advantage of indigenous knowledge and activities of different groups. In addition, it must be kept in mind that the flexibility of forest-based climate change social intervention programs facilitate adaptation to new conditions and uncertainties. It can also help turn these challenges and uncertainties into opportunities to reduce the impacts of climate change on forests and forest communities. This had been forgotten in many traditional forest-based and climate change related interventions. The reason for this forgetfulness is due to the linear, top-down, and instrumental nature of traditional interventions (Bhandari 2009).

- *Institutional analysis in social interventions*

At the design stage of forest-based social intervention programs, a very detailed and comprehensive assessment of institutional environment should be done. The institutional context of the intervention plan has a tremendous impact on its success and sustainability of forest exploitation under climate change. Interventions that take the advantage of an accurate institutional analysis are more successful in achieving goals such as reducing poverty and supporting the participation of various institutions. Determining the strengths of institutions and organizations involved in sustainable forest management under climate change circumstances contributes to the better focus of efforts. In addition, institutional analysis results in determination of the economic capacity of institutions, development of financial planning of intervention, and sustainability of intervention program (Bhandari 2009).

- *Considering intervention cycle as a long-term process*

Participatory problem analysis is one of the key factors in development of forest-based climate change social intervention programs. Because it contributes to the achievement of long-term goals of sustainable forest development. However, it should be noted that these participatory analyzes require long-term program planning cycles. In addition, institutional strengthening and capacity building, natural resources (such as forests) management, and achieving equality usually require a long-term time cycles. Therefore, it should not be expected that the participation of different groups involved in forest-based climate change social interventions be occur in a short period of time. Participation is a time-consuming process, and social intervention planners and policy makers need to be aware of this fact. Because any hasty (short-term) action may have long-term negative and irreversible side effects. Social interventions, if they want to be participatory, must be implemented with a long-term vision (Bhandari 2009).

- *Risk assessment in forest-based climate change social interventions*

Any forest-based climate change social intervention needs to strengthen the risk management capabilities of different groups. The ability to manage risk contributes to the long-term sustainability of forest management programs. Risk assessment may involve identifying potential risks that threaten practitioners of intervention. By identifying such risks, risk mitigation strategies should also be considered in the plan. Risk management strategies must be applicable and enforceable, especially in terms of being responsive to social risks. For example, insurance of products of forest dwellers and diversification of their income sources are among the risk reduction strategies currently being used in forest-based climate change social interventions (Bhandari 2009).

- *Being focused on existing assets and knowledge*

Many social interventions in the management of forest sustainability do not achieve desired and satisfactory results. Because they consider indigenous and local people as individuals without knowledge. In other words, most of them are based on predetermined goals, policies, and knowledge content. At the same time, social participation and commitment to intervention programs increase when policymakers, practitioners, and managers of intervener agencies engage in and pay attention to the local practices and knowledge. In some cases, community institutions in forested areas can even be used instead of new structures and mechanisms. By focusing on existing community assets and knowledge, intervening organizations can eliminate the distrust of individuals and local organizations. They can create a positive social attitude towards participation and joint decision-making. In addition, paying attention to existing capital and knowledge can help increase organizational and social cohesion by strengthening relationships between external and internal organizations (Bhandari 2009).

- *Top-down intervention as an inappropriate approach for sustainable forest and land management*

Although the disadvantages of social interventions have been briefly discussed in the previous sections, they are explained in more detail here again due to their great importance. Top-down social interventions have many drawbacks. One of the main problems with these interventions is that they are biased against the innovations and ideas produced by scientists. In other words, innovations and ideas produced "by scientists and policymakers" are considered "useful innovations for the local community". However, in many cases, such judgment is inconsistent with the realities of society and the local context. In other words, the need for these innovations and ideas among locals is less than the need for them among scientists and policymakers. Another problem with top-down interventions in forest and land management programs is that they ignore social and ecological impacts. In addition, most of these interventions have used psychological and economic models to analyze the decision-making of local people. At the same time, they have ignored the social, political, cultural, and historical contexts of the forest and land management. Accordingly, interventions in the field of forest and land management need to avoid pursuing merely a top-down approach. They should use participatory, bottom-up, and integrated approaches. Of course, this does not mean romanticizing a particular approach and recommending a one-size-fits-all approach. Because, holding a biased idea of the application of participatory approaches is again trapped in dogmatism (Vanclay 2004).

- *Forest as a socio-cultural phenomenon and forestry as a socio-cultural activity*

Many policymakers, decision makers, and forest management interveners view the forest as a technical and static phenomenon. However, forest is a socio-cultural phenomenon. Similarly, forestry is a socio-cultural activity. Each forest area is located in a larger area or environment, which has different social, cultural, and environmental elements. The presence of different elements indicates the need for their dynamic communication. Forest can never be imagined apart from the people who live around it. Similar connections can be seen between the forest and other elements such as organisms, foresters, and the culture of local people where the forest is located. From this perspective, forests are a source of income, a place of residence, and a culture-oriented concept. Being aware of this fact contributes to design and implement comprehensive and multidimensional forest-based climate change social interventions and development of forest management (Vanclay 2004).

- *Profit cannot be the main motivational force for participation of all stakeholders in intervention*

Many forest-based climate change social interventions use only economic incentives to attract the participation of different stakeholders. However, for some stakeholders, profit is not the most important thing. For example, many forest dwellers are willing to pay to protect the forests and reduce the effects of climate change. Although in many interventions, participation is profit-oriented, it should not be generalized to all interventions as a one-size-fits-all principle. Paying attention to this issue, forest practitioners should try to develop forest-based climate change social interventions based on a variety of incentives to encourage different groups (Vanclay 2004).

- *Social learning in interventions*

Each intervention requires at least two actors (intervener and target group). However, it should be mentioned that the number of actors in interventions, especially in forest-based climate change social interventions is usually more than two. In general, linear or top-down interventions in the field of forests and lands suggest that the process of adopting ideas should include the stages of awareness, interest, evaluation, testing, and acceptance. However, it must be kept in mind that considering a technology and innovation as a fixed and pre-determined object is not correct. Ideas, innovations, and policies have different characteristics and dimensions. Their design in interventions should not be one-sided and controlled. Rather, designing must take place synergistically within processes of network building, social learning, and negotiation between different groups. In the process of social learning, different patterns, methods, and activities are developed. Steps need to be taken to create a suitable environment for social learning in forest-based climate change social interventions (Leeuwis 2013):

- Increasing awareness about the problem situation instead of the problem itself;
- Aligning the interests of different actors in forest-based climate change social interventions;
- Social designing or redesigning the environments of creating innovation in order to increase social learning among actors and facilitate the negotiation between them;
- Gradually preparing and improving of supportive requirements for innovations of the intervention practice;
- Redesigning and evolution of innovations to be diffused by forest-based social interventions.

24.5 Constraining Factors in Forest-Based Climate Change Social Interventions

Just as in any forest-based social intervention program there are factors that contribute to their sustainability and their positive impacts, there are factors that lead to the unsustainability of these interventions. In other words, these factors are constrains that prevent social intervention from achieving those goals. These factors are summarized in Table 24.2.

Table 24.2 Constraining factors in forest-based climate change social interventions

1	Vague goals
2	Poor adaptation of goals to the real situation of intervention
3	Ineffective linkage of different sectors
4	Insufficient implementation support for interventions
5	Inappropriate risk assessment measures
6	Insufficient technical support for interventions
7	Insufficient and ineffective capacity-building
8	Lack of monitoring system
9	Non-existence of holistic perspective towards networking
10	Focusing on quantitative outcomes of interventions
11	Insufficient community participation
12	Inappropriate follow up system in interventions
13	Ignoring local knowledge on climate change
14	Insufficient trust building programs
15	Non-proficient interveners and change agents
16	One-dimensional goal setting approach
17	Defining one-size-fits-all approach for interventions
18	Lack of attention to the real needs of target group
19	Conflict of interests
20	Lack of collaboration and networking
21	Lack of formal approval for interventions
22	In appropriate entrance methods
23	Inappropriate exit strategies
24	Lack of funding sources and financial supports
25	Parallel and overlapped interventions by different institutions
26	Inappropriate education methods
27	Short term program planning cycle
28	Geographical remoteness of intervention site
29	Lack of coproduction of knowledge
30	Self-serving of intervention staff
31	Disregarding the differences between target group members and stakeholders of intervention

24.6 The Uses of Forest-Based Climate Change Social Interventions

By studying the literature on social interventions, we can identify specific trends in terms of their "use" and "nature". The interesting fact is that these trends in social interventions are consistent with the different periods of evolution of rural development theories in the world. Comparing the trends of social interventions with the evolutionary periods of rural and social development theories can help identify the reasons why a trend is becoming more general in a particular historical period. In this regard, in this section, these trends of social interventions are firstly introduced from the perspective of their use, and in the next stage, the relationship between each of these trends and the evolution of rural development theory is discussed. These trends also apply to forest-based climate change social interventions. It should be noted that the uses introduced for social interventions in this section are based on the uses of evaluative intervention research, which have been introduced by Shadish et al. (1991) in detail. These uses have also been reflected by Valizadeh and Bijani (2019) as the investigative and research interventions' uses. Given that forest-based climate change social interventions are also considered as an intervening research process, in this regard, the applications mentioned to research interventions can also be generalized to these interventions.

Thinkers and researchers have been debating the concept of "use" in social interventions since the 1970s. In the initial definitions of use, the scope was very limited and included only "immediate", "concrete", and "observable" effects of interventions on a particular decision or policy. However, in the following decades, thinkers focused on the fact that social interventions may not necessarily lead to rapid consequences and changes. In other word, they may change people's minds. Accordingly, in the early 1980s, researchers began to expand the concept to include measures such as learning from the process of conducting social interventions, increasing awareness of research audiences and stakeholders about the effects and consequences of intervention, and thinking about evaluation (Shadish et al. 1991). Patton (2008) defines it as "how real people in the real world apply intervention research findings and experience". Henry and Mark (2003) have recently introduced a broader concept of use in evaluative interventions research that emphasizes the importance of "influence". Such thinking led to a paradigm shift in attitudes towards the use of the results of social interventions, which highlights the long-term use instead of focusing on short-term uses. To facilitate this, it was suggested that each interventionist researcher should try to explain the potential use of his/her results. Because, the research results themselves are not impelling and the intervenors themselves must work to make better use of the results.

With the passage of time and the development of different types of social interventions, different uses have been proposed. In general, three main types of uses can be named for interventions which include instrumental, conceptual, and persuasive uses. Instrumental use occurs when decision makers use the social intervention findings to modify or revise programs (Fleischer and Christie 2009). This type of use is what Patton (Patton 2008) points out. In this type of use, the results and information collected in the social intervention process are used "directly" and "objectively" and for a specific decision.

Conceptual use occurs when the results of intended social intervention help key stakeholders to understand social intervention and its context properly. Conceptual use may be used for social interventions performed by novice practitioners and intervenors, or for intervening situations where there is insufficient knowledge of them. Because instead of direct and instrumental use, it leads to a better or new understanding of social intervention and the context of intervention (Shadish et al. 1991). A crystal-clear example of this is forest-based climate change social interventions. Suppose a social intervention is being carried out to increase the adaptability and resilience of forest dwellers to climate change; the results of this intervention show that residents of forested areas do not have the ability to adapt to climate change. Based on the results, it is suggested that changes be made in farmers' attitudes toward risk mitigation methods. In this case, the conceptual use of forest-based social intervention has been considered. But if, following the results of social intervention, strategies such as changes in techniques and methods of forest and land management are undertaken by the government, the results have instrumentally been used. In this regard, conceptual use can be called enlightenment, organizational learning, and cognitive processing.

Persuasive use, sometimes referred to as the political use of social interventions, is often not considered a positive use. Examples of the use of negative persuasion include the use of the results of social interventions to justify, approve, and legitimize decisions made by the informant and executive decision makers (Fleisher and Christie 2009). Of course, in some cases, political use is not intentional. In other words, they are not necessarily used to justify a particular decision or action, but they are applied for directing and using the results of social interventions (Patton 2008). This form of use is called the "positive persuasive use". However, it should be noted that the ability of persuasive use to be positive and negative is one of the fundamental features of this type of use. In general, any type of forest-based climate change social intervention may have at least one instrumental, conceptual, and persuasive use. This means that practitioners, policymakers, decision makers, and intervenors should not try to fix one of the instrumental, conceptual, and persuasive uses as a one-size-fits-all use. Because, as mentioned in the previous sections on the top-down approaches to social intervention, extreme adherence to a particular approach in social interventions may be misleading. Predetermined bias toward a particular approach of social intervention can have many social, economic, and environmental impacts in forested areas.

Prior to the 1970s, most interventions were instrumental. From the 1970s to 1990s, however, most social interventions had conceptual uses. In addition, since 1990s, interventions have generally had a combination of uses (instrumental, conceptual, and persuasive) with an emphasis on persuasive use. Therefore, a regular and clear trend of the uses of social interventions in different periods of time is understandable. It should be mentioned that most of the social interventions are aimed at the sustainable development of forest communities and rural areas located in forested areas. These interventions follow at least one social dimension, even if they are not implemented with such a goal in mind. Therefore, it can be seen that the goals and uses of social interventions in different time periods are affected by the goals of social and rural development programs. Given that there are different periods for the evolution of rural and social development theories in the literature, so these periods are examined here. Explaining these courses will help to better understand the reasons for the prevalence of various uses for forest-based climate change social interventions in different time periods.

By examining the opinions of experts in the field of social and rural development, three generations of definitions can be identified. At each stage, the concept of development pays attention to the current problems of rural and social interventions alongside its evolution. In the first generation of development theories, the concept of "gross national product" was introduced as the most important "criteria" of development. This view was very common before the 1970s. This period is also called the "period of optimism and trust". According to the assumptions of these theories, with the increase in gross national product, not only all issues and problems can be solved, but the argument was that development would only take place in these circumstances. Due to the economic nature of the concept of development in this period, indicators such as poverty, improvement in literacy, education, health and services, housing, unemployment, and even balanced income distribution were not very important (Zamanipour 2013). It should be noted that in this period, top-down and centralized planning by the governments was considered as the best approach. Social and rural interventions in this period were inspired by the economic concept of development. Therefore, the intervenors tried to increase the income of the target group. Accordingly, economic growth and economic growth-based interventions were considered as an approach to social and rural development. In general, top-down and centralized planning used rural development programs as a means of increasing economic growth. With such an approach to rural and social development, it can be understood that social interventions were an instrument for economic growth.

The second generation of development theories, (which were dominant from 1970 to 1990s), stemmed from the experimental results of the late 1960s. These experiences have shown that despite many efforts, development has not taken place in many third world countries or its pace has been very slow. Few countries that have been able to achieve acceptable economic growth have faced many problems, such as increasing class divisions, poverty, and other unresolved social issues. It was on this basis that the issue of eradicating poverty and hunger and meeting the basic needs of human beings (rather than relying solely on economic growth) played an important role in defining development. One of the most important reasons for this change in thinking in defining the concept of development was that the most of needy people (usually villagers) who were considered the poorest people, were among those who benefited the least from development (Zamanipour 2013). This period is also called the "period of doubt or revision". During this period, doubts arose about the idea of economic growth as the only criteria and instrument for development. Most of the rural and social interventions had a conceptual use in this period. Rural development approaches sought to explore the concept of development in its context and to understand it more deeply. Accordingly, most of them were carried out in order to identify intervening contexts. Instead of direct and instrumental use for government policymakers and planners, they led to a better or new understanding of the concept of development and its various dimensions. This concept was well known in previous interventions: economic growth. However, in order to recognize the various dimensions of social and rural development, the interventions were mostly conceptual in this period.

The third generation of development theories emerged in the late 1990s. The various experiences gained in different countries and the World Bank research emphasized the ineffectiveness of second-generation slogans. Therefore, the ideologues and experts revisited the idea of economic growth (which was emphasized in the first generation of development theories). Revisions were carried out through change and renewal in the economic and social structures of the society. Therefore, comprehensive reforms were made with full attention to human dignity and cultural values. This was the same concept that was later called "self-dependent development" or "endogenous development". Endogenous development in this period meant development that could produce the necessary ability to reproduce and advance itself (Zamanipour 2013). Inspired by this approach, social interventions also tried to focus on bottom-up and endogenous planning instead of top-down planning. This approach shift was the first step in putting the last first. Social interventions during this period sought to involve local people in social development programs. In other words, social interventionists tried to encourage people to participate in planning and implementing interventions. Accordingly, the persuasive use of interventions were encouraged. Although persuasive interventions have negative applications in some cases, they were the first step to prioritize the views of indigenous peoples. For this reason, they are considered as a turning point in the history of social interventions.

24.7 Typology of Forest-Based Climate Change Social Interventions

Interventions to combat land use change are made by land users, governments, civil society organizations, and market participants. Analysis by Agrawal et al. (2014) represents a combination of changes in "resource rights to agricultural land and forests (which are created by political and institutional changes related to socio-political choices)", "incentives and rewards for changing land use attitudes", and "technological mechanisms such as agricultural intensification". It can be understood that these three types of influences embodied in interventions represent political, economic, agro-economic, and agro-ecological logics (Agrawal et al. 2014). Of course, there may be other logics for interventions, and the above mentioned ones do not mean denying other types of intervention logics. In general, according to Agrawal et al. (2014), forest-based climate change interventions can be divided into three categories.

24.7.1 Interventions Related to Resource Rights

Formal resource rights interventions can be derived from legislative action, policy reform, or management actions. Policy reform can create new rights to resources including the right to development and generation of management programs and the right to buy or sell lands. Policy reforms can directly change land use by changing the types of rights. For example, taking serious steps in line with "conservation-oriented" or "extraction-oriented" measures for forests and reallocation of rights using limiting the rights of some users is one of the policy reforms. These reforms through changes in political, economic, and social relations between actors, relative access to capital, and choices over resource use lead to deforestation. Policy changes can encompass mechanisms such as new incentives or drivers to adopt new technologies that result in forestry. In general, these rights and measures are reflected in their impact on forest use, forest cover, and land use outcomes. The main factor of rights-based interventions is redistribution of resource rights from a group/user to another group/user. This is true for all types of rights-based interventions such as controlling the technical dimensions of land use interventions, changes in zoning, and carbon sequencing mechanisms. For example, forest land use may change by improving conflict management skills, enforcing/dictating rules or increasing the ability to monitor and impose sanctions. Interventions related to resource rights can also be including forest and agriculture policy reforms, titling/land tenure, protected and conserved forests, zoning and spatial planning, decentralization and CBNRM, and logging bans and moratoria (Agrawal et al. 2014).

24.7.2 Incentives and Rewards-Based Interventions

Some interventions use economic incentives to support/maintain sustainable land and forest use. Examples of these interventions include payments for carbon sequestration/forest cover maintenance/watershed and access to finance for implementing forestry projects. In all of these interventions, the party interested in an environmental value rewards other sectors for protecting forests (Calvo-Alvaradoet al. 2009). Payment for environmental services (PES), voluntary standards and certification, and sustainable commodity supply chain interventions are some of the most important rewards-based climate change interventions in the field of forest and land management (Agrawal et al. 2014).

24.7.3 Technological Interventions

Interventions based on technical approaches to improve forest and land outcomes through carbon storage, tree cover, and forest conditions have a long history. These interventions must be carried out in such a way that they do not have a negative impact on the communities and their livelihoods. These interventions include agricultural intensification and tree planting. In general, technical interventions and social and institutional changes to support these interventions need to be undertaken together (Agrawal et al. 2014).

24.8 Summary and Conclusion

The main goal of this study was to provide a theoretical framework for forest-based climate change social interventions. Given that each social intervention program has three stages of design, implementation, and evaluation, in this regard, these three stages should be the basis of any framework in the field of interventions (Fig. 24.1). However, there are many constraining and enabling factors that must be taken into account in any forest-based climate change social intervention. The most important constraining factors to forest-based social interventions include structural, political, organizational, economic, executive, collaborative, network building, and follow-up barriers. Enabling factors consist flexible designing, institutional analysis, long-term intervention, risk assessment, prioritizing local knowledge, site-specific intervention, socio-cultural forestry, non-profit incentives, social learning, and participation. The basis of the framework provided for forest-based climate change social interventions is related to the different types of these interventions. This typology is based on the work of Agrawal et al. (2014). This section shows the framework of different areas of forest-based social intervention. In general, forest-based climate change social interventions include interventions related to resource rights, incentives and rewards-based interventions, and technological interventions. This classification can be very insightful in (re)directing forest-based climate change social interventions. Because it allows decision makers, policymakers, managers, and ecosystem practitioners to be able to focus their interventions.

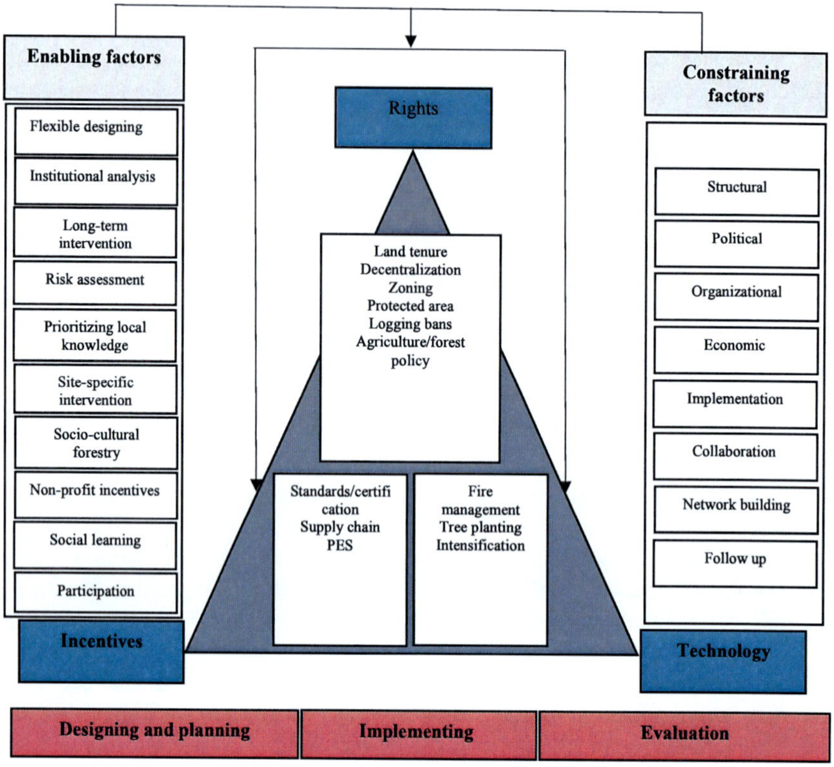

Fig. 24.1 A theoretical framework for forest-based climate change social interventions

References

Agrawal A, Wollenberg E, Persha L (2014) Governing agriculture-forest landscapes to achieve climate change mitigation. Glob Environ Change 29:270–280. https://doi.org/10.1016/j.gloenvcha.2014.10.001

Baumgartner RJ (2019) Sustainable development goals and the forest sector—a complex relationship. Forests 10(2):152. https://doi.org/10.3390/f10020152

Bernier P, Schoene D (2009) Adapting forests and their management to climate change: an overiew. Information Service of FAO

Bhandari H (2009) Sustainability of rural development projects. The eighth in a series of discussion papers produced by the Asia and the Pacific Division, IFAD

Blate GM, Joyce LA, Littell JS, McNulty SG, Millar CI, Moser SC, Neilson RP, Halloran K, Peterson DL (2009) Adapting to climate change in United States national forests. Information Service of FAO

Bösch M, Elsasser P, Rock J, Weimar H, Dieter M (2019) Extent and costs of forest-based climate change mitigation in Germany: accounting for substitution. Carbon Manag 10(2):127–134. https://doi.org/10.1080/17583004.2018.1560194

Bukoski JJ, Drazen E, Johnson WR, Swamy L (2018) Tropical forests for sustainable development: shaping the 2030 Agenda for Sustainable Development with knowledge from the field. https://doi.org/10.1080/10549811.2018.1418255

Calder I, Hofer T, Vermont S, Warren P (2008) Towards a new understanding of forests and water. UNASYLVA-FAO 229:3
Calvo J, McLennan B, Sánchez- A, Garvin T (2009) Deforestation and forest restoration in Guanacaste, Costa Rica: putting conservation policies in context. Forest Ecol Manag 258(6):931–940. https://doi.org/10.1016/j.foreco.2008.10.035
Chia EL, Somorin OA, Sonwa DJ, Tiani AM (2013) Local vulnerability, forest communities and forest-carbon conservation: case of southern Cameroon. Int J Biodiv Conserv 5(8):498–507. https://doi.org/10.5897/IJBC2012.0533
Davis EJ, Hajjar R, Charnley S, Moseley C, Wendel K, Jacobson M (2020) Community-based forestry on federal lands in the western United States: a synthesis and call for renewed research. Forest Policy Econ 111:102042. https://doi.org/10.1016/j.forpol.2019.102042
Dixon RK, Solomon AM, Brown S, Houghton RA, Trexier MC, Wisniewski J (1994) Carbon pools and flux of global forest ecosystems. Science 263(5144):85–190. https://doi.org/10.1126/science.263.5144.185
Dlamini CS (2014) African forests, people and climate change project: forest and climate changes policies, strategies and programmes in the SADC and COMESA regions. In: African forest forum, working paper series, vol. 2, no. 17. African Forest Forum, Nairobi, pp 1–17
Duong NT, De Groot WT (2020) The impact of payment for forest environmental services (PFES) on community-level forest management in Vietnam. Forest Policy Econ 113:102135. https://doi.org/10.1016/j.forpol.2020.102135
FAO (2016) Global forest resources assessment 2015. How are the world's forests changing?, 2nd edn. FAO, Rome
FAO (2015) Forests and climate change: working with countries to mitigate and adapt to climate change through sustainable forest management. Food and Agriculture Organisation of the United Nations, Rome, Italy
Fleischer DN, Christie CA (2009) Evaluation use: results from a survey of US American evaluation association members. Am J Eval 30(2):158–175. https://doi.org/10.1177/1098214008331009
Hajost S, Zerbock O (2013) Lessons learned from community forestry and their relevance for REDD+. United States Agency for International Development, Arlington, VA 22209
Hengeveld GM, Didion M, Clerkx S, Elkin C, Nabuurs GJ, Schelhaas MJ (2015) The landscape-level effect of individual-owner adaptation to climate change in Dutch forests. Reg Environ Change 15(8):1515–1529. https://doi.org/10.1007/s10113-014-0718-5
Henry GT, Mark MM (2003) Toward an agenda for research on evaluation. New Direct Eval 97:69–80. https://doi.org/10.1002/ev.77
Holden E, Linnerud K, Banister D, Schwanitz VJ, Wierling A (2017) The imperatives of sustainable development: needs, justice, limits. Routledge
Humphries S, Holmes T, de Andrade DFC, McGrath D, Dantas JB (2020) Searching for win-win forest outcomes: learning-by-doing, financial viability, and income growth for a community-based forest management cooperative in the Brazilian Amazon. World Dev 125:104336. https://doi.org/10.1016/j.worlddev.2018.06.005
Jindal R, Kerr JM, Carter S (2012) Reducing poverty through carbon forestry? Impacts of the N'hambita community carbon project in Mozambique. World Dev 40(10):2123–2135. https://doi.org/10.1016/j.worlddev.2012.05.003
Katila P, Jong W, Colfer C (2019) Galloway. Sustainable Development Goals, Their Impacts on Forests and People
Leeuwis C (2013) Communication for rural innovation: rethinking agricultural extension. Wiley
Madani K (2014) Water management in Iran: what is causing the looming crisis? J Environ Stud Sci 4(4):315–328. https://doi.org/10.1007/s13412-014-0182-z
Mader S (2019) What drives climate change? Doctoral dissertation. Faculty of Business, Economics and Social Sciences, Institute of Sociology, University of Bern, Switzerland
Maraseni TN, Bhattarai N, Karky BS, Cadman T, Timalsina N, Bhandari TS, San SM (2019) An assessment of governance quality for community-based forest management systems in Asia:

prioritisation of governance indicators at various scales. Land Use Policy 81:750–761. https://doi.org/10.1016/j.landusepol.2018.11.044

Nhem S, Jin Lee Y, Phin S (2018) The impact of forest resource decline: analyzing forest-related income supplements to reduce income inequality and poverty of the Kouy indigenous people living in Kampong Thom province, Cambodia. J Sustain Forest 37(2):97–119. https://doi.org/10.1080/10549811.2017.1369887

Ofoegbu C, Chirwa PW, Francis J, Babalola FD (2017) Socio-economic factors influencing household dependence on forests and its implication for forest-based climate change interventions. South Forest J Forest Sci 79(2):109–116. https://doi.org/10.2989/20702620.2016.1255420

Patton MQ (2008) Utilization focused evaluation, 4th edn. Sage Publications, Thousand Oaks

Paudel G, Adhikari S, Bhusal P (2019) Integration of forest and climate change policies in Nepal. J Forest Nat Res Manag 1(1):1–13. https://doi.org/10.3126/jfnrm.v1i1.22647

Purnomo EP, Ramdani R, Salsabila L, Choi JW (2020) Challenges of community-based forest management with local institutional differences between South Korea and Indonesia. Dev Pract, 1–12. https://doi.org/10.1080/09614524.2020.1749561

Ray D, Petr M, Mullett M, Bathgate S, Marchi M, Beauchamp K (2019) A simulation-based approach to assess forest policy options under biotic and abiotic climate change impacts: a case study on Scotland's national forest estate. Forest Policy Econ 103:17–27. https://doi.org/10.1016/j.forpol.2017.10.010

Rennaud JP, Ruitenbeek J, Tennigkeit T (2013) Challenges of community-forestry based carbon projects: process, participation, performance. Field Actions Science Reports. J Field Actions 7:21–34

Shadish WR, Cook TD, Leviton LC (1991) Foundations of program evaluation: Theories of practice. Sage

Shiba PK, Michael GJ (2012) NTFP income contribution to household economy and related socio-economic factors: lessons from Bangladesh. Forest Policy Econ 14(1):136–142. https://doi.org/10.1016/j.forpol.2011.08.003

Silva E (1994) Thinking politically about sustainable development in the tropical forests of Latin America. Dev Change 25(4):697–721. https://doi.org/10.1111/j.1467-7660.1994.tb00533.x

Torkar G, Verlič A, Vilhar U (2014) Importance of forest ecosystem services to secondary school students: a case from the North-West Slovenia. South-East Europ Forest 5(1):35–43. https://doi.org/10.15177/seefor.14-02

USDA (2008) Watershed forestry resource guide center for watershed protection and US forest service

USDA (2013) Watershed forestry. USDA Forest Service 1400 Independence Ave. SW Washington, DC

Valizadeh N., and Bijani M. (2019). Agricultural Research: Applications and Future Orientations. In: Leal Filho W., Azul A., Brandli L., Özuyar P., Wall T. (eds) Zero Hunger. Encyclopedia of the UN Sustainable Development Goals. Springer, Cham. pp: 1–8. https://doi.org/10.1007/978-3-319-69626-3_5-1

Vanclay F (2004) Social principles for agricultural extension to assist in the promotion of natural resource management. Australian J Exper Agric 44(3):213–222. https://doi.org/10.1071/EA02139

Walsh ES, Vierling KT, Strand E, Bartowitz K, Hudiburg TW (2019) Climate change, woodpeckers, and forests: current trends and future modeling needs. Ecol Evol 9(4):2305–2319. https://doi.org/10.1002/ece3.4876

World Bank (2008) Forests sourcebook: practical guidance for sustaining forests in development cooperation. World Bank, Washington, DC. https://doi.org/10.1007/s11069-008-9278-7

Zamanipour A (2013) Agricultural extension in development process. Birjand University Press, Birjand, Iran

Chapter 25
Conversion of Land Use Land Cover and Its Impact on Ecosystem Services in a Tropical Forest

Soumen Bisui, Sambhunath Roy, Debashish Sengupta, Gouri Sankar Bhunia, and Pravat Kumar Shit

Abstract The tropical forest ecosystem provides society-wide range of ecosystem services, but increasing human activity changes the land use/land cover dynamics (LULC) with significant change of the ecosystem services values. The present study is considered to evaluate and monitoring of LULC change along with the ecosystem services of a tropical forest region in Jhargram block West Bengal in India during 1972–2019. LANDSAT satellite imageries (1972, 1987, 1992, 2002, 2012, and 2019) have been analyzed to identify the LULC change and to determine the ecosystem service values using the value transfer method. Based on the Maximum likelihood algorithm image classification techniques using Arc-GIS software v.10.3, the study area is divided into eight land use land cover (LULC) categories like forest, cultivated land, grassland, water bodies, settlement/built-up land, barren land, and sand. We found that forest land continuously decreased by 42.3% between 1972 and 2019 and cultivated land is also decreased by 31.6% from 1972 to 2019. Settlement and Barren land are increased progressively 1209.6%, 394.3% respectively between 1972 and 2019 respectively. The total ecosystem services in Jhargram block have been calculated as $30.52, $29.69, $27.03, $34.1, $29.2, and $21 million in the year of 1972, 1987, 1992, 2002, 2012 and 2019 respectively. The total ecosystem services have reduced by 31.2% due to deforestation. Variations can also be seen in the value of individual ecosystem service functions viz. food production, raw material, erosion control, nutrient cycling, recreation, and climate regulation, which are important

S. Bisui (✉) · S. Roy · P. K. Shit
PG Department of Geography, Raja N. L. Khan Women's College (Autonomous), Gope Palace, Midnapore 721102, West Bengal, India
e-mail: bisui.soumengeo@gmail.com

D. Sengupta
Department of Geology and Geophysics, Indian Institute of Technology (IIT), Kharagpur 721302, West Bengal, India
e-mail: dsgg@gg.iitkgp.ernet.in

G. S. Bhunia
TPF Gentisa Euroestudio SL, Gurgaon, India
e-mail: rsgis6gouri@gmail.com

P. K. Shit et al. (eds.), *Spatial Modeling in Forest Resources Management*, Environmental Science and Engineering,
https://doi.org/10.1007/978-3-030-56542-8_25

contributions in total ecosystem services. The sensitivity analysis indicates that our estimated ecosystem service value for our present study is reasonable and robustness concerning the value coefficient. This information will be helpful for sustainable land use management strategies and policy-making processes in the tropical forest region.

Keywords Landsat data · Ecosystem service values · LULC · GIS techniques · Sensitivity analysis · Deforestation

25.1 Introduction

Ecosystem through their function to provides directly or indirectly multi-services, which are important to the living organism of the earth and also important for human survival (MEA 2005; Schagner et al. 2013; Costanza et al. 2014a, b). Different ecosystem service provides an extensive range of services like provisioning (food production, raw material, water), regulating (climate regulation, erosion control, water regulation), cultural services (cultural and recreation) and supporting services (soil formation and nutrient cycling) which differ in quantity and as well as quality (MEA 2005). These services help to maintain the ecological process and sustaining a life of the earth. The ecosystem service directly linked to the type of ecosystems, e.g., LULC type in a specified region (de Groot et al. 2002; Styres et al. 2010). Due to huge population growth, economic development, and urbanization, LULC have experienced massive change throughout the world in the last decades, especially from forest to agriculture land, built-up area and pasture land (Lambin and Meyfroidt 2011; Chen et al. 2014). Dynamic change of LULC has been recognized that one of the important driver of ES loss (Costanza et al. 2014a, b; Kubiszewski et al. 2017). However, the evidence is showing about 60% of the ecosystem service has degraded since few decades (MEA 2005) and also results showed that global forest has faced an environmental crisis on account of heavy deforestation (Tsegaye et al. 2010; Sloan and Sayer 2015). Presently, the particularly topical forest is quickly diminishing the cause of natural landscape convert to farm pasture and harvest timer for fuel and construction (Miyamato et al. 2014) as a result ecosystem services and their individual functions are also rapidly declining. Henceforth, for sustainable land utilization, the assessment of ecosystem service (ESV) is a significant decision support tool (Forster et al. 2015) and provides an effective approach to policy-maker and land planner for appropriate land management (De Groot et al. 2012; Iverson et al. 2014; Yi et al. 2017). However, numbers of researches have conducted several research works on the assessment and impact of ecosystem services at a global scale (Costanza et al. 1997, 2014a, b; Zang 2011; Kindu et al. 2016; Yi et al. 2017; Polasky et al. 2011).

Likewise, the tropical land of India has been faced challenges for deforestation, expansion of agriculture land, increase of settlement and also demand fuel wood and construction material for the few decades. Till date, very few studies have been conducted with little attention on assessment and change of ESV and dynamics of LULC, (Sinha and Mishra 2015; Everard et al. 2019; Mondal et al. 2018; Kumar and Chaudhry 2015). Further, a recent study indicates approximately 7% of national GDP, 57% of the income of rural people in India which have been contributed from the Indian forest (Sukhdev 2009).

Now a day, much of the rural people particularly in indigenous tribal area are being suffered from deforestation and forest degradation, as they depend on forest recourses for their food fuel, livelihood. About 42.3% of forest is degraded during the period from 1972 to 2019. As a result, tribal people's livelihoods have a frightening condition; many newspapers have reported this situation (www.anandabazar.com, 2017; the national green tribunal, 2019). Therefore, the present study attempts to analyse LULC dynamic in Jhargram block in West Bengal (India) and to assess variation of ecosystem service values during the period between 1972 and 2019.

25.2 Study Area

The study was conducted in Jhargram block which is part of Chottonagpur plateau with undulating topography. Geographically Jhargram block is located between 22° 15′ 18″ N latitude to 22° 30′ 02″ N latitude and 86° 54′ 22″ E longitude to 87° 15′ 33″ E longitude (Fig. 25.1). The total land area covers 55,365 ha. The geological formation of the area is mainly lateritic and the general slope of the area is forming northwest to southwest.

This area experiences very high temperature 40–45 °C in May–June month and receives rainfalls mainly during monsoon season (June–October). The climate is warm humid tropical nature falling under Koppen "AW" type of climate classification. Jhargram block is one of the draught prone block of Jhargram district; the total forest cover of the block covers 6168 ha.

Administratively Jhargram block has one Panchayat samity and 13 gram panchayats. As per the census 2011, the total population of this block is 170,097, out of which 22.7% and 14.83% population belongs to ST and SC community respectively. The different primitive tribes living in this block are Bhumij (11.6%) Lodha (3.85%), Munda (6.10%), Koras and Mahalis (Paschim Midnapore District Human Development report, 2016).

The main economic activities are farming and livestock rearing and the dominant crop is paddy. About 49.02% of peoples are living below poverty line. Many of them are depending upon the natural recourses for food, fodder and fuel. This region is also very attractive for tourism.

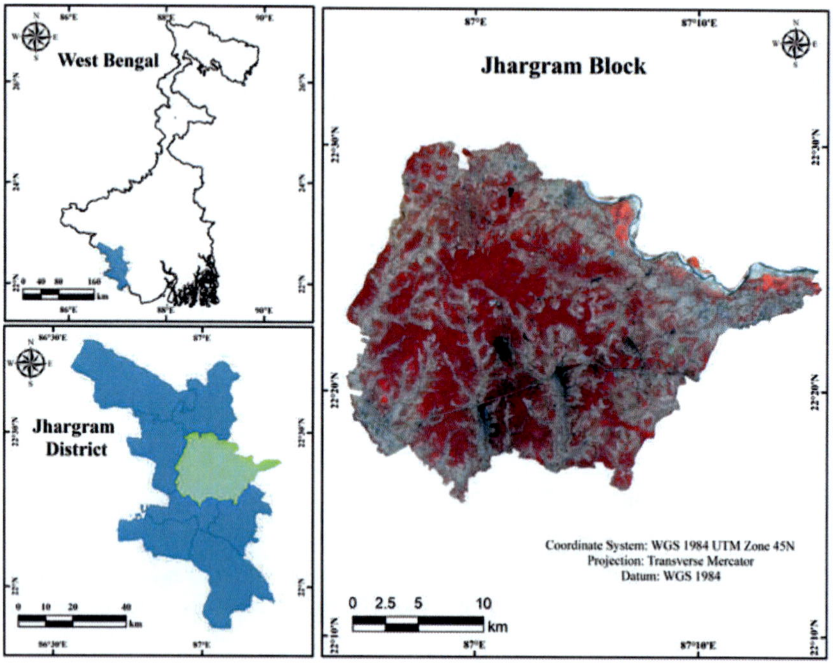

Fig. 25.1 Location map of the study area

25.3 Material and Methods

25.3.1 LULC Classification

In this study we used Landsat MSS (1972), Landsat TM (1987 and 1992), Landsat ETM + (2002) and Landsat OLI (2012, 2019) satellite images for LULC change analysis with different resolution. The years are selected on the basis of major economic and political changes and data availability which images are acquired from United States Geological Survey Earth Explorer Community (USGS) (Table 25.1).

Remotely sensed images were pre-processed including image enhancement and image classification using by EDRAS Imagine 2014 for preparation of land use/land cover map. The images were classified into eight LULC classes (Table 25.2) using maximum likelihood classification and calculating the area statistics for subsequently used to calculate the ESV for different LULC indifferent year (Fig. 25.2).

Table 25.1 Description of imagery data used for land cover change study in Jhargram block

Imagery date	Imagery type	Resolution (m)	Path and row	Source
12–12–1972	Landsat MSS	57 × 57	149/44, 149/45	USGS
15–12–1987	Landsat TM	30 × 30	139/44, 139/45	USGS
13–11–1992	Landsat TM	30 × 30	139/44, 139/45	USGS
16–12–2002	Landsat ETM +	30 × 30	139/44, 139/45	USGS
11–04–2012	Landsat OLI	30 × 30	139/44, 139/45	USGS
17–10–2019	Landsat OLI	30 × 30	139/44, 139/45	USGS

Table 25.2 Description of land covers type in the study area

Land cover type	Description
Forest	Land covered by trees mainly tropical forest
Cultivated land/cropland	Land under cultivation of agriculture
Grassland	Land dominated by grass
Water bodies	Pond, river, reservoir and stream
Settlement/built-up land	Land covered by building and infrastructure
Barren land	No vegetation covered including rock, sand
Fallow land	Land without crop at few season
Sand	Land covered by sand

25.3.2 LULC Change

LULC changes important role to changes of ecosystem service and their function (Hao et al. 2012). Therefore we calculate percentage change (PC) and annual percentage change (AAC) of LULC by the following formula:

$$PC = \{(Y_k - Y_1)\} \times 100$$

Y_1 is taken the base year, i.e., 1972 and Y_k is the successive year taken for study i.e., 1987, 1992, 2002, 2012, 2019.

AAC = PC/difference between the year of study and base year.

25.3.3 Assignment of ESV

In the research work, Costanza et al. (1997, 2014a, b) estimated ecosystem service value coefficients are adopted for quantifying the ESV values of different LULC. Most of each LULC use as a proxy of representative biome (Table 25.3). Costanza et al. (1997) assessed the economic value of 16 types of biome and 17 ecosystem

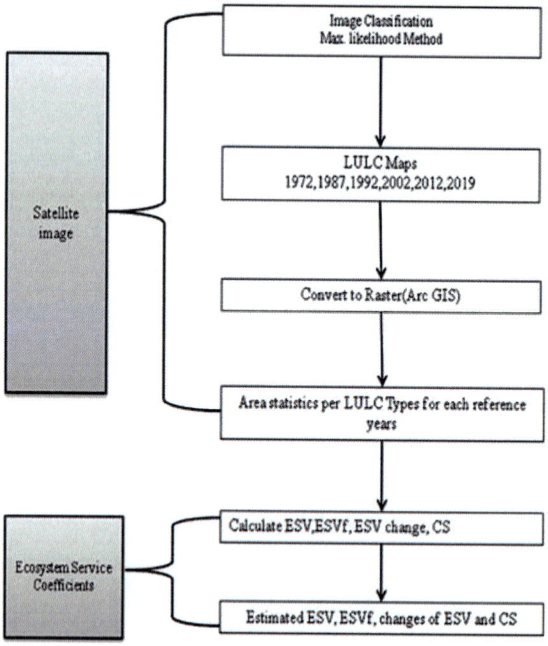

Fig. 25.2 Methodological framework on the study landscape. ESV: ecosystem service value of each LULC in reference year, ESV; f: estimated value of each individual function, CS: coefficient sensitivity

Table 25.3 Summary of land use land cover type with equivalent biome and value coefficient adopted from costanza et al. (1997, 2014a, b)

LULC category	Equivalent biomes	Ecosystem service value coefficient (US$ ha^{-1} yr^{-1})
Forest	Tropical forest	2007
Cultivated land/cropland	Cropland	92
Grassland	Grass/rangeland	232
Water bodies	Lake/river	8498
Settlement/built-up land	Urban	0
Barren land	Urban	0
Fallow land	Urban	0
Sand	Desert	0

functions (Table 25.4). Ecosystem service values are calculated using the method followed by Li et al. (2007). The total ESV value of selected landscape in different study years e.g. 1972, 1987, 1992, 2002, 2012, and 2019 were estimated as follows:

Table 25.4 The annual value coefficients for ecosystem service in (US\$ ha^{-1} $year^{-1}$) adopted from Costanza et al. (1997, 2014a, b)

Ecosystem service type	Sub type	Forest	Cultivated land/cropland	Grassland	Water bodies	Settlement/built-up land	Barren land	Fallow land	sand	Total
Provisioning service	Food production	32	54	67	41	0	0	0	0	194
	Raw materials	315	0			0	0	0	0	315
	Genetic resources	41	0	0	0	0	0	0	0	41
	Water supply	8	0		2117	0	0	0	0	2125
Regulating services	Gas regulation	0	0	7		0	0	0	0	7
	Climatic regulation	223	0	0	0	0	0	0	0	223
	Disturbance regulation	5	0	0	0	0	0	0	0	5
	Waste treatment	87	0	87	665	0	0	0	0	839
	Erosion control	245	0	29		0	0	0	0	274
	Pollination		14	25		0	0	0	0	39
	Water regulation	6		3	5445	0	0	0	0	5454
	Biological control		24	23	0	0	0	0	0	47
Supporting services	Nutrient cycling	922	0	0	0	0	0	0	0	922
	Soil formation	10	0	1	0	0	0	0	0	11
	Habitat/refugia		0	0	0	0	0	0	0	
Cultural services	Cultural	2	0	0	0	0	0	0	0	2
	Recreation	112	0	2	230	0	0	0	0	344
Sum of ES for LULC		2007	92	232	8498		0	0	0	0

$$ESV = \sum (A_k \times VC_k)$$

where ESV is the estimated ecosystem service value; A_K the area (ha) of LULC category k and VC_K value coefficient (US \$ ha^{-1} $year^{-1}$) for LULC category k. The changes of ESV value were calculated by the output estimated values of difference for each LULC Category reference year. The percentage of ESV was calculated through the following formula:

$$\text{Percentage of ESV change} = \left(\frac{ESV_{\text{final year}} - ESV_{\text{initial year}}}{ESV_{\text{initial year}}} \right) \times 100$$

where ESV signifies total estimated ecosystem service value. Positive values indicate an increase and negative values mean a decrease of quantity. Furthermore, we also calculate the individual ecosystem function of study landscape using following formula:

$$ESV = \sum (A_k \times VC_{fk})$$

$ESV_{f=}$ Ecosystem service value of ecosystem function 'f', A_k is the area (ha) and VC_{fk} is the value coefficient of function 'f' (US \$ ha^{-1} year^{-1}) for LULC category 'k'.

25.3.4 *Sensitivity Analysis*

As suspicion of value coefficient, we use sensitivity analysis to the following equation (Li et al. 2007; kindu et al. 2016; Mamat et al. 2018) by adjusting 50% of value coefficient and then calculate the corresponding change in ESV of each LULC, which is similar to grade economic concept of elasticity By using the following formula.

$$CS = \frac{(ESV_j - ESV_i)/ESV_i}{(VC_{jk} - VC_{ik})/VC_{ik}}$$

where ESV_iandESV_j = initial and adjusted total estimated ecosystem service value,VC_{jk} and VC_{ik} is initial and adjusted value coefficient (US\$ ha^{-1} $year^{-1}$) of LULC category 'k'. When CS value greater than one then ESV is considered elastic relative to value coefficient, but if the CS value is less than one then the ESV considered as to be inelastic.

25.4 Result and Discussion

25.4.1 Land Use and Land Cover Change

Based on the six different (1972–2019) year remote sensing images, eight LULC classes were identified of the Jhargram block (Fig. 25.3) and area under different LULC presented in Table 25.5. There was a significant increase in the settlement of 217.44 ha in 1972 to 2847.72 ha in 2019, overall increase about 1209.6% because this area were converted from the forest land and cultivated land. The forest cover of this region has been decreased continuously from 1972 to 2019, with an overall decrease by about 4.3%. The downward forest cover indicates the subsistence agriculture for livelihood and increases the demand for fuel wood. The maximum loss of forest cover rate in the period of 2002–2019 due to anti-social activities.

Significant changes of water body have been found in the period of 1972–2012 after 2012. Rapid loss of forest cover has been found after 2002 with an overall decrease of about 24.6%. The fallow land and barren land have been increased by 57.06% and 394.3% respectively in the study area. Grassland or scrubland have been decreased continuously from 1972 to 2012 but slightly increased after 2012; however, slight increase has been observed after 2019. Cultivated land slightly increases from 1972 to 1992, after that it was recorded decreasing trend up to 2019 (Table 25.5).

25.4.2 Spatio-temporal Changing Pattern of Ecosystem Service Values

25.4.2.1 Ecosystem Service Value Changes During 1972 and 2019

The ecosystem service values and changes in total ESV were calculated using the coefficient value (Table 25.3). The total ecosystem service during the study period was US$30.52, US$29.69, US$27.03, US$34.1, US$29.2, and US$21 million in the year of 1972, 1987, 1992, 2002, 2012 and 2019 respectively. In this study, the net ecosystem service value was reduced by 31.2% over the four decades of the study period, from US$ 30.52 million in 1972 to US$ 21 million in 2019 (Table 25.6). The highest loss of ESV from the forest was 42.28% and cultivated land, Grassland, water bodies were 31.81%, 33.02%, and 24.48% respectably. While ESV of eight LULC type have changed in different reference years across our study landscape (Table 25.6). The ESV for forest land is calculated as maximum; however, the lowest value is calculated for the cultivated land. The last year ESV of the forest is US$12.37 million while cultivated land, grassland, water bodies accounted about US$1.877 million, US$2.86 million, US$8.29 million of the total ESV across entire the block.

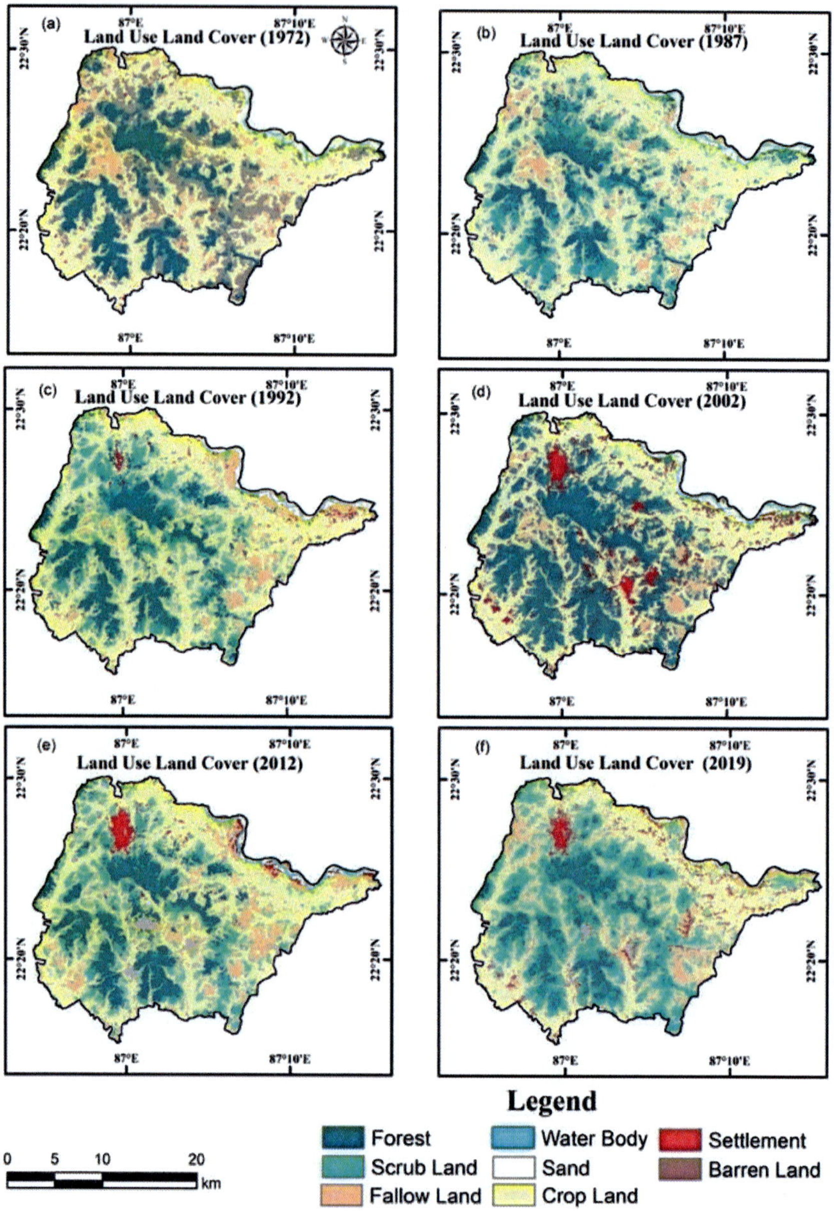

Fig. 25.3 LULC of the study area in 1972, 1987, 1992, 2002, 2012, 2019 respectively

Table 25.5 Land use and land cover change from 1972 to 2019

Year	Forest			Cultivated land/cropland			Grassland/scrub land			Water bodies		
	LU (ha)	PC	AAC (%)	LU (ha)	PC	AAC (%)	LU (ha)	PC	AAC (%)	LU (ha)	PC	AAC (%)
1972	10,680.48	–	–	21,536.28	–	–	14,101.2	–	–	452.88	–	–
1987	9572.751	−10.4	−0.7	23,777.64	10.4	0.7	9904.77	−29.7	−2.0	708.21	56.4	3.8
1992	9949.05	−6.8	−1.4	23,477.04	9.0	1.8	9931.5	−29.6	−5.9	308.61	−31.8	−6.36
2002	10,503.99	−1.7	−0.2	20,406.49	−5.2	−0.5	12,335.22	−12.3	−1.2	976.68	115.7	11.6
2012	9315.09	−12.8	−1.28	18,332.46	−14.8	−1.5	9983.52	−29.2	−3.0	768.24	69.6	7.0
2019	6168.12	−42.3	−6.0	14,741.18	−31.6	−4.5	18,787.66	33.2	4.7	341.26	−24.6	−3.5
Year	Settlement/Built-up land			Barren land			Fallow land			Sand		
	LU (ha)	PC	AAC (%)	LU (ha)	PC	AAC (%)	LU (ha)	PC	AAC (%)	LU (ha)	PC	AAC (%)
1972	217.44	–	–	100.16	–	–	7506.72	–	–	770.04	–	–
1987	556.9788	156.1	10.4	18.71	−81.3	−5.4	10,109.92	34.7	2.3	716.22	−6.9	−0.5
1992	599.48	175.7	35.1	1266.3	1164.3	232.9	9376.02	24.9	5.0	457.2	−40.6	−8.1
2002	1769.847	713.9	71.4	58.76	−41.3	−4.1	8690.51	15.8	1.6	623.7	−19.0	−2.0
2012	2354.86	982.9	98.2	2756.78	2652.4	265.2	11,319.74	50.8	5.0	534.51	−30.6	−3.0
2019	2847.72	1209.6	172.8	495.1	394.3	56.3	11,790.25	57.06	8.1	193.91	−74.9	−10.7

LU- Land use, PC- Percentage of change, AAC-Annual average% change, PC and AAC calculated by taking 1972 as base year

Table 25.6 Total ecosystem service values estimated for each LULC category changes from 1972 to 2019 in the study area following Costanza et al. (1997, 2014a, b) and Li et al. (2007) valuation coefficients

LULC class	ESV(US$ million)						ESV (US$ million) change					
	1972	1987	1992	2002	2012	2019	1972–1987	1987–1992	1992–2002	2002–2012	2012–2019	1972–2019
Forest	21.43	19.21	19.96	21.08	18.69	12.37	−2.22 −10.36%	0.75 −3.9%	1.12 5.61%	−2.39 −11.33%	−6.32 −33.81%	−9.06 −42.27%
Cultivated land/cropland	1.98	2.18	2.15	1.87	1.68	1.35	0.2 10.1%	−0.03 −1.37%	−0.28 −13.02%	−0.19 −10.16%	−0.33 −19.64%	−0.63 −31.81%
Grassland	3.27	2.29	2.30	2.86	2.31	4.35	−0.98 −29.9%	0.01 0.43%	0.56 24.34%	−0.55 −19.23%	2.04 88.31%	1.08 33.02%
Water bodies	3.84	6.01	2.62	8.29	6.52	2.90	2.17 56.51%	−3.39 −56.4%	5.67 216.40%	−1.77 −21.35%	−3.62 −55.52%	−0.94 −24.27%
Settlement/built–up land	0	0	0	0	0	0	0	0	0	0	0	0
Barren land	0	0	0	0	0	0	0	0	0	0	0	0
Fallow land	0	0	0	0	0	0	0	0	0	0	0	0
Sand	0	0	0	0	0	0	0	0	0	0	0	0
Total	30.52	29.69	27.03	34.1	29.2	21	−0.83 −2.7%	−2.66 −8.9%	7.07 20.15%	−4.9 −14.36%	−8.2 −28.18%	−9.52 −31.2%

25.4.2.2 Change in Ecosystem Function

The estimation of annual values of ecosystem function and their changes from 1972 to 2019 in the study block is presented in Table 25.7. This order of contribution changed with change the time due to change of LULC. The collective contribution of seven ecosystem function represented about US$24.38 million, US$23.58 million, US$21.69 million, US$27.08 million, US$23.23 million, US$16.39 million in 1972, 1986, 1992, 2002, 2012, 2019, respectively (Table 25.7). Only two functions increase their value during study period gas regulation and pollination, the

Table 25.7 Annual estimated value of ecosystem function (ESV_f in US$ million US$ year^{-1}) under each service category in different year and their changes (1972–2019) in the study landscape

Ecosystem service	Sub type	Using global coefficient (Adopted from costanza et al. (1997, 2014a, b)						
		1972	1986	1992	2002	2012	2019	Overall change
Provisioning service	Food production	2.46	2.27	2.26	2.3	1.98	2.26	−0.2
	Raw materials	3.36	3.01	3.13	3.3	2.93	1.94	−1.42
	Genetic resources	0.43	0.39	0.4	0.43	0.38	0.25	−0.18
	Water supply	1.04	1.57	0.73	2.15	1.7	0.77	−0.27
Regulating services	Gas regulation	0.09	0.06	0.06	0.08	0.06	0.13	0.04
	Climatic regulation	2.38	2.13	2.21	2.34	2.07	1.37	−1.01
	Disturbance regulation	0.05	0.04	0.04	0.05	0.04	0.03	−0.02
	Waste treatment	2.45	2.16	1.93	2.63	2.18	2.39	−0.06
	Erosion control	3.02	2.63	2.72	2.93	2.57	2.05	−0.97
	Pollination	0.65	0.58	0.57	0.59	0.5	0.67	0.02
	Water regulation	2.57	3.94	1.76	5.41	4.26	1.95	−0.62
	Biological control	0.84	0.79	0.79	0.77	0.66	0.78	−0.06
Supporting services	Nutrient cycling	9.84	8.82	9.17	9.68	8.58	5.68	−4.16
	Soil formation	0.12	0.1	0.1	0.11	0.1	0.08	−0.04
	Habitat/refugia	0	0	0	0	0	0	0
Cultural	Cultural	0.02	0.01	0.01	0.02	0.01	0.01	−0.01
	Recreation	1.32	1.25	1.23	1.42	1.23	0.8	−0.52
Sum		30.52	29.69	27.03	34.1	29.2	21.0	−9.48

rate of change is US$0.04 million, US$0.02 million respectively. All other function decreased their contribution with nutrient cycling (−US$4.16 million), raw materials (−US$1.42 million), food production (−US$0.2 million), water regulation (−US$0.62 million), climatic regulation (−US$1.01 million), most to the over decrease (−US$9.52 million) (Table 25.7). The comparative result indicates that the highest value calculated for the group of regulating service US$12.05 million, supporting service US$9.96 million, provisioning service US$6.25 million and cultural service US$1.33 million presented in the year 2019 about regulating service US$9.37 million, supporting service US$5.76 million, provisioning service US$5.22 million and cultural service US$ 0.81 million. Therefore, the order of involvement remains the similar over the study period but individual service decline throughout the year except for gas regulation and pollination (Table 25.7). It can be implicit from this study the dynamic change of LULC can have a significant effect on the overall ESV of the study landscape.

25.4.3 Ecosystem Sensitivity Analysis

To confirm the strength and reliability of our research using alternative coefficient (50% adjusting in the value coefficient) the total ESV over the study period are represented in Table 25.8. The coefficient of sensitivity (CS) of this analysis was less than one in all LULC revealed that the total estimated ESV of study landscape was inelastic, concerning modified value of coefficient. If the CS value was less than one, then the estimated ecosystem value is regarded to be inflexible but if the CS value is greater than one then ecosystem value considered changeable with esteem to that coefficient. The CS range from 0.05 to 0.07 for cultivated land and high CS values (0.58–0.73) is estimated for forest land when the value coefficient for these LULC types was adjusted by 50%. The sensitivity analysis point out that estimation was reasonable and strength and the used ESV index was a well fit for our research landscape.

25.4.4 Discussion

LULC change is dynamic and non-linear nature, and it depends upon different natural and anthropogenic factor (Meshesta et al. 2014). In this study, we observed forest land, cultivated land, and water bodies have been continuously loss but settlement and fallow land have been constantly increase between the study period 1972 and 2019. Many studies point out that the forest land and agricultural land are being decreased by different human activity (Tripathi et al. 2019; Kreuter et al. 2001).

LULC change has been a significant factor of regular change in the ecosystem and its services. Moreover, ecosystem services of the study area were found to be significantly reduced over the years due to the decline of important components of the

Table 25.8 Percentage change in estimated total ecosystem service values and coefficient sensitivity (CS) after a 50% adjustment of service valuation coefficients (VC)

	1972		1987		1992		2002		2012		2019	
	(%)	CS	(%)	CS	(%)	CS	(%)	CS	(%)	CS	(%)	CS
Forest VC ± 50%	35.11	0.70	32.35	0.64	36.92	0.73	30.91	0.61	32.00	0.64	29.49	0.58
Cultivated land/cropland VC ± 50%	3.24	0.06	3.67	0.07	3.98	0.07	2.74	0.05	2.88	0.06	3.22	0.06
Grassland VC ± 50%	5.36	0.10	3.86	0.07	4.25	0.08	4.19	0.08	3.96	0.08	10.37	0.20
Water bodies VC ± 50%	6.29	0.12	10.12	0.20	4.85	0.09	12.16	0.24	11.16	0.22	6.91	0.13

landscape especially forest. The similar results are also corroborated with the earlier research work conducted by Kindu et al. (2016) and Tolessa et al. (2016). Moreover, the ecosystem individual function like nutrient cycling, raw material, recreation, erosion control was also reduced in our study area which is also earlier reported by Hao et al. (2012) and Sharma et al. (2019). The methods used in this study will also help to assess the magnitude and spatial distribution of changes in the ecosystem services with respect to LULC changes over the study period 1972 to 2019. Our study landscape ESV was found to be reliable by sensitivity analysis that proves to be vigour of our estimation, which was similar to the finding of kindu et.al. (2016) and Tolessa (2016).

The loss of natural forest in Jhargram block is an important role to the decline of values from several ecosystem services and individual ecosystem function, (Hu et al. 2008; Kreuter et al. 2001). Most developing countries lose the ecosystem service from day to day. Thus, based on trustworthy information the deep action is needed to make the trade-off between land uses and ecosystem service that can optimize current and future ecosystem service requirements (Tolessa 2016).

Our study will help the local level assessment for other data scare region of the West Bengal particularly tropical region in India. The quantitative valuation of total and individual assessment of ESV changes shown in the study can give an important possibility for effective land use management resolution and optimizing deliverance of important ecosystem services.

25.5 Conclusion

This study describes the significant relation between LULC changes and impact on the function and structure of the ecosystem services. It is necessary to explore how much service value is lost responding to LULC transformation on a spatial and temporal scale. We believe this study will help to decision-maker for proper planning to achieve the sustainable development goal and it is also important for further research in the Jhargram block.

We found that the ecosystem service and their function reduced due to loss of forest cover could have an impact on the livelihood of local tribal community people. Therefore, it is necessary to adopt proper ecological protection to achieve the trade-off land use and ecosystem services. Beyond the completion of these finding, we recommend that a further challenge will continuously put forward specific alternative strategies and future planning for improving the ecological environment and ecological services and increase the involvement of local people for managing forest resources and services.

Acknowledgements We are grateful to the PG Department of Geography, Raja N. L. Khan Women's College (Autonomous), affiliated to Vidyasagar University, Midnapore, West Bengal, India for supporting this research. The author (P. K. Shit) grateful acknowledges West Bengal

DSTBT for financial support through R&D Research Project Memo no. 104(Sanc.)/ST/P/S&T/10G-5/2018.

References

Chen J, Sun BM, Chen D, Wu X, Guo LZ, Wang G (2014) Land use changes and their effects on the value of ecosystem services in the small Sanjiang Plain in China. World J Sci. https://doi.org/10.1155/2014/752846

Costanza R, d'Arge R, de Groot R, Farberk S, Grasso M, Hannon B, Limburg K, Naeem S, O'Neill RV, Paruelo J, Raskin RG, Suttonkk P, van den Belt M (1997) The value of the world's ecosystem services and natural capital. Nature 387:253–260

Costanza R, de Groot R, Sutton P, van der Ploeg S, Anderson SJ, Kubiszewski I, Farber S, Turner RK (2014a) Changes in the global value of ecosystem services. Glob Environ Change 26:152–158

Costanza R, Kubiszewski I, Giovannini E, Lovins H, McGlade J, Pickett KE, Ragnarsdottir KV, Roberts D, De Vogli R, Wilkinson R (2014b) Time to leave GDP behind. Nature 505:283–285

de Groot R, Brander L, Ploeg S, Costanza R, Bernard F, Braat L, Christie M, Crossman N, Ghermandi A, Hein L, Hussain S, Kumar P, McVittie A, Portela R, Rodriguez LC, Brink P, van Beukering P (2012) Global estimates of the value of ecosystems and their services in monetary units. Ecosyst Serv 1:50–61

de Groot RS, Wilson MA, Boumans RM (2002) A typology for the classification, description and valuation of ecosystem functions, goods and services. Ecol Econ 4:393–408

Everard M, Kangabam R, Tiwari MK, McInnes R, Kumar R, Talukdar GH, Das L et al. (2019) Ecosystem service assessment of selected wetlands of Kolkata and the Indian Gangetic Delta: multi-beneficial systems under differentiated management stress Wetlands. Ecol Manag 27(2–3):405–426

Förster J, Barkmann J, Fricke R, Hotes S, Kleyer M, Kobbe S, Kübler D, Rumbaur C, Siegmund-Schultze M, Seppelt R, Settele J, Spangenberg JH, Tekken V, Václavík T, Wittmer H (2015) Assessing ecosystem services for informing land-use decisions: a problem-oriented approach. Ecol Soc 20(3):31

Hao F, Lai X, Ouyang W, Xu Y, Wei X, Song K (2012) Effects of land use changes on the ecosystem service values of a reclamation farm in Northeast China. Environ Manage 50(5):888–899

https://www.anandabazar.com/district/midnapore/green-is-being-destroyed-says-report-1.730578?ref=midnapore-new-stry

https://www.indiaenvironmentportal.org.in/files/file/Jhargram-forest-NGT-order.pdf

Hu H, Liu W, Cao M (2008) Impact of land use and land cover changes on ecosystem services in Menglun, Xishuangbanna Southwest China. Environ Monitor Assess 146(1–3):147–156

Iverson L, Echeverria C, Nahuelhual L, Luque S (2014) Ecosystem services in changing landscapes: an introduction. Landsc Ecol 29:181–186

Kindu M, Schneider T, Teketay D, Knoke T (2016) Changes of ecosystem service values in response to land use/land cover dynamics in Munessa-Shashemene landscape of the Ethiopian highlands. Sci Total Environ 547:137–147

Kreuter UP, Harris HG, Matlock MD, Lacey RE (2001) Change in ecosystem service values in the San Antonio area Texas. Ecol Econ 39(3):333–346

Kubiszewski I, Costanza R, Anderson S, Sutton P (2017) The future value of ecosystem services: global scenarios and national implications. Ecosyst Serv 26:289–301

Kumar S, Chaudhry P (2015) Ecosystem services valuation of the forests of Arunachal Pradesh State India. Brazil J Biol Sci 2(4):369–375

Lambin EF, Meyfroidt P (2011) Global land use change, economic globalization, and the looming land scarcity. Proc Natl Acad Sci USA 108:3465–3472

Li RQ, Dong M, Cui JY, Zhang LL, Cui QG, He WM (2007) Quantification of the impact of land-use changes on ecosystem services: a case study in Pingbian County China. Environ Monitor Assess 128(1–3):503–510

Mamat A, Halik Ü, Rouzi A (2018) Variations of ecosystem service value in response to land-use change in the Kashgar region, northwest China. Sustainability 10(1):200

Meshesha DT, Tsunekawa A, Tsubo M, Ali SA, Haregeweyn N (2014) Land-use change and its socio-environmental impact in Eastern Ethiopia's highland. Reg Environ Change 14(2):757–768

Millennium Ecosystem Assessment (2005) Ecosystems and human well-being: a framework for assessment. Island Press, Washington

Miyamoto M, Parid MM, Aini ZN, Michinaka T (2014) Proximate and underlying causes of forest cover change in Peninsular Malaysia. Forest Policy Econ 44:18–25

Mondal PP, Zhang Y (2018) Research progress on changes in land use and land cover in the western Himalayas (India) and effects on ecosystem services. Sustainability 10(12):4504

Polasky S, Nelson E, Pennington D, Johnson KA (2011) The impact of land-use changeon ecosystem services, biodiversity and returns to landowners: a case study in the of Minnesota. Environ Resour Econ 48(2):219–242

Schägner JP, Brander L, Maes J, Hartje V (2013) Mapping ecosystem services' values: current practice and future prospects. Ecosyst Serv 4:33–46

Sharma R, Rimal B, Baral H, Nehren U, Paudyal K, Sharma S, Kandel P et al. (2019) Impact of land cover change on ecosystem services in a tropical forested landscape. Resources 8(1):18

Sloan S, Sayer JA (2015) Review and synthesis: forest resources assessment of 2015 shows positive global trends but forest loss and degradation persist in poor tropical countries. Forest Ecol Manage 352:134–145

Sinha B, Mishra S (2015) Ecosystem services valuation for enhancing conservation and livelihoods in a sacred landscape of the Indian Himalayas. Int J Biodiv Sci Ecosyst Serv Manag 11(2):156–167. https://doi.org/10.1080/21513732.2015.1030693

Styers DM, Chappelka AH, Marzen LJ, Somers GL (2010) Developing a land-cover classification to select indicators of forest ecosystem health in a rapidly urbanizing landscape. Landsc Urban Plann 94:158–165

Sukhdev P (2009) Costing the earth. Nature 462(7271):277

Tolessa T, Senbeta F, Abebe T (2017) Land use/land cover analysis and ecosystem services valuation in the central highlands of Ethiopia. Forests, Trees and Livelihoods 26(2):111–123

Tripathi R, Moharana KC, Nayak AD, Dhal B, Shahid M, Mondal B, Shukla AK et al. (2019) Ecosystem services in different agro-climatic zones in eastern India: impact of land use and land cover change. Environ Monit Assess 191(2):98

Tsegaye D, Moe SR, Vedeld P, Aynekulu E (2010) Land-use/cover dynamics in Northern Afar rangelands Ethiopia. Agric Ecosyst Environ 139:174–180

Yi H, Güneralp B, Filippi AM, Kreuter UP, Güneralp İ (2017) Impacts of land change on ecosystem services in the San Antonio River basin, Texas, from 1984 to 2010. Ecol Econ 135:125–135

Zang SY, Wu CS, Liu H, Na XD (2011) Impact of urbanization on natural ecosystem service values: a comparative study. Environ Monit Assess 179:575–588

Chapter 26
From Genesis to Awaited Success of Joint Forest Management in India

T. K. Mishra, Sudipta Kumar Maiti, Saikat Banerjee, and S. K. Banerjee

Abstract In most of the developing countries of the world including India local people are the chief users and guardians of the different types of ecosystems, and they make the vast majority of daily environmental decisions with their land use and investment choices. They have used their traditional knowledge, since days unknown, to manage natural resources, conserve ecosystems, and adapt to environmental changes. Community management is always based on variety of reasons like resource enhancement, religious and cultural purposes, and many other needs. In India the onset of community management in a large scale during 1990s was mostly for the purpose of resource enhancement, livelihood and biodiversity conservation. But the movement to protect forests did not get its deserved share. Empowerment of people to manage forest resources remained as a far cry. Thereafter, there was no further development in the process of community management as it was envisaged. In the present time, in view of increased need of carbon sequestration, sustainable forest management has beecome a need. But this can not be compromised with the subsistence need of the forest dependent people. Therefore, a new generation community management needs to be deviced to sustain our forest resources for the need of the forest dwellers in the one hand and global need of carbon sink in the other.

T. K. Mishra
Ex Guest Faculty, Department of Environmental Science, University of Kolkata, Kolkata, India

S. K. Maiti
Department of Botany, Raja N. L. Khan Women's College (Autonomous), Midnapore, West Bengal, India
e-mail: maitisudiptakumar@gmail.com

S. Banerjee · S. K. Banerjee (✉)
Forest Ecology and Climate Change Division, Tropical Forest Research Institute, P.O. RFRC, Mandla Road, Jabalpur, M.P. 482021, India
e-mail: banerjeeskdr@gmail.com

S. Banerjee
e-mail: saikatantara@gmail.com

P. K. Shit et al. (eds.), *Spatial Modeling in Forest Resources Management*, Environmental Science and Engineering,
https://doi.org/10.1007/978-3-030-56542-8_26

Keywords Joint forest management · Sustainability · Ecological impact · Productivity enhancement · Carbon sequestration

26.1 Introduction

Peoplels' participation in managing forest resources has never been a new idea in India. Though in some other form, it has been a part of our heritage since time unknown. In ancient ages, life was mostly forest oriented. Then came the idea of domestication of plants and animals. As the society evolved, people became more and more detached from their forest bases. But the rulers gradually took hold of forest resources and during the first half of British era brought almost all good forests under state control without considering about the forest dwellers. Lele (1998) writes "In a country whose last 150 years are a history of overwhelming state control over forests, one could ask the question: Why should there be joint management of forests? A simple-minded answer would be that joint management is required because the current system of management has failed to arrest the degradation of forests in India, the basic objective of forest policy".

Forests and forest resources have since long been supporting the livelihood and overall subsistence of humans. Its importance is much magnified for the forest dwelling communities and villagers around the forests. Forests are home to at least 80% of terrestrial biodiversity, and are a major carbon sink for regulating global climate (Lugo and Brown 1992; Brown et al. 1993; Brown 1996). Over 1.6 billion people worldwide depend on forest resources for their livelihood and many rely on forests for food, shelter and water (Arnold 1998; Byron and Arnold 1999). In India, the poverty-stricken areas coincide with the natural forest areas, in which nearly 400 million people derive livelihood benefits from forests (Lynch and Talboot 1995; FAO 1998). Forests and forest resources have also long been meeting the needs of communities in urban areas in the form of timber, raw material, forest based industries and recreation (Mathur and Sachdeva 2003). The varied user profile and further inequity of distribution and the growing demand has created a pressure on the forests. The combined effects of over exploitation of timber, deforestation of arable land and pastures, industrialization, construction, urbanization etc. has resulted in the dwindling and degradation of forest resources and many of the plant resources are threatened today (Davidar et al. 2010; Aggarwal et al. 2009; Ravindranath et al. 2012).

Thus, the idea of participatory management evolved as a movement and not as a government programme. Even now the majority of the forest areas are under state control as reserve forest, sanctuaries and national parks. Only protected and unclassed open forests (less than 40% of canopy) and degraded forests (less that 10% forest cover) were allowed to protect by the villagers.

The protection of forests by people's participation and the empowerment of people in taking management decision started with the intervention of Government or non-Government organization and the participation started only when the forest administration failed to tackle the situation. Few individual successful efforts made in the states of West Bengal (Arabari village 1972), Uttaranchal (Chipko protest movement 1973), Haryana (Sukhomajri village 1976), and Rajasthan (Gopalpura village 1986) were testimony to the people's movement to reestablish a relationship between the man and nature in respect of the management of forests and develop rural livelihood and regenerate forests through co-management which caught the attention of foresters, India's politicians as well as international environmental activists.

Sundar (2017) stated that joint forest management (JFM) in India has been instrumental in changing the socioeconomic lives of forest dependent communities (FDCs) from being forest-centered to one based on the money economy. JFM has resulted in the elite capture of forest resources, and enabled regeneration of degraded forests though the evidence on the quality of regeneration is mixed. The government records suggest that with the drying up of donor funds for the JFM program, FDCs are losing interest and many JFM committees are now defunct in many states. This may be due to losing of interest of forest villages in protecting forests as no responsibility of management has been bestowed on them.

It has been rightly expressed by some authors that there has been fall of community oriented forest protection and management in our country. But there is plenty of scope in developing the joint protection of forests to joint management. In spite of all the pros and cons, the total number of JFMCs in the country as on 2016 are 118,213 and the forest area brought under it is 22,938,814 ha (28.17% of the forests of the country). The first promising step in the direction towards the development of new participatory management systems was taken by the Government of West Bengal through their orders of 12 July 1989 to constitute Forest Protection Committees (FPCs) who were allowed to take 25% of the timber produced as a measure of incentive. The order contained a detailed resolution on composition of FPCs, their duties, functions and usufructuary benefits to be provided to them. Some amendments were made to the resolution in July 1990 so as to make it more democratic and effective.

Clean Development Mechanism (CDM) of the Kyoto Protocol provides an opportunity to promote JFM, to increase carbon stock in vegetation and soil, to build capacity and enable communities to participate effectively in forest protection and management and increase income or flow of benefits, creating long term stake for the communities. CDM has two main goals: (1) to assist countries without emission targets (in developing countries) in achieving sustainable development, (2) help those countries without emission reduction targets under Kyoto (i.e. developed countries) in achieving compliance by allowing them to purchase offsets created by CDM Projects. Recent development has recognized afforestation and reforestation as activities, which could be undertaken under Clean Development Mechanism (CDM) of the Kyoto Protocol. In a related development, forest management in India experienced a dramatic change in 1990 when a community based approach (Joint Forest Management) was adopted as one of the strategies for protection and management of state forest lands.

The present day effort of protecting forests by forest fringe people who are ecosystem people as well as ecological refugees do have antecedents in systems such as "Ban Panchyats" of Kumaon, many of which have been functioning successfully since the late 1920s. Results of CHIPKO movement in early 1970s inspired the idea of protecting forests by village people. In the Northern India mostly in Tehri Garwal Himalayas many Van Suraksha Samities (VSS) were formed by the local youths to protect their own forest resources from the external exploiters. Now-a-days a large number of community forest management groups are functioning throughout India. These include legally sanctioned Swayam Forest Groups of Kumaon, the Forest Protection Committees of West Bengal, and the Hill Resource Management Societies of Haryana, the Village Forest Committees of Uttar Pradesh, Karnataka and the informal indigenous groups operating in Orissa, Karnataka and Bihar.

As mentioned earlier recent developments have recognized afforestation and reforestation as activities which could be undertaken under CDM and this could easily include JFM activities. In all likelihood, a JFM project would qualify any substantial development criteria, as it promotes biodiversity, improves ecology and enhances availability of various forest products to the local communities.

The regenerating JFM forests would sequester carbon dioxide and the carbon credits would be accumulated by the Forest Development Agency (FDA). The FDA, in turn would provide the necessary guarantees to the private entity interested in purchasing these credits and transfer the certified emission reduction (CER) in lieu of agreed price.

JFM projects under CDM have the potential of providing the communities with carbon credits for regenerating degraded forest lands. These projects could attract funds for adopting innovative institutional interventions to enhance regeneration of forests and biodiversity conservation ultimately enhancing forest productivity. This would not only help the local communities in obtaining benefits in the form of enhanced flow of forest products and watershed protection through enhanced vegetation cover, but would also help them in getting financial returns for every ton of carbon sequestered. The importance of increasing current income levels as well as providing alternative income sources to forest communities is finally being recognized as an essential process to reducing pressure on forest. This provides an opportunity to facilitate village development activities with the help of carbon credits (Singh 2004). A study was carried out to estimate the creation of carbon sinks and sequestration achieved in community projected forests of Sambalpur Forest Division, Orissa. The results have shown that 1.53–3.01 tonnes of carbon is being sequestered per ha per year with only protection, which can be enhanced through proper implementation of the management prescriptions (Gera et al. 2002).

26.2 Initiation and After

It is difficult to trace back the genesis of participatory forest management. No where in history of ecology and that of natural resource management one can find that the state power has allowed people to manage forest with their own acumen and needs. But one can illustrate the case of 'Arabari socio-economic experiment' model from early 1970s to late 1970s as the beginning of present concept of JFM. So, much before 1988, the forest fringe villagers in southwest Bengal with the help of a newly elected Panchayat system protected their own forest with the apprehension that they will get something in liu of their effort. West Bengal government sanctioned something more than non-timber resources in 1984–85 to the tribal people of the forest fringe people for their immediate requirements for agriculture and house repair. This movement did not confine to the borders of the state, rather spread over most of the states of central India.

The Arabari experience demonstrated that local people would effectively protect the degraded forest if their basic needs of fuel wood, fodder, and small wood are fulfilled, if they are provided exclusive rights to non-wood forest products, wage-paid employment, and are assured of substantial cash benefits from the final harvest. However, the experience also indicated that more comprehensive discussions with the local communities having legitimate claims to benefits from the forest need to be held before forest protection, management responsibilities and system of distribution of the benefits from joint management are determined and finalised. Based on the encouraging results of the Arabari experiment, the Government of West Bengal has prepared an ambitious programme to regenerate some 259,000 ha of sal (*Shorearobusta*) forest under the joint forest protection scheme.

The success of the Arabari experiment is attributed mainly to the political commitment of the state government for a better forest management, substantial and immediate benefits to the participating villagers. They were allowed to get 25% of the timber production in every ten years, wages during felling etc. and non-timber forest produce as per 1989 Assembly Legislation. This was a clear-cut policy for sharing of benefits from forest. It needs to be mentioned here that people were hardly involved in forest management activities other than protection of the forest.

Even prior to JFM movement the Van Panchayat movement in the foot hills of Himalayas set the road map. Van Panchayats in the northern part of erstwhile undivided Uttar Pradesh, now Uttaranchal, born out of conflicts and compromises that followed the settlements and reservations of forests in the hills during 1920s. The first government approved Van Panchayat was thus formed in 1921. According to estimates, there are 6069 Van Panchayats managing 405,426 hectares of forests (13.63% of total forest area) in the state (Mukherjee 2003). Most of these have been carved out of civil (protected) forests under the jurisdiction of the Revenue Department. The area under each Van Panchayat ranges from a fraction of a hectare up to over 2,000 hectares. Under JFM regime i.e. after 1988 there has been some fresh inputs to this effort.

Besides these two much talked experiments there are many other such efforts within different states of India. All those experiments and efforts were mainly people centered and can never be compared with 'Social Forestry' which started as a government sponsored programme after the recommendation of National Agriculture Commission, 1976. But it could not associate people with the protection and management of forest resources around them.

In the view of success of Joint Forest Management (JFM) experiments and failure of government mechanism to protect forests in 1988, New Forest Policy came into existance. During 1990 the operative part of such policy was detailed out in black and white. Till then to the Forest Rights Act (FRA) of 2006 that has extended rights over forestland to local communities, the devolution of authority in the case of India has only been increasing. Here it needs to be mentioned that the Panchayat Extension to Scheduled Areas Act (PESA) of 1996 gave power over minor forest products to gram sabha (village assemblies) in scheduled areas, thus taking 'decentralization' beyond Forest-Department-managed territory. It made gram sabha "competent to safeguard and preserve the traditions and customs of the people, their cultural identity, community resources, and the customary mode of dispute resolution" [section 4 (d)]. 2006 FRA went much ahead of that allowing Community Forestry rights to the real forest dwellers. The Biodiversity Conservation Act (2002), while taking within its purview all forms of local governance like gram panchayat (village level), panchayat samiti (sub-district level), zilla parishad (district level), and even municipal corporations in urban areas, authorized local level institutions to grant or refuse permission to outsiders to use products of biotic resources, and to charge fees for their use. Although the implications of PESA are not evident at the ground level, and rules of the Biodiversity Conservation Act are yet to be finalized by the National Biodiversity Authority (Gadgil 2008), the two Acts reflected the Government's commitment to decentralization. But till date there is a wide gap between what is there in acts and policies and what is happening in the field. Delay in implementation creates skepticism that the rights bestowed might be withdrawn through other windows in course of time.

In July 1990, the Govt. of India issued an order which stressed that the right holders should identify themselves with the development and protection of forests from which they benefits. The order asked the state forest departments to work out modalities to ensure participation of communities in afforestation programmes. It recommended that NGOs and voluntary agencies should act as interface between the forest communities for the revival, restoration and development of degraded forests. The institutions through which this could be achieved could be a Panchayat, a village co-operative or a village forest committee. The village community would get a share of usufructs, lops and tops and if protection has been successfully carried out, a share of the final revenue or crop. However, with regard to the share of Village Forest Committees (VFCs) in timber produce, the share ranges from 25% to VFC members in Kerala to 100% in Andhra Pradesh. It is 50% in many states including Gujarat, Maharashtra, Orissa and Tripura. In most of the states, 25% of the timber revenue is to be deposited in the village development fund. For access to NTFPs, barring a few nationalized products, in most states all NTFPs are available to the people free of royalty. There is a range of specifications for collection and sale of NTFP in different states. In Karnataka, 50% of NTFP, fruits, timber and final harvest are to be sold to local villagers at Forest Department rates, while the remaining may be auctioned and proceeds shared between the government, VFC members and village development fund. With regard to grazing, while some states have banned grazing completely, some have rotational grazing which has helped the regeneration of vegetation in forests. In spite of all these odds and apprehensions, Poffenberger and McGean (1998) enthusiastically stated that JFM would bring in a "reversal of the alienation of forest people's rights, of institutional conflict, and of ecological patterns of forest degradation".

In majority of the states, JFM is implemented through government order/guidelines/resolution or instruction, except the state of U.P. where States Legislative Assembly has passed JFM Rules (TERI 1998). On analysing the government order/resolution passed by the various State Governments, it is clear that no state government was in a position to confer giving ownership to community over forest before 2006. State governments were only able to provide mainly limited rights to use or share benefit of the forest resources ranging from NTFPs to timbers.

Forests being in the concurrent list of the Constitution of India, it is clear that no State has confered any kind of ownership rights over forest. Various State Governments have only developed a mechanism of benefit sharing which is also conditional. Jindal et al. (2003) gave a detailed comparison of the JFM regulation of some leading states of India which is provided Table 26.1.

Rao et al. (2005) reviewed the Joint Forest management studies in India pertaining to their monitoring and evaluation. They found that there were no nationwide monitoring studies done by any agency. There are few regional studies only and mostly based on the objectives of the donor agencies.

Agarwal et al. (2005) made some case studies on social aspects of JFMC in India. Till then as per MoEF data there were 84,642 JFMCs in India protecting 17.33 million ha of forest throughout the country. They studied JFMCs like Barror in Karnataka, Sijua-Murakata-Kuanburi of Paschim Medinipur, West Bengal. Both of these JFMCs were found to have developed as independent institutions and confidently implement their management decisions.

Mishra et al. (2005) made detailed study of spread performance and impact of JFM in West Bengal. They studied 200 FPCs of West Bengal. Their study states that there is much dependence of local people on the forest resources for food, medicine, building materials and others. As both overstory and understory plant species of the forest, provide them the resources for their livelihood, more active community-based conservation is essential for sustainable availability and use of such resources. They recommend that the community has to ensure least human interference in the natural regeneration of the forest in order to maintain its phyto-diversity. Their study also show that, women are richer in local ecological knowledge than men. But inspite of women's greater knowledge about forest ecosystem and involvement in forest protection activities, their participation in institutional activities for JFM in West Bengal is not satisfactory. It is, therefore, the liability of the JFM Executive Committee (JFMC) members to moralize more women members towards forest conservation and management activities. It is also required that any operation of unsustainable extraction by private agencies within the forests should be thoroughly monitored and checked by the joint effort of foresters and JFMC members. They also studied the nutrient status of the degraded forest soil managed by JFMC and found that the nutrient status has increased considerably. This is because of increased vegetation cover. The results are in agreement with those obtained by Singh et al. (1987) who

Table 26.1 Status of ownership right under JFM in various Indian states

State	Status of right provided to community	Legal status
Andhra Pradesh	Vana Samrakshana Samities (forest conservation committees) are entitled to 100% share in timber and bamboo harvested from the regenerated degraded forests in approved micro plan	Servitude
Arunachal Pradesh	25% revenue accruing from thinning and felling to be distributed amongst member	Servitude
Bihar	Income earned will be divided into three shares and one share will be deposited in village development fund, one in forest development fund and the third in working fund	Servitude
Gujarat	All benefits deriving form sale of timber will be utilize by the village committee in a planned manner	Servitude
Haryana	30% benefit will flow to village forest committee and 70% to government	Servitude
Himachal Pradesh	25% of benefit will go to village forest development committee	Servitude
J&K	Benefit sharing is not decided	No right
Karnataka	50% community and 50% government	Servitude
Kerala	Village Samrakshana Samities will get 10% of net revenue harvested forest produce and 100% benefit revenue from NTFP	Servitude
Madhya Pradesh	Funds allotted for protection against fire, felling etc. will be deposited in village-resource development plan fund in the even if people successfully carry out these protection measures. Work to be done as per approved micro plan	Conditional servitude
Maharashtra	Yet to be decided	Right undecided
Nagaland	No provision for flow of revenue to community	No right
Orissa	Benefit sharing is to be decided by government expenditure incurred for implementation of JFM will be provided by government	Right undecided
Punjab	No provision for flow of revenue to community	No right
Rajasthan	50% benefit to village committee	Servitude
Tamil Nadu	Pruned material to given to head loader and land less household free of cost 90% profit obtained by sale of NTFP will be shared equitability among the members of VFC 50% profit to VFPC in case of bamboo & small timber	Servitude
Tripura	No provision for flow of revenue to community fund	No right

(continued)

Table 26.1 (continued)

State	Status of right provided to community	Legal status
Uttar Pradesh	50% to be distributed to village community members and remaining 50% t be spent on community work including recycling of funds for management of village forest	Servitude
West Bengal	No provision for flow of revenue to community	No right

studied the effect of vegetation covers on the nutrient status of soils of a degraded land. Their results indicate that available average nitrogen content of the soil of barren land was only 38.0 ppm but in the JFMC managed sites it was 122.0 ppm. The enrichment of nitrogen is not only due to the presence of the large number of ground vegetation but also due to the fixation of atmospheric nitrogen by increased presence of nitrogen fixing species in the herb and shurb level. A large number of ground vegetation also directly or indirectly contributed to the enrichment of nutrients to some extent in the soil.

26.3 Issues Before JFM

Joint Forest Management is mainly a forest management activity rather than a socio-economic activity. It is essential to know the different management strategies for different forest regeneration patterns in respective state agro-ecological zones. Any sort of participation in natural resource management has a multidimentional implication. It has always a positive impact on the conservation of the resources and it also creates space for the participants to receive something for their livelihood. In this give and take environment there always have some important issues to address. But it is a general finding throughout the country that the positive trends that enriched the participatory forest management process during late twentieth century started to fade out in early part of twenty-first century though there are supporting government policy and guidelines and massive fund support. In this case both forest department and the community have responsibility to resolve this issue and strategies should be framed for its revival and JFMC may take effective measures towards forest conservation for enhancing productivity and livelihod opportunities for future. Malhotra et al. (1992) demonstrated that compared to plantation approach, natural regeneration of degraded forests is not only cost effective but also socially relevant and ecologically sound. A natural forest production system will benefit the FPC members much more than a plantation production system and is, therefore, more sustainable.

26.4 Transforming Users to Managers

In India forests play a vital role in the rural economy. In many areas, forests and its resources provide biomass for livelihood to the rural poor. They provide different kinds of benefits like jobs and incomes often needed to supplement inadequate returns from agriculture produce such as fuel wood, food, fodder, NTFPs, and building poles for the home and a range of environmental benefits, without which the other activities such as agriculture might be impossible. In India forest sector is the second largest land use after agriculture. The forest dwellers has always been users of the forest resources. For about last one hundred years or more these forest users were labelled as intruders. But in the JFM regime there developed some scope for involving villagers in the process of management of forest resources. Mishra (1999) stated that in West Bengal the forest legislation of 1989 and the amendments of 1990 have made possible the integration of the indigenous knowledge of local communities with the 'scientific' principles involved in modern forestry management. Quoting case studies from Midnapore district he showed how this has led not only to better management of forests, but to a richer anthropogenic activity and better biomass yields. Sarkar and Das (2006) did an extensive study on the economic outcome of Joint Forest Management in West Bengal which is based on the strategic decision making in between government and forest fringe community using evaluation of cost and revenue in randomly selected FPC. Their empirical study suggests that economic outcome of the JFM programme has been beneficial for both community belonging to marginal landholding, small landholding and landless agricultural households and government and this is due to the strict dominant cooperative strategy of community. But the earlier forest policy of the government was oriented with the commercial need of the government disregarding the traditional right and benefit of the forest fringe communities. Kumar (2005) has opined that Joint forest management represents a progressive shift towards state recognition of the interdependence between the well-being of the forests and well-being of the women and men dependent on them for subsistence and livelihood needs. Although it does not transfer the ownership or title of the land to the local community, the administrative legitimacy that state joint forest management resolutions lend to benefit showing arrangement, with community institutions have motivated thousands of villagers in the three states to initiate or participate in forest regeneration through community controlled protection.

Sustainability of JFM initiatives is being viewed as a matter of concern as also a challenge because sustainability can be assured only when there is genuine involvement of the community at each level. At present, the communities are performing their roles of protection of forest effectively. However, their roles in management, decision-making, access to information, etc. need to be further strengthened. The foundation of participatory management of forests has been laid. It is time to consolidate and build upon these initiatives so that the long-term security and development of forests can be ensured. Switching over from control to support system of management, treating forests as part of the resources of the communities, modernization of its nurseries, plantation and seed technology, promoting biodiversity conservation through eco-development and introduction of information technology as a management tool are some of the major initiatives.

Forests are expected to serve as a more secure source of meeting basic needs related to fodder, fuel wood and other minor forest products. While regeneration efforts can increase wage employment opportunities for the poor, biomass increase can enhance the scope for additional employment and income generation through the collection of NTFPs. Positive implications of the successfully implemented JFM for the community and the society would be preservation of the biodiversity through regeneration of local and resident species of the region and by sustaining local wild life, low runoff of rain water, reduction of soil erosion, higher recharges leading to augmentation of the ground water level, employment generation, improvement in income, an increase in team building capacity, women empowerment etc. All the stated advantages can be commanded only when vegetation cover is successfully established and sustained by the local communities. Therefore, it is imperative that the forest development, management and exploitation have to be collaborative effort between the government and local community.

26.5 Ecological Impact

Several studies have documented the extensive ecological knowledge regarding forests that many local and indigenous populations maintain, and forest management practices that are ecologically sound. There were various studies to assess the ecological impact of JFM throughout India. Their study noted that there are significant strides made in promoting JFM, but the program still needs to address a lot of inadequacies. Though it is estimated that nearly 23.0% of forests are brought under JFM, covering nearly 50% of the open forests in India, how much of it has developed into good forests is not really known. They also noted that there are many lacunae in implementation of the program and there are also gaps in policies to promote JFM.As stated before, West Bengal forest department in India was a pioneer in initiating Joint Forest Management (JFM) involving participatory forest management by both foresters and local communities in order to protect degraded forests. Gupta and Mishra (2019) did a case study in Jhargaram district of West Bengal. The research objective was qualitative and quantitative study of tree communities by quadrat method along with identification of major impacts of environmental factors affecting regeneration of five locations under JFM in natural coppicing Sal (*Shorea robusta*) dominated tropical dry deciduous forests of West Bengal. 23 families, 33 genera and 36 tree species were identified. Dominant families were *Anacardiaceae* and *Combretaceae*. There were statistically significant differences in stand diversity, dominance, richness and evenness. Generalized Linear Model predictors such as site categories, seasons, grazing intensity and invasive species frequency had significant impacts on seedling diversity, dominance, richness and density. Surface soil potassium and soil texture were best predictors of seedling abundance. Species with "poor", "no" and "new" regeneration status necessitate proper attention in forest management plans involving regulation of exotic invasive species populations, grazing and browsing, lopping, fire, over-extraction of non-timber forest produce and prevention of illegal felling through vigilance and more active participation of Forest Protection Committees.

Forest management is an interdisciplinary subject involving both social scientists and research workers in the field of forestry. The scientifically sound silvicultural systems can be introduced based on increased level of consciousness about the ecology of forest (Mishra et al. 1994). Mishra et al. (1997) studied the phytosociological and soil physico-chemical aspects of a highly degraded lateritic land of Midnapore Forest Division, West Bengal and observed that proper management and silvicultural practices have dramatically changed the ecology of the region within five years. The impact of different management systems (FPC managed forests, plantation stand managed by State Forest Department and degraded forest managed by Forest Department) on biodiversity conservation in Amlachati Range of West Midnapore, W. Bengal has been assessed by Mishra et al. (2008). The study was conducted covering the four seasons (summer, monsoon, post-monsoon and winter). Joint Forest Management involving local people both for protection and management of forests have brought about visible quantitative and qualitative changes in the structure of forest community in this region. The results also indicate that the nutrient status was higher in FPC managed forest soil than that of the other two sites. The results reveal that different management practices like thinning, burning, weeding, collection of regulated fuel wood as well as other non-timber forest produces etc. within a sustainable limit may create a possible avenue for the invasion of new species and these practices can be assured through joint forest management by FPCs leading to fulfillment of the criteria of sustainability of forest ecosystem.

26.6 Productivity Enhancement

A forest consists of many components that can be broadly divided into two categories that are biotic (living) and abiotic (non-living). The net primary production (NPP) of a forest is a well suited indicator of forest productivity. It consists of the accumulation of stem wood in standing trees plus the growth of all the other tissues or components including those that are short lived or roots. Forest productivity is affected by numerous important ecological, environmental and social factors. Enhancing productivity of each type of forest ecosystem through appropriate management (technological and material inputs) would lead to fulfillment of the criteria of sustainability of forest ecosystem. The productivity of an ecosystem can be enhanced to a certain level which could be defined as "optimum level of productivity". The productivity could be enhanced above the optimum level temporarily by providing material inputs (irrigation, soil amendments, use of pesticides, insecticides, fertilizers etc.) more intensively. Such a level of higher productivity cannot be sustained over a long period of time as it causes excessive strains on the system and leads to nutrient deficiency in due course which ultimately may have a serious adverse effect on the site conditions causing irreparable damage. Sometimes, the ecosystem may lose its capacity to recover to its original state and collapse. It is always advisable to manage the forest ecosystem at an optimal level of productivity. Therefore, one of the pre-requisite of the sustainable forest development is to assess the optimum level of productivity of each forest ecosystem based on its site factors, prevailing micro-climatic conditions and socio-economic and environmental conditions. Total biomass production at an optimum productivity level would be an indicator to determine the sustainability of the forest ecosystem.

Mishra (2001) made a detailed study of species wise biomass of non-timber species in both disturbed and undisturbed condition in few FPCs of West Midnapur Forest Division. During study on ecological impact of JFM in Bhagabati Chak Forest of Midnapore (West Bengal) managed by FPC since 1986 found that the community has evolved some management practices of their own which have positive role on the ecosystem. These practices include (1) restricting fuel wood collection by cutting green shrubs etc. to four times a year which is fixed by the FPC, (2) identifying six gregarious shrub species for fuel wood, (3) banning total sweeping of litter from the forest floor etc. The FPC identified six species viz. *Combretum roxburghii, Holarrhena antidysenterica, Lantana camara, Ehretia laevis, Antidesma acidum* and *Helicsteris isora* which are the major associates of coppice sal (*Shorea robusta*) dominated forest in the study site. In fenced condition the annual biomass production of these six species is 5812.95 kg/ha. In unfenced condition where there is regular exploitation of fuel wood, the annual productivity is 3611.25 kg/ha. Major plant species of ethno-botanical use of the region are *Hemidesmus indica, Ichnocarpus frutescens, Smilax macrophylla, Andrographies paniculata* (all of medicinal use), *Basella alba* (whole plant used as vegetable) and *Dioscorea alata* (rhizome used as vegetable) (Table 26.2).

Budhikhamari is situated in the Mayurbhanj district of orissa had a rich coppice sal forest prior to 1970. All kinds of biotic pressure such as illicit felling, grazing, collection of leaves and NTFPs caused destruction of sal forest almost completely by late 1970s. The whole forest range was degraded to almost nil forest resource which

Table 26.2 Annual productivity of fuel wood producing species

Species	Biomass (kg/ha)	
	Fenced condition	Unfenced condition
C.. roxburghii	5340.17	2933.78
H. antidysenterica	182.50	431.73
E. laevis	94.08	22.76
L. camara	96.00	78.20
A. acidum	82.00	100.40
H. isora	9.99	24.38
Total	5812.95	3611.25

Table 26.3 Annual productivity of NTFPs with ethno-botanical importance

Species	Biomass (kg/ha)	
	Fenced condition	Unfenced condition
Hemidesmus indicus	8.96	8.20
Ichnocarpus frutescens	15.66	14.50
Smilax macrophylla	10.36	11.61
Andrographis paniculata	8.28	3.70
Dioscorea alata	0.98	4.99
Basella alba	13.50	–
Total	57.74	43.00

evoked a sense of deprivation of a natural resource. The tribals such as Santals, Bhunyas, Bhumijas, Gondas etc. of the entire range with an initiative of Forest Department decided to protect their forest resources. At present there are 42 Village Forest Protection Committees (VFPCs) protecting nearly 3500 hectare of forest in the whole range. In the late 80s the Forest Department Officials of the area took special initiatives to organize, resolving conflicts and framing their own regulations to protect, manage and extract forest resources within sustainable limits. All the FPCs here in general have accepted the principle that fuel wood, fodder and NTFP collection without felling trees is allowed to the members protecting forests. Some nominal fees are levied by certain FPCs for certain items of NTFP (Table 26.3). It was observed that procurement of certain NTFPs has increased due to protection. Sal seed collection record shows that in 1991, 150 qtl of sal seed was collected whereas in 1992, the figure goes up to 250 qtl. Sharing the minor forest produce and fuel wood is decided by the villagers. In Bolangir and Mayurbhanj district villagers are free to collect sal leaves (for making plates), and mahua flowers, hill broom plants and tamarind. In Mayurbhanj the money collected resulted in an institution called Purti Society, which now leads forest protection activities in the village. The forest provides employment to villagers. Agricultural incomes of the villages have gone up. Land productivity has improved.

26.7 Towards Sustainability

One of the primary objectives of forest management in India outlined in the National Forest Policy 1988 is to help meet the resource and livelihood needs of forest dependent communities vis-a-vis ecological restoration. Brundtland Commission (World Commission on Environment and Development) defined it as development that "seeks to meet the needs and aspirations of the present without compromising the ability to meet those of the future (Smith and Rees 1988)". Sustainable condition is an ecosystem condition in which biodiversity, renewability and resource productivity are maintained over time. Sustainability can, thus be considered in three conceivable dimensions.

1. Ecological dimension (avoiding any damage to the environment)
2. Economic or production dimension (where a constant flow of goods can be assured)
3. Socio-cultural dimension (preserving all socio-cultural diversities).

Present day forests occupy about a quarter of the world's ice-free land with about half of these occurring in the tropics (WRI 1998; Groombridge and Jenkins 2002). In another estimate (FAO 2006), the global forest cover is about 3952 million ha which is about 30% of the world's land area. In the more recent assessment (FAO 2015), the forest cover in the world is 3999 million hectare which is 31% of the land area. The results indicated that total forest area declined by 3% from 4128 Mha in 1990 to 3999 Mha in 2015. According to FAO (2006), globally net carbon stocks in forest biomass decreased by about 4000 $MtCO_2$ annually between 1990 and 2005. In temperate and boreal regions forest area is gradually increasing but deforestation in the tropics is of major concern (Groombridge and Jenkins 2002). Loss of biodiversity occurs largely from the habitat loss and fragmentation produced by the human appropriation of land for development, forestry and agriculture as natural capital is progressively converted to man-made capital (Tilman et al. 2001; Gibbs et al. 2009).

Under such conditions, it is felt necessary by the global community that a management system for forests should be endured so that it not only meets the needs of the communities now, but also is perpetual to satiate the demand of the future generation. As compared to many other industries sectors, it is relatively easier for the forest community to expand its scope from sustained yield to sustainable development, which requires a shift from forest management to forest ecosystem management (Makela et al. 2000). Forest ecosystem management (FEM) approaches forest conservation, utilization, administration and regulation on the basis that the forest is a highly integrated, complex, generally resilient, multivalue biophysical system that has thresholds of tolerance for disturbance (either too much or too little) beyond which its resilience and certain values and environmental services are changed, and often reduced. FEM is the management of forest ecosystem processes and disturbance regimes to sustain the desired values and ecosystem services from a shifting mosaic of different ecosystem conditions across the landscape, and a non-declining pattern of change over time in the values and services provided by each stand in that landscape. It is also the management of the human use of, and interactions with the forest, because humans are part of forest ecosystems.

The concept "Sustainable Development" has been defined by various people as: Sustainable development is the management and the conservation of the natural resource base and the orientation of technological and institutional change in such a manner so as to ensure the attainment and continued satisfaction of human needs of present and future generations. Such sustainable development (in the agriculture, forestry and fisheries sectors) conserve land, water, plant and animal genetic resources, is environmentally non-degrading, technically appropriate, economically viable and socially acceptable. Sustainable development can thus be elaborated in the following terms:

- It requires elimination of poverty and deprivation
- It requires conservation and enhancement of the resource base
- It requires a broadening of the concept of development so that it should cover economic growth and cultural development, and
- It requires unification of ecology and economics at all levels.

Sustainable forest management (SFM) provides great opportunities for adapting to climate change by increasing the resilience of people and ecosystems. Forests are a major mitigation option over the next 30–40 years and can play a key role in the necessary transition towards low-carbon economy (GCEC 2014; IPCC 2014). Efforts to tackle climate change are thus becoming increasingly linked with efforts to conserve forests (Dixon et al. 1993). Climate change signifies the deforestation issue as a major development challenge. It is one of the most important global environmental problems facing humanity. FAO (2010) estimates that 13 million hectares of the world's forests are disappearing annually and that accounts for 20% of all global greenhouse gas emission. If deforestation continues at the present rate then the tropical forests may be lost by 2050. Thus, substantially reducing deforestation is a critical part of avoiding dangerous climate change. As previously mentioned, Kyoto Protocol's Clean Development Mechanism (CDM), allows investment from developed countries to compensate for their greenhouse gas emission through forestry schemes for developing countries. The Kyoto Protocols make provision to take into account Afforestation, Reforestation and Deforestation and other agreed Land Use, Land Use Change and Forestry (LULUF) in meeting their commitments. One of the ways suggested is to make project based activities aiming at reducing greenhouse gas or enhancing carbon stocks on a specified period through carbon sequestration. Such mechanism could offer potential benefits to farmers and pulp and paper industry by way of farm forestry/industrial plantations initiatives as carbon offsets.

Maintenance, conservation and enhancement of biodiversity are one of the criteria selected for sustainable forest management (SFM) in India under India-Bhopal Process (Prasad 2000) The process has proposed eight criteria and a total of 50 indicators associated with differen criteria. The proposed criteria are:

- Forest resource security
- Ecosystem function and vitality
- Biodiversity conservation
- Soil and water conservation
- Forest resource productivity
- Forest resource utilization
- Social, cultural and spiritual needs
- Policy and legal frame work.

The ultimate aim of criteria and indicators is to promote improved forest management practices over time, and to further the development of a healthier and more productive forest estate, taking into consideration the social, economic, environment, cultural and spiritual needs of the full range of stakeholder groups in countries concerned.

In order to monitor the progress of conservation of biodiversity towards SFM, following indicators were suggested under India-Bhopal process (Prasad 2000).

- Area of protected ecosystem
- Area of fragmented ecosystem
- Number of Rare, Endangered, Threatened and Endemic species
- Level of species richness and diversity in related area
- Availability of medicinal and aromatic plants in various forest area
- Status of non-destructive harvest of NWFPs
- Number of Keystone and Flagship species in various forest types.

There is a need for field-testing of these indicators followed by benchmark studies to establish site-specific minimum acceptable standards for monitoring progress towards SFM in the large tract of India.

Tropical forest ecosystem is one of the most biodiversity rich terrestrial ecosystems and it stores approximately half of the world's living terrestrial carbon and a very significant proportion is fixed in above ground biomass. Thus, it plays an important role in global carbon cycle and is regulating the biospheric climate (Ketterings et al. 2001). A study on estimation of forest carbon in four villages of South Balaghat Forest Division (M.P.) was conducted where Joint Forest Management as an institutional frame work could claim the benefits of REDD+, using conservation finance to reward and compensate the communities engaged in protecting and enhancing forest cover. In this case study, an economic value of net carbon stored comprising carbon storage and sink values, has been estimated using the market price approach based on IPCC methods to estimate carbon storage services by forests. The study demonstrates that wherever carbon stock exist in the forests, financial incentives can help to build resilience towards climate change. Moreover, REDD+ is the mechanism with focus on communities and has the potential to deliver lots of benefits including biodiversity conservation, improving livelihoods and building capacities for reducing poverty associated with climate change (Khera 2017).

Sahu and Rath (2011) made a study on the socio-economic impacts of JFM in some villages of Orissa. The findings of their study established that the improvement in forest cover under JFM has positive impact on the reduction of stress migration in the study area. The environmental stress migration, which was severe in these villages before the implementation of JFM, has come down significantly over the years. There results have also established that the health of local labour market has improved a lot due to the proper and sustainable utilization of the common property resources through JFM.

26.8 Making JFM Meaningful

In India, forestry development paradigms, other than community based ones, have become anachronistic and should be rejected in order to make further headway. The present method of forestry development, referred to as a scientific management system has been practiced for the last 130 years. However this system is unable to sustain the quantity and quality of India's forests and it thus fails to meet the demands of the rising local population for industrial wood, fuelwood and needs for non-wood products and recreation.

Not much work has been done to quantify the development of forest ecosystems under protection of joint forest management. At present there is an urgent need to monitor a comparative study of forest managed by community's vis-à-vis forest department to have ideas on impact of management options on ecosystem development. In various parts of India the communities have their own indigenous management practices which certainly have some positive impacts on ecosystem development. But there have some dark side also. Therefore, monitoring of JFM is important from the point of view of its successful implementation. The success of JFM would not only depend upon the extent of people's participation, but the management regime and technology adopted.

To maintain ecological balance and the intricate web of life, it is not just essential to have a mixed forest with diverse species but at the same time one that meets the varied needs of the people depending on it. Community forest management practices have shown to be giving equal importance to ecological requirements as well as cultural and livelihood needs that a forest ecosystem ensures, thereby well integrating forests into the social fabric of the community. There are several areas of forest management like flora and fauna interactions, regulation of yield, biodiversity inter-linkages etc. that requires scientific research. Institutions and agencies having such knowledge in forestry science have a great role to play in extending the knowledge base to the local communities, thereby strengthening communities' practices (Sarangi 2002).

26.9 Conclusion

JFM, being a social movement, cannot solely focus on the development of forests. Thus, developing the NTFP market, adding value to various forest products, agricultural development, watershed development, afforestation, prevention of grazing on forest land, fire control and such programs related to natural resource development need to be promoted along with forest regeneration and development. Thus, several related developmental departments need to work together in a given village or a cluster of villages for overall development. Also, it is necessary to work towards an integrated development approach rather than isolated efforts. Sustainable natural resource management is the key to sustainable development of our society. Thus, there has to be harmony between conservation of forests, carbon sequestration and development of communities through livelihood security. There has to be an effective partnership among all the stakeholders for capacity building, monitoring and evaluation of JFM to achieve the ultimate goal of planning and development i.e. self-reliance. Forest right act has created a scope to go forward in this direction. But its real implementation is Payment for environmental services (PES) and Reducing Emissions from Deforestation and Degradation (REDD+) is being experimented worldwide. Its impacts on the forest conservation and meeting livelihood needs of people in developing countries, REDD+ is a mystery. Nobody is sure that in a market driven economy how the poor little or uneducated villagers going to negotiate with global carbon market. It is essential that in the global mechanism that aims at sustainable forest management through protecting forests and enhancing carbon sequestration, devolution of power to local communities is one of the important components. It is also want of the time that the REDD+ mechanism has to meet the livelihood needs of forest-dependent communities by adding value to the collected forest produce. Then only we can say that a sustainable forest management system has developed in India.

Forest management has the potential to increase the terrestrial C pool. Forest and forest soil are the sink for greenhouse gases that can contribute to meeting the national commitment to emission reduction. The goal of reducing carbon sources and increasing the carbon sink can be achieved efficiently by protecting and conserving pools in existing forests (Brown et al. 1996). The practices such as the rehabilitation of degraded forest land, agroforestry, regeneration, afforestation, prevention of grazing on forest land and fire control are some of the suitable options for carbon management in Indian soils. Therefore, sustainable forest management through participatory approach will bring in its wake all round economic development of the forest dependent communities, besides conservation of biodiversity and watershed protection as well as wood biomass for energy for carbon mitigation (Banerjee et al. 2017).

References

Agarwal A, Mishra TK, Nagraj BC, Maiti SK, Maity DK, Ramprasad V, Jagannath Rao R, Bhat PR, Nagendra MDV, Khan H (2005) Joint forest management: lessons from case studies. In: Ravindranath NH, Sudha P (eds) Joint forest management in India, spread, performance and impact, centre for ecological sciences. IISc, Bangalore

Aggarwal A et al (2009) Forest resources: degradation, livelihoods and climate change. In: Dattaand D, Nischal S (eds) Looking back to change track. TERI, New Delhi

Arnold JEM (1998) Forestry and sustainable livelihoods. In: Corney D (ed) Sustainable rural livelihoods: what contribution can we make? Department of International Development, London, UK

Banerjee S, Singh S, Banerjee SK (2017) Biodiversity conservation and enhancement of forest productivity through participatory approach for adaptation and mitigation to climate change—Indian scenario. In: Murlikrishna CH et al. (eds) Climate change and role of communities in adaptation and mitigation. Madhya Pradesh State Forest Research Institute Publication, pp 215–238

Brown S (1996) Present and potential roles of forests in the global climate change debate. Unasylva 47:3–9
Brown S, Hall CAS, Knale W, Raich J, Trexler MC, Woomer P (1993) Tropical forests: their past, present and potential future role in terrestrial carbon budget. Water Air Soil Pollut 70:71–94
Byron N, Arnold JEM (1999) What futures for the people of the tropical forests? World Dev 27:789–805
Davidar P et al (2010) Assessing the extent and causes of forest degradation in India: where do we stand? Biol Conser 43:2937–2944
Dixon RK, Winjum JK, Schroeder PE (1993) Conservation and sequestration of carbon: the potential of forest and agroforest management practices. Global Environ Change 3:159–171
FAO (1998) Asia and Pacific forestry towards 2010—report of the Asia and Pacific forestry outlook study. Food and Agricultural Organization of the United Nations, Rome, Italy
FAO (2006) Global forest resource assessment 2005. Progress towards sustainable forest management. FAO forestry paper 147, 320 pp
FAO (2015) Forest restoration and rehabilitation. Food and Agricultural Organization f the United Nations, Rome, Italy
Gadgil M (2008) Baharude Hakkachi Vanraji! [Marathi], Dhananjayrao Gadgil Janashtabdi Prabodhanmala: 22. Pune, Lokayat Prakashan, p 41
GCEC (2014) Better growth, better climate: new climate economy report. The Global Commission on the Economy and Climate, Washington DC
Gera M, Bist NS, Gera N (2002) Carbon sequestration through community based forest management—a case study from Sambalpur Forest division, Orissa. Indian Forest 129:735–740
Gibbs HK, Ruesch AS, Achard F, Clayton MK, Holmgren P, Ramankutty N, Foley JA (2009) Tropical forests were the primary sources of new agricultural land in the 1980s and 1990s. Proc Nat Acad Sci 107:16732–16737
Groombridge B, Jenkins MP (2002) World atlas of biodiversity. University of California, Barkley, USA
Gupta B, Mishra TK (2019) Analysis of tree diversity and factors affecting natural regeneration in fragmented dry deciduous forests of lateritic West Bengal. Trop Ecol 60:405–414
IPCC (2014) Intergovernmental panel on climate change: WORKING group III. Summary for policy makers
Jindal A, Upadhyaya RK, Varma VK (2003) Joint forest management in India: the unaddressed legal issues. Canada, World Forestry Congress
Ketterings QM, Coe R, Van Noordwijk M, Ambagau Y, Palm CA (2001) Reducing uncertainty in the use of allometric biomass equations for predicting above ground tree biomass in mixed secondary forests. Forest Ecol Manag 146:199–209
Khera A (2017) A case study on estimation of forest carbon in South Balaghat Forest division and benefit sharing by JFM Committees under REDD+. In: Murlikrishna CH et al. (eds) Climate change and role of communities in adaptation and mitigation. Madhya Pradesh State Forest Research Institute Publication, pp 132–136
Kumar N (2005) Tribals and forest: emerging symbiosis. IASSI Quarterly 23(4):92–103
Lele S (1998) Why, who, and how of jointness in joint forest management: theoretical considerations and empirical insights from the Western Ghats of Karnataka. Paper presented at the International Workshop on Shared Resource Management in South Asia
Lugo AE, Brown S (1992) Tropical forests as sink of atmospheric carbon. Forest Ecol Manag 54:239–255
Lynch OJ, Talboot K (1995) Balancing acts: community based forest management and national law in Asia and Pacific. World Resource Institute, Washington D.C.
Makela A et al (2000) Process based models for forest ecosystem management: current state of the art and challenges for practical implication. Tree Physiol 20:289–298
Malhotra KC, Dev D, Dutta M, Vasulu TS, Yadav G, Adhikary M (1992) Role of NTFP in village economy. IBRAD, Calcutta

Mathur A, Sachdeva A (2003) Towards an economic approach for sustainable forest development. Working Paper Series No. 2/2003—planning Commission, Govt. of India
Mishra TK (1999) Users become managers: indigenous knowledge and modern forestry. Econ Polit Weekly XXXIII(6):262–263
Mishra TK, Maity S, Mandal D (2005) Joint forest management in West Bengal: its spread, performance and impact. In: Ravindranath NH, Sudha P (eds) Joint forest management in India: its spread performance and impact. Universities Press, Hyderabad pp 160–179
Mishra TK (2001) Assessing impact of extraction on sustainability of forests under joint forest management—a case study. Annal Forest 9:181–187
Mishra TK, Banerjee SK (1997) An ecological reconnaissance of lateritic forest of South West Bengal. Adv Forest Res India 16:1–43
Mishra TK, Nandi A, Banerjee SK (1994) Management of coppice sal forests in lateritic region of West Bengal. Indian J Forest 17:212–217
Mishra TK, Sahoo TK, Mishra PN, Banerjee SK (2008) Species diversity under different management systems—a case study in Amlachati Range of Midnapore West Forest Division. Econ Environ Cons 14:101–110
Mukherjee P (2003) Community forest management in India: the Van Panchayats of Uttaranchal. XII World Forestry Congress, Canada
Poffenberger M, McGean B (1998) Village voices, forest choices: joint forest management in India. Oxford University Press, Delhi
Prasad R (2000) Report of the national task force on SFM. Submitted to the Govt. of India (MOEF). Indian Institute of Forest Management, Bhopal
Rao JR, Murali SK, Murthy IK (2005) Joint forest management studies in India: review of the monitoring and evaluation methods. In: Ravindranath NH, Sudha P (eds) Joint forest management in India, spread, performance and impact. Centre for Ecological Sciences. IISc, Bangalore
Ravindranath NH et al (2012) Deforestation and forest degradation in India—implication of REDD+. Curr Sci 102:1–9
Sahu NC, Rath B (2011) Impact of joint forest management (JFM) on environmental stress migration: evidence from Orissa. Int J Rural Manag
Sarangi S (2002) Working paper 6, community forest management and ecological balance: experience from the state of Orissa
Sarkar D, Das N (2006) A. study of economic outcome of joint forest management programme in West Bengal: the strategic decisions between government and forest fringe community. Indian Econ Rev 43:17–45
Singh SB, Das PK, Banerjee SK, Singh B (1987) Effect of vegetation cover on nutrient status of soils. J Indian Soc Soil Sci 35:232–237
Singh TP (2004) Joint forest management in India: a potential CDM activity. Environ Sci. http://www.mendeley.com
Smith C, Rees G (1988) Economic development, 2nd edn. Basingstoke, Macmillon
Sundar B (2017) Joint forest management in India—an assessment. Int Forest Rev 19(4):495–511
TERI (1998) Looking back to think ahead-green India 2947. Tata Energy Research Institute, Delhi, pp 21–25
Tilman D et al (2001) Forecasting agriculturally driven environmental change. Science 292:281–284
WRI (1998) World resource institute: world resources, 1998–99. Oxford University Press, Oxford

Chapter 27
Google Earth Engine and Its Application in Forest Sciences

Mojtaba Naghdyzadegan Jahromi, Maryam Naghdizadegan Jahromi, Babak Zolghadr-Asli, Hamid Reza Pourghasemi, and Seyed Kazem Alavipanah

Abstract In order to support sustainable forest management, it is essential to estimate the extent and change of forest cover and to evaluate the environmental and socio-economic impacts of forest dynamics. It is challenging, however, to calculate forest area on a large scale using traditional statistical survey methods. Access to satellite images make it feasible to monitor the Earth's forest at different spatial and temporal resolutions. The Google Earth Engine (GEE) is a cloud computing platform, which provides data analysis toolkits to access and to handle remote sensing datasets easily and freely. GEE has been used to analyze environmental changes with the emphasis of forest monitoring. Through GEE's platform, the user can monitor forest cover by investigating satellite images in different spatial and temporal resolutions with acceptable accuracy. Moreover, this platform's impressive satellite image archives, coupled with sophisticated in-built processing and analyzing toolkits, immensely help remote sensing-based studies. Hence, a systematic review

M. N. Jahromi
Department of Water Engineering, School of Agriculture, Shiraz University, 714416-5186 Shiraz, Iran
e-mail: Naghdyzadegan@gmail.com

M. N. Jahromi (✉) · S. K. Alavipanah
Department of Remote Sensing and GIS, Faculty of Geography, University of Tehran, Tehran, Iran
e-mail: naghdizadegan.m@ut.ac.ir

S. K. Alavipanah
e-mail: salavipa@ut.ac.ir

B. Zolghadr-Asli
Department of Irrigation and Reclamation Engineering, Faculty of Agricultural Engineering and Technology, College of Agriculture and Natural Resources, University of Tehran, 3158777871 Karaj, Iran
e-mail: ZolghadrBabak@ut.ac.ir

H. R. Pourghasemi
Department of Natural Resources and Environmental Engineering, School of Agriculture, Shiraz University, Shiraz, Iran
e-mail: hamidreza.pourghasemi@yahoo.com

P. K. Shit et al. (eds.), *Spatial Modeling in Forest Resources Management*, Environmental Science and Engineering,
https://doi.org/10.1007/978-3-030-56542-8_27

has been conducted here to survey those studies that have employed this platform for forest monitoring. According to the analysis, when it comes to forest monitoring, the GEE's platform has been mainly used for two objectives, namely, classification and change detection. Random forest has been identified as the most popular classification method and spectral index difference has been the most efficient method for forest change detection while considering GEE limitation for image preprocessing. Overall, the survey's result revealed how applying this platform for forest monitoring is trending.

Keywords Forest · Google earth engine · Classification · Change detection · Spectral indices

27.1 Introduction

Forests are considered to be major terrestrial carbon sinks due to *photosynthesis*, which is a crucial part of the Earth's carbon cycle for stabilizing the global climate (Lausch et al. 2016). Arguably, the increase in the concentration of greenhouse gases (GHG), such as CO_2, in the atmosphere has resulted in human-derived climate change (Landwirtschft 2015), and forests can play a pivotal role to buffer the adverse effect of GHG emissions. Naturally, the capacity of forests' carbon storage is directly linked to the forest biomass (Amidon et al. 2008; Schmid 2017). Additionally, forests can potentially reduce stormwater runoff and even enhance its water quality by filtering the pollutants of local waters, which in turn, can lead to *healthier* watersheds (Schlautman and Smink 2008). As such, forests can also be seen as a reliable water resources supplier. Reportedly, for instance, forest lands supply more than half of the annual water in the United States (Sun and Vose 2016). Overall, it can be seen that biodiversity richness, climate regulation, water supplies, and, in turn, any credible attempt for sustainable development are critically dependent on the forest cover status (Foody et al. 2003; Lu et al. 2016). Therefore, by integrative forest biomass management and optimizing environmental utilization in local, regional, and even global scales, it will be possible to mitigate the GHG and balance the environment and energy security (Bartuska 2006). In the last few decades, however, human interferences in the environment has caused notable changes innatural landscapes, including forest status (Zhang et al. 2019). This, in turn, has had a significant influence on the forest canopy cover and consequently carbon storage, hydrological system, climate regulation, soil erosion, land degradation, biodiversity, and other ecosystem services (D'Almeida et al. 2007; Liu et al. 2012).

Over the years, land conversion from forest to urban and rural areas has resulted in massive ecosystem degradation (Hur et al. 2008; Schlautman and Smink 2008). Given the prominent role of forests as indispensable natural resources (Foley et al. 2005), over the past few years, the scientific community and environmental advocates have attempted to raise public awareness about the alarming rate of forest degradation around the globe. In fact, the idea behind *forest mapping* is to achieve reliable information on different scales for regional and global management of these natural resources (Hansen et al. 2013).

Monitoring the changes in forests in the long-term horizon can significantly increase the understanding of land cover dynamic effects on the structure and functionality of ecosystems (Gong 2012; Zurqani et al. 2018). These findings could, in turn, help achieve integrated land–water management (Dronova et al. 2015; Foster et al. 2003).

From the 1900s till a few decades ago, forest monitoring was mainly done through on-site surveys, which needless to say, were costly, time- and energy-consuming endeavors, even where high accuracy was not much of a concern (Global Forest Atlas 2017). Back in the days, field surveys were limited to measuring the heights of trees, the numbers of trees, and canopy cover percentage using permanent plots (Paneque-Gálvez et al. 2014). A milestone in forest monitoring was the introduction of aerial photographs (Osei and Andam-Akorful 2019). However, these could not foster any significant improvements, mainly due to the low accuracy of cameras used in the past.

Presently, thanks to recent technological advancements in improving remote sensing (RS), state-of-the-art satellite images can be used to monitor any gradual or abrupt changes in forest cover. As such, having access to high-resolution images of such satellites inconvenient time-lapses made RS the most important and efficient methods for forest monitoring and change detection (Belgiu and Drăguţ 2016; Rogan and Chen 2004).

The first Earth observation satellite of the Landsat program of the National Aeronautics and Space Administration (NASA) was launched in 1972. This satellite started a new era in forest monitoring on an unprecedented scale (Global Forest Atlas 2017). In the following years, seven additional satellites were launched by the United States Geological Survey (USGS) and NASA, two of which have started their missions as late as 2017 (USGS 2017). These long-term satellite images, of course, help researchers to gain a better understanding of land cover dynamics and forest change detection (Gong 2012). Generally, these satellites are tuned to monitor specific frequency bands of sunlight reflections from Earth's surface area (Tempfli et al. 2009). By applying pre- and post-processing procedures on the obtained data sets, the gathered information could then be used for forest monitoring purposes.

During the last decade, thanks to the enhancement of satellite images and their readily available datasets, detecting and monitoring changes in forests has been facilitated immensely (Fadli et al. 2019). This is mainly due to the fact that forest monitoring is a complex and multi-dimensional task that requires a variety of different datasets. A wide variety of tools, ranging from simple cameras to advanced sensors can be implemented as an RS instrument. Furthermore, all satellite images for environmental monitoring, regardless of their spatial and temporal accuracy, can convey some information of value and, as such, can be interpreted and analyzed using RS and geospatial techniques. In fact, one of the benefits of RS is that users can achieve more comprehensive results by consideration and integration of different data gathered using a variety of parameters such as surface elevation, slope, canopy cover (Osei and Andam-Akorful 2019). That being said, RS can indeed be seen as a helpful instrument for forest monitoring. Consequently, both public and private institutions are actively trying to promote RS, not just in terms of applying its data, but rather enhancing, installing, and upgrading the physical infrastructures and computational platforms required for using RS. As such, for instance, in addition to the institutes that are responsible for the Landsat programs (USGS), there are other agencies, such as National Oceanic and Atmospheric Administration (NOAA), USGS, International Union of Forest Research Organizations (IUFRO), and European Space Agency (ESA) that are actively monitoring the Earth's environment via their satellites images. Such continuous monitoring of the Earth's land changes can indeed carry vital information that is applicable in both regional and global scale.

It should be noted that prior to the point when the USGS made their Landsat archive freely available for the public in 2008, the most common practice in RS-based forest monitoring was to perform the analysis of satellite images that were captured in a five to ten years intervals (e.g., Griffiths et al. 2014; Masek et al. 2008; Teferi et al. 2013). However, after the Landsat archive became readily and freely available, which contained more than three decades of data, many researchers were able to focus on the long-term forest change monitoring at higher spatial and temporal resolutions (Huang et al. 2017; Loveland and Dwyer 2012). (Schmidt et al. 2015), for instance, have used Landsat and Moderate Resolution Imaging Spectroradiometer (MODIS) datasets to detect forest distribution using the Normalized Difference Vegetation Index (NDVI) time series. In fact, over the last few years, several studies have implemented an RS-based framework for forest monitoring (e.g., Desclée et al. 2006; Guo et al. 2020; Hansen et al. 2013; Huang et al. 2010; Kennedy et al. 2010; Loveland and Dwyer 2012; Osei and Andam-Akorful 2019; Zhang et al. 2019; Zurqani et al. 2018). If RS-based studies are to be applicable on a global scale, however, satellite images of different sensors are to be readily, freely, and easily available for each region in acceptable spatial and temporal resolutions. This, however, is a challenging and costly task, for it requires massive computational and storage capacity. In response, Google Earth Engine (GEE), a revolutionary cloud-based platform with high-performance computing capabilities, has been introduced to handle online RS datasets processing (Tempfli et al. 2009). As such, in this chapter, we have surveyed more than 30 forest science articles in which the GEE platform has been used. The main idea here is to evaluate various methods, satellite images and the subject matter of these studies.

27.2 Google Earth Engine

Nowadays, satellite image archives of the United States government agencies, such as NASA, NOAA, and USGS (Loveland and Dwyer 2012; Nemani et al. 2011; Woodcock et al. 2008), and also of the European Space Agency (Copernicus Data Access Policy 2016) are available freely in different spatial and temporal scales. This, in turn, has made a lot of raw RS-related data available, though storing and processing these data to gather useful, practical information has turned into a real challenge for researchers. As such, over the years several platforms and computer software have been developed and promoted to handle and process this overwhelming amount of available RS-based data [i.e., TerraLib (Camara et al. 2000), Hadoop (Whitman et al. 2014), GeoSpark (Yu et al. 2015), and GeoMesa (Hughes et al. 2015), ERDAS Imagine (Geosystems 2004), ENVI (Guide 2008), and Arc-GIS (Johnston et al. 2001)]. Despite the explicit privileges of having access to this massive readily available datasets and huge leaps provided by the computational advancements oft he platforms mentioned above, to this day, retrieving, selecting, and downloading satellite images in larger scales remains as a challenge since these are massive datasets, by nature, which makes handling and analyzing these mega datasets a time and energy-consuming process (Huang et al. 2017). Furthermore, such analysis commonly requires high-tech supercomputing machines for pre-processing the data, and as such, applying these methods for most real-world practical cases would demand a highly experienced analyst to implement and interpret these data in a meaningful way. These problems inspired the Google Company to look for a solution that, in essence, addresses all the above-mentioned challenges (Gorelick et al. 2017). The main concern here was to facilitate handling mega datasets by reducing the computational requirements to import and process the satellite images for individual users, and they came up with the GEE for online remote-sensing datasets processing (Gorelick et al. 2017).

Perhaps one of the most notable features of this online geospatial analysis platform is that it has access to a large dataset of satellite images and cloud computing (Gorelick et al. 2017), which in turn can make it possible for the users to analyze changes in the Earth's surface in seemingly real-time continuously, dynamically, and freely (Housman et al. 2015; Zurqani et al. 2018, 2019a). Thanks to the GEE online geospatial analysis platform, users all around the world cannot only have access to vast satellite image datasets but, more importantly, can efficiently process these datasets with minimum on-site effort and equipment (Zhang et al. 2019). GEE is also efficient in data acquisition and storage, file pattern analysis, database management and equipment distribution for large regions, thus enabling desktop processing and by applying algorithms that combine data from multiple sensors (Gorelick et al. 2017).

In addition to being equipped with visualizing functions for geospatial data sets and high-performance computing resources, GEE enables the users to share their findings and projects easily. More interestingly, GEE's platform makes it possible for users to apply and develop a new computational algorithm with little orno programming experience (Gorelick et al. 2017). Users can simply log into the GEE home page (https://earthengine.google.com) to freely gain access into four decades satellite and aerial images that covers varies datasets, including but not limited to, land cover and land use, climate and weather data, and other environmental-related variables (Housman et al. 2015). The GEE archive supports many of the most well-known Earth-observing satellite images, including Landsat, Sentinel 1 and 2, MODIS, ASTER (see Table 27.1). GEE actively and continuously updates its database on a daily basis, which amounts to approximately 6000 new scenes per day (Gorelick et al. 2017). The platform also enables each user to upload and implement personalized information from external sources (e.g., pictures taken by personal drones) to

Table 27.1 Summary of selected articles for forest mapping using GEE

Serial no	Authors	Purpose	Satellite	Year	Area	Method	Journal
1	Hansen et al. (2013)	Forest change detection	Landsat	2000–2012	Global scale		Science
2	Johansen et al. (2015)	Woody vegetation mapping	Landsat	2004–2010	Australia	CART, Random forest, NDVI	Society and Environment
3	Riitters et al. (2016)	Forest dynamic detection	Landsat	2000–2012	Global scale	Hansen method	Landscape Ecology
4	Chen et al. (2017)	Mangrove forest mapping	Landsat and Sentinel-1A	2014–2016	China	Spectral indices	ISPRS Journal of Photogrammetry and Remote Sensing
5	Huang et al. (2017)	Land cover change detection	Landsat	1985–2015	China	Random forest, spectral index time series	Remote Sensing of Environment
6	Schmid (2017)	Forest Change detection	Landsat 5 and 8	1984–2013	Germany	NDVI time series	
7	Shelestov et al. (2017)	Crop mapping	Landsat 8	2013	Ukraine	Random forest, CART, SVM, Naïve Bayes	Frontiers in Earth Science
8	Portengen (2017)	Mangrove forest classification	Sentinel-1 and Sentinel-2	2017	Vietnam	k-means clustering, Random forest	
9	Thomas et al. (2017)	Mangrove forest mapping	Japanese Earth Resources Satellite (JERS-1)	1996–2010	Japan	Manually interpretation	PloS One

(continued)

Table 27.1 (continued)

Serial no	Authors	Purpose	Satellite	Year	Area	Method	Journal
10	Zurqani et al. (2018)	Land use change	Landsat	1999, 2005, 2009 and 2015	USA	Random forest classification	International Journal of Applied Earth Observation and Geoinformation
11	Pimple et al. (2018)	Mangrove forest mapping	Landsat	1987–2017	Thailand	Random forest classification	Journal of Computer and Communications
12	Tsai et al. (2018)	Forest cover mapping	Landsat	2011 and 2016	China	Classification	Remote Sensing
13	Wu et al. (2018)	Global forest pattern	MODIS and ESA Landcover	2003–2013	Global scale	SOM	IEEE
14	Yang et al. (2019)	Forest biomass estimation		2011–2012	USA	Pixel level AGB estimation algorithm	International Journal of Digital Earth
15	Arai et al. (2019)	Forest degradation by fire	Landsat 8	2017–2018	Brazilian	Using temporal series of imageries	IEEE International Geoscience and Remote Sensing Symposium
16	Alaoui et al. (2019)	Forest degradation by fire	Landsat	1997–2016	Morocco	Spectral bands and spectral indices	Biodiversidade Brasileira
17	Ceccherini et al. (2019)	Effect of forest cover on terrestrial carbon cycle	Landsat	2000–2017	Global scale	present a retrieval methodology	Geophysical Research
18	Fadli et al. (2019)	Forest cover change	Landsat	2000–2016	Indonesia	Hansen method	Journal of Physics

(continued)

Table 27.1 (continued)

Serial no	Authors	Purpose	Satellite	Year	Area	Method	Journal
19	Duan et al. (2019)	Urban forest distribution	Sentinel-2	2016	China	Random forest classification	Forests
20	Koskinen et al. (2019)	Forest mapping	Landsat and Sentinel-1 and Sentinel-2	2013–2015	Tanzania	Random forest, CART, SVM	ISPRS Journal of Photogrammetry and Remote Sensing
21	Osei and Andam-Akorful (2019)	Forest monitoring	Landsat, MODIS and NOAA AVHRR	1985–2018	Ghana	k-means clustering, Random forest	
22	Tieng et al. 2019	Mangrove forest mapping	Landsat 8 and Sentinel-2	2014 and 2015	Cambodia	Random forest classification	Earth and Environmental Science
23	Zhang et al. (2019)	Global forest cover mapping	Landsat 8	2018	global scale	Random forest classification	IEEE
24	Zurqani et al. (2019a,b)	Change from forest to urban area	Landsat	2013, 2015, and 2017	USA	k- means clustering, random forest	Remote Sensing in Earth Systems Sciences
25	Jena and Pradhan (2019)	Forest change detection due to mining	ETM + , MODIS	2012–2017	Indonesia	spectral indices time series	IEEE
26	Anchang et al. (2020)	Woody canopy cover mapping	Sentinel-1 and Sentinel-2	2015–2017	Senegal	Random forest classification	Frontiers in Environmental Science
27	Guo et al. (2020)	Forest cover detection	Landsat	2018	China	Random forest classification	science
28	Martín-Ortega et al. (2020)	Forest change detection	Landsat	1984–2017	Costa Rica	Correlation with illuminate	Remote Sensing

complement this massive dataset. With the help of Google's computing infrastructure for parallel processing of massive datasets and millions of servers around the world, GEE makes swift analysis of trillions of images possible (Dong et al. 2016). GEE's revered elastic computing and parallel processing are mainly rooted in its efficiency in importing sophisticated datasets, as well as its built-in internet-accessible application programming interface (API) (Yang et al. 2019).GEE currently provides APIs for two of the most popular programming languages, namely *JavaScript* and *Python*, though to this day they are still in active development (Wu et al. 2018).

From a practical point of view, perhaps one of the most advantageous features of GEE is that it can enable the users to easily sort and filter the RS data fora specific sensor for different spatial and temporal scales. In other words, users can select specific images in a "collection," using specific temporal, spatial, cloud-cover filters or other criteria. This feature, in turn, allows users to analyze relatively massive RS datasets promptly. For instance, if users want to select summer collection of Landsat 8 images to analyze specific changes during this timeframe, they can easily do so using the following syntax:

Collection = ee.ImageryCollection("Landsat8")
Summer = Collection.filter(ee.Filter.calenderRange(6, 8, "month"))

GEE also provides specific and targeted features to handle time series. As such, time-series analysis in GEE is relatively easy and fast. GEE is also embedded with a lot of useful and time-saving built-in functions. For instance, it is possible in GEE to remove cloudy scenes automatically using *SimpleCloudScore* function. The functions in GEE ranges from simple mathematical functions to complex imagery processing.

These features and many more, have made GEE a revered platform to conduct RS-based researches. As such, it has been used to cope with a large variety of topics including but not limited to crop yield production (e.g., Lobell et al. 2015; Xiong et al. 2017), surface water change (e.g., Pekel et al. 2016; Tang et al. 2016; Wang et al. 2018), forest change (e.g., Chen et al. 2017; Fadli et al. 2019; Hansen et al. 2013; Wu et al. 2018; Yang et al. 2019; Zhang et al. 2019; Zurqani et al. 2019b), land use (e.g., Zurqani 2019), urban mapping (e.g., Goldblatt et al. 2016; Liu et al. 2018b; Sun et al. 2019; Zhang et al. 2015), flood mapping (e.g., Coltin et al. 2016; Liu et al. 2018a; Zurqani et al. 2019a) and fire mapping (e.g., Long et al. 2019; Parks et al. 2018).

Over the years applying GEE for forest monitoring has gained momentum among researchers. Here, a systematic review method was employed for searching the relevant papers, and the results have been summarized in Table 27.1. This review was limited to the publications from peer-reviewed scientific journals (*in English*) which they have focuses on forest in GEE platform from January 2013 to March 2020. While this is a thorough review, it might not be an exhaustive one. The gathered database contained 31 articles. As shown in Fig. 27.1, the survey reveal an increasing trend in the number of articles that use GEE's platform for forest monitoring. The survey

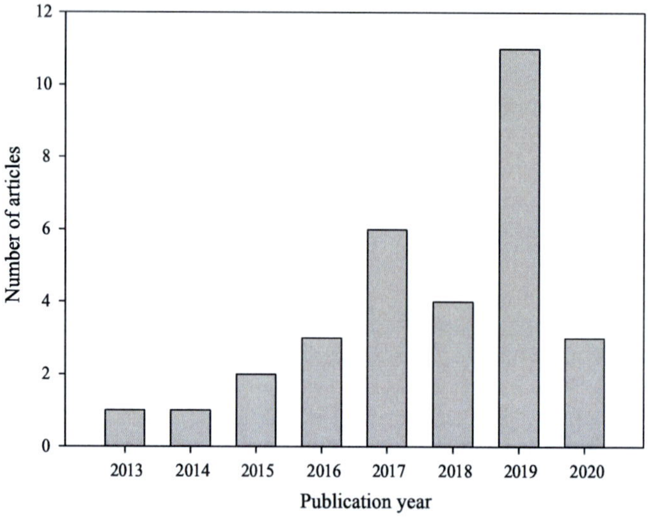

Fig. 27.1 The number of articles in forest science that used GEE platform

also monitored the location of case studies in these articles. Figure 27.2 illustrates how frequently these methods have been used to tackle forest-related problems in different continents. As can be seen, studies on Asia have employed this platform most often, while the researches that conducted on Australia contributed the least

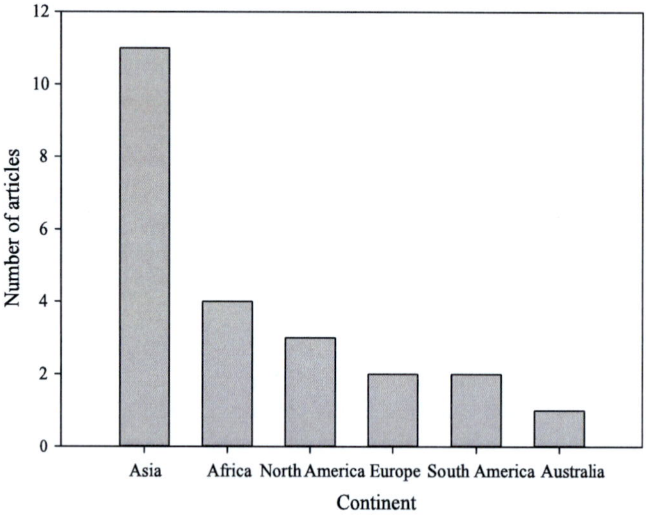

Fig. 27.2 The number of articles in forest science that employed GEE platform in each given continent

number of articles. The following section explores the survey's findings and how GEE has been used for forest monitoring.

27.3 Results and Discussion

According to the conducted survey, the researches in the field of forest science could be divided into two main categories, namely classification and change detection. In the former group, GEE' platform has been used to segregate the forest section from other land cover in mostly large scale study areas. In the latter category, however, the forests' characteristics have been monitored and evaluated through time to detect any abrupt or gradual changes in these characteristics. Figure 27.3 demonstrates the progression of each category over time, where it can be seen that classification is the most frequently tackled subject matter. However, in the past years, it seems that change detection has also been received more attention by the researchers in the field.

Furthermore, nearly half of the studies in our survey were conducted in Asia alone, 64% of which used some sort of classification method. More specifically, the review revealed that approximately 25% of the reviewed articles are based in China, and that 63% of them were aimed at forest classification. According to the conducted review, the United States followed China in the number of papers that employed GEE's platform for forest detection studies (14% of all the reviewed articles). The following sections explores both categories a bit further.

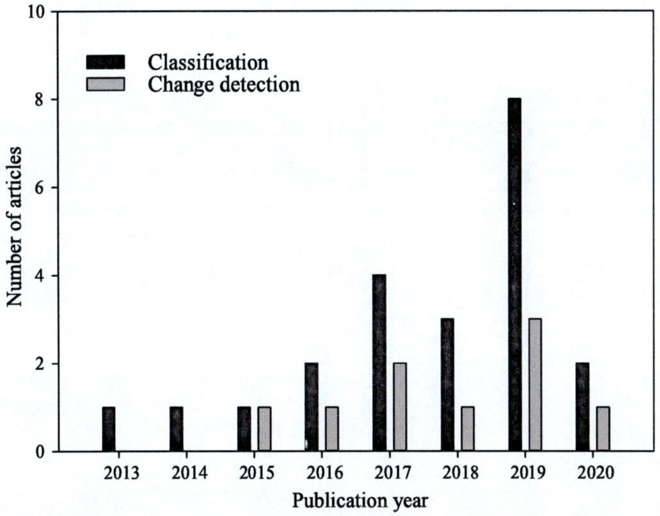

Fig. 27.3 The number of forest-related articles that employed the GEE platform and their primary objective, namely classification and change detection since 2013

27.3.1 Imagery Classification in GEE

While most studies have employed GEE for land cover classification (e.g., Huang et al. 2017; Koskinen et al. 2019; Pimple et al. 2018; Tsai et al. 2018; Zurqani et al. 2019b; Zurqani et al. 2018)), some have solely focused on using the application for forest and non-forest classification (e.g., Chen et al. 2017; Duan et al. 2019; Fadli et al. 2019; Guo et al. 2020; Hansen et al. 2013; Martín-Ortega et al. 2020; Osei and Andam-Akorful 2019; Schmid 2017; Zhang et al. 2019). The classification of forest via GEE consists of five main steps: (1) data selection, (2) feature selection, (3) model training, (4) classification, and (5) accuracy evaluation.

27.3.1.1 Data Selection

As the first step of classification, one needs to select the appropriate satellite images for classification. While there are a wide variety of images with different spatial and temporal resolution and bands available, the survey revealed that when it comes to implementing GEE's platform for forest classification, MODIS, Landsat, NOAA AVHRR, and Sentinel-2 were the most commonly cited satellites (Table 27.2).In fact, according to the survey's results, approximately 85% of the studies that employed GEE's platform have used the Landsat free database (Alaoui et al. 2019; Arai et al.

Table 27.2 Frequently used datasets in GEE-based studies for forest classification

Satellite	Image collection	Spatial resolution (m)	Temporal resolution (Day)	Data availability	Provider
Landsat	Landsat 1–5, 7, 8 (MSS, TM, ETM + , OLI, TIRS)	15, 30, 60	16	1972–Present	USGS
Sentinel	Sentinel 1, 2	10, 20, 60	3, 5	2014–Present	European Union/ESA/Copernicus
MODIS	MOD09, 11, 12, 13, 14, 15, 17, 43, 44, 45	250, 500, 1000	1–8, 16, 30	2000–Present	NASA LP DAAC at the USGS EROS Center
ASTER	L1T radiance, Global emissivity	15, 30, 90, 100	1, Once	2000–Present	NASA LP DAAC at the USGS EROS Center
NOAA AVHRR		$\approx$1.09 km (Different resolutions in different products)	1	1981–Present	NOAA

2019; Ceccherini et al. 2019; Fadli et al. 2019; Guo et al. 2020; Martín-Ortega et al. 2020), which is perhaps due to the length of Landsat archive which dates back to 1972.

27.3.1.2 Feature Selection

Different combinations of spectral bands must be used to distinguish forests from other land covers. Conventionally, different vegetation indices have been proposed and implemented for forest classification. It should be noted that the indices mentioned above could also be used as supplementary data for the classification of forests from non-forest areas. Furthermore, some have employed the time series of such indices for forest change evaluation (e.g., Martín-Ortega et al. 2020; Schmid 2017). In the context of GEE's platform, some of the most cited indices have been summarized in Table 27.3.

Vegetation indices have also been used for studying *forest loss and gain*, and as such, it has been implemented to promote a framework to estimate *biomass* on a large scale (Hansen et al. 2013; Kim et al. 2014). Moreover, GEE's platform can integrate seasonal vegetation indices' time-series and textural features to improve classification accuracy. As such, GEE's platform can help minimize the risk of misclassification of different elements, most commonly cloud shadows and water, which in turn improve the segregation of different vegetation types.

Naturally, different vegetation types and geographic conditions will cause the spectral characteristics of the pixels to vary from one another. However, some external factors, such as the slope or the topography, can make some pixels with different vegetation attributes appear somewhat similar to each other. This could, in fact, turn into a major issue in large-scale studies, where there are a lot of different factors that could vary significantly throughout the study area. Using GEE's platform, however, some studies have implemented and combined texture, DEM datasets, and spectral indices to overcome this issue (Zhang et al. 2019).

27.3.1.3 Training Data

The main idea here is to label each given pixel of the obtained images according to its spectral similarity to a set of pre-analyzed pixels known as training data. Training data could be obtained from different resources, such as field surveys and high spatial resolution images. Training samples could also be collected from previous GPS-based field research.. GEE's platform also helps users in this step by providing a variety of options to obtain training data that are online, such as Google Map, Map World, and OpenStreetMap, as well as land cover services such as Geo-Wiki (Gong 2012).

Table 27.3 Frequently used indices in the GEE-based studies for forest classification

Index	Acronym	Description	Examples
Normalized Difference Vegetation Index	NDVI	$\frac{NIR-Red}{NIR+Red}$	Duan et al. (2019), Jena and Pradhan (2019), Osei and Andam-Akorful (2019), Pimple et al. (2018), Schmid (2017), Tsai et al. (2018), Zhang et al. (2019)
Enhanced Vegetation Index	EVI	$2.5 \times \frac{NIR-Red}{NIR+(6\times Red)-(7.5\times Blue)+1}$	Chen et al. (2017), Martín-Ortega et al. (2020), Zhang et al. (2019)
Green Ratio Vegetation Index	GRVI	$\frac{NIR}{Green}$	Zhang et al. (2019)
Soil Adjusted Vegetation Index	SAVI	$\frac{1.5\times(NIR-Red)}{NIR+Red+0.5}$	Guo et al. (2020)
Modified Soil-Adjusted Vegetation Index	MSAVI	$\frac{2\times NIR+1-\sqrt{[(2\times NIR+1)^2-8\times(2\times NIR+1)]}}{2}$	Tsai et al. (2018), Zhang et al. (2019)
Normalized Difference Water Index	NDWI	$\frac{(SWIR+Red)+(NIR+Blue)}{(SWIR+Red)-(NIR+Blue)}$	Chen et al. (2017), Duan et al. (2019), Guo et al. (2020), Zhang et al. (2019)
Normalized Difference Built-up Index	NDBI	$\frac{SWIR-NIR}{SWIR+NIR}$	Duan et al. (2019), Guo et al. (2020)
Normalized Difference Moisture Index	NDMI	$\frac{NIR-SWIR}{NIR+SWIR}$	Guo et al. (2020)

Note NIR, Red, Blue, Green, and SWIR are the near infra-red, red, blue, green and short wave infra-red bands, respectively

27.3.1.4 Classification Methods

A wide variety of supervised and unsupervised classification algorithms have been used to detect land cover/land use changes from remotely sensed data (Butt et al. 2015; Lam 2008). Although unsupervised classification algorithms are trendy for land cover mapping for large scale area as they do not need training data for classification, the survey revealed that they had been used less in comparison to supervised classification methods (see Fig. 27.3). GEE also provides a variety of computational

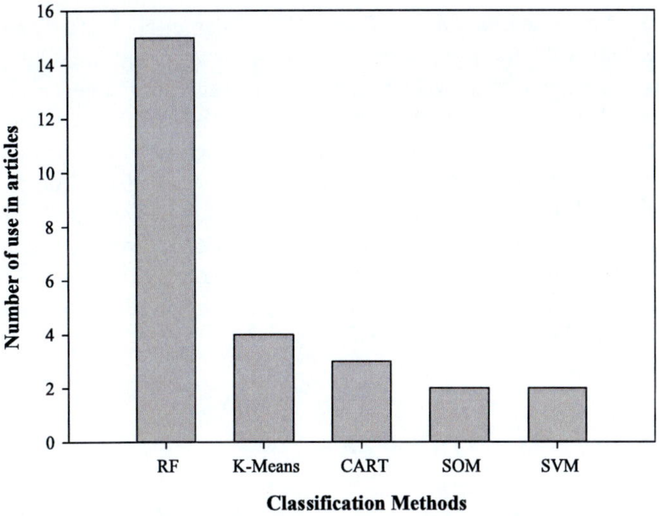

Fig. 27.4 The frequency of using differnt classification methods in surveyed articles

algorithms for both supervised and unsupervised classification. As for the supervised classification algorithms, the survey revealed that machine learning-based algorithms are more exclusively used when it comes to the GEE platform. K-means clustering (e.g., Osei and Andam-Akorful 2019; Portengen 2017; Zurqani et al. 2019b) and self-organizing map (SOM) (e.g., Wu et al. 2018) are the most notable supervised algorithms embedded in GEE's platform. Random forest (RF) (e.g., Anchang et al. 2020; Duan et al. 2019; Huang et al. 2017; Pimple et al. 2018), classification and regression trees (CART) (e.g., Johansen et al. 2015; Koskinen et al. 2019), support vector machine (SVM) (e.g., Koskinen et al. 2019; Shelestov et al. 2017) are the supervised algorithms for classification that are available in GEE's platform where RF is the most used methods in surveyed article (Fig. 27.4).

27.3.1.5 Accuracy Evaluation

Accuracy assessments are used to determine the classification efficiency in segregating various land cover (Congalton 1991; Sader et al. 1995; Tsutsumida and Comber 2015). The classification accuracy could be evaluated using stratified random sampling. Using the Google fusion table, both the training samples and validation samples were uploaded to the GEE, where the accuracy of the classification algorithms would be computed using cross-validation (Huang et al. 2017). Here, each pixel is compared to its corresponding labeled are using visual evaluation of applied images and high-resolution images (Zhang et al. 2019). As such, the accuracy of the model in terms of correctly labeling the pixels can be computed. Training and validation data are commonly collected from *Ground data* or very high-resolution

(VHR) images provided via Google Earth (GE) (Duan et al. 2019). Overall accuracy, kappa statistics, producer's accuracy, and user's accuracy could be calculated from this confusion matrix to evaluate classification accuracy (Congalton and Green 2019; Deus 2016). According to the survey, when it comes to assessing the accuracy of the classified images for both supervised and unsupervised classification, overall accuracy is the most commonly cited measure (Chen et al. 2017; Guo et al. 2020; Huang et al. 2017; Koskinen et al. 2019; Shelestov et al. 2017). Satellite images in GEE are not atmospherically corrected, so they may cause parameter acquisition to encounter some difficulties (Huang et al. 2017; Sun et al. 2019; Tian et al. 2019). Mostly the top-of-atmosphere (TOA) reflectance data are used for forest extraction, which may affect the accuracy of classification.

27.3.2 *Forest Change Detection in GEE*

In the context of forest change detection, some of the most cited methods are image overlay, change vector analysis, image ratioing, and principal component analysis (Pelletier et al. 2016). The post-classification comparison method, however, is perhaps one of the most popular methods for change detection (Jackson et al. 2004). By employing a technique called *mosaic plot*, a visual representation of land use losses and gains be computed (Comber et al. 2016). Statistical summary of the progression of losses and gains over time can also be presented by the mosaic plot. The main advantage of the post-classification method is that, in contrast to more conventional methods, not only can it detect the occurring changes, but it can also verify the type of transition from one land cover to another (Foody 2002; Singh 1989). As such, using this method, one can calculate the magnitude of the Land Use/Land Cover (LULC) changes over time (Pimple et al. 2018). However, it has been suggested that when it comes to implementing GEE's platform for forest monitoring, a more efficient approach is to just resort to a *single change index*, as that is more efficient (Zhu et al. 2012). The most common practice here is to use a technique called *imagery difference*, in which the fluctuation of the single change index, say NDVI, would be computed and monitored for consecutive time-lapses by subtracting the current value of the index from its predecessor. Multi-date differencing is another known differencing practice, where a higher order of differencing is used (e.g. NDVI bands for the Previous Year (RPNDVI) (Yang et al. 2011). Here, pixels with change will be flagged as "probable change" (Zhu et al. 2012). This method is used for the evaluation of changes from forest to urban or bare areas (Zurqani et al. 2019b), deforestation due to fire (Arai et al. 2019), and mining (Jena and Pradhan 2019).

27.4 Conclusion

The reviewed articles could clearly demonstrate the influence of GEE as a powerful cloud platform for geo-spatial data processing used for forest monitoring. In these articles, the spatial and temporal changes in forest attributes and cover, which occurred as a consequence of land-use change over time were identified. According to the survey, when it comes to forest monitoring, GEE's platform has been used for two main objectives that are classification and forest change detection. While both categories are trending, the results suggest that classification-type studies are more commonly practiced via GEE's platform. One should note, however, that despite the acceptable accuracy of classification methods for forest detection, there are some limitations such as the lack of atmospheric correction and the need for more supplementary data in large scale, areas which one needs to consider. Forest analysis using GEE could be useful to assist decision making in evaluation of ecological services in better estimation of carbon stock in forest areas, planning ecosystem restoration activities, evaluating the rates of biodiversity and enumerating the relationships between forest changes and social welfare, human health, governance and policy actions, and other relevant regional-to-global-scale applications regarding human issues.

References

Alaoui HM, Assali F, Hajji H, Aadel T, Lahssini S (2019) Big remotely Sensed data application on forest fires in morocco. Biodiversidade Brasileira (1):182

Amidon TE et al (2008) Biorefinery: conversion of woody biomass to chemicals, energy and materials. J Biobased Mater Bioenergy 2(2):100–120

Anchang JY et al (2020) Toward operational mapping of woody canopy cover in tropical savannas using Google Earth Engine. Front Environ Sci

Arai E, Shimabukuro YE, Dutra AC, Duarte V (2019) Detection and analysis of forest degradation by fire using Landsat/Oli images in Google Earth Engine. In: IGARSS 2019–2019 IEEE international geoscience and remote sensing symposium. IEEE, pp 1649–1652

Bartuska A (2006) Why biomass is important: the role of the USDA forest service in managing and using biomass for energy and other uses. Citeseer

Belgiu M, Drăguţ L (2016) Random forest in remote sensing: a review of applications and future directions. ISPRS J Photogram Remote Sens 114:24–31

Butt A, Shabbir R, Ahmad SS, Aziz N (2015) Land use change mapping and analysis using remote sensing and GIS: a case study of Simly watershed, Islamabad, Pakistan. Egypt J Remote Sens Space Sci 18(2):251–259

Camara G et al (2000) TerraLib: Technology in support of GIS innovation, II Workshop Brasileiro de Geoinformática, GeoInfo2000. São Paulo, pp. 1–8.

Ceccherini G et al (2019) Combining global Biomass and forest cover to generate dynamic maps of above ground forest biomass for the period 2000–2017. Geophys Res Abstracts

Chen B et al (2017) A mangrove forest map of China in 2015: analysis of time series Landsat 7/8 and Sentinel-1A imagery in Google Earth Engine cloud computing platform. ISPRS J Photogramm Remote Sens 131:104–120

Coltin B, McMichael S, Smith T, Fong T (2016) Automatic boosted flood mapping from satellite data. Int J Remote Sens 37(5):993–1015

Comber A et al (2016) Methods to quantify regional differences in land cover change. Remote Sens 8(3):176
Congalton RG (1991) A review of assessing the accuracy of classifications of remotely sensed data. Remote Sens Environ 37(1):35–46
Congalton RG, Green K (2019) Assessing the accuracy of remotely sensed data: principles and practices. CRC press
D'Almeida C et al (2007) The effects of deforestation on the hydrological cycle in Amazonia: a review on scale and resolution. Int J Climatol J R Meteorol Soc 27(5):633–647
Desclée B, Bogaert P, Defourny P (2006) Forest change detection by statistical object-based method. Remote Sens Environ 102(1–2):1–11
Deus D (2016) Integration of ALOS PALSAR and landsat data for land cover and forest mapping in northern Tanzania. Land 5(4):43
Dong J et al (2016) Mapping paddy rice planting area in northeastern Asia with Landsat 8 images, phenology-based algorithm and Google Earth Engine. Remote Sens Environ 185:142–154
Dronova I, Gong P, Wang L, Zhong L (2015) Mapping dynamic cover types in a large seasonally flooded wetland using extended principal component analysis and object-based classification. Remote Sens Environ 158:193–206
Duan Q, Tan M, Guo Y, Wang X, Xin L (2019) Understanding the spatial distribution of urban forests in China using Sentinel-2 images with Google Earth Engine. Forests 10(9):729
Fadli A, Kosugo, A., Ichii, K., Ramli, R., 2019. Satellite-based monitoring of forest cover change in indonesia using google earth engine from 2000 to 2016, Journal of Physics: Conference Series. IOP Publishing, pp. 012046.
Foley JA et al (2005) Global consequences of land use. Science 309(5734):570–574
Foody GM (2002) Status of land cover classification accuracy assessment. Remote Sens Environ 80(1):185–201
Foody GM, Boyd DS, Cutler ME (2003) Predictive relations of tropical forest biomass from Landsat TM data and their transferability between regions. Remote Sens Environ 85(4):463–474
Foster D et al (2003) The importance of land-use legacies to ecology and conservation. Bioscience 53(1):77–88
Geosystems L (2004) ERDAS imagine. Atlanta, Georgia 7(12):3209–3241
Global Forest Atlas (2017) Global Forest Atlas, 1–2, viewed 21 March 2020. https://globalforestatlas.yale.edu/temperate-zone
Goldblatt R, You W, Hanson G, Khandelwal AK (2016) Detecting the boundaries of urban areas in India: a dataset for pixel-based image classification in Google Earth Engine. Remote Sens 8(8):634
Gong P (2012) Remote sensing of environmental change over China: a review. Chin Sci Bull 57(22):2793–2801
Gorelick N et al (2017) Google Earth Engine: planetary-scale geospatial analysis for everyone. Remote Sens Environ 202:18–27
Griffiths P et al (2014) Forest disturbances, forest recovery, and changes in forest types across the Carpathian ecoregion from 1985 to 2010 based on Landsat image composites. Remote Sens Environ 151:72–88
Guide E.U.s. (2008) ENVI on-line software user's manual. ITT Visual Information Solutions
Guo Y et al (2020) China forest cover extraction based on Google Earth Engine. Int Arch Photogram Remote Sens Spat Inf Sci 42:855–862
Hansen MC et al (2013) High-resolution global maps of 21st-century forest cover change. Science 342(6160):850–853
Housman I et al (2015) Monitoring forest change in southeast Asia: case studies for USAID Lowering Emissions in Asia's Forests. US Department of Agriculture, Forest Service, Remote Sensing Applications Center: Salt Lake City, Utah, pp 1–16
Huang C et al (2010) An automated approach for reconstructing recent forest disturbance history using dense Landsat time series stacks. Remote Sens Environ 114(1):183–198

Huang H et al (2017) Mapping major land cover dynamics in Beijing using all Landsat images in Google Earth Engine. Remote Sens Environ 202:166–176

Hughes JN et al (2015) Geomesa: a distributed architecture for spatio-temporal fusion, Geospatial Informatics, Fusion, and Motion Video Analytics V. International Society for Optics and Photonics, p 94730F.

Hur J et al (2008) Does current management of storm water runoff adequately protect water resources in developing catchments? J Soil Water Conserv 63(2):77–90

Jackson TJ et al (2004) Vegetation water content mapping using Landsat data derived normalized difference water index for corn and soybeans. Remote Sens Environ 92(4):475–482

Jena R, Pradhan B (2019) A model to detect forest change relating To mining using Google Earth Engine application in Belitung Island, Indonesia. In: 2019 6th international conference on space science and communication (IconSpace). IEEE, pp 47–52

Johansen K, Phinn S, Taylor M (2015) Mapping woody vegetation clearing in Queensland, Australia from Landsat imagery using the Google Earth Engine. Remote Sens Appl Soc Environ 1:36–49

Johnston K, Ver Hoef JM, Krivoruchko K, Lucas N (2001) Using ArcGIS geostatistical analyst, 380. Esri Redlands

Kennedy RE, Yang Z, Cohen WB (2010) Detecting trends in forest disturbance and recovery using yearly Landsat time series: 1. LandTrendr—Temporal segmentation algorithms. Remote Sens Environ 114(12):2897–2910

Kim D-H et al (2014) Global, Landsat-based forest-cover change from 1990 to 2000. Remote Sens Environ 155:178–193

Koskinen J et al (2019) Participatory mapping of forest plantations with Open Foris and Google Earth Engine. ISPRS J Photogram Remote Sens 148:63–74

Lam NS-N (2008) Methodologies for mapping land cover/land use and its change, Advances in land remote sensing. Springer, pp 341–367

Landwirtschft BfEu (2015) The forests in Germany-selected results of the third national forest inventory

Lausch A, Erasmi S, King DJ, Magdon P, Heurich M (2016) Understanding forest health with remote sensing-part I—a review of spectral traits, processes and remote-sensing characteristics. Remote Sens 8(12):1029

Liu Y et al (2012) The long-term effects of reforestation on soil microbial biomass carbon in sub-tropic severe red soil degradation areas. For Ecol Manage 285:77–84

Liu X et al (2018) High-resolution multi-temporal mapping of global urban land using Landsat images based on the Google Earth Engine Platform. Remote Sens Environ 209:227–239

Liu C-C, Shieh M-C, Ke M-S, Wang K-H (2018) Flood prevention and emergency response system powered by Google Earth Engine. Remote Sens 10(8):1283

Lobell DB, Thau D, Seifert C, Engle E, Little B (2015) A scalable satellite-based crop yield mapper. Remote Sens Environ 164:324–333

Long T et al (2019) 30 m resolution global annual burned area mapping based on Landsat images and Google Earth Engine. Remote Sensing 11(5):489

Loveland TR, Dwyer JL (2012) Landsat: building a strong future. Remote Sens Environ 122:22–29

Lu D et al (2016) A survey of remote sensing-based aboveground biomass estimation methods in forest ecosystems. Int J Digit Earth 9(1):63–105

Martín-Ortega P, García-Montero LG, Sibelet N (2020) Temporal patterns in illumination conditions and its effect on vegetation indices using Landsat on Google Earth Engine. Remote Sens 12(2):211

Masek JG et al (2008) North American forest disturbance mapped from a decadal Landsat record. Remote Sens Environ 112(6):2914–2926

Nemani R, Votava P, Michaelis A, Melton F, Milesi C (2011) Collaborative supercomputing for global change science. Eos, Trans Am Geophys Union 92(13):109–110

Osei JD, Andam-Akorful S (2019) Long term monitoring of Ghana's forest reserves Using Google Earth Engine

Paneque-Gálvez J, McCall MK, Napoletano BM, Wich SA, Koh LP (2014) Small drones for community-based forest monitoring: an assessment of their feasibility and potential in tropical areas. Forests 5(6):1481–1507

Parks SA, Holsinger LM, Voss MA, Loehman RA, Robinson NP (2018) Mean composite fire severity metrics computed with Google Earth Engine offer improved accuracy and expanded mapping potential. Remote Sens 10(6):879

Pekel J-F, Cottam A, Gorelick N, Belward AS (2016) High-resolution mapping of global surface water and its long-term changes. Nature 540(7633):418–422

Pelletier C, Valero S, Inglada J, Champion N, Dedieu G (2016) Assessing the robustness of Random Forests to map land cover with high resolution satellite image time series over large areas. Remote Sens Environ 187:156–168

Pimple U et al (2018) Google Earth Engine based three decadal landsat imagery analysis for mapping of mangrove forests and its surroundings in the trat province of Thailand. J Comput Commun 6(1)

Portengen L (2017) Classifying mangroves in Vietnam using radar and optical satellite remote sensing: processing Sentinel-1 and Sentinel-2 imagery in Google Earth Engine

Riitters K, Wickham J, Costanza JK, Vogt P (2016) A global evaluation of forest interior area dynamics using tree cover data from 2000 to 2012. Landscape Ecol 31(1):137–148

Rogan J, Chen D (2004) Remote sensing technology for mapping and monitoring land-cover and land-use change. Progress Plan 61(4):301–325

Sader SA, Ahl D, Liou W-S (1995) Accuracy of Landsat-TM and GIS rule-based methods for forest wetland classification in Maine. Remote Sens Environ 53(3):133–144

Schlautman MA, Smink JA (2008) Evaluating the collective performance of best management practices in catchments undergoing active land development. J Soil Water Conserv 63(2):54A–55A

Schmid J (2017) Using Google Earth Engine for Landsat NDVI time series analysis to indicate the present status of forest stands, Thesis submitted in B. Sc. Geography at the Faculty of Geoscience and …

Schmidt M, Lucas R, Bunting P, Verbesselt J, Armston J (2015) Multi-resolution time series imagery for forest disturbance and regrowth monitoring in Queensland, Australia. Remote Sens Environ 158:156–168

Shelestov A, Lavreniuk M, Kussul N, Novikov A, Skakun S (2017) Exploring Google earth engine platform for big data processing: classification of multi-temporal satellite imagery for crop mapping. Front Earth Sci 5:17

Singh A (1989) Review article digital change detection techniques using remotely-sensed data. Int J Remote Sens 10(6):989–1003

Sun G, Vose JM (2016) Forest management challenges for sustaining water resources in the Anthropocene. Forests 7(3):68

Sun Z, Xu R, Du W, Wang L, Lu D (2019) High-resolution urban land mapping in China from sentinel 1A/2 imagery based on Google Earth Engine. Remote Sens 11(7):752

Tang Z et al (2016) Assessing Nebraska playa wetland inundation status during 1985–2015 using Landsat data and Google Earth Engine. Environ Monit Assess 188(12):654

Teferi E, Bewket W, Uhlenbrook S, Wenninger J (2013) Understanding recent land use and land cover dynamics in the source region of the Upper Blue Nile, Ethiopia: spatially explicit statistical modeling of systematic transitions. Agr Ecosyst Environ 165:98–117

Tempfli K et al (2009) Principles of remote sensing: an introductory textbook. International Institute for Geo-Information Science and Earth Observation

Thomas N et al (2017) Distribution and drivers of global mangrove forest change, 1996–2010. PloS One 12(6)

Tian F, Wu B, Zeng H, Zhang X, Xu J (2019) Efficient Identification of corn cultivation area with multitemporal synthetic aperture radar and optical images in the Google Earth Engine cloud platform. Remote Sens 11(6):629

Tieng T et al (2019) Mapping mangrove forest cover using Landsat-8 imagery, Sentinel-2, Very High Resolution Images and Google Earth Engine algorithm for entire Cambodia, IOP Conference Series: Earth and Environmental Science. IOP Publishing, p 012010

Tsai YH et al (2018) Mapping vegetation and land use types in Fanjingshan national nature reserve using Google Earth Engine. Remote Sens 10(6):927

Tsutsumida N, Comber AJ (2015) Measures of spatio-temporal accuracy for time series land cover data. Int J Appl Earth Obs Geoinf 41:46–55

USGS (2017) Landsat missions timeline, viewed 25 March 2020. https://Landsat.usgs.gov/Landsat-missions-timeline

Copernicus Data Access Policy (2016) Viewed 10 March 2020. https://www.copernicus.eu/main/data-access

Wang C, Jia M, Chen N, Wang W (2018) Long-term surface water dynamics analysis based on Landsat imagery and the Google Earth Engine platform: a case study in the middle Yangtze River Basin. Remote Sensing 10(10):1635

Whitman RT, Park MB, Ambrose SM, Hoel EG (2014) Spatial indexing and analytics on Hadoop. In: Proceedings of the 22nd ACM SIGSPATIAL international conference on advances in geographic information systems, pp 73–82

Woodcock CE et al (2008) Free access to Landsat imagery. Science 320(5879):1011–1011

Wu W, Zhao X, Gong C, Li X (2018) Obtain the patterns of global forest NPP and its influence factors with Google Earth Engine. In: IGARSS 2018–2018 IEEE international geoscience and remote sensing symposium. IEEE, pp 2898–2901

Xiong J et al (2017) Automated cropland mapping of continental Africa using Google Earth Engine cloud computing. ISPRS J Photogram Remote Sens 126:225–244

Yang Z et al (2019) A scalable cyberinfrastructure and cloud computing platform for forest above-ground biomass estimation based on the Google Earth Engine. Int J Digit Earth 12(9):995–1012

Yang Z, Di L, Yu G, Chen Z (2011) Vegetation condition indices for crop vegetation condition monitoring. In: 2011 IEEE international geoscience and remote sensing symposium. IEEE, pp 3534–3537

Yu J, Wu J, Sarwat M (2015) Geospark: a cluster computing framework for processing large-scale spatial data. In: Proceedings of the 23rd SIGSPATIAL international conference on advances in geographic information systems, pp 1–4

Zhang F et al (2015) The influence of natural and human factors in the shrinking of the Ebinur Lake, Xinjiang, China, during the 1972–2013 period. Environ Monit Assess 187(1):4128

Zhang X, Long T, He G, Guo Y (2019) Gobal forest cover mapping using Landsat and Google Earth Engine cloud computing. In: 2019 8th international conference on agro-geoinformatics (agro-geoinformatics). IEEE, pp 1–5

Zhu Z, Woodcock CE, Olofsson P (2012) Continuous monitoring of forest disturbance using all available Landsat imagery. Remote Sens Environ 122:75–91

Zurqani H (2019) Application of geospatial technologies for land use analysis and soil science education

Zurqani HA, Post CJ, Mikhailova EA, Schlautman MA, Sharp JL (2018) Geospatial analysis of land use change in the Savannah River Basin using Google Earth Engine. Int J Appl Earth Obs Geoinf 69:175–185

Zurqani H, Post C, Mikhailova E, Ozalas K, Allen J (2019a) Geospatial analysis of flooding from hurricane Florence in the coastal South Carolina using Google Earth Engine. Graduate Research and Discovery Symposium (GRADS), Clemson, SC, USA

Zurqani HA, Post CJ, Mikhailova EA, Allen JS (2019b) Mapping urbanization trends in a forested landscape using Google Earth Engine. Remote Sens Earth Syst Sci 2(4):173–182

Chapter 28
Free-Open Access Geospatial Data and Tools for Forest Resources Management

Gouri Sankar Bhunia, Pravat Kumar Shit, and Debashish Sengupta

Abstract This chapter presents a description of the various data and techniques used by open-source geospatial foundations (OSGeo) to gather geo specific knowledge and forestry ecosystem services. Remote sensing plays a key role in the estimation of forest parameters and in the detection and reconfiguration of forest cover. The new management of the forest strengthens geospatial tools, approaches and inventions. Crowdsourcing modes range from the collection of data passively passed on to large groups on the web to the active participation of the crowd in the production of data via special mobile apps and web platforms. We searched the Google Scholar and the Science Web for peer-reviewed articles using a combination of the terminology crowd sourcing, social media, volunteered geographical knowledge and landscape and perception, interest and ecosystem resources in forestry to provide an overview of current crowding for the collection of earth observation. For this analysis, the application focus studies either describe active open access geospatial data or research projects using actively crowded geo-information. System-focused research—the second group—offered the framework for addressing participants' participation in relevant forestry applications. The use of open-source geospatial data and software to collect real-time, locational data to explain forestry application and forestry management is also promising. Actionary open source geospatial technologies and data face a number of challenges: the involvement of adequate participants and sample representatives in order to ensure the utility of the outcome from an approach to public policy—e.g. through access to technology and/or a specific interest in nature—must be discussed more specifically.

G. S. Bhunia
Department of Geography, Nalini Prabha Dev Roy College, Bilaspur, Chhattisgarh, India

P. K. Shit
PG Department of Geography, Raja N.L. Khan Women's College (Autonomous), Gope Palace, Vidyasagar University, Paschim Medinipur, West Bengal 721102, India

D. Sengupta (✉)
Department of Geology & Geophysics, Indian Institute of Technology (IIT), Kharagpur 721302, West Bengal, India
e-mail: dsgg@gg.iitkgp.ac.in

P. K. Shit et al. (eds.), *Spatial Modeling in Forest Resources Management*, Environmental Science and Engineering,
https://doi.org/10.1007/978-3-030-56542-8_28

Keywords Open source · Geospatial data · Crowd sourcing · Remote sensing

28.1 Introduction

Forest is an area of at least 0.05–1.0 ha with an area of over 10–30% of tree-sized crown cover (or equivalent stock level), of trees that may achieve a minimum height of 2-5 m at in situ maturity. A forest can be either closed forest structure with trees covering a large portion of the land and open forest in different stories and undergrowth (UNFCCC 2001). Forests are an important natural resource base that needs to be used, administered, and protected in an informed way from a local to a global scale. Forests supply goods of materials, such as fuelwood, trade wood and other non-wood products. Forests are inventoried in many ways. For example, knowledge about forest resources is gathered for political, logistical, and organizational preparation and management of forests. National forest inventories (NFIs) are examples of inventories conducted to collect national forest resource information and facilitate strategic planning and policy development at national level. Information of interest in this case may include land cover, growing stock volume, biomass, carbon balance, and potential for large-scale procurement of timber. NFI's was based on field studies in several countries (Tomppo et al. 2010). Forest attribute maps may be created but the maps' accuracy is usually insufficient for operational preparation of forest management. Sustainable forest management is a tricky balance between the needs of an ever-increasing human population and the preserving of balanced land habitat ecological services (MacDicken et al. 2015). Forest inventories have historically been built to support the timber harvesting-related knowledge needs. For certain countries, such inventories are focused on field sample plots or, for some cases, on systematic analysis of aerial imagery for stand delineation and attributes such as species distribution, stand height, breast height diameter (DBH), and basal area, which are then supplemented with field plot data collected at comparatively few locations to reflect the forest landscape (Tomppo et al. 2010).

Inherent in the word "sustainable" is a multidimensional forest management strategy that addresses a wide variety of considerations, particularly ecology, forest health, and disease and fire tolerance (Gauthier et al. 2014). Over the last 10 years, open source geo-space tech has taken significant strides in the fields of efficiency and user-friendly graphical interfaces. Forestry is the world's greatest land habitat and is an important contributor to natural, economic and social benefits (Davis et al. 2001). Forests with rich composition and complex structure have seen to be much more productive and wood-producing relative to forests with a basic structure (Meng et al. 2014). In this background of competitive demands on forest capital and growing global rivalry among fiber suppliers, value chain optimization has become a core driver of the forest industry of today (Shahi and Pulkki 2015). Forest management intelligence demands are becoming increasingly dynamic and wide-ranging, raising new barriers to forest resource programmes. As such, forest knowledge needs to be precise, spatially specific, up-to-date, and must describe the morphology, structure, and ultimately qualities of wood supply (Groot et al. 2015). Considering the above current scenario, present chapter analyze the open source geospatial data and tools for forest resource mapping and monitoring for sustainable conservation, focusing on the Open Source Geospatial Foundation (OSGeo) data and techniques in forest resource applications.

28.2 Remote Sensing Technology in Forestry Application

Remote sensing offers a systematic, consistent view at regular intervals of earth cover and is useful for land cover changes and exposes dimension of biodiversity (Cohen and Goward 2004; Kumar et al. 2010) directly. The steady stresses on the base of forest resources have given rise to much discussion about how they can best handle their future. Myers (1998) indicates that 'The gradual loss of many of the world's forests is one of the world's greatest challenges and opportunities.' The rates and drivers of deforestation in the world's hot spots of biodiversity and tropical habitats are commonly used for the identification of satellite image detection, change analysis (Armenteras et al. 2004). However, a number of high spatial and spectral resolution airborne and satellite sensors have been available for the study of forest changes over the last decades.

The MODIS sensor on the NASA's Aqua and Terra satellite provides an overall mapping of a percent tree cover, fires of vegetation and shift of land cover (Savtchenko et al. 2004). However, MODIS data has global forestry monitoring capability and its spatial resolution is limited to 250 m. ASTER is a multispectral variable spatial resolution (15–90 m) imager with track-long stereos capabilities (Abrams 2000). ASTER has high-level spatial resolution (15 to 90 m). The Advanced Wide Field Sensor (AWiFS) Indian multi-sensor ResourceSat uses double cameras to provide forest cover loss with a space resolution (56 m). The high-resolution SPOT (European sensor) can be used for mapping changes in the ground cover in spatial range of 5–20 m (King 2002). As a consequence, Loveland et al. (2000), at about 1 km of space resolution, created the pan-continental map from the NOAA-AVHRR sensor data at the end of the 1990s. The image data of AVHRR have 2 spatial resolutions: 1.1 km local area coverage (LAC) and 5 km global area coverage (GAC) for a global area coverage. The NOAA–AVHRR Land Pathfinder Standardize Vegetation Vegetation Index (NDVI) data (Zeng and Rao 2003) also extracted a global 8 km fractional vegetation coverage dataset for 1982–2000. They are commonly used to study and monitor ecosystem vegetation conditions, forestry, tundra, grasslands, land, land cover mapping and large-scale maps for this subject. They are also very popular. The lower cost and high likelihood for cloud-free views on land surfaces are one of the obvious advantages of AVHRR. More recently, TERRA-MODIS (Hansen et al. 2003) or ENVISAT-MERIS (Arino et al. 2007) have been producing new global land cover datasets in a finer resolution (250–500 m) (Table 28.1). Landsat data have been used to carry out detailed assessments of changes in global tropical forests (Foody et al. 2003). The three Landsat sensors, MSS, TM and ETM+, combine to

Table 28.1 Open source satellite-sensor and global forest cover mapping

Map title	Spatial resolution	Satellite/sensor	References
IGBP	1 km (global)	NOAA-AVHRR	Loveland et al. (1999)
University of Maryland (UMD)	1 km (global)	NOAA-AVHRR	Hansen et al. (2003)
Trees	1 km (tropics)	NOAA-AVHRR	Achard et al. (2001)
FRA 2000	1 km (global)	NOAA-AVHRR	FAO (2001)
MODIS Land cover	1 km (global)	TERRA MODIS	Friedl et al. (2002a, b)
Global land cover	1 km (global)	SPOT-VEG	Bartholome and Belword (2005)
Vegetation continuous filed	1 km (global)	NOAA-AVHRR	De-Fries et al. (2000)
Vegetation continuous filed	500 m (global)	TERRA MODIS	Hansen et al. (2003)
GLOBECOVER	300 m (global)	Envisat MERIS	Arino et al. (2007)

Source Kumar et al. (2010)

create the longest series of images that can be used to track changes in vegetation in the world at a high spatial resolution.

Remote sensing plays a crucial role in estimating forest parameters and detecting forest cover destruction and change. Geospatial tools, methods, and innovations improve current forest management. Advances in Geographic Information Systems (GIS), Global Positioning Systems (GPS), and remote sensing in recent decades have created new ways to conduct forest productivity evaluations (Turcotte 2003), forest inventories (Tomppo et al. 1999), harvest planning (Laamanen and Kangas 2011), infrastructure scheduling (Abdi et al. 2009), log transport (Devlin et al. 2008), emissions monitoring (Patenaude), classification of the ecosystems (Rieman et al. 2000), monitoring and optimization of processing technologies (Turcotte 2003), and control of forest health (Coops et al. 2006). The application of remote sensing and GIS in forestry application are summarized in Table 28.2. Such techniques include airborne laser scanning (ALS), terrestrial laser scanning (TLS), digital aerial photogrammetry (DAP), and satellite optical imagery with high spatial resolution (HSR) and very high spatial resolution (VHSR). In the field of forest surveillance, various satellite sensor forms play specific and compatible roles, e.g. optical, hyperspectral, LiDAR or radar at various spatial and temporal resolutions. VHR satellite data also offer potential forest monitoring possibilities (Solano-Correa et al. 2018). Multitemporal and multispectral VHR data are obtained by passive sensors (e.g., QuickBird, WorldView-2, GeoEye, Pleiades) (Solano-Correa et al. 2018) and are economically transmitted through DigitalGlobe, Geo-Airbus and Earth (Planet 2018). Time series of low resolution images (MODIS, MERIS, etc.) were widely used to generate vegetation, forest cover and landscape-type maps (Friedl et al. 2010), forest cover changes, tree density (DiMiceli et al. 2011), vegetation indexes, primary gross and net production (Running et al. 2004) and forest disturbances (Justice et al. 2002). Also, RADAR and LiDAR

Table 28.2 Current remote sensing approaches in forest resource applications

	Method										LiDAR		Optical								SAR						
Mapping Approach	Dense time series tracking	Change detection	Vegetation indices	Data transforms	Spectral Mixture Analysis	Classification	Interferometry	Modelling	Data Fusion	Visual Interpretation	ICESat GLAS	Airborne LiDAR	MODIS	CBERS	Landsat	SPOT	Sentinel-2	RapidEye	Quickbird	IKONOS	ALOS 1/2; PALSAR 1/2	ENVISAT ASAR	SRTM	TerraSAR-X	TanDEM-X	Consmo-SkyMed	GeoSAR
Forest Disturbance Mapping	√		√												√												
	√		√												√												
	√	√				√									√												
	√		√												√												
		√		√											√												
					√					√				√	√												
					√								√														
					√													√									
					√	√									√												
	√		√										√														
			√			√																					
		√																		√							
						√									√					√				√			√
Identification of canopy gaps					√														√								
			√																								√
		√																						√			
	√	√																				√		√			
						√															√	√		√			
						√																		√			
						√						√															
								√		√		√															
Proxies						√				√					√					√							
										√														√			
										√																	√
						√															√			√			
										√		√															
and degraded	√				√									√													
					√									√						√							
					√		√				√																
Canopy height change		√					√				√																
							√			√																	
						√														√							
						√																√			√		
						√																√		√			
						√																√		√			
AGB Change							√				√																
							√				√																
							√				√																
							√				√																
							√	√		√																	
							√	√		√																	
							√													√							
						√	√															√		√			
						√	√															√		√			

Modified after Mitchell et al. (2017)

data were successfully used to measure biomass and canopy height (Santoro et al. 2015). With the launch of the Landsat archive in 2008 (Wulder et al. 2012) with a more than 40-year time span, it became easy to track long-term forest shifts at a greater resolution. The European Space Agency's (ESA) open data strategy with regard to the Sentinel satellites ensures higher-resolution Earth coverage more regularly than Landsat, that is especially important for cloud-dominated forest areas. Sentinel 1 can also supplement forest architecture surveillance by using RADAR data at a higher spatial and temporal resolution than other radar items such as ENVISAT ASAR. Despite synergistic advances in geospatial technology and education, it is not shocking that 81% of recent graduates in forestry and natural resource management use GIS in their work in the United States of America at least once a month (Merry et al. 2007). Hyperspectral imagery includes, relative to multiple imaging that contains just a dozen spectral bands, hundreds of spectral bands. Hyperspectral sensors are well suited for plant study, as both the spectral signatures for reflectance and absorption of individual species can be better distinguished from much wider spectral hyperspectral imaging bands (Varshney and Arora 2004). In the earth remote sensing area, for example, hyper spectral pictures of AVIRIS are widely used. In 224 contiguous spectral channels with wavelengths between 400 and 2500 nm, AVIRIS is a unique visual sensor which provides calibrated images of upwelling spectral radiance. A similar research was also carried out by Rosso et al. in 2005, in which vegetation dynamics were monitored to suggest sustainable wetlands management in San Francisco Bay, CA.

28.3 Open-Source Geospatial Tools and Technologies

Open source generally means transparency, free and unlimited access to information, and inclusive decision-making based on consensus (Brovelli et al. 2020). Digital architects from the twenty-first century claim to be the fundamental principle of transparency. Geo-spatial techniques vary from conventional imaging and electronic data collection tools to a broad variety of applications. The introduction of open source geospatial applications in a short period has shifted the use of these innovations from skilled geospatial to everyday users. Functionality of the open-source project:

- Data conversion using command line (like GDAL library)
- Spatial databases (like Postgres)
- Desktop GIS (like QGIS)
- Web GIS (like OpenLayers)
- Mobile GIS (like Android with OpenLayers)
- Geo processing libraries (GDAL) and much more.

The GDAL is an open source platform which provides an abstract data model for reading and writing different GeoTIFF, ERDAS Imagine, JPEG2000 geospatial raster formats. The app library is a permissive online resource for raster geospatial data reading and writing. The retrieval of data can be done using command line software. The utility can be built into a library on different platforms such as .net and can be developed using languages such as C#. Through fixing the bugs and adding the new functionality, the user will upgrade the source code that is available. GDAL version 2.0.2 is available for use at this time.

28.4 Open Source Satellite Data Used in Forest Mapping

The free and free availability of global Earth monitoring data obtained and stored in a systematic manner by space agencies such as the ESA sequence of Sentinel and the United States. The advanced AVHRR radiometer (NASA), National Aerospace Science Administration (NASA) and the Landsat system of NASA and the USA (USGS) have allowed the rapid growth of private- and public data commons for the most advanced high-definition radiometers. Landsat (1972) and NOAA AVHRR (1978) have been the first terrestrial remote sensing data. Since the beginning, large areas (i.e. national to global coverage) have been created using AVHRR data since gross space resolutions as data are free and the small amount of rough spatial resolutions (approximately 1.1 km of nadir data) associated with these data has resulted in data volumes that can be stored and exploited over large areas (Franklin and Wulder 2002). The advent of the MODIS Terra and Aqua sensors with AVHRR and Landsat architecture heritage (Justice et al. 2002) has contributed to the creation and upgrade of the systemically annual overhead commodity (Friedl et al. 2002a, b) of the 500 m (500 m) world-wide. Several countries have undertaken forest cover mapping using satellite data and developed their own national forest cover products. Whilst some countries have high-quality national forest cover products, considerable differences exist in terms of geographical coverage, their spatial and temporal resolution as well as thematic depth (Table 28.3).

During the past 10 years, rapid growth in computation power, combined with free and open Satellite data, and significant decreases in the associated cost have resulted in the migration of programs that facilitate the mapping of broad-based land areas solely under the control of public agencies and a small number of well-funded research groups. Cloud computing services are becoming increasingly possible for users to process data and create products for land cover without significant investment in computing infrastructure (Gorelick et al. 2017). At the same time, this data (and related products), in particular Landsat, are gradually being used in national and international programs and programmes (Pettorelli et al. 2016). Since the launch of the Earth Resources Engineering Satellite in 1972 (re-named Landsat 1), remotely sensed imagery has increasingly been used to track Earth's ecosystems (Gillespie et al. 2008) by quantifying land cover change (Asner et al. 2010), deforestation (Asner et al. 2002), carbon stocks and pollution (Asner et al. 2010), habitat degradation and disease (Tang et al. 2010), species composition. However, satellite imagery products differ in their spatial and spectral resolution, geographic and temporal exposure, cloud coverage, security regulations and price factors, which can impede their clear

Table 28.3 Global forest cover platform for remote sensing data

Product	Spatial resolution (m)	Coverage of years	Contents/overall reported accuracy	Source
GlobeLand30	30	2000, 2010	10 classes/80.3%	https://www.globeLand30.com
Global tree cover	30	Annual (2000–)	One class (forest but with percentage cover)/unknown	https://glcf.umd.edu/data/landsatTreecover/ FCC: https://glcf.umd.edu/data/landsatFCC/
Global forest change	30	Annual (2000–)	Forest canopy cover %/gains/losses	https://earthenginepartners.appspot.com/science-2013-globalforest/download_-v1.5.html
Copernicus land service: dynamic land cover	100	Annual (2015–)	10 classes/74% (2015)	https://land.copernicus.eu/global/products/lc
Forest and non-forest global map	25	Every year 1993–1998 2007–2010 2015–2016	Two classes (forest/non-forest)/84% accuracy/L-badn SAR	https://www.eorc.jaxa.jp/ALOS/en/palsar_fnf/fnf_index.htm
UNEP environmental data explorer			Includes global forest cover, global potential evapotranspiration	https://en.wikipedia.org/wiki/UNEP_Environmental_Data_Explorer
Pilot analysis of global ecosystems: forest ecosystems		Every year	Data from the World Resources Institute includes: percentage tree-cover, population density and tree cover, share of wood in fuel consumption, etc.	https://www.wri.org/publication/pilot-analysis-global-ecosystems-forest-ecosystems#data

application in conservation (Loarie et al. 2007). Satellite imagery used in the conservation work varies in resolution from 1000 to <1 m (Xie et al. 2008). Monitoring of forest clearance on a global to biome scale is mostly performed at spatial resolutions of 250–1000 m (Hansen et al. 2013). MODIS images, including 36 spectral bands with a range of 250–1 km, have been collected primarily for mapping the dynamics of vegetation and large scale processes. The potential for classifying phenological vegetation coverage of the Albemarle-Pamlico Estuarine Network using 250 m 16-day composite data has been explored by Knight et al. (2006). Landsat imagery (30 m of multispectral resolution) has been an integral part of scientific research since 1972 (Wallace et al. 2006), particularly in mapping and assessing land cover change, and is available at no cost at present. In order to identify and characterize continuous changes in the early succession of forestry, for example, almost 20-year continuous Landsat TM/ETM+ image datasets (19 images) covering the West of Oregon are used (Schroeder et al. 2006). Because of the limited spatial resolution, Landsat products are usually used to map community-level vegetation. It is difficult to use Landsat images to map them in a heterogeneous setting at species level. However, it becomes possible to map species when integrated with other auxiliary data.

28.5 Spectral Vegetation Indices in Forestry Applications

Space-borne observations are now commonly available to provide information on various thematic challenges on a variety of frontiers and on many levels, which is important and harmonized (Turner et al. 2007). They provide an economical and coherent source of data suitable for landscape change mapping and monitoring (Roy et al. 2014). But the knowledge of a variety of landscape attributes such as structure, architecture and connectivity are needed for the understanding of the changes observed (Chetkiewicz et al. 2006; Lausch et al. 2015). The reflectance on the surface of organ and leaf is determined by chemical and morphological characteristics of vegetation from vegetation to electromagnetic spectrum (spectral reflections or vegetation emission properties) (Zhang and Kovacs 2012). The major applications for remote vegetation detection include: (i) the ultraviolet region (UV), which goes from 10 to 380 nm; (ii) the visible spectra, which are composed of the blue (450–495 nm), green (495–570 nm), and red (620–750 nm) wavelength regions; and (iii) the near and mid infrared band (850–1700 nm) (Cruden et al. 2012). During the visible and near/middle infrared regions, different applications are based on peaks or overtones for the reflectivity of particular compounds in light spectrum. Vegetation information from remote sensed images is interpreted primarily by variations, changes in green plant leaves and spectral canopy characteristics. The most frequent method of validation includes direct or indirect associations between VIs and plant value characteristics measured on the land, such as coverage of foliage, LAI, biomass, growth and vigor valuation. More defined methods for the assessment of VIs are used to track sentinel plants using Direct and georeferenced methods in comparison with VIs from the same plants for calibrations (Table 28.4).

Table 28.4 Summary of important vegetation indices used for forestry application

Index	Equation	Application	References
AGVI	$GVI - (1 + 0.018GVI) * YVI - \frac{NSI}{2}$		Gitelson et al. (2001)
Atmospherically resistant vegetation index (ARVI)	$\frac{(NIR-RB)}{(NIR+RB)}$ where RB is the difference between Red and Blue	Commonly used in atmospheric aerosol removal	Kaufman and Tanré (1992)
Ratio Vegetation Index (RVI)	$\frac{R}{NIR}$	This indicator is highly vegetation sensitive and has a strong association with the biomass of plants	Jordan (1969)
Difference vegetation index (DVI)	NIR–R	The DVI can be extended to the monitoring of the atmosphere in vegetation and is highly sensitive to changes in soil history	Richardson and Weigand (1977)
Perpendicular vegetation index (PVI)	$\sqrt{(\rho_{soil} - \rho_{veg})^2 - (\rho_{soil} - \rho_{veg})^2 NIR}$	Efficient PVI effects in the earth's environment; PVI also has less exposure to atmospheric effects and is used primarily to reverse surface vegetation (grass yields, chlorophyll content) parameters	Richardson and Weigand (1977)

(continued)

Table 28.4 (continued)

Index	Equation	Application	References
Soil-adjusted vegetation index (SAVI)	$\frac{(\rho_n - \rho_r)(1+L)}{(\rho_n + \rho_r + L)}$	In most common environmental conditions, the value of L is about 0.5. If L is near 0, the value of SAVI is NDVI. In order to get the optimum soil impact modification, L should therefore differ with the amount of vegetation present	Huete (1988)
Modified SAVI (MSAVI2)	$0.5 * [(2NIR + 1) - \sqrt{(2NIR +)^2 - 8(NIR - R)}$	This method is mainly used in plant growth analysis, desertification research, the assessment of grasslands yield, LAI assessment, the organic soil matter analysis, the monitoring of droughts, and the analysis of soil erosion. MSAVI2 does not rely on the soilline principle and has a simpler algorithm	Richardson and Wiegand (1977)

(continued)

Table 28.4 (continued)

Index	Equation	Application	References
Optimized soil-adjusted vegetation index (OSAVI)	$\frac{(NIR-R)}{(NIR+R+X)}$	OSAVI is not soil line based and can effectively eradicate the impact of the soil history. It is used mainly to determine the aboveground biomass, the content of leaf-nitrogen and chlorophyll	Baret et al. (1993)
Enhanced vegetation index (EVI)	$\frac{(TM_4-TM_3)(1+L)}{TM_4-C_1TM_3+C_2TM+L}$	The feedback mechanism has been implemented by establishing a soil and atmospheric output parameter simultaneously. The values of the atmosphere correction NIR, R, and B, L, and its value equal 1 are the parameters of soil modification, with parameters equivalent to 6 and 7.5	Liu and Huete (1995)
Normalized difference vegetation index	$\frac{NIR-RED}{NIR+RED}$	This index is used to detect plant greenery, due to the high NIR chlorophyll reflectance	Zarco-Tejada et al. (2012)

(continued)

Table 28.4 (continued)

Index	Equation	Application	References
Visible-band difference vegetation index (VDVI)	$\frac{(2*\rho_{green}-\rho_{red}-\rho_{blue})}{(2*\rho_{green}+\rho_{red}+\rho_{blue})}$	Vegetation extraction accuracy dependent on VDVI is higher than that of other VI and Green band visible lights. In addition, more than 90% of VDVI accuracy has been recorded	Wang et al. (2015)
Wide dynamic range vegetation index (WDRVI)	$\frac{\propto\rho_{nir}-\rho_{red}}{\alpha\rho_{nir}+\rho_{red}}$	If α equals 1, WDRVI is equivalent to NDVI. If αis equal to the ratio, WDRVI is zero. WDRVI offers a simple way to increase a dynamic range for environments with high biomass (LAI > 2). NDVI is, however, the best choice for classification of plants when biomass is small (LAI < 1)	Gitelson (2004)
Modified chlorophyll absorption ratio index (MCARI)	$\frac{1.5*[2.5(R_{800}-R_{670})-1.3(R_{800}-R_{550})]}{\sqrt{(2R_{800}+1)^2-(6R_{800}-5R_{670})-0.5}}$	Leaf chlorophyll concentrations are more susceptible to MCARI	Daughtry et al. (2000)

28.6 Open Source Software for Forest Data Mapping and Analyzing

Open source software originates from the beginning of computation, when problems of computing were solved through scientific collaboration. Each programmer introduced a new dimension to existing information and software was shared. It has grown into a software creation and licensing strategy to make the source code available and to cooperate through a range of copyright safeguards (Coetzee et al. 2020). The idea of a developer community participating in a software product is part of open source software development. In the past, there has often been no legal arrangement or financial reward, but many software developers now contribute to open-source geospatial applications. This approach may encourage innovation by reducing obstacles to own computer operating systems and software products, such as software licensing cost (Yeung and Hall 2007). Many of these characteristics are always obvious to the human eye, but computational methods and resources are required to use these data for the accurate analysis, quantification and evaluation, across large regions, of the effect of human actions on our environment (Riitters et al. 2000). Reconnaissance of shifting trends and not just the shifting locations is important for safe, multifunctional management of forest ecosystems as well as the creation of efficient forest policy (Vogt and Riitters 2017). Similar to those given in other free software, such as QGIS [https://qgis.org/en/site/], GRASS [https://grass.osgeo.org/], GDAL [https://www.gdal.org/], R [https://www.r-project.org/], Conefor (Saura and Torné 2009) and FRAGSTATS (McGarigal et al. 2012) were fine-tuned to provide detailed answers to end users in a combination of thematic fields of application, requiring the analysis of imaging. An example of the standard tasks that environmental specialists conduct with GIS are provided by a short survey of the recent forestry ecology (FE) publication is illustrated in Fig. 28.1. We also identified a number of free open source desktop applications to create, edit and analyze data based on an Internet survey. Table 28.5 explains the comprehensive features of the software.

28.7 Open-Source Tools for Forestry Application

Many technical problems around storage capacities and computing capacity have been addressed recently by advancement in cloud based remote sensing technology. Open source software application helps experts with limited remote sensing experience to perforate comprehensive ground evaluations using free and open source resources to view Extremely High Resolution Satellite Imagery (VHR) for all regions. The ***Orfeo Toolbox*** (OTB) is an open source (Grizonnet et al. 2017), high-tech remote sensing project with a quick picture view, Python or QGIS commandline applications, and a versatile C++ API. It is based on the Insight Toolkit (ITK) and is written in C++. The OTB depends for raster and vector size on the Geospatial Data Abstraction Library (GDAl); geometric sensor models OSSIM; machine-learning libSVM,

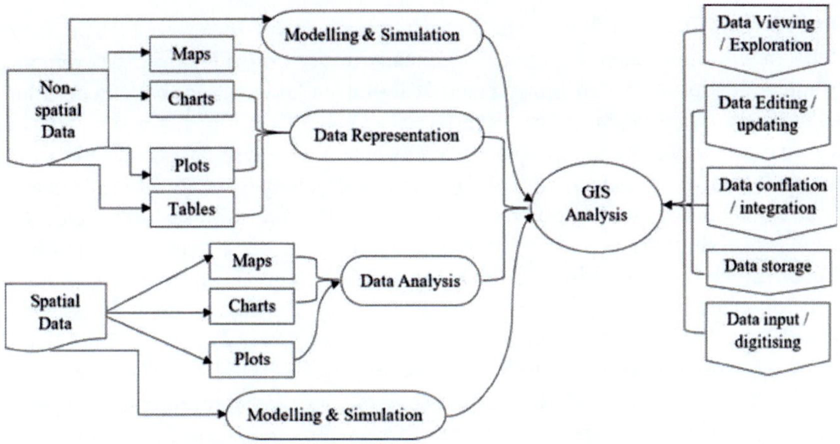

Fig. 28.1 Tasks that can be accomplished with Open Source software. Modified after Steinigera and Hay (2009)

Table 28.5 Free and open source desktop GIS suitable for Forest ecology tasks

Software	Year	Webpage	User focus[a]	Data focus (raster/vector)[b]	Platform	
					Operating system	Language
GRASS	1982	grass.osgeo.org	Experienced, …, research	More raster	Ms-Windows, Linux, MacOSX	C, Tcl/Tk, Python
ILWIS	1984	ilwis.org	Novice ,…, research	Raster and vector	Ms-Windows, Linux	MS Visual C
MAP WINDOW	1998	mapwindow.org	Novice, …, research	Raster and vector	Ms-Windows	MS Visual Studio .Net
SAGA	2001	saga-gis.org	Novice, …, research	More raster	Ms-Windows, Linux	MS Visual C
Quantum GIS	2002	qgis.org	Novice, …, research	More vector	MS-Windows, Linux, MacOSX	C++, Qt4, Python
gvSIG	2003	gvsig.gva.es	Novice, …, research	More vector	MS-Windows, Linux, MaxOSX	JAVA

[a]User levels: (i) novice (viewing), (ii) experienced (editing, simple analysis), (iii) expert (analysis), (iv) research (scripting, programming)
[b]Data focus: subjective evaluation with respect to (i) software history and (ii) number of functions for raster and vector data editing and analysis

OpenCV and Shark; and mathematical parsing MuParser for example. The cross platform with more than 90 applications, ready-to-use OTB kit ships, for example, orthorectification, output mapping, external digital model elevation chart, mode interpolation, etc. The Graphical User Interface for Object Classification and its Shapes Toolbox (***GuidosToolbox***) offers user-friendly access to a range of specialized spatial morphological analysis techniques (Vogt et al. 2007), which are not included in any other applications. GuidosToolbox is four-pillar structure. The main modules are Batch Processing, Pre-processing and other modules for Image Analysis. Providing GuidosToolbox as a stand-alone device in a self-contained directory ensures full portability: the device can be conveniently accessed from an external USB drive. "***Open Foris***" applications were released at the 24th World Congress of the International Union of Forest Research Organizations (IUFRO) in Salt Lake City, USA, with open software tools to help countries compile vast forest inventories and allow nations to consider the importance of their forests (Fig. 28.2). These instruments are intended to transform the way nations track the status of their forests and enhance the data needed to establish deforestation management policies and successful climate mitigation plans. ***Collect Earth*** (https://collect.earth) was developed by SERVIR, a joint venture of the National Aeronautics and Space Administration (NASA) with the United States, as a standard, open source, high resolution satellite image processing and visualization program. Internal Development Agency (USAID). Users are able to access lower definition and high resolution satellite images for a range of purposes in conjunction with Google Earth, Bing Maps and Google Earth Software, including: multistage support including national forest inventories, Land Use, Land Use Change and Forestry (LULUCF) assessments, confirmation of current maps and so on (Saah

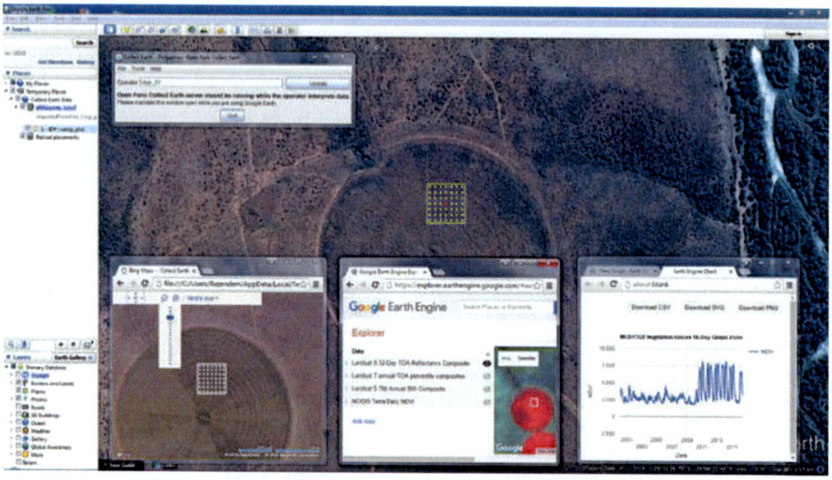

Fig. 28.2 FAO-developed Open Foris Collect Earth. The Collect Earth package provided easy access to multi-temporal and high-resolution satellite imagery along with innovative tools for rapidly sampling changes to land use and land cover over time. *Source* https://www.openforis.org/tools/collect-earth/case-study.html

Fig. 28.3 Google Earth Engine cloud platform offers direct access to Google's. *Source* https://www.earthengine.google.com

et al. 2019). ***Google Earth Engine (GEE)*** lets users analyze cloud satellite imagery. Collect Earth provides users with free access to both high-resolution GEE images and high-resolution (VHR) images from Google Earth and Bing maps (Fig. 28.3). It brings together images from different sources stored for a range of time periods in these three repositories, and from 30 cm to 30 m of spatial resolution. ***Collect Mobile*** is a simple, intuitive and versatile resource for data collection on the field. The software helps comprehensive datasets, such as biophysical, socio-economic and biodiversity assessments, to be performed. It provides several features including: data quality validation on-the-fly; embedded GPS location; integrating with data collection; analyzing and exporting into widely used formats; input processes; and estimation of the qualitative control attributes in the field. ***CLASlite*** has been developed to enable high-resolution mapping and monitoring of forests with satellite images by governments, non-governmental organizations and academic institutions. It can also be a daily practice for non-experts, who advocate climate conservation, forest governance and growth policy for resource management, to track tropical deforestation and satellite forest destruction.

28.8 Open Geospatial Platform and Forest Resource Mapping and Monitoring

Information Communication Technology (ICT) is often marketed as a tool for strengthening evidence base and governance quality in developing countries (Bertot et al. 2010). It is, however, also criticized for being deployed top-down so that the needs of the local authorities and communities have not been addressed (Huang 2013). The geospatial service System (GSW) is a virtual, internet-based geospatial infrastructure that integrates geospatial resources such as sensor, data resources, computing resources, information resources, machine resources, network resources, and storage resources for data management and data extraction, and information gain in geospatial communications. In specific, geographic information systems were attacked for selective enhancement and exploitation of the local interests of a techno community (Dunn et al. 1997). Mapping strategies have developed to enhance local participatory planning processes and collective decision-making to tackle the inherently disempowering dimensions of traditional GIS (McCall and Dunn 2012). Many of these participatory GIS approaches have centred on low-tech methods to allow communities spatially. The potential for decentralized use of emerging low-cost digital space tools and data for improving governance is also growing in interest (Fisher and Myers 2011). The availability of free geospatial data and software has been increasing rapidly over the last ten years (2012). GSW's mission includes:

- acquire global spatial data for all seasons, every day and all the directions using any type of satellite, aviation and ground surface sensors.
- seamlessly connect the whole process from sensors to applied services via unified satellite communication networks and network data relays and the use of cable and wireful communications.
- register sensors, computing resources, storage resources, internet resources, manipulate software and spatial data on the internet, and process spatial data online quantitatively, automatically, intelligently, and in real time; and
- provide geospatial services, compose virtual service chains and transmit user-required information in the most effective and efficient ways. Using GSW for real-time environmental data manage will help describe, organize, manage, manipulate, interchange, search, and release environmental data in a unified framework.

28.8.1 OpenLayers API'S

OpenLayers is an open source, web-based map-generation packed library. OpenLayers 3 v3.15.1 is currently being developed. For building a thin 2D map viewers, OpenLayers 3 (OL3) is being used. The OL3 API create the fundamental GIS functions such as zoom-in, zoom-out, pan, recognition, mapping distance and measurement calculation. In the map viewer the raster data (WMS) along with the baseline layers is applied with OL3 API. The current ability of open-source geo-source technology and innovations have been used to build a solution for the aggregation of spatial information sets at an end-to-end level. The program was developed to satisfy flood control needs in real time.

28.8.2 MapX

MapX (https://app.mapx.org) supports an online platform which offers reputable local, national and global spatial information, an authentication data integrity framework through a scorecard and a series of online tools for viewing, analyzing and accessing geospatial data (Lacroix et al. 2019). It has a graphical user interface and the ability to no program geospatial data, analyze data and create story maps is required. In particular, MapX was created to collect data through existing structured web services such as Open Geospatial Consortium (OGC) Web Map Services (WMS), currently available in four languages (French, English, Dari and Pashto).

28.8.3 GFW Map Builder

Forest Atlases are provided using Map Builder, a tool recently released publicly by GFW. It helps everyone to build their own forest surveillance website using their own data coupled with GFW geospatial analysis tools. GFW partnered with Blue Raster to build this software application framework based on ArcGIS online, and it is released under the MIT license so that any company can create its own forest monitoring platform with custom data, free of charge at any geographic scale. Users can deploy Map Builder for server on both ArcGIS online and ArcGIS portal (https://www.globalforestwatch.org/howto/tags/map-builder/). Through modifying the logo and branding, plugging in custom data, focusing the map on a specific area of interest or style data for a particular subject, users can customize the web applications.

28.8.4 ArcGIS Open Data

Forest ministries depend on ArcGIS Open Data (https://www.esri.com/en-us/arcgis/products/arcgis-open-data) to distribute data publicly via the Forest Atlas. Ministry using a Forest Atlas has its own Open Data Platform, where it publishes its public access geospatial data as well as additional tools, such as contracts for forestry, reports and other publications. Data distributed through the Open Data Sites was published along with extensive metadata in their original projections. Data are compiled in English and French using independent, open data pages. Users can search and access the desired dataset by keyword, full text, or location. Web developers can find the API endpoints available for use in their applications.

28.8.5 GEO and GEOSS

The Global Earth Observation System of Systems (GEOSS) is organized by the Group on Earth Observations (GEO), which includes more than 40 international organizations, 62 nations and the European Union. GEOSS is structured into nine societal benefit areas (SBAs), with forestry included explicitly within the agricultural benefit area (Bastin et al. 2012). There is clearly a huge theoretical potential for forest data from a variety of scales and disciplines to be collated and aggregated into fuller, richer datasets which permit consistency checking, better analysis of multi-scale phenomena and filling of important data gaps. Forest data and statistics are typically compiled by forest authorities through national forest inventory (NFI) programs, which collect in-situ information, including estimates of forest area, species composition and growing stock (Tomppo et al. 2010). These data are used for strategic planning and production forecasting at national and regional levels, but they are also used to generate indicators for compliance with international reporting requirements. The data gathered at an international scale is typically managed by bodies such as the United Nations Food and Agriculture Organization (FAO) Forest Resource Assessment, the Ministerial Conference on the Protection of Forests in Europe and the United Nations Framework Convention on Climate Change (UNFCCC) Land use, Land-use Change and Forestry (LULUCF).

28.9 Conclusion

Geospatial tools for the visualization, tracking and simulation of forest resources include key data underpinning Natural Resources Management Planning (NRM). The current generation of geospatial tools and open data Free and Open Source (FOS) has eliminated the need for licensing and data acquisition, which was a significant barrier to broader participation in forest resource mapping, tracking and modelling for local governments. In addition, the availability of free satellite imagery, elevation data and raster-based spatial analysis tools offers community members new opportunities to combine quantitative analysis of 'hard' data with qualitative local knowledge. Satellite imagery information can also be shared with environmental and conservation organisations and education leaders in user-friendly formats. We suggest enhancing interdisciplinary development and collaboration between ecology, environment and remote sensing fields, and to increase accessibility to high resolution imagery at low (or no) costs to ensure the prevalence of satellite image analysis for ecology purposes. Although online available free data sets and free satellite images, many of these services are accessible with smaller resolutions (e.g. MODIS,

Landsat), restricted to accessible dates (e.g. 4-m multispectral OrbView-3 data are freely available from 2003 to 2007, but with no global complete coverage), and/or uninitialized (e.g. Google Earth). When lower resolution images are used to evaluate land cover without checking results of modelling in the field, connectivity estimates and correct corridors can be inaccurate and/or misleading. Given those potential benefits, it could be argued that funding for Free Open Source GIS applications, collaborative mapping and monitoring should be included in all rural development and forest resource management initiatives creating new spatial data in less developed countries. Continued support to use Free Open Source GIS capabilities could:

- Creating Performance Legitimacy by participation.
- Foster equity, integrity and responsibility by skills development.
- Contribute to a shared global educational content capital pool;
- legitimize the use of Free open source GIS tools; and
- help build a larger legal geospatial industry in the developing world.

Acknowledgements We are grateful to the PG Department of Geography, Raja N. L. Khan Women's College (Autonomous), affiliated to Vidyasagar University, Midnapore, West Bengal, India for supporting this research. The author (P. K. Shit) grateful acknowledges West Bengal DSTBT for financial support through R&D Research Project Memo no. 104(Sanc.)/ST/P/S&T/10G-5/2018).

References

Abdi E, Majnounian B, Darvishsefat A, Mashayekhi Z, Sessions J (2009) A GIS-MCE based model for forest road planning. J For Sci 55(4):171–176

Abrams M (2000) The advanced spaceborne thermal emission and reflection radiometer (ASTER): data products for the high spatial resolution imager on NASA's Terra Platform. Int J Remote Sens 21:847–859

Achard F, Eva HD, Mayaux P (2001) Tropical forest mapping from coarse spatial resolution satellite data: production and accuracy assessment issues. Int J Remote Sens 22:2741–2762

Arino O, Gross D, Ranera F, Gross D, Bicheron P et al (2007) GlobCover: ESA service for global land cover from MERIS. https://www.gofc-gold.uni-jena.de/documents/boston/Arino_Globcover_ENVISATSYMPOSIUM.pdf

Armenteras D, Rudas G, Rodriguez N, Sua S, Romero M (2004) Pattern and cause of deforestation in Colmbian Amazon. Ecol Indic 6:353–368

Asner GP, Keller M, Pereira R, Zweede JC (2002) Remote sensing of selective logging in Amazonia: assessing limitations based on detailed field observations, Landsat ETM+, and textural analysis. Remote Sens Environ 80:483–496

Asner GP, Powell GVN, Mascaro J, Knapp DE, Clark JK et al (2010) High-resolution forest carbon stocks and emissions in the Amazon. Proc Nat Acad Sci USA 107:16738–16742

Baret F, Jacquemoud S, Hanocq JF (1993) The soil line concept in remote sensing. Remote Sens Rev 7(1):65–82

Bartholome E, Belword AS (2005) GLC 2000: A new approach to global land cover mapping from earth observation data. Int J Remote Sens 26:1959–1977

Bastin L, McInerney D, Revez G, Figueiredo C, Simonetti D, Barredo J, Achard F, San-Miguel-Ayanz J (2012) Web services for forest data, analysis and monitoring: developments from EuroGEOSS. Earthzine

Bertot J, Jaeger P, Grimes J (2010) Using ICTs to create a culture of transparency: E-government and social media as openness and anti-corruption tools for societies. Gov Inf Q 27(3):264–271

Brovelli M, Ilie CM, Coetzee S (2020) Openness and community geospatial science for monitoring SDGs—an example from Tanzania. In: Rajabifard A (ed) Sustainable development goals connectivity dilemma: land and geospatial information for urban and rural resilience. CRC Press, Boca Raton, pp 313–324. ISBN 978-0-429-29062-6

Chetkiewicz C-LB, St. Clair CC, Boyce MS (2006) Corridors for conservation: integrating pattern and process. Ann Rev Ecol Evol Syst 37:317–342

Coetzee S, Ivánová I, Mitasova H, Brovelli MA (2020) Open geospatial software and data: a review of the current state and a perspective into the future. ISPRS Int J Geo-Inf 9:90. https://doi.org/10.3390/ijgi9020090

Cohen WB, Goward SN (2004) Landsat role in ecological applications of remote sensing. Bioscience 54:535–545

Coops NC, Johnson M, Wulder MA, White JC (2006) Assessment of QuickBird high spatial resolution imagery to detect red attack damage due to mountain pine beetle infestation. Remote Sens Environ 103(1):67–80

Cruden BA, Prabhu D, Martinez R (2012) Absolute radiation measurement in venus and mars entry conditions. J Spacecr Rockets 49(6):1069–1079

Daughtry CST, Walthall CL, Kim MS, De Colstoun EB, McMurtrey JE (2000) Estimating corn leaf chlorophyll concentration from leaf and canopy reflectance. Remote Sens Environ 74(2):229–239

Davis LS, Norman Johnson K, Bettinger PS, Howard TE, Alván Encinas L, Salazar M, Gretzinger S, Lange G, Schmithusen F, Hyde W et al (2001) Forest management: to sustain ecological, economic, and social values. Universidad Nacional Agraria La Molina: Lima, Perú

De-Fries R, Hansen M, TownShend JRG, Janetos AC, Loveland TR (2000) A new global 1 km datasets of per cent tree cover derived from remote sensing. Global Change Biol 6:247–254

Devlin GJ, McDonnell K, Ward S (2008) Timber haulage routing in Ireland: an analysis using GIS and GPS. J Transp Geogr 16(1):63–72

DiMiceli CM, Carroll ML, Sohlberg RA et al (2011) Annual global automated MODIS vegetation continuous fields (MOD44B) at 250 m spatial resolution for data years beginning Day 65, 2000–2010

Dunn CE, Atkins PJ, Townsend JG (1997) GIS for development: a contradiction in terms? Area 29:151–159

FAO (2001) Pan-tropical survey of forest cover changes 1980–2000. FRA Working Paper No. 49, https://www.fao.org/docrep/004/y1997e/y1997e1f.htm

Fisher RP, Myers BA (2011) Free and simple GIS as appropriate for health mapping in a low resource setting: a case study in eastern Indonesia. Int J Health Geogr 10(11). https://doi.org/10.1186/1476-072X-10-15.

Foody GM, Boyd DS, Cutler MEJ (2003) Predictive relations of tropical forest biomass from landsat TM data and their transferability between regions. Remote Sens Environ 85(4):463–474

Franklin SE, Wulder MA (2002) Remote sensing methods in medium spatial resolution satellite data land cover classification of large areas. Prog Phys Geogr 26(2):173–205

Friedl MA, McIver DK, Hodges JCF, Zhang XY, Muchoney D et al (2002) Global land cover mapping from MODIS: algorithms and early results. Remote Sens Environ 83:287–302

Friedl MA, McIver DK, Hodges JC, Zhang XY, Muchoney D, Strahler AH, Woodcock CE et al (2002) Global land cover mapping from MODIS: algorithms and early results. Remote Sens Environ 83(1):287–302

Friedl MA, Sulla-Menashe D, Tan B et al (2010) MODIS collection 5 global land cover: algorithm refinements and characterization of new datasets. Remote Sens Environ 114:168–182

Gauthier S, Bernier P, Burton PJ, Edwards J, Isaac K, Isabel N, Jayen K, Le Goff H, Nelson EA (2014) Climate change vulnerability and adaptation in the managed Canadian boreal forest. Environ Rev 30:256–285

Gillespie TW, Foody GM, Rocchini D, Giorgi AP, Saatchi S (2008) Measuring and modelling biodiversity from space. Prog Phys Geog 32:203–221

Gitelson AA (2004) Wide dynamic range vegetation index for remote quantification of biophysical characteristics of vegetation. J Plant Physiol 161(2):165–173

Gitelson AA, Merzlyak MN, Chivkunova OB (2001) Optical properties and nondestructive estimation of anthocyanin content in plant leaves. Photochem Photobiol 74(1):38–45

Gorelick N, Hancher M, Dixon M, Ilyushchenko S, Thau D, Moore R (2017) Google Earth engine: planetary-scale geospatial analysis for everyone. Remote Sens Environ 202:18–27

Grizonnet M, Michel J, Poughon V et al (2017) Orfeo ToolBox: open source processing of remote sensing images. Open Geospat Data Softw Stand 2:15. https://doi.org/10.1186/s40965-017-0031-6

Groot A, Cortini F, Wulder MA (2015) Crown-fibre attribute relationships for enhanced forest inventory: Progress and prospects. For Chron 91:266–279

Hansen MC, DeFries RS, Townshend JRG, Carroll M, Dimiceli C, Sohlberg RA (2003) Global percent tree cover at a spatial resolution of 500 meters: First results of the MODIS vegetation continuous fields algorithm. Earth Interact 7:1–15

Hansen MC, Potapov PV, Moore R, Hancher M, Turubanova SA et al (2013) High-resolution global maps of 21st-century forest cover change. Science 342:850–853

Huang YE (2013) The future's fix: using information and communications technology to reduce global poverty. Available at: https://hdl.handle.net/11021/24185. Accessed 13 Nov 2016

Huete AR (1988) A soil-adjusted vegetation index (SAVI). Remote Sens Environ 25(3):295–309

Jordan CF (1969) Derivation of leaf-area index from quality of light on the forest floor. Ecology 50(4):663–666

Justice CO, Giglio L, Korontzi S et al (2002) The MODIS fire products. Remote Sens Environ 83:244–262

Kaufman YJ, Tanré D (1992) Atmospherically resistant vegetation index (ARVI) for EOS-MODIS. IEEE Trans Geosci Remote Sens 30(2):261–270

King R (2002) Land cover mapping principles: a return to interpretation fundamentals. Int J Remote Sens 18:3525–3545

Knight JF, Lunetta RS, Ediriwickrema J et al (2006) Regional scale land cover characterization using MODIS-NDVI 250 m multi-temporal imagery: a phenology-based approach, GIScience Remote Sens 43:1–23

Kumar D, Borah S, Sarma H, Shankar U (2010) Monitoring and assessment of land use and land cover changes in Kamrup district of Assam, India. Communicated data

Laamanen R, Kangas A (2011) Large-scale forest owner's information needs in operational planning of timber harvesting—some practical views in Metsähallitus. Finnish State-Owned Enterprise. Silva Fennica 45(4):711–727

Lacroix P, Moser F, Benvenuti A, Piller T, Jensen D, Petersen I, Planque M, Ray N (2019) MapX: An open geospatial platform to manage, analyze and visualize data on natural resources and the environment. SoftwareX 9:77–84

Lausch A, Blaschke T, Haase D, Herzog F, Syrbe R-U, Tischendorf L, Walz U (2015) Understanding and quantifying landscape structure—a review on relevant process characteristics, data models and landscape metrics. Ecol Model 295:31–41

Liu HQ, Huete A (1995) Feedback based modification of the NDVI to minimize canopy background and atmospheric noise. IEEE Trans Geosci Remote Sens 33(2):457–465

Loarie SR, Joppa LN, Pimm SL (2007) Satellites miss environmental priorities. Trends Ecol Evol 22:630–632

Loveland TR, Zhu Z, Ohlen DO, Brown JF, Reed C, Yang L (1999) An analysis of the IGBP global land-cover characterization process. Photogramm Eng Remote Sens 65:1021–1032

Loveland TR, Reed BC, Brown JF, Ohlen DO, Zhu Z, Yang L, Merchant JW (2000) Development of a global land cover characteristics database and IGBP DISCover from 1 km AVHRR data. Int J Remote Sens 21:1303–1330

MacDicken KG, Sola P, Hall J, Sabogal C, Tadoum M, de Wasseige C (2015) Global progress toward sustainable forest management. For Ecol Manage 352:47–56

McCall MK, Dunn CE (2012) Geo-information tools for participatory spatial planning: fulfilling the criteria for 'good' governance? Geoforum 43(1):81–94

McGarigal K, Cushman SA, Ene E (2012) FRAGSTATS v4: spatial pattern analysis program for categorical and continuous maps. Computer software program produced by the authors at the University of Massachusetts, Amherst. Retrieved from https://www.umass.edu/landeco/research/fragstats/fragstats.html

Meng J, Lu Y, Zeng J (2014) Transformation of a degraded *Pinus massoniana* plantation into a mixed-species irregular forest: Impacts on stand structure and growth in southern China. Forests 5:3199–3221

Merry KL, Bettinger P, Clutter M, Hepinstall J, Nibbelink NP (2007) An assessment of geographic information system skills used by field-level natural resource managers. J For 105(7):364–370

Mitchell AL, Rosenqvist A, Mora B (2017) Current remote sensing approaches to monitoring forest degradation in support of countries measurement, reporting and verification (MRV) systems for REDD+. Carbon Balance Manage 12:9. https://doi.org/10.1186/s13021-017-0078-9

Myers N (1988) Threatened biotas: hot spots in tropical forests. Environmentalist 8:187–208

Pettorelli N, Wegmann M, Skidmore A, Mücher S, Dawson TP, Fernandez M, Lucas R et al (2016) Framing the concept of satellite remote sensing essential biodiversity variables: challenges and future directions. Remote Sens Ecol Conserv 2(3):122–131

Planet (2018) High resolution monitoring. In: Planet. https://planet.com/products/hi-res-monitoring/. Accessed 11 May 2018

Rieman BE, Lee DC, Thurow RF, Hessburg PF, Sedell JR (2000) Toward an integrated classification of ecosystems: defining opportunities for managing fish and forest health. Environ Manage 25(4):425–444. https://doi.org/10.1007/s002679910034

Riitters KH, Wickham JD, Vogelmann JE, Jones KB (2000) National land-cover pattern data. Ecology 81:604

Richardson AJ, Wiegand CL (1977) Distinguishing vegetation from soil background information. Photogram Eng Remote Sens 43:1541–1552

Rosso PH, Ustin SL, Hastings A (2005) Mapping marshland vegetation of San Francisco Bay, California, using hyperspectral data. Int J Remote Sens 26:5169–5191

Roy DP, Wulder MA, Loveland TR, Woodcock CE, Allen RG, Anderson MC, Helder D, Irons JR, Johnson DM, Kennedy R, Scambos TA, Schaaf CB, Schott JR, Sheng Y, Vermote EF, Belward AS, Bindschadler R, Cohen WB, Gao F, Hipple JD, Hostert P, Huntington J, Justice CO, Kilic A, Kovalskyy V, Lee ZP, Lymburner L, Masek JG, McCorkel J, Shuai Y, Trezza R, Vogelmann J, Wynne RH, Zhu Z (2014) Landsat-8: science and product vision for terrestrial global change research. Remote Sens Environ 145:154–172

Running SW, Nemani RR, Heinsch FA et al (2004) A continuous satellite-derived measure of global terrestrial primary production. Bioscience 54:547–560

Saah D, Johnson G, Ashmall B, Tondapu G, Tenneson K, Poortinga A, Markert K, Quyen NH, Aung KS, Schlichting L, Matin M, Uddin K, Arya RR, Dilger J, Ellenburg WL, Flores-Anderson AI, Wiell D, Chishtie F (2019) Collect Earth: an online tool for systematic reference data collection in land cover and use applications. Environ Model Softw 118:166–171

Santoro M, Beaudoin A, Beer C, Cartus O, Fransson JES, Hall RJ, Pathe C, Schmullius C, Schepaschenko D, Shvidenko A et al (2015) Forest growing stock volume of the northern hemisphere: spatially explicit estimates for 2010 derived from Envisat ASAR. Remote Sens Environ 168:316–334

Saura S, Torné J (2009) Conefor Sensinode 2.2: a software package for quantifying the importance of habitat patches for landscape connectivity. Environ Model Softw 24:135–139. https://doi.org/10.1016/j.envsoft.2008.05.005

Savtchenko A, Ouzounov D, Ahmad S, Acker J, Leptoukh G, Koziana J, Nickless D (2004) Terra and Aqua MODIS products available from NASA GES DAAC. Adv Space Res 34:710–714
Schroeder TA, Canty MJ, Yang Z et al (2006) Radiometric correction of multi-temporal Landsat data for characterization of early successional forest patterns in western Oregon. Remote Sens Environ 103:16–26
Shahi S, Pulkki R (2015) A simulation-based optimization approach to integrated inventory management of a sawlog supply chain with demand uncertainty. Can J For Res 45:1313–1326
Solano-Correa YT, Bovolo F, Bruzzone L (2018) An approach for unsupervised change detection in multitemporal VHR Images acquired by different multispectral sensors. Remote Sens 10:533
Steinigera S, Haya GJ (2009) Free and open source geographic information tools for landscape ecology. Ecological Informatics, Version 4th July 2009. Available at: https://www.geo.uzh.ch/~sstein/manuscripts/sstein_freegitools_ecoinf2009.pdf
Tang L, Shao G, Piao Z, Dai L, Jenkins MA et al (2010) Forest degradation deepens around and within protected areas in East Asia. BiolConserv 143:1295–1298
Tomppo E, Goulding C, Katila M (1999) Adapting finnish multi-source forest inventory techniques to the New Zealand preharvest inventory. Scand J For Res 14(2):182–192
Tomppo E, Schdauer K, McRoberts R, Gschwantner T, Gabler K, Stl̀´hl G (2010) Introduction. In: Tomppo E, Gschwanter T, Lawrence M, McRoberts R (eds) National forest inventories pathways for common reporting. Springer, Heidelberg, 607pp
Turcotte P (2003) Multidata and opti-grade: two innovative solutions to better manage forestry operations. Paper presented at the proceedings of the 2nd international precision forestry symposium: Seattle, WA
Turner BL II, Lambin EF, Reenberg A (2007) The emergence of land change science for global environmental change and sustainability. Proc Natl Acad Sci 104:20666–20671
UNFCCC (2001) Seventh conference of the parties to the UN framework convention on climate change. 29 Oct–9 Nov, Marrakesh, Morocco. https://www.iisd.ca/climate/cop7/mon29.html.
Varshney PK, Arora MK (2004) Advanced image processing techniques for remotely sensed hyperspectral data. Springer, Berlin
Vogt P, Riitters K (2017) GuidosToolbox: universal digital image object analysis. Eur J Remote Sens 50(1):352–361
Vogt P, Riitters KH, Iwanowski M, Estreguil C, Kozak J, Soille P (2007) Mapping landscape corridors. Ecol Ind 7:481–488
Wallace J, Behn G, Furby S (2006) Vegetation condition assessment and monitoring from sequences of satellite imagery. Ecol Manage Restor 7:S31–S36
Wang X, Wang M, Wang S, Wu Y (2015) Extraction of vegetation information from visible unmanned aerial vehicle images. NongyeGongchengXuebao/Trans Chin Soc Agric Eng 31(5):152–159
Wulder MA, Masek JG, Cohen WB et al (2012) Opening the archive: how free data has enabled the science and monitoring promise of Landsat. Remote Sens Environ 122:2–10
Xie Y, Sha Z, Yu M (2008) Remote sensing imagery in vegetation mapping: a review. J Plant Ecol 1:9–23
Yeung AKW, Hall GB (2007) Spatial database systems: design, implementation and project management. GeoJournal library. Springer, Dordrecht. ISBN 978-1-4020-5391-7.
Zarco-Tejada PJ, González-Dugo V, Berni JAJ (2012) Fluorescence, temperature and narrow-band indices acquired from a UAV platform for water stress detection using a micro-hyperspectral imager and a thermal camera. Remote Sens Environ 117:322–337
Zeng X, Rao P (2003) Interannual variability and decadal trend of global fractional vegetation cover from 1982 to 2000. J Appl Meteorol 42:1525–1530
Zhang CH, Kovacs JM (2012) The application of small unmanned aerial systems for precision agriculture: a review. Precision Agric 13:693–712. https://doi.org/10.1007/s11119-012-9274-5